A Sustainable Future for the Mediterranean

A Sustainable Future for the Mediterranean

The Blue Plan's Environment
and Development Outlook

Edited by
Guillaume Benoit and Aline Comeau

Preface by
Lucien Chabason

London • Sterling, VA

First published by Earthscan in the UK and USA in 2005

Copyright © Plan Bleu 2005

All rights reserved

ISBN–13: 978-1-84407-259-0 paperback
978-1-84407-258-3 hardback

ISBN–10: 1-84407-259-2 paperback
1-84407-258-4 hardback

Typesetting by Fish Books
Printed and bound by Gutenberg Press Ltd, Malta
Cover design by Andrew Corbett
Page design by Danny Gillespie

For a full list of publications please contact:

Earthscan
8–12 Camden High Street
London, NW1 0JH, UK
Tel: +44 (0)20 7387 8558
Fax: +44 (0)20 7387 8998
Email: earthinfo@earthscan.co.uk
Web: **www.earthscan.co.uk**

22883 Quicksilver Drive, Sterling, VA 20166-2012, USA

Earthscan is an imprint of James and James (Science Publishers) Ltd and publishes in association with the International Institute for Environment and Development

Plan Bleu – Regional Activity Centre
15, rue Beethoven
Sophia Antipolis – 06560 Valbonne
France
Tel: +33 (0)4 92 38 71 30
Fax: +33 (0)4 92 38 71 31
Web: **www.planbleu.org**

DISCLAIMER
The designations employed and the related data presentation in this report do not imply the expression of any opinion whatsoever on the part of MAP/Blue Plan or the cooperating institutions concerning the legal status of any country, territory, region or city, or of its authorities, or of the delineation of its frontiers or boundaries.
The elaboration of this report and related data collection took place from 2002 to mid-2004. Some information elements subsequent to this date have been integrated before achieving the manuscript on March 2005.
This document has been realized with the financial support of the European Community. The views expressed are those of the Blue Plan and in no way represent the official opinion of the European Commission or the countries bordering the Mediterranean.

A catalogue record for this book is available from the British Library

Library of Congress Cataloging-in-Publication Data has been applied for

Printed on elemental chlorine-free paper

Contents

Preface	viii
Acknowledgements	ix
Introduction	xi

Part 1 The Mediterranean and its Development Dynamics — 1

Section 1 The Mediterranean Region: A Unique But Neglected Heritage — 3
A unique heritage — 3
Signs of unsustainable development — 12

Section 2 Determining Factors of the Mediterranean Future — 29
Climate change and its possible impacts — 29
Towards north–south demographic convergence — 30
The Mediterranean between globalization and regional integration — 34
A deficit of governance in the face of environment/development challenges — 50
The SEMC faced with multiple development challenges — 57
Summary of the baseline scenario assumptions — 63

Part 2 Six Sustainability Issues — 67

1 Water — 69
Continuing growth in pressures on vulnerable water resources — 72
Paths to more sustainable water management — 89

2 Energy — 109
Unsustainable energy development — 111
Towards more sustainable energy systems — 123

3 Transport — 147
A boom in transport benefiting the road sector — 149
The large impacts of land and air transport — 164
Reducing pollution and the risks of maritime transport — 179
Towards a more sustainable transport system — 183

4 Urban Areas — 197
Widespread urbanization, once in the north, now in the south — 199
Increased expansion and mobility, vulnerable and fragmented towns — 206
Urban environment: Air, water, waste — 223
What possibilities for action do cities have? — 231
Mediterranean and Euro-Mediterranean cooperation: Beyond the challenges — 242

5 Rural Areas — 247
    The originality, fragility and diversity of Mediterranean rural areas — 249
    Rural and agricultural population: Divergent and unsustainable developments — 253
    Land dynamics and the extent of environmental degradation — 261
    Food and agriculture outlooks — 273
    Territorial challenges in the Mediterranean — 282
    Alternatives for rural areas — 289

6 Coastal Areas — 303
    The coastal zone: A description — 305
    An increasingly built-up coast: Megapolization? — 309
    Pursuing the fight against coastal pollution — 318
    Protecting coastal ecosystems and their biodiversity — 325
    Ensuring sustainable management of fisheries and aquaculture — 330
    Mobilizing for an integrated coastal area management — 335
    Directing tourism towards sustainable development — 341
    A sustainable coastal management scenario — 348
    Legal References — 354

Part 3 Summary and Call for Action — 357
    The Mediterranean in motion — 359
    The risks in the baseline scenario — 361
    Facing sustainable development challenges — 364
    Striving for more sustainable development — 373
    Concluding remarks — 386

Statistical Annex — 389
List of Figures, Tables and Boxes — 433
List of Acronyms and Abbreviations — 439
Index — 444

Preface

This Blue Plan report brings together three topics: possible scenarios for the future, sustainable development and the Mediterranean. This preface would, and rightly should, have been signed by Michel Batisse, chairman of the Blue Plan till 2004. He guided the development of the entire project, but sadly died before the report could be published. Until his final days, he put all his talent and immense international experience into this task. He deserves our gratitude and admiration for his invaluable contribution to the analysis of Mediterranean issues.

I would also like to thank all the members of the steering committee and everyone who has contributed to this report.

As a prospective exercise, the project was faced with the same methodological questions that were present when the Blue Plan was developing various trend and alternative scenarios in 1989. But a different method has been adopted, which takes into account the lessons learned from past prospective exercises.

Far from being limited to a simple, mechanical projection of past trends, a *prospective* analysis aims to anticipate the feedback and adjustments that may be generated over time by the projected developments. For example, it is easy to imagine that a continuation of current consumption trends for fossil energy sources will produce tensions in energy markets that will result in increased prices, which, in turn, will put pressure on demand and generate multiple responses. These could include a greater competitiveness of alternative energy sources, a search for improved energy efficiency or even a reduction in the rate of economic growth because of increasing energy costs as a production factor. The same process could be imagined in the field of water resources.

The methodological difficulties associated with prospective analysis result from the need to achieve a precise introduction of economic variables and social compromises, together with physical projections, and to model the system's evolution over time. And it is precisely because of the lack of appropriate ways of achieving this that so much previous prospective work has encountered problems and produced results that were not particularly perceptive.

The Blue Plan has sidestepped these obstacles by presenting an alternative sustainable development scenario in this report, based not on mechanical market reactions or future socio-economic changes but on the assumption that proactive sustainable development policies are implemented; this scenario is then compared with a baseline scenario that assumes a continuation of the trends in the relationships between economic growth and the growth in environmental effects that have been observed over the past 30 years. From this point of view, it reflects the approach of the Johannesburg Earth Summit, which relies on political engagement by nations and actors for a more sustainable development.

In this way, the report tackles the issue of sustainable development head-on. The Blue Plan's past work on the environment and development had drawn attention, well before the notion of sustainable development even appeared, to the interactions between population, economic development, and pressures on the coasts and natural resources such as water and the natural environment. This approach, which is systemic as well as forward-looking, corresponds in practice with the concept of sustainability with its three pillars: economic, social and environmental, a concept that has particular meaning for the Mediterranean.

Sustainable development in the Mediterranean has specific features. Particular pressures come from tourism, urban concentration in coastal areas, the development of irrigated and intensive agriculture, the trend to abandon or poorly manage mountain regions, overfishing and intercontinental (Asia/Europe) maritime transport. All these pressures are exerted on particularly limited and fragile resources: water, natural coastal areas and the marine environment.

Furthermore, the economy/environment interface cannot be separated from social issues. Among the social questions covered by the Millennium Development Goals adopted within the United Nations framework, unemployment among the young is the most burning issue in the Mediterranean area. In the southern-shore countries, 30 per cent of young people were out of work in 2003, a 'record' compared with the world's major regions (unemployment among the young is also significant in some European Mediterranean countries), while significant progress has been made in access to education, gender equality and public health.

As in every region of the world, the Mediterranean presents specific sustainable development problems.

A Sustainable Future for the Mediterranean

The topics covered in this report illustrate the degree to which sustainability problems are of growing importance and the number of crucial natural resource issues.

In this regard, it is important to stress the difference between 'strong' and 'weak' sustainability, the latter allowing for the substitution of natural capital by human or built capital. This distinction is also relevant in the Mediterranean, for example, in the water sector. Today, with large scale desalinization of sea water in some countries of the region, we are seeing the gradual substitution of some natural capital (the fossil fuels used for desalinization) and built capital (the desalinization plants) for the natural capital that is being used up (freshwater in its catchment area). The often-mentioned perspective of water as a limiting factor for economic development is perhaps becoming more distant while there is a growing use of fossil fuel energy, which itself could soon reach its physical and economic limits. In this regard, prospective work is all the more important for examining possible future developments.

The adoption of a weak sustainability approach based on these mechanisms of substitution may therefore eventually turn out to be dangerous. The Precautionary Principle would suggest that, without veering from this path, the alternative approach of strong sustainability should be explored, initially through a more optimal use of water and energy resources rather than a relentless pursuit of permanently increasing supply. In other words, the questions hanging over the future of water- and energy-related matters call for the implementation of a precautionary policy, which is exactly what is needed in uncertain situations such as those presented in this report.

This situation of relative uncertainty linked to assumptions on substitution may be highly relevant to the management of renewable natural resources, but not to that of limited resources such as natural coastal areas, coastal agricultural plains and island landscapes; these are doomed to vanish forever at the present rate of consumption of coastal space by urban sprawl, roads, tourism and harbours. Here what is lost is lost for ever. And we cannot rely on self-regulation and substitution in these fields as long as the market works poorly for matters such as the irreversible use of heritage 'goods'. Ethical principles such as the rights of future generations to have worthwhile natural heritage such as sand dunes, beaches and lagoons, should be implemented by public authorities exercising their responsibilities. The draft protocol on integrated management of coastal areas responds to this concern, a protocol that the Contracting Parties to the Barcelona Convention need to adopt and implement as soon as possible.

The Barcelona Convention plays an important role in the Mediterranean region as a forum for sustainable development, as well as a framework for cooperation in the management of common goods such as the sea. It is vital that this Convention continues and develops its actions and that it receives essential support from the European Union. It is also vital that the governance of sustainable development issues is improved and benefits from the necessary resolution of the remaining conflicts occurring around the Mediterranean Basin, that an atmosphere of detente and cooperation follows, and that increased human and financial resources are devoted to more active policies for managing cities, public transport, rural facilities, education and health, and to reducing pollution.

If this report contributes to progress in raising awareness of sustainable development issues in the Mediterranean and the adoption of the necessary actions, the Blue Plan will have successfully fulfilled its mission.

Lucien Chabason
Chairman of the Blue Plan
Sophia Antipolis, July 2005

Acknowledgements

A *Sustainable Future for the Mediterranean* has been edited by the Blue Plan Regional Activity Centre of the Mediterranean Action Plan.

The work is the fruit of a collective expertise based on numerous studies and workshops organized in the last decade by the different components of the Mediterranean Action Plan, other institutions, Mediterranean networks of experts and NGOs, which has mobilized more than 300 experts from all Mediterranean countries as well as from some European countries not bordering the Mediterranean sea. Much of the report owes to analyses produced within the framework of the Mediterranean Commission on Sustainable Development since 1996 and to the *Fascicules du Plan Bleu* series. The original version was in French.

The Blue Plan thanks the many individuals and institutions that have contributed to this collective work.

The production team

Editors: Guillaume Benoit (Director of the Blue Plan) and Aline Comeau (Scientific Director)

Coordination of the editions: Silvia Laría. *Editing of the English version*: Peter Saunders and Mirjam Schomaker. *Translation*: Timothy Fox

Main authors: Aline Comeau (Part 1, water, energy, land-based pollution), Guillaume Benoit (Part 3, rural areas, coastal areas), Silvia Laría (urban areas, environmental policies), Patrice Miran (transport, household waste), Elisabeth Coudert (prospects for coastal areas, tourism). *Contributors*: Vito Cistulli, Abdeljaouad Jorio, Stéphane Quéfélec (economics, regional integration), Luc Dassonville (woodlands, desertification)

Advisory panel: Michel Batisse †, Serge Antoine, Lucien Chabason, Jean Margat, Jean de Montgolfier, Bernard Glass

Statistical annex: Jean Pierre Giraud. *Data collection*: Jean Louis Couture, Ingrid Deo, François Larini, Jerôme Lartigue, Nicolas Rivier

Maps: François Ibanez

Documentation and bibliography: Hélène Rousseaux, Catherine Martin

Secretariat: Bassima Saïdi, Hakima Kerri, Christiane Bourdeau, Isabelle Jöhr, Brigitte Février

† deceased.

Steering Committee

Michel Batisse † (President of the Blue Plan and President of the Committee), Lucien Chabason (former Coordinator of the Mediterranean Action Plan, Athens), Serge Antoine (Comité 21, France), Georges Corm (Economist and historian of the contemporary Mediterranean, Lebanon), Maria Dalla Costa (Agenzia per la Protezione dell'Ambiente et per i Servizi Tecnici, Italy), Djamel Echirk (Inspection générale de l'environnement, Algeria), Mohamed Ennabli (Hydrologist, Tunisia), Emilio Fontela (University of Madrid, Spain), Magdi Ibrahim (ENDA Maghreb, Morocco), Thierry Lavoux (Institut français de l'environnement, France), Jacques Lesourne (President of Futuribles International, France), Ronan Uhel (European Environment Agency, Denmark), with contributions from Arab Hoballah (Deputy Coordinator of the Mediterranean Action Plan, Athens), Anne Burrill (European Commission), George Abu Jawdeh (former UNDP resident Coordinator, Lebanon), Azzam Mahjoub (Economist, Tunisia)

Other components of the Mediterranean Action Plan

Programme for the assessment and control of pollution in the Mediterranean region (MEDPOL, Greece), Specially Protected Areas Regional Activity Centre (SPA RAC, Tunisia), Regional Marine Pollution Emergency Response Centre for the Mediterranean (REMPEC, Malta), Regional Activity Centre for Cleaner Production (CP RAC, Spain), Priority Actions Programme Regional Activity Centre (PAP RAC, Croatia), Programme for the protection of coastal historic sites (France), Environment Remote Sensing Regional Activity Centre (ERS RAC, Italy).

Expert's contributions

Background papers. *Population*: Youssef Courbage, Isabelle Attané (Institut national d'études démographiques, France). *Economics and Geopolitics*: Emilio Fontela (Spain), Azzam Mahjoub (Tunisia), Rudolf El Kareh (Lebanon). *Regional environmental cooperation*: Lucien Chabason (former MAP coordinator). *Financing for* *sustainable development*: Georges Corm (economist, Lebanon). *Climate change*: Gérard Begni (Medias France). *Water*: Jean Margat. *Energy*: Michel Chatelus † (University of Grenoble), Habib El Andaloussi, Houda Allal, Michel Grenon (Observatoire Méditerranéen de l'Energie). *Land transport*: Christian Reynaud (Nouveaux Espaces de Transports en Europe Applications de recherche). *Maritime transport*: Jean-Pierre Dobler (Ecomar, France). *Urban areas*: Lionel Urdy (Euro-Mediterranean Network of Economic Institutes), the MOLAND team (European Joint Research Centre, Ispra, Italy), Xavier Godard (CODATU), Francis Papon (Inrets, France). *Rural areas*: Michel Labonne, Martine Padilla (CIHEAM, IAM Montpellier), Michel Dubost (Icalpe), Grigori Lazarev, Luca Fé d'Ostiani, Jean de Montgolfier, Bernard Roux (INRA, France). *Tourism*: Ghislain Dubois (TEC-Conseil, France) *Millennium Objectives*: Meriem Houzir

Peer review of the chapters. *Energy*: Samir Allal (University of Versailles, France), Mohamed Berdai (Centre de Développement des Énergies Renouvelables, Maroc), Laurent Dittrick (International Energy Agency), Rabia Ferroukhi (OME, France), Fatiha Habbèche (Agence de Promotion et de Rationalisation de l'énergie, Algeria), Manfred Hafner (OME, France), Stéphane Pouffary, Philippe Beutin, Bernard Cornut, Nicolas Dyèvre and Robert Angioletti (Ademe, France), Néji Amaimia (Agence Nationale des Énergies Renouvelables, Tunisia), Adel Mourtada (Association Libanaise pour la Maîtrise de l'Énergie et de l'Environnement, Lebanon), M. Soliman (Centre d'efficacité énergétique, Egypt). *Water*: Mohamed Ennabli (Tunisia), Jean Margat (France). *Transport*: Lilia Khodjet El Khil (REMPEC, Malta), Jacques Molinari (GIR Maralpin), Farès Boubakour (Université de Batna, Algeria), Rami Semaan (Sitram-Consultants, Lebanon), Yücel Candemir (Istanbul Technical University). *Urban areas*: Serge Antoine (Comité 21, France), Claude Chaline (Institut d'Urbanisme de Paris), Omar El-Hosseiny (Ain Shams University, Cairo), Henda Gafsi (Programme de gestion urbaine, Tunisie), Rușen Keleș (Ankara University), Joan Parpal (Medcities, Barcelone), Jacques Theys (Centre de Prospective du Ministère français de l'Equipement), Farès Boubakour (Université de Batna), Gabriel Jourdan (GIR Maralpin), Chebbi Morched (Urbaconsult, Tunis). *Rural areas and agriculture*: Gérard Ghersi, Mahmoud Allaya, Omar Bessaoud (CIHEAM, IAMM), Frédéric Devé (FAO), Denis Groëné, Ali Mhiri (INATunis). *Coastal areas*: Lucien Chabason (MAP, Athens), the directors of MAP components, the focal points of PAP and Blue Plan, Fouad Abousamra (MEDPOL, Athens), Chedly Rais (SPA RAC, Tunisia), Pierre Bougeant (Conservatoire de l'Espace Littoral et des Rivages Lacustres, France), Mahfoud Ghezali (Université de Lille, France), Mohamed Larid (Institut des Sciences de la Mer et de l'Aménagement du Littoral, Algeria), Adalberto Vallega (University of Genoa, Italy), Alain Bonzon (FAO/CGPM), Denis Lacroix (IFREMER, France). *Reading of the English draft*: Erdal Özhan (Medcoast), M. Scoullos (MIO-ECSDE)

Blue Plan Focal Points

The Blue Plan Focal Points in Albania, Algeria, Bosnia-Herzegovina, Croatia, Cyprus, Egypt, France, Greece, Israel, Italy, Lebanon, Libya, Malta, Monaco, Morocco, Serbia and Montenegro, Slovenia, Spain, Syria, Tunisia, Turkey and the European Commission were consulted on the draft report.

Photo credits

Bernard Glass, Florence Pintus (Blue Plan), Christian Perrochon, PMA's collection (www.marseille-port.fr)

Project funding

Mediterranean Action Plan (the bordering countries and the European Commission), European Commission (GD EuropeAid), European Environment Agency, France (Ministries in charge of the Environment and of Agriculture).

Introduction

The Mediterranean is an original and unique ecoregion because of its geographical and historical characteristics, its natural and cultural heritage and the feeling shared by its peoples of belonging to the 'Mediterranean world'. It cannot be simply defined. Fernand Braudel described it as '…a thousand things at the same time. Not just a landscape, but countless landscapes. Not just a sea, but a string of seas. Not just a civilisation, but many civilisations… The Mediterranean is an age-old crossroads. For thousands of years, everything has converged on this sea, disturbing and enriching its history…'

At the crossroads of three continents, the Mediterranean is also a north–south fracture zone, an arena for multiple international exchanges of strategic importance. Because of its special characteristics – a pattern of development highly conditioned by the natural environment, a region that brings together countries at different levels of economic and social development that share a joint heritage – it is a perfect illustration of the global problems of sustainable development. Will the region be able to show the way to a pattern of development that brings people together, which is more balanced and respectful of a heritage to be passed on to future generations? Or will it fall into an inequitable and short-term pattern that squanders the resources it has inherited? It all depends on whether it is destined to become a model for the regional regulation of globalization or to reinforce global instability.

The Blue Plan[1] in 1989 already highlighted the risks of a growing divide between the north and the south of the Basin and of continuing and sometimes irreversible environmental degradation. It pointed the way to a more equitable pattern of development, one more respectful of the environment, which would integrate development with the environment, and strengthen the institutional capabilities of states and cooperation between north and south and between southern countries. It already contained the principles of sustainable development, a concept that subsequently emerged as the search for a developmental mode that tries 'to meet present needs without compromising the ability of future generations to meet their own needs'. Today, has this long-term vision been realized? Have we followed the paths it mapped out?

This report constitutes an indispensable tool for each and every coastal country, the European Union and all those trying to build a Euro-Mediterranean zone of stability and shared prosperity.[2] It presents a new analysis of the dynamics at work in the Mediterranean area, linked as far as possible with the social dimension. The approach highlights the relationships between development and the environment and focuses on the strategic priorities for the region. Six issues are analysed: water, energy, transport, urban areas, agriculture and rural areas, and coastal areas. All are subject to public policies and social practices that call for major changes if we want to maintain our vast natural capital, reduce risks and disparities, and get a genuine economic development process under way.

The scarcity and irregularity of *water* resources and the wide range of *energy* resources in the region require particular attention if the needs of a growing population and economy are to be met while preserving resources and avoiding crises.

The *transport* sector, which is inseparable from energy issues, urban sprawl and the spatial distribution of activities, is growing more rapidly than gross domestic product. How can this demand for mobility, which is being increased by economic liberalization and changing lifestyles, be met while minimizing the expected growth of environmental and social impacts?

By 2025, three of every four Mediterranean inhabitants will live in *urban areas*. Lifestyles and consumption patterns change rapidly with increasing urbanization, and urban sprawl and car use become ubiquitous. With shortages of space and financial, human and natural resources in cities, will we find ways and means of avoiding major impacts on the environment and human health?

The very future of *rural areas* is intimately tied to that of cities. These areas, which often suffer from human and economic abandonment, are highly dependent on appropriate management and development to ensure the conservation of natural resources (water, soil and biodiversity) and landscapes, the reduction of some risks (floods and fires) and territorial disparities. In some countries agriculture, vulnerable to the shock of globalization, plays a major role. Will we be able to reinvigorate these areas and avoid their irreversible degradation?

Finally, *coastal areas*, the interface between land and sea, a unique natural and cultural space, are the fundamental and symbolic issue of the Mediterranean's future.

A Sustainable Future for the Mediterranean

They are full of potential for economic development, mainly for tourism, but also subject to every kind of pressure. Will we discover how to enhance them sustainably and succeed, better than in the recent past, in achieving the necessary reconciliation between development and the environment?

To tackle these six issues, the report is structured as follows:

- Part 1 defines the prospective framework of how the Mediterranean area may evolve between now and 2025, by building a 'baseline scenario', which extrapolates the strong current trends while taking into account the major determinants of the future: climate, population, geo-political and economic factors, and regional and national governance.
- Part 2 analyses, for each of the six key issues, the possible environmental and social impacts of the baseline scenario up to 2025, some responses underway in the Mediterranean countries, and paths to alternative scenarios.
- Part 3 summarizes the main findings of the analyses, highlights the overall impacts and risks of ongoing trends up to 2025, and suggests alternative directions for shifting policies and actions towards more sustainable development.

The overall approach is firmly action-oriented. The magnitude of the possible changes is of more interest than the accuracy of the projections. The challenge is to enhance the Mediterranean assets and find a better balance between the improvement of living standards and increased demand for motorized transport, water, energy and land. Although the future is not predictable, the report shows that progress should be possible through approaches that have already proved their worth in the Mediterranean.

The great diversity of the region and its multiple interactions with other regions lead to define the Mediterranean area in different ways, depending on the theme of each chapter. The political reference point remains the countries bordering the Mediterranean that have signed the Barcelona Convention, but, as far as possible, the characteristics of the eco-region will be analysed on scales closer to the bio-climatic region and the coasts – coastal regions, catchment areas, agro-climatic region and coastal settlements.

The time horizon for analysing the past will also vary according to the issues tackled. Changes that have occurred over the past 20 years will be described and related to the long-term trends. Better understanding of what is happening may, however, be gained from looking further back in time. The development of international tourism, for example, has occurred over more than two centuries: starting in Italy towards the end of the 18th century; developing on the French Riviera in the early years of the 19th century; and

Figure 0 Structure of the report

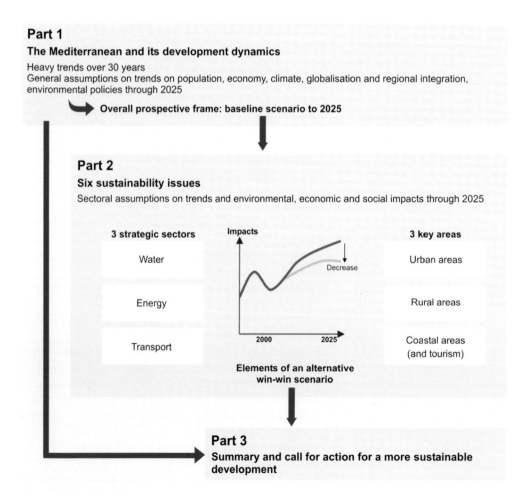

reaching a huge scale on the Spanish, Greek and Croatian coasts in the 1960s. Regions where tourism is in full growth (Turkey, Tunisia, Malta, Cyprus and Egypt) or just emerging (Syria, Libya, Algeria and Albania) could profit from the lessons learned from this long history in order to avoid some of the past mistakes of other Mediterranean regions or benefit from the positive experiences.

The time horizon for exploring the future is 2025. This falls between the long time scales of global population and climate change and the shorter time scales of changes in consumption, production and distribution patterns. It takes a century to grow a forest, dozens of years to eliminate the commonest forms of soil or water pollution, but only a few years to destroy a landscape.

Exploring this future, which is not very far away, can point to paths that can be taken without delay, for the benefit of the Mediterranean of today and tomorrow. The imagination shown by the Mediterranean peoples at all stages of their history proves that such 'changes of scenario' are far from impossible.

Notes

1. Grenon, M. and Batisse, M. (eds) (1989) *Futures of the Mediterranean Basin. The Blue Plan*, Oxford University Press (shortened to *Blue Plan 89* for the rest of this book), also published in French by Economica, in Arabic by Edifra, in Spanish by the Spanish Ministry of Public Works, and in Turkish by the Turkish Ministry of Environment.
2. It responds to a request from all the Mediterranean coastal countries and the European Union, that is the Contracting Parties to the Barcelona Convention 'for the Protection of the Marine Environment and the Coastal Region of the Mediterranean'.

Part 1

The Mediterranean and its Development Dynamics

Section 1
The Mediterranean Region: A Unique But Neglected Heritage

The Mediterranean area, because of its geography and its history, is an exceptional region. Its features, uniqueness and permanencies (the sea, the climate, the relief, biodiversity, its populations and its landscapes), which have been described so many times in literature, together encapsulate the 'Mediterranean world'. They are briefly reviewed here.

Following on from this, the most evident signs of unsustainable development in the Mediterranean will be dealt with, by identifying the major geo political, socio-economic, spatial and environmental trends in the last 20 to 30 years.

A unique heritage

Definitions of the Mediterranean area

The Mediterranean area can be defined taking into account several dimensions (climate, vegetation, bio-diversity, culture, etc.). According to the dimensions dealt with in this report, it will be described at the various levels shown in Figure 1.

The first level (given the code N1 in the illustrations) comprises the 22 countries and territories bordering the Mediterranean (Table 1). Although, strictly speaking, this is a wider area than the Mediterranean bio-geographical region (defined on the basis of climate and vegetation), it is at this level that the institutional framework, the sectoral and economic policies and the directions of regional cooperation are defined, with all the many consequences for the region. Information about current trends is more easily accessible (long term series of statistics). At this first level (N1), the Mediterranean countries and territories occupy 8.8 million km², or 5.7 per cent of the land area of the globe, and have 427 million inhabitants, 7 per cent of the world's population in 2000 (see Statistical Annex). Four demographic 'heavyweight' countries contain 58 per cent of the total: Turkey and Egypt (66 million inhabitants each), France (59 million) and Italy (57 million). For the purposes of analysis, the countries will occasionally be brought together in continents: the north Mediterranean countries (NMC) in Europe, and the southern and eastern Mediterranean countries (SEMC) of Africa and Asia, respectively, with regional subsets if necessary (Table 1). These groupings are by nature arguable. On the economic level, for example, Israel would be part of the northern Mediterranean countries and Albania of the 'southern' countries. Other groupings will be used if necessary. Turkey, for example, will sometimes be attached to the countries of the northern shore.

But whenever possible, the Mediterranean specificities are better illustrated at a second level (given the N3 code), closer to the Mediterranean eco-region and defined by the 234 coastal regions of the Mediterranean (administrative units of the NUTS[1] 3 level or equivalent of 'départements', 'willayas' or provinces, see Statistical Annex). The coastal region level (N3) thus defined represents 12 per cent of the total surface area of the countries and contains 33 per cent of their total population, 143 million permanent inhabitants in 2000. Some countries, such as Libya, Israel, Lebanon, Greece, Monaco, Cyprus and Malta, are very 'Mediterranean' from this point of view because their populations and their activities are concentrated in the Mediterranean coastal regions. This is less the case for Italy where the economic heart, the valley of the Po, is turned more towards Europe than the Mediterranean, and for Morocco, Spain and Turkey, which are continental countries and Mediterranean mainly because of their

A Sustainable Future for the Mediterranean

Figure 1 **A multi-dimensional Mediterranean region**

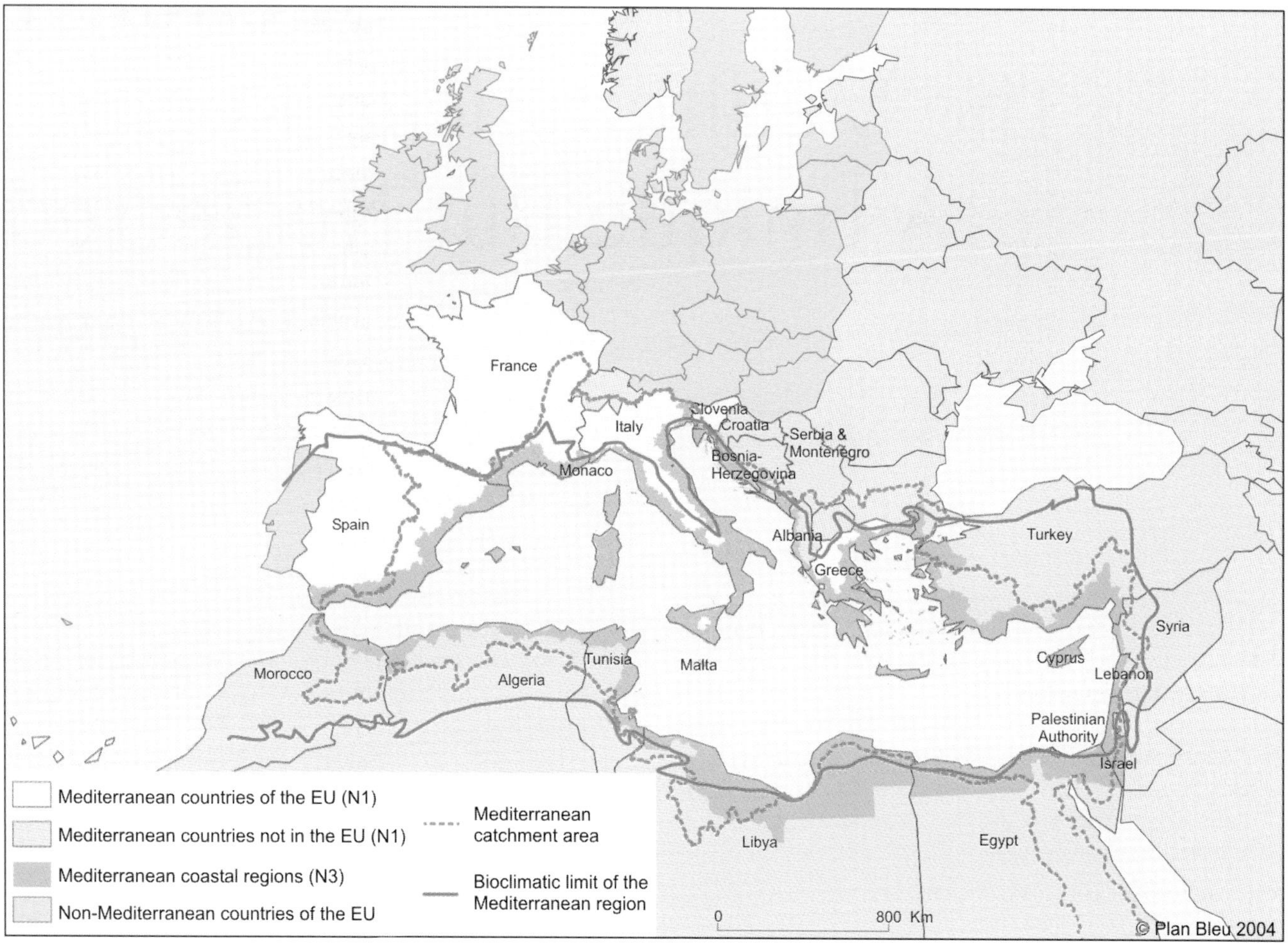

- Mediterranean countries of the EU (N1)
- Mediterranean countries not in the EU (N1)
- Mediterranean coastal regions (N3)
- Non-Mediterranean countries of the EU
- ---- Mediterranean catchment area
- — Bioclimatic limit of the Mediterranean region

0   800 Km

© Plan Bleu 2004

Source: Plan Bleu

climate and vegetation, but open to other seas (the Atlantic, the Marmara Sea, the Black Sea). It is also less true of Egypt and Syria, with huge arid areas where human settlements are mainly organized around fertile valleys, oases and continental routes. It is even less true of France and Croatia (despite the importance of their Mediterranean coasts) and Slovenia, Bosnia-Herzegovina and Serbia-Montenegro, which to a large extent belong to non-Mediterranean temperate Europe.

For the issues dealt with in Part 2 of the report, other geographic levels will be used:

- Chapter 5 on Rural areas deals with an area close to the bio-climatic region (extended to arid regions);
- Chapter 1 on Water refers to catchment areas formed by rivers watersheds (level NV) in the region;
- a more accurate approach is needed for a better understanding of the changes on the terrestrial and maritime Mediterranean coast, focused on a narrow coastal strip.

Box 1 Share of the 'Mediterranean coastal regions' population in the countries, 2000 (N3, N1)

The proportion of inhabitants in the coastal regions (N3/N1) varies from country to country:

- some countries have more than 80 per cent of their total population in the coastal regions: Greece, Israel, Libya, Malta, Cyprus, Lebanon and Monaco;
- others have between 60 and 70 per cent: Tunisia and Italy (which alone has a quarter of the total Mediterranean coastal population);
- around 40 per cent: Spain, Algeria, Croatia, Egypt, Albania and Palestinian Territories;
- less than 20 per cent: Turkey, Morocco, Syria, Slovenia, Serbia-Montenegro, Bosnia-Herzegovina and France.

The report will also take into account the interactions of the Mediterranean with other areas such as the European Union (EU), with which interactions are extensive

The Mediterranean Region: A Unique But Neglected Heritage

Table 1 List of Mediterranean countries and their abbreviations, N1

Group	Shore	Sub-group	Country	ISO code
Mediterranean (MED)	Northern Mediterranean Countries (NMC)	4 EU-Med[a]	Spain	ES
			France	FR
			Italy	IT
			Greece	GR
	Northern shore	Monaco	Monaco	MC
		Islands	Malta	MT
			Cyprus	CY
			Slovenia	SI
			Croatia	HR
		Eastern Adriatic Countries (EAC)	Bosnia-Herzegovina	BA
			Serbia and Montenegro	CS
			Albania	AL
	Eastern shore	Turkey	Turkey	TR
	Southern and Eastern Mediterranean Countries (SEMC)	Israel and Palestinian Territories	Israel	IL
			Palestinian Territories	PS
		Mashrek	Syria	SY
			Lebanon	LB
			Egypt	EG
	Southern shore	Maghreb	Libya	LY
			Tunisia	TN
			Algeria	DZ
			Morocco	MA

Note: a The four riparian countries of the European Union before the integration in 2004 of Cyprus, Malta and Slovenia.

and the destiny closely linked. Figure 1 shows the superposition of these multiple areas (N1, N3, NV and bio-climatic region), which together make up, or interact with the Mediterranean area.

Population growth in the south and east

At the N1 country level (all countries together), the area has seen spectacular growth in population, from 285 million in 1970 to 428 million in 2000. The average growth rate was 1.36 per cent per year, equivalent to adding a country the current size of Bosnia-Herzegovina each year (Figure 2 and Statistical Annex). This is slightly less than the average world annual growth rate in the same period (1.7 per cent), so the share of the Mediterranean in the world population remains quite stable (7 per cent in 2000). Population growth is essentially in the SEMC (all the riparian countries from Morocco to Turkey) where, with 3.9 million more inhabitants per year, there was a record growth rate of 2.35 per cent per year between 1970 and 2000. This was five times higher than in NMC (0.45 per cent per year on average) during the same period. Thus, since 1990, the population in the SEMC has overtaken that in the NMC. Growth in the south was overestimated by *Plan Bleu* 89 (especially in Lebanon and Syria) since there has been a faster than predicted fall in fertility rate. The population growth forecasts, updated by *Plan Bleu*[2] in 2001 show that the total population in the Mediterranean countries could reach 523 million by 2025, an average increase of 1.32 per cent per year, or 96 million more in 25 years (of which 92 million would be in the SEMC).

At the N3 coastal region level, the population increased from 95 million in 1970 to 143 million in 2000. The increase (48 million in 30 years) was mostly (80 per cent) in the SEMC. The changes were underestimated by *Plan Bleu* 89. The population growth rates in the Mediterranean coastal regions between 1985 and 2000 were, in most countries, higher than the maximum of the ranges in the 1989 scenarios. The forecasts for the coastal regions, updated in 2001, show that the population of these regions could reach 174 million by 2025, an average increase of 0.8 per cent per year (31 million in 25 years), mainly in the SEMC.

A Sustainable Future for the Mediterranean

Figure 2 Population of countries and coastal regions, 1970–2025

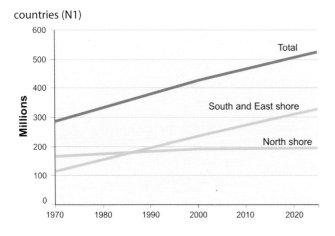

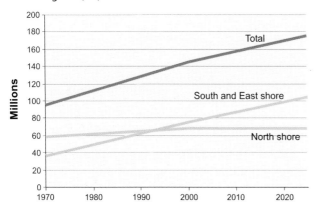

Source: *Plan Bleu*, Attané and Courbage, 2001

The sea and trade

The sea is at the heart of the Mediterranean eco-region. It belongs to all the Mediterranean people; it has fashioned their history and is a natural link between them.

Partly enclosed by the straits of Gibraltar and the Dardanelles, which make possible water renewal because of heavy evaporation, the Mediterranean sea covers only 26 million km², 0.8 per cent of the total surface area of the oceans. With small tides, the sea is as deep as an ocean (3700m in the Tyrrhenian sea and 4900m in the Ionian sea). It is fragmented into a 'complexity of seas' each possessing different biocenose and histories. The Sicily shelf that links Sicily to Tunisia at less than 400m in depth divides it into two and separates the western from the eastern basin. From the Gibraltar strait, the main natural entrance for water in the Mediterranean, the marine circulation makes a large cyclonic movement to the east that is divided into autonomous circuits and returns in deep currents to the west at the end of a 100-year water renewal cycle. These straits are the third busiest in the world (with 240 ships a day), and are therefore particularly at risk of pollution.

The sea shelters a very varied living world. The marine fauna, including 600 species of fish, benefits from the diversity of the nature at the bottom of the sea. However, it is not very abundant because of poor water productivity (narrowness of the continental shelves, low input of organic matter). The coastal zone, which is the main area of primary production for the food chain, concentrates fauna and flora in a limited and particularly vulnerable area. Human pressures on the coasts threaten many species (turtles, monk seals, gruper) and very valuable habitats. The Mediterranean, and especially the narrow areas between the northern and southern shores (Gibraltar, Sardinia, the strait of Sicily, Crete, Cyprus, the Dardanelles), is on one of the main migration routes for terrestrial avifauna between Europe and Africa.

The sea has not always been the natural link between land and people, although it has often been described as such. The Mediterranean region, despite the building of great empires, was for many centuries divided into autonomous areas and it took a long time for navigation to become safe. However, maritime links have made the Mediterranean Basin become an historical area that has been essential for human trade to such an extent that Fernand Braudel described it as the prototype of a 'world economy', with its simultaneously diverse and unified nature and its heritage of civilization. This 'world economy' culminated in the 16th century by associating, in a relationship of conflictual complementarity, the great empires and powers of the time as well as their populations. Since that time, the emergence of the pole position of the Atlantic, and later Asia, have largely removed that advantage.

The sea remains a fundamental basis for trade at both regional and world levels, especially since the opening of the Suez Canal in 1869. As well as manufactured products and food for the southern economy, mainly hydrocarbons are traded. Some countries export oil and gas (Algeria and Libya), while others import them. But the main flows are from the Gulf countries and Asia to Europe and America. This transport of hydrocarbons gives a geo-strategic dimension to the region and also represents a danger to the marine and coastal environment.

An exceptional, but fragile, natural area

The Mediterranean bio-climatic area: Drought and floods

The Mediterranean is also a specific bio-climatic area characterized by a summer water deficit. The 'Mediterranean' climate is defined by warm and dry summers, forcing a phase of xeric stress to the vegetation lasting between 40 days (sub-Mediterranean limit) and 200 days

The Mediterranean Region: A Unique But Neglected Heritage

Figure 3 The bio-climatic features of the Mediterranean

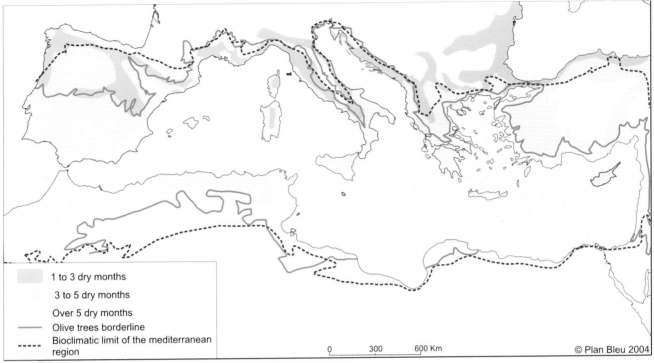

Source: Lemée, G. (1967) *Précis de Biogéographie*. Masson, Paris; Desfontaines, P. (1975) *La Méditerranée, Géographie Régionale*. Gallimard, Paris; Quézel, P. and Medail, F. (2003) *Ecologie et Biogéographie des forêts du Bassin Méditerranéen*. Elsevier, Paris

(sub-desert limit). This drought obliges the vegetation to adapt (xerophytic characteristics). A delimitation of the Mediterranean bio-climatic area based on climate (temperature, precipitation) and vegetation criteria is suggested in Figure 3. In the south, it is generally based on the 100mm isohyet (locally 150mm), below which Saharan conditions are found. In the north and east, the boundaries are set according to bio-geographic criteria and can vary depending on the author. Defined this way, the area is narrow in the east Adriatic and leaves room for the desert in most of the south-east quadrant of the Basin (Libya and Egypt) and for a non-Mediterranean temperate climate from Ancona to Trieste. However, the area covers most of the countries such as Morocco, Spain, Greece and Lebanon. It includes the natural area of olive trees, the emblematic species of the Mediterranean.

Summer drought is accentuated in the south by the great irregularity of rainfall and by aridity, which may increase as a result of global warming. It is a major problem for vegetation, agriculture and societies, and explains the fundamental importance of water in the region and the scale of the efforts by successive generations for the storage, transport and use of water. The Mediterranean countries have 7 per cent of the world's population but only 3 per cent of the world's water resources and more than half of the population of the planet that lacks water, defined as having less than 1000m^3 of natural renewable water resources per capita per year.

Maximum rainfall is recorded in the winter and autumn in the north-west of the basin. The irregularity of rainfall is also the cause of violent downpours in the whole basin (except in Egypt and Libya) and floods that are often sudden and catastrophic.

A *rugged relief*

The Mediterranean, at the crossroads of the African and European tectonic plates, has a young relief (alpine folds) displaying two contrasting forms. In the southeast of the basin, in Libya and Egypt, the Sahara table reaches the sea directly, interrupted only by the Nile valley and delta, the Nile being the only tropical river that flows into the Mediterranean. Everywhere else, from Mount Lebanon to the Taurus and to the Pindus, the Dinaric Alps to the Apennines and the southern Alps, the Cevennes to the Pyrenees and the Sierra Nevada, and the Rif to Kabylia, the coastal regions are dominated by mountains.

The Mediterranean area is composed of multiple cramped, steep and contrasting topographic units. Its interpenetration with the sea explains the large number of islands (Cyprus, Malta, the Greek islands, the islands

A Sustainable Future for the Mediterranean

Figure 4 The Mediterranean, a fragmented area

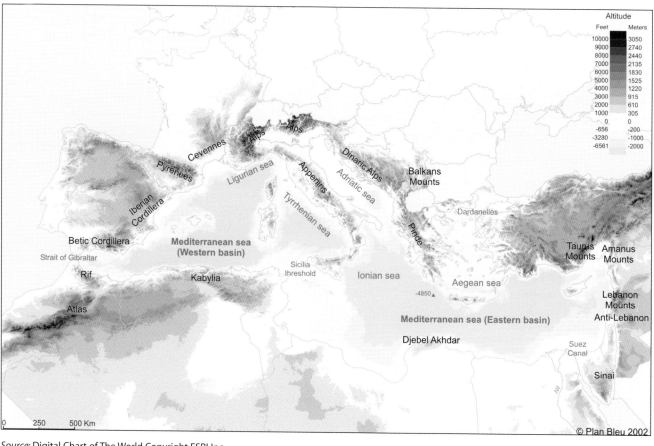

Source: Digital Chart of The World Copyright ESRI Inc

Note: The Mediterranean is a fragmented area: a complex of seas, three maritime gateways, 46,000km of coastline and many islands; an area 'eaten up' by mountains.

of the East Adriatic and of the western basin) and the indented morphology of many coasts. A multitude of harbours and ports are narrowly inserted between water and rock. The isolation of the islands worsens the scarcity of resources such as water. Some islands that survive with difficulty on fishing and agriculture have managed to find economic and strategic 'openings' and become privileged tourist destinations.

The domination of the mountains going down into the sea reduces the size of the watersheds. Only four large rivers, the Po, the Ebro, the Rhone and the Nile go far inland. Terrestrial transport is therefore difficult. This partly explains the low rate of industrial development in the region, although the difficulty in circulation imposed by the relief is gradually being overcome by costly infrastructures such as tunnels and bridges.

Except for the four large river basins, the plains are generally narrow, and vast agricultural regions are scarce. Once they have been developed, rehabilitated and irrigated, they are of great value to agriculture, but their conquest was long and difficult. This agricultural land is now threatened by the sprawl of the towns that it used to supply.

Mountains play a fundamental role in the equilibrium of the region. They receive rainfall and feed the rivers. Their vegetation and development (terraces, small hydraulic systems) retain the soil and regulate the water cycle. Many communities have thrived in these areas, but only after extreme efforts to develop them. Indeed, it is these mountain communities that have supplied manpower to the cities and the coasts. Mountain dwellers live in difficult natural conditions and get few rewards for their labour. These once worked areas are therefore often abandoned, leading to an increase in forest fires and a widening of regional imbalances between coastal zones and inland areas. But when there are few prospects for emigration, and poverty and strategies for survival prevail, mountain populations can be the source of overexploitation of resources and degradation of land and vegetation (erosion and desertification), of which they are the first victims, and which increases the extent of downstream floods.

The Mediterranean Region: A Unique But Neglected Heritage

The fertility of Mediterranean soils is often limited by dryness, shallowness and sensitivity to erosion (Figure 5). Many ways of preventing erosion have been developed in the region. Since antiquity, steep land has been arranged in terraces to plant crops. Because of these constraints, only 13 per cent of Mediterranean land is considered fit for agricultural purposes. The land is rich in bases: fluviosoils (young alluvial soils) and luviosoils (terra rossa on hard limestone). However, some of this fertile land is threatened by urban sprawl, infrastructures and the growing risk of degradation (erosion, salinization, pollution, etc.).

At the complicated crossroads of the Eurasian and African tectonic plates, geology also explains the strong seismic activity in Algeria, Italy, the East Adriatic, Greece and Turkey (Figure 6), and the strong volcanic activity in Italy and Greece with disastrous and recurring consequences for human life (Figure 7).

An exceptionally diverse eco-region

A *major worldwide reservoir of biodiversity*

The geographical (climate, land, split-up relief, variable frontiers over time between tropical and temperate zones) and historical specificities of the Mediterranean region make it one of the most original bio-geographical regions in the world from the point of view of biodiversity, but also one of the most threatened.

Its *terrestrial* biodiversity is of great value. It constitutes one of the main reservoir of *plant* biodiversity in the world. Mediterranean flora, which is more or less adapted to drought, has a wealth of about 25,000 species of higher plants, accounting for 10 per cent of known species in the biosphere (on less than 1.6 per cent of total land area). And more than half of these species is *endemic*, that is, peculiar to the region (Figure 8). Thus, the Mediterranean contains 4.3 per cent of the nearly 300,000 species of vascular plants known in the world; thanks to the number of endemic species, the Mediterranean lags just behind the tropical Andes (6.7 per cent of the total of the world's endemic species) and the Sundaland (5 per cent). It is also a major world area for migratory birds: about two thousand million birds, of 150 species, stop over in the Mediterranean wetlands or live there periodically. One of the reasons is the high compartmentalization of habitats because of the relief and the many islands. Moreover, there are many '*relict*' plant associations, established under past climatic and ecological conditions that have not persisted. These plant species are particularly sensitive to any kind of degradation, since they cannot regenerate once they have died out. A*nimal* biodiversity is often equally

Figure 5 Constraints to Mediterranean soil fertility

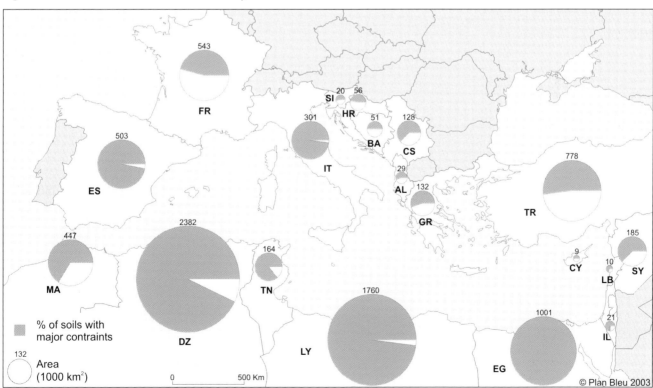

Source: FAO (2000) *Land Resource Potential and Constraints at Regional and Country Levels.* FAO, Rome (World Soil Resources Report 90)

Figure 6 Seisms, probable maximal intensity

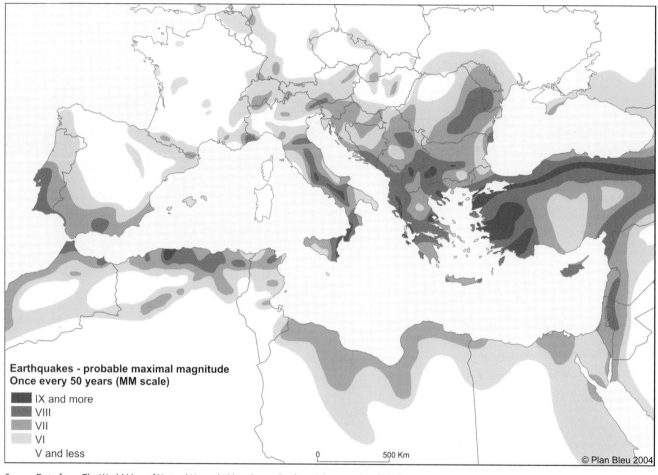

Source: Data from *The World Map of Natural Hazards*; Munchener Ruckversicherung Gesellshaf

important (35 amphibian species of the 62 living in the Mediterranean, are endemic, and 111 of the 179 reptile species),[3] although much less is known about the fauna. The species richness of invertebrates, particularly insects, is significant since this ecosystem is outside the inter-tropical zone. In most cases, areas with high plant and animal endemism appear to coincide.

In this region, civilizations have 'domesticated', or transformed the milieu and 'shaped' landscapes and the environment significantly over a prolonged period. Almost everywhere, the primary vegetation has been replaced by landscapes affected by humans, in some cases degraded, but in others improved, abandoned or re-conquered. Because of these changes, a number of animal and vegetable species have disappeared or are under threat (some Felidae, certain antelopes, a number birds such as birds of prey and limicolous birds). *Agricultural biodiversity*, which has been enriched over the ages (with many varieties of cereal, vegetables and fruit, plus horned cattle, sheep and goats) has turned the Mediterranean into one of the world's eight most important dispersion centres for cultivated plants. This rich genetic heritage is experiencing a remarkable change and is now facing a serious threat as a result of the abandonment of traditional practices.

Marine biodiversity is also both especially rich and endangered. The Mediterranean contains 7 per cent of the world's known marine species in 0.8 per cent of the oceans (in terms of surface area). The ratio of endemic marine species is also very high, often more than 20 per cent (for algae, sponges, Echinodermata, and 50 per cent for the ascidians). This biodiversity is highly concentrated in the sublittoral zone.

The current *threats* to this heritage (climate change, destruction of habitats and species by urban development, infrastructures, pollution and practices or agricultural abandonment, invasive species, tourism and leisure, etc.), combined with its exceptional wealth have made the Mediterranean region one of the world's main *critical hot spots* for terrestrial biodiversity.[4]

The Mediterranean Region: A Unique But Neglected Heritage

Figure 7 Volcanic activity and earthquakes in Mediterranean cities

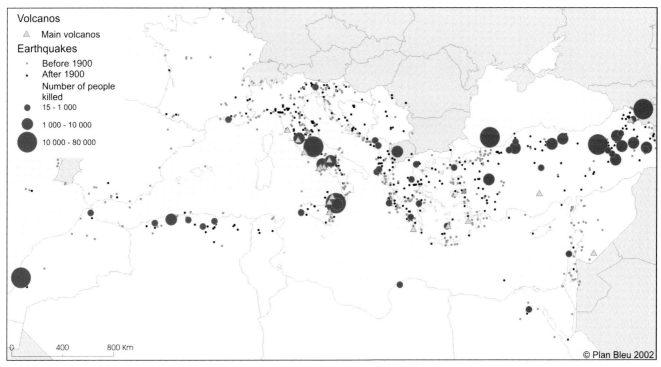

Source: Data from NOAA (www.ngdc.noaa.gov); Weber, C. (1990) *Sismicité en Méditerranée*. BRGM, Paris (modified)

Figure 8 Zones with high levels of endemic plant biodiversity in the Mediterranean bio-climatic area

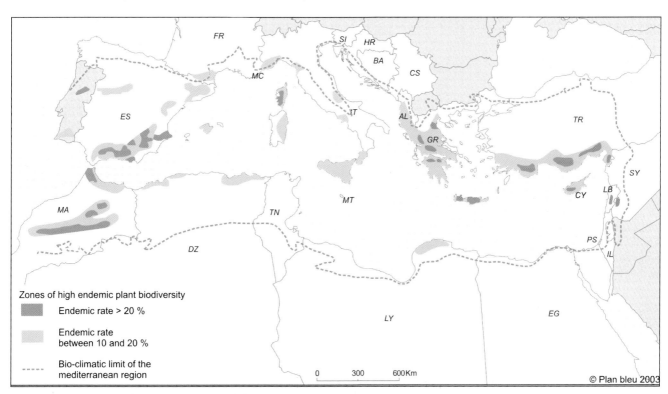

Source: Adapted from Médail and Quezel, in *Annals of the Missouri Botanical Garden*, 84 (1997)

A Sustainable Future for the Mediterranean

Landscapes, a cultural heritage and specific lifestyles

The Mediterranean countries are characterized by a valuable cultural, historical and landscape heritage.

The great majority of Mediterranean towns are extremely old and rich in *historical sites* of exceptional architectural value. The old cities matched the landscape and were adapted to the environment and the climate. Quiet narrow streets and residential areas are combined with public places where animated crowds can gather. Shade and the circulation of air are enhanced. Fountains are often an important component. Today's building, however, seems to be turning its back on what were once recognized as the qualities of towns. Now, sprawling towns, horizontal constructions and shopping centres are gradually altering the Mediterranean area, with serious repercussions in the north as well as in the south.

More generally, all the valuable *Mediterranean landscapes*, coastal, urban or rural landscapes formed over centuries, are now *under threat*. Sometimes mythical and idealized in painting, literature and films, Mediterranean scenery offers inhabitants and visitors the beauty of the sea, the coastlines and the ancient towns, an exceptional luminosity that has inspired many painters and film-directors, the vast horizons of the sea and the mountains, the wise mixture of culture and nature, variety of the relief, and vegetation that differs from that of temperate Europe. To this can be added an incomparable *historical and archaeological heritage*. In 2003, around 191 sites in the Mediterranean countries (N1) were on the UNESCO World Heritage List, of which 86 are in the coastal regions (N3).

Finally, Mediterranean *lifestyles*, characterized by great conviviality and 'sociability' and a special tradition of quality food (taste, health), also appear to be undergoing major change.

Signs of unsustainable development

The Mediterranean region encountered much disruption during the 20th century, which lead to questions about the 'sustainability' of its development. Sustainable development is a worldwide objective. According to a widely agreed definition, it aims to 'meet current needs without compromising the ability of future generations to meet their own needs'.[5]

The concept of sustainable development, already present in *Plan Bleu 89*, requires development to be considered from several points of view. According to the OECD, it refers to promoting a process to facilitate 'the reconciliation of economic, social and environmental objectives of society or, if necessary, to arbitrate between them'. One major challenge is to guarantee the *rights of future generations* to develop their potential at least to the level that we enjoy today, which leads to questions about the *future* impacts of *current* development, particularly when they are irreversible. Another major challenge relates to equity for *current generations*, which leads to the complex questions of poverty, social justice, health and education.

Many signs of the poor sustainability of Mediterranean development, already identified in *Plan Bleu 89*, have been confirmed and are increasing. They will be detailed in Part 2 of this report, for each one of the selected issues (water, energy, transport, urban areas, rural areas, coastal areas). But the general past trends and the most evident signs of poor sustainability are reviewed[6] here: the persistence of conflicts, the poor performance of economies, the continuation of social and territorial disparities, and the increases in pressure on the environment (an essential element of support for economic development in this region).

Persistence of conflicts

First of all, it is difficult to talk in a reasonable way about progress in terms of sustainable development when one considers the huge conflicts that have caused bloodshed in the region during recent decades. The collapse of the former Yugoslavia in the 1990s and the resumption of the Israeli–Palestinian conflict in 2000 have marked the recent history of the region. Conflicts in the Mediterranean are solidly anchored in history and culture. Social and economic differences between countries and increasing difficulties of access to natural strategic resources such as water and energy could perpetuate the risk of conflicts.

To the instability in the Near East, which to a large extent conditions the future of the region, many other tensions can be added, for example those between communities within the various countries (separatist factions in Corsica for France, in the Basque country for Spain, Kurdish ones in Turkey, de facto partition in Cyprus). The region has also been indirectly affected by the Gulf wars and the embargo imposed in Iraq (reduction in trade with Syria, Lebanon and Turkey, Mafia-style activities, smuggling, etc.).

These conflicts involve considerable *human costs*. In 20 years, more than 500,000 people have been killed in the Mediterranean countries. Millions have been displaced. Direct environmental impacts (bombs launched and accumulated in the Adriatic sea, pollution linked to the destruction of industrial sites, the destruction of forested areas, etc.) are high and often persistent, even if often ignored. Conflicts destroy societies for long periods: the displacement of populations that increases migration flows, the capture of land, the destruction of local solidarity, long-term malfunctioning of institutions, and the justification of exceptionally authoritarian regimes, etc.

The Mediterranean Region: A Unique But Neglected Heritage

These conflicts also act to monopolise important financial resources, particularly in countries on the eastern shore where *military expenditure* accounted for 5.9 per cent of GDP, on average, between 1988 and 2002, compared with 2.7 per cent for all Mediterranean countries. It reached more than 9 per cent of GDP in Israel, 6.8 per cent in Syria, more than 5 per cent in Croatia (1992–2002) and in the Lebanon, and more than 4 per cent in Turkey and Greece. In comparison, it accounted for 2.3 per cent of GDP in East and Pacific Asia, 1.2 per cent in Latin America and 2.4 per cent worldwide. Even if total Mediterranean military expenditure has decreased slightly in absolute value since the second half of the 1990s and at the beginning of the new millennium, some countries still devote a significant part of their public funds to it: one-quarter in Syria in 1999, around one-sixth in Greece in 1998 and in Israel in 2001.[7] Such a mobilization of public (national and international) funds linked to the many conflicts in the Mediterranean put a severe strain on budgets that could be allocated to action to promote sustainable development.[8]

Finally, these conflicts slow progress towards regional cooperation, the importance of which will be seen throughout this report.

Poor economic performance, persistent north–south gap and high unemployment

The Mediterranean has stabilized its macro-economic balances, but shows slow growth and has not managed to reduce the gaps in wealth between its shores.

Slower growth

The Mediterranean recorded slower growth than the world average between 1970 and 2002. Indeed, during that period, real GDP in the Mediterranean[9] increased from nearly US$1,900 thousand million to just over US$4,500 thousand million ($1995), an average of 2.7 per cent per year. This is below the rate of growth in other emerging regions (7.2 per cent in East and Pacific Asia, 3.2 per cent in Latin America, 3.1 per cent worldwide). Thus, the Mediterranean share of global GDP has been falling steadily for 30 years.

Mediterranean growth itself fell steadily between 1970 and 2002, from 3.7 per cent on average during the 1970s, to 2.6 per cent during the 1980s and 2.1 per cent between 1990 and 2002. This is in contrast with southern Asia, in particular, where growth has been accelerating from one decade to the next for the past 30 years.

The decrease in economic growth is a common feature of the four EU-Med countries and the SEMC. However, these overall results hide great differences (see Statistical Annex). The four EU-Med countries, with 90 per cent of the total GDP in the Mediterranean region, experienced rather lower annual growth between 1985 and 2000 (2.3 per cent per year) than the Plan Bleu 89 projections (Spain stands out with 3.3 per cent). But GDP growth in the SEMC was greater (4.1 per cent per year), at the higher end of projections of the 1989 Plan Bleu scenarios. However, for the SEMC, as a result of population growth, GDP growth *per capita* fell between 1985 and 2002 compared with 1970–1985 (Figure 9). Thus, over the 1985–2002 period, average GDP growth per capita was higher in the four EU-Med than in the SEMC.

Figure 9 GDP growth rates, 1970–2002

Total GDP growth rates per period with constant prices (annual average rates)

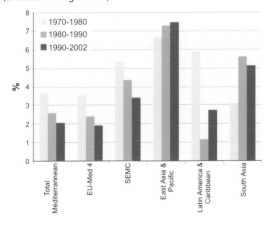

GDP growth rates per capita (PPP constant) per period

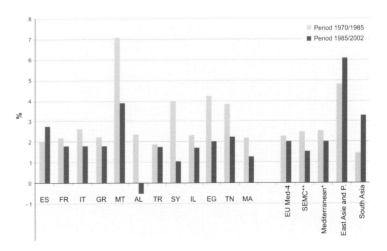

Source: World Bank, 2004

*   Cyprus, Monaco, Palestinian Territories, Lebanon, Libya and East Adriatic countries not included.
** Palestinian Territories, Lebanon and Libya not included.

A Sustainable Future for the Mediterranean

Figure 10 Income gaps: SEMC income per capita compared with four EU-Med average income (GDP per capita PPP)

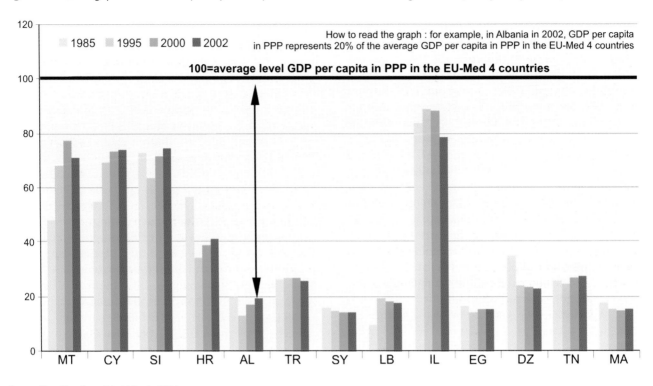

Source: Blue Plan from World Bank, 2004

Note: Croatia and Slovenia: 1990 and 1993 instead of 1985. Lebanon: 1989 instead of 1985. Cyprus: 2001 instead of 2002.

North–south income gaps not reducing

Thus the gaps in per capita income between the north, and the south and east have not been reduced. Only four countries got near to the average level of the four EU-Med countries: Cyprus, Israel (decreasing since 2000), Malta and Slovenia, but these make up only 2 per cent of the population of the Mediterranean countries. When comparing average income in all the Mediterranean countries (Figure 10) with average income in the four EU-Med countries over the past 15 years, little progress is seen, indeed the reverse trend appears, with the gaps increasing slightly in some countries. Turkey and Tunisia have a standard of living equivalent to one-third of that in the four EU-Med countries; in Egypt, Lebanon, Morocco, Albania and Syria, it remains less than a fifth. A particularly worrying increase in the income gap between 1985 and 2002 can be seen in Algeria. On the other hand, after strong falls, Albania (now enjoying once again a rank previously achieved 15 years ago) and Croatia are improving.

Very high unemployment rates

Unemployment has become a major concern for societies and governments, in the north as well as in the south. It represents a strong disparity between those with access to a job and those who are excluded. All the Mediterranean countries have experienced severe increases in unemployment since 1980 (see Statistical Annex).

In the four EU-Med countries (Spain, France, Italy, Greece), the unemployed accounted for more than 9 per cent of the active population in 2002. Spain was most badly affected with 11.4 per cent in 2002, and a peak of 24.1 per cent in 1994.

In the SEMC, unemployment grew faster than in any other region of the world, due to a combination of several factors: slower growth in public employment, privatization, increase in population, agricultural modernization, a decrease in opportunities for migration, and insufficient economic growth. Unemployment rates in this area are among the highest in the world, with an average of 15–20 per cent of the working population out of work.

In 2002, unemployment reached 18 per cent of the active population in Morocco, and 12 per cent in Syria. It reached 14 per cent in Tunisia in 2003. Algeria and the Palestinian Territories experienced the highest rate with almost 30 per cent in 2001; Egypt went from 5 to 9 per cent unemployed between 1980 and 2001. The figure is

The Mediterranean Region: A Unique But Neglected Heritage

more than 24 per cent in Serbia-Montenegro and approached 15 per cent in Croatia and Albania in 2002. Only Turkey experienced a fall in unemployment between 1980 and 2000 but it increased again in 2002 to 10 per cent of the active population.

Where work precariousness (insecurity) exists, a disparity occurs between workers who profit from welfare benefits and those in the informal sector who are excluded. In the SEMC, the informal sector is estimated to account for between 30 and 40 per cent of the urban work force and has been the main supplier of jobs in recent years.

Exclusion from the world of work often leads to social exclusion and generates poverty. In this respect, *long-term unemployment* (more than 12 months), is especially worrying: in 2002, 58 per cent of the unemployed were long-term unemployed in Italy, 50 per cent in Greece, 40 per cent in Spain and 33 per cent in France.

It is the least skilled persons, the young and women, who are most badly affected. The youth (between 15 and 20 years old) unemployment rate is, everywhere in the region, 15–40 per cent of the active population. It is more than 20 per cent in Spain, Italy and Greece. In many eastern and southern countries, almost half of the unemployed have never had any experience of work.

Economic development based on the exploitation of declining natural resources

The Mediterranean economy remains highly dependent on natural resources (agriculture/water, tourism/coastal areas, residential economy/space, energy/hydrocarbons) that it paradoxically tends to overexploit.

Despite a relative decline, *agriculture* is still a driving sector for the economy in several countries (Figure 11). Agriculture accounts for between 10 and 16 per cent of GDP in all Maghreb countries, 12 per cent in Lebanon, 13 per cent in Turkey, 17 per cent in Egypt, 18 per cent in Bosnia-Herzegovina and up to 23 per cent in Syria and 25 per cent in Albania in 2002 (against 2 per cent for the 15 EU countries in 2001). Being the main water consumption sector (but also a sector consuming marginal lands in the poor countries), agriculture always makes a large contribution to employment (48 per cent in Albania, 46 per cent in Turkey, 36 per cent in Morocco, 33 per cent in Egypt, 28 per cent in Syria, 25 per cent in Tunisia against roughly 4 per cent in the EU-15 in 2000). The agro-food network represents no less than 10 per cent of manufactured production. By far the first sector in terms of water consumption, agriculture is, in several countries, at the root of serious problems of water resource overexploitation and soil degradation.

Figure 11 The social importance of agriculture in 2001

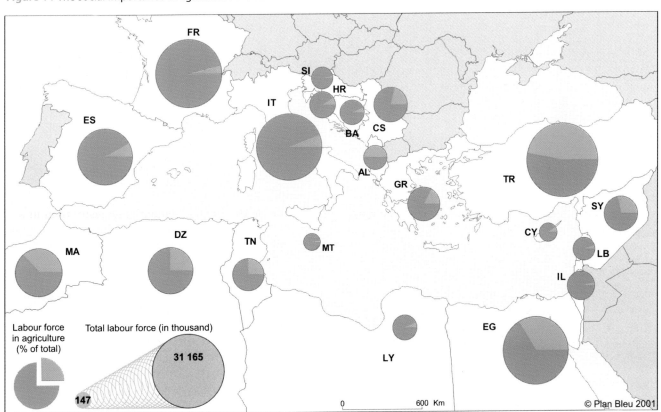

Source: FAO (http://faostat.fao.org)

The share of the *industrial sector* in national economies has tended to increase in the SEMC but has fallen in the four EU-Med countries. It generally accounts for between one-third and one-fifth of GDP and shows signs of gradual diversification. The share of 'low-technology industry based on resources'[10] is still high in the SEMC, and accounts for 61 per cent of the added value in the manufactured sector (Egypt) and 81 per cent (Tunisia) compared with 46 per cent (France) and 69 per cent (Greece).

Four Mediterranean countries remain highly dependent on the *extraction* industry: Algeria with 23 per cent of GDP, Libya with 18 per cent, Syria with 12 per cent and Egypt with 6 per cent.

Overall, in the SEMC, the substitution of technical capital for natural capital in the production process remains rather slow compared with the NMC. All these features clearly affect the export structure of a country. The proportion of *energy product* sales account for more than 95 per cent of Algerian and Libyan export income and 32 per cent of Algerian GDP. Agricultural products retain an important position in total foreign trade (imports + exports): 9 per cent in the Maghreb countries, 19 per cent in the Mashrek countries in 2001. A recent Femise study indicates that several SEMC are specialized at the international level in intensive non-skilled products (particularly the textile and clothing industry) or in natural resources exploitation.[11]

In almost all the Mediterranean countries, the *service* sector has become the largest (Figure 12) as a result of

Figure 12 GDP structure and services structure

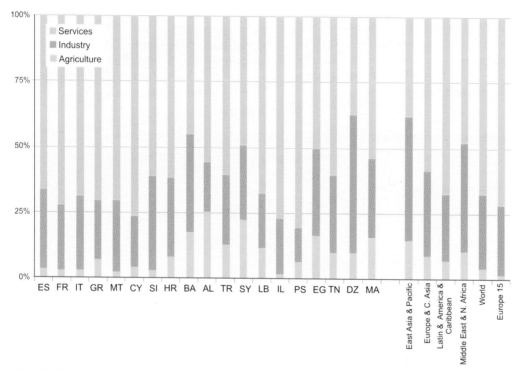

Source: World Bank, 2004

Services structure in 2001 (as % of GDP)							
	TN	SY	MA	LY	LB	EG	DZ
Commerce, rest. and hotels	15.9	12.5	21.1	11.4	28.8	1.5	21.8
Transport, communications	8.4	12.3	7.2	8.2	3.1	8.4	
Finance, insurance, banking	4.1	3.9		2.2	6.2	20.3	
Domestic services	3.3			3.1	7.1	1.9	
Government services	13.0	7.9	13.9	19.8	21.0	7.4	11.4
Other services	1.6	2.3	11.3	2.6	6.7	7.5	

Source: Arab Monetary Fund

The Mediterranean Region: A Unique But Neglected Heritage

tourism-connected activities, which particularly exploit shared public assets such as the sea and the coasts. The services sector generally accounted for more than 50 per cent of GDP in the SEMC in 2002 and more than 60 per cent in the NMC. However, in the SEMC this sector has a dual economic nature with formal major added-value activities (banking, finance, mass marketing) coexisting with informal activities, often at the survival level (door-to-door salesmen, shoe-shiners, car minders, servants, etc.).

Apart from providing considerable employment, the tourism sector constitutes an important source of currency for Mediterranean countries. Between 1990 and 2002, it accounted for approximately 27 per cent of world tourism receipts. For the Mediterranean countries, this is equivalent to 12 per cent of total receipts from the export of goods and services (compared with a 6 per cent world average) (Figure 13). However, receipts from tourism require major investments, often from public sources, and tourism can have negative impacts on the environment.

Tourism receipts have been particularly important for the economic growth of some east Adriatic countries, Lebanon and Cyprus where they were equivalent to between 35 and 53 per cent of the export value of goods and services in 2002. The contribution was also significant in Syria, Malta, Greece, Egypt, Tunisia, Morocco, Turkey and Spain, where the ratio is around 20 per cent on average.

Exploitation and, sometimes, waste of natural resources

A strong dependence on all sorts of natural resources (water and soil for agriculture, coastal zones for tourism, lands for residential economies, hydrocarbons for energy, etc.) contributes to the 'rent economy' and a 'mining-like' exploitation nature of the Mediterranean economy, which is often mentioned as a reason for its weak performance. It should lead to attempts to reconcile development and the environment and result in considerable attention being paid to resource management. However, urban, tourism and agricultural development is often poorly controlled, with uneconomical consumption or use of water, energy, land and coastal zones, as will seen in Part 2 of this report. Strictly speaking, the 'ecological footprint' is not an indicator of sustainable development, but it appraises, in a synthetic way, the pressure exerted by human activities on the environment and natural resources. According to this indicator, the environmental capital of each Mediterranean country is being spent more quickly than it is being renewed (Box 2).

Figure 13 Share of the receipts from international tourists as a percentage of goods and services exports

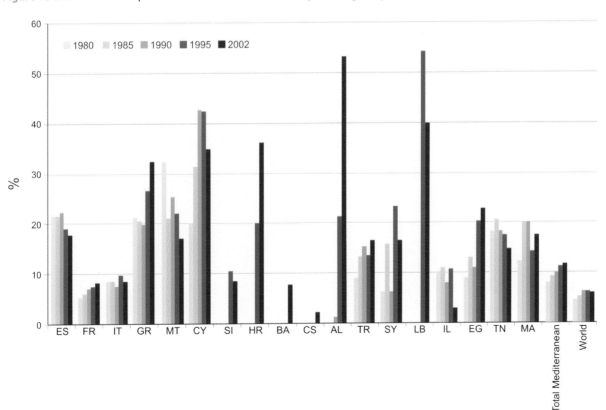

Source: World Tourism Organization, World Bank, 2004

A Sustainable Future for the Mediterranean

Box 2 — Ecological footprint of Mediterranean populations, (N1)

The **ecological footprint** is a measure of the theoretical size of the biologically productive area (of land and water) required by an individual, a city, a country, a region or humankind to produce the resources consumed and absorb the wastes generated, while using technologies and existing management modes.

The ecological footprint gives information on human demand pressures on natural resources. The method of calculation of footprints may be complicated, since it assumes that one can take all the resources and intermediate consumptions involved in the production of goods into account and convert them into area-equivalents. For this reason, the information must be used with care since the methodology is still under debate and collecting reliable environmental data remains difficult.

According to the WWF, the global ecological footprint (for all the world's population) in 2001 was 13.5 thousand million hectares, or 2.2 hectares per capita. This can be compared with the Earth's total bio-capacity, estimated at 11.3 thousand million hectares, one quarter of the Earth's land surface, or 1.8 hectare per capita. In 2001, the human ecological footprint was therefore 21 per cent, or 0.4 hectare per capita, larger than the Earth's bio-capacity.

The **ecological deficit** is the extent to which a population's ecological footprint exceeds the bio-capacity of its territory. When compared with bio-capacity, the ecological deficit is obtained. All the Mediterranean countries, without exception, had an ecological deficit in 2001, which means that the environmental capital of each country is spent faster than it is renewed.

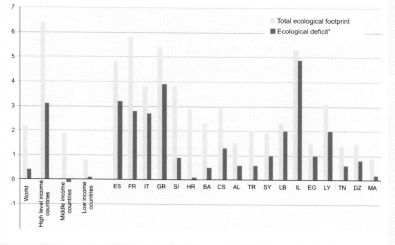

Source: WWF, Living Planet Report, 2004

* difference between the ecological footprint and the bio-capacity.

Insufficient decoupling between economic growth and pressures on environment

The ecological footprint concept takes into account the *eco-efficiency* of development patterns, that is, the consumption of natural resources or emission of pollutants per unit of created wealth.

Indeed, when countries' ecological footprints are compared with their Human Development Index (HDI), which takes account of GDP per capita, life expectancy and education level (Figure 14), the highest HDIs are generally in the countries with the largest ecological footprint per capita.

A given increase in development results in a parallel increase in HDI and ecological footprint per capita. This illustrates that there is no or little decoupling between economic growth and pressures on the environment. The high income countries (the four EU-Med countries, Slovenia and Israel) are further from sustainable development in terms of ecological impact and those countries with intermediate income (SEMC and East-Adriatic except Slovenia) are further away in terms of human development.

Figure 14 Human Development Index (HDI) and ecological footprint

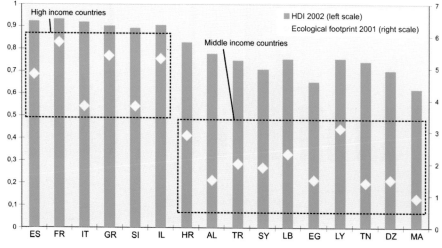

Source: PNUD, WWF

The Mediterranean Region: A Unique But Neglected Heritage

> **Box 3** The concept of decoupling between economic growth and pressure on the environment
>
> Decoupling occurs when the growth rate of environmental pressures is below that of economic activity. Decoupling is absolute when environmental pressure is steady while economic activity grows. It is relative when pressure on the environment is increasing but less rapidly than economic activity.
>
> Decoupling can be measured by comparing an environmental pressure variable (such as CO_2 emissions) with an economic variable (such as GDP). Comparison between an environmental pressure and population change or other variables may also be made.
>
> *Source:* OECD, 2002a

The challenge for the northern countries is therefore to decrease their ecological impact while maintaining a high HDI level, and for southern countries to increase their HDI while retaining a relatively low ecological impact. Examples of successful decoupling are given in Part 2.

Environmental pressure on regional commons: a shared but differentiated responsibility

The NMC, because of their economic activity, consumption patterns and lifestyles, contribute even more to pressures on the environment than the SEMC. For example they are responsible for almost 70 per cent of total CO_2 emissions from the energy and transport sectors in the Mediterranean. The industries of the four EU-Med countries (Spain, France, Italy, Greece) emit 1500 tonnes of liquid discharges per day, measured in terms of Biologic Oxygen Demand, one and a half times more than the 18 other riparian countries together (1000 tonnes per day).

This differentiated responsibility appears even more clearly when comparing indicators per capita. The four EU-Med countries emit twice as much industrial CO_2 and nitrogen oxide and consume three times more energy per capita than the other Mediterranean countries.

Persistant social disparities, despite progress

Other signs of unsustainable development come under the social sphere. In absolute terms, despite progress in the fields of health, education and access to basic facilities, many social inequalities remain both between and within countries (particularly between rural and urban areas). These gaps could widen further in the future and move the region away from sustainable development. Even if the average standard of living, measured in terms of GDP per capita, has improved in nearly all Mediterranean countries, its conversion into 'social well-being' is conspicuously unequal. In the absence of a global 'development' index (Box 4), some large indicators have been chosen to show this on the national scale. Additional analyses of internal differences within the countries would no doubt be extremely useful but remain to be done.

Continuing poverty

Poverty affects all Mediterranean countries. It makes a large part of the population vulnerable by depriving them of access to essential basic services and equipment. It is difficult to measure it because it is a 'relative' concept, and reliable and comparable data are scarce.

In *monetary* terms, *poverty* affects almost 10 per cent of the population of the Mediterranean countries, including the high income ones. The proportion of the population living below the *threshold of absolute poverty* (US$14.4 PPP per capita per day in the developed countries) is 12 per cent in France, 21 per cent in Spain and up to 23 per cent in southern areas of Italy. In the other countries, scarce official statistics indicate between 6 per cent and 18 per cent of the population living on less than US$2 per day (Turkey, Tunisia, Algeria, Morocco) and up to 44 per cent in Egypt. In terms of *relative poverty thresholds* (the value of which, for each country, depends on the standard of living) the proportion is between 8 and 19 per cent. As regards income disparities, for the countries for which UNDP publishes information, there is a distinction between countries with *low income inequality* (Egypt, Italy, France, Spain, Slovenia, Croatia) and those with *large income inequality* (Algeria, Israel, Tunisia, Turkey, Morocco). However, these disparities do not reach the extreme values of some countries in South America, Central America and even subtropical Africa.

In the broadest sense of *human poverty* (a concept that integrates deprivation in health, education, living conditions, etc.), poverty rates are significant. The UNDP human poverty index, the definition of which varies according to the level of development, indicates that between 10 and 12 per cent of the population in Italy, Spain and France, a total of 17 million people, are affected by human poverty in these three countries. In the other countries, the proportion varies between 10 per cent (Lebanon) and 35 per cent (Morocco).

In the absence of long time-sets of data, it is not possible to measure changes over time. However, most analysts agree that the level of poverty in the Mediterranean is not as high as the alarming levels in other parts of the world. Family solidarity helps to lighten it (Box 5), even though there are *locally* very high levels, especially in rural areas. The inequality between social groups, the rise in poverty (linked to the degradation of social services) and the emergence of new forms of urban poverty are more and more worrying for national and international institutions. *Impoverishment* generates both social and political tension. Governments are led to rede-

A Sustainable Future for the Mediterranean

Box 4 — Sustainable development indicators. The example of HDI

GDP and even GDP ppp (in purchasing power parity) is not an adequate way of measuring development level. For example, it does not take into account the cost of much environmental and social damage or social progress. Initial studies have tried to assess the cost of environmental and social degradation and collect various types of information that better account for overall progress in a society, by defining new synthetic indicators, without systematic valuation of the variables in monetary terms. The best known is the UNDP's Human Development Index (HDI), which, in addition to GDP PPP per capita, takes into account the expected life span at birth and the level of education. An HDI of more than 0.8 is deemed as being high, less than 0.5 as low.

This indicator shows that the Mediterranean region is globally quite well placed compared with the rest of the world, with an average HDI, estimated at 0.8 with, nevertheless, many differences between countries:

- nine countries have a high HDI, above 0.8: France (16th out of 177 on a world scale), Spain, Italy, Israel, Greece, Cyprus, Slovenia, Malta and Croatia (48th on a world scale).
- nine have an HDI between 0.7 and 0.8: Libya (58th on a world scale), Albania, Bosnia-Herzegovina, Lebanon, Turkey, Tunisia, the Palestinian Territories, Syria and Algeria (108th on a world scale).
- five have an HDI between 0.6 and 0.7: Egypt and Morocco (120th and 125th on a world scale).

HDI has been increased continuously in absolute terms since 1975 in all Mediterranean countries. However, countries with a lower HDI have fallen by between 10 and 20 places in the world ranking since 1994 and seem to be less able to convert economic growth into social well-being than other regions of the world.

Human Development Index (HDI) of some Mediterranean countries

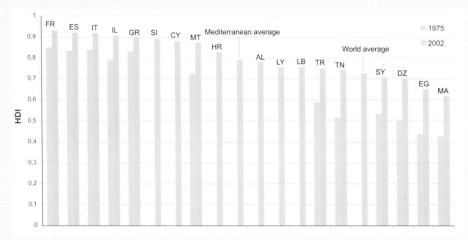

Source: PNUD, 2004

More recently, other indices that integrate social and/or environmental variables have been proposed and calculated in some developed countries. However, there are currently no synthetic indicators of the 'Social Health' type in the Mediterranean that would group together approximately 15 variables such as drug use and suicide among young people, unemployment and health insurance, child poverty, fatal road accidents or violent offences; there are also no indices of 'real progress' or 'sustainable well-being'. This would certainly be a direction to be explored for the Mediterranean countries that have already adopted a common set of 130 indicators for sustainable development in 1999.

---

fine their social and economic policies: education and training, primary and secondary redistribution systems, and more precisely targeted social protection networks. Without this, the 'social contract' will certainly fail, bringing with it more migration to towns or abroad, or, in the south, a strengthening of the power of fundamentalist movements.

Box 5 — Resilience of Mediterranean societies towards impoverishment

The impoverishment observed may lead to less social exclusion than in the north of Europe thanks to the presence of family solidarity networks that are still very powerful. Indeed, the family solidarity networks play a major role and confer a particular 'resilience' to Mediterranean societies.

Thus, in the Mediterranean, poverty is not experienced in loneliness. The proportion of poor persons who live alone is 3.9 per cent in Spain, 6.1 per cent in Italy and 10.5 per cent in Greece, compared with 18 per cent in the United Kingdom and 46 per cent in Denmark. The long-term unemployed are more integrated into social life and benefit from a stronger family solidarity.

This is probably inherited from a certain 'view' of the poor. In the Catholic and Muslim religion, unlike the Protestant one, the poor are considered more as victims than as persons to be blamed. However, this resilience could be affected by a growing pressure due to a possible increase in the ratio of dependents to workers.

Source: Reiffers, 2000

The Mediterranean Region: A Unique But Neglected Heritage

Health, food security and education

In the field of health, signs of degradation with regard to sustainable development can also be seen, even though there has been considerable progress in the SEMC since 1960. *Life expectancy at birth* has increased from 50 to more than 70 years in 2002 and *infant mortality* has been nearly halved in many countries. It has been a remarkable achievement in the context of high population growth.

With regard to *health facilities*, gaps of four to one persist between the four EU-Med and the other Mediterranean countries, for simple indicators such as the number of hospital beds (about eight per 1000 inhabitants compared with two) or the number of doctors per inhabitant. Many SEMC still have less than one doctor per 1000 inhabitants. The lowest health cover is in Morocco and some areas in Algeria.

Another worrying sign is the low level of *public* health expenditure in the SEMC, despite growing needs due to population growth. Public health expenditure was overall less than 5 per cent of GDP in 2002 and the per capita figure is actually falling in some countries (Algeria, Egypt). In Albania, Bosnia-Herzegovina, Morocco and Egypt, expenditure is less than 100 euros/per capita/year. On the other hand, *private* health expenditure has gone up considerably in most Mediterranean countries in the past ten years. In Lebanon there are record levels of spending (more than 8 per cent of GDP in 2002) and also in Algeria. There may be a risk of a slow-down in the progress achieved up to now, as a result of reductions in public health spending and the growing difficulties of access to the health system by poorer people. The return of diseases that had been suppressed (typhoid, cholera, the plague) could be the first signs.

In the very different demographic context of ageing, the NMC are having difficulty controlling public health expenditure, which accounts for a growing proportion of GDP (from 5 per cent to 7 per cent of GDP in 2002). For example, the cost per capita reached US$2000 per year in France in 2002. The emergence of new illnesses linked to changes in lifestyles (sedentary) and food habits, for example obesity and cardio-vascular disease, is a public health concern. Some cancer specialists are becoming alarmed by the doubling of the death rate from cancer since 1945 and attribute it mainly to degradation of the environment. Air pollution may have been responsible for the death of 6500–9500 persons in France in 2002.[12] These concerns, which are also beginning to affect the SEMC, confirm the need to assess lifestyles and production methods, not only from an economic perspective but also in terms of health and environment.

Overall improvements in *food security*, for example average daily caloric input in the Mediterranean increasing to 3340 cal/day[13] (a value close to that of the OECD countries), masks many differences linked to poverty.

In Egypt, Algeria and Morocco, about 6 million people are undernourished, 3–7 per cent of the population. According to the FAO, in Turkey and in the East Adriatic countries there were 1.2 and 1.7 million undernourished people respectively in 2000. In most of the SEMC, between 5 and 20 per cent of children under five years of age still show clinical signs of malnutrition, even if infantile malnutrition has dropped considerably in the past 15 years. The share of food in total household consumption exceeded 40 per cent in Morocco and Egypt, 35 per cent in Tunisia, 25 per cent in Greece, while it was less than 15 per cent in France, Slovenia and Italy.[14] In this context some fringes of the population are especially vulnerable to drought and to increases in the price of basic foodstuffs.

Access to *education* is one requirement for sustainable development. Here too, the SEMC have made considerable efforts to generalize school education at the primary level, within a context of high population growth and budgetary constraints. On the other hand, there are still many differences between countries at higher levels of education. In 2000, some countries had a rate of schooling at secondary level very much lower than the world average, for example Morocco (31 per cent) and Syria (39 per cent). This rate deteriorated considerably (by 16 per cent) between 1990 and 2000. At the higher education level the gap between countries with a high schooling level (more than 50 per cent in the four EU-Med countries) and others such as Albania, Algeria, Morocco and Syria (less than 15 per cent) has grown since the middle of the 1980s.

Public expenditure on education has increased significantly since 1960. This reached an average of US$700 per capita in 2000, 5 per cent of Mediterranean GDP (varying in 2000 between 2.7 per cent in Libya and 7.3 per cent in Israel). The four EU-Med countries, the islands and Israel increased public spending on education per capita threefold between 1980 and 2000, and reached US$1000 per capita in 2000 (current $), while such spending remained, at best, stagnant in Albania and Morocco, or even decreased in Syria and Algeria. Although this indicator does not necessarily reflect the quality of teaching, there is a high risk of an increase in the gaps between countries. The education system in many SEMC is deteriorating (status and remuneration of teachers, staff, etc.) and it is often considered as not well adapted to the job market. According to the World Bank, even though the *illiteracy rate* of people above 15 years of age was halved in the SEMC between 1970 and 2000, from 60 to 30 per cent, the East Asia and Pacific region cut their illiteracy rate by more than four, from 45 to 10 per cent, at a lower cost per capita, showing a relatively greater efficiency.

Shortage of housing and basic facilities in many SEMC

Due to high population growth in the south and the east of the basin, changes in family lifestyles and migration to cities, a large part of the Mediterranean population remains excluded from decent accommodation. The shortage of accommodation is still high in Algeria (a shortage of 6.9 million dwellings), Morocco (5.9 million) and Egypt (9.8 million) (1997–1998 figures). Public efforts, as well as more and more private ones, are insufficient to deal with the growing needs (there is coverage of less than 60 per cent in some countries). Accommodation is often dilapidated because of bad maintenance and 'over-occupation'. Informal housing is continuing to develop in cities (see Chapter 4).

Households with modest incomes are hard hit by the accommodation crisis, since housing for this category is becoming scarce. Accommodation takes a growing part of family savings in the south and east (70 per cent in Tunisia). Although not as acute, a growing number of people in the four EU-Med countries are also experiencing problems with access to accommodation.

Basic facilities (drinking water, sanitation, waste collection, electricity) are often lacking in rural zones and in the unregulated urban housing areas of the SEMC. As many as 27 million Mediterranean people are still without adequate sanitation, 30 million lack access to drinking water and 16 million people have no electricity.

The condition of women

The difference of rights and conditions between men and women is a fundamental aspect of sustainable development.

During the past 40 years there has been undoubted progress in the condition of women, thanks to education and birth control. Women have increasingly important roles in *economic activity*. Everywhere in the northern Mediterranean countries, as well as in Turkey and in Israel, there are more than 66 economically active women for every 100 economically active men compared with fewer than 40 in 1960. In the other Mediterranean countries, the proportion of women in the working population has tended to rise too. However, in the Maghreb and Mashrek countries, this percentage remains lower than in other regions of the world. As the number of people on the job market increases, women become more vulnerable to *unemployment*. The differences between the unemployment rates for men and women are often high: by more than 17 percentage points in Egypt in 2001; in the four EU-Med countries the gap (between 2 and 8 points in 2002–2003) has not stopped rising since 1980. For example, in Spain, in 1980, the unemployment rate for women was 2 points higher than for men; the gap reached 7.7 points in 2003. Women, on average, are also less well paid than men, despite some progress. In France salary gaps between men and women were 25 per cent in 2001. The gap was 36 per cent in 1950, but at this rate women will need another century to find equality. Women are also the most vulnerable within the poor, especially single women.

As far as *education* is concerned, while there has long been no gap between the schooling of girls and boys in the north, large differences remain in most of the SEMC. Despite considerable progress, in 2001–2002 there was still a difference of 13 points between the percentages of girls and boys in primary education in Morocco, more than 8 points in Algeria and 6 points in Egypt. Concerning *literacy*, gaps between men and women can rapidly decrease: from 29 points in Syria in 1999 to no more than 8 points in 2002. In 2002 in Morocco only 38 per cent of female adults were illiterate compared with 61 per cent of the male adult population. Women still play a small role in *political institutions* in all Mediterranean countries. In the NMC, only 9–29 per cent of members of parliament are women (compared with an average of 25 per cent in the OECD countries and 35 per cent in northern Europe); in the SEMC only between 2 and 12 per cent are women; the east-Adriatic countries stand out with more than 15 per cent and Spain with 29 per cent.

Despite undeniable progress, the daily reality for a great number of women belonging to poor classes is still that they are very vulnerable; their rights are often restrained by a public opinion divided by unchanged old traditions or by the more or less radical views of religion. Reforms that would improve their status are found in legal texts that are sometimes ambiguous and slow to enter into force.

Some indicators of social changes

This brief overview of social changes shows real progress in terms of coverage of basic social needs and facilities, resulting from proactive public policies. This is confirmed by an analysis of the change in some millennium follow-up indicators (Box 6). However, many social differences persist between the Mediterranean countries and within countries.

Box 6   The Mediterranean Region and the Millennium Development Goals

The Millennium Development Goals were adopted in September 2000 by the 191 UN Member States. They set targets to 2015 compared with 1990 for improving standards of living. Analysing some of these indicators provides information about strong and weak points in the Mediterranean.

Poverty: There is much less extreme poverty (the percentage of the population living on less than $1 a day) in the Mediterranean (from 0 to 3 per cent depending on the country) than in Asia (16 per cent) and sub-Saharan Africa (45 per cent), but

The Mediterranean Region: A Unique But Neglected Heritage

the percentages shoot up as soon as the $2-a-day threshold is passed (44 per cent of the Egyptians, comparable with southeast Asia). Poverty fell in Tunisia and Turkey but rose in Egypt and Morocco, whereas several regions in Asia have already reached the target of reducing poverty by half between now and 2015.

Youth unemployment: The rate (14.6 per cent) is higher than the average in developed regions. It has risen so much since 1990 that it has now reached record levels in the SEMC (29 per cent on average in northern Africa compared with 16 and 17 per cent in Latin America and southeast Asia respectively).

Primary education and literacy: The net percentages of primary enrolment are good and not far from those in southeast Asia (91 per cent). The country farthest from the target (100 per cent by 2015) is Serbia-Montenegro (75 per cent). The literacy rate among young adults (15–24 years old), despite progress in all developing and transition countries, is still often less than in southeast Asia and Latin America. It is low in Morocco and Egypt, where only 70 young people in 100 are literate.

Gender equality and the empowerment of women: The region still suffers an appreciable lag. The percentage of women in salaried, non-agricultural employment in the SEMC is between 15 and 30 per cent, depending on the country, compared with 40–50 per cent in the countries of Europe, Latin America and southeast Asia. The ratio between the number of girls and boys enrolled in primary and secondary education shows that the differences have been reduced considerably in the SEMC since 1990, although in Turkey and Morocco significant disparities remain (ratios of 85 and 84 per cent respectively). Ratios for the other countries are greater than 92 per cent, and in 2000 Tunisia had already reached the 2015 target of equality.

Child mortality: Bosnia-Herzegovina, Serbia-Montenegro, Albania, Libya, Tunisia, Syria, Lebanon and the Palestinian Territories register child mortality rates between those of emerging regions and those of developed regions (between 15 and 28 per 1000). Rates in Turkey, Egypt, Algeria and Morocco, despite sharp falls, are still high and comparable with those in Asia and Latin America (between 28 and 39 per 1000). The countries that come closest to the target of reducing infant mortality by two-thirds by 2015 are Libya, Serbia-Montenegro, Tunisia and Egypt.

Maternal health: Maternal mortality rates in the Mediterranean countries have fallen considerably but remain high (more than 120 per 100,000 live births) in Maghreb, Lebanon and Syria, which is comparable to the Latin American and southeast Asian rates. The brake on progress for reaching the target of a 75 per cent reduction in maternal mortality by 2015 is due mainly to the lack of access to health services in rural areas and social support for unwanted pregnancies.

CO_2 emissions: Annual CO_2 emissions per capita in the Mediterranean were close to the world average in 1990 (4.8 tonnes per year compared with 4.3) but considerably greater by 2000 (5.4 tonnes compared with 4), which is a sign of development that is not husbanding natural resources. Emission levels vary markedly from one country to another, ranging in 2000 from 0.9 tonnes per capita in Albania to 10.9 in Libya. Libya, the EU Member States (including those that joined in 2004) and Israel are the largest emitters, while the Maghreb countries (excluding Libya) and Egypt emit relatively little (between 3.1 and 1.4 tonnes per capita). Average per capita emissions in the Mediterranean in 2000 were nearly half those in the EU-15 (5.4 tonnes per year compared with 9) and almost four times less than in the US (21 tonnes).

Access to drinking water: By 2000 the percentage of people with sustainable access to an improved water source was greater than 80 per cent in most Mediterranean countries. The overall situation is nearly on a par with that in southeast Asia and a little less good than that in Latin America. However Morocco and Syria clearly stand out by having wide disparities in access to drinking water between their urban (92–98 per cent) and rural areas (54–56 per cent). Rural access rates in 2000 in these countries remained greater than those of sub-Saharan Africa (45 per cent) but are much lower than the average in southeast Asia and Latin America (about 70 per cent). By 2000 only Egypt had reached the 2015 objective of halving the percentage of the population without access to drinking water.

Access to sanitation: The percentage of the population in 2000 with access to improved sanitation facilities (connection to a wastewater disposal system) was greater than 90 per cent for the countries with available data, except Morocco and Tunisia (68 and 84 per cent), which have nevertheless made progress since 1990 (58 and 76 per cent). These rates are markedly higher than in Latin America (75 per cent in 2002) and southeast Asia (61 per cent in 2002). Although access for urban dwellers in 2000 was nearly 100 per cent in all countries (except Morocco at 86 per cent), rural dwellers were at 44 per cent in Morocco, 62 per cent in Tunisia and 70 per cent in Turkey.

Official Development Assistance: ODA as a percentage of the Gross National Income (GNI) in 2003 was 0.41 per cent for France, 0.21 per cent for Greece, 0.17 per cent for Italy and 0.23 per cent for Spain (compared with 0.15 per cent for the United States). The rates tended to fall between 1990 and 2000 (0.6 per cent for France in 1990, 0.31 per cent for Italy and 0.21 per cent for the US). Although they tended to rise between 2000 and 2003, they still remained beneath the 0.7 per cent target of the Action Plan of the Johannesburg Summit and the Monterrey Conference. Total ODA contributed by the EU Member States and the US (the main donors to the SEMC) dropped a little if one compares the annual averages for 1991–1992 and 2001–2002 at constant 2001 prices and exchange rates, whereas ODA had increased perceptibly between 1981–1982 and 1991–1992. Since 1990 the percentage of aid given by the US to the Mediterranean countries (the total net flow, in current dollars, of ODA and development aid) has tended to fall (from more than 40 per cent in 1990 to 15 per cent in 2002). In Euro-Mediterranean relationships the reverse trend is seen, with the percentage for the Mediterranean increasing from 8 per cent in 1990 to 13 per cent in 2001–2002.

Debt: Debt servicing as a percentage of exports of goods and services in the 1990s fell in Maghreb, Egypt and Syria but increased significantly in the eastern Adriatic countries, Turkey and Lebanon, where the rate in 2002 (51 per cent in Lebanon and 47 per cent in Turkey) was much higher than in Latin America, the region for which debt repayment as a percentage of trade is the highest (31 per cent in 2002). Servicing the debt in Morocco and Croatia (about one-quarter of their export income) is relatively costly compared with Tunisia or Egypt (10 and 13 per cent), which is comparable to East Asia (12 per cent).

Internet access: Progress since 1995 has been fairly strong. Half of the population in Monaco uses the Internet, with about 33 per cent in Slovenia, Italy, France, Malta and Israel and about 12–18 per cent in Lebanon, Greece, Spain and Croatia. Tunisia, Turkey and Serbia-Montenegro (between 5 and 7 per cent) have a comparable level to southeast Asia (4.6 per cent) but less than Latin America (8.2 per cent). Less than 3 per cent of the population in Albania, Algeria, Morocco, Bosnia-Herzegovina and Egypt uses the Internet.

Territorial disparities and concentration on coastal zones

The question of unsustainable development is also raised by the extreme disparities between territories within the Mediterranean countries. The wealth and populations of a country are concentrated more and more in limited areas such as *coastal areas* or large cities. This increases vulnerability to natural hazards. The earthquakes in the region of Izmit in Turkey, in 2001, highlighted how hazardous it can be to choose a way of development that concentrates more than 50 per cent of a country's economic activity in just a few areas. It accentuates the pressure on the most valuable natural resources (built-up coasts, loss of agricultural land and valuable natural habitats, pollution and degradation of ecosystems and coastal landscapes) and the difficulty of managing them. On the other hand, *rural areas*, with a reduction in population and economic activity, are faced by other types of development problems (poverty, difficult access to basic services, desertification, environmental degradation, forest fires, degradation of agrarian landscapes, etc.), but which have the same cause: the concentration of development in specific areas.

'Coastal overdevelopment' ('*littoralisation*' in French), the trend to concentrate activities and people along the coast, is a worldwide phenomenon that is particularly noticeable in the Mediterranean. The process was accentuated in the 20th century with: the building of infrastructures, mosquito control (Languedoc), improvement works in plains, drainage of wetlands (Marais pontins), industrial investment (refineries, petro-chemicals, cement, iron and steel industries) focused on a few coastal areas (Barcelona, Valence, Fos-Marseilles, Genoa, Tarente, Venice, Izmir Bay, coast of Alexandria, and ports in the Maghreb countries) and benefited for a long time from the full support of public authorities. And the tourism boom was solely concentrated on the seaside.

At the same time the hinterland, often mountainous, entered a crisis: the hardship of life in rural areas, which had been abandoned or had inadequate aid and was suffering competition to the advantage of more favoured regions, and the lack of dynamism in the rural economy reinforced the attraction of cities and coastal zones, whether well-founded or not.

All this has generated strong internal disparities within countries to the detriment of inland areas and a coastal overdevelopment that has affected the whole region and resulted in major changes across the catchment areas (Figure 15).

The magnitude of changes over the past century in the northwest of the basin are illustrated in Figure 16. The trend can be assessed at various levels. For example, the population density in the coastal regions is almost three times the average for the countries (see Statistical Annex). Nevertheless, the trend can best be seen very near the coast (a few km). It can also be seen roughly for the whole basin in Figure 17, where the lights of the main agglomerations exactly follow the coast. Part 2 will return to the impact of this phenomenon in terms of sustainable development, which is highly differentiated between the coastal zones and the hinterland.

The high cost of environmental degradation

The poverty that characterizes many rural and peri-urban areas of developing countries, coastal over-development and urbanization where they have been poorly controlled, and a resource-hungry pattern of economic development can together explain the extent of environmental degradation, in the north as well as in the south and east.

Assessment of the costs of environmental degradation provides an estimate of the value of environmental assets and services lost by excessive pressures. The availability of data is poor and assessments have to go through three complex phases: (i) identification of the main pressures on the environment; (ii) their link with

The Mediterranean Region: A Unique But Neglected Heritage

Figure 15 The coastal over-development

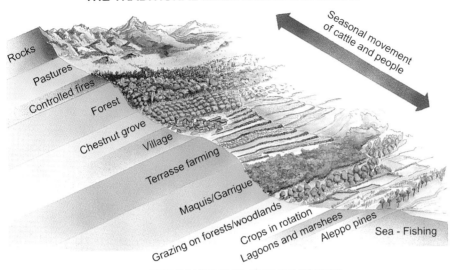

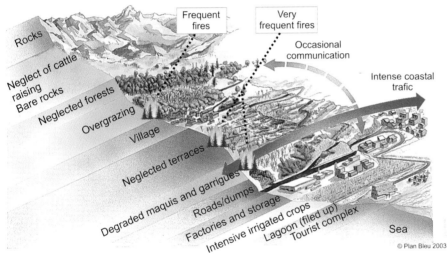

Source: Plan Bleu

the impacts; and (iii) a monetary assessment of these impacts. This last step is particularly sensitive. The use of a natural asset (everything that can be obtained from it – forest timber, recreation, etc.), is relatively easy to value. However, this is not the same for the value of non-use, which is the value that an individual or a society attaches to the existence of the asset and the necessity of transmitting it to future generations (value of legacy or values of existence).

Despite all the limitations (methodological and ethical) of an *economic cost* assessment of degradation, an approximate calculation has been made in some Mediterranean countries (Table 2). For the major environmental variables (water, air, land, coastal zones, etc.) it could reach 5 per cent of GDP in Algeria and 5.4 per cent in Egypt. Generally, these costs are paid for collectively. From a monetary point of view, the effects of the degradation of the living environment mainly affect poor people, who, deprived of fundamental rights such as access to healthy water, live in a bad or polluted environment.

The thematic chapters in Part 2 will show that other costs of degradation, which are also very high, can be assessed for the more developed countries of the northern shore, particularly those related to urban congestion and pollution, and increased vulnerability to risks (fires, floods, etc.).

A Sustainable Future for the Mediterranean

Figure 16 Evolution of the population in the coastal zone cities of the Latin Arc, 1910/1911–1999/2000/2001 (according to the various censuses)

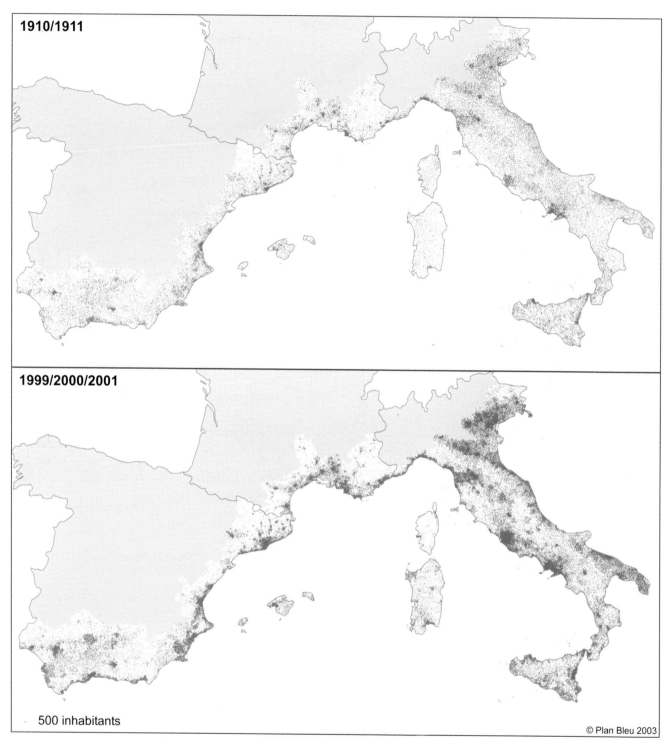

Source: INE, INSEE, ISTAT

The Mediterranean Region: A Unique But Neglected Heritage

Figure 17 The Mediterranean area by night

Source: NOAA (www.nqdc.noaa.gov)

Table 2 Average annual damage cost of environmental degradation in some Mediterranean countries (as a percentage of GDP)

	Algeria	Egypt	Lebanon	Morocco	Syria	Tunisia
	1999	1999	2000	2000	2001	1999
Air pollution	1	2.1	1.0	1.0	1.3	0.6
Lack of access to water supply and sanitation	0.8	1.0	1.1	1.2	0.9	0.6
Land degradation	1.2	1.2	0.6	0.4	1.0	0.5
Coastal zone degradation	0.6	0.3	0.7	0.5	0.1	0.3
Waste management	0.1	0.2	0.1	0.5	0.1	0.1
Subtotal	3.6	4.8	3.4	3.7	3.3	2.1
Global environment (CO_2 emissions)	1.2	0.6	0.5	09	1.3	0.6
Total	4.8	5.4	3.9	4,6	4.6	2.7

Source: World Bank estimates, 2004

Section 2
Determining Factors of the Mediterranean Future

After this overview on the past trends and the most evident signs of unsustainable development, the future of the Mediterranean will be explored, with all its uncertainties and hazards, while assessing the capacity of the eco-region to confront these and adapt to them, or its particular vulnerability.

For exploring the future, a trend scenario is built. The 'baseline scenario' extrapolates the major trends and their possible effects to 2025, while taking account of some signs of growing change. It is a *realistic* scenario that does not pretend to predict the future (which is never predictable), but describes the frame within which the Mediterranean could evolve in the next 25 years if no deep change, or break, or surprise events, take place for changing the observed trends.[15]

As all scenarios do, the baseline scenario is founded on several assumptions. Assumptions need to be explicit, whether they be *general* (for example concerning population growth, or changes in the global economic context, or climate change) or more *sectoral* and *detailed* (on future energy demand, on growth in land and maritime transport, on tourism changes, etc.).

This section formulates the general assumptions of the baseline scenario concerning five major determinants of the Mediterranean future: impacts of climate change, population, globalization and regional integration, taking the environment and sustainable development into account in national policy making, and the constraints that burden economic development in the southern and eastern Mediterranean countries. All the general assumptions are summarized at the end of Part 1 (Table 5 p63), and the detailed sectoral assumptions on water, energy, transport, urban areas, rural areas and coastal areas, will be formulated in the thematic chapters of Part 2.

Climate change and its possible impacts

One of the upheavals that could affect the Mediterranean area is climate change. An increasing body of observations is confirming the perception of a warming world and other climate changes linked to increased emissions of greenhouse gases (such as CO_2) from human activities (energy, transport, agriculture, etc.). The Intergovernmental Panel on Climate Change (IPCC)[16] states that during the 20th century, average global surface temperature has increased by 0.6°C and average sea level by 1–2 mm per year, while CO_2 concentration increased by 31 per cent between the pre-industrial period and 2000. The 1990s were undoubtedly the warmest decade of the last millennium in the northern hemisphere. The IPCC extreme global scenarios project global warming ranging from 1.4°C to 5.8°C, by 2100, an increase in average annual rainfall ranging from 5 to 20 per cent and a rise in sea level of between 9cm and 88cm, with some regional differences.

There is still uncertainty about the size and the speed of this warming and even more about its impacts on the planet during the 21st century. One major and additional uncertainty concerns the repercussions of *global* warming for *regions* such as the Mediterranean. The marine currents (horizontal and thermohaline circulation) might be altered at the global level and could in turn influence local climate towards a cooling trend.

A recent study[17] cautiously concluded that, according to some current models used by climate experts, a hypothetical average global warming of 1°C is projected to have the following effects in the *Mediterranean*:

- A *warming* ranging from 0.7 to 1.6°C, depending on the area. A warming has already been recorded in temperate Europe where annual temperatures have been increasing by 0.1 to 0.4°C per decade. Warming would be highest in the Mediterranean Basin with a rise in summer temperature twice that in northern Europe: harsh winters would disappear by 2080 and hot summers would be more and more frequent.
- *Changes in rainfall*: in winter and spring, an increase in rainfall in the north and a decrease in the south, all with a very wide uncertainty range, from −2 per cent to +26 per cent; in summer, a decrease in rainfall in the north and the south; in autumn, a reduction in rainfall in the west and an increase in the east and centre. A decrease in total rainfall has already been observed during the 20th century in some Mediterranean areas and in North Africa.
- An increase in the frequency, intensity and duration of *extreme meteorological events* (heatwaves, summer drought, winter floods and mudslides in the north of the basin).

Since action by the international community may not be able to reduce greenhouse gas emissions enough to reverse the strong trend, *the **baseline scenario** assumes that climate change will lead to an intensification of extreme climatic events and a warming of less than 1°C by 2025.*

The *multiple impacts* of such changes to 2025 mainly concern:

- The *sea* with possible initial changes in temperature, salinity, content of organic matter, CO_2, nitrates, phosphates, etc., which could in turn affect the thermohaline circulation (linked to differences in temperature and salinity between various layers of sea water). Changes in salinity and temperature have already been observed, but the modelling of water exchange in the Mediterranean with other seas and with the atmosphere is particularly complex. Scientists are very prudent when assessing the risk of a rise in the Mediterranean sea level on this time horizon. However, locally, the subsidence of the south of Europe linked to post ice-age tectonic changes (5cm of subsidence by 2080) and that in some deltas (Rhone, Nile) would aggravate a hypothetical rise in sea level.
- *Rainfall pattern*: in the south, there is a risk of an increase in the occurrence of droughts; irrigation water requirements are likely to increase. In the north, an increase in torrential rainfall in autumn and spring would increase the frequency of mudslides and floods.
- *Biodiversity and marine and terrestrial ecosystems* may be altered. The risk of extinction of vulnerable species in the Mediterranean is considered by several studies to be serious. Increased perturbation of ecosystems is also projected, for example those due to forest fires, droughts, parasitic attacks, invasive species and storms.

Climate change calls for preventive, curative and adaptive responses, which will be analysed in Part 2. With the United Nations Framework Convention on Climate Change and the Kyoto Protocol, the international community has tried, but not yet successfully, to achieve a global preventative response to reduce gas emissions. On the time scale of this report (to 2025), this response may have little impact on climate change. Nevertheless, the mechanisms set up through this process (clean development mechanisms) to differentiate efforts between countries at various stages of development may well be applicable in the Mediterranean eco-region.

Towards north–south demographic convergence

Demography is an essential component of the baseline scenario. The growth of the permanent and tourist population, their poorly controlled concentration in cities and on coastal zones is projected to continue. This, together with the widespread change in living and consumption patterns, could increase the pressures on the Mediterranean environment and territories.

Reduced fertility

One of the main determining factors for the Mediterranean countries in the decades to come (and one of the surprises compared with the demographic projections by the United Nations and Plan Bleu 89) is the spectacular fall in fertility[18] in the SEMC (Figures 18 and 19). Fertility in these countries had already fallen from 5–7 children per woman in the 1970s to 2–4 children per woman in 2000, a steady fall of 2–3 per cent per year for 30 years. This fall in fertility is continuing in the developed countries of the northern shore, and is below the population renewal threshold of 2.1 children per woman, with values among the lowest in the world in Italy (1.28 children per woman after a fall of 2.8 per cent per year between 1970 and 1990) and in Spain (1.27 children per woman, with a fall of 4 per cent per year between 1970 and 1990).

The education of girls, transformations of lifestyles that accompany urbanization, family planning policies, the use of contraceptive methods, changes in the role of religion in society, the generalization of the model of smaller families, the influence of television and migration on mentality, all help to explain the extent of the fall in fertility in the SEMC. In the north, these same

Determining Factors of the Mediterranean Future

factors, accentuated by the rise in individualism and the weakening of family bonds, led to a spectacular drop in fertility, which in the long term could put national identities in jeopardy.

The baseline scenario assumes a continuation of this trend, with fertility rates around the Mediterranean converging, with a slight rise in the north.

This convergence is of vital importance for the future Mediterranean population because of its long-term effects: bringing closer together a model of society for both shores (size of households, way of life, etc.), re-balancing the active and inactive age brackets, and opportunities for exerting less pressure on the environment and the land.

By 2025, another 100 million inhabitants and 40 million households

If the demographic transition in the SEMC accelerates, with a fall in average population growth rate from 2.3 per cent in 1975–2000 to 1.3 per cent in 2000–2025), growth will nevertheless remain high until 2025 and beyond.

Figure 18 Fertility rate (per woman), N1, 1950–2025

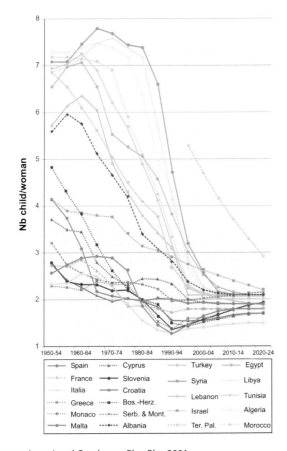

Source: Attané and Courbage ; *Plan Bleu* 2001

Note: Convergence of north–south fertility rate: around two children per woman in most Mediterranean countries.

In the baseline scenario the population of the riparian countries is projected to be stable in the north and to increase by 40 per cent in the south and east. The Mediterranean countries will have 523 million inhabitants in 2025, 96 million more between 2000 and 2025, of which 31 million would be in the Mediterranean coastal regions. Most of this growth will be in the south and east where there will be 3.7 million more inhabitants each year between 2000 and 2025, in absolute terms almost as many as between 1970 and 2000. This population will remain concentrated on the coastal zones and in cities.

In the Mediterranean, as elsewhere, the family unit is undergoing major change. In practically all countries, the size of households is falling. On average, it fell from 4 people per household in 1985 to 3.7 in 2000 and could reach 3.3 by 2025 (Figure 20 and Statistical Annex). With urbanization and the reduction in fertility, relationships with children are changing and the role of women is evolving. Co-habitation of several generations, which is still frequent in the south and east, is gradually giving way to nuclear households or even one-parent households in the north. This decrease in household size is particularly strong in the NMC (northern Mediterranean countries) but also in Turkey (from 5.3 in 1985 to 4.2 in 2000 and projected to be only 3.3 in 2025), and Algeria (from 7 to 6.1 in 15 years, and to 4.9 in 2025). In 2025, the SEMC would have between 4 and 5 persons per household except in Libya (more than 7) and Israel (3.1), with, however, some internal differences between rural and urban areas.

In the baseline scenario, these trends towards smaller households size continue, with the average size possibly going down to 3.3 people in 2025.

Ageing in the north; working-age population increasing by 3 million per year in the south and east

As fertility continued to fall and life expectancy to rise, it is the number of people aged 65 years or over that has increased the fastest in most Mediterranean countries since 1970. This ageing has been particularly noticeable in *the north*. Italy, for example, with fewer than 15 per cent of people aged under 15 and 18 per cent over 65 in 2000, is among the first countries in the world to be confronted with this new phenomenon (Figure 21 and Statistical Annex).

In the baseline scenario, ageing will be amplified in the north. People over 65 years of age currently represent 16 per cent of the total population; this could reach 22 per cent by 2025.

With the ageing population and the long-term widening of demographic deficit, the NMC are faced with new problems that could lead them to look again at their social 'model', particularly at pensions and the distribution of health costs. In France and Italy the number of active people per retired person will fall by

A Sustainable Future for the Mediterranean

Figure 19 Convergence of demographic parameters

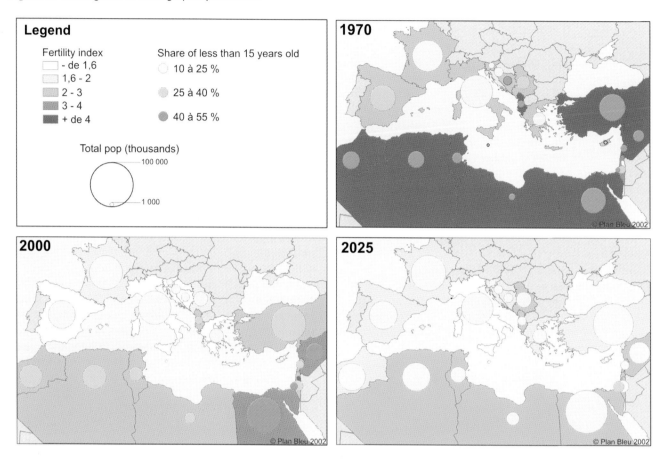

Source: Maps after an idea in *Méditerranée* no. 3.4, 2001; data from Attané and Courbage, *Plan Bleu* 2001.

Figure 20 Number and size of households, 1985–2025, all Mediterranean countries together (N1)

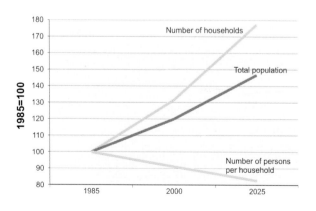

Source: UN-Habitat (2001) Global Urban Observatory. *Statistical Annexes to the Global Report on Human Settlements* 2001 (www.unhabitat.org)

Note: Between 2000 and 2025, the number of households will grow faster (32 per cent) than the total population (20 per cent). This trend could be pursued in the future. There could be 40 million more households in the Mediterranean by 2025.

half in the next 50 years, which could put solidarity among generations at risk. Ageing will disrupt social and economic behaviour (economic growth, consumption, number and structure of households, savings, mobility, ageing of the workforce).

The SEMC are very different from the NMC having, still, a high proportion of people under the age of 15 (32 per cent in 2000 compared with 17 per cent in the north) but they are already at the start of an ageing process. The proportion of the population under 15 will probably continue to fall, to 22 per cent in 2025, while the number over 65 years will double, from 6 per cent of the total population in 2000 to more than 10 per cent in 2025. This could create a new opportunity for a rebalancing between the active and inactive (young and old) population. It would reduce the costs to the active population and result in a stabilization of the numbers to be educated. However, it presents a *major challenge for the SEMC in terms of employment*. The age distribution in 2000 brought a yearly average of 3.7 million into the working age bracket of the population (1.4 million in Turkey and Egypt). The demand for new jobs is therefore

Figure 21 Population age distribution, N1, 1970–2025

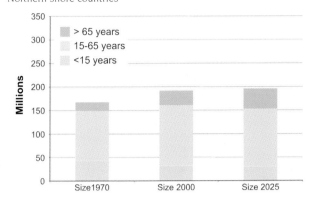

Source: Plan Bleu; Courbage and Attané, 2001

projected to increase before declining slowly after 2030. Between 2 and 4 million new entries into the labour market are expected per year in the SEMC. In total 55 million more net entries into the working age population are expected between 2005 and 2020 and 24 million more between 2020 and 2030 (Figure 22). Between 2000 and 2020 alone, if the activity rate[19] and employment rate[20] remain unchanged, 34 *million jobs would need to be created within 20 years*.

Conversely, by 2010, the NMC will have more people coming out the age bracket of the working population (2.3 million) than entering it (2.2 million). The deficit will be more than 300,000 from 2020–2024. Time differences will be seen, however, between countries: the Italian job market will be in deficit before 2010 while France will only lack active people after 2020.

Greater urban populations

To the phenomenon of population concentrations on the coast can be added that of urban development, which has been very rapid in the Mediterranean. During the last 30 years, most population growth has been in cities: of 143 million additional inhabitants in Mediterranean countries between 1970 and 2000, 84 per cent, or 120 million, were in towns.

In all Mediterranean countries together (N1), the urban population (living in towns of more than 10,000 inhabitants) increased from 153 million in 1970 to 273 million in 2000, an increase of 1.9 per cent per year, which represents a doubling in 30 years, or 4 million more town-dwellers per year. More than 80 per cent of this urban growth is in the south and east where urban growth has been 3.6 per cent per year on average since

Figure 22 Net entries and withdrawals of the 20–64 year-old bracket (in 1000 inhabitants)

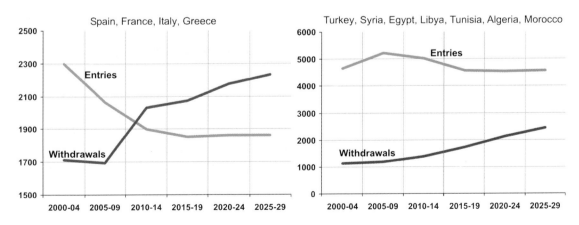

Source: Plan Bleu; Courbage and Attané, 2001

Note: Entries: 1/5 of the 20–24 year olds.

Withdrawals: 1/5 of the 60–64 year olds.

A Sustainable Future for the Mediterranean

1970, and could go beyond 4 per cent per year in many countries (Libya, Syria, Turkey). The south and east Mediterranean is urbanizing more rapidly than the rest of the world (3.6 per cent per year compared with 2.5 per cent). Urbanization has been faster than Plan Bleu 89 had projected; for example there were about 43 million urban-dwellers in Egypt and the same number in Turkey in 2000, which was far beyond the ranges of all the 1989 scenarios.

All the Mediterranean countries have registered a growth in *urbanization rates*. The average on the northern shore increased from 62 to 67 per cent between 1970 and 2000, and from 43 to 62 per cent in the SEMC. For all Mediterranean countries, urbanization increased on average from 54 to 64 per cent between 1970 and 2000.

This strong trend towards urbanization is projected to continue. Projections of urban population, obtained by extrapolating past trends country by country, show that the *total urban population* in the Mediterranean (273 million in 2000), could reach 378 *million by* 2025. Again, most of this growth will be in the SEMC where there will be an average of 3.9 million more urban-dwellers per year (2.08 per cent per year, above the annual average rate of total population growth in these countries).

At this rate, there will be a total upheaval in lifestyles of people in the SEMC since, in only 50 years, what were essentially rural countries (with an average urbanization of 40 per cent in 1970) will become urban ones (74 per cent by 2025).

In *coastal regions*, the urbanization process ends in overdevelopment. The urban population of coastal regions may increase by 33 million between 2000 and 2025, of which 30 million would be in the SEMC (see Statistical Annex).

To summarize, the determining population factors for the baseline scenario show a demographic transition that is accelerating in the south, slightly reducing the needs for education, but resulting in a high demand for employment. In absolute terms, population growth remains high and its impacts on the environment are likely to become more accentuated. The north is faced with an ageing phenomenon, and even a deficit in its working population. Concentration of population in cities and on the coasts is accentuated.

The Mediterranean between globalization and regional integration

The region's future is part of an international geo-political and economic context that, since Plan Bleu 89, has been profoundly changed by globalization and the creation of major regional centres. In this context, the Mediterranean's future is ever more linked to that of the European Union, which is gradually enlarging over the entire northern shore and developing cooperation with the south and east through increasingly intensive relationships.

Globalization and major regional economic poles

The worldwide geo-political context has been overturned with the end of the 'East–West' poles and the boom in globalization. Since the collapse of the USSR, the US is now without a major rival and has assumed a hegemonic role in the world. Its model of society – democratic functioning of institutions, respect of individual freedom, economic liberalism and a consumer society – is being disseminated all over the planet. There is a vast movement towards deregulation and *liberalization* aimed at strengthening the functioning of markets everywhere. The international financial organizations (World Bank, IMF) and trade organizations (WTO) disseminate the principles of a 'healthy economy',[21] which have inspired the structural adjustment recommendations to transition and developing countries. In many cases, welfare states that have tried to find a way of reconciling socialism, planning and market liberalization, are being challenged. 'Healthy economy' principles are, however, difficult to apply and do not always meet with the expected success. They are being debated even in the

Figure 23 Urban and rural populations in the Mediterranean countries, N1, 1970–2025

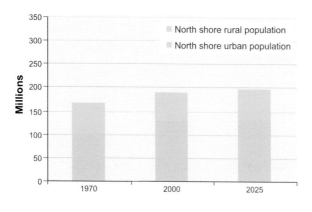

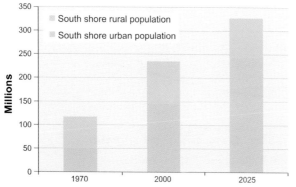

Source: Plan Bleu; Courbage and Attané, 2001

Determining Factors of the Mediterranean Future

Bretton Woods institutions and pose problems for the pattern of development and, *a fortiori*, sustainable development.

These influences, combined with major *technological and scientific progress* in the fields of computer science, communication and telecommunication, have contributed to a process of globalization that is particularly active in financial markets. In globalizing economies, enterprises that compete with one another are no longer subject to fragmentation of markets and their strategies can become worldwide. The new trans-national corporations are adopting strategies that are adapted to the global functioning of finance, with less and less attention being paid to national contexts.

In the baseline scenario globalization continues with an increasing influence on the market economy and information technology. Global production and trade grows.

But with globalization we also see a *polarization* of the economy and greater competition between three major poles, each of which is trying to expand its sphere of economic and geo-political influence: the Americas, Europe and Asia, bolstered by an emerging China.

The *European pole* (in the broad sense, including the Mediterranean) accounts for 28 per cent of world GDP and about 30 per cent of trade (Box 7). At the core of this pole, the European Union plays a structural and strategic role through trade in goods and services, circulation of capital and regulation policies (cohesion, solidarity and agriculture). Through its success, it is showing a possible way of positioning itself in globalization through a regional type of path.

Within this pole, the *Mediterranean* (all countries together) has witnessed a fall in its ranking in the world economy: about 13 per cent of world GDP and 14 per cent of external trade in goods and services (Table 3). The four EU-Med countries contribute the lion's share of this percentage since they account for 11–12 per cent of world GDP and trade, with the rest of the Mediterranean countries accounting for only 2 per cent.

Box 7 The Euro-Mediterranean pole in two world economy scenarios to 2050

In a first trend scenario, called '**Chronicle of a predicted decline**', the European pole loses economic weight to profit Asia, while the American pole retains its place. Such a decline would essentially be due to demographic change (a fall in fertility and restrictive migration policy in the EU) and to the absence of EU political will to co-develop with its southern and eastern shores. The EU enlargement to the east is not sufficient to guarantee parity with the US.

An alternative scenario, called '**Europe-Russia-Mediterranean**' is based on the assumption that the EU undertakes a voluntary integrated development policy (economic, technical and political) with the south and east of the Mediterranean and with Russia. In this scenario, the EU launches a technical cooperation programme in order to develop confidence and accelerate reforms in the southern and eastern Mediterranean economies, resulting in improvements in productivity. At the same time, the EU energizes its demography (increases its labour force) by a policy to increase the birth rate and a more open immigration policy towards its neighbours. In this scenario, the wider European pole achieves an increased weight in the world economy compared with NAFTA and the Asian pole. The world remains tri-polar.

Share of the three major poles in the world economy in 2000, and two scenarios for 2050
(in % of the world total)

| | World GDP | | | World exports of goods | | | World imports of goods | | |
| | | 2050 | | | 2050 | | | 2050 | |
	2000	Trend scenario	Alternative scenario	2000	Trend scenario	Alternative scenario	2000	Trend scenario	Alternative scenario
Europe-Russia-Mediterranean pole	28	20	32	30	24	47	29	25	48
American pole (NAFTA, MERCOSUR)	31	31	26	19	17	12	32	37	25
Asia-Pacific pole (China, Japan-South Korea, South-Asia, ASEAN countries)	35	45	39	41	51	29	32	35	21

Source: IFRI, 2002

Note: GDP values in US$90 ppp.

Trade excluding internal trade in each pole.

A Sustainable Future for the Mediterranean

Table 3	The Mediterranean share in the world's economy, 1980–2002		
	1980	1990	2002
GDP	14.0%	13.5%	13.0%
Exports	15.2%	15.5%	14.0%
Imports	16.5%	16.5%	13.9%
World GDP (thousand million US$ 1995)	19,400	26,400	35,300

Source: World Bank, 2004

Note: Some countries are not included in 1980 and/or 1990 (east-Adriatic Countries, Palestinian Territories, Lebanon).

In *geo-strategic* terms, greater rapprochement between the EU and the Mediterranean would make it possible to strengthen a 'Euro-Mediterranean' pole and give it a better ranking in the world's economy. This is what emerges from a recent prospective study,[22] dealing with the EU's place and role in the world up to 2050; it highlights two major risks for the future of Europe in the 21st century: a widening gap (economic, technological, cultural and military) between Europe and the US, and instability related to disparities in development with its neighbouring regions. Both risks rely on the same observations of demographic variables and technological progress: the EU is showing a certain technological tardiness and is gradually entering demographic 'hibernation'. In contrast, the southern and eastern Mediterranean countries have a tremendous labour force but do not have the material and human resources to absorb foreign technology. From this position, two extreme scenarios are envisaged (Box 7).

The following paragraphs analyse the intensity of the links and the level of political integration between the Mediterranean shores, and define the assumptions of the baseline scenario concerning the strategic rapprochement of the EU with its southern- and eastern-shore neighbours. This will enable us to discern whether the future is of an EU that is drawing closer to its 'neighbours' or moving further away from them.

Intensifying exchange between Europe and the south and east of the Mediterranean

The intensity of the exchange of goods and people between the Mediterranean countries and the EU is increasing the interdependence of their respective destinies, which are already linked by geographical proximity and a joint natural heritage. It should encourage rapprochement between the shores. According to the baseline scenario, how might these relationships evolve?

An intense intermixing of peoples: Long-standing and persistent migration between the SEMC and the EU

The Mediterranean identity has been forged through age-old periods of migration and intermixing. More recently, the period between 1950 and 1970 saw considerable migratory flows towards Europe and the Gulf countries to meet the need for predominantly unskilled manual labour. This work force came from the Maghreb countries and Turkey, and from Mashrek and Egypt. Since 1973, when the EU closed its borders, the large projects of the 1980s in the Gulf countries came to an end, and the first Gulf War occurred, migratory flows have taken other forms: families coming together, refugee flows, political asylum seekers and illegal immigration.

Despite the difficulty in measuring migration flows with reliable statistics, it is estimated that 10 million foreigners, 5 million of whom are from other Mediterranean countries, are living in the Mediterranean countries. Foreign immigrants of Mediterranean origin accounted on average for 4 per cent of the total population in the Mediterranean countries, without counting people who have been naturalized, who are the descendants of immigrants, or who are illegal immigrants (Figure 24).

Historical links (colonization, political influence, language) are a deciding factor in the choice of emigrant destinations: Turks go mainly to Germany, the people of the Maghreb go to France, Albanians go to Italy or Greece and Egyptians go to Libya or the Gulf countries.

The 'fortress' EU strategy is faced by the difficulty of monitoring borders and coordinating the operations of the various countries. Illegal migratory flows continue to grow, maintaining criminal organizations that endanger the migrants' most basic human rights (Box 8).

Box 8 Illegal immigration to the EU

According to estimates, between 400,000 and 500,000 illegal immigrants enter the European Union each year where they have unofficial employment. Requests for official recognition help to measure the extent of this immigration: in France 300,000 requests were recorded in 1999; in Spain, 400,000 in 2000; in Italy, nearly 1 million between 1987 and 2000; in Greece 450,000 in 2000. Despite these various recognitions, there are between 3 and 4 million illegal migrants in the EU.

The Mediterranean Basin contains the main routes for clandestine immigration into western Europe. Highly lucrative illegal networks charge several thousand euros per transfer, often in dramatic conditions for the illegal migrants. The number of people interrogated per day on the southern coast of Spain and the Adriatic coast in Italy, and the number of victims found drowned in the Straits of Gibraltar, only reveal part of the true scale of this phenomenon.

Determining Factors of the Mediterranean Future

Figure 24 Foreign and Mediterranean emigrant peoples

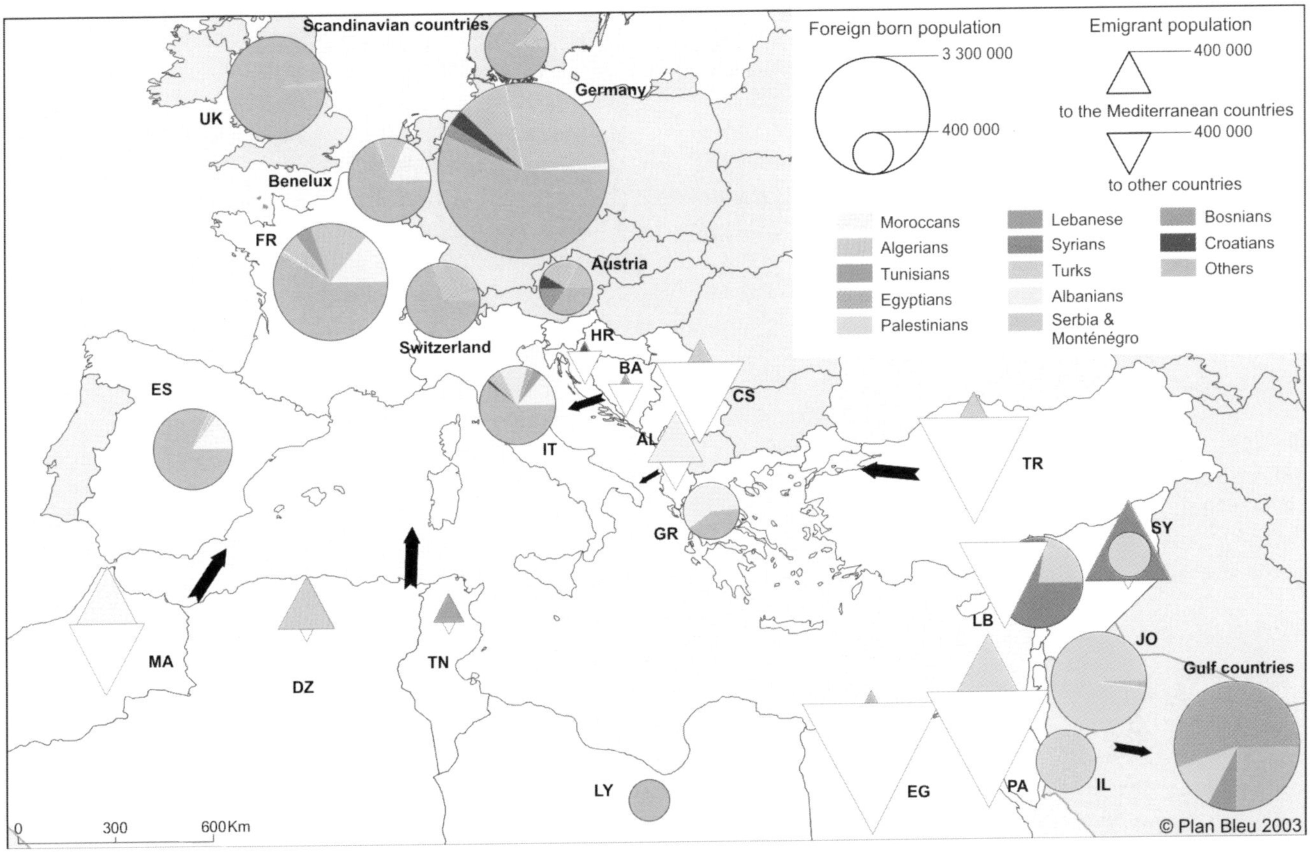

Source: Plan Bleu, 2003, compilation from Eurostat New Cronos (http://epp.eurostat.cec.eu.int)

Note: In the four EU-Med countries, foreigners of Mediterranean origin are between 3 and 7 per cent of the total population (7 per cent in Greece). If the naturalized populations or descendants of migrants are taken into account, the proportions are higher: in France, there are 1.3 million foreigners from the Mediterranean (2.2 per cent of the total population) but also 5.1 million descendants of Mediterranean migrants. Lebanon, a country of great diaspora, is home to 500,000 Syrians and 430,000 Palestinians. Libya has 410,000 foreigners, 7 per cent of its total population. During the past decade, Israel has welcomed and naturalized 900,000 foreigners (15 per cent of its total population) from the ex-Soviet Union and Ethiopia.

The baseline scenario assumes that the migratory policies of the EU countries remain very restrictive, although monitoring illegal immigration remains problematic. In future, migratory flows from all over the world (legal or otherwise), encouraged by globalization, are most unlikely to dry up. The propensity in the SEMC to emigrate could remain high because of the gap between the number of jobs created and the number of people entering the job market.

These past and future migratory flows are therefore expected to strengthen the economic, social and cultural *interdependency* between Mediterranean countries. Migrants provide manpower and income transfers between countries.

With migrants, there is also a transfer of knowledge and learning, often seen as a loss of human capital (brain-drain) when it is not followed by a return to the source country. Approximately 250,000 graduates have left the Maghreb since Independence; during the year 1996–1997, 1200 assistant professors and professors left Algeria, mainly for Canada, the US and Europe. About 7.5 per cent of graduates from Egypt live in OECD countries. From the macro-economic point of view, there is the question of 'return on investment' for the long years of study needed to train skilled labour, which then benefits other countries that can offer more attractive salaries.[23]

Moreover, migrants can prove to be potential vectors of rapprochement and change in both their home and their adopted societies because their outlooks are enriched by different cultures. For example their influence on the rapid fall in fertility rates in Morocco has been demonstrated.[24] Migrant flows can also introduce new lifestyles and consumption patterns in their home countries.

By 2025, an additional 178 million international tourists

The intermixing of peoples and the interdependency between Mediterranean countries is also strengthened by the development of tourism, a key economic sector. The Mediterranean remains the first destination for

A Sustainable Future for the Mediterranean

European tourists; 85 per cent of the 218 million international tourists visiting Mediterranean countries each year are Europeans. A vast majority of Europeans make up the clientele in the NMC, as well as in Turkey, Tunisia, Egypt, Morocco and Israel. A less-known fact is that tourism in Syria and Lebanon is mostly Arabic (60 per cent and 37 per cent respectively). Many Tunisians travel to Libya and many Saudis and Libyans vacation in Egypt. Also worth noting is the importance of holidays taken by nationals residing abroad, accounting for 40 per cent of visits to Morocco and 81 per cent to Algeria. Mediterranean tourism thus remains mostly a Euro-Mediterranean affair, which creates economic interdependence and a vector of additional cultural exchange.

The sector has seen strong growth for several decades. In the Mediterranean-rim countries (N1) international tourist arrivals increased 3.7-fold between 1970 (58 million visits) and 2000 (more than 218 million visitors). These values are at the upper end of the ranges of the *Plan Bleu 89* scenarios.

So much so that international visits to the Mediterranean countries (N1) in the baseline scenario will be roughly 396 million visitors by 2025, or 178 million additional visitors compared with the year 2000 (see Statistical Annex). The emergence of central and eastern Europe as a source of tourists to the Mediterranean countries change tourist flows in the basin. Some of the factors that may influence the future changes are the hard-to-predict behaviour of an ageing northern European population (the main source of tourists), the outcome of the Middle East conflicts and the future of international terrorism.

According to the World Tourism Organization's projections to 2010 and 2020, extrapolated to 2025, international tourist flows towards the Mediterranean should increase. Annual growth could be greater than 10 per cent in Libya and between 5 and 10 per cent in Croatia, Bosnia-Herzegovina, Serbia-Montenegro, Albania, Slovenia, Turkey, Syria, Lebanon and Egypt. Three of the four EU-Med countries (France, Italy and Spain) will still accommodate 65 per cent of the

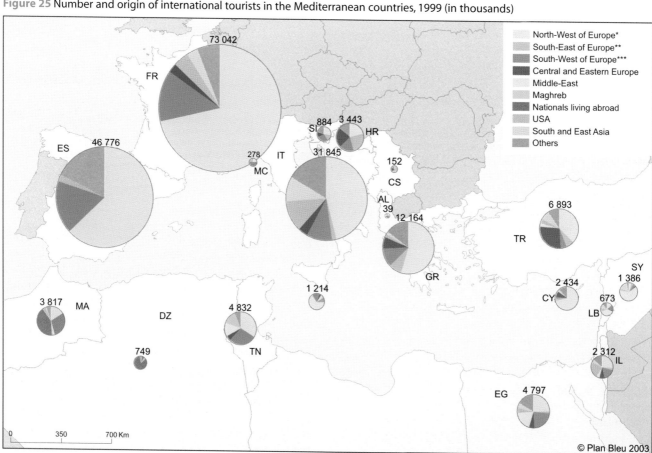

Figure 25 Number and origin of international tourists in the Mediterranean countries, 1999 (in thousands)

Source: World Tourism Organization, 2001

*   North-west Europe: Belgium, Germany, Netherlands, Sweden, Switzerland, United Kingdom.
** South-east Europe: Albania, Bosnia-Herzegovina, Croatia, Cyprus, Greece, Slovenia, Turkey and Serbia-Montenegro.
*** South-west Europe: France, Italy, Malta, Monaco and Spain.

Determining Factors of the Mediterranean Future

international tourists to the Mediterranean (more than 75 per cent in 2000). By 2025 Turkey would become the fourth most visited country in the Mediterranean with 34 million international tourist arrivals, and Egypt the fifth with 24 million.

In addition there are *domestic tourists*, the numbers of which will grow with the SEMC's changing demographics, extended life expectancy and rise in living standards. *Plan Bleu* has estimated the additional flows of domestic tourists between now and 2025 in the countries and coastal regions.[25]

The baseline scenario thus projects a considerable development of tourism with its parade of economic, social and cultural exchanges. By 2025 the Mediterranean countries will accommodate an additional 273 million domestic and international tourists (a total of 637 million), of which an additional 136 million will be in the coastal regions (Figure 26 and Statistical Annex). The possible opportunities and effects of such growth will be analysed in Chapter 6 on coastal areas.

The intensification of trade and financial exchanges

Trade in goods

If the intermingling of peoples brings the destinies of the two shores of the Mediterranean closer together, their economies are also intertwined with a certain asymmetry in favour of the EU. In 2002 the EU was the leading trading partner of the SEMC, accounting for about 50 per cent of their total trade compared with 43 per cent in 1990.

The Maghreb countries are the most integrated into the European area with Tunisia at 75 per cent, Morocco at 64 per cent and Algeria at 62 per cent. In contrast, the SEMC have accounted for less than 8 per cent of the EU's trade since 1990 (Box 9). Trade between the SEMC and the EU is *polarized* between three SEMC (Egypt, Israel and Turkey, together having 55 per cent of the SEMC share) and four EU countries (France, Greece, Italy and Spain, with roughly half of total EU/SEMC trade). Generally, SEMC have a trade deficit (except Algeria, Libya and Syria, who are oil exporting countries), with the world as well as with the EU.

With the exception of Israel and Turkey, and to a lesser extent, Tunisia, which are diversifying their industrial fabric, and export manufactured articles with a high capital-intensive content to the EU (vehicles in Turkey and telecommunication devices in Israel), the SEMC are experiencing *traditional specialization*. They focus on producing and exporting labour-intensive industrial products such as clothing and shoes, and energy products (Algeria, Syria, Egypt). They import mainly machinery and transport goods and manufactured products.

The imbalance of trade is also seen in *weak intra-SEMC trade*; it accounts for barely 5 per cent of their total trade (3

Figure 26 Tourist visits (domestic and international), N1, N3, trends and projections to 2025

Countries (N1)

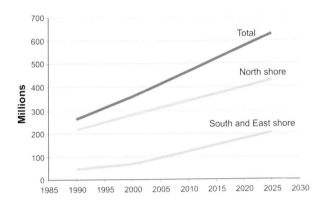

Coastal regions (N3)

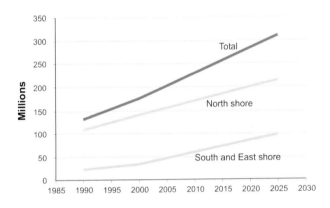

Source: Plan Bleu estimates

per cent in 1990), even if it is increasing faster than trade with the EU (by 10 per cent per year on average), which may be evidence of the beginnings of some intra-SEMC specialization. Thus Turkey (accounting for nearly 40 per cent of intra-SEMC trade) imports raw materials and sends about three-quarters of its exports to the SEMC in the form of high value-added products (manufactured products, machines and transport materials).

The *eastern Adriatic countries* also enjoy growing but asymmetric trade with the EU. They are developing more than half (58 per cent) of their trade with the EU while they account for only 0.5 per cent of EU imports and 1.3 per cent of exports.[26] Association and stabilization agreements signed with the EU in 1999 grant these countries non-reciprocal preferential trade, allowing them to increase the EU's share of their exports between 1995 and 2002 from 57 to 60 per cent. All have trade deficits, especially with the EU (50–60 per cent of the total deficit).

A Sustainable Future for the Mediterranean

Figure 27 SEMC and EAC trade in goods with the EU

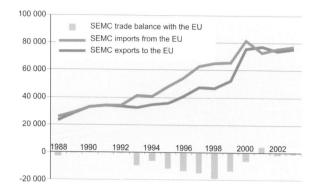

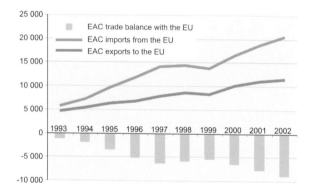

Source: Eurostat (EU declared figures)

Trade in services

Although the value of trade in services in 2002 was lower than that of goods (about five times less for SEMC/EU trade), it has seen a higher rise since 1992. As with trade in goods, the EU is a major partner of the SEMC and exchanges between the two groups of countries have more than doubled in ten years. Almost all the SEMC and EAC record a surplus with the rest of the world and with the EU-15 (Israel being a notable exception).

With the prospect of a Euro-Mediterranean free trade zone (objective date: 2010), the baseline scenario is therefore one of increased trade in goods and services in the Mediterranean with increasing polarization between the EU and the SEMC and eastern Adriatic countries.

Exchange of capital:[27] Official development assistance, direct foreign investments

Over the last decade the non-EU Med countries received total *net capital flows* (official and private) of an average of US$16.8 thousand million ($1995) a year, an increase on the two previous decades (US$11 thousand million). But their relative share in all international financing fell (10 per cent between 1991 and 2000, compared with 17 per cent between 1971 and 1980) to the benefit of southeast Asia and South America. With reference to GDP, net capital flows have also fallen, accounting for 3 per cent of their GDP over the last decade (compared with 6 per cent in the previous decade, see Statistical Annex).

Of these capital flows, the net *official* flows (essentially Official Development Assistance), although decreasing in the total flows over the past 30 years, have remained predominant (57 per cent of total flows). They are mostly in the form of increasing contributions from the EU, in particular the European Commission and are highly polarized. Nearly 85 per cent of these contributions are still composed of bilateral funds. Between 1991 and 2000, some 60 per cent of the official contributions came from just five countries (the US 32 per cent, France 9 per cent, Germany 8 per cent, Italy and Japan 6 per cent each). More than 90 per cent of the net official American contribution focused on three countries (Egypt, Israel and Turkey). EU-sourced aid, with better geographical distribution, has recently been redirected towards the rebuilding of conflict zones (the eastern Adriatic countries and the Palestinian Territories) and the countries joining the EU in 2004 (Malta, Cyprus and Slovenia).

At the same time, the percentage of net *private* flows – basically foreign direct investments (FDI) and portfolio investments – in total contributions increased (from an average of 28 per cent between 1981 and 1990 to 43 per cent between 1991 and 2000). Excluding the four EU-Med countries, the Mediterranean is nevertheless characterized by a low capacity to attract FDI. It attracts barely 1.3 per cent of the world's total FDI and little more than 5 per cent of FDI destined for developing countries (Table 4).

Table 4 FDI in the Mediterranean countries (as a percentage of total world FDI)

	Average 1991–96 (%)	Average 1997–2002 (%)
Four EU-Med countries	14.2	10.2
Islands	0.1	0.1
EAC	0.2	0.3
SEMC	1.3	0.9
- of which Israel	0.3	0.3
Total of world FDI (million $)	254,326	852,499

Source: UNCTAD, Report on Investment, 2003, participation in enterprise capital is considered as FDI

Note: In absolute value the total of FDI in the SEMC has certainly increased (from less than US$3 thousand million in the early 1990s to nearly US$9 thousand million in 2000), their share of world FDI remains low (0.9 per cent of the total, or the equivalent for 12 countries of the FDI received by Poland alone) and has fallen since 1990. The observed increase in portfolio investments, moreover, could dry up with the end of privatization programmes, except for Israel and Turkey.

Box 9 Trade in Goods between Mediterranean countries, 1998–2000

Trade balance

Food products	82
Drink and tobacco	−257
Non-food raw material	−411
Energy	23,893
Mineral fuel, etc	−240
Chemical products	−8060
Manufactured articles	−6949
Machinery and vehicles	−19,418
Various manufactured articles	8885

Note: In 2002, SEMC contributed 19 per cent of energy supplies to the EU (expressed in value).

Supplies in textiles/clothing make up about 15 per cent (mainly intermediate goods imports+clothing export). SEMC absorb 7 per cent of EU exports of vehicle sector and nearly 9 per cent of agricultural product sector.

Imports (1998–2000 in million US$)

Exports (1998–2000 in million US$)

Source: Eurostat – COMEXT

A Sustainable Future for the Mediterranean

Box 10 Trade in services between the SEMC, the EAC and the European Union

As with the trade in goods, EU/SEMC trade in services concentrates on some of the 15-EU (the UK, France, Germany, and Italy, at 59% in 2002). The distribution between the Maghreb, the Mashrek, Turkey and Israel is relatively balanced (between 19 and 30% of total trade for each country). A great part of the SEMC and EAC surplus regarding the EU comes from 'travel' and 'transport' activities linked to tourism. But the SEMC are in deficit regarding services for enterprises and building trades as well as civil engineering. Among the SEMC, Israel is an exception since it is in overall deficit towards the EU and has a more diversified trade in services. The surplus recorded by Israel towards the EU, especially concerning the trade in services for enterprises, information technology and communication services, does not counterbalance its deficit regarding insurance services and travel and transport services (many Israeli tourists go to Europe).

SEMC Trade of services with the EU in million euros, in 2002

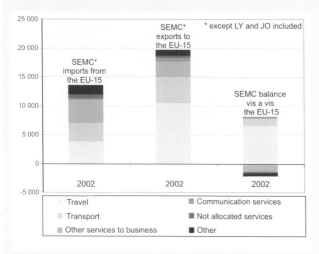

EAC Trade of services with the EU in million euros, in 2002

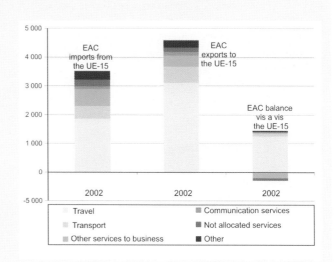

Note: LY not included, JO and IL included (this explains the difference observed between total imports and exports in 2002 compared with the previous graph).

SEMC Trade of services with the EU in million euros

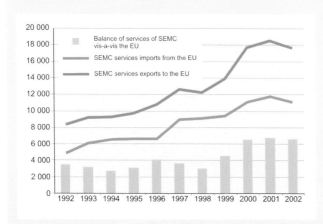

Note: IL and LY not included, JO included.

Source: Eurostat (EU declared figures).

Determining Factors of the Mediterranean Future

Highly polarized, FDI comes mainly from the EU (France and the UK) and the US, along historic links. Between 1991 and 2000, two countries, Israel (22 per cent) and Turkey (31 per cent), received more than half of the FDI received in the Mediterranean (the four EU-Med countries excluded). FDI per capita in Israel, Cyprus, Malta, Slovenia and Croatia is between 5 and 15 times greater than the Mediterranean average (see Statistical Annex).

The transfer of funds from emigrant Mediterranean workers represented more than US$20 thousand million in 1999 with, at the top, Turkey, Egypt, Lebanon and Morocco, but also two former European emigration countries, Spain and Greece, for the development of which these funds played an essential role. As a fraction of GNP, these funds are very important for some countries such as Lebanon (16 per cent), Albania (10 per cent) and Morocco (6 per cent), Egypt and Tunisia (4 per cent) even if they are drying up with time. They represent a large share of total revenue *in foreign currency* for the SEMC (57 per cent for Albania, 31 per cent for Egypt, 24 per cent for Morocco, 16 per cent for Greece, 11 per cent for Syria, 10 per cent for Turkey, 9 per cent for Tunisia). The total of these transfers is often two to three times higher than the Official Development Assistance for the Maghreb and the Mashrek countries.

We are seeing an increase in capital flows from the EU to the SEMC and the eastern Adriatic countries, where the official flows play a major and growing role. Financial flows are still polarized by historical bilateral links between countries, although we are seeing a certain rising importance from the European Commission. Except for Turkey and Israel, the SEMC only marginally attract direct foreign investments. This trend defines the assumption of the baseline scenario.

Multi-speed regional integration and cooperation

Although trade in goods, services, capital and exchanges of people is increasing and indicates a certain rapprochement between the EU and the Mediterranean, what about the convergence of policies and regional integration on which the formation of a genuine pole of stability depends?

Already, in 1989, development of north–south and south–south cooperation was one of the basic assumptions of *Plan Bleu* scenarios.[28] Since then, some progress in regional cooperation has been shown by the progressive expansion of the EU on the Mediterranean's northern shore, the launch of the Euro-Mediterranean partnership, and regional cooperation on the environment and sustainable development. However, this cooperation is still insufficient in the context of the challenges that must be met by the SEMC.

In the north, EU enlargement is overtaking deeper sustainable development

More than just a single market, the European Union is a political project of historic reconciliation, affirmation of shared values and principles, and building one of the world's major poles of stability and prosperity. This project is being built on successive enlargement and deepening.

The *deepening dynamics* is gradually leading the Member States to transfer a growing part of their sovereignty to the Union (foreign trade, environment, currency), even if in terms of foreign policy and defence, the EU is still seeking a unified project. Economic integration goes hand in hand with financial support for agriculture and the regions whose development is lagging, in order to reduce the negative effects of liberalization, encourage the upgrading of infrastructures and thus strengthen the internal cohesion of the Union. Agricultural subsidies accounted for approximately 51 per cent of total EU operating expenditure in 2002, and structural operations (structural and cohesion funds) for around 27 per cent. Regulatory tools encourage convergence, for example European Directives that formulate sometimes very ambitious objectives for member and accession countries.

This tripod (single market, agriculture and structural policies, and European legislation) makes the EU a real example of an attempt to *reconcile free trade and sustainable development*. Despite this, certain trends show only small signs of greater sustainability (transport, agriculture, unemployment and poverty, the environment, internal regional imbalances). The EU Strategy for Sustainable Development adopted in Göteborg in 2001 reiterates the need to integrate the objective of sustainable development into all European policies and to evaluate their social and environmental effects.

Moreover, the integration model has to encompass a great diversity of situations and does not imply standardization. The principle of mutual recognition[29] promotes subsidiarity and shows that modern economic management can rely on common principles, while respecting a great variety of national cultures and strategies. The model is becoming more and more attractive to countries neighbouring the EU the northern shore of the Mediterranean. The latest wave of *enlargement* in 2004 brought in ten new countries, including Cyprus, Malta and Slovenia. Turkey, as well as the other eastern Adriatic countries (Croatia, Albania, Bosnia-Herzegovina, Serbia-Montenegro) are seeking entry into the EU in the longer term.

To join the EU, candidate countries, by accepting the 'Community *acquis*' (the body of common rights and obligations which binds all the Member States together within the EU), must fulfil the political, economic and institutional accession criteria. In return they benefit

1 A Sustainable Future for the Mediterranean

Box 11 Origin and destination of capital flows to the Mediterranean

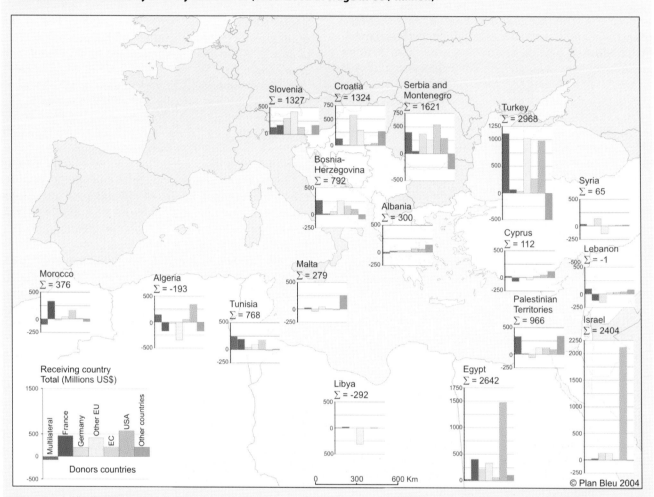

For example, Turkey received an annual average of US$2968 million in 2000–2002, and registered a negative contribution of US$500 million from 'Other countries'.

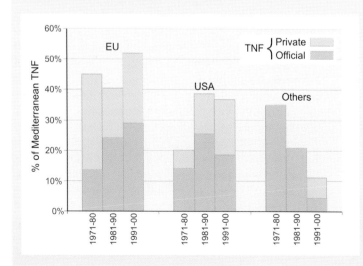

Source of capital destined to the Mediterranean (excl. the EU) –
as a percentage of the net total contributions for the Mediterranean

Note: TNF: Total Net Flows.

EU includes Member States and the European Commission.

The funds come mostly from the EU and its Member States and the US. The EU share of these contributions is higher and increasing (with a large increase in official contributions). The US provides 36 per cent of the total (with an increase in private contributions).

Source: OECD Development Aid Committee (DAC)

from advantages of the single market and the EU's institutional and financial accompanying measures for compensating for the social, territorial and environmental effects of the reforms and help in upgrading their economies and infrastructures. During the transition periods, upgrading is facilitated by financial transfers.[30] This EU support has been decisive for the success of the European integration model, even if it has not always been enough to prevent a certain negative impact as in Greece and Spain (Box 12).

Box 12 Spain and Greece in the EU: A success story clouded by environmental impacts

The successful integration into the EU of Greece in 1981, and Spain and Portugal in 1986 (democratic stability, the relative convergence of their economies with the rest of the EU and the stopping of emigration) is most certainly due to the domestic reforms carried out by these countries but also, in large part, to the extent of the support from EU community funds.

In the agricultural sector, for example, integration led to important structural changes (specialization and reducing the working population and the number of farms) the social and environmental effects of which were reduced because of EU support (aid in upgrading facilities and infrastructures and reconversion to other activities).

Without this support and the environmental directives, just liberalizing trade would have had much higher social and environmental **costs** than those actually seen, even if some could not have been avoided.

Thanks to the agro-environmental programmes, **Greece**'s joining the EU enabled the use of chemical fertilizers to remain fairly limited. On the other hand, there was a greater polarization of activities around the Athens urban area and increased tourist pressures on the most sensitive coasts.

Despite an active policy and European support, the pressures on the environment in **Spain** grew more rapidly than the economy: water consumption, the dumping of industrial waste and polluted water into the Mediterranean, gaseous emissions because of increased road traffic and deforestation related to the large expansion of tourism activities on the coasts. Modernization and agricultural specialization are leading to overexploitation of water resources, the growing use of chemical products and the pollution of water tables. Desertification has also increased since membership. Finally, in Spain, the richest country in the EU in terms of biodiversity, the development of its economic activities threatens that biodiversity.

Source: Plan Bleu, based on Kuik, O. J. and Oosterhuis, F. H. (2000) Free Trade and the Environment in the Euro-Mediterranean Context: Lessons Learned from Spain, Portugal, Greece and Poland. Paper prepared for the Mediterranean Commission on Sustainable Development. Plan Bleu, Sophia Antipolis

With EU enlargement, solidarity is being reoriented in favour of the less rich accession countries. In the coming years, structural funds, one of the biggest items in the Community's budget, should gradually be redistributed to the regions of these new member countries and 'leave' the slowly developing regions of the four EU-Med. However, the structural funds for the new members will probably never reach the levels received by the four EU-Med countries in the past. In 2002 net financial transfers from the EU to Greece and Spain (as farm subsidies and structural support), were about 5–6 times higher than those for the ten new member countries (Figure 28, p47), and equivalent to about 1.2 per cent of Spanish and 2.3 per cent of Greek GDP.

At the same time, in view of their accession in the longer term, the EU has, since 2000, carried out a stabilization policy towards the other eastern Adriatic countries that aims to consolidate peace and ensure democracy and prosperity through 'stabilization and association agreements' and the CARDS programme.[31]

The EU enlargement process, as it spreads gradually over all the Mediterranean's northern shore, perhaps even to Turkey, currently seems to have carried the day over that of deepening. The speed of the expansion contrasts with the slowness of decision-making reforms for making the European project more democratic and popular, for strengthening the Union's budget, and for building a Community policy on external affairs and defence.

In the south and east, cooperation limited by conflicts

The regional integration model that is being built in the north has no equivalent in the southern- and eastern-rim countries. Despite several initiatives since 1945, cooperation between SEMC exists more on paper than in reality. The region is still characterized by continuing conflicts and a lack of structured cooperation. Although the League of Arab States plays an important consultative role between its member countries, regional development in the Near East and the Arab Maghreb Union is still being blocked by the difficulty of making the transition from a logic of rivalry (or conflict) to one of cooperation.

From the solely economic point of view, the low intensity of trade among the SEMC bears witness to their very weak integration with each another. Intra-SEMC trade accounts for less than 5 per cent of their external trade.

With only a few exceptions, few major regional projects or programmes (industry, transport or energy) have been implemented that could exploit economies of scale. Some initiatives have been launched for *liberalizing trade*. In addition to many bilateral agreements between the SEMC, several regional initiatives have been launched. The project of a Great Arab Free-Trade Zone agreed in 1997 by 17 signatory countries, eight of which are Mediterranean, includes dismantling of tariffs by 10 per cent per year over a ten-year period so as to stimulate inter-Arab trade. A free-trade zone of the Mediterranean Arab States (Jordan, Tunisia, Morocco and Egypt) has also been launched by the Agadir process. These initiatives are superimposed on the project of a Euro-Mediterranean Free-Trade Zone, initiated by the EU (presented below). Finally, a Middle East Partnership Initiative (MEPI) was proposed by the US in 2002. MEPI relies initially on bilateral agreements between the US and Middle East partners (three have been signed with Israel, Jordan and Morocco), which depend on satisfying political and economic criteria. They are supposed to evolve towards sub-regional accords, then to a free-trade area (MEFTA) by 2013. Financial support is planned for projects in three sectors (education, political and economic reform) but remains limited considering the area covered: from Morocco to Iran (US$29 million in 2002 and US$100 million in 2003).

All these 'variable geometry' initiatives do not make it possible to form a coherent and mutually supporting ensemble. We are still a very long way from the alternative scenarios of Plan Bleu 89. Without such cooperation, the southern and eastern countries remain divided. Their positions at the regional level and in international organizations such as the World Trade Organization (WTO) are often contradictory and are not based on any shared vision.

North–south cooperation: An ambitious Euro-Mediterranean partnership with limited means

North–south cooperation was for a long period dominated by bilateral approaches that still account for most Official Development Assistance in the SEMC and benefit from the often-remarkable know-how of powerful cooperation agencies, particularly in France, Germany, the US and Japan.

Box 13 The Euro-Mediterranean Partnership

Established in 1995 in Barcelona, the Euro-Mediterranean Partnership has three components:

- A political and security partnership to 'define a common area for peace and stability'.
- An economic and financial partnership for 'building an area of shared prosperity'; this aims to realize a Euro-Mediterranean free trade zone (EMFTZ) by 2010 with total suppression of tariffs on industrial products* from the EU to the partner countries (the EU having already removed tariffs on industrial products from its partners). The partnership has star-shaped bilateral association agreements between the EU and each Mediterranean partner, replacing the cooperation agreements made in the 1970s. By mid-2004, all partner countries, with the exception of Syria, had signed an association agreement.
- A partnership in the social, cultural and human sectors for developing human resources, encouraging inter-cultural understanding and exchanges between civil societies.

The EMP has two financial-aid instruments for partner countries to accompany the liberalization of their economies: a MEDA fund and European Investment Bank (EIB) loans.

The MEDA instrument, now in its second phase (MEDA II from 2000 to 2006), has an average annual budget of 764 million euros, an 11 per cent increase compared with MEDA I (1995–1999). The part of MEDA II devoted to regional projects accounts for 28 per cent of the total. Credit is mostly (70 per cent) meant for supporting economic transition and the private sector. Under MEDA II, funds for social aspects were reduced to 20 per cent while those for the environment increased to more than 9 per cent.

Loans from the EIB complete the financial set-up with an average of 880 million euros per year (1995–1999), increasing to 1.25 thousand million euros per year (2000–2003). The new Facility for Euro-Mediterranean Investment and Partnership (FEMIP) should make it possible to increase the EIB annual commitments to partner countries from 1.4 to 2 thousand million euros per year for 2002–2006. This facility is basically meant for the private sector and infrastructures. The financial programmes are mostly bilateral and cover four sectors: energy, water supply and sanitation, industries and transport/ storage.

* And a near status quo on agriculture and services, with agricultural products having to be renegotiated after 2010.

A basic stage of structuring regional integration began in 1995 with the launch of the Euro-Mediterranean Partnership (EMP). Strengthening a rapprochement process that has been readjusted on several occasions since the 1970s, the EMP brings the EU together with ten 'partners' from the southern and eastern countries.[32] It aims simultaneously to re-equilibrate EU policies in a priority area that affects its security and create a counterweight to the big economic poles of Asia and North America. It provides some form of response to the 1994 US initiative of the North American Free Trade Agreement (NAFTA) signed between the US, Canada and Mexico. It is the first expression of a genuine regional policy between the EU and its partner countries, covering all their relations, and willing to use the economy for the benefit of more ambitious political, cultural and security objectives (Box 13).

Despite its progress and ambitions, the Partnership is still criticized for its implementation and resources. Although it has been slowed by the continuing conflict in the Near East, its lack of real *political impetus* is also criticized. Economic and security considerations predominate over political and cultural ones, while the partner countries have limited possibilities to influence the overall process.[33]

Determining Factors of the Mediterranean Future

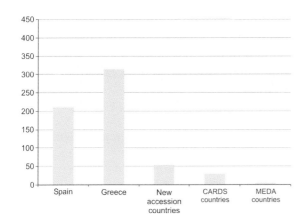

Figure 28 Net amounts of the EU's main financial transfers in 2002 (euros per capita)

Source: Plan Bleu from available statistics

Note: Spain and Greece: net operating budget for 2002 per inhabitant. This includes the amounts received from the EU for agriculture, structural actions and internal policy minus contributions of the country to the EU budget.

Amounts from PHARE, ISPA and SAPARD programmes for AC 10 countries, amounts of the CARDS programme for Balkan countries and MEDA funds for Mediterranean countries: average for 2000–2006.

Another concern relates to the Partnership's *insufficient accompanying resources* in terms of both staff and funding (MEDA and EIB), which seem to be on too small a scale in view of the risks and costs of the reforms needed in the SEMC. Despite recent increases, MEDA credits, excluding loans from the EIB, amounted to only 5 euro per capita per year in 2002, or, as a comparison, one-sixth of what people in the eastern Adriatic countries received from the EU (30 euro per capita per year), one-tenth of what the people of the ten new member countries receive (53 euros per capita per year) and nearly 60 times less than what some four EU-Med countries receive for structural and agricultural aid (Figure 28). These figures reflect the different political and economic relationships that link the EU to its member countries, to accession countries, and to partner countries.

Some recent changes, such as the setting-up of a Parliamentary Assembly of the Euro-Mediterranean Partnership and the Euro-Mediterranean Foundation for dialogue between cultures, constitute a response to these criticisms. The Facility for Euro-Mediterranean Investment and Partnership (FEMIP), created within the EIB, could be transformed into a Euro-Mediterranean bank (Box 13).

The Partnership is also criticized for poor real political commitment from the SEMC other than for tariff removals (economic and environmental policies). More generally, the question of its global vision is raised: the Partnership is being built brick by brick mainly in the form of star-shaped bilateral relations. Thus, when a country signs an association agreement with the EU, it commits to making a free-trade agreement simultaneously with all the other signatories within a maximum time frame of five years, but the mode of negotiation is on a country by country basis, and the development of south-south sub-regional cooperation is not favoured by MEDA funding (mostly bilateral). This is particularly the case in the current negotiations on agriculture, undertaken without a common view by the partners on the future of agriculture, especially of its multiple roles and possible contribution to sustainable development in the region.

To sum up, the *objectives of sustainable development* would not be sufficiently taken into account by the Partnership, even if several recent conferences have reiterated the primary objective (Box 14).

Box 14 The Euro-Mediterranean Partnership and sustainable development

Although the Barcelona Declaration giving rise to the EMP in 1995 explicitly aimed at 'reconciling economic development with protecting the environment and including environmental concerns in relevant aspects of economic policy', it has to be acknowledged that the environmental and social aspects of development are often eclipsed by the economic aspects.

The EMP has a priority programme for the **environment**, SMAP (Short and Medium Term Action Programme), defined in 1997 around five priorities: integrated water management, waste management, hot spots (pollution and biodiversity), integrated coastal area management and combating desertification.

But resources remain weak in view of the challenges. Between 1995 and 2000 the EMP funded environmental projects to the tune of 260 million euros, more than 86 per cent of which was in the form of interest rebates on EIB loans in the water and sanitation sectors. Between 2001 and 2002, 133 million euros supported projects in water and sanitation: 70 million euros of regional projects (SMAP II and Local Water Management) and 63 million euros of bilateral projects, mostly interest rebates on EIB loans. In all, between 1995 and 2002, nearly 393 million euros were invested in the environment (including water and sanitation) or 8 per cent of all commitments and an average of 49 million euros per year.

The EMP is not succeeding in truly influencing policies on the environment, urban and rural development and coastal zone management. Until recently, the other sectoral programmes of the EMP (energy, water, transport, the information society and industry), funding by the EIB and the Partnership's economic surveys have seldom included environmental and sustainability aspects despite the very strong interrelationship between the environment and development in the region.

Analysing the contents of the various **bilateral association**

agreements shows few references to protecting the environment. Some agreements mention a rapprochement with European environmental standards (Algeria and Tunisia) or refer to regional cooperation in environmental matters, but nothing is specified about the objectives and resources, for example, strengthening capabilities in environmental management or monitoring/assessing the social and environmental impacts of free trade.

The EMP, unlike NAFTA, does not have a multilateral legal framework or regional coordination mechanism for the environment, the equivalent of the North American Commission for Environmental Cooperation.

These observations suggest the need for **readjustment** of the EMP. The European Union Strategy For Sustainable Development, adopted in Göteborg in 2001, by affirming that the internal and external policies of the Union should support efforts made by non-EU countries to achieve sustainable development, expressed a desire for an evaluation of the impact of the agreements made with the Community and monitoring through indicators. The Conference of Euro-Mediterranean ministers of Foreign affairs (Valencia, 2002) reaffirmed sustainable development as the EMP's guiding principle; it requested a survey of the impact of the EMFTA in terms of sustainable development (launched in 2004) and demanded a strengthening of the technological and institutional capacities in the region in environmental and sustainable development matters. Finally the Conference of the Euro-Mediterranean ministers of the Environment (Athens, 2002), acknowledging the imperatives of integrating the environment into the EMP, requested the establishment of integration strategies for each sector. Greater synergy is also sought with other regional programmes such as the Mediterranean Action Plan. However, it is too early to know how these intentions will be implemented so as to integrate the different sustainable development dimensions into the EMP.

In future, Euro-Mediterranean relations should be part of the new '*European neighbourhood policy*', introduced in 2003, which opens up perspectives of strengthening economic and political integration for the near neighbours of a wider EU.[34] The objective is 'to work with the partners to reduce poverty and create an area of prosperity and shared values, based on free trade, increased economic integration, intensified political and cultural relations, strengthened cross-border cooperation and sharing responsibility for conflict prevention and resolution'. In the long term this means facilitating a move towards a framework where the EU and its neighbours will maintain relations that are 'comparable with the close political and economic ties that are currently characteristic of the European economic area'. This relates to increasing participation in the internal European market, continuing liberalization and integration with a view to extending the free circulation of people, goods, services and capital. Although the new neighbourhood policy recalls that free trade cannot be considered as an end in itself, the question still remains as to how the implementation will make trade and sustainable development more compatible among countries with large gaps in development.

The implementation phase began in 2004–2005, through National Action Plans agreed with the countries that have already ratified the association agreements – Israel, Jordan, Morocco, Tunisia and the Palestinian Territories. These plans seek to integrate sustainable development principles, and their environmental sections put an emphasis on cooperation for better governance, legislation and implementation as well as stronger regional and global involvement. The process with similar plans will be extended to Egypt and Lebanon.

These new directions introduce the differentiation principle by virtue of which the offer of real advantages and preferential relations is subordinated to progress being made by neighbours with political (democracy, human rights), economic (macro-economic reforms, developing the private sector) and regulatory reforms (rapprochement of laws, protection of health and the environment). *Under these conditions, the EMP could well evolve towards a 'variable geometry' integration depending on the countries and their efforts.*

Mediterranean cooperation on the environment

Although the EU, through the EMP, has a major responsibility for searching for a reduction in the risks related to the liberalization of the SEMC economies and more generally for sustainable development in the region, other regional partners have also been working, sometimes for a long time, to promote economic development that 'respects the Mediterranean environment'.

In the first place, the *Mediterranean Action Plan* (MAP) has played an important role for over 25 years in regional cooperation on the environment with a heightened interest in sustainable development. Implemented in Barcelona in 1975 under the aegis of the UNEP (United Nations Environment Programme), MAP comes under the authority of the Contracting Parties of the 1976 *Barcelona Convention* for the Protection of the Mediterranean Sea against Pollution, covering the riparian states and the European Community (Box 15). Although its primary purpose was to protect the Mediterranean Sea through the progressive implementation of legal instruments, its brief was expanded in 1995 to the sustainable development of the Mediterranean and its coastal region. MAP has developed coastal management projects and, with GEF (Global Environment Facility) support, programmes for curbing pollution and protecting coastal and marine biodiversity. In 1996 it instituted the *Mediterranean Commission for Sustainable Development* (MCSD) and plans to propose a 'Mediterranean strategy for sustainable development', to which this report is meant to contribute.

The Barcelona Convention, its protocols and MAP provide a legal framework and vital instruments in the service of intergovernmental cooperation on the environment and sustainable development in the Mediterranean. They constitute the only institutional framework that gathers all Mediterranean countries and the European Commission together at regular meetings between Mediterranean decision-makers.

Despite the efforts carried out by MAP to build a joint vision and draw up rules for all Mediterranean countries to protect and manage the Mediterranean region's environmental public goods in a sustainable way, there are disparities in the objectives, standards and resources of policies implemented in this field as big as those in the socio-economic sector. In the north of the basin effective *implementation* of the Barcelona Convention is powerfully strengthened through European laws (which are often more restrictive[35]), through enforcement measures of a jurisdictional nature and also through the EU structural funds. In contrast, the southern and eastern countries face serious difficulties, particularly financial ones, in trying to tackle all issues related to protecting the marine environment. There is, therefore, a high risk of a widening 'fracture' between the shores as regards environmental regulations (maritime transport security, protection of marine biodiversity, combating land-based pollution) and diverting or displacing pollution to the south and the east.

Other regional initiatives also deserve a mention. In 1989 the World Bank, in conjunction with UNDP, the EIB and the European Commission, launched a Mediterranean Environment Technical Assistance Programme (METAP), which gave support to 15 of the region's countries for identifying projects, strengthening environmental management capabilities, formulating environmental policies and mobilizing funds for investment. However, the programme has suffered from only limited resources compared with its ambitions and needs.

The major international UN agencies are also involved but without distinguishing in their approaches the specificities of the Mediterranean eco-region, which is always fragmented into three 'regions' (Africa, Europe and Western Asia).

The presence of multiple *networks* and *actors* in eco-regional cooperation, too many to mention, bears witness to a sometimes very old awareness of the importance of collectively preserving such specific resources and eco-systems: *Silva Mediterranea* for the forest sector, the International Commission for the Scientific Exploration of the Mediterranean Sea (CIESM), the General Fisheries Commission for the Mediterranean (GFCM), ICAMAS (International Centre for Advanced Mediterranean Agronomic Studies) for agriculture, the

---

Box 15 The Barcelona system (Barcelona Convention, MAP, MCSD)

In 1976 the Mediterranean countries and the EEC adopted the Convention for the Protection of the Mediterranean Sea against Pollution, the 'Barcelona Convention', which entered into force in 1978.

Since then, six protocols have been adopted by the Mediterranean countries.

1. **Dumping** Protocol for the prevention and elimination of pollution in the Mediterranean Sea by dumping from ships and aircraft or incineration at sea (1976, amended in 1995).
2. **Prevention and Emergency** Protocol, for preventing pollution from ships and, in cases of emergency, combating pollution of the Mediterranean Sea (1976, re-examined in 2002).
3. **LBS** Protocol for the protection of the Mediterranean Sea against pollution from land-based sources (1980, amended in 1996).
4. **SPA and Biodiversity** Protocol concerning Specially Protected Areas and biological diversity in the Mediterranean (1995, replacing the 1982 SPA Protocol).
5. **Offshore** Protocol on sea protection against pollution resulting from exploration and exploitation of the continental shelf and the seabed and its subsoil (1994).
6. **Hazardous Wastes** Protocol, on preventing pollution of the Mediterranean Sea by trans-boundary movements of hazardous wastes and their disposal (1996).

The Barcelona Convention was amended in 1995, and became the Convention for the Protection of the Marine Environment and the Coastal Region of the Mediterranean, which entered into force in 2004. Its modified protocols are in the process of being ratified. By mid-2004, the new *SPA and Biodiversity* Protocol and the *Prevention and Emergency* Protocol had also entered into force.

The Mediterranean Action Plan (MAP) ensures the follow-up and implementation of the whole system. Including right from the start three constituents (legal, scientific and economic), it operates with a secretariat based in Athens and six centres located in various Mediterranean countries. The *Plan Bleu* Regional Activity Centre, the MAP systemic and prospective centre, acting as a Mediterranean environment and development observatory, is the author of the present report.

The Mediterranean Commission on Sustainable Development (MCSD), set up in 1996, is composed of representatives of the coastal countries and the European Commission, as well as co-opted representatives of environmental NGOs, local authorities and socio-economic actors. With support from MAP's Centres, in particular *Plan Bleu*, MCSD mobilizes specialists and officers, organizes forums and has already produced a number of strategic papers on water, tourism, the impact of the Euro-Mediterranean Free Trade Zone, urban management, etc. Within this same framework, a first set of 130 Mediterranean indicators for sustainable development was adopted in 1999.

Centre for Environment and Development for the Arab Region and Europe (CEDARE), the Medwet programme for the conservation of wetlands, and the Mediterranean programme of World Conservation Union. In parallel to the development of Mediterranean intergovernmental cooperation, many regional NGOs and Mediterranean networks with very uneven capabilities have been formed. They cover highly varying fields: water, forests, cities, natural hazards, energy, biodiversity and the protection of nature, the environment, coastal areas, academic networks, consular institutions and many others. The recent development of decentralized cooperation between regions and other local authorities is also the source of important dynamics for the region's future.

However, all these initiatives and networks have only limited resources compared with the challenges to be met.

What assumptions for regional construction?

At the end of this analysis, although interactions seem to be increasing in the Mediterranean, the political building of a regional pole remains embryonic. Nothing can guarantee that the rapprochement of objectives announced by the EU in its new neighbourhood policies will be accompanied by enough resources to help the SEMC meet their challenges; there is a high risk that the EU will give short-term priority to the success of its enlargement. In the Mediterranean, it seems that political solidarity wins out over geographical solidarity. The Mediterranean region remains more a communications buzzword than a political reality.

The baseline scenario assumes that all the northern shore countries are gradually integrated into the EU and experience a strengthening of democracy, as well as a certain convergence of socio-economic and environmental standards. They form a unified, pacified and modernized whole where cohesion is restored despite the upheavals caused by liberalization. However, the production and consumption patterns follow those of the rest of the EU and often remain not very sustainable. Lacking concomitant political deepening, this dynamic of enlargement is accompanied by a certain weakening of the European project that is essentially reduced to a single market with common values and rules.

Despite awareness that its fate is linked to that of the south and east, the EU hardly manages to play the historic role that would be justified by the importance of its interdependencies (economic, social and environmental) and its well understood interests. South–south cooperation and eco-regional cooperation remain insufficient, limited by continuing conflicts; despite its progress, Euro-Mediterranean integration continues with variable geometry and few resources, at least in the initial period (to 2015) during which most resources are mobilized for the new member countries.

The Mediterranean then finds itself in danger of an accentuated fracture between a northern shore where development towards the single market is accompanied by powerful political, financial and regulatory commitments, vital for strengthening cohesion and sustainable development, and a southern and eastern shore where economic liberalization is continuing without an equivalent level of support and political commitment.

This baseline scenario intentionally excludes other more pessimistic assumptions, such as an extension of conflicts or a decline in the European project that would reduce the chances of Mediterranean regional integration.

A deficit of governance in the face of environment/development challenges

If the future of the Mediterranean comes within an international and regional framework, it also depends especially on the way in which national policies and actors perceive the objective of sustainable development and, in particular, on their ability to integrate the environment into development. As long ago as 1989, the *Plan Bleu* scenarios showed that protecting the Mediterranean could not be accomplished by actions carried out on the sea and the coastal regions alone, but that it depended largely on national policies on development, the environment and regional planning. The sustainable development scenarios were then typified not only by strengthened regional cooperation but also by a different approach to environmental problems at the whole country level: internalizing the costs of protection, taking account of the environment in decision-making mechanisms, reducing centralization, implementing better coordination and associating the public with decision-making and management.[36]

Since 1992 the United Nations Conference on the Environment and Development (UNCED) has added international political impetus to environment/development issues by approving Agenda 21 and the Rio Declaration. By returning people to the heart of sustainable development concerns, the Rio Declaration stresses the rights of each individual to access to environment-related information, to taking part in the decision- making process and to access to legal and administrative actions. Governments are invited to work out nationwide laws relating to responsibility for pollution and other damage to the environment, to promote the internalization of costs for protecting the environment, and to use economic instruments (based on the polluter-pays principle). Many international and European conferences have since highlighted the large practical difficulties in achieving this. They advocate the passing of more targeted objectives, centred on priorities. The 2002 Johannesburg summit, while recalling the importance of 'good governance' in the process, offered a targeted statement based on three priorities: (i) the eradication of poverty; (ii) adapting consumption and production patterns; and (iii) protecting and sustainably managing the stock of natural resources.

In the Mediterranean, despite considerable efforts to strengthen environmental administrations and legislation, it has been extremely difficult to prevent environmental degradation by means of effective upstream action on the development process.

The recent expansion in environmental policies

Over the last two decades, the Mediterranean countries, like many others, have mainly strengthened environmental policies and institutions as a sector in its own right, without really finding complementary intervention mechanisms upstream of consumption and production activities or other economic and social policies. While physical and land planning policies registered a certain decline, unprecedented efforts have been devoted to structuring this new sector of state intervention, the protection of the environment. In the four EU-Med countries, impetus from the European Union has been decisive, as has that from the major cooperation programmes (World Bank, MAP) in the other Mediterranean countries.

Environmental administrations and agencies

Being a recent area for public intervention, environmental policies have run up against two problems: the definition of their field of application and their institutional position and the types of action to be taken by the administrations in charge of implementation. The cross-sectoral nature of the environment, the management of which interacts with the activities of many ministries, makes the question of its integration into conventional policies and administrations especially tricky.

Environmental policies have generally been structured around three major *fields of competence*: (i) pollution prevention and control; (ii) the protection of nature; and (iii) the preservation of renewable natural resources, with notable differences from country to country. Generally speaking, the missions and modes of action of administrations in charge of the environment cover technical coordination, impetus, laws and regulations, and sometimes research.

The *institutional positioning* of administrations in charge of the environment that have appeared in the Mediterranean countries is generally typified by a certain instability. Such administrations have often been created within other older ministries and then experienced frequent changes of supervisory ministries.

Despite this instability, central environmental administrations have been set up in all the riparian countries in the form of a ministry, a 'sub'-ministry or an independent agency. In recent years the linking between the Environment and Spatial Planning in several countries (Italy, Greece, Slovenia, Croatia, Bosnia-Herzegovina, Algeria and Morocco) has represented a significant development in the search for consistency. Elsewhere, environmental policies come from an independent ministry (Spain, France, Malta, Albania, Turkey, Lebanon, Israel and Egypt) or coexist with agriculture (Cyprus and, more recently, Tunisia) or other sectoral authorities (Syria and Libya).

Institutional uncertainties contribute to a certain *weakness* in environment-related administrations, which is accompanied by a weakness of human and financial resources with budgets amounting in general to less than 0.1 per cent of government budgets,[37] and, above all, weak inter-ministerial power. In the Mediterranean countries of the EU, environmental matters have benefited over the past few years from increased attention, which has been partly expressed by an increase in staff and budgets of the lead ministries. In the countries joining the EU in 2004, progress has been made in reinforcing capabilities in implementing the 'Community *acquis*'.

In addition to the central environmental administrations, *technical agencies* have often been set up to support and implement environmental policies and allow them more autonomy (particularly financial autonomy), flexibility and a greater capacity for mobilizing various stakeholders. This applies to generalist agencies in Italy, Egypt and Tunisia or analogous agencies that existed in other countries (for example Albania and Algeria), transformed later into ministerial divisions (Box 16).

To inadequacies in financial and staffing levels can be added the critical weakness of authorities in administrative policing matters. More generally, in the Mediterranean as elsewhere, the critical factor is the *low priority given to environmental issues* by governments through their inter-ministerial coordinating body, since many environmental issues derive from more powerful sectoral policies such as public works, transport, agriculture and industry, on which environmental administrations have only limited influence.

Policies focused on regulating the environment

Most efforts in this new environmental protection sector have focused on regulations.

At the *international* world and regional level, the 1980s saw a veritable outburst of *environmental law* with nearly 300 documents, treaties or conventions. A first wave of regulations focused on protecting the major components of the environment: the sea, continental waters, the atmosphere and biological diversity, a second dealt more with the control of polluting substances, and a third, in the 1990s, with standards that were aimed

Box 16   Agencies for environmental protection

In **Egypt**, the Egyptian Environmental Affairs Agency (EEAA) was established in 1982. Until 1990, EEAA evolved slowly with little impact on the development of environmental policy, and little presence in the public arena as a result of limited staffing and resources, and having no regulatory authority. The mandates of the EEAA which also include the recommendation of financial mechanisms for encouraging the different partners to undertake environmental protection activities, were extended in 1994 to authority for regulating air pollution and controlling hazardous waste management and discharges into maritime water. At the regional level, five Regional Branch Offices represent EEAA and address urgent issues.

In **Tunisia** the National Agency for the Protection of the Environment (ANPE), established in 1988, is in charge of monitoring the state of the environment and pollution control. It carries out environment impact assessments of new agricultural, commercial and industrial projects, analyses the documents submitted for approval prior to investments with a polluting potential, and controls polluting establishments.

In **Italy** the National Agency for the Protection of the Environment (ANPA) was established in 1994 to serve as a scientific auxiliary to the Ministry of the Environment by dealing with the functions of pollution prevention and reduction previously handled by local health units. Its initial activities (monitoring the state of the environment, preventing and treating pollution, natural and technological risks and working out integrated strategies) have been gradually expanded to *ex ante* economic analysis of environmental projects, the promotion of clean technologies and sustainable urban development. Becoming an agency for protecting the environment and technical services (APAT), its jurisdiction has recently been extended to hydro-geological risks, the preservation of water resources, and soil protection. APAT is part of a federal system of 19 regional agencies (ARPA) and two provincial environmental agencies. With nearly 6000 agents, the ARPAs are becoming the main environmental inspection body; they do not operate as a regional section of APAT but are under the direct leadership of regional authorities.

More **specialized** national agencies have sometimes been entrusted with protecting a specific environment (the *Conservatoire du littoral* in France, Tunisia's Agency for the protection and planning of coastal zones, etc.) or the management of water resources. In terms of protecting nature, national parks in most countries and regional natural reserves have often been created by Ministries of Agriculture and/or Forests. Other agencies operate under the shared supervision of several ministries. Examples include: in France, the Agency for the Environment and Energy Management (ADEME), ONF (National Forestry Office) and the National Agency for Radioactive Waste Management (ANDRA), and in Italy the National Agency for New Technologies, Energy and the Environment (ENEA) and the Central Institute for Scientific and Technological Research as applied to the sea (ICRAM).

increasingly often at human activities themselves. The new rules did not, however, supplant the previous legislative developments; the three kinds of regulations coexist and advance simultaneously. Integrated protection of the environment is the most recent phase in this development. In order to be more effective, it no longer simply aims at protecting the main environments or combating the potentially noxious effects of specific substances but at all human activities that might have unfavourable impacts on the environment as a whole.[38]

Environment-related texts have existed in the *countries*, sometimes for a long time, on matters such as forest management and the protection of natural sites or water resources. But those relating to pollution control, and especially those setting up an overall framework for protecting the environment, have generally been lacking. They have, however, proliferated over the past 15 years, but little effort has been made to codify them with a view to harmonizing the many national legislative documents.

For EU members and the accession countries, some 200 legislative documents on the environment, including more than 140 directives, contribute to harmonizing environmental policies and legislation around three principles: the precautionary and preventive action principle, the principle of prioritizing corrective action at the source of damage to the environment, and the polluter-pays principle. Although the application of community law varies markedly from one country to another, this has enabled enormous progress to be made, for example on water quality and waste treatment. At the beginning of 2004, before integrating into the EU, the 10 accession countries had already transposed an average of about 90 per cent of the Community *acquis*, mainly the directives on air quality and waste management, and strengthened their administrative capabilities for implementation.

Above and beyond the arsenal of specific laws for each environmental milieu or sector (water, soils, biodiversity, waste, etc.) in each Mediterranean country, it is worth noting the adoption of *framework laws for environmental protection* that include the major legal principles for protecting the environment such as the precautionary principle and the polluter-pays principle. These framework laws less frequently affirm the existence of general pollution violations (Box 17).

Box 17   Framework laws for protecting the environment

Greece was one of the first countries in the world to refer to the environment in its 1975 Constitution; in 1986 it passed a framework law on environmental protection that affirms the polluter-pays principle and provides for levying taxes on waste and charges for water consumption. In **France**, although the legal regime for protecting the environment is very old, it was

Determining Factors of the Mediterranean Future

only in 1995, with the Barnier law, that the precautionary and polluter-pays principles were formally adopted; in 2004, a constitutional law with a Charter for the environment was adopted, which includes the precautionary principle. A long effort to harmonize the many laws and decrees finally resulted in the publication of the Environmental Code in 2000. In **Spain** and **Italy** the development of environmental policies relies on legislation and planning developed by the national and, largely, regional authorities. At the national level, the main efforts are devoted to transposing European legislation into national law and ratifying international conventions; the legal framework, however, remains extremely fragmented with the lack of a comprehensive underlying law on the environment.

In **Slovenia**, the Environmental Protection Act of 1993 was the first comprehensive collection of aims, principles and rules of environmental management since the country's independence. It is based on Agenda 21, environment-related EU directives and Slovene experience with environmental management. In **Croatia**, the Law on Environmental Protection was adopted in 1994. This 'umbrella law' sets out the rights and responsibilities of actors in the environmental sector, stipulates who is responsible for pollution and who has to clean it up and defines environmental inspections. In **Albania**, the law setting a comprehensive framework for Environmental Protection was adopted by the Parliament in 2002.

In **Turkey** the 1983 law on the environment includes the polluter-pays principle and sets out a regime of objective liability. Its provisions make impact studies mandatory, ban certain polluting operations and the disposal of dangerous chemical substances and waste by making them punishable with penalties, provide incentives for less pollution, and create a fund for the environment.

In **Lebanon** the draft National Code for the Environment was put before parliament in 1999 but has not yet been promulgated. In **Egypt**, Law 4/1994 for the Environment enlarged the EEAA mandate and provided an array of implementation instruments including regulatory standards, inspection, enforcement, the review of environmental impact assessments and the implementation of economic instruments.

In **Algeria** a new environmental protection law was passed in 2003, based on the principles emerging from the Rio Summit (prevention and precaution, polluter-pays, integration and participation) and lays down the foundations of integrated pollution control. **Morocco** passed an umbrella law in 2003 for protecting the environment; it includes the polluter-pays principle while adding that of the user-pays, legislates on polluting facilities, institutes an environmental protection fund, and defines a regime of civil liability for actions harmful to the environment.

Sources: UNECE *Environmental Performance Reviews*, Slovenia 1997, Croatia 1999, Albania 2002. OECD *Environmental Performance Reviews*, Turkey 1999. Algeria. *Journal officiel* no 43 of 20 July 2003. Morocco. *Bulletin officiel* no 5118 of 19 June 2003

Efforts in planning and environmental assessment

In addition to these entirely legislative efforts, which are always under development, most Mediterranean countries have developed *national plans for environmental protection*: indicative, multi-annual planning aimed at facilitating the allocation of funds.

Italy, with its three-year plans in 1989 and 1994, France, with its ten-year plan in 1990, and Greece, with its five-year plan from 1994 to 1999, have tried to set objectives for environmental policies according to observed performance or counter-performance. Then the European community's policies in environmental planning and regulation gained greater importance. The European Union is on its sixth environment action programme for the 2002–2012 period. Spanish and Italian regions continue to deal with environmental planning.

In the southern and eastern Mediterranean countries the preparation of plans for protecting the environment has often benefited from technical and financial support from the World Bank, UNDP and sometimes German bilateral cooperation. Technical support in the eastern Adriatic countries has also come from the European Programme of Pre-Accession Aid for Central and Eastern Europe (PHARE). National Environmental Action Plans (NEAP) have been developed in Slovenia (1997), Albania (1994, updated in 2002), Croatia (2002), Turkey (ten-year NEAP in 1998), Syria (1998–2007 NEAP), Egypt (first NEAP in 1992, a new one for 2002–2017), Algeria (2001–2011 ten-year plan) and Morocco (1995 strategy with targets up to 2005 and 2020).

While they sometimes refer to sustainable development, these plans are basically 'green' with highly variable scope and objectives. They do not always enable a clear-cut determination of priorities or a re-orientation of other sectoral policies to reduce their pressures on the environment.

Moreover, in order to improve *knowledge on the state of the environment* and sometimes evaluate the implementation of 'green plans', considerable efforts have been made to implement monitoring and evaluation systems. The creation of the European Environmental Agency in 1993 led European countries to reinforce their national capacities for producing environmental information and give themselves specialized institutions (IFEN in France, APAT in Italy) that have since been widely recognized. In the SEMC, the setting-up of environmental observatories began in 1994 in Morocco (ONEM), when the first assessments by the World Bank of the annual costs of environmental degradation showed that these costs accounted for 8 per cent of the country's GDP and it was judged necessary to develop information for helping policymakers to reorient the pattern of development. Tunisia then set up its national observatory (OTEDD), while other bordering countries gradually implemented similar facilities. These bodies met together at several Mediterranean workshops organized by *Plan Bleu* in order to share experiences, particularly in the field of indicators for sustainable development. In short, while several Mediterranean countries first turned critical attention to their environment and policies at the 1992 Rio summit, today nearly all regularly publish *Reports on the State of the Environment*. However, the monitoring/ evaluation facilities in several SEMC remain fragile, and still depend on the expertise of external or international consultants.

A certain powerlessness in the face of pressures on the environment

Efforts for implementing environmental policies have therefore been considerable, but implementation has experienced and continues to experience formidable problems: many international agreements, community directives, national laws and environmental action plans often remain unimplemented. The so-called *implementation gap* includes highly variable realities, anything from a certain disjointedness in a few countries to a real gulf in others, and always all kinds of serious delays in implementation.

There are many signs of these failures, beginning with the continuing degradation described in a number of reports on the environment published by individual countries, OECD, UNECE, NGOs and the EEA, as well as the present report, which questions the *effectiveness* of environmental policies and calls for more integration into economic and sectoral policies.

In the EU *countries*, the growth of disputes about the environment bears witness to this difficulty of implementation. It simultaneously shows a positive trend to mobilize citizens to protect the environment and the importance of recourse to tribunals for controlling and monitoring environmental policies (Box 18).

In the *other Mediterranean countries*, similar failures can be mentioned, with the aggravating factor of a chronic weakness of environmental administrative staff (especially the regional services in the big countries), an often lesser awareness of ecological issues and more limited possibilities of using tribunals. A few examples of pollution prevention and control will show these implementation problems.

In the *eastern Adriatic countries*, environmental laws and regulations have been passed at a rapid pace since 1990, leading to delays in adopting the required application documents and considerable delays in implementation.[39] For example *Albania* included the principle of sustainable development in its new Constitution of 1998, but the environment is not yet considered to be a priority because of the difficult economic and financial situation. Only in few cases have the environmental permit procedures been enforced, mainly for new

Box 18 Disputes over applying Community environmental law

The number of complaints sent to the European Commission more than tripled between 1996 and 2000 in the environment sector. The environment alone accounts for more than a third of infractions for non-compliance with community law.

Environment-related infractions opened by EU Member States

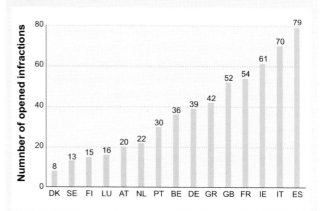

Source: European Commission, situation as of 4 November, 2003

By the end of 2003, of the more than 500 infractions dealing with environment being considered, Spain and Italy headed the list of countries with infractions, followed closely by France and Greece. This may result from non-communication to the Commission of implementation measures being taken by a Member State, non-conformity of national legislation to European directives, or a poor implementation of directives.

Infractions of community law give rise to procedures that vary from a notice-to-comply to referring the case to the European Court of Justice, which can impose penalties on the Member State involved, calculated according to the duration and seriousness of the infraction. Generally, dissuasive judgements have been handed down, for example, against Spain in 1998 and 2003 for poor implementation of the directive on the quality of swimming water; against France in 2001 for the same reason and in 2002 for non-compliance with the directive on water pollution by nitrates from agricultural sources; against Italy in 2001 for the lack of action and monitoring programmes as set out in the nitrates directive and in 2002 for non-compliance with waste management (Sicily and Basilicata); against Greece in 2002 for a lack of substantial measures for protecting the *Caretta caretta* turtle on Zante Island, and for the insufficient transposition of the directive on integrated pollution prevention and control.

Source: European Commission. *Examen de la politique de l'environnement 2003*. COM(2003) 745 final/2; European Commission. *XXème Rapport sur le contrôle de l'application du droit communautaire*. COM(2003)669

facilities and state-owned industries that are being privatized.[40]

In I*srael*, the Ministry of the Environment supervises all land-based sources of marine pollution within the framework of an inter-ministerial permit committee. When a permit is granted, it is strictly regulated; courts may impose fines or even imprison the offender. Marine and Coastal Environment Division inspectors ensure law enforcement. Nevertheless, in 2001 it was reported that it would take two to three years to regulate all industrial emissions to the marine environment.[41]

In E*gypt*, environmental permitting is used for the discharge of effluents. The EEAA performs multimedia inspections as part of an 'environmental register' for all discharges, while other inspection bodies implement medium-specific or issue-specific laws; multiple inspections therefore place an unnecessary burden on the inspected facilities. To partially overcome these problems, several governors have gathered together all the entities concerned with the inspection of large facilities. Inspections are carried out following public complaints, and periodically according to current legislation. However, because of the huge number of facilities to be inspected, the frequency of inspections is insufficient, qualified inspection personnel at the national level is limited, and there is a high rate of non-compliance with environmental legislation.[42]

In A*lgeria* the general inspections of the environment and the recently organised *wilaya* inspections are in charge of impact assessments, granting permits, making inventories of polluting facilities (more than 50,000), recovering the tax on polluting activities, inspecting treatment plants, and consulting with local governments and industries. On the practical level, technical prescriptions concerning emissions into the air by fixed facilities, wastewater treatment and the conditions for discharging industrial wastewater have remained somewhat unclear for 20 years. Despite this lack of clarity in the legislative framework (at least until the 2003 framework law), some 6700 annual inspections are carried out, sometimes leading to sanctions for non-compliance (about 2000 notices to comply, 400 temporary closures and 66 cases taken before the courts). On the incentive side, existing facilities can benefit from financial aid through the National Fund for the Environment.[43]

The weakness of the administrations in charge of the environment (lack of human and financial resources devoted to implementation and monitoring, lack of power and credibility with other sectoral administrations) is often cited among the explanatory *factors for difficulties* encountered in all riparian countries. The complexity of environmental legislation and the difficulty in funding the high costs of implementation also contributes. But in addition to the question of insufficient resources for implementation (however real), five factors are likely to influence this lack of effectiveness of environmental policies:

1. Their logic of action, generally characterized by **a top-down approach** where a state authority defines the problem to be dealt with and entrusts its being carried out to administrations with no clear identification of the main actors at the source of the damage (qualitative and quantitative) of the various milieus. Despite the flowering of ecologically sensitive associations almost everywhere, the environment still seems to be a state affair in almost all Mediterranean countries. Regional and local authorities are hardly involved, except in decentralized countries such as Spain and Italy and, to a lesser degree, France.

2. The basically **curative approach** of environmental policies which, because of their often emergency nature, emphasize the treatment of symptoms and deal with repairing *end-of-the-pipe* degradation rather than attacking the root causes of the problems. Despite the passing of *ex ante* procedures for assessing impacts, the logic of prevention does not seem to count for much compared with the logic of cure. In the SEMC the generalization of impact studies is not accompanied by risk management or in-depth changes on the way the involved actors act (private and public industrialists, small and medium companies, farmers, not to mention the informal sector).

3. The **complex mix of instruments** needed for implementation. Priority is given to the regulatory command-and-control approach and the search for compliance with procedures, whereas it is increasingly agreed that legislation by itself cannot improve the environment. The question is seldom asked in the SEMC as to how to combine the various instruments for attaining *quality* environmental objectives at the lowest administrative, human and financial cost. As a complement to the required regulatory instruments, economic instruments (water pricing, taxes on pollution, subsidies for clean technologies, etc.) have not been noticeably developed. In the EU-Med countries, at the same time as the regulatory instruments, there has been an increased use of incentive mechanisms (income tax, tax and other charge rebates, a reduction of harmful subsidies to the environment, etc.), voluntary initiatives and the promotion of eco-technologies. However, evaluation of the Fifth European Environment Programme (1994–2000) showed that despite the incentives that had been implemented (EMAS management and environmental audit system,

A Sustainable Future for the Mediterranean

> **Box 19   Company commitment to sustainable development**
>
> Companies play a decisive role in the resource-production-consumption process, and some of them have for the past ten years and sometimes longer changed their behaviour and discipline regarding the environment (annual reports, notations, changes in processes, etc.). Sustainable development leads them to recognize their local and global responsibilities. More than 330 companies have joined the Global Compact initiative of the UN Secretariat, which aims at promoting civic responsibility for human rights, labour standards and the environment. However, a whole set of companies and, especially, the small and medium-sized enterprises are on the fringe of this movement. Some companies still, for example, abandon polluted sites, and others look after the image for the sake of their brand's good name without concrete action. But there is certainly more recognition, within companies and sometimes their staff, of the importance of environmental labels and standards (EMAS, ISO 14001).
>
> In the Mediterranean, some 7380 companies, still mostly on the northern shore, conform to exercises of the ISO 14001 type, sometimes in groups as part of professional branches (work with UNEP in 2002–2003) or with encouragement from chambers of commerce.
>
>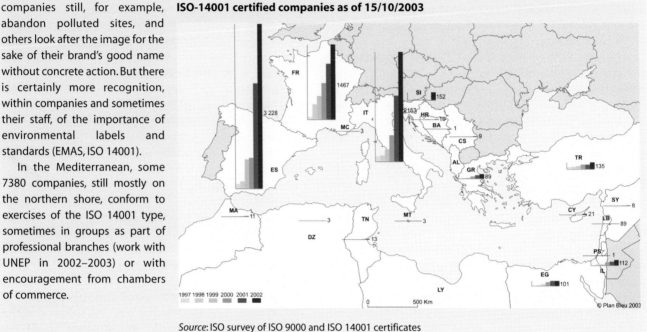
>
> *Source*: ISO survey of ISO 9000 and ISO 14001 certificates

ecological labels, environmental agreements, etc.), the objectives of the programme were not reached, and the targeted activity sectors (transport, energy and agriculture) did not manage to consider environmental concerns as their own.[44]

4   The **unresolved issue of integrating the environment and development**. Despite the cross-sectoral nature of environmental issues, there may be a trend towards sectorizing environmental policies as the lead administration gets stronger and acquires more autonomy. This sectorization is expressed by an increase of necessary plans dealing with various issues (dangerous waste, contaminated soils, wastewater treatment, etc.), but these plans do not lead to a reduction in pressures upstream. This explains the need for effective inter-ministerial mobilization mechanisms to counter possible conflicts of interest (economic development/environmental protection), and to ensure that the environment permeates the administrative culture of economically oriented ministries and sectoral policies. The recent implementation of National Commissions for Sustainable Development and analogous inter-ministerial agencies in most Mediterranean countries does not yet seem to have succeeded in changing the way in which economic, social and environmental development objectives are defined to make them compatible. When national 'sustainable development' strategies are developed, they do not get beyond the formulation stage or remain focused on repairing environmental damage.

5   Finally, **concerns about spatial solidarity** seem to be weakened today in a number of countries despite the considerable efforts of the European Union's cohesion policies and the relative power of land-planning administrations in the SEMC. In this era of globalization and the search for economic competitiveness, sectoral policies and approaches are being increasingly preferred. It would be extremely worrying if the regional development approaches were also devoted to strengthening competitive poles rather than working for the reduction of regional imbalances, with the concomitant risk of weakening integrated environmental and physical planning approaches that require cross-sectoral teams able to create a 'bridge' between administrations.

Determining Factors of the Mediterranean Future

All these findings on progress and failures of environmental policies in the Mediterranean countries differ little from those in other countries and regions. Perhaps this is a sign of a certain 'globalization' of environmental policies and approaches and, at the same time, of little account being taken of specific eco-region features in the national policies of the riparian countries. An indicator of this is the absence of a clearly defined policy for the protection of the Mediterranean coastal zones in several riparian countries (see Chapter 6).

Ultimately, the issue of effective governance of development that is respectful of the environment remains. Undeniable progress has been recorded in reducing pollution from point sources such as the big polluting industries of the northern shore. But the growing importance of non-localized pollution (agriculture, transport, energy, urban growth patterns), irreversible degradation, withdrawals of non-renewable or non-substitutable natural capital, and increased territorial imbalances because of coastal overdevelopment and urban development, raise even more difficult problems that call for an examination of the lifestyles and consumption patterns of entire societies (Part 2 of this report) and much more significant efforts in policy integration and the accountability of all those involved.

This analysis leads to the following assumptions for the 'environmental governance' determinant of the baseline scenario:

- *Faced with the growing rise in economic and short-term interests, environmental policies remain polarized around curative and catch-up actions while land planning and territorial cohesion get weaker in national policy agendas.*
- *Despite a broader mobilization of regional/local authorities and civil societies, the environment essentially remains a matter for the state in most countries.*
- *Given the political weakness of environmental administrations, integrating environmental concerns into the economic and sectoral policies remains difficult. The legitimacy of these administrations for taking care of broad sustainable development issues is in doubt.*
- *Environmental governance, although strengthened, proves increasingly impotent in the face of growing pressures on the environment and sustainable development challenges, which also include social issues.*

The SEMC faced with multiple development challenges

Although, as we have seen, all Mediterranean countries are finding it difficult to implement governance for the environment and sustainable development, the SEMC are faced with the additional challenge of ensuring strong economic and social development to meet the needs of a growing population. This means pursuing major reforms to overcome certain handicaps and take advantage of the opportunities of globalization rather than weather the risks. Will they manage to carry these reforms out? This is the determinant of the baseline scenario analysed here.

Economic and financial handicaps limit the room for manoeuvre

The SEMC have stabilized their large macro-economic balances during the past two decades. But economics are still fragile, which may cast a shadow over future growth and limit room for manoeuvre. This fragility can be seen by analysing some macro-economic, budgetary and fiscal indicators and the balances of external accounts.

Financial resources earmarked for investment in the SEMC are relatively low. Investment, which accounted for 24 per cent of GDP in 1988 (excluding Libya, Lebanon, Israel and Palestinian Territories) amounted to just 21 per cent of GDP in 2002 (excluding Israel and Libya) because of the low per capita incomes and weakness of *foreign investments* (only 1 per cent of world FDI was directed towards the SEMC in 1997–2002, the end of privatization programmes).

Despite major progress resulting from the stabilization of public financing, *budget deficits* in some countries remain large. After deducting donations from abroad, deficits returned to below 5 per cent of GDP in most of SEMC at the end of the 1990s, from more than 15 per cent of GDP in Egypt and more than 30 per cent of GDP in Israel over the 1980–1985 period.[45] But in Lebanon and Turkey, and to a lesser extent in Israel and Morocco, the deficit remains particularly high. Reducing the public deficit has often been at the expense of slowing down public expenditure (especially in Egypt, and Syria).

The budget deficit is projected to increase, under the combined effects of increases in expenditure (or the impossibility of decreases) and falls in income.

- *Expenditures* may *increase* in the medium term because of population growth. After the efforts made over the past two decades, there is already a resumption of public expenditure in Morocco and Egypt, with a maximum in Turkey (nearly an 18 per cent increase between 1995 and 1999). Reducing investment expenses could endanger what is already weak economic growth. A drop in current expenditure seems improbable because of the level of unemployment and public debt.[46]
- *Income* may *fall* with the dismantling of import duties and, in the longer term, with the fall in non-fiscal income (revenues from public enterprises) following disengagement of the state. This income carries considerable weight in total government tax revenue.

On average, import duties account for 17 per cent of public receipts in the SEMC (averaged over recent years): between 16 and 27 per cent in the Maghreb countries and between 11 and 45 per cent in the Mashrek countries (less than 0.06 per cent in the four EU-Med). According to an IMF study,[47] the lowering of import duties in the SEMC as part of EMFTA could in the short and medium term mean a drop in budgetary income of 1 to 4 per cent of GDP, depending on the country. Only Israel and Turkey are free of this risk since import duty accounts respectively for only 0.4 per cent and 2.9 per cent of government income. Government income arising from public companies represents an important proportion of State revenue (more than 15 per cent in Tunisia, Egypt and Syria). Finally, income from privatization is a one-off so cannot be repeated.

Tax revenue as a percentage of GDP is often relatively low in the SEMC (except in Israel and Algeria where oil tax systems exaggerate the figures). It varied between a minimum of 14 per cent in Lebanon and a maximum of 26 per cent in Tunisia in 1999 (it reached about 40 per cent in Italy and France at the end of the 1990s). The fiscal context of the SEMC is also characterized by a relatively *low* proportion of *direct taxation*, varying between 1994 and 1998 between 3 per cent of GDP in Lebanon and 10 per cent in Tunisia, with two exceptions: Algeria (20 per cent) and Israel (21 per cent). In three of the four EU-Med countries (Spain, France and Italy), this figure varies between 20 and 30 per cent.

Debt (public and private) leads to a net outflow of money. The present value of the external debt related to Gross National Income (GNI) can exceed 60 per cent in some countries. Consequently debt service expressed in relation to the GNI has continued to increase since 1985 and is not tending to decrease in most of the Mediterranean countries except Morocco and Egypt. In 2002, debt service accounted for about 10 per cent of SEMC GNI (from 2.3 per cent in Egypt to 15.2 per cent in Turkey). This is a relatively high figure compared with countries with intermediate income (6.8 per cent in 2002) and even in comparison with Latin America (8.5 per cent). In absolute terms, debt service in the SEMC reached almost US$40 thousand million in 2002: the equivalent of more than twice the net total flows (public and private, US$13.3 thousand million) that they received in 1991–2000 (see Figure 29).

The SEMC regularly record a *trade deficit* (with the exception of hydrocarbon-exporting countries). This is partly explained by the fact that their exports are dominated by low value-added goods, subjecting them to a deterioration of exchange terms, which has affected oil as well as agriculture products. This deterioration could lead to increased environmental degradation since countries that are highly dependent on natural resources might seek to counterbalance decreases in relative export rates through increases in export volumes.

Mainly as a result of trade deficits, (excepting Syria, Algeria and Libya), most of the SEMC record a *deficit in their current accounts* (Figure 30). Although this is sometimes relatively weaker than in the EU-Med countries, its financing mode is vulnerable. Actually, the SEMC balance their current deficits mainly through receipts linked to the trade in services (mainly tourist-related income[48]) and through transfers[49] (essentially from emigrants). However, these receipts are highly volatile since they fluctuate significantly according to the economic and political situation. Terrorist attacks and armed conflicts have major repercussions on tourist activity.

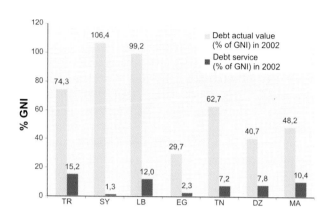

Figure 29 Debt and debt service

Present value of debt and debt service, 2001

Source: World Bank, 2004

Total Debt service 2002 (current million $)	
Turkey	27,604
Syria	258
Lebanon	2188
Egypt	2066
Tunisia	1438
Algeria	4166
Morocco	3691

Source: World Bank, 2004

Determining Factors of the Mediterranean Future

Figure 30 Current account balance (as % of GDP)

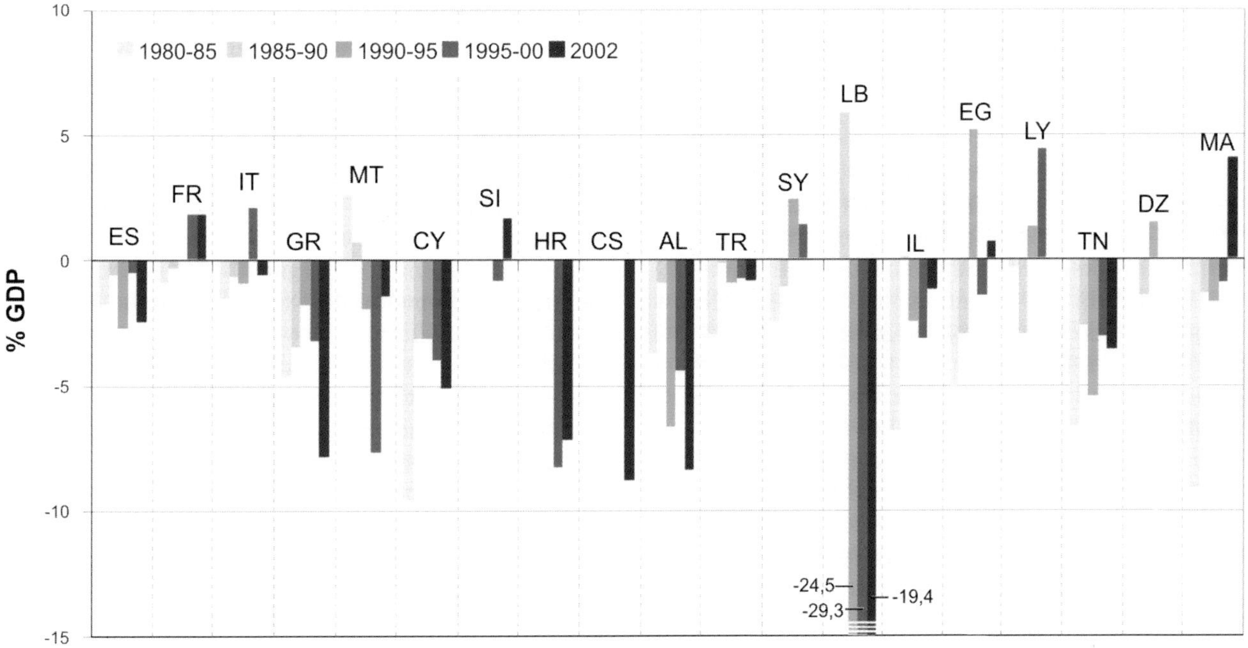

Source: World Bank, 2004

The *exposure* of the SEMC economies is also linked to their *weak diversification* and their high *dependence* in relation to:

- The European economy, which exposes them to any unfavourable turn of events by affecting their growth and in consequence their government income. According to FEMISE, for a 3 per cent decrease in European growth, the expected decrease in GDP in SEMC through the evolution of trade could reach 3.3 per cent in Algeria, 6.2 per cent in Tunisia and 1.2 per cent in Turkey. If a decrease of 3 per cent should occur in the US, its potential repercussions on GDP would be a reduction of 1.5 per cent in Algeria, 1.3 per cent in Israel, 0.9 per cent in Tunisia and 0.4 per cent in Turkey.
- Natural resources and climate hazards. In Morocco the major droughts (1994, 1996 and 1999) explain to a great degree the economic recession of the past few years and the difficulties in balancing the state budget. In Syria, Tunisia and Algeria climate hazards engender wide fluctuations in agricultural production, a still important sector of the economy and employment.
- Fluctuations in world prices. Economic growth in Algeria is closely linked to the world price of oil, a sector that on its own ensures 62 per cent of the country's fiscal income.

All of these *structural weaknesses* (debt, tax system, weak diversification, deterioration in terms of trade, volatile receipts) reduce the room for manoeuvre for investing in *physical and human capital*, which is indispensable for the potential of future SEMC development.[50] Some signs show that falling behind other competing regions could become more marked. The level of *gross formation of fixed capital* in the SEMC is close to the world average, but much lower than that in the southeast Asian countries. Their educational system, although recording much quantitative progress, is still typified by a low ability to adapt to socio-economic needs and produces few students in advanced studies. Their expenditure in the field of *research and development*, for the countries that publish data, has a hard time reaching 0.6 per cent of GDP (compared with more than 1 per cent in the Eastern Adriatic countries and 2.1 per cent on average worldwide). The weakness of their efforts in terms of *innovation* is confirmed by other indicators such as the number of patents accorded to residents or the number of engineers and scientists in R&D per 1000 people. Their telecommunications network and access to *new communication and information technologies* (NCIT) show a certain tardiness. Turkey and Egypt thus devoted less than 4 per cent of their GDP in 2001 to expenditure on NCIT compared with more than 7 per cent in Israel and 9 per cent in France. The SEMC (excluding Israel) account for 53 per cent of the Med-

environnementale; cas des PME du secteur Textile-Habillement au Maroc, Plan Bleu, Sophia Antipolis

PNUD. (2004) Rapport mondial sur le développement humain. De Boeck Université, Paris

PNUE. PAM. Plan Bleu. (2002) Libre-échange et environnement dans le contexte Euro- Méditerranéen, MAP Technical Report Series no 137, Montpellier/Mèze, France, 5–8 octobre 2000. 4 vol. Athènes: PAM

Poulin, R., Salama, P. (dir.) (1998) L'insoutenable misère du monde: économie et sociologie de la pauvreté. Hull: Vents d'Ouest

Ramade, F. et al (1997) Conservation des écosystèmes méditerranéens. Enjeux et prospective. 2nd ed. Les Fascicules du Plan Bleu, no 3, Economica, Paris

Ravenel, B. (1995) Méditerranée, l'impossible mur. L'Harmattan, Paris

Regnault, H. (2003) Intégration euro-méditerranéenne et stratégies économiques. L'Harmattan, Paris

Reiffers, J.-L. (dir.) Institut de la Méditerranée. (2000) Méditerranée: Vingt ans pour réussir. Economica, Paris

Reynaud, C. et al (1996) Transports et environnement en Méditerranée. Enjeux et prospective. Les Fascicules du Plan Bleu, no 9, Economica, Paris

Tapinos, G. P. (2000) Mondialisation, intégration régionale, migrations internationales. Revue internationale des sciences sociales, Septembre 2000.

UNCHS. (2001) Global report on human settlements. Earthscan, London, Nairobi

UNCTAD. (2002) Trade and Development Report. UNCTAD, Genève

UNDP (2000) Human Development Report. UNDP, New York

UNDP. Regional Bureau for Arab States, Arab Fund for Economic and Social Development. (2002 & 2003) Arab Human Development Report. UNDP, New York

UN Economic Commission for Europe (UNECE) (2003) Politique del'environnement dans les pays en transition: Bilan de 10 années d'études de performance environnementale dans le cadre de la CEE. Fifth ministerial conference, Kiev (Ukraine), 21–23 May

UNEP/MAP/MED POL. (2001) Meeting of the Informal Network on Regulation Compliance and Enforcement. Sorrento, 15–17 March 2001

Villevieille, A. et al (1997) Les risques naturels en Méditerranée. Situation et perspectives. Les Fascicules du Plan Bleu, no 10, Economica, Paris

Wackernagel, M., Rees, W. (1999) Notre empreinte écologique: comment réduire les conséquences de l'activité humaine sur la Terre. Ecosociété, Montréal

World Bank. (2001, 2002, 2003, 2004) World Development Indicators. World Bank, Washington

World Commission for Environment and Development (WCED) (1987) Our common future (Bruntland report), WCED

World Tourism Organization. (2001) Tourism 2020 Vision. Vol. 7: Global forecasts and profiles of market segments. WTO, Madrid

WWF. (2000 & 2004) Living Planet Report. WWF, Glan

Part 2

Six Sustainability Issues

Chapter 1

Water

From the beginning of time, water has shaped the Mediterranean environment and its economies and societies. At the dawn of the 21st century it is taking on a new importance. With just 3 per cent of the world's freshwater resources and more than half of the world's 'water poor' population,[1] the Mediterranean countries are faced with the vital issue of access to this source of wealth for their populations and ecosystems.

This chapter deals with continental freshwater; coastal seawater will be discussed in Chapter 6, Coastal areas. It contains an additional geographic level to N1 (the Mediterranean coastal countries and territories): the Mediterranean Catchment Basin (NV level); this level is particularly relevant for studying water resources. Its area is defined as the entire catchment areas of the small and large rivers that empty into the Mediterranean that are included in the territorial limits of the Mediterranean countries (Figure 1.1). So as to avoid covering too large an area that is not part of the Mediterranean eco-region, the watersheds that are outside the Mediterranean countries have been excluded from the conventional Mediterranean catchment area. In particular, the designation excludes Egypt's upstream Nile watershed, which stretches into Tanzania and covers a total of 2,726,000km². The surface area of the catchment area included (restricted to the Mediterranean countries) is 1,753,850km². To avoid confusion in the statistics used in this chapter, the illustrations will refer to the geographical level concerned by coding titles (i.e. N1 or NV).

The countries that have similar hydrological profiles are grouped by continent:

- Northern Mediterranean countries (NMC or northern shore): Spain, France, Italy, Greece, Monaco, Slovenia, Croatia, Bosnia-Herzegovina, Serbia-Montenegro, Albania, Cyprus and Malta
- Southern and eastern Mediterranean countries (SEMC):
  - the Eastern shore: Turkey, Syria, Lebanon, Israel and the Palestinian Territories;
  - the Southern shore: Egypt, Libya, Tunisia, Algeria and Morocco.

Figure 1.1 **Mediterranean catchment area, NV**

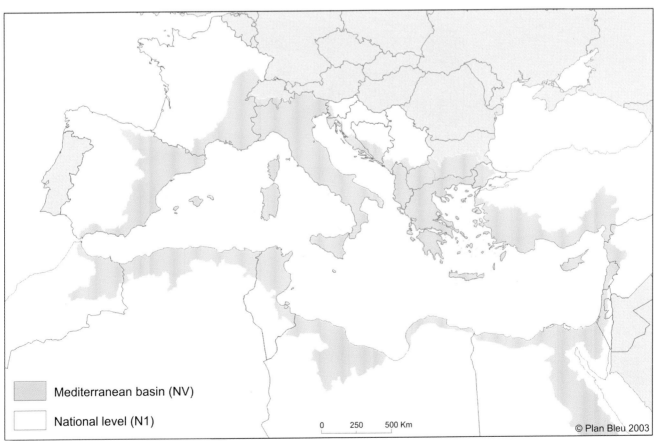

Note: This definition of the NV level excludes the part of the Nile upstream of Aswan, and Andorra, Bulgaria, Macedonia and Switzerland.

2 A Sustainable Future for the Mediterranean

In the southern and eastern shore areas of the Mediterranean, withdrawals from aquifers and surface waters, and the disposal of effluents, both resulting from economic and demographic growth, are threatening the quantitative and qualitative regenerative capacities of freshwater. In the north, withdrawals are stable or even falling. However, decision-makers and public opinion are deeply concerned about degradation of the quality of freshwater and ecosystems and the growing costs of water supply. On both shores, urbanization and population concentrations on the coasts are continuing to increase local pressures on resources.

These developments, and the associated economic, strategic and environmental risk factors, will be analysed in the first part of this chapter in the context of the *Plan Bleu* baseline scenario. However, this increased demand could be dealt with by limiting the impact on the environment through a significant improvement in water management. This is the preferred way forward and will be considered in the second part of the chapter in the context of an alternative scenario where the possibilities, problems and ways of achieving 'sustainable' water management on the basis of experience accumulated in the Mediterranean region are analysed.

Continuing growth in pressures on vulnerable water resources

It is important to keep in mind the extreme diversity of the situations in different Mediterranean countries and the large disparity in terms of space and time in the distribution of water resources, which may well become more pronounced.

Water in the Mediterranean area is an irregular, rare and fragile resource

The Mediterranean climate is typified by an enormous irregularity in the distribution of rainfall in space and time (Figure 1.2).

Although the entire Mediterranean catchment area (NV) receives a total average volume of precipitation estimated at 1100km³ (1000 million cubic metres) per year, nearly two-thirds of this volume is concentrated in one-

Figure 1.2 Average rainfall distribution in the Mediterranean basin

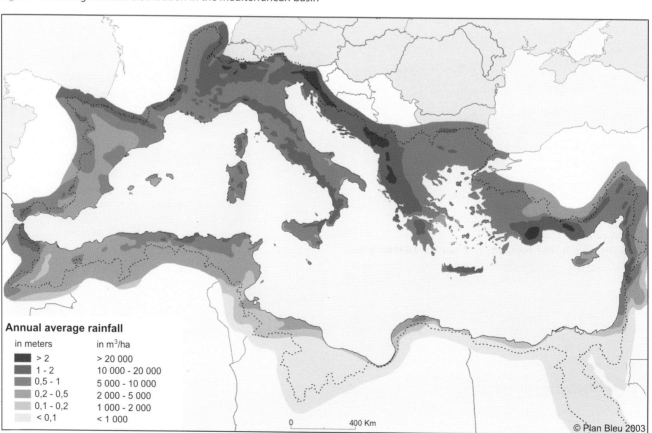

Source: *Plan Bleu*; Margat, 2004

fifth of the Basin's surface area. While three countries, France, Italy and Turkey, receive half of the total precipitation, the southern countries receive only 13 per cent.

Part of this precipitation is used directly by vegetation; the other part produces surface or underground flow, which constitutes an 'internal inflow' of precipitation to freshwater resources. Adding this interior inflow to the Basin's external inflow^2 gives the total of 'natural, renewable freshwater resources' (underground and surface) of the Mediterranean catchment area, which amounts, on a yearly average, to about 600km^3 per year (NV see Table A.33 in Statistical Annex). These resources are the maximum potential of water resources 'given by nature' on average every year. This potential can be exploited more or less intensively without affecting the rights of future generations since it is renewed every year by rain, at least insofar as this is not altered by climate change.

This potential is very unevenly distributed in the Mediterranean since it is concentrated (more than 85 per cent) in the NMC and Turkey, which also supply 90 per cent of the total water flow into the Mediterranean, estimated at about 475km^3 per year.

These average renewable natural freshwater resource values, calculated over long periods, are characteristic of the Basin's hydrology and should be considered as relatively constant in time over the period in which we are interested (to 2025) (Box 1.1), even if they mask a wide variability of input from one year to another.

Box 1.1 What is the outlook for renewable natural resources?

For the sake of simplicity, we assume that the average annual volume of **renewable natural resources** will remain constant between now and 2025. This is probably optimistic given two important risk factors: climate change and modifications in the hydrological regime in the catchment areas (not to mention increased pollution, which affects the quality of the resource).

Climate change could increase still further the occurrence of droughts and, in the southern and eastern Mediterranean, diminish average precipitation, but there are very serious uncertainties. Indeed available research on the possible effects of global warming on water resources in the Mediterranean are very sketchy and often contradictory. Although temperature rise appears to be more and more likely, the scenarios for the evolution of precipitation and, even more, run-off, in the various available models do not allow any conclusions to be drawn. Experts, however, seem to agree that there will be a risk of greater contrasts between different seasons and regions and an increase in the occurrence of random droughts, although they cannot specify their severity or when and for how long they might occur.

Changes in hydrological regime and desertification in the catchment areas attributable to human activities are more immediate and likely risks than global warming. Water resources are and will be affected by changes in land use in the catchment areas, mainly the sealing of soils, as a result of increasing urban development and transport infrastructures, which increases surface run-off (amplifying the irregularity of input) and make control of this problem even more necessary. Deforestation (fires and logging) and the drying of wetlands also change the water regime. Erosion due to some agricultural practices (abandoning terraced cultivation, overgrazing) amplify the phenomenon even more. In contrast, tree replanting, particularly that related to the abandonment of farmland in the northern rim Mediterranean countries, could control the temporal patterns of the flows, even if it contributes to reducing average run-off.

Source: Margat, 2004

Although the average renewable natural water resources may be considered as unchanging for the 2000–2025 period, the inequality in natural water availability per inhabitant between the north and the south and east of the Mediterranean will grow as a result of demographic factors alone. Although the average 'share' of natural renewable water resources per capita in the Mediterranean's southern countries was on average 2.5 times lower than that of people in the north in 1950, it was about 6 times lower in 2000 and is expected to be 8 times lower by 2025.

In absolute terms, the *water poor* Mediterranean population (NV), those with less than 1000m^3 per capita per year, was 108 million in 2000 and could reach 165 million by 2025 in nine SEMC. Of these, 63 million people would be 'in shortage', that is, with less than 500m^3 per capita per year (compared with 45 million in 2000). These already troubling figures mask local or temporary shortages in many catchment areas.

Most of the renewable natural resources in the Mediterranean Basin are from surface run-off (about three-quarters of the water that runs off within the basin and 90 per cent of the run-off that leaves the basin). Six major rivers have an average natural flow greater than 10km^3 per year: the Nile in Egypt, the Rhone in France, the Po in Italy, the Drin in Albania–Serbia-Montenegro, the Ebre in Spain and the Neretva in Bosnia-Herzegovina and Croatia. The hydrographical structures are highly scattered since only 21 basins have areas of more than 10,000km^2 and cover only 42 per cent of the entire Mediterranean catchment area.

The total of underground flows in the Mediterranean Basin would be about 150km^3 per year out of the 600km^3 per year of total inflow. Underground waters have an essential function in regulating flows. In the northern countries, they contribute to most of the river-creating run-off and are the main factor in their permanence. In contrast in the south, they are fed by the often temporary flooding of rivers, and many flow into evaporation fields (if they are not already harnessed), particularly in enclosed low-lying areas. Where aquifer reservoirs are sufficiently developed, they improve resistance to drought.

There are three types of aquifers prevalent in the Mediterranean Basin:

1   karstic carbonate aquifers, consisting of water 'towers' with mainly perennial run-off;
2   localized alluvial aquifers in the valleys and deltas of major rivers, strongly linked and with high rates of interchange with their respective rivers;
3   aquifers formed mainly of detrital sedimentation on coastal plains in contact with the sea or extensive basins, especially in the south (Egypt and Libya) in the Sahara, largely beyond the Mediterranean Basin. They contain deep-lying water tables with considerable reserves but are currently seeing very little renewal ('fossil' water), and are practically independent of surface water.

In addition to this uneven spatial distribution of the water resources, there is a very great irregularity in time, both within and between years.

The intra-annual variability between seasons of the same year is characterized by a concentration of rain into relatively few days (an average of 50–100 days per year) and by summer droughts that correspond with the highest peaks of water demand (for irrigation and tourism). To measure this variability, the *regular* resource is defined as the run-off assured for 90 per cent of the time during an average year, in practice the 'basic' effective run-off for 11 months out of 12. Of the total volume of renewable resources in the Mediterranean Basin in an average year, only a small proportion, about 30 per cent, is regular in this sense.

Added to this variability within the year is a very strong inter-annual variability between succeeding years. Droughts[3] are very frequent and may get worse with climate change. This irregularity considerably limits the exploitation of renewable surface water resources in the basin and justifies the construction of many projects designed to regulate intra- or inter-annual flows (Figure 1.3).

To this irregularity, which confers a certain 'quantitative' vulnerability to freshwater supplies, one has to add a 'qualitative' vulnerability.

Freshwater resources in the Mediterranean Basin often have a natural quality (for example natural salinity or high hardness levels) that limits their use. For example, in Tunisia 26 per cent of the surface water, 90 per cent of the underground waters drawn from water

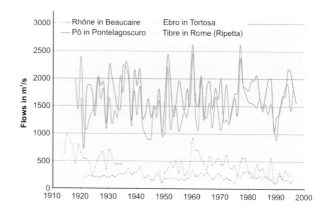

Figure 1.3 Inter-annual variation in river discharge

Source: Med-Hycos Project (www.medhycos.com or http://medhycos.mpl.ird.fr)

tables and 80 per cent of the water pumped from deep water tables has a salinity of more than 1.5g per litre.

The resources are also particularly vulnerable to human activity. This is the case with many shallow alluvial water tables. Mediterranean river beds are often dry in summer and, unlike rivers in more temperate zones, are therefore unable to cleanse themselves. High summer temperatures diminish this ability even more in river beds and perennial waters. Coastal water tables are in fragile balance and can easily be contaminated by sea water intrusion when too much water is taken from them.

Climate change may contribute to increasing this natural irregularity and vulnerability of water resources by increasing the frequency and magnitude of droughts and reducing precipitation in the southern region of the Mediterranean (Box 1.1).

Water shortages and frequent droughts have a particular effect on the SEMC, which are expected to experience the greatest water needs in coming years.

Growing demands in the south and east

Between now and 2025 water demands in the south and east are expected to increase by 25 per cent

Water is 'used' by people for vital functions (drinking and hygiene), and many economic activities (irrigation, industry, energy, shipping and leisure). W*ater demand* is defined here as the total volume of water needed to satisfy the different water users, including volumes 'lost' during transport, for example leaks from pipes and evaporation between the withdrawal point (or production site) and the point of final consumption. Demand thus includes withdrawals from natural resources, water imports and the unconventional production of water such as desalination and re-use. It differs from final consumption of water by users, in that it includes all related losses and non-use, for example water transport (see below).

The main forces that drive Mediterranean water demand are irrigation (to make up for the deficit in precipitation and a growing supply or export demand in the SEMC), domestic requirements (which increase with urbanization) and rapidly expanding tourism. It is difficult to obtain reliable data about water demands that are comparable over time and between countries. The following data are therefore valid only to an order of magnitude. Retrospective data sets over long periods are only available at the whole country level and show that during the second half of the 20th century, the total water demand of the Mediterranean countries doubled, and has now reached about 290km^3 per year for all the Mediterranean countries (N1). The countries experiencing the greatest growth (more than 2 per cent per year) are Turkey, Syria and France. Over the past decade only a few countries either stabilized (Israel) or even reduced their total water demand (Italy, Malta and Cyprus, Figure 1.4). On the smaller scale of the various Mediterranean catchment basins (NV), the available time series are less extensive. The total demand for water was 190km^3 per year in 2000 (see Table A.34 in Statistical Annex), or about a third of the 600km^3 per year of renewable natural resources.

Projections of these demands by country and sector up to 2025 have been made[4] on the basis of analyses of available national strategy and planning documents, reviewed in the light of major prospective regional and worldwide projections[5] and certain consistency tests. These projections form the 'baseline scenario' for water by Plan Bleu. The projections, carried out country by country at the two N1 and NV scales, are summarized in Figure 1.4 (and Table A.34 in the Statistical Annex). In relation to Plan Bleu previous scenario exercises[6] for water, the future demands for 2025 have been significantly downgraded. The national projections have been readjusted from their previous levels, which generally had a tendency to overestimate the projected growth in water demand. According to this baseline scenario, the total water demand in the basin (NV) would increase by 20km^3 between 2000 and 2025, or 0.4 per cent per year. Most of this growth would be due to the southern rim countries (0.7 per cent per year and 25 per cent for the whole period) and above all the eastern rim countries (1.5 per cent per year, mainly in Turkey and Syria). The demand may stabilize or even drop in certain northern rim countries such as Italy.

Irrigation, the main consumer, is growing rapidly

Analysis of demand by sectors shows that the main user in volume terms in most countries remains agriculture

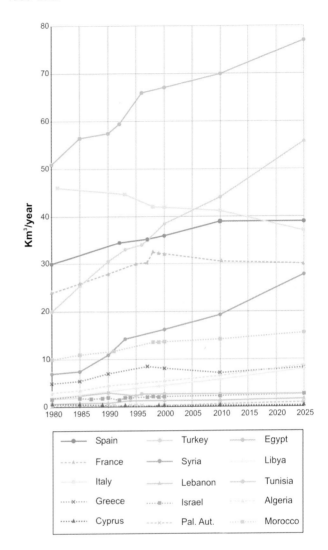

Figure 1.4 Total demand per country, baseline scenario, N1, 1980–2025

Source: Plan Bleu

for irrigation (except in the eastern Adriatic countries and France), followed by drinking water supply and, in third place, use by industry and the energy sector.[7]

Agricultural water demand includes the quantities of irrigation water brought 'artificially' to plants. It also includes losses in distribution networks through infiltration and evaporation but excludes the rainwater directly captured by plants, known as 'green water'. It is the largest item in water demand, representing 65 per cent of total demand in the Mediterranean Basin, 48 per cent in the north and 82 per cent in the south. In the *north* the area under irrigation is becoming more or less constant (reducing in Italy) and is expected to lead to a more or less constant demand for agricultural water, in both absolute and relative terms. On the other hand,

Figure 1.5 The structure of the water demand per sector and per country groups, NV, 2000–2025

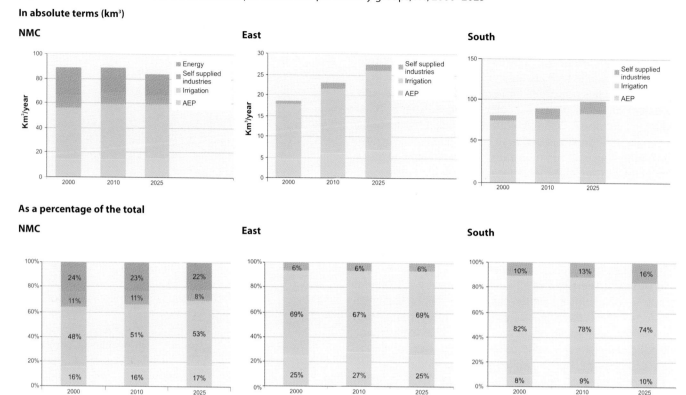

Source: Plan Bleu, J. Margat

AEP: domestic water supply

forecasts project a large increase in agricultural water demand in the *south and especially in the eastern* part of the Basin. Agricultural development policies in most of the SEMC (Turkey, Syria, Egypt, Algeria and Morocco) are planning both to extend the areas under irrigation and to increase farming intensity[8]. According to a FAO study[9], the irrigated area could increase by 38 per cent in the south to reach 9.1 million ha and 58 per cent in the east to reach 7.7 million ha by 2030. However, some hoped-for efficiency gains in water-use for irrigation and a relatively larger increase in demand for drinking water could result in the share of agriculture in total demand in the eastern countries remaining more or less constant. The share may even drop in the south, from 82 to 74 per cent of total demand by 2025, to the benefit of the supply of drinking water.

Under the combined effect of the rise in living standards and, most importantly, population changes, the baseline scenario projects a continuing large increase in *drinking water* demand in the SEMC and a stabilization in the NMC. The drinking water demand includes the water required to supply the tourism sector, which can lead to large seasonal spikes, although it has, generally speaking, no significant effect on average annual water demands countrywide. Locally, tourist demand can lead (on coastlines and islands) to over-sizing of water works or serious supply difficulties, even if these are usually quite easy to overcome by the capacity of the sector. In the baseline scenario the share of drinking water in total demand will continue to increase at the expense of the shares of the energy and industrial sectors in the north and the agricultural sector in the south and east. By 2025 it will reach an average of 15 per cent of total demand in the Mediterranean Basin (NV). In certain water-poor countries, given its priority in allocations, it may constitute more than a third of total demand (Cyprus 34 per cent, Israel and Algeria 45 per cent, the Palestinian Territories 58 per cent and Malta 87 per cent). In the water-rich countries, as in most of the eastern Adriatic countries, domestic water represents most of the demand, as irrigation is useless or marginal.

Water demand by the *energy and industrial* sectors (which consume small quantities but are often polluting) is expected to fall in absolute terms (efficiency gains and energy transition) in the NMC. In the SEMC, on the other hand, the industrial sector is expected to increase its water demand significantly in absolute terms, reaching up to 16 per cent of total demand in the southern countries by 2025.

As the limits set for preserving natural ecosystems are approached, the need for an *'environmental* water demand' is becoming increasingly accepted; this would include the water required for the operations of these ecosystems. Some countries have included in their legislative arsenal the respect of a minimum flow in their rivers, set for the survival of species (France), or have included, even more explicitly, an environmental demand (Spain); others are considering such measures (Italy), but this demand is usually not yet quantified in supply and demand balance sheets and is seen more as an attempt to limit resource exploitation.

Pressures on natural water resources

Water supply strategies to meet these demands vary according to the different situations in the different Mediterranean countries. The forecasts of demand growth are highest in the generally water-poor SEMC. On the other hand, demand seems to have stabilized in most of the water-richer NMC, where efforts are directed more at reducing some disparities between regions or ensuring the quantity and quality of water supplies. According to the baseline scenario, there would be across the whole region a continuation of large construction programmes for securing supply, the extraction of a larger share of renewable natural resources or, in countries with more limited natural resources, 'producing' water from so-called 'unconventional' sources such as sea-water desalination and the re-use of wastewater.

Increasing withdrawals and waterworks

The level of pressures on resources can be gauged in a very approximate but indicative way by the *exploitation index* of renewable natural resources. This is defined as the ratio of 'withdrawals from renewable natural water resources to average renewable natural water resources', expressed as a percentage. Calculation of this index for the Mediterranean catchment basin (NV) for 2000 and 2025 highlights the variety of situations, as shown in Figure 1.6.

- A first group of countries, where water withdrawals are close to or exceed the average annual volume of renewable natural resources (exploitation index equal to or greater than 75 per cent): Egypt, Israel and Libya, to be joined by 2025 by the Palestinian Territories and Spain's Mediterranean basins. The natural resources in all these countries are already very highly stressed and they will have to meet a growing part of their demand from other 'unconventional' sources.
- A second group of countries, where total demand represents a growing share of the average annual volume of renewable natural resources, but where the exploitation index will stay between 50 and 75 per cent until 2025: Malta, Syria and Tunisia.
- A third group of countries, where the exploitation index lies between 25 and 50 per cent, may nevertheless experience local or exceptional stress: Lebanon, Cyprus and Morocco, joined by Turkey and Algeria by 2025.
- A fourth group of countries, where the exploitation index is less than 25 per cent: Greece and the eastern Adriatic, France and Italy, where total demand is dropping.

The countries in the first group are drastically limited by the *availability* of their renewable natural freshwater resources and they must either re-use it several times or consider other sources. For the other countries, the

Figure 1.6 Exploitation indices per basin, NV, 2000–2025

2000

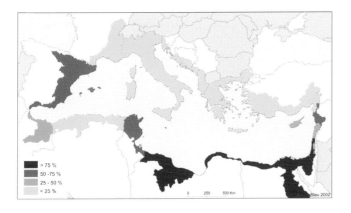

2025

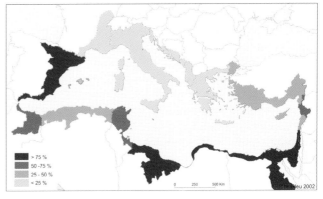

Note: Exploitation index: ratio of annual volume extracted from natural resources/annual average volume of renewable natural resources, expressed as a percentage

limitations on water supply are due less to the resource's *availability* than to the *capacity* to exploit it, that is to carry out and manage the exploitation and the transport infrastructures for the water from its source to its end use.

The exploitation index is increasing in all the SEMC and in Spain's Mediterranean basins, indicating a growing pressure on the already scarce resources, whereas it is decreasing in the northern-rim countries, essentially for demographic reasons. The overall index can mask stronger pressures locally or in dry years.

Withdrawal is mostly from surface waters (80 per cent of total withdrawals of average renewable natural resources in the Mediterranean, 87 per cent in the southern-rim countries) and only 20 per cent from underground waters.

This stress on resources appears even greater when it is realized that not all renewable natural resources are necessarily 'exploitable' by mankind (Box 1.2). According to the assessments available in the various countries,[10] only about 60 per cent of the renewable natural water resource of the entire Mediterranean catchment area (NV, 360km^3 per year[11]) would be 'exploitable' – 56 per cent in the north, 60 per cent in the east and 79 per cent in the south. Thus, only a little over half of the Mediterranean's natural resources (between 50 per cent in Algeria and 93 per cent in Libya) were considered 'exploitable' in 2000. The exploitation indices given above would be even higher if calculated on the basis of exploitable (and not total) resources.

> **Box 1.2 What is the 'exploitable part' of renewable natural resources?**
>
> Not all renewable natural water resources are directly **exploitable**. A first limitation is **technical-economic** factors. Beyond a cost so high as to be considered prohibitive, the resources are considered to be non-exploitable. A second limitation is **environmental** and aims at preserving the proper functioning of water-related ecosystems and the self-cleansing capacity of rivers and landscapes. A third limitation is **geopolitical** and takes account of the right of access to other people's water where resources are shared.
>
> These limitations vary according to the country and they change over time, thus changing the proportion of resources considered as 'exploitable'.
>
> By reducing exploitation costs, progress (both technological and organizational) in exploiting resources increases the exploitable proportion of natural resources, for example through improvements in exploitation yields (deep drilling techniques, harnessing tapped water from undersea sources, reducing losses from evaporation or the silting up of dams). The most striking example in the Basin is the possible development upstream on the Nile in Sudan, which would contribute to reducing losses by evaporation, to the joint benefit of Sudan and Egypt, and would increase the exploitable external renewable resources in Egypt by 9km^3 per year.
>
> In contrast, there are many factors that reduce the exploitable proportion of natural resources, for example the increasing distances between the sources of supply and the points of use, increasing environmental considerations in developing new infrastructures, and the silting-up of dam reservoirs, not to mention the growing pollution of resources that further increases the price of exploitation. Reducing the costs of alternative water supply solutions (such as sea water desalination, recycling of wastewater, improving demand management) lowers the threshold that defines the acceptable cost of 'exploiting' natural resources and contributes to reducing the 'exploitable' proportion.
>
> *Source:* Margat, 2004

Three-quarters of these 'exploitable' renewable natural resources are *irregular* and require the construction of control structures to enable a year's water to be stored, either for use in the summer (for irrigation and tourism) or from one year to the next. Water demand in the Mediterranean catchment basins of very many countries is already more than twice the regular natural renewable resource and could not be met without such structures. This is the case for Spain, France, Syria, Israel, Egypt, Libya, Tunisia, Algeria and Morocco. The investments required for infrastructures are generally out of the reach of individual users and have long justified collective action. The Mediterranean peoples have always developed a high degree of know-how in managing the scarcity of water and much ingenuity in building such structures. The last century was typified by growing activity in countries in contracting such work, which became increasingly larger in scale. More than 500 large-scale dams were built during the last century in the Mediterranean catchment basin (NV), providing a total of more than 230km^3 of storage.

In the baseline scenario, national strategies are still largely dominated by efforts to increase water *supply* through the building of large *infrastructures* intended to exploit a growing proportion of renewable resources, and no value is given to the large potential for saving water. The policy of building new dams will probably slow in the north, given the exhaustion of possible construction sites, but will be pursued in the SEMC to take greater advantage of surface waters. Underground waters will increasingly have to be extracted, including water from massive overuse for irrigation (Egypt) or even water from non-renewed aquifers (fossil water). Such reinforcing of water infrastructures will be linked to the development of water transfers between basins of the same country (e.g. Spain, Greece, Egypt) or between countries (e.g. France–Spain, Albania–Italy). Box 1.3 lists some of the major projects contained in the hydrological planning documents of the Mediterranean countries.

Box 1.3 Supply policies, still dominated by major project programming

In **Spain** the new national hydrological plan, adopted in 2002, sets out a programme of 119 new dams, storing 2.5km³, linked to the project of transferring water from the northern rivers, mainly the Ebre (1km³ per year at first, then 3.35km³ per year in the long term) to the Mediterranean coastal basins. Catalonia is also studying the feasibility of transferring water from the Rhone River in France to Barcelona. This project has launched a major debate about questions of water demand management as opposed to supply management. One of the main criticisms of the project concerns the lack of studies dealing with evaluation of future water demand by Barcelona and socio-economic and environmental analysis of the different possible options for meeting or changing demand. These criticisms have lead to the cancellation of the transfer project from the Ebre.

In **Greece** the project to develop the Acheloos River to enable the transfer of 1.1 to 1.3km³ of water per year to Thessaly includes 4 dams.

In **Turkey** the 'GAP' (SE Anatolia Project) in the upper basins of the Tigris and the Euphrates will contain 22 dams in its final stage, storing 60km³ of water (including 48km³ by the already-completed Ataturk Dam).

In **Cyprus** ten additional dams are planned (1999) for storing 85hm³ (hectometer) and regulating 25hm³ per year.

In **Syria** several dams are under construction or are planned in the Mediterranean Basin (Oronte and coastal basins) and the Steppe Basin in the master plans.

In **Egypt** the extension of the use of Nile water outside its basin, already begun by the Peace Canal (towards the northern Sinai) and the Tuska Canal (towards the Kharga oasis and the 'New Valley'), will be continued and completed by increasing the use of the Nubian aquifer (non-renewable resources). From an optimistic point of view, this extension relies on the discharge gains that would result from loss-reducing improvements in the Nile Basin in Southern Sudan (the Jonglei Canal, already begun), with a predicted proportional share for Egypt of about 4km³ per year in the initial phase, and from 9 to 10km³ in the long term.

In **Libya** the already advanced project of transferring water taken from the Saharan fossil aquifers ('GMR') will carry 2.2km³ of water per year to the coastal areas in its final phase by 2007.

In **Tunisia**, it is planned to fully use by 2010 the exploitable water resources (4.03km³ per year on average) by exploiting 0.835km³ per year more than in 1995, of which 0.67km³ will be of surface water, by increasing the storage capacity of dams to 1.9km³ (from 1.48 in 2000).

In **Algeria** the national water plan projects to exploit an additional 5.4km³ between now and 2020, of which 3.1km³ will be from surface waters, in particular by building 50 additional dams.

In **Morocco** the building plan foresees the construction of 60 major dams between now and 2020, controlling 14km³ per year, plus some 100 small or medium-sized dams (2 to 3 per year), and an increase in the exploitation of underground water to 3km³ per year.

Figure 1.7 Main water transfers, NV

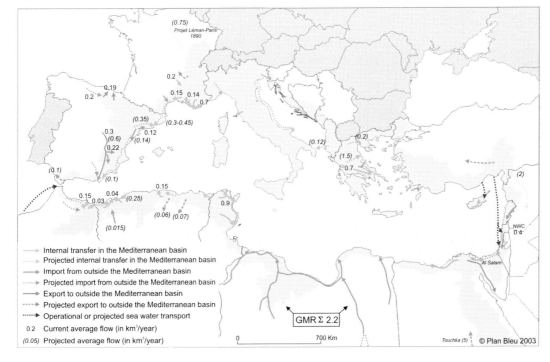

Source: Margat, 2004

A Sustainable Future for the Mediterranean

Unsustainable exploitation of an increasing proportion of water resources

With the more intensive use of renewable water resources, a growing proportion of water supply in some countries is being met in an unsustainable way.

With increasing water demand, the underground water tables are sometimes *overexploited*, which lowers their levels excessively. Here again we are, in a way, borrowing the 'natural capital of water' from future generations. Such 'overexploitation' has been diagnosed and inventoried in most Mediterranean countries, even if it has not always been defined according to standardized criteria.

Box 1.4 Overexploitation of Mediterranean coastal aquifers

In **Spain,** the condition of 'overexploitation' has received legal recognition with regulatory consequences since the law on water was passed in 1985. Sixty-one 'hydro-geological units' in the Mediterranean Basin, covering 7222km^3, are considered to be overexploited with an overall exceedence of withdrawals that are in excess of the natural recharge of 404hm^3 per year, or 41 per cent of the total pumped volume in these aquifers. This corresponds to an overall exploitation index of 172 per cent, mainly concentrated in the Segura Basin. Nine of these units have been 'temporarily declared as over-exploited' according to the law.

In **Israel** withdrawals from the coastal plain aquifer (using about 1000 boreholes) have been greater than the 'Safe Yield' (estimated at 240–280hm^3 per year) since the 1960s, since when they have been reduced progressively, from 480hm^3 to 330hm^3 per year.

In **Libya** excessive exploitation of the Jeffara aquifer to the south of Tripoli has created a regional depression of the water table, the levels of which have been lowered by several dozen metres, up to 100m in places, with accelerated falls in level of 1–5m per year. It has been calculated that if all withdrawals were to stop, it would take 75 years for the aquifer to replenish to its original level.

In **Tunisia,** out of 196 water table aquifers, 54 are being extracted in excess of their average recharge.

Overexploitation has particularly damaging effects in the case of coastal aquifers where the balance between freshwater and sea water is fragile and can easily be disturbed, causing almost irreversible *salt water intrusion into aquifers*. This has already occurred in many coastal regions of the Mediterranean countries, resulting in the abandonment of some wells. Underground coastal water tables have been reduced below sea level by excessive pumping in *Spain* (Campo de Dalia, near Almeria, in the 1980s), in *Italy* (Ravenna, up to 40m below sea level), in *Sicily* near Augusta, up to 50m below sea level, in *Sardinia* at Iglesias (as a result of mine pumping), in *Greece* (the Argolis Plain where the water table, exploited by 6000 wells and boreholes, has been lowered by 80m to 150m and is now below sea level over 100km^2), in *Cyprus*, in *Turkey* (the region of Cesme, Marmaris and Bodrum), in *Israel* (in the coastal plain aquifer where several 2–15m depressions have been found), in the *Nile* delta, where sea-water intrusion is estimated at 2km^3 per year, in *Libya* (where overexploitation of the Jeffara aquifer has caused a sea-water invasion over a 50km front in the Tripoli region, estimated at 166hm per year) and in *Tunisia* (in the Sfax region).

The proportion of the total quantities of underground water withdrawn that can be attributed to overexploitation (in excess of their average natural recharge, and thus unsustainable) is already considerable in several Mediterranean countries: 20 per cent in Spain (25 per cent in the Jacar Basin, 4 per cent in the Balearic islands), 13 per cent in Cyprus, 24 per cent in Malta (in 1990), 29 per cent in Gaza, 32 per cent in Israel (in 1994, where overexploitation is, however, compensated for to a large extent by artificial recharge), and 8–10 per cent in Tunisia.

Source: Margat, 2004

An *indicator of unsustainable water production* (calculated on the catchment basin level or NV, Table 1.1) shows that more than 10 per cent of water supply is probably already being taken from unsustainable sources in Libya, Israel, the Palestinian Authority, Cyprus and Malta, and also in some regions of Spain.

The situation at the N1 level (all Mediterranean countries) is even less 'sustainable', as some water-poor countries (group 1) are exploiting *fossil water* stocks,[12] thus irreversibly depriving future generations of this resource, as in all 'mining' undertakings. Overexploitation is added to fossil-water withdrawals to bring the unsustainable production indices in Libya, Tunisia and Algeria to 84, 22 and 35 per cent respectively.

In some ways, irregular resources created by dams could also be considered as unsustainable. Indeed, the *silting-up of retained water* in many dams gives them a limited life span while simultaneously reducing the number of possible sites for building new dams, so that future generations with need to develop other sources (it should be remembered that 70 per cent of the renewable natural resources in the Mediterranean catchment basin are *irregular*).

Box 1.5 Water regulated by dams – an unsustainable resource?

The high sediment load in run-off water in the Mediterranean countries makes the 'silting-up' of dammed water a particularly serious issue; it reduces the regulatory function of dams despite the planned provision of large volumes of 'dead storage'.

In **Spain**'s Mediterranean basins some 50 dams examined in 1996 had lost 6 per cent overall of their original capacity; the annual loss of capacity, although small on average (about

Water

Table 1.1   Unsustainable production indices, NV, 2000

Country	Value date	Overexploitation of renewable resources (in km³/year) (¹)	Demand in water (in km³/year) (2)	Index of unsustainable water production in % (1)/(2)
Spain	1997	0.70	18.20	**4**
Malta	1997–1998	0.02	0.05	**31**
Cyprus	1998	0.04	0.33	**12**
Israel	1999–2000	0.19	1.80	**10**
Palestinian Territories	1995	0.03	0.13	**23**
Egypt	1995–1996	0.00	66.00	**0**
Libya	1999	0.77	2.24	**34**
Tunisia	2000	0.18	2.27	**8**
Algeria	2000	0.00	2.90	**0**

Source: *Plan Bleu*, Margat, 2004

0.1–0.5 per cent), nevertheless amounts in a few extreme cases to more than 1 per cent, even 2.8 per cent in the Jucar basin dam, which is already 84 per cent silted-up.

In the southern countries, the loss of useful reservoir capacity is typically between 0.5 and 1 per cent per year, sometimes even more: 1–2.5 per cent in **Tunisia** (28hm³ per year on average at present); 0.5–2 per cent in **Algeria** (0.8 per cent on average at 19 dams studied, i.e. 15hm³ per year); in **Morocco** these losses are generally of the order of 0.5 per cent (70hm³ per year), leading to a reduction in regulatory capacity equivalent to the loss of potential irrigation of 6000–8000 ha per year.

The dams in **Algeria** have already lost more than a quarter of their original capacity (over 1km³), and six are more than half silted-up; the height of five dams has already had to be increased, and several large reservoirs are irreplaceable. In **Morocco** the reservoirs had, by 1990, lost 8 per cent of their original capacity (800hm³) some of which were already half filled. In **Tunisia** the reservoirs of seven dams studied had, by 1976, lost 15 per cent of their original capacity. The total reservoir capacity should reach a maximum (1.9km³) by 2010, then reduce and stabilize at around 1.8km³ between 2020 and 2030 with the consequent reduction of exploitable irregular resources. In **Egypt** the Aswan dam is on an entirely different scale: a new Nile delta is forming in its upper reaches where sediment is accumulating at 60–70hm³ per year (about 2km³ since 1970), but the 31km³ of dead storage is designed to be enough for five centuries.

Small dam reservoirs are no less exposed to these problems, and their life expectancy can be even shorter, about 20 years in **Tunisia**, for example, for reservoirs of from 50,000 to 100,000m³. Many Mediterranean reservoirs will doubtless fill up during the 21st century. Moreover, as the number of sites suitable for building dams is limited, there is an *irreversible* loss of the exploitable irregular resources that make up most of the exploitable volumes in the Mediterranean. Preventative measures (such as re-foresting basins, sediment trapping) can at best delay matters but cannot indefinitely extend the life expectancy of these dams. This will inevitably lead to a diminution of the water resources that are manageable through regulation. The 'post-dam' period will begin in the 21st century.

Source: *Plan Bleu*, Margat, 2004

Other freshwater sources? The limited potential for re-use and desalination

To limit their dependence on 'unsustainable' withdrawals, countries faced by a limit to their exploitable natural resources (mostly group 1 countries: Libya, Malta, Israel, Syria, Spain and Cyprus) are committed to developing water production from other sources: successive water-recycling or sea-water desalination.

Recycling agricultural drainage water currently represents about 12.6km³ per year at the NV level, mostly in Egypt, making it possible to meet annual demands (73km³ per year) that are greater than the average annual natural resources (57km³ per year) by recycling the same water for several successive irrigations. Such recycling could double between 1990 and 2025 (an increase of 8km³ per year between 2000 and 2025). However, it does not come without major public health problems since the drains are very often used for waste (in the Nile delta, for example, where market gardens are irrigated by water of doubtful quality). Moreover, there is a serious risk of soil salination, since drainage water concentrates salts.

Re-using wastewater is also being developed for irrigation. It represents about 1.1km³ in the entire Mediterranean Basin (Spain, Israel, Cyprus, Egypt and Tunisia). In Cyprus the quantities re-used could triple or quadruple between now and 2010. In Egypt re-use could increase by a factor of ten. In Israel this potential is being intensively explored. However, the development of wastewater re-use at the basin level may remain technically and economically limited by the volume of wastewater produced, the places of discharge (usually far from places of re-use), and the need to store it and provide preliminary treatment before re-use, as otherwise the risks for animal and human health or soil contamination are significant. The potential for wastewater re-use may reach a total of 5.7km³ in the NV catchment areas by 2025, or 3 per cent of total demand in the baseline scenario (from less than 1 per cent in 2000).

The industrial production of freshwater through *desalination* of sea water or brackish water is another and growing non-conventional resource. Total production in the entire Mediterranean Basin (Spain, Malta, Cyprus, Syria, Israel, Egypt, Libya and Tunisia) approached 0.4km³ per year in 2000. However, in spite of a regular reduction in costs,[13] and its particular adaptability to meeting the needs of islands and tourists, its development remains limited (0.1 per cent of the 360km³ per year of exploitable natural resources in the NV basin and 0.2 per cent of total demand). It is reserved for domestic or industrial uses. Desalination represents a large fraction of supply in only a few countries (Malta and Libya). In Malta all additional demand can only be met by desalinated water. The impacts of desalination on the cost of supply, energy consumption and the local environment must also be remembered, even if these should be put into perspective at the Basin level, given the small volumes involved compared with those of conventional water.

The *direct use of brackish water* for industry or even agriculture is also being developed (in Israel, 166hm³ in 1999, and in Tunisia).

Thus a growing percentage of water supply is coming from sources other than withdrawals from renewable natural resources. This trend should continue, mainly where it has already begun: increased use of non-renewable resources (Libya, and if necessary Egypt, Tunisia and Algeria), re-using drainage water and wastewater (Egypt, Israel, Syria and Tunisia), desalination (Malta, Cyprus, and in certain areas in Spain and Israel) and the recourse to imports from other basins (inter-regional or international transfers: Spain, Israel). Progress will be determined by the relative evolution of the costs and technological feasibility of these sources compared with those of withdrawals from natural resources. It may be limited by their impacts on the environment and health. It is important not to be misled by this apparent diversification of supply sources. In all Mediterranean countries, withdrawals of natural renewable resources are expected to remain the main, if not the only source of supply for a long time and by a large margin. On the basis of the trend projections to 2025, as a proportion of total demand, withdrawals would remain at:

- 100 per cent or thereabouts in most of the northern-rim countries, as well as in Turkey, Syria, Lebanon, Algeria and Morocco; about 95 per cent in Spain, Libya and Tunisia; 87 per cent in Cyprus;
- about 70 per cent in Israel, the Palestinian Territories and Egypt;
- 33 per cent in Malta.

By 2025 the demand met other than by withdrawals from natural resources may reach a maximum of about 25km³ per year (90 per cent of which will be in Egypt, by using return water from agricultural drainage) and represent less than 8 per cent of total water demand in the Mediterranean region (and less than 1 per cent ignoring the particular case of return water in Egypt).

Water policies and planning, in the various Mediterranean countries considered by the P*lan* B*leu* baseline scenario, rely mainly on an extension of water supply in order to keep pace with increasing demand that is too often considered to be irreducible. First, the plan is to pursue major hydraulic projects aimed at exploiting an increasing proportion of renewable natural resources. Secondly, other non-conventional resources are envisaged, but in fairly limited quantities and mostly in water-poor countries. These strategies for a considerable increase in water supply are expected to increase environmental as well as social impacts and the cost of water supply. The two following sections attempt to analyse these impacts. Without claiming to be exhaustive (the landscape and social aspects, and the risks related to non-conventional water production are not considered), they stress, above all, their sometimes irreversible effects on the water resources themselves.

Resource and ecosystem degradation

Man-made degradation that modifies the regime and quality of natural water resources adds to the stresses on them. Although not a recent problem, such degradation increased to a significant level during the 20th century as a result of demographic growth, major development projects and industrialization.

Changes to the water regime

I*ncreasing withdrawal* and the infrastructures that go with it modify the natural water regime. The increase in water withdrawals has inevitably reduced the flow of many Mediterranean rivers, especially during low-water periods. Generally speaking, about 80 thousand million m³ per year are currently being taken from Mediterranean Basin rivers, which cannot fail to have an impact on the regimes and the functionalities of many rivers, especially in the south.

When these rivers feed aquifers, the reduced flow can in turn have harmful effects on underground water resources. In Italy, for example, in the lower Piave valley to the north of Venice, upstream diversions of water for irrigation and hydroelectric plants have reduced the levels of the free or captive alluvial water table. Conservation measures of low water flows of the Piave were vital for preserving the supply to the water table and limiting the sinking of Venice.

Intensifying underground water withdrawal has also had inevitable and major effects on surface run-off. The most visible have been the drying-up of springs in the Po plain in Italy or in Tunisia's Djeffara coastal plain. The 'basic flows', the regular components of river flows, have consequently been affected. In Spain the basic flow of the Rio Jucar in the Albacete region has been reduced by a factor of four in roughly ten years as a result of intensive exploitation of the eastern Mancha aquifer.

Water projects also disturb water regimes. Reservoirs may well be useful for regulating water, but these installations and their operations also have negative effects on downstream regimes: reducing average and low waterflows; interrupting the continuity of and degrading aquatic ecosystems; impacts of water release and decennial emptying (for example impacts in the Rhone Valley on local wells and boreholes from plugging caused by reservoir emptying); attenuating the positive effects of flooding, disrupting and blocking the movement of sediments (reservoirs are silted-up but deltas are depleted and made even more vulnerable to erosion by the sea); and loss of the retained water through evaporation.

The case of the Nile in Egypt is the most evident. The regulation of water flows by the Aswan dam resulted in a considerable increase in the quantity of water used and the perpetuity of irrigation. But 97 per cent of the 134 million tonnes (Mt) of sediment transported annually by the river, some of which (about 16Mt) once fertilized the irrigated land, are now held back in Lake Nasser. To compensate for this, the land now requires 13,500t per year of nitrate fertilizers. For want of inputs of sediment, the delta's coastline has already receded noticeably. Moreover, it is estimated that one of every 10m^3 of water regulated by the dam is lost through evaporation. The Ebre delta also now receives only 3Mt of sediment a year compared with 20Mt formerly and has begun to recede by nearly 50m a year.

River-bed water developments can also cause considerable lowering of alluvial surface-water levels; this is the case with the 'run-off canals' downstream of hydroelectric plants, for example in the Rhone Valley in France. *Poor irrigation practices* (especially in alluvial plains where drainage is difficult) can be the cause of an increase in salinity and be responsible for the irreversible loss of millions of hectares of soil (see Chapter 5, Rural areas).

These regime changes are aggravated by other factors such as changes in *land use* in the catchment areas or the *exploitation of gravel in active river beds*, which leads to considerable drops in river levels and, in consequence, in the levels of associated alluvial surface waters. In France's southern Var valley, intensive gravel diggings in the 1960s near the city of Nice – satisfing large demands from the building sector – caused the alluvial surface water level to drop several metres, harming farmers and the harnessing of drinking water.

Degradation of the quality of natural water and ecosystems

Changes of a *qualitative* nature that alter the quality of water resources that are already naturally vulnerable have to be added to the changes in the *quantities* of the water resource regimes described above and may further limit the possibilities of using them.

The increase in pollutant emissions

The *increase in withdrawals* in the SEMC projected in the baseline scenario threatens to degrade the quality of water resources. The examples already given in Box 1.4 provide a good picture of such overexploitation, which can sometimes have irreversible effects. Moreover, the expected increase in withdrawals of water for drinking, which demands top quality, poses a threat to the best-quality resources. The example of the Milan region in Italy illustrates the interaction between water regime and quality: after an industrialization phase that drastically lowered the level of the water tables, changes to the industrial structure (an evolution towards the financial sector) resulted in the water-table levels rising again, bringing to the surface poor-quality water that had been affected by years of various polluting activities.

But water quality is also threatened by the many *pollutants discharged into freshwater* either locally (for example untreated industrial or domestic waste, uncontrolled discharges, drains used as conduits for household waste) or in a diffuse manner (fertilizers and pesticides for intensive agriculture, solid waste, sludge from treatment plants). The threat is hard to quantify, as there is little information on these pollutant flows (particularly for diffuse pollution), so one can only use rough estimates that have been made indirectly.

Wastewater discharges by industry and households, although stabilized in volume in the NMC (or even falling) as a result of demographic changes and progress made in industrial processes, still constitutes a very considerable pollution load (in terms of the concentrations and the nature of the products it contains, which is becoming more and more complex). The amount of pollution will certainly increase in the SEMC as a result of demographic growth, rising living standards and urbanization (which often leads to an increase in water consumption per capita), and the arrival of more than 133 million additional tourists per year by 2025 (Part 1). Given the evolution of sectoral demands, it is estimated that, globally speaking, emissions of urban and industrial wastewater in the Mediterranean countries could continue to increase by 1 per cent per year (30 per cent in total) between 2000 and 2025, from 37km³ to 47km³ per year (NV). The increase is expected to be mainly in the SEMC, which may experience a doubling of wastewater volumes during this period. More than 60 per cent (80 per cent in the SEMC) of this wastewater could affect freshwater, the estimated remainder being emptied into the sea.

Industrial wastewater in the NMC, when expressed in terms of the volume of dissolved Biological Oxygen Demand (BOD) per year, has stabilized (or even decreased slightly) over the past 20 years, but is growing rapidly in the SEMC. At the N1 level, an increase of 1 per cent per year between 1980 and 1997 has been recorded[14] in Turkey, 2 per cent in Egypt, 5 per cent in Tunisia and 7 per cent in Morocco.

Figure 1.8 Urban and industrial wastewater, NV, 2000–2025

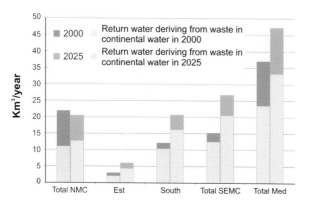

Source: Plan Bleu, J. Margat; from sectoral water demands

Note: Urban and industrial emissions should decrease in volume from now to 2025 in the NMC, but not in pollutant load.

Urban and industrial emission volumes could double by 2025 in the SEMC.

Total commercial *fertilizer* consumption in the agricultural sector in the NMC has stabilized, or even slowed, since 1990, after a period of high growth between 1960 and 1990. Many SEMC still have high growth in the use of fertilizers (4 per cent per year between 1985 and 1997 in Syria, 2 per cent in Turkey and Israel, 1.3 per cent in Egypt) which will probably continue (see Chapter 5). *Pesticide* consumption is less well known; it has stabilized in France (the world's second-biggest user).

This information provides only very indirect information on discharges of pollutants into the environment. It must be complemented by information (that is still hard to come by) about the level of processing pollutant flows prior to their release into the natural environment (see below, the section entitled 'Combating pollution') and about the capacity of the environment to absorb the flows without being altered.

Degradation in water and ecosystem quality

If pressure on the water resource increases, the resulting degradation in water quality is impossible to assess on the too-vast scale of the Mediterranean catchment area. One of the main problems is the variety of parameters that define quality and its variability in space and time. Added to this is the serious inadequacy of the monitoring networks in most Mediterranean countries, which makes it very hard to establish a summary overview of the situation. Data are noticeably incomplete (in time and space and in terms of the parameters covered). Nevertheless, attempts have been made to map rivers in terms of quality (identifying those whose 'poor quality' is attributable to pollution) in a few countries: Spain, France and Algeria. Figure 1.9 provides an incomplete summary, probably not standardized, and limited by the state of the available information.

Despite inadequacies in monitoring, many *signs of degradation* are observed locally. Degradation of water quality is becoming a major concern in many countries. Surface waters very frequently have high biochemical dissolved-oxygen demand (BOD), high phosphate, nitrates and heavy metals content, and local bacteriological pollution. The wadis that flow through many Mediterranean cities are often transformed into literal open-air sewers (for example Tirana, Damascus and Beirut). Underground waters are the most vulnerable because the pollution takes much longer to reverse (Box 1.6). Natural lakes and dam reservoirs are also under threat from climate-induced eutrophication.

Since 1940 the EU-Med countries have experienced considerable degradation of their water resources. More recently, with improvements in the level of treatment of domestic and industrial waste,[15] they have registered some improvements in surface water quality (concentrations of BOD, phosphates and heavy metals in rivers have been falling for the past 10 years). On the other hand, non-localized agricultural pollution, particularly by nitrates, is continuing to inflict lasting degradation on resource quality.[16]

Data on the presence of dangerous substances such as pesticides in the environment are very scarce, even in the European countries, where the European Environment Agency has deplored the continuing inadequacies of the monitoring networks.[17] Many compounds escape detection, despite most of them being very persistent in the environment and their possible effects on health and ecosystems being under study. In the few countries that are beginning to monitor them, traces of phytosanitary products are found in natural water, drinking water and food. In France, of the 3000 monitoring points in 2000, 90 per cent of those located in surface water and 58 per cent of those in underground water are affected by the presence of pesticides.[18]

Figure 1.9 Main rivers subjected to chronic pollution, NV

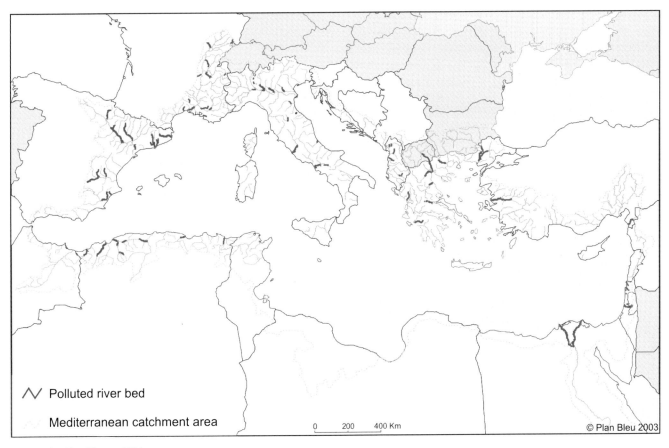

Source: Plan Bleu, Margat, 2004

Note: Attempt at a summary based on a compilation of water-quality maps from various countries, which do not always specify the quality-defining criteria.

Box 1.6 Examples of underground water pollution in the Mediterranean Basin

Many types of pollution from agriculture (fertilizers, pesticides), industry (wastes) and cities (untreated sewage, wastes) threaten the quality of the underground water in plains with intensive land use: urban areas or intensive farming. However, such pollution is no more marked in the Mediterranean Basin than elsewhere.

In **France**, more than half of the country is already classified as zones suffering from excessive nitrate pollution.

In **Italy**'s Turin province alone, for example, where 381 collecting wells for producing drinking water (5m³ per second in 1985) are concentrated in an area of some thousands of square kilometres, 252 rubbish dumps (mostly urban), 159 quarries and 840 boreholes for injecting polluting industrial effluents have been recorded, which leads to significant interactions between the areas drained by the withdrawals and those subject to the effects of sources of pollution.

Results showing a high nitrate content are beginning to appear in various areas. For example, more than 100mg per litre are found in several **Spanish** aquifers (Levant, the Balearic Islands, Catalonia – even up to 600mg per litre in places) where 25 (out of 206) polluted hydro-geological units are located in the Mediterranean basins. In **Italy**, aquifers are often contaminated by nitrates, heavy metals and persistent organic chemical compounds, making it necessary to draw water from deeper springs, the natural quality of which is not always adequate (because of the presence of manganese, iron and sulphates). In **Turkey**, occasional checks show the infiltration of wastewater or the seepage of discharges, and the contamination of water tables by toxic substances from industry (cyanide in the Kelmalpsa Valley and fertilizers and pesticides in the Cukurova, Bursa and Bornova Valleys). In **Israel**, the average nitrate content of water extracted from the coastal aquifer has exceeded 40–50mg per litre since 1975. In **Malta**, as a result of the use of fertilizers over the past 20 years, the nitrate content in the aquifer high on the Rabat plateau has increased significantly, often exceeding 100, and sometimes even 300mg per litre.

Source: Margat, 2004

The intensive exploitation of underground water can also affect *aquatic ecosystems*, especially local 'wetlands' associated with emerging sites, the conservation of which is desirable for many reasons. This is particularly the case for the brackish water of coastal wetlands, the quality of which is very sensitive to the balance established between emerging underground water and sea water.

Is access to water becoming more difficult?

Stress on water resources and the increase in pollution thus have considerable environmental impacts, which, in turn, limit the quantities of water physically available to meet any additional increases in demand. But they also increase the vulnerability of water supplies by increasing costs, health risks and conflicts between users.

The rising cost of water supply

The 'easiest' investments for exploiting water resources were the first to be made. With dam sites progressively being used up, the lowering of water tables and, more generally, the increasing distances between the points of abstraction and water use (longer and longer transport of water), the costs of exploitation and supply are increasing. For example, the intensive exploitation of many Mediterranean aquifers (even without disturbing the balance between use and replenishment) has had the direct effect of lowering the levels of water tables, as observed during recent decades (by several dozen, even hundreds of metres), which has increased the pumping costs.

Box 1.7 Examples of lower levels observed in water tables

In **Italy**'s Milan region, the intensive and increasing exploitation of the alluvial aquifer of the Po plain has lowered levels by more than 25m, even up to 40m in places, in 80 years. The rate of reduction increased during the 1970s from 1 to 2m per year. In the karstic aquifer of the Sierra de Crevillente, near Alicante in **Spain**, exploitation, initially through galleries starting in 1967, then by drilling and pumping, lowered the levels by 250m in 20 years; 75 per cent of the extracted volume corresponded to a reduction in stocks. In the Messara Valley in Crete (**Greece**) 30m falls in ten years were observed. In **Libya**'s Jeffara plain, increased pumping has lowered the water table to the south of Tripoli by several dozen metres (up to 70m in places), with accelerated reductions in level varying from 1 to 5m per year. In the **Tunisian** Djeffara, the reductions in the level of the intensively exploited water table have accelerated, going from 0.5 to 0.7m per year in 1970 to 1 to 2m per year by 1987.

Costs are also being pushed up by the growing need to *treat* water before using it (and the increased stringency of drinking water standards in Europe). The more a resource is degraded, the more expensive is the treatment. Moreover, this rising cost of water exploitation and treatment is expected to be increasingly passed to the end-user. Budgetary limitations make it no longer possible for states to pay, as they often did in the past, the huge costs of major hydraulic facilities for distribution or treatment. The contribution of users to the funding and operating of facilities is increasing with a more general acceptance of the 'polluter- or user-pays' principle. For all these reasons, satisfying water demands by increasing the supply (baseline scenario) is therefore expected to result in increasing costs to the community and users for access to water; this will probably not be defrayed by technological progress in the sector. This trend is seen very clearly in the EU-Med countries.

Today 30 million people in the Mediterranean without clean drinking water, and tomorrow?

'Water is vital for life and health. The *right to water* is indispensable for enabling a healthy life in human dignity. It is a preliminary condition for arriving at other human rights'.[19] But this basic right may increasingly be put to the test with the rising stress on ever-more polluted water resources.

While drinking water is still allocated priority over other uses such as agriculture or industry[20] and still represents much smaller volumes than the other sectors, the limiting factor for access to drinking water is very rarely the quantitative availability of the resource but much more the *quality of service* of water distribution and the *access conditions* to these services for the poorest. There is now a risk that the degradation of the quality of natural water resources and the predictable increase in supply costs will increase the difficulty of access by the poorest populations to drinking water services, the frequency of accidents, and the risks of a breakdown in distribution services. For these very different reasons, a growing part of the population may see itself excluded from access to water that fulfils the conditions of drinkability.

The scarcity of available statistics about access to drinking water often makes it impossible to distinguish between the causes that limit access to 'safe' water. Nor do they allow comparisons in time or between countries with different standards when trying to define what is meant by 'access' and 'drinking' water. Caution is therefore needed when using these statistics. There were nearly 30 million Mediterranean inhabitants[21] who officially did not have permanent access to drinking water in 2002 (Figure 1.10). The largest numbers were in Turkey, Algeria, Morocco, Egypt and Syria. Rural dwellers and the poorest in suburban neighbourhoods are often the worst affected.

Water

Figure 1.10 People not having access to an improved source of drinking water or improved sanitation, N1

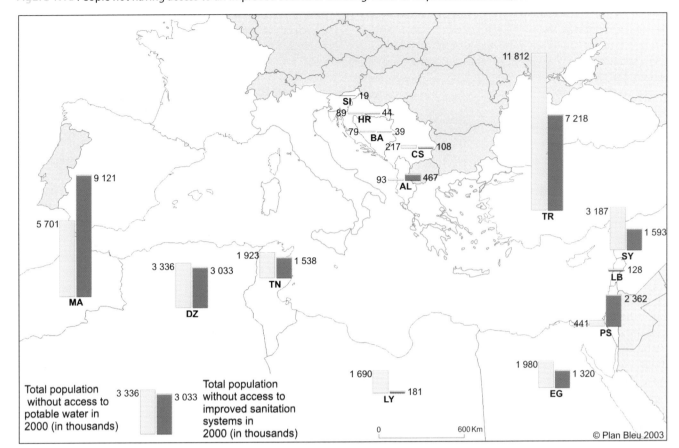

Source: World Health Organization and United Nations Children's Fund. Water Supply and Sanitation Collaborative Council. Global Water Supply and Sanitation Assessment (http://unstats.un.org/unsd/mi/mi_goals.asp or www.unicef.org/wes/mdgreport)

Note: 27 million Mediterranean inhabitants may be deprived of an adequate sanitation system, and 30 million of access to safe water.

In the sense of the World Health Organization, 'improved' water supply technologies are household connections, public standpipes, boreholes, protected dug wells, protected springs and rainwater collection. 'Not improved' are unprotected wells, unprotected springs, vendor-provided water, bottled water (due to concerns about the quantity of supplied water, not the water quality) and tanker truck-provided water. 'Access' implies a source would be likely to provide 20 litres per capita per day at a distance no greater than 1000 metres.

In the sense of the World Health Organization, 'improved' sanitation technologies are connection to a public sewer, connection to a septic system, a pour-flush latrine, a simple pit latrine and a ventilated, improved pit latrine. The excreta disposal system is considered adequate if it is private or shared (but not public) and if it hygienically separates human excreta from human contact. 'Not improved' are service or bucket latrines (where excreta are manually removed), public latrines and open-pit latrines.

The *quality of distributed water* is a growing concern in all Mediterranean countries. In 1997 in Turkey, quality was not acceptable in 12 per cent of the samples taken at the national level. In France in 1998, 8 per cent of the population was served by water with a non-conformity rate of over 5 per cent from the bacteriological point of view.[22] These figures conceal problems of management capacity[23] and equipment condition, and the pollution of springs.

With an increase in chemical substances poured into freshwater, *uncertainties* persist about the long-term effects on health and ecosystems. Knowledge about the effects on health of many substances included in water (but not regularly measured, such as plant-protection products) is still sketchy, especially if one takes the possible synergistic effects of the different compounds into account. Epidemiological studies that show the effects on health of the poor quality of distributed drinking water are lengthy, scarce and difficult to access. A study begun by the EU in 1991, which was to test the health effect of 900 active compounds, was, by 2003, able to reach conclusions on only 5 per cent of these products. It is estimated that in France 10 million citizens drink water that is contaminated by pesticides

2 A Sustainable Future for the Mediterranean

without, however, being able to identify the immediate risks to health. However, some studies are beginning to show the possible effects of plant protection products on the health of rural and farming populations, linked in particular to the handling of the products and the accumulation of very persistent substances in water and food (fertility, cancer, development anomalies and headaches).[24]

As a result of such concerns, the people of the Mediterranean are increasingly turning away from tap water to bottled mineral water. A survey[25] has shown that, in France between 1989 and 2000, the proportion of people saying that they drink tap water fell sharply from 72 to 58 per cent, and that 39 per cent of them drink only bottled water.

The poorest populations are the hardest hit by these shortcomings in terms of both resources and management facilities. They are often the first affected by disease (e.g. gastro-enteritis, hepatitis diseases and diarrhoea) or forced to buy high-priced water (in comparison with their incomes) from tankers or in bottles.

The scope for progress in improving access to safe water lies far more in protecting upstream resources and improving management than exploiting new resources.

It is also about the implementation of compensation systems to allow the poorest to have access to water distribution services. These paths to progress will be explored in the second part of the chapter.

Conflicts over water

Tensions over water resources may also increase conflicts:

- between users of the same resources: for example falling water-table levels resulting from their intensive exploitation have already caused violent conflicts between the best-equipped irrigators (with deep-drilling technology) and those using more traditional techniques;
- between the main categories of water users (drinking water/industry/agriculture), with arbitration often being at the expense of the agriculture sector;
- between regions or even countries (Spain, the Balkans, the Near East, Egypt). For example, the issue of sharing underground resources (or those in the Jordan Basin) between Israelis and Palestinians is indissociable from that of the territorial conflict and its resolution. Generally speaking, although the diffi-

Figure 1.11 Shared Mediterranean basins

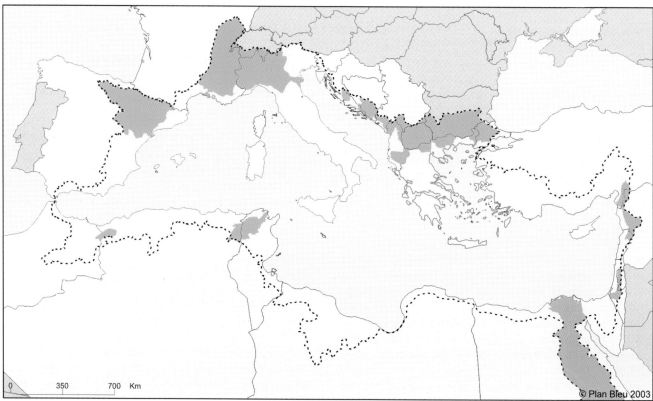

Source: Plan Bleu, Margat, 2004

culty of access to water is rarely the only cause of conflicts, it feeds them by generating stress, resentment and even some violence. Some do not hesitate to call water the 'oil' of the 21st century.

Countries become more vulnerable to external issues, for example as a result of an increased sensitivity to droughts or an increase in the frequency of shortages, with increased *dependence* on others for their water supply (cross-border water transfers or increasing recourse to imported energy to operate desalination plants). From this point of view, many Mediterranean countries depend to varying degrees on inputs from outside their national territories. Apart from the specific case of Egypt (which depends for 98 per cent of its renewable resources on the Nile from up-stream countries), the other countries have dependency indices[26] ranging from 0 to 38 per cent. The most dependent countries are Croatia, Israel and Albania. Trans-border basins are common in the Mediterranean, with some 20 covering about a third of the Mediterranean catchment area, representing 46 per cent of the Basin's overall run-off. This essentially comprises the Nile, the basins of the eastern Adriatic countries (Drin, Vardar, Struma, Maritza) and the basin of the Orontes in the Near East. These basins may be sources of conflict over access to resources, hydroelectric power plants and the distribution of their effects (cross-border pollution). More optimistically, they can also be special opportunities for cooperation strengthened by necessity.

The projections of water demands and resources thus show a highly contrasting situation depending on the country. The situation is particularly worrying for most of the SEMC, expected to experience the largest growth in water demand. Some of them are already in water shortage situations and exploit their resources in an unsustainable way. The predictable increase in tensions over water resources is expected to result in geo-political, environmental, social and economic risks that could limit the prospects for future generations by altering some resources in an irreversible way. The NMC have less stressful 'supply and demand' balance sheets. On the other hand, they are more and more concerned about degradation of water quality, which boosts supply prices and the effects of which on health and ecosystems are still poorly known.

In any case, classic 'supply' policies, which rely on natural resources to meet all the forecast increases in demand, are being stretched to the maximum. Although the limits in some countries are physical, they are also above all socio-economic and environmental. Such policies will have to be justified more and more on the basis of economic, social and environmental criteria.[27] They will also have to be compared with other policies, such as those that involve, as a priority, managing the water we already have in a better way before asking Mother Nature to cough up more, as discussed in the next section.

Paths to more sustainable water management

Compared with the baseline scenario, which projects increasing exploitation of natural resources, resulting in greater vulnerability, and increasing effects of such exploitation, which is already reaching its limits, an alternative path of *greater efficiency* in water management provides many advantages from three points of view: social, economic and environmental. Such an alternative path is already being explored in many Mediterranean countries.

The goal is the *integrated management* of water resources, 'a process which promotes the coordinated development of water, land and related resources to maximize the resultant economic and social welfare in an equitable manner, without compromising the durability of vital ecosystems'.[28] This concept is increasingly being used in water policies, even if effective implementation is slow. It is at the heart of the recent European Framework Directive on Water.

> **Box 1.8** The Framework Directive on water, a major stimulus for integrated management
>
> The European Union has, after more than five years of negotiations, adopted a new framework (Directive 2000/60 EC) for a common integrated policy in the water sector (water supply and the disposal of wastewater).
>
> Its dual goal is coherence and a search for effectiveness in national measures for preserving or restoring the quality of the resource. It is a major stimulus for the integrated management of water resources in the EU and future member countries. It acknowledges the needs for hydrographic basin planning and management and confirms the principle of cost recovery and the importance of public participation in management. It introduces a new approach based on objectives, with Member States in principle responsible for the results.
>
> It should be transposed to national legislation during 2004. Plans of action are to be drawn up basin by basin, relying on shared diagnoses and targets for 'good ecological status' appropriate to each hydrologic unit, to be achieved by 2015. The European Commission will publish a first assessment report in 2016, to be updated thereafter every six years.

For the sake of clarity, this section will discuss first the management of the *resource*, then the management of water *demand*. However, it is important not to lose sight of the basic interdependence underlying the very concept of integrated management, since the boundary between resource and demand is sometimes difficult to define. Nevertheless particular stress will be placed on the less-frequently explored path of possible progress in managing *water demand*. The *Plan Bleu* and the Mediterranean Commission on Sustainable Development have shown in recent studies how such an approach could lead to more sustainable development in the Mediterranean.

Given the many political, sectoral and geographical interactions between the actors, the greatest scope for progress in water management is in *water governance*, since water crises are often crises of governance. This concept refers to the broad panoply of political, social, economic, administrative and cultural systems that govern the development and management of resources and water services. It involves a multitude of actors and interests, including public actors, who, because of water strategic, social and long-term implications, have played a particularly important role in the water sector in recent decades.

Improving resource management

One of the first areas where progress in water management may be possible is the management and protection of water resources. It is dealt with only briefly here, because it comes up again in Chapter 6, in the section entitled 'High-impact land-based pollution, improvements remain insufficient'. Policies for preserving and conserving water resources rely on a wide range of legal instruments; these are often less expensive than restoring the quality of the resources, but they are not yet sufficiently effective.

Combating pollution

Most countries have introduced legislation to protect their water resources but are encountering great difficulty in applying their laws, mainly because of the spread of responsibilities between the various institutions involved and their limited resources. Effective monitoring of withdrawals and waste disposal in most Mediterranean countries remains limited.

Treating polluting wastewater

The scope for improvements in processing wastewater can be appreciated by considering the expected increase between now and 2025 in the flows of pollutants in the SEMC and the inadequacy of the financial resources that have been allocated for treating them. Sewage is defined here as the *collection* of wastewater (essentially a health-oriented service) followed by *treatment* of the water collected (essentially an environment-oriented service).

Information on the rates and levels of collection and treatment of wastewater are highly variable and difficult

Table 1.2 Percentage of population connected to a sewage collection network and benefiting from treatment, N1

	Year	Resident population (thousands)	Percentage of the population connected to public collecting sewer (1)	Percentage of the population connected to public treatment plant (2)	Percentage of the population connected to a collecting system and benefiting from treatment (2)/(1)
Spain	1995	39,433		48	
France	1998	58,497		77	
Italy	1995	57,268		63	
Greece	1997	10,507		56	
Malta	2000	391	100	13	13
Cyprus	2000	669	34	34	100
Slovenia	1999			30	
Turkey	1998	62,810	59	23	39
Syria	1994	13,782	59	10	17
Israel	1999		100	89	89
Tunisia	2000	9618	47	42	89
Algeria		29,100	66	4	6
Morocco	1996	26,848		3	

Source: Eurostat. New Cronos 2003 (http://epp.eurostat.cec.eu.int/); Eurostat (2003) Statistiques de l'environnement dans les pays Méditerranéens. Compendium 2002. Office des publications officielles des Communautés Européenes, Luxembourg

Note: The last column is estimated as the ratio of the percentages of population concerned, not in pollutant loads.

Water

to compare. According to the WHO, for the countries publishing such data, the proportion of the population with access to an adequate sanitation system in 2000 was more than 85 per cent everywhere except in Morocco (68 per cent) and the Palestinian Territories (25 per cent). Between 1990 and 2000, an additional 10 per cent of the population in Morocco, Egypt and Tunisia gained access to an adequate sanitation system. However, 27 million Mediterranean inhabitants still do not have adequate sanitation systems (Figure 1.10 p87).

The percentage of the population not connected to a *treatment plant*, on the other hand, remains high (between 30 and 70 per cent depending on the SEMC.[29] (Table 1.2) In Turkey, for example, in 1998, 69 per cent of the population was connected to a collection system but only 23 per cent to a joint collection *and* treatment plant.

The levels and efficiencies of *effluent treatment* are even less well known; the efficiency of treatment plants is rarely measured and published. It is normally very low in many SEMC because of the lack of maintenance and the poor training of maintenance staff. Although treatment rates in the industrial sector are generally higher, they seldom exceed 50 per cent (see Chapter 6).

In the *northern countries*, there is more pollution, which is coming more and more effectively under control, but the NMC still have some way to go to meet the requirements of the 1991 European Waste Water Directive.[30] Although the EEA has welcomed the improvements in urban wastewater treatment in Europe, it has stressed the low treatment rates of southern European wastewater, where less than 50 per cent of the population is connected to a treatment plant and only between 30 and 40 per cent to a secondary or tertiary treatment system.[31] In 1995, 64 per cent of the population in the four EU-Med countries were connected to a system for purifying urban wastewater with primary treatment.[32] But when connection to more elaborate tertiary treatment is considered, the figures for Spain, Italy and Greece fall to 3, 24 and 10 per cent respectively. In France only 40 per cent of pollution flows receive satisfactory treatment.[33] Increasing treatment raises the question of treating sludge, which is less and less easy to spread on agricultural land and requires ever more expensive treatment (dumping, incineration).

In the *southern and eastern countries*, the levels of collection and (in particular) treatment are very low, not to say non-existent. Many large cities still have no treatment plants. Investments in pollution control are very low. In Turkey these are very modest compared with those devoted to developing distribution infrastructures and irrigation networks.[34] Given the large growth expected in the urban populations of the SEMC (98 million between 2000 and 2025) and the low current level of water treatment, there is an urgent need to develop simple treatment technologies that are not too expensive, such as lagooning. Israel is the exception, with 85 per cent of the population connected to tertiary treatment in 1999.

Prevention rather than cure

Steps can be taken upstream to limit the polluting water discharges. They and their impacts are more economical than curative treatments, which are always very expensive. Preventive steps against pollution include facilities that regulate discharges, pollution monitoring and, further upstream, the creation of protective boundaries for harnessing drinking water, reasonable use of fertilizers in agriculture and the promotion of clean production processes in industry. Water demand management (see below) reduces emission volumes by limiting the volumes required for a given use, therefore the volumes to be purified, and is thus again beneficial from this point of view.

There are many examples that demonstrate the considerable potentials for savings in the *industrial* sector in the Mediterranean that result from the introduction of *clean technologies*. Major savings in natural and financial resources in industrial establishments can be made just by raising awareness during an environmental audit and providing relatively limited investments. These are often amortized in only a few months (Box 1.9). In the SEMC the scope for progress by introducing such clean techniques is even greater since production processes are generally poor compared with worldwide standards. Foreign direct investment can be the vector of such modernization, but the low current levels in the SEMC justifies massive public action in this area (subsidies and public aid for development).

Box 1.9   Examples of savings made by clean production techniques

Much experience has been gained in the Mediterranean countries that demonstrates the effectiveness of industrial action to introduce 'clean' production processes that save resources and limit polluting emissions. Many case studies in the Mediterranean registered by RAC/PP show the magnitude of the savings that are made possible by such processes, with return-on-investment times seldom exceeding two years.

In **Turkey**, for example, a tyre-making factory in the Izmit region reduced its water consumption by nearly three-quarters, from 900,000 litres to 250,000 litres per day, thus reducing its discharges into the community sewers. A detailed analysis made it possible to replace a cooling system with a closed-circuit system for an investment cost of US$5000 and a return time of two years.

In **Egypt**, one of the largest tinned food manufacturers (Montazah near Alexandria) underwent an eco-audit and introduced measures to reduce energy consumption: insulating steam pipes, replacing leaky parts, fitting a pressure regulator to the sterilizers, and improving the recuperation system and boiler efficiency. Water consumption was reduced by implementing water-consumption hydrometer monitors, installing sprinklers (so that water flows only when needed) and improving the water collection and recycling system. The savings made in water, steam and energy (nearly a 40 per cent saving in fuel consumption) made it possible to reduce discharges and amortize investments in between 1 and 44 months.

In **Croatia,** one of the biggest dairy companies, LURA in Zagreb-Lurat, undertook measures such as employee training programmes, a reduction in the diameter of cleaning pipes, and changes to the hot-water circuit, that resulted in a reduction of 286,000m^3 per year (or 27 per cent) in the quantity of effluents, and savings of 280,000m^3 per year of drinking water. As a result of these simple measures, which involved the employees, and at a low cost (a total investment of 31,000 euros), the factory made significant savings in water and energy, equivalent to 328,000 euros per year, with an amortization of the investment in less than one month and a reduction in its effluents.

Source: Regional Activity Centre for Clean Production, Barcelona; www.cema-sa.org

Increasing the exploitable potential in a sustainable way

There is also a wide range of resource management methods – not yet sufficiently explored – that would make it possible to increase the exploitable potential of renewable natural resources (a definition was given in Box 1.2) at the lowest possible environmental and economic cost. For the arid countries, some options are described in Box 1.10 and merit consideration in detail in all countries, although they often call for difficult innovations covering a wide spectrum (hydro-geology, hydrology, agronomy, pedology (study of soils), geology and socio-economics) and a large number of actors. Other options such as harnessing coastal or underwater freshwaters or mildly brackish springs are beginning to be explored in the Mediterranean.

Box 1.10 Some options for better resource management in arid climates

It may be possible to better manage water in an arid country like Tunisia with very scarce and vulnerable water resources. The following options may enable progress towards more sustainable and integrated management:

Artificially recharging the water tables: in arid areas, unlike in wetter climates, the immense regulatory capacity of the water tables could be better exploited by using water from dams when they are at their highest level for recharging water tables through infiltration (in infiltration zones or wadi beds), thus achieving underground water storage. The advantage of this is to transfer part of the irregular surface water (high water run-off) to regular water (in the aquifers), which is more easily exploitable. It would also make it possible to sustain dry-weather flows and limit losses through evaporation. For example in Tunisia in 1996, 65 million m^3 of water were transferred through the artificial refilling of aquifers. By 2030 this transfer is planned to reach 200 million m^3.

Dividing up regulatory works upstream of catchment areas by creating many hillside lakes above big dams would help to limit the silting-up of reservoirs by reducing the erosive energy of run-off and reinforcing the effects of artificial water-table charging.

Interconnecting water networks would make it possible to re-establish regional balances and optimize resource allocation.

A more balanced exploitation of aquifers with poor renewable resources could be achieved by matching withdrawals to practicable reductions in 'natural loss' (by evaporation in endoreic depressions), thus limiting withdrawals to achieve such a better balance. Depending on the situation, this strategy would be opposite to a more intensive exploitation of reserves (which would be more productive but unsustainable) or else could replace the final phase of such exploitation.

More active management of renewable-resource aquifers, making more use of their reserves when there is large capacity, could artificially amplify their natural 'regulatory reserve'.

Water and soil conservation: some steps (re-vegetation, cultivation practices, soil conservation, biological processes) make it possible to increase storage capacity for rainwater via the ground and thus limit irrigation needs while also limiting erosion and the resulting silting-up of downstream reservoirs.

All these measures, which predominantly aim to increase rainwater infiltration and ground storage, would enable a re-balancing of the distribution of water resources in the regions. It might lead to a rethinking of the function of big dams around preventing floods and refilling water tables as a main goal. We can see with these examples that it should be possible to achieve better management and distribution of water resources by an **integrated** approach, by understanding the interactions between the different components of the water system (and the economic, social and environmental effects of the different possible options) and through:

- understanding the physical conditions that govern productivity and the long-term behaviour of the water resource 'system';
- implementing exploitation strategies that ensure acceptable levels, duration, production costs and environmental impacts.

Source: Plan Bleu, based on a personal communication from Mr Mohammed Ennabli, former Minister of the Environment in Tunisia

Water

Improving demand management

In addition to improving resource management, there is huge scope for improving water demand management (WDM).[35] This includes *all actions and organizational systems intended to expand technological, social, economic, institutional and environmental efficiencies in the various uses of water*. It means making water consumption doubly efficient, by increasing both the efficiency of the ways by which water needs are met, and those by which water is allocated to various uses.

There is an analogy here with the energy sector. The rational use of energy (developed in Chapter 2) involves the same issues, limits and leverage as WDM (which could also be called 'the rational use of water'). The energy sector has experience in implementing efficiency measures that could probably be useful in the water sector.

The challenge: Saving nearly a quarter of the demand

There is scope for improvement in both the allocation between uses and each particular use of water.

It is impossible to quantify the possible gains for the entire Mediterranean from a more *efficient allocation* of the different uses from the threefold economic, social and environmental points of view. Such gains can only be evaluated locally by 'cost effectiveness' studies of the various options, by including the costs and benefits of the environmental and social externalities. Such studies are seldom undertaken, particularly on the questions of optimizing allocations on the basis of the various relevant criteria. In practice, arbitration of allocation issues remains dominated by power struggles and influenced by lobbying or is based on essentially economic considerations. Some Mediterranean countries are beginning to determine their allocation arbitrations on the basis of an optimization criterion of the 'more value per drop' type. This promotes considerable improvements in the technical and economic efficiency of water usage, but *social and environmental* impacts are still rarely factored into these decisions.

It is easier to quantify the physical *efficiency improvements* in the various sectoral uses than in the choice of allocations. Plan Bleu has tried to estimate very approximately the magnitude of water losses and 'poor use' in each sector and for each Mediterranean country, losses that artificially increase water demand in the various national planning documents. Many estimates of these average losses are found in the literature; losses in the distribution networks and by the users themselves often add up to a volume as large as the actual uses. Plan Bleu has estimated these losses for the entire Mediterranean from a set of hypotheses (ambitious but 'feasible') for the various sectors (for example losses in networks, leaks at the point of use), as summarized in Table 1.3.

The *feasible potential savings* at the Mediterranean catchment area level have been estimated at about 24 *per cent of current demand* in 2000, or 46km^3 of total demand of 190km^3.[36] This estimate should be treated with extreme caution because of the scarcity and poor reliability of the available statistics; it provides only the order of magnitude of the possible improvement in the purely physical efficiency of use.

This savings potential in the Mediterranean is therefore far from insignificant. The main quantitative potential is in *irrigation for agriculture*: in the northern countries the losses occur in the major networks while in the south and east they may also be in more fragmented irrigation practices. The potential savings in the agriculture sector are five times higher than in the domestic sector. Industry can contribute effectively through recycling, as shown by the example of French industry, and which will be promoted by the European Framework Directive. The drinking water sector would contribute only a small fraction of the total potential savings, but may be easier to achieve in the mid-term in the south and the north and to justify economically, *given the present price of water*.

Table 1.3 Estimate of recoverable losses by Mediterranean Basin regions, NV, 2000 (in km^3/yr)

Sub-regions of the Mediterranean NV	Domestic Rates of loss reduced to 15%, user leaks reduced to 10%	Industries Recycling generalized to 50%	Agriculture/Irrigation Transport losses reduced to 10% and efficiency raised everywhere to 80%	Total Total	And percentage of sub-regions
North	3.0	4.9	11.9	19.8	43%
East	1.1	0.5	3.5	5.1	11%
South	2.3	4.0	14.6	20.9	46%
Total	6.4	9.3	30	45.7	100%

Source: Plan Bleu, Margat, 2004

Note: 'Recoverable losses' from the point of view only of available technology, without allowing for opposition and social difficulties.

2 A Sustainable Future for the Mediterranean

Figure 1.12 Total demand, baseline and alternative scenarios, NV, 2000–2025

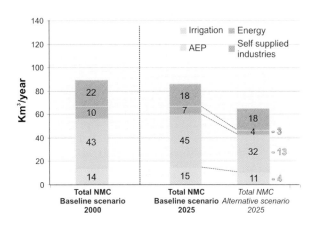
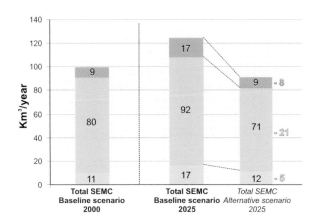

Source: Plan Bleu, Margat, 2004

These possibilities suggest a *variation* to the baseline scenario that would make more use of these potential water savings. In such an *alternative scenario*,[37] the losses mentioned above would be progressively reduced and finally eliminated entirely by 2025. Figure 1.12 shows the savings to be made in 2025 in relation to the baseline scenario by sectors and sub-regions. Compared with the baseline scenario, there would be a total saving of 54km³ by 2025, or 24 per cent of the baseline scenario demand by 2025. Most the potential (34km³) would be in the SEMC (in the irrigation sector), where it would actually amount to more than the expected growth in total demand between 2000 and 2025.

By reducing the demand without affecting the activity for which it is used, water demand management can reduce all the impacts and risks described above. The *benefits* are economic, social and environmental, all at the same time.

Over 25 years, the potential savings could reach a total of 675km³[38] assuming a progressive (linear) exploitation of the potential. If the water 'wasted' in this way is considered as a cost (exploitation, supply, processing), the possible financial savings are huge. Given an average cost of supply per cubic metre of water of 0.4 euros (a cost that is near to that of irrigation water, without the treatment to turn it into drinking water) the *savings over 25 years* would be nearly 270 thousand million euros, or 11 thousand million euros per year. Of course, the costs of exploiting this potential (such as subsidies, awareness campaigns, training) have to be subtracted from these figures, but they are much lower than the resulting benefits. These (obviously theoretical) figures show only the scale of what could be achieved by better management of water demand in the Mediterranean and thus provide an indication of what is at stake.

For some countries, a strong commitment to WDM could prevent predictable water demand crises and even remove the need for some investments with large costs and impact while making it possible to meet needs and keep up with demographic developments.

The benefits would also be *social* by contributing to a rise in agricultural incomes, job creation (for the maintenance and management of infrastructures) and access to water for the poor (by reducing the supply costs).

Finally, there would also be *environmental* benefits from reducing or stabilizing withdrawals from ecosystems and resources and decreasing the number of new infrastructures needed (from the hundreds planned, shown in Box 1.3). This objective is, however, seldom the first in the WDM processes, and quite often the water savings made are re-used at once, so much so that total demand is not even reduced. Careful management does not necessarily mean allocating more water to nature, unless that is clearly the aim.

Benefits could also be seen in *energy* savings. Considering that it takes nearly 1kWh to produce, treat and distribute 1m³ of water, drinking water savings alone could represent nearly 8 thousand million kWh by 2025. Spread over 25 years, the energy savings would be about 100TWh for the production of drinking water alone, more than the output of two 500MW power plants for a year.

On the Mediterranean Basin scale, these global savings are theoretical, but they are exemplified by the experience of cities like Rabat-Casablanca in Morocco (Box 1.11). WDM made it possible to avoid building many water infrastructures that had been planned in previous planning documents, which may have been too ambitious in terms of efficiency.

Water

Box 1.11 Infrastructure savings through WDM, Rabat-Casablanca

The area of Rabat-Casablanca in Morocco has experienced a substantial reduction in the growth rate of water demand during the past 12 years, in spite of high and growing urbanization. This was achieved as a result of measures to find and repair leaks in the system; setting up a progressive pricing system to make consumers, including public users, more responsible but with welfare provision for the poor; systematic metering of the water supply; and a major public awareness campaign about water savings. These actions were made possible by an appropriate administrative framework that combines private business with government agencies and local authorities in a 'delegated water service management' supported by a charter involving several communities. These measures have allowed costly investments (dams, water transfer channels) formerly planned in the 'Plan Directeur' in 1980, to be deferred. These investments, which would have been difficult to finance without extra debts, may prove not to be necessary in the long term.

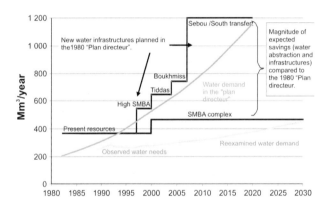

Source: DGH Rabat (2002) Analysis of the Case Study on the Drinking Water Supply in the Rabat-Casablanca Coastal Area. Final Report. In *Plan Bleu*, CMDD (2002b); www.gwpforum.org

The WDM issues vary according to the country. In the *northern countries*, rather better endowed in water and with falling demand, it is the *qualitative* aspects of the resources that dominate, as well as the interest in maintaining, even restoring, ecosystems and simultaneously *lowering water supply costs*. For the SEMC, with the squeeze coming from finite water resources and rapidly increasing demand, it is the *quantitative* issue that dominates.

Nevertheless, despite their immense potential, these issues are not yet fully appreciated in the Mediterranean countries, and even less included in policies. Even if WDM seems to be a winner from many sustainable development points of view, it still comes up against much resistance, which must be analysed in order to be overcome.

Too few Mediterranean countries commit themselves to WDM

Many private and local initiatives combine to achieve better WDM, but they are generally not enough to exploit the vast savings potential mentioned above that quite often requires much voluntary *public involvement*. More and more Mediterranean countries, often the most water-poor, such as Israel, Malta, Cyprus, Spain, Tunisia and Morocco, are making commitments to WDM. They are developing official national WDM strategies, announcing quantified goals that combine economic, technical and institutional instruments that mobilize the actors through the refurbishment of networks, the promotion of economic irrigation techniques, progressive pricing, bolstering water policies and raising the awareness of users. Some devolution of water management (such as catchment areas at the local level), a growing participation by users, and redefining the role of the state can favour such strategies.

The Tunisian example is without doubt one of the most advanced in this field, with the setting up of institutions officially responsible for WDM and real impacts on demand (Box 1.12). With a few exceptions, however, few Mediterranean countries have formally committed themselves to the WDM path; efforts are still mostly concentrated on exploiting new resources. Although WDM is an increasingly shared concern, it is only rarely included in the official objectives of water planning and even more rarely expressed in terms of targeted and quantified objectives. Very few Mediterranean countries have specific institutions to manage demand or define an environmental demand. In this regard, the EU Framework Directive on water and the Johannesburg objectives could act as a driver of reform in many Mediterranean countries.

Barriers to WDM

One of the main impediments to improvement in WDM is the lack of *understanding of the importance of the issue and its potential benefits by the various actors*, in particular the public actors. All too often, because of professional training, decision-makers in the water or agricultural areas rely too much on technology for increasing supply, underestimate the resulting impacts, and play down the credibility of alternative options. Thus dams, desalination or water transfers are seen as ready-made responses to the assumed increase in demand without factoring in their costs, the sum of their impacts (emissions, energy consumption, increased vulnerability) and, most importantly, with no adequate examination of alternative approaches and analysis of the demand structure and the possibility of achieving savings.

Systematic *evaluations* of cost effectiveness, comparing several options (increasing supply or using WDM) are still rarely carried out and published. Such evaluations,

Box 1.12 Tunisian national strategy for managing water demand

Tunisia was one of the first countries to adopt a **national water-saving strategy** for both urban and agricultural demand. By doing this, it confirmed a cultural 'oasis' tradition of a frugal and patrimonial management of water, such a rare resource in Tunisia. Thanks to this policy, water demand for irrigation has stabilized in spite of increasing agricultural development, the size of peak demands and the unfavourable climatic conditions (droughts). The water demands of both tourism (a source of foreign currency) and cities (a source of social stability) have been assured.

The underlying principles of the Tunisian WDM strategy are:

- moving from isolated technical measures to an integrated approach;
- a participatory approach, which makes the users more responsible (960 water user associations were created covering 60 per cent of the public irrigated area);
- a gradual introduction of reforms and adaptation to local situations;
- introduction of financial incentives systems aiming at promoting water conservation-based equipment and technologies (60 per cent grant for purchasing such equipment);
- support to farmers' income, allowing them to plan for and secure agricultural investment and labour;
- a pricing system that combines transparency with flexibility, compatible with the national goals of food security and allowing a gradual recovery of costs.

As shown in the following diagram, thanks to these measures, the demand for irrigation water has stabilized while the added value of irrigated production has increased.

Water consumption and added value of irrigation, Tunisia, 1990–2000

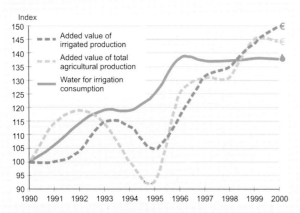

Source: M. Hamdane, 2002, in *Plan Bleu*, CMDD (2002b)

which would include more detailed estimates of the potential for savings on the basis of precise analysis, and internalization, as far as possible, of the costs of the environmental impacts of the various options, would increase awareness of the opportunities and feasibility of WDM among decision-makers. Carried out prior to investments, they should make it possible to compare increases in supply with WDM or to optimize allocations. Broadly speaking, the question is whether it is less 'costly', after taking accumulated long-term effects into account, for the community to reduce losses than to exploit new resources. Very often WDM appears to be economically much more advantageous than increasing supply, even if it has a higher 'governance' cost. The few studies available that make this kind of comparison show differences of 1–3, even 1–10, between the cost per cubic metre of water 'saved' and 'supplied',[39] depending on the case considered.

But before being better managed, the demand has to be *known* (social, economic and environmental demand), which leads to defining the efficiency potentials of highest priority or the most 'profitable' for exploitation. It turns out that the largest potentials for savings are often in agriculture.

As well as the need for greater awareness, WDM also requires a profound change in practices and attitudes, even a questioning of production and consumption patterns. The challenge is how best to *combine approaches* of the 'hard' type (stressing supply technology and infrastructures) with those of the 'soft' type (trying to involve all actors in the search for the best possible enhanced value of each cubic metre of water while keeping in mind the needs of nature). More generally speaking, this means repositioning people at the heart of concerns, since people are not just users or clients but also 'citizens', and are therefore responsible for the patrimonial management of water.

Beyond this 'quasi-cultural' change of perspective, the problem for WDM is the need to integrate a mass of sectoral policies and actors that have sometimes antagonistic objectives, with the objective of improved water management. *Integrating* WDM objectives into *sectoral* policies (agriculture, energy, industry, commerce) raises first institutional difficulties. Indeed many policies conflict with water demand, in particular commercial agricultural policies, given the importance of irrigation for water demand. Integrating WDM objectives into the various *actors* activities in a country presents an additional level of difficulty. To overcome it, some Mediterranean countries have set up *coordination, cooperation or arbitration agencies* to facilitate analysis and cooperation.

At the *national* level, coordination of government action is the responsibility of inter-ministerial committees on water (Algeria). National water councils (Tunisia and Morocco) have a more consultative role but make it possible to carry out analyses, raise awareness of actors and formulate proposals (such as developments of the regulatory and legislative framework).

At a more *local* level, *deconcentrated management units*, for a catchment basin or an aquifer, also prove to be favourable institutional frameworks for promoting subsidiarity. Water agencies (water or basin agencies) can be the preferred forum for political mediation in water management if they are based on strong legitimacy with the users (for example, willingness to listen and a sensitivity to social needs, independence, transparency and the authority to manage and to impose sanctions). User associations (irrigators, users, the local economic base) are also bodies that can cooperate, define and apply effective rules for WDM. Experience shows the advantage of *involving users* in managing common resources. It is often a matter of restoring local regulatory systems for water withdrawal that might have been destroyed by modern 'nation-states' without being replaced by more effective systems. The poor performance of the policing of water in most Mediterranean countries (a lack of resources, persistent illicit practices) encourages the use of more local and often more effective self-monitoring by means of concerted management processes. Groundwater or river contracts that are being developed in the Mediterranean illustrate the benefits of these processes (Box 1.13).

Box 1.13 The benefits of processes that involve users

Faced with the growing overexploitation of water tables associated with a number of failures in policing water, some contractual processes linking the multiplicity of actors have been shown to be effective in the Mediterranean. 'Water-table contracts', derived from voluntary processes that make users responsible (drilling declarations followed by metering) have made it possible in a number of cases to prevent water degradation and even restore environments:

- The high-basin region of the river Guadiana in **Spain** has seen a rise in alluvial ground-water and spring rehabilitation in spite of a large increase in irrigation. This was possible because new regulations were introduced for pumping from boreholes (including various measures for protection, warning, real sanctions and permits) together with compensation for loss of income for the irrigators and financial incentives for water-efficient technologies or for planting crops that need less water. These agri-environmental measures have played a part in making the plan to restore ground-water levels feasible and also to protect the natural park which is the region's major tourist attraction.
- In the coastal zone of the Languedoc region in **France**, the Astien water table is at present viewed by the 20 towns and villages that depend on it as a natural resource to be protected after 100 years of intensive exploitation (700 wells) that had led to its level being lowered and the penetration of sea water into its depths. In this tourist area, matters were made worse after 1960 by the development of camp grounds and holiday homes. In association with the Rhône-Méditerranée-Corse basin agency, a mixed interdistrict study and project agency was formed. Through a water-table management plan, it took a series of measures: an inventory of resources, information (on the rights and duties of well-owners and public awareness raising), advice on the digging and maintaining of wells, a standardization of municipal drinking water and water savings by the gating of artesian wells, etc. A strong technical team and a system of aid for restoring wells and metering withdrawals made it possible to reduce withdrawals by 25 per cent, or a million cubic metres a year, without altering needs. In ten years, the quality of the water table stopped degrading, and the average piezometric level of the water table was raised.
- **Egypt** has also developed a participative approach to the management of irrigation and modernization of technology in the irrigated areas of the Nile valley. The technology is based on modern equipment and bottom-up irrigation management, including a centralized control, simplification of the network, and above all the involvement of user organizations in decision-making, management and maintenance through intensive training.

Source: Centro de Estudios Hidrograficos del CEDEX (Spain); Syndicat Mixte d'Etudes et de Travaux de l'Astien (SMETA), Agence de l'eau Rhône Méditerranée Corse; Ministry of Water Resources and Irrigation Nile Water Sector. In *Plan Bleu*, MCSD. *Case Studies*. Forum on Progress Towards Water Demand Management in the Mediterranean Region. City of Fiuggi Terme, 3–5 October 2002

Nevertheless, improving local management capacity cannot be done without legal and financial strengthening of its legitimacy and its decision-making power, along with increased accountability (penal and financial) of leaders and transparency of transactions. It also implies a clear-cut separation of monitoring and management functions. The large increase in the number of irrigation organizations that are emerging in the Mediterranean (as elsewhere) under pressure from the major donors (e.g. World Bank) can only provide real improvement in water management with a reform of their statutes and funding methods, which are often hard to implement (for example in Turkey where a reform project is still under way).

In the absence of cooperating agencies and political will at the highest level, more often than not the specific interests of pressure groups carry the day over the general interest. There can be powerful *resistance* (more or less intentional) to some WDM measures among actors. For example some major *donors* have difficulty in adapting their projects to WDM requirements (small-scale, soft-type actions); the *civil servants* in water administrations, having acquired power and fame at the time of major projects, may be afraid of losing some of their power or feel uncomfortable with participatory processes; *local authorities* often do not have the resources to match their ambitions; and *water distributors*, trying above all to balance their operating income, may be afraid of water-saving measures. For example some WDM measures introduced in Madrid were compromised by the budgetary imbalance that they caused (a drop in income and cash flows in water companies and imbalances in municipal agencies). And, finally, *consumers* (industry, farmers, tourists and residents) seek to minimize the immediate cost of their water supply without realizing that measures that improve the efficiency of water use very often allow them to modernize their technology and increase their income (examples include the agriculture sector in Tunisia, see Box 1.12, and the introduction of cleaner processes in the industrial sector, see Box 1.9). Interest and lobby groups (farmers, builders and politicians) remain very powerful in favouring policies that aim to increase supply and seek to discredit the potential of WDM.

But the main cause of resistance is very often ignorance of the issues and opportunities for progress. This is why the efforts to *raise awareness* and understanding are as vital as political firmness. Often missing from academic curricula, the *training* of water professionals and technicians in the methods and issues of WDM could be a major contributor to the emergence of new, more integrated and economic water strategies.

Water demand management thus results from a *combination of instruments* (laws, economics, voluntary agreements, awareness-raising) and determination. It can provide considerable benefits, especially for irrigation. But to do so requires *progressive approaches* that are adapted to each local situation, with a greater involvement of consumers and a greater awareness by the decision-makers of issues at stake. For this almost 'cultural' change to extend to several Mediterranean countries, it must overcome much resistance. Support at the highest state level is needed to create a coherent strategic framework that is essential for coordinating actions and a continuing, long lasting involvement.

The difficult problem of funding

A lack of financial and human resources is often given as an obstacle to water management. Although substantial savings in infrastructures and reductions in environmental impact can be achieved in the medium-term, implementation of sustainable water management requires adequate resources for:

- the building and sustainable management of water supply and treatment facilities;
- the improvement of institutional capacities, in particular the drawing-up of national strategies, the perfecting of institutional, legal and financial frameworks, and the training and hiring of qualified staff.

Without these necessary (but not sufficient) financial resources,[40] it is illusory to try to improve water management. The financing needs of the water supply and sanitation sector for improving services to increasing numbers of people and activities are considerable. At a global level, the financial requirement for bringing drinking water to a thousand million people and providing sanitation and cleaning up pollution for two thousand million has been estimated at US$180 thousand million per year compared to the US$80 thousand million per year spent today.

The *northern Mediterranean countries* are soon to be faced with the necessity of changing and standardizing their old infrastructures to bring them in line with ever stricter European standards. These costs are considerable and, moreover, fully passed on to users.

The *southern and eastern Mediterranean countries* are faced with a double challenge: (i) the need to *upgrade* their infrastructures in accordance with minimum standards (as mentioned above, this objective is far from being achieved since 27 million people are currently not connected to sanitation systems and most polluting emissions are not processed); and (ii) the expected and very significant *population growth* (98 million additional city dwellers in the SEMC by 2025). To meet these challenges, the financing requirements will be huge. For domestic wastewater alone, applying the minimum standards of the European Waste Water Directive (91/271/EEC) in the SEMC would require the investment of a minimum of 40 thousand million euros in the SEMC up to 2010 and 60 thousand million euros up to 2025, that is around 10 per cent of their current GDP, an average of 2.5 thousand million euros per year for 25 years.[41] To that must be added operating costs of at least 0.3–1 thousand million euros per year.

These projections require a rethinking of the *financing modes* of water management and a definition of new roles for the several actors (states, local authorities, users, firms, banks, etc.).

To meet this funding need, *public budgets* are being cut back. Balancing budgets on a large macro-economic scale imposes increasing difficulties on most of the SEMC in making choices about the allocation of public expenditure. States, which have long been the main funders of large water infrastructures, are looking for other sources of funding. The *users* of water services are therefore being asked to contribute increasing amounts, but there are major difficulties in ensuring that all the investment and operating costs of water services are fully and rapidly covered. As we will see in the next sub-section, this can only be achieved gradually (to factor in the ability of consumers to pay the real costs of water and the acceptability to them of having to make such payments), especially if the costs and very long time scales for amortizing the costs of water infrastructures are taken into account. It is worth noting that it has taken European countries more than a century to balance their sewage budgets with consumers, through the very recent 'polluter pays' principle.

Given the difficulty of recovering costs from consumers and the low profitability of water investments (long return times, high costs of infrastructures, commercial and political risks), *private capital* is slow to invest in this sector. 'Public–private' partnerships, an approach that appeared during the 1980s, have so far had rather mixed results. Several of these sorts of approaches have been tried, successfully in Rabat-Casablanca (Box 1.11), but many others throughout the world have failed. Generally speaking they have been unable to meet the challenge of attracting massive private funding in a sector that is risky and not profitable enough. Moreover, these approaches are more appropriate for big cities and pose problems for contractors (local authorities, states) in supervising and regulating them (social and environmental clauses).

This is what has led many specialists in international cooperation in the water sector to propose an overhaul of the funding system[42] with the intention of increasing the amount of investment available through a combination of local (micro-credit, regional banks), national, international, public and private funding. The proposals aim to invert the present funding architecture by promoting a *decentralization* of funding and decision-making in order to improve the efficiency and monitoring of infrastructures and services.

Better application of economic instruments

Economic instruments are often considered the preferred instruments for integrated water management. Often given as *the* solution to water management problems, they deserve a closer analysis of their advantages and limitations in the light of their practical use in the Mediterranean.

The debate on applying economic instruments for water management inevitably leads to the question of the legal status of water and its ethical, cultural and even religious value. Far from being clear-cut, ideas on the nature of water are very controversial: it has status as an essential global public good (seen as a human right), an intermediary public good, a common good, an economic good or one that is both economic and social. Without going into this debate, it is important to remember that access to drinking water is not freely available to all, despite being a fundamental human right. The Dublin Conference of 1992 recognized water as an economic good too.

Table 1.4 Use of economic instruments for WDM in a few Mediterranean countries

Country	Group(*)	Importance of economic instruments	Incentive to save water — Agriculture	Incentive to save water — Drinking water	Other measures
Spain	1	Moderate	Low	Moderate to high	Seasonal restrictions
Israel	1	High	Moderate to high	Very high	Possible restrictions
Egypt	1	Low	Nil	Low	
Malta	2	Moderate to high	NA	High	
Tunisia	2	Low	Low	Low	
Cyprus	3	Moderate	Moderate	NA	
Turkey	3	Moderate	Low	Very high to low	
Lebanon	3	Low	Low	Very low	
Morocco	3	Moderate	Low	Moderate to high	
France	4	Moderate to high	Low to high	Very high	
Italy	4	Moderate	Low	High	
Greece	4	Moderate	Low	High	User awareness / Possible restrictions & bans
Albania	4	Low	Nil	NA	
Slovenia	4	Low	NA	Very low	
Croatia	4	Low	NA	Very low	

Source: Cemagref, 2002, Fiuggi forum, in Chohin-Kuper et al (2002)

Note: Assessments of the encouragement to water savings rely on the combination of criteria of price levels and progressiveness.
NA = 'Not Available'

(*) the groups refer to the exploitation index, Figure 1.6 p77.

2 A Sustainable Future for the Mediterranean

Table 1.5 Cost recovery and agricultural water prices in some Mediterranean countries

		Cost recovery		Change in prices	Forecasts
	Type of irrigation scheme	Operation and maintenance	Capital		
Spain	Small scale IPP	Low (to partial?)	Nil to low		1985 Law: principle of covering full costs
	Individual (underground water)	Total	Total		
France	RDS ASA	Total	Partial	Stable	
Italy		Partial	Nil		
Greece	Collective (LLEC)	Partial	Nil		Introduction of a volume-related pricing system for new networks
	Individual	Total	Total		
Malta	Poor water quality				Grants for micro-irrigation equipment
Cyprus	Small scale IPP	Partial		+30 to +80% between 1990 and 1999	a 80% ↗ in prices by 2003, to cover 38% of the cost of water
	Private perimeters	Total	Partial		
Albania			Subsidized		Pricing system to be defined
Turkey	Large perimeters	Partial	Partial		
Lebanon		Low to partial			Prices: + 20 to 30% in 2002
Israel	Irrigation water (1999) Mekorot	Partial		↗ in price between 1986 and 1996, ↘ between 1996 and 1999. Lowering of subsidy	
Egypt		Nil	Nil		
Tunisia	Private area (210,000 ha) Small IPP (160,000 ha)	Total Partial	Partial Nil	+12% in nominal terms since 1983	Pricing system by hydro-agro-ecological areas. Capital cost in pricing system
Morocco	ORMVA	Partial to total	Partial	+ 0.001 to 0.004US$/m³/year	Objective 100% for O&M and 40% for investment

Source: Plan Bleu, Cemagref, 2002, Forum de Fiuggi, both in Chohin-Kuper et al (2002)

Notes: Irrigated Public Perimeters (IPP), Rural Development Societies (RDS), Authorised Syndicated Associations (ASA), Local Land-Enhancement Commissions (LLEC), Regional Offices for Farming Enhancement ORMVA.

Despite its scarcity, in practice water does not have an intrinsic commercial value: there is (as yet) no market for water as a trade in goods (in the sense of the market for raw materials), but we are gradually moving towards a services market linked to its availability (and its recovery after use through treatment).

Economic instruments are seldom used

Economic instruments (e.g. pricing, quotas, subsidies, taxation) can make a considerable contribution to the more efficient allocation of resources at the sectoral and inter-sectoral levels by improving access to water for the poorest and factoring in environmental concerns. On the one hand, they can result in behavioural changes by the various users and, on the other, contribute to the funding essential for water management.

A brief overview of the economic instruments in use in the Mediterranean[43] shows that they are still seldom used. Moreover, there is no clear relationship between the level of water scarcity in countries and the level of recourse to economic instruments (Table 1.4).

In the Mediterranean, economic instruments are generally used for *recovering the costs* of water-distribution services and more rarely for research into the better allocation of resources or environmental protection (for example by integrating water scarcity into the price). Consequently it is the various forms of *pricing* (contractual or by volume) of irrigation and drinking water (block systems) that are by far the most used. Other instruments, such as quotas or subsidies, are much less common or else are used in conjunction with pricing.

In terms of *results*, the objective, even limited, of cost recovery is seldom achieved. It is above all in *irrigation*, where the savings potential is the greatest, that the prices are the lowest and the operational costs, as well as the investment costs, are furthest from being covered by consumers, as shown in Table 1.5. Lobbying and resistance from the agricultural sector explain most of this gap.

For drinking water, however, there is a general trend of increasing prices aimed at recovering an increasing proportion of the real supply and treatment costs (France, Spain, Morocco, Tunisia and Egypt). Thus in France, full recovery of costs has become a legal obligation.[44] This trend of an increasing recovery of water costs and thus an increase in the price of water is expected to continue. The price of drinking water has therefore risen considerably over the past decade: by 130 per cent in Egypt in 1995, and by 70 per cent (in current prices) in France between 1991 and 2000, from US$1.44 to US$2.45 per m^3 (in US$ 2000).

Enforcing the Framework Directive on Water on current and future EU-Med countries should result, increasingly, in the introduction of *environmental concerns* (such as resource scarcity, treatment) into prices, but there is considerable variation in these concerns from country to country. In France, for example, collection and treatment are systematically invoiced whereas in Spain and Greece only the biggest cities charge for treatment. In SEMC, in Morocco for example, invoicing for treatment was introduced in 1995, but some countries, such as Lebanon, continue to exclude treatment from their invoicing of drinking water. For the countries that have included treatment in the price of water (France, Spain, Greece, Italy, Slovenia, Turkey, Israel, Egypt, Tunisia and Morocco), its share of the total price of water per cubic metre is about 30–50 per cent and varied in 2000 from US$0.02 (Egypt) to US$0.8 per cubic metre in France. In addition to treatment charges, some countries have begun to introduce pollution or resource taxes that increase the price of drinking water and encourage resource saving while enabling the funding of pollution abatement and new resource development (in France in 2000 the resource tax was US$0.42 per m^3 and the pollution abatement tax US$0.38 per m^3.

Conditions for the use of economic instruments

Although more use of economic instruments should ensure better water demand management, certain conditions[45] must be satisfied for them to function properly.

Coherent goals and compensation mechanisms

First, there has to be a political will that is expressed by defining *clear and understandable objectives*: cost recovery and/or water savings; financial and/or social and environmental concerns (the protection of water resources); factoring in the cost of treatment. The selection of appropriate instruments and price levels depends on those objectives.

The effectiveness of economic instruments also depends on the *economic and institutional environment* within which they are implemented: proper performance of the regulatory and legal systems and complementarity with other non-economic instruments are indispensable conditions for their proper operation. Thus, it has been observed in the Mediterranean (Tunisia for example) that the imposition of taxes or quotas on water distributed by public networks can turn consumers away from the network and have the perverse effect of increasing withdrawals from aquifers, which are seldom monitored.

Once defined, these objectives must be made *consistent* with other national objectives that could influence water demand (for example respecting the right of access to water, regional balances, agricultural independence or support of farmers' incomes). Such interference is very often cited to justify the total lack or the low level of current pricing of irrigation water in the Mediterranean. Some sectoral measures (subsidies for irrigated crops in Europe, Egypt and Turkey) even conflict with the goal of managing the water resource. Policies to freeze the prices of certain agricultural products, as in Morocco and Egypt, limit the room for manoeuvre in implementing incentive pricing for water savings.

But, in most cases, this apparent contradiction could be resolved through well-targeted implementation of *compensation mechanisms*. To restore some *intra-national imbalances*, a tax could be imposed on the volumes of drinking water distributed throughout a country, making it possible to finance the transfer of drinking water to the public network for the poorest districts (as in France and Morocco). The double objective of territorial equity and cost recovery would then be satisfied. Similarly, many

Box 1.14 Tariffs compatible with social goals

Pricing reforms in all Mediterranean countries have to face the problem of making WDM goals (e.g. reduction of wastage, recovery of costs) compatible with the aims of sustaining farming incomes or ensuring access to drinking water for everyone.

For **irrigation**, which has the largest water-saving potential, one can identify four types of policy in the Mediterranean that seek to ensure acceptable incomes for farmers:

1. Free allowances of water for farmers (as in Egypt); although these do not encourage water saving, the costs of withdrawal (formerly using animals, now pumps) in Egypt are increasing with the increase of salination problems and are increasingly being met by the irrigators, although they are far from marginal.
2. Price increases that are lower than actually required. In Lebanon, for example, prices are limited to 'what the users can pay'; a partial adjustment in the price of water is planned in those deficit-ridden areas of Morocco where irrigation water would be too expensive (adjusted tarif of UD$0.07 as opposed to US$0.02 per m³ in the other areas).
3. Implementation of specific tariff structures with bonuses for water savings (a single volumetric tariff or block tariffs) that enable the impacts on the incomes of the producers (for a given saving of water) to be reduced. In the Valencia region of Spain, the reduction in income could go from 70 per cent with proportional pricing, to 30 per cent with proportional pricing plus bonus, to only 15 per cent with block tariffs plus bonus.
4. Recourse to other instruments such as quotas. The quantities of water used can be limited even if there are no economic instruments for encouraging water saving. In Egypt, for example, community water management has for centuries imposed taking turns in the use of water, the time allowed for each turn being a kind of quota.

For **drinking water** various approaches are used to maintain the right of access to drinking water for everyone, including the most deprived (see also Chapter 4, Urban areas):

1. *A block tariff*: a single tariff structure is offered to all domestic users with a first, especially low-level block acting as a socially based tariff. In the block tariff structure, the balance between goals (the social goal and water-savings goals) can be adjusted by varying: (1) the volume of the first block; (2) the price level of the social block; and (3) the price step between the first 'social' block and the higher blocks.

Too many Mediterranean countries have tariff systems that are not sufficiently targeted, with too large a segment of their population not having to pay any water charges. These systems do not have the required effect on water demand. The volumes of the first consumption block are more often than not larger than the 'social' volume (estimated at about 15m³ per quarter). In Egypt and Greece (Dyonisos), the volume of the first block is large and the initial tariff level and the progressive price levels are low. Even when the volume of the first block is small, the step in price between the two first blocks is often too small to discourage wastage (Greece and Israel).

On the other hand, in Malta, Morocco and Rome the volume of the first block is limited to 20–25m³, and there is a large difference in price between the first two consumption blocks. These tariff structures therefore reflect a real social concern but remain compatible with the goal of reducing water demand. In Turkey, the large difference in prices between cities results from this social policy, with low prices for the first consumption block and a larger first-block volume for poorer cities such as Diyarbakir (compared with richer cities such as Izmir).

Drinking water prices that are more or less favourable for saving water

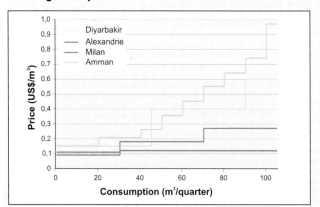

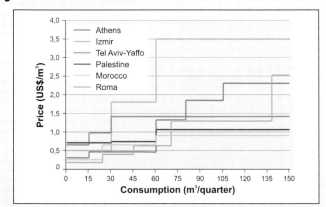

Tariff structure favouring social goals: low price level, little progression between blocks

Tariff structure combining social and environmental goals: high average price and large progression between blocks

Box 1.14 Tariffs compatible with social goals (continued)

2   *Tariffs with allowances that depend on certain criteria*: in Seville (Spain) there is a two-level block tariff with an allowance (50 per cent price reduction) for domestic consumption below a certain level and for families with more than five children (first-block tariff irrespective of water consumption); in Athens, the two-level block tariff is closely linked to the social objective by putting a ceiling on prices for large families (those with more than three children) and providing free water allowance for the most underprivileged.

3   *Paying the bill depending on social position*: in France, there is no social pricing as such, but the law of 1998, bans the cutting off of water for any person or family living in penury, and the Charte Solidarité Eau (Water Solidarity Charter) recommends cancelling the debt of households unable to pay.

Source: *Plan Bleu*, Cemagref, Fiuggi Forum, 2002

Mediterranean countries have implemented economic instruments that make it possible to reconcile *social* imperatives (access rights to water for everyone), *economic* imperatives (cost recovery) and *environmental* imperatives (reduction of water wastage). Box 1.14 shows some of the approaches adopted by the Mediterranean countries. However, poor design of these compensatory mechanisms (for example too large a tranche of low socially based prices), can nullify the desired effect on water savings or, indeed, make the poorest even poorer. The system must be adjusted as necessary by measuring its effects.

Relevance and social acceptability

The economic instruments must be chosen on the basis of whether they are likely to achieve the goals that have been set. *Tariffs*, for example, seem particularly appropriate for drinking water consumption or water for irrigation. But *subsidies* may be more appropriate when it comes to targeting water-saving equipment (for example water-saving irrigation equipment is subsidized in Tunisia). When there is strong pressure on the resource, *quotas* seem to be the most appropriate instrument for regulating demand, provided they are set in collaboration with all consumers and include environmental needs (e.g. low-water plans as in France and Spain).

But the implementation of economic instruments also has to take account of *transactional costs*, which should not exceed the expected benefits. These are the costs of implementing the system, for example installation of meters, awareness-raising campaigns, loss of well-being for some actors, the costs of implementing compensation mechanisms, etc.

And, finally, a proper understanding of the water supply and demand situation makes it possible to adapt the economic instruments to the desired goal in the best possible way. The choice of pricing structures requires knowledge of: the volumes consumed; the reaction of consumers to prices (Box 1.15) and income (Box 1.14); and the possible existence of other water resources (that may enable consumers to avoid economic measures).

Box 1.15 Large variability in consumer reactions to water prices

The **price elasticity of demand** provides an indication of the influence of water price on consumer demand. Many studies show that this is extremely variable and depends on many criteria.

Among these, it is important to consider whether or not there are any **replacement solutions**. The elasticity is greater when there are many water supply options available, for example households that have access to sufficient quantities of good-quality underground water stop using distributed water when there is even a small price rise (in Cyprus, for example). The same thing is seen in irrigation. In irrigated areas of Tunisia (Jebel Ammar), where a fourfold increase in the price of water resulted in the amount being consumed falling by a factor of three, while in other situations (Morocco and Tunisia), a price rise resulted in irrigators leaving the network, which sometimes led to an overexploitation of aquifers. In agriculture, the price elasticity of demand is influenced by the availability of accessible underground water, crop substitution, the flexibility of irrigation systems and the added value of irrigated crops. On the other hand, because it is so indispensable, drinking water demand is relatively insensitive to the price for low levels of consumption (basic needs), while it is more sensitive for 'comfort' needs (swimming pools, garden watering).

Elasticity also depends on the **price level** of water and the **size of the increase**. It also changes with **time**: the effect of a price rise on demand is short-lived. In France, for example, 'the large price rise towards 1971, due to inclusion of the treatment tax (30–40 per cent) resulted in a slight bend in the consumption curves that disappeared 2 or 3 years later.'

Source: *Plan Bleu*, Cemagref, Fiuggi Forum, 2002

The implementation and adaptation of economic measures thus implies a system of *monitoring and evaluation* based on audits and informed performance indicators that is rarely available in most countries.

Many failures of public–private partnerships in the water field result from a failure to consider social behaviour when implementing economic instruments and include consumers in the choice of systems. Just as in the energy sector, the acceptability of measures for WDM is fundamental and remains closely linked to the perception that users have of their utility, which can be promoted through dialogue, awareness-raising, confidence in the instruments proposed (transparency, simplicity) and in the institutions that are in charge. The decentralization of decision-making and 'crisis' periods (droughts) facilitate the acceptance of measures that would otherwise only be accepted as a result of much awareness-raising and progressiveness.

This overview of economic instruments has shown that they are still barely (albeit more commonly) used in the Mediterranean, especially in the irrigation sector, the main potential area for water savings. They could prove to be more effective for improving water management, provided they are not seen as a pre-packaged and single answer to the extreme complexity of the situations encountered. For economic instruments to work well, many conditions (discussed above) have to be met, including, most importantly, the definition of a clear objective, a coherent framework, and an indispensable combination with other instruments (regulatory, awareness-raising, etc.).

Consolidating Mediterranean cooperation

A firm priority for 15 years

International cooperation is also being reoriented more and more clearly towards the objective of integrated water management. The many international conventions and conferences on water (from the Dublin Conference to the Kyoto Forum 30 years later) have gradually redirected the priority objectives of cooperation towards the promotion of integrated water management and access for everyone to adequate drinking water and sanitation services. In Johannesburg in 2002, the priority objectives of the millennium were to:

- implement plans for water efficiency and integrated management between now and 2005;
- reduce by half the number of people with no access to drinking water or sewage systems, between now and 2015;
- achieve better integration of sustainable agriculture and better consumption and production patterns.

But the fact that nearly a thousand million human beings are still deprived of drinking water and proper treatment facilities[46] shows that these objectives are clearly not being met. The reasons for these failures have often been analysed, including (in simple terms) failures in governance and weaknesses in financial and human resources for better water management. International cooperation should help to better target these objectives and improve the effectiveness of actions by promoting synergies with the private sector and the actors of civil society.

In the field of *regional cooperation for water management*, the Mediterranean can boast of something of a head start. Long-standing close networks between water professionals and actors have resulted in a valuable exchange of experiences. Without returning to the systems of the Mediterranean Action Plan, which, since 1975, indirectly involve upstream soft-water management (land-based emissions into the sea,[47] and clean industrial production, promoted by the Barcelona CAR/PP), it should be remembered that the issue of water demand management has been highlighted as one of the major Mediterranean priorities since 1996. Stimulated by *Plan Bleu*, many Mediterranean fora (Marseilles, Fréjus and Fiuggi) have enabled the exchange of experiences on WDM and developed assessments and recommendations that have been adopted by the Mediterranean Commission for Sustainable Development and the Contracting Parties to the Barcelona Convention in 1997, aimed at better integration of water demand management in national strategies and the regular monitoring of progress.[48]

Twinning between water companies,[49] and the increased number of professional *networks* and consultations (basin agencies and irrigation associations) illustrate a promising cooperation that should enable progress in WDM. These exchange networks, sometimes quite long-standing in the Mediterranean, give it a certain head start in WDM compared with other regions of the world. Evidence for this is the emergence of an organization, the Global Water Partnership (GWP-med), that brings together most of the Mediterranean networks around a shared objective of promoting integrated water resource management. However, the real range of these exchanges in the field of integrated water management remains hard to evaluate, since they often rely on individual or local involvement, even fragmented partnerships. These should be intensified and consolidated.

The *Euro-Mediterranean partnership* is also a special arena for cooperation in the water sector. In founding the partnership, the 1995 Barcelona Declaration recognizes the rational management of water resources as a priority and recommends the strengthening of regional cooperation. Several European sub-programmes include water among their priorities as a result. The integrated management of water is one of the five priorities of the Euro-Mediterranean Regional Programme for the Environment.[50] A specific Euro-Mediterranean *regional* programme for strengthening local water management capacities was adopted in Turin in 1999. With 40 million euros (for five years), it reinforces the EMWIS, a regional information programme on know-how in the integrated water-management sector. However, although it may open up interesting possibilities, its range is nevertheless limited by its modest budget when faced with the magnitude of its ambitions, but also when faced with the very great diversity of the subjects covered and the relative complexity of its procedures. B*ilateral* programmes (about 85 per cent of the MEDA budgets) might provide an occasion for stronger stimulation for implementing national integrated water management strategies in the most willing of the partner countries.

In the context of the EU Water Initiative launched in the Johannesburg Summit, the European Commission has pursued in partnership with Greece (lead country) the Mediterranean component of the EU Water Initiative (MED EUWI). With the objective of enhancing regional cooperation in the water sector in the Mediterranean, emphasis is put on Mediterranean and south-eastern European priorities.

Official Development Assistance, a necessary new stimulus

Among the systems for funding water management, Official Development Assistance (ODA) needs to play a stronger role in factoring in the long-term issues and the vital commonality between consumers, zones and generations, and the sector's limited attractiveness for private international capital. There is no doubt that the implementation of internal national mechanisms that make it possible to cover the investment and operating costs will, in the end, be the only guarantee of sustainable water management, but without international public aid, much of the necessary investment will not be forthcoming.

The ODA in the Mediterranean is experiencing the same serious reductions as in the rest of the world, and this also affects the water sector. Although it may be difficult to measure the flows of ODA *in the water sector of the Mediterranean* (since the region is split between the different areas of activity of international agencies), some trends can nevertheless be identified on the basis of the statistical data bank of the OECD Development Aid Committee (DAC).

Despite its importance, water[51] received less than 8 per cent of the total ODA attributed to the Mediterranean countries between 1973 and 2001, with a cumulative total over the period of about 9.8 thousand million dollars (or a yearly average of US$350 million). The funds allocated to the region in this sector underwent large fluctuations during the period, with a peak in 1996. On average, between 1990 and 1999, the region would have received about 700 million dollars per year, and aid would have halved between 1996 and 1998.

Multilateral *donors*[52] are well behind bilateral cooperation. More than 90 per cent of the cumulative aid to Mediterranean countries since 1973 comes from the following six countries: the US (US$2.7 thousand million) and Japan (US$2.3 thousand million), followed by Germany (US$2.3 thousand million), France (US$0.9 thousand million), then Italy and the UK. Funding from the Gulf Arab countries, remains to be assessed in detail, but an initial assessment shows only US$0.44 thousand million in the period 1961–2002. The EU makes a very modest contribution, mainly through its MEDA programme (a few tens of million of dollars per year[53]). On the other hand, it is very active through loans from the European Investment Bank with only the interest rebates being considered as public aid for development.[54]

Egypt is the biggest *beneficiary* of ODA for the Mediterranean countries in the water sector, followed by Turkey, Morocco, Tunisia, and more recently, the Palestinian Territories. The eastern Adriatic countries have so far received only a small amount and percentage of international aid, spread among many countries, despite post-war reconstruction efforts. Large countries such as Algeria and Syria remain the poor relations of the aid system. Flows of aid are still strongly influenced by geo-political and historical ties.

Figure 1.13 Official Development Assistance in the water sector in the Mediterranean, cumulative 1973–2001 (in million US dollars)

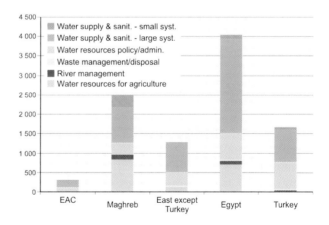

Source: DAC/OECD database

More than half (53 per cent) of the ODA aid received by Mediterranean countries in the water sector is still for major drinking water and treatment systems. However, the second item (21 per cent) involves support for water policy and reform of the management system, which gives hope that there will be improvements in water governance, even if three-quarters of the aid under this heading in the region focuses on three countries: Turkey, Egypt and Morocco. Next (18 per cent) comes actions for the development of irrigation water for agriculture. Aid for small-scale drinking water and treatment systems[55] (5 per cent), essentially from Germany, goes only to Maghreb (12 per cent of the aid). Although some national funding methods may be such as to distort the analysis,[56] one can nevertheless get some measure of the scale of effort needed to reach the millennium goals, especially in Turkey and Morocco, which still have a high percentage of rural dwellers. Projects for 'resource protection', 'effluent management', 'basin management' and 'training and education' are poorly funded.

The donor countries have their sectoral priorities, but certain turnarounds are striking. Germany, for example, is disengaging from irrigation investment, as had the EIB.

ODA in the water sector in the Mediterranean, while falling, is still largely influenced by geo-political considerations, and continues to be absorbed mainly by the building of large infrastructures, with, however, a move towards support for water policies and management in a few countries.

If the states honour the commitments they made in Monterrey in 2002 to increase their development aid by 25 per cent by 2025, this could create new resources for ODA, the effectiveness of which could be reinforced by better *targeting* of integrated water management. Aid should be oriented towards tangible results in water demand efficiency matters and focus on the organizational and institutional aspects of demand management, the poor relation of national water policies. ODA would be more effective if it were seen as a 'catalyst' for reinforcing sustainable water-management policies in the SEMC, and if it were used more as start-up capital for private funding than for funding infrastructures. It seems essential to *reform the international aid system* to adapt it to the specific requirements of WDM (low investment costs compared with operational and transaction costs, dispersed activities). The mechanisms for monitoring and evaluating ODA also need to be strengthened (mandatory reporting, result indicators, monitoring implementation, etc.).

Some *priority paths* for better targeting of ODA in the Mediterranean water sector are:

- *Strategic support*: aid for establishing national strategic frameworks for sustainable water management, for example by:
  - twinning or providing technological expertise in the institutional sector; aid to decision-making; south–south exchanges (many SEMC have highly advanced experience in WDM that could benefit other SEMC), and closer relations with energy-efficiency experts;
  - the general use and publication of 'cost effectiveness' studies, comparing alternatives (improvement of management compared with exploitation of new resources);
  - support for the implementation of more lasting financial mechanisms, for example tax reforms; setting up cash flow funding for certain WDM activities that often have an immediate additional cost (for example water-saving equipment) but delayed benefits (cumulative water savings resulting from the use of such equipment);
  - the development of socio-economic and environmental indicators for planning integrated management at the various relevant levels.

- *Technology transfer*: promoting water-saving equipment, audits and aid for implementing clean-production procedures.

- *Awareness-raising and the strengthening of capacity* in the WDM sector:
  - awareness-raising of actors in Mediterranean water; training programmes and exchange fora;
  - the development of regional networks: basin agencies; sectoral networks (managers, agricultural consumers, consumer associations, water distributors, etc.);
  - the creation of support cells (legal and technological) for local authorities and administrations for specifying the terms of public–private partnership contracts, and follow-up and monitoring;
  - multi-disciplinary research to minimize consumption; the development of low-cost water-treatment technologies.

***

The Mediterranean countries are beginning to make noteworthy progress towards better water management but this is still inadequate given the outstanding issues and delays, particularly in the treatment sector. Progress in ensuring supply without increasing pressure on already degraded resources must be significantly speeded up if several SEMC are to avoid predictable crises. The economic, environmental and social challenges of such progress are considerable.

It is vital that reforms of all policies and moves to implement them (such as establishing sustainable financial systems) clearly focus on the goal of integrated water management. Understanding the vital issues and strengthening local management capacities will be just as critical as providing the financial resources. Regional cooperation, which benefits from a long tradition in the water sector in the Mediterranean, has a basic role to play in catalysing and accelerating the necessary changes.

Thus water, that precious resource in the Mediterranean, could, as a result either of necessity or proper foresight, become one of the prime drivers of sustainable development policies. The path to a complete change of approach is clear – aim always to prevent rather than cure; take the long-term view, not the short; manage demand sustainably rather than increase supply.

Notes

1. In the Falkenmark sense, i.e. populations possessing on average under 1000m³ per year and per capita of natural, renewable freshwater resources.
2. Run-off (surface or underground) from other, non-Mediterranean countries such as the Nile at Aswan.
3. Insufficient precipitation compared to averages.
4. Margat, 2004.
5. FAO, Global Water Partnership, Plan Bleu.
6. Plan Bleu, MEDTAC, 2000.
7. Mainly in France, for power plant cooling.
8. I.e. the number of harvests per ha and per year.
9. FAO, 2000.
10. Each according to criteria proper to the country or the evaluator.
11. 605km³ per year in all Mediterranean countries (N1).
12. I.e. generally deep aquifer reservoirs fed by an annual volume of input less than 1 per cent of their reserves.
13. With technological advances, desalination is becoming less expensive: it varies depending on the technology, the degree of original salinity and, above all, the price of energy, from 0.5 to 1.8 euros per m³. (Hoffman and Zfati, 2003, p11).
14. World Bank, 2002.
15. Application of the Directive on urban wastewater treatment.
16. EEA, 2003.
17. EEA, 2003.
18. IFEN, 2002.
19. The United Nations Committee of Economic, Social and Cultural Rights, 2002.
20. Except in conflict zones; the Palestinian Territories.
21. These fairly subjective statistics provide little information about the variations in quality that can occur in a year, for example for populations connected to a network. They certainly underestimate the populations not having access to drinking water by underestimating the number of people in many Mediterranean cities subjected to frequent water cuts (see Chapter 4).
22. IFEN, 2002.
23. In France small municipalities with limited management capacities are generally the most affected.
24. Le Monde, 23 April 2003.
25. IFEN, 2000.
26. Report: input from outside/(outside + inside input).
27. As shown by the very large demonstrations in 2001 in Spain while preparing the National Hydrological Plan, which brought more than 100,000 demonstrators onto the street.
28. GWP, 2000.
29. With the exception of Israel, 10%.
30. Directive 91/271/EEC (see Chapter 6 for a map of EU-Med large urban areas complying with the Directive).
31. EEA, 2003.
32. Eurostat and Table 1.2.
33. Laimé, 2003.
34. OECD, 1999.
35. Water demand is here defined as the sum of the water volumes consumed and lost during production, transport and use (water use); it in fact corresponds to water supply (water withdrawal + water production + imports/exports).
36. These figures concern the Mediterranean catchment area (NV) as defined at the beginning of the chapter. Estimated on a scale of entire countries (N1), they amounted to 70km³ per year in 2000.
37. With several possible variants.
38. $25 \times 54/2 = 675km^3$.
39. Louhichi-Flichman-Comeau, 'Amélioration de l'efficience de l'eau sur un périmètre irrigué en Tunisie'; 2000
40. There is also need for a good governance framework, i.e. a legal framework securing long-term investment, monitoring operators, anti-corruption measures, etc.
41. On the basis of an average cost of 100 euros/Equivalent habitant (EH) for upgrading and 400 euros/EH for new infrastructures. European Topic Center-IW, Implementation of the urban waste water treatment directive in the 10 accession countries. Final draft report, April 1999 and ADEME. Europe estimated between 3 and 10% of GDP as the cost of the necessary investments only for the upgrading of treatment infrastructures to meet the waste water directive in the ten accession countries (although their population is stable or even decreasing).
42. Camdessus, 2003.
43. Cemagref, June 2002, The economic instruments for WDM in the Mediterranean, Fiuggi Forum, SEE Chohin-Kuper et al, 2002
44. Accounting instruction MP49, which obliges municipalities to balance their water department budgets with consumers.
45. These conditions are derived from real experiences presented at the Plan Bleu–MCSD-GWP forum in Fiuggi in 2002.
46. Including 30 million in the Mediterranean.
47. Described in Chapter 6.
48. For the full text of the recommendations, see www.planbleu.org
49. Groupe des Eaux de Marseille and the Etablissement Public des Eaux d'Alger.
50. Short and Medium Term Environmental Action Programme for the Mediterranean (SMAP), with a total of 30 million euros for the MEDA II programme (2000–2006).
51. The Water and Agriculture section (the water and soil subsection).
52. International Development Association, International Fund for Agricultural Development.
53. Between 1995 and 2004, there were 34 projects supported by various programmes of the MEDA, including 8 regional and 26 bilateral ones (see above for the various Euromed Partnership programmes). The total of these projects were worth 450 million euros or about 50 million euros per year.
54. All the loans granted by the EIB to the SEMCs in the water sector for the 1995–2003 period totalled 1.76 thousand million euros, or an average of 220 million euros per year, which represents about 19% of the total of the EIB loans in the region.
55. Systems using reduced-cost technologies.
56. For example, in Turkey, international aid generally reinforces the national budget and not the local projects that are usually funded by local communities' borrowing from the Provinces Bank, Iler Bankasi.

References

Camdessus, M. (2003) Financing Water for all. Report of the world Panel on Financing Water Infrastructure. 3rd World Water Forum, Kyoto, 16–23 March 2003 (www.worldwatercouncil.org)

Chohin-Kuper, A., Rieu, Th., Montginoul, M.; CEMAGREF. (2002a) Les outils économiques pour la gestion de la demande en eau en Méditerranée. In Plan Bleu, CMDD

EEA. (2003) Europe's Environment: The third assessment. Luxembourg: Office for Official Publications of the European Communities

European Topic Center for Inland Waters (ETC-IW). (1999) Implementation of the Urban Waste Water Treatment Directive in the Ten Accession Countries. Final draft report (under contract to the European Environment Agency)

FAO. (2000) Agriculture: toward 2015/2030, Global perspective studies unit, April, FAO

GWP. (2000) Integrated Water Resources Management. Stockholm: GWP

Hoffman, D., Zfati, A. (2003) Reducing Middle East water and power station costs. Water & Waste Water International, vol 18, no 4

IFEN. (2000) La préoccupation des Français pour la qualité de l'eau. Données de l'environnement no 57, Orléans: IFEN

IFEN. (2002a) L'environnement en France. Paris: La Découverte

IFEN. (2002b) *Les pesticides dans les eaux. Bilan annuel 2002.* Orléans: IFEN
Laimé, M. (2003) *Le dossier de l'eau. Pénurie, Pollution et corruption.* Paris: Ed. du Seuil
Louhichi, K. Plan Bleu, CIHEAM. (1999) *L'amélioration de l'efficience de l'irrigation pour une économie d'eau: cas d'un périmètre irrigué en Tunisie. Rapport final.* www.planbleu.org
Margat, J. (1992) *L'eau dans le bassin méditerranéen. Situation et prospective.* Les Fascicules du Plan Bleu, no 6. Paris: Economica
Margat, J.; Plan Bleu. (2004) *L'eau des Méditerranéens: situation et perspectives.* Athènes. MAP Technical Report Series no 158. PAM, www.unepmap.gr
OECD. (1999) *Examens des performances environnementales: Turquie.* Paris: OCDE
Plan Bleu. (1997) *L'eau en région méditerranéenne. Situations, perspectives et stratégies pour une gestion durable de la ressource.* Sophia Antipolis: Plan Bleu
Plan Bleu, MEDTAC. (2000) *Vision méditerranéenne sur l'eau, la population et l'environnement au XXIème siècle.* Sophia Antipolis: Plan Bleu. www.planbleu.org
Plan Bleu, CMDD. (2002a) *Études thématiques.* Forum 'Avancées de la gestion de la demande en eau en Méditerranée', Fiuggi (Italie), 3–5 Octobre 2002
Plan Bleu, CMDD. (2002b) *Études de cas.* Forum 'Avancées de la gestion de la demande en eau en Méditerranée', Fiuggi (Italie), 3–5 Octobre 2002
Plan Bleu, CMDD. (2003) *Résultats du Forum de Fiuggi sur les 'Avancées de la gestion de la demande en eau en Méditerranée'. Constats et propositions.* Sophia Antipolis: Plan Bleu. www.planbleu.org
World Bank. (2002) *World Development Indicators,* Washington, DC: World Bank

Chapter 2

Energy

Energy is at the heart of many concerns about sustainable development. It comes from natural resources (minerals, plants, the wind, the sun) and, like water, it constitutes one of the essential pillars of all human activity.

The use of energy results in short- and long-term effects on the environment and the climate (the greenhouse effect). The investments required for energy supply are so large that energy systems have a certain inertia. The long-term effects of energy choices on future generations, therefore, need to be particularly carefully considered. There are many prospective studies in this field, which raise questions about sharing responsibilities and costs between generations and people in different parts of the world.

The situation becomes more complicated when several geographical scales have to be considered: local, national, regional and global. This chapter will focus on the N1 country level since energy policies are implemented at the national level. At all scales, a growing share of energy consumed comes from trade, and impacts on the environment now have to be assessed at the global level. That is why the energy consumption of each *individual* affects the future of *all*.

The Mediterranean region is a perfect illustration of these interdependencies, and is therefore an ideal *laboratory* for changing interdependencies into complementarities that benefit everyone. Grouped around a commonly shared sea, the Mediterranean countries are unequally endowed with energy resources. Nevertheless, all of them have room for manoeuvre to improve the efficiency of their energy use and strengthen the security of their supply while contributing to more sustainable energy development.

Unsustainable energy development

The energy situation in the Mediterranean region is very diverse, with four hydrocarbon exporting countries (Algeria, Egypt, Libya and Syria); all the others are energy importers (including Tunisia, a small producer). The northern-rim countries (NMC), poorly endowed with fossil energy resources, consume two-thirds of the energy used in the Mediterranean region. Faced with a certain environmental impasse (in particular their responsibility for global warming), they are seeking to diversify their energy supply while trying to reduce the environmental impacts of their energy consumption. For the southern and eastern Mediterranean countries (SEMC), with rapid energy growth, the main challenge is to plan the future in such a way that their development patterns do not inevitably lead to the same impasse as in the northern countries and so that wasteful costs are avoided.

General growth in energy demand

Rapid growth in energy demand

Energy demand includes consumption by the users and the losses that occur between the sources and the users, which can sometimes be substantial. The demand for *primary energy* has increased rapidly in all Mediterranean countries during the past 30 years. Primary energy is that delivered by nature: the chemical energy contained in fossil fuels or biomass, the mechanical energy of wind and water, the thermal energy of subterranean water and solar radiation, and nuclear energy.

In most countries, most of the energy consumed is *primary commercial energy*. But in some very rural countries, *non-commercial energy*, directly from biomass for domestic needs, still represents a large part of energy supplies that is not included in the statistics. In Morocco, for example, this is estimated at nearly 30 per cent of overall energy consumption, about 15 per cent in Tunisia and 10 per cent in Turkey.

Demand for *primary commercial energy* in the whole Mediterranean region has more than doubled over the past three decades (an increase of 2.7 per cent per year) and in 2000 stood at 820 Mtoe (million tonnes of oil equivalent). The NMC, with an average annual increase of 1.9 per cent per year since 1971, accounts for three-quarters of the total, but the SEMC, with an increase of 6.1 per cent per year since 1971, accounts for a growing fraction (Figure 2.1). Details of demand by country and sub-region are given in the Statistical Annex. These figures match the highest values of the various scenarios considered by the *Plan Bleu* in 1989.

The Observatoire Méditerranéen de l'Energie (Mediterranean Energy Observatory, OME) has developed a *baseline scenario* for the whole Mediterranean basin up to 2025. This scenario, based on the main orientations defined in energy strategies (of both the Mediterranean countries and the major companies operating in the region) enables country-by-country projections of primary commercial energy demand and the contributions of the different energy sources (see the Statistical Annex). The driving force of energy growth in this scenario remains population and economic growth. Energy policies are dominated by the expectation of considerable increases in energy demand. The scenario assumes that there will be no significant shift towards priority being given to energy restraint, but that the current trends in technological progress will continue (a reduction of about 0.9 per cent per year in energy intensity[1]).

According to these projections, *total demand for primary energy* in the whole Mediterranean Basin could reach 1365Mtoe in 2025. Compared with 2000, this would be an increase of 65 per cent over the period (544Mtoe), or 2.1 per cent on average per year, compared with an average annual GDP growth of 2.7 per cent per year. This scenario is in line with a global reference scenario drawn up by the International Energy Agency (IEA) for 2020,[2] which projects an average annual increase in the world's primary energy demand of 2 per cent between 1997 and 2020.

It is projected that energy demand in the SEMC will increase at four times the rate in the NMC between 2000 and 2025, an increase of 340Mtoe, or 3.8 per cent per year, compared with 205Mtoe, or 1.2 per cent per year in the NMC. This growth will be the result of economic development in the SEMC and from the needs of an increasing population (16 million Mediterranean people still do not have access to electricity). By 2025, Turkey is projected to become the second largest consumer of energy in the Mediterranean (Figure 2.1). The relative share of the SEMC in total energy consumption in the Mediterranean will increase from 10 per cent in 1970 to 40 per cent by 2025.

These growth rates are equivalent to a doubling of total Mediterranean demand during the next 30 years and a tripling in the SEMC.

Figure 2.1 Primary energy demand, baseline scenario to 2025

Trends and projections

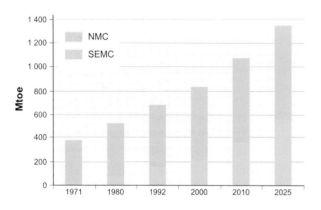

The six main energy consumers in the Mediterranean

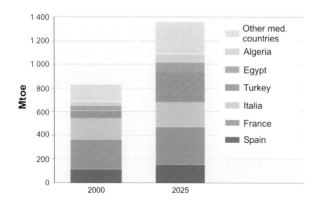

Source: OME, 2002

Given the demographic changes that have occurred, total energy demand *per capita* in the Mediterranean countries has increased continuously, but at the slower rate of 1.2 per cent per year since 1970, to reach an average of about 1930koe (kg of oil equivalent) per capita. In 2000, consumption of primary commercial energy per capita in the NMC was about 3100koe per year, half that in the US[3] but three times that in the SEMC. In terms of per capita energy consumption in the Mediterranean, the highest consuming country (France with more than 4000koe per capita) consumes over ten times more than the lowest (Morocco with less than 400koe per capita). By 2025, primary energy consumption could exceed 4000koe per capita in the NMC and 1700koe per capita in the SEMC.

Spectacular growth in electricity demand

Electricity production is the largest use of primary commercial energy (34 per cent of the total on average, possibly reaching 40 per cent by 2025). Electricity is the energy form that has seen the largest growth over the past few years with developments in the industrial sector (for example new processes and automation) and improvements in living standards in the residential sector (for example appliances and air-conditioning).

Total electricity demand in the Mediterranean countries has more than tripled over the past three decades and by 2000 stood at nearly 1500 terawatt hours (TWh). The average annual growth rate (AAGR) between 1971 and 2000 was 4.5 per cent, well above that of primary energy consumption and even that of GDP. In the baseline scenario, developed by *Plan Bleu* from OME studies, rapid growth in electricity consumption is projected to continue and may reach 2770TWh by 2025 (an increase of 2.5 per cent per year), with a possible tripling of electricity consumption in the SEMC by 2025.

Total electricity consumption (including all economic and domestic activities) per capita has increased since 1971 at a rate of 3.1 per cent per year and has now reached an average of nearly 3500kWh per capita per year in the Mediterranean. In 2000, northern-rim countries consumed an average of 6000kWh per capita, or about four times more than in the south and east. The largest consumer (France with 9000kWh per capita) uses nearly 20 times as much electricity per capita as the smallest (Morocco with 500kWh per capita). In the baseline scenario, per capita consumption in all countries will increase significantly and reach an average by 2025 of more than 8500kWh per capita in the NMC, but only 3400kWh per capita in the SEMC.

The importance of the transport and residential sectors

In the NMC, the *share of transport* in final energy consumption[4] has grown significantly and now stands at 32 per cent of the total, compared with only 21 per cent in 1971 (Figure 2.2). Growth in traffic and mobility has been well above economic growth (see Chapter 3) and has not been compensated by technological improvements. Car manufacturers have made much technological progress (lower consumption for a given power, lower pollutant emission per car), but the size of the fleet has increased even more rapidly, increasing the total consumption by the sector in a way that seems to be out of control.

In the SEMC, transport-related energy consumption is increasing at an average of 4 per cent per year and this could accelerate with reductions in trade restrictions, the growth of the automobile fleet and urban sprawl (see Chapters 3 and 4). However, the most spectacular growth in final energy consumption in the SEMC has been in the *residential* sector (an increase of more than 5 per cent per year between 1974 and 1999). Because of population growth, smaller families, very rapid urbanization and the spread of new lifestyles, the residential and service sectors have become the leading energy consumers (approaching 40 per cent).

Figure 2.2 Final consumption per sector, 1971–2000 (% in 2000)

NMC

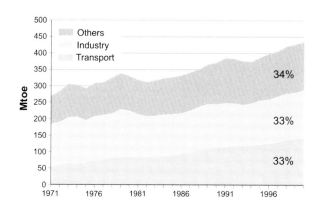

SEMC

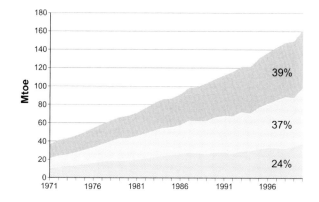

Source: IEA, 2001c and OECD

The dominance of hydrocarbons and the breakthrough of natural gas

To appreciate the possible impacts of such growth in primary energy demand, the possible evolution of the energy supply structure needs to be analysed (Figure 2.3). In 2000, fossil fuels (oil, coal and gas) dominated energy supply in the Mediterranean: more than 75 per cent of consumption in the north and 96 per cent in the south and east, the rest being mainly nuclear and hydroelectric. During the past few decades, it appears that the share of coal has remained constant, that of nuclear has stabilized and that of *natural gas* has increased rapidly at the expense of oil. This is projected to continue over the coming years, confirming the predominance of fossil fuels, which will still meet 87 per cent of energy needs in 2025 with oil still representing 40 per cent of the total. Renewable energy remains marginal, especially as far as primary *commercial* energy is concerned.

Since more than a third of primary commercial energy is used to produce electricity, the diversity of the supply facilities in the Mediterranean countries reflects the diversity of their power production systems (Figure 2.4). The choice of fuel types depends on the country's resources: gas in Algeria and Egypt, and coal in Greece, Turkey and in the eastern Adriatic countries. Nuclear energy plays a substantial role in France, important in Spain and Slovenia. Hydroelectric power is only significant in the eastern shore countries (Syria and Turkey) and in the eastern Adriatic countries (Albania, Bosnia-Herzegovina, Croatia and Slovenia, Serbia-Montenegro). The NMC intend to build coal and natural gas-burning power plants (except in France where the nuclear option is being maintained). In the SEMC, natural gas plants are expected to predominate; only Turkey, Israel and Morocco plan more coal-burning plants.

According to the baseline scenario, based on national forecasts, electricity production in 2025 could require 75 million tonnes more coal and 160Gm^3 (thousand million m^3) more natural gas than in 2000, a significant increase in primary energy consumption for all Mediterranean countries. There is as yet no co-generation of 'electricity and heat', but this might be developed, in particular to produce desalinated water in the coastal areas of water-poor countries (from Algeria to Israel) and some islands.

In 2000 *oil* was the primary energy source, meeting half of primary energy demand both in the north and the south. Consumption increased by 40 per cent between 1971 and 2000, although its share of energy consumption fell from 70 per cent to 28 per cent following the oil crisis. Although in the Mediterranean, as in the rest of the world, oil continues to dominate the energy market, it is increasingly used exclusively for purposes where

Figure 2.3 Primary energy demands by source, baseline scenario, 1971–2025

Figure 2.4 Electricity production by source, baseline scenario, 1971–2025

NMC

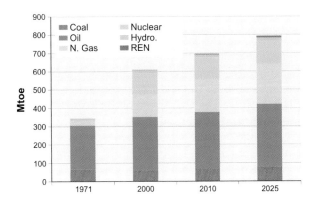

NMC

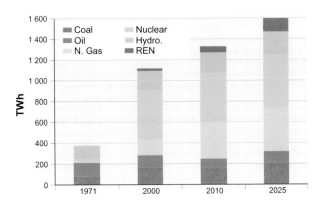

SEMC

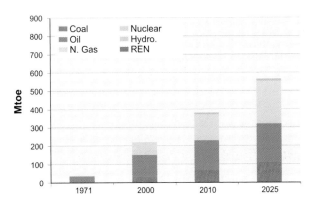

SEMC

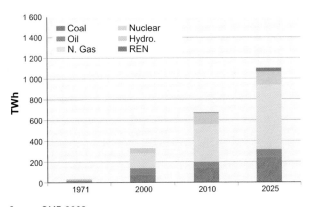

Source: OME, 2002

Note: REN – geothermal, wind and solar energy

Source: OME, 2002

there are no alternatives (such as transportation fuel, lubricants); since the 1970s it has been used as a source of heat only to meet shortfalls in global energy needs (as seen, for example, in Italy). In future this stop-gap role may be reduced as a result of increasing competition from natural gas. To sum up, oil is projected to remain the main energy source for the Mediterranean, but its share in overall supply is projected to continue to fall, from 48 per cent in 2000 to 40 per cent by 2025.

This *relative* reduction in the share of oil is explained mainly by competition from gas in the residential and tertiary markets and even more for electricity generation. Oil demand will probably increase in *absolute* terms, from 409Mt in 2000 to 550Mt by 2025. This growth (2.3 per cent per year) is related to that of transport (petrol and diesel) and mainly concerns the SEMC.

Natural gas has seen the largest and steadiest increase over the past three decades, contributing to a diversity of supply (particularly if nuclear energy is excluded, the rapid growth of which is due to the massive development of the French nuclear programme during the 1970s). The share of gas in the energy balance, insignificant in 1971, reached 21 per cent in 2000 in all Mediterranean countries and 27 per cent in the SEMC.

Because its transport requires large investment, natural gas was first developed in the countries that had large reserves (Algeria and Italy). The most abundant source of energy in the region and among the least polluting, gas is rapidly overtaking other sources and is projected to meet most of the increase in demand between now and 2010–2025 (see Figure 2.10 p122). The preferred fuel for new electricity power plants in both the north and south, gas is projected to penetrate the residential and tertiary markets of the SEMC. This will be facilitated by the increasing concentration of populations in towns and along coasts. For the whole of the Mediterranean, natural gas consumption is projected to increase rapidly between 2000 and 2025, from 176Mtoe to 445Mtoe (an increase in absolute terms of more than 3.7 per cent per year). Electricity companies will be the main purchasers, with their consumption increasing more than 3.5 times by 2025, by 260Mtoe. The largest gas-consuming countries will be Italy at 92Mtoe by 2025, followed by Turkey (75Mtoe), France (61Mtoe), Algeria (47Mtoe), Egypt and Spain at 43Mtoe. The share of gas in the Mediterranean energy balance is projected to increase from 21 per cent in 2000 to 33 per cent by 2025 (and its share in electricity production from 20 per cent in 2000 to 38 per cent by 2025).

The share of *coal* in energy supply has remained stable (at 10–15 per cent) although, in absolute terms, its total consumption has increased sharply, especially in the SEMC (by 6.5 per cent per year). The largest amount (two-thirds of the total) is consumed by the northern-rim countries. It is used in electricity power plants (Greece, the eastern Adriatic countries and Spain where 30 per cent of electricity is supplied by coal) and the metalworking industries. In the northern countries, coal consumption is projected either to remain stable or even to decrease under pressure from greenhouse gas-related agreements. The remaining third is used in the SEMC and more particularly in the countries without hydrocarbon resources such as Turkey, Israel and Morocco. These countries are planning new coal-burning power plants and their total consumption could double by 2025.

Present projections still show strong continued use of coal at the regional level, related to concerns about diversifying energy supply and cost considerations. Its share is projected to remain at around 14 per cent of the overall total despite the increase in the use of natural gas, with consumption doubling from 103Mtoe in 2000 to 190Mtoe by 2025, mainly because of increased use in Turkey (by 72Mtoe between 2000 and 2025).

Although nearly all southern and eastern-rim countries have *nuclear* research installations of varying sizes, the use of nuclear energy remains concentrated in the north, with a total installed capacity of 72GW. This nuclear power plant capacity is all in France (63.2GW, representing 78 per cent of the electricity produced), Spain (7.6GW or 28 per cent) and Slovenia (0.7GW or 40 per cent). France, in its long-term multi-year planning, is intending to keep its nuclear options open, with a recent decision to build a new EPR-type plant, which includes some technical improvements but is still based on pressurized water reactors.

In the baseline scenario, the share of nuclear power in meeting energy demand in the countries investigated is projected to fall from 15 to 9 per cent between 2000 and 2025, stabilizing at around 126Mtoe. Over the time period examined (to 2025), it is difficult to envisage a significant resumption in the construction of nuclear power plants in the northern countries as long as the major problem of treating nuclear waste is not resolved. In Europe, Finland is building a new EPR-type reactor, but Germany is committed to abandoning nuclear power, and Italy has shut down its four production units. Between now and 2025, the SEMC will probably stay away from this energy source. For the moment, Turkey has abandoned its project of building a nuclear plant. In the longer term, the use of nuclear power in some of these countries, using improved designs, cannot, of course, be ruled out.

Renewable energy (REs) sources considered here include hydro (small and large) and the other renewable energies (REn: geothermal, wind, solar; biomass and waste are excluded because of a lack of available statistics). Renewables contribute about 3 per cent of commercial energy supply in the Mediterranean countries (but more, at least 6.6 per cent of the total, if biomass, which is often not marketed, is included). Most of the electricity produced from renewable sources is from hydropower. There is a reawakening of interest in renewable energy sources, which have increased significantly (by 3.6 per cent per year on average between 1992 and 2000 in the Mediterranean region) and the costs of which are falling very rapidly. Renewable energy may well play an increased role up to 2025, especially in the NMC, because of momentum in Europe. The baseline scenario therefore projects an increase of 5.8 per cent per year in the use of these energy sources. In absolute terms, this corresponds to a quadrupling between now and 2025. But, because of the growth in total demand, the share of renewable energy sources (excluding biomass) in total primary energy in 2025 is projected to be less than 4 per cent. If biomass is included, this figure may increase to more than 10 per cent.

To sum up, the baseline scenario, which envisages a continuation of 'traditional' national and international strategies, projects a very large increase in primary commercial energy and electricity demands in the Mediterranean in 2025 (2 per cent per year), concentrated in the SEMC (4 per cent per year). Of these demands, 87 per cent is projected to be met by fossil fuel sources with a growing share being taken by natural gas without, however, allowing the use of oil and coal to be reduced. By 2025[5] renewable energies are projected to provide no more than 4 per cent of energy supply.

What are the possible impacts of such a scenario in terms of sustainable development up to 2025?

2 A Sustainable Future for the Mediterranean

Increasing risks

The very strong growth in demand projected in the baseline scenario suggests the need to consider two main kinds of risks for the achievement of sustainable development that may lead to a reorientation of current energy choices:

- the first, which is of a geo-political and socio-economic nature, concerns the growing insecurity of energy developments in the Mediterranean countries, linked to the problem of access to energy by present and future generations;
- the second is linked to the predictable worsening of impacts on the environment and human health.

The increased vulnerability of energy developments

The strong growth in energy demand naturally leads to questions about the ability of Mediterranean countries to ensure their supplies and about the increase in their energy dependence. The situation in this regard is clearly different for the oil-producing countries.

Oil producers, transition to the post-oil period

The five oil-producing Mediterranean countries (Algeria, Egypt, Libya, Syria and Tunisia) are in a special situation. In these countries, which represent 6 per cent of the world's oil production and nearly 5 per cent of the world's gas production, the risk of very rapid energy development must be analysed from two points of view: (i) the eventual exhaustion of their fossil reserves; and (ii) the funding of planned ambitious extensions to energy production capacity on which a large part of their economies depend.

The risk of hydrocarbon *reserves* being exhausted needs to be analysed with much care, since the Mediterranean is one of the regions of the world where there has been relatively little prospecting. Every new exploratory campaign and new technology has resulted in an increase in the proven reserves in Algeria (oil and gas), Libya (oil) and Egypt (gas). Twenty years of prospecting have doubled the known gas reserves in the region.

At the present rate of exploitation (about 200Mt per year), and in the absence of major discoveries, *oil* is likely to have the shortest exploitation period of the known reserves: approximately 30 years. With the exception of Libya, these reserves would be exhausted by 2025, a possibility that weighs heavily on the prospects for economic development in the mid-term. Egypt may become a net importer of oil (but an exporter of gas) well before this date, as Tunisia has already done. On the other hand, the known and probable reserves of *natural gas* appear to be much larger in volume than the oil reserves and should last beyond the end of the 21st century given the present rate of exploitation of the major known fields, which are concentrated predominantly in Algeria, Egypt and Libya.

Given the strong growth in domestic energy demands, these countries will need to ensure a large growth in production capacity if they are to export a total of about 170Mt per year of oil and 100 to 170Gm^3 per year of natural gas by 2025 (Figure 2.5). The challenge for these countries is therefore to attract and secure the necessary investments for expanding their production capacities in the oil and gas sectors. Because of the inertia of energy systems, there is also a need to get ready for the 'post-oil' period and the transition of oil-income economies to more diversified economies over the coming decades. Furthermore, the importance of the income from energy exports makes these countries especially vulnerable to the volatility of hydrocarbon prices. Hydrocarbons in Algeria, Libya and Egypt are a major source of income (98 per cent of export income and 32 per cent of GDP in Algeria, and 95 per cent of Libya's export income) and that income fluctuates with the global market.

Energy dependence is increasing

The Mediterranean producer countries are in danger of seeing their export capacity reduced by growth in internal demand, and the other Mediterranean countries face a growing *energy deficit* (increasing by 1–2 per cent per year between 2000 and 2025). Indeed, the planned increase in national primary energy production (nuclear in France, coal in Turkey[6] and renewable energies) will not be enough to meet the growing demand. The Mediterranean basin as a whole may thus see a doubling of its hydrocarbon imports (oil and, in particular, gas into the NMC), increasing from 290Mtoe in 2000 to 530Mtoe in 2025.

Despite some diversification of supply resulting from the increased use of gas (Figure 2.3), the energy dependence of most Mediterranean countries is projected to increase, with imported fossil fuels still having the dominant share of the total (87 per cent in 2025). This growing energy dependence can be illustrated by a *dependency index*, defined as the ratio '(Demand – Production)/Demand' for primary energy, expressed as a percentage. The index is increasing in most countries. In the NMC it is projected to increase from 68 per cent in 2000 to 71 per cent by 2025. Dependence on oil in the NMC is likely to remain stable (at 97 per cent), but is projected to increase for gas. It is projected to increase slightly in the SEMC,[7] so that the average dependency index of all Mediterranean countries will have doubled between 1970 and 2025, from 21 to 38 per cent. The most dependent countries are the smallest (various islands, Israel, the Palestinian Territories and Lebanon) but also Morocco, Italy, Spain, Tunisia, Greece and above all Turkey, where the index may go from 35 per cent in 1971 to 70 per cent in 2025.

Energy

Figure 2.5 Primary energy production by NMC and SEMC source, baseline scenario

NMC: little increase in production

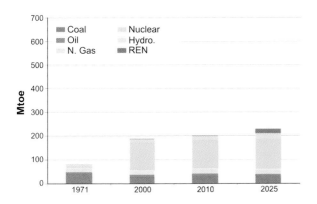

SEMC: increase concentrated in four countries: Algeria, Egypt, Syria, Libya

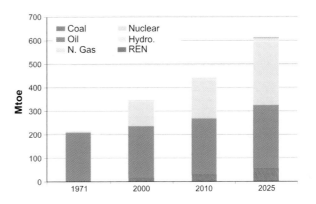

Source: OME, 2002

Note: REN – geothermal, wind and solar energy

A brief overview of the world's energy outlook (Box 2.1) provides a better indication of the vulnerability that can result from such dependence and enables a better analysis of the possible ways in which growing Mediterranean import needs may be met. It shows that, essentially for geo-political reasons, with the increasing concentration of the world's oil reserves, supplies of hydrocarbons on the world market are likely to become more and more expensive and risky (with price volatility persisting).

Figure 2.6 Growing energy-dependence index of a number of Mediterranean countries

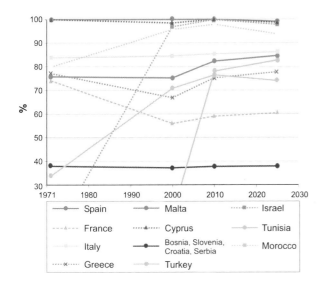

Source: Plan Bleu, OME, 2002

Note: The dependence index is defined as (Demand – Production)/ Demand for primary energy ratio.

Box 2.1 The risks to the worldwide supply of hydrocarbons

Many forecasts of world energy supply try to be reassuring about the global availability of energy resources. The reserves are generally physically sufficient to cover the growing worldwide demand, particularly up to 2025. Indeed the known world reserves are estimated to be good for about 40 to 50 years of production for oil, 60 to 100 years for gas and more than 200 years for coal and uranium, time scales well beyond the time scale of this report. Moreover, these reserves are certainly widely underestimated, since knowledge of fossil resources is still partial, and the role of non-conventional oil may be considerable.

The main risks cited by energy experts for the development of worldwide supplies of hydrocarbons therefore result less from exhaustion of reserves than from geo-political factors. The progressive substitution of oil by gas and the increase in the world's production capacity requires a level and phasing of investment (in production and transport infrastructures) that is much higher and longer than for oil and requires long-range planning. With the growing geographic concentration of oil reserves, the geo-strategic importance of the Middle East is expected become even greater (the five major oil producers: Saudi Arabia, the United Arab Emirates, Iraq, Iran and Kuwait will probably produce more than 50 per cent of the world's oil by 2050 and play a growing role in the world's exports of gas). If the situation in the major producing countries does not improve, the capital invested in energy production by the world's oil and gas operators could prove to be insufficient to meet world demand, which could eventually lead to considerable tension in world energy markets, even as early as 2025. Even in the mid-term, therefore, the risk of a disruption in hydrocarbon supplies cannot be excluded. This is bound to affect the price of energy (risk coverage).

Few forecasters venture to make very long-term predictions on fuel *prices*, which govern energy bills, or the development of alternative energies. The price of oil plays a crucial role in energy markets and is characterized by large fluctuations related to the volatility of financial markets. In such a context the largest uncertainty is on supply, which depends on geo-political balances and the future level of investment in production. The few studies available predict a moderate *rise* in oil prices, linked less to the exhaustion of reserves than to the economic, political and environmental limitations of its exploitation and on continuing high price volatility. A 2001 reference scenario from the IEA[8] starts at $20 a barrel in 2000 rising to $21 in 2010 and $28 in 2020. The price of gas is still closely linked to the price of oil, although it may gradually free itself from oil. However, the increasing distances between production and consumption areas may lead to a price increase. The question is whether this will be compensated for by technological progress in the industry or a liberalization of the energy sector. Some experts forecast an increase in the price of natural gas of nearly 20 per cent by 2010,[9] but here too there remains much doubt.

Ever more expensive access to energy?

In the context of increasing prices and volatility, the massive growth in energy imports may well increase the costs of fuel for Mediterranean countries and households.

The economic impacts on the NMC may remain fairly mild, since the increase in their demand is moderate. These countries have diversified their supplies, gradually excluding oil from electricity production and they have undergone structural changes in their economies that reduce their dependence on erratic variations in the price of a barrel of oil. The cost of energy now represents only a small and decreasing share of GDP (2 per cent on a worldwide scale).

On the other hand, because of their growing trade deficits and possible inflation, the SEMC could face serious problems in balancing their macro-economic budgets. The combination of increasingly volatile prices and uncertainties about funding the investment needed for basic energy infrastructures could result in a larger energy bill and greater difficulty in providing essential energy services for a growing proportion of the population. In Morocco, for example, the cost of energy in 2002 was US$1.5 thousand million dollars (or about 4 per cent of GDP). A 20 per cent increase in oil consumption in a year would lead to a considerable increase in this cost, about eight times the annual average ODA received by Morocco in the energy sector.

It should be noted that rural electrification infrastructures in some countries are still inadequate, which calls for considerable investment. Access to modern forms of energy in many rural areas remains one of the essential requirements for improving living conditions (lighting, telecommunications, refrigeration and basic health services) and is an important factor in reducing poverty, raising educational levels, limiting rural–urban migration and a vital driving force for economic activity. In 2000, nearly 16 million people in the SEMC did not have access to electricity,[10] mainly in Syria, Egypt and Morocco (see Chapter 5). Since then, important efforts have been carried out, mainly in Syria, for serving 99 per cent of the population.[11]

With the ending of the welfare state, recovery of the costs of investment may affect the final price of energy, and thus consumers' budgets, with not insignificant social impacts. Although price increases would hit the SEMC harder, they are also of concern in the northern-rim countries where steep rises in petrol prices can sometimes result in social conflict. In this context, looking for ways to reduce the costs of energy supply seems to be one major challenge for sustainable development that concerns all countries, especially the SEMC.

Consequences for the environment and the increase in risks

In addition to the geo-political and socio-economic criteria already discussed, environmental criteria are playing an increasingly important role in national energy choices. Awareness of the global effects of their energy choices involve countries in international conventions such as the Climate Change Convention and the Kyoto Protocol, which are having a growing influence on national strategies. This Protocol commits the industrialized countries (Annex 1 countries – in the Mediterranean, the four EU-Med countries and Monaco, Croatia and Slovenia) to limit their greenhouse gas emissions (GHG) by 2012, but does not commit the SEMC.

Without going into all the possible risks and impacts on the environment of the various energy options (Table 2.1), it should be remembered that 'traditional' fuels consumed for energy (firewood, charcoal), for which the statistics are poor, can have considerable environmental impacts (Chapter 5). In Morocco, for example, firewood consumption in rural areas is blamed for the over-exploitation of forests, which contributes to the loss of about 30,000 ha of forest per year and thereby to desertification and soil erosion.[12] The impacts, in the baseline scenario, may get worse, especially if access to commercial fuels becomes more expensive and more sustainable management of woodland resources is not implemented.

The environmental impacts of the increase in maritime transport of hydrocarbons, with the increase in world trade in energy in and through the Mediterranean, are also considerable and are analysed in Chapter 3.

In this chapter, two especially sensitive aspects of the baseline scenario that concern urban areas and Mediterranean coastal regions will be analysed in more

Table 2.1   Risks and effects on the environment related to energy consumption and production

Energy consumption

Air	Emissions of CO_2, CO, SO_2, NO_x, COV, dust, heavy metals, PAH (combustion)
	The release of CFC and fluorinated gas (air-conditioning and refrigeration systems)
Waste	Ash, oils, sludge (combustion engines)
Risks	Fires, explosions, electrocutions, intoxications

Energy production and transport

Air	Emissions of SO_2, NO_x, CO, COV (coal extraction)
	Emissions of organic compounds, dusts and CH_4 (oil and gas extraction, oil refining, gas storage and distribution)
	Emissions of COV (storage, transport and distribution of oil products, especially fuels)
Water	Effluents that can contaminate river water (electric power plants, coal mines, refineries)
	Thermal pollution that can alter the ecosystem of a river (the cooling of traditional and nuclear thermal power stations)
	Change in the regimen of rivers (hydraulic equipment)
	Fuel dumping at sea (freighters)
	Fertilizer and pesticide irrigation and pollution (biofuels)
Accidental risks Nuclear accident risks (electro-nuclear power station, reprocessing plant, transport of nuclear materials)	Fires (oil-drilling platforms, storage or refining sites, hydrocarbon transport networks) Oil spills (freighter break-up) Explosions, land collapse, landslides (coal mines) Breaks (hydroelectric dams)
Risks of vandalism and terrorism	Attacks on production tools, transport or storage (nuclear power stations, repository sites for nuclear materials, a methane terminal, hydroelectric dams, and so forth)
	Misappropriation of dangerous materials
Waste	Coal mining waste
	Radioactive waste (nuclear)
	Residue (refining)
Footprint on the area and nature	The destruction or modification of ecosystems (hydraulic, wind, solar)
	In particular destruction of coastlines for production (land takeover of production and supply infrastructures by fuel-burning power plants and coal harbours)
	Fragmentation or alteration of landscape (pipelines, gas pipelines, high-tension wires)
Chemical substances and radioactive matter	Liquid radioactive emissions (nuclear plants, spent fuel reprocessing plants) Toxic substances contained in unburned solids (ash) or dust from burning fuel
Noise	Thermal installations (particularly engines)
Soil	Contamination (coal or uranium extraction sites, former gas factories)
	Abandoned industrial sites

Source: based on IFEN, 2002 p401

detail: emissions of polluting gases and the impacts of infrastructures.

The increase in gas emissions

Fossil fuels dominate Mediterranean energy provision. Any burning of fossil fuels produces large quantities of *atmospheric pollutants*, in particular nitrogen oxides and sulphur dioxide, which are responsible *locally* for the degradation of air quality and *globally* for acid rain.

Before discussing long-term global impacts, it is important to remember that the burning of fossil fuels is responsible for degradation of air quality in large cities and around industrial areas. The effects of this pollution on human health are beginning to be evident in epidemiological studies that demonstrate higher mortality and many respiratory diseases (see Chapters 3 and 4).

Such atmospheric pollution can be carried over smaller or larger distances and alter the chemical composition of the atmosphere. A recent study[13] has found very high pollution levels over the Mediterranean Sea, mostly from Europe but also from North America and even Asia; the presence of aerosols (in particular particulates of sulphur soot) in the troposphere over the Mediterranean could be the cause of a reduction in evaporation and in the amount of rain in the Middle East and North Africa. Concentrations of ozone higher than European standards have also been recorded, with possible impacts on human health, ecosystems and agricultural production.

Still more generally, the energy sector is a major contributor to the *greenhouse effect* and to global climate change. The OECD estimates that more than half of greenhouse gas emissions (GHG), most importantly CO_2[14] originate from the use of energy. CO_2 emissions from energy use will therefore be used as the main indicator of the possible effects of gaseous emissions in the baseline scenario. The emission level depends on the fuel used and its conversion efficiency. Natural gas, composed mostly of methane (CH_4), is considered to be the 'cleanest' fossil fuel, since it emits less carbon[15] than oil or coal and enables electricity production technologies with better efficiencies (co-generation, combined cycle power plants). The CO_2 emissions related to primary energy consumption are estimated here by applying average GHG emission coefficients for the contribution of each fuel to the energy supply of each Mediterranean country (current and those projected in the baseline scenario).[16] The estimates to 2025 are given country by country in the Statistical Annex. The 1990 value is particularly important since it acts as the reference level for the Kyoto protocol commitments.

According to these estimates, total emissions from Mediterranean countries from energy use have increased steadily and amounted to 1900 million tonnes of CO_2[17] in 2000 (Figure 2.7). They represent more than 7 per cent of total worldwide emissions and this may reach 9 per cent by 2025, in the baseline scenario. The transport sector is one of the main contributors (between a quarter and a half depending on the country).

In 2000 the NMC accounted for 70 per cent of total energy-related CO_2 emissions in the Mediterranean, with an average increase of 1 per cent per year between 1971 and 2000. Given the current projections for the energy sector, the four EU-Med countries (the main countries officially committed to reducing their emissions through the Kyoto Protocol) may find it difficult to reach their reduction targets (Figure 2.8), especially Greece and Spain. Because of the delays in achieving their goals, these countries should agree to make greater efforts, given the short time available between now and 2008 to 2012.[18]

Figure 2.7 CO_2 emissions related to energy activities, baseline scenario, 1971–2025

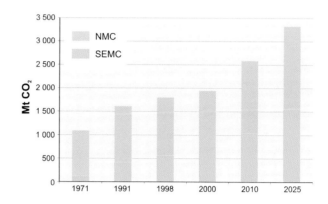

Source: OME, 2002

Figure 2.8 CO_2 emissions from the energy sector in the four EU-Med countries, 1970–2025

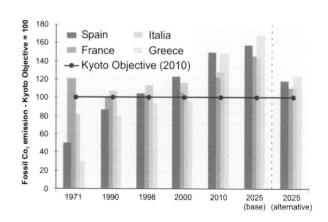

Source: Plan Bleu and OME, 2002

Note: Baseline and alternative scenarios.

At the same time, the SEMC are experiencing spectacular growth in their emissions (6 per cent per year between 1971 and 2000), and may increase their emissions still further, by a factor of 2.5, by 2025. There will probably be a huge increase of more than 500 per cent in Turkey between 1990 and 2025.

Thus in the *baseline scenario*, the gradual replacement of oil by gas, although beneficial in terms of specific emissions, does not seem to compensate at all for the increase in emissions resulting from the overall increase in energy consumption. Although a modest contributor to the increase in the global greenhouse effect, the Mediterranean Basin may well, however, be particularly vulnerable to the possible impacts of climate change (see Part I).

Impact and risks of energy infrastructures

The production, distribution and transport of energy, all of which – particularly electricity production – are growing rapidly, also call for massive infrastructures that have impacts in terms of land use and destruction of landscapes and coastal ecosystems. Some infrastructures, such as nuclear reactors, pose problems of dismantling at the end of their useful life.

Drilling and oil and gas production installations, refineries, power stations and high-voltage electric cables all *occupy ground* areas that may appear small in comparison with other activities,[19] but pose problems of competition for highly coveted coastal areas and may contribute to the degradation of coastal ecosystems. They also have considerable impacts in terms of landscapes, *fragmentation of land* (linear enclosure by electricity cables) and increased risk of death for some bird species.

Expansion of these infrastructures also increases the risks of *accidents and pollution*. The production and transformation of hydrocarbons (refining, liquefaction and petro-chemistry) are hazardous (having led to the development of the European 'Seveso' directive), highly polluting (gaseous emissions, damage to coastlines and liquid waste), and are not adequately assessed. Some indication of the risks can be obtained by analysing the planned increase in infrastructures related to gas and electricity developments in the Mediterranean.

Gas infrastructures

To meet the forecasted doubling of intra-Mediterranean trade between now and 2025, a large number of infrastructures are projected that will have considerable impacts in terms of land use, especially on coastlines (Figures 2.9 and 2.10).

Natural gas is carried as a gas by pipelines or as a liquid at very low temperatures (LNG) by methane carriers. Liquefaction plants (at the points of production) and re-gasification terminals (at the points of arrival) are needed for LNG, as well as a fleet of methane or LNG carriers. International gas transfers should ideally be by sea because of the technical, economic and even political limitations of long-distance land routes. LNG is therefore expected to take a growing share of international gas trade, already standing at 25 per cent.

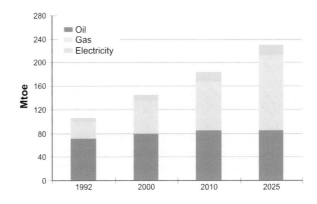

Figure 2.9 The doubling of intra-Mediterranean gas trade, 2000–2025

Source: OME, 2002

In general the transport of natural gas does not present pollution problems comparable with those of oil, but it does carry risks of major technological accidents, especially when the delivery points are in populated industrial areas. Permanent infrastructures are potentially very vulnerable to terrorism and vandalism. It is important to note that although accidents have so far been rare, the consequences could be serious.

Electricity infrastructures

The capacity of the electricity *generation system* has increased 2.5 times over the past 20 years in the Mediterranean Basin and installed capacity is now 350GW. Growth is projected to continue, with a further 120GW planned between 2000 and 2010, including 70GW in the SEMC, a near doubling of capacity between 2000 and 2010. This will lead to an increase in infrastructures (power plants, supply networks feeding the power plants, for example coal ports and transport, and electricity distribution). To meet the planned growth to 2025 (to 1270TWh per year in 2025), about 400 new power stations (of 500MW each) would have to be built[20] (beyond the 440 existing plants). These energy infrastructures will accentuate the *'littoralization'* or coastal overdevelopment effect, that is the process of population and activities being concentrated on the coasts. Power plants are often located on the coast, which enables them to be closer to the areas of consumption and facilitates the supply of fuel by sea and water for cooling. Many oil and natural gas fields are located near the coast (the western Egyptian desert, the eastern areas in Libya) or offshore. Inland power plants use water that is precious in arid environments and produce large amounts of waste and emissions, some of which are directly disposed of on site.

2 A Sustainable Future for the Mediterranean

Figure 2.10 The development of gas infrastructures in the Mediterranean

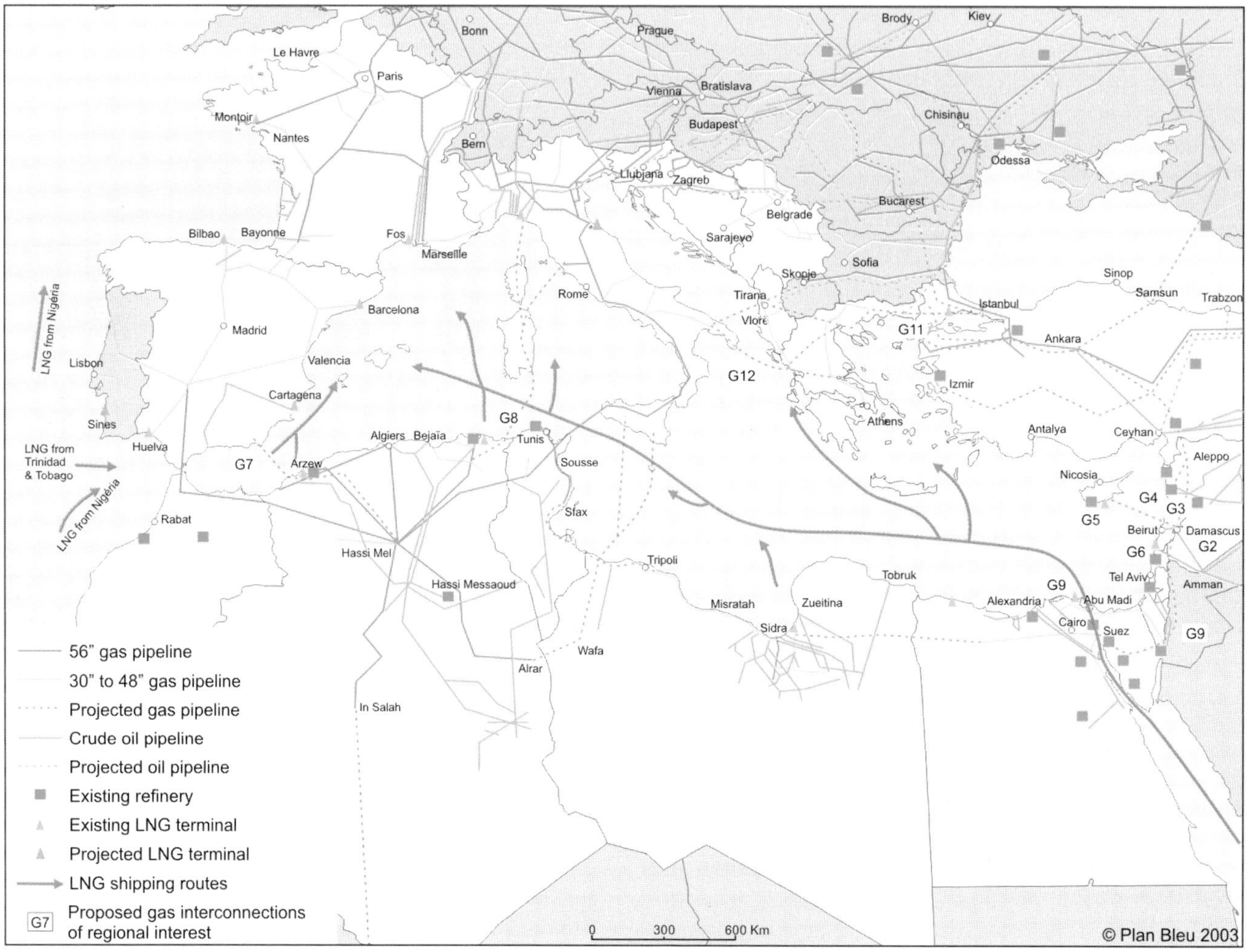

Legend:
- 56" gas pipeline
- 30" to 48" gas pipeline
- Projected gas pipeline
- Crude oil pipeline
- Projected oil pipeline
- ■ Existing refinery
- ▲ Existing LNG terminal
- ▲ Projected LNG terminal
- → LNG shipping routes
- G7 Proposed gas interconnections of regional interest

© Plan Bleu 2003

Source: OME, 2002

Distribution networks and electricity interconnectivity are also projected to increase and will have a growing impact on landscapes and the fragmentation of land. The corridor of land affected by a high-voltage electricity line can be between 30 and 120m or up to 400m wide. In the NMC, projects for building overhead lines are facing increasingly hostile public opinion because of their environmental impact. People living near such lines also have growing concerns about the possible impact of electromagnetic fields on their health, although this has never been established by epidemiological studies, more of which are needed. These public reactions sometimes result in a rethinking of the proposed technical options and a shift towards solutions that combine decentralized power production and energy management, making it possible to defer (or even cancel) the building of some infrastructures (see the preliminary studies for reinforcing the 400kV Boutre-Broc Carros line in France's Provence–Côte d'Azur region).

Box 2.2 The main electricity interconnections in the Mediterranean, 2002

Before 1990 there were only a few, low-capacity electricity interconnections between Maghreb countries. Of the 17 interconnections built between 1952 and 2003, half came into service during the past decade, which shows the efforts made in electric interconnection in recent years. Between now and 2010, 16 new links are planned for construction and the North Africa network will be strengthened to 400kV.

In addition to the existing interconnections in Maghreb (Algeria–Morocco (2 links) and Algeria–Tunisia (4 links)), and recently inaugurated links (like those connecting Spain to Morocco, Libya to Egypt, Egypt to Jordan and Syria to Jordan

122

Figure 2.11 Electricity interconnections in the Mediterranean

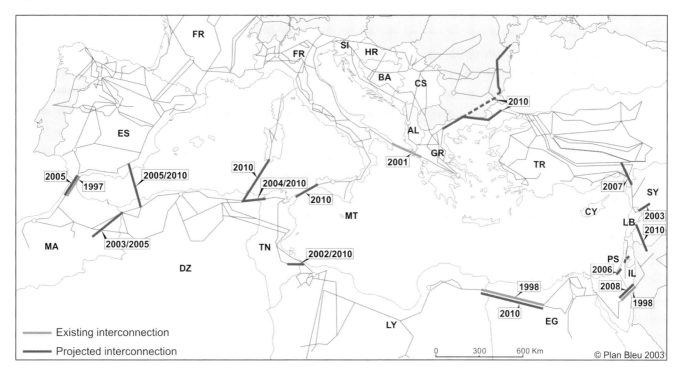

Source: OME, 2002 and Systmed report, 2003

and Lebanon), there are a number of projects and lines being boosted to 400kV (the 400kV links of the northern African countries and the interconnections between Tunisia and Libya (3rd link), Tunisia to Algeria (5th link), Algeria to Morocco (3rd link) and Syria to Turkey).

When the Spain to Morocco interconnection was put into service in 1997, the two shores of the Mediterranean were joined, and a doubling of this link is now planned for 2005. It will be boosted between now and 2010 with interconnection projects joining Algeria to Spain (direct undersea link), Algeria to Italy via Sardinia, Tunisia to Italy, and Turkey to Greece.

According to the preliminary estimates of the MedRing project (a study of an electricity loop around the Mediterranean, which is underway), plans are being made for the trade in electricity between the Mediterranean countries to exceed 75TWh per year between now and 2010, compared with only 45TWh in 2000, of which 5TWh was between the SEMC.

These interconnections create technical but also political links between the Mediterranean shores.

In conclusion, under the baseline scenario, the projected strong growth in energy demand will *increase the level of risk in the Mediterranean*. Energy supplies will be ever more dependent on today's uncertain geo-political situation. The energy bill may well increase, and all countries (both importers and exporters) may suffer from highly volatile prices. Environmental effects are likely to be considerable, some may be irreversible, and will affect the worldwide and regional heritage through an increasing contribution to global warming and the destruction by infrastructures of some of the coastline and landscapes. These risks and effects will differ between countries but will be most serious in the SEMC. These countries will urgently need to find answers to the spectacular increase in demand to meet the basic needs of the population today and in the future. They will also be exposed to a significant share of the environmental risks of transporting hydrocarbons to supply energy to the northern-rim countries.

Towards more sustainable energy systems

Faced with geo-political, socio-economic and environmental risks, there are alternative paths to more sustainable energy development.

The first is that of *energy efficiency* (EE) or *energy demand management*. This aims to increase the efficiency of current energy systems, breaking the link between economic growth and energy growth. The aim is to provide the same services while using less energy. This is a winning strategy from all points of view: it reduces the dependence, the costs and the impacts connected with energy supply.

A Sustainable Future for the Mediterranean

The second is to ensure supply while reducing the related environmental impacts, by *diversifying energy supply and promoting 'clean' sources of energy*. Many countries have already committed themselves to this path, particularly through increasing use of natural gas and renewable energy sources that reduce geo-political and environmental risks and, under some circumstances, costs. There has also been progress in reducing the environmental risks of conventional energy sources (e.g. research into nuclear waste processing, coal technologies using fluidized beds, regulating the transport of hydrocarbons), and measures aimed at optimizing supply (for example electrical interconnections). Such projects are already well underway, but it is impossible to list them all here.[21] In contrast to the energy efficiency path, it is not always possible to minimize all types of risk *simultaneously*. Choice, governed by societal factors, is called for. For example, some countries (e.g. Turkey and France) are still subsidizing nationally produced coal, despite local and worldwide pollution, with the objective of maintaining jobs in the mining sector or limiting energy dependence. These strategies may sustain economic development or social cohesion, but at the expense of deprived populations who are exposed to air pollution, or future generations being directly affected by the effects of climate change. There is no such thing as an impact-free, neutral energy source. Energy decisions require choices between various options, choices that increasingly take environmental and long-term considerations into account, as is clear from the debate about the nuclear option.

In the vast field of possible energy strategies, this chapter explores energy efficiency and renewable energy, which, in combination, seem to promise more sustainable energy supplies, without, however, obviously excluding the use of other sources for which technological improvements can be expected.

The challenges of energy efficiency and renewable energy

Advantages

As noted at the Johannesburg Summit on sustainable development in 2002 (Box 2.9), energy efficiency and renewable energy are considered the most promising paths for ensuring access to energy for the greatest number and contributing to the development of the least advanced countries, while facing up to the planet's major environmental challenges. Indeed, in the light of the risks highlighted by the baseline scenario, these solutions provide many advantages:

- Reducing national energy *dependence* through reducing the need to import energy.
- Reducing the *vulnerability* of supply systems, since renewables are not subject to geo-political risks or those of reserve depletion, and thus contribute to security and diversity of supply. Well adapted to the production of decentralized energy, they reduce some risks of supply breakdown (accidents or terrorist attacks on distribution networks or power plants). In remote or lightly populated areas, such as the many Mediterranean islands, they remove the need for large connection infrastructures with the mainland. In such areas, they can provide the energy needed for local development and satisfy basic needs at the lowest price. Renewables may be phased in gradually, which gives them an economic advantage in terms of flexibility in adapting to demand, since the investment is also phased.
- *Savings in the costs* of energy, from offsetting the investment costs of large infrastructures, energy efficiency and the negligible operational costs of renewable energies (or lower investment costs than for traditional energy in remote areas).
- *Local development and job creation*: developing renewables and energy efficiency may create more jobs than in the energy production sector (for example the insulation of dwellings, solar water-heating, wood-burning heaters).
- Renewables may also provide *economic export opportunities* for some SEMC (exporting 'green' fuels, green certificates or within the framework of new 'clean' funding measures).
- Indirect effects on the *efficiency of use* of other economic items (water, labour, waste) and fall-out in the industrial sector (benefits related to advanced technology) in the high-growth sectors that are being developed in preparation for the 'post-oil' period.
- *Reducing the risks and environmental impact* through energy efficiency that reduces the need for infrastructures and fuel transport. Energy efficiency and renewables also result in considerable reductions in gaseous emissions thus reducing local and *global* air pollution and impacts on climate change. Like all energy sources, renewables are not without *local* impact on the environment. The most troublesome in the Mediterranean are wind farms in the landscape, which call for preliminary studies. Micro hydropower plants are also often opposed because of their potential impact on wildlife. This calls then for choices between various types of impact, although renewables have undeniable advantages in terms of gaseous emissions and the minimization of environmental risks.

The benefits of energy efficiency and renewables are quantified on the Mediterranean basin scale in an alternative scenario in the section entitled 'The challenge: an alternative scenario to 2025', below (p143).

Reasons for committing to more rational use of energy

Seeking a more rational use of energy aims to accelerate the decoupling of economic growth from energy growth and to make a significant break from the trend observed over the past 30 years in the Mediterranean, which is the basis of the baseline scenario. Currently, a 1 per cent increase in GDP means, on average and for all Mediterranean countries, an 0.8 per cent increase in primary commercial energy consumption.

This weak decoupling between energy and the economy can also be illustrated by the *energy intensity* indicator, which compares primary commercial energy consumption with GDP. Diagrammatically it is an 'inverted' indicator of energy efficiency, in that a large and growing value of energy intensity reflects a highly energy 'consuming' mode of economic growth and suggests areas for possible savings. Table A.18 in the Statistical Annex shows that the NMC have reduced their energy intensity only slightly over the past 20 years.[22] Greece and Spain have even increased it. Moreover, the reduction seems to have been the result of structural changes in economies that are 'dematerializing' rather than of active efforts to improve energy efficiency.[23] In *contrast* all the SEMC, except Israel and Tunisia, have seen an increase, sometimes a very large increase, in energy intensity (Turkey, Lebanon, Syria and Libya).

Generally, energy intensity is observed to fall as income levels rise, a trend that should be accentuated through energy efficiency. The *challenges* and room for manoeuvre vary between countries.

The SEMC still need to build their essential industries and infrastructures. They have not yet reached the dematerialization phase of their economic development. If they manage to achieve their share of globalization by technological 'leap-frogging' while remaining in control of their development, they will shorten their learning time and may well be able to choose paths that are more energy-saving and economically more efficient. They have before them an exceptional opportunity to manage their energy needs for the coming 25 years through judicious choices of infrastructures. Depending on whether the enormous investments planned for transport, housing, industry and energy supply do or do not include energy-efficiency objectives, the structure of energy demand by future generations may be completely different and this may indirectly affect their ability to develop their economies. The challenge of *anticipation* remains considerable and is also one for the energy-producing countries which may be able to increase the amounts available for export.

For the NMC the question is less one of anticipating what infrastructures to build than one of reducing their contributions to the growth of global environmental risks and to inequalities in the distribution of resources.

It is also a question of fulfilling their international commitments and improving their economic performance through better performance of their energy systems. And there is a need to prepare for the 'post-oil' period by promoting the emergence of more flexible and decentralized energy systems.

The potential for energy efficiency in the Mediterranean

The objective of energy efficiency is to optimize energy systems through a broad range of actions.

First it means *improving the efficiency* of the energy chain (energy production, distribution and consumption) through actions such as improving the management of energy infrastructures, house insulation, reducing the electricity consumption of household appliances and light bulbs, and reducing wasteful consumption.

It is also an opportunity to generally *rethink energy supply systems* so as to reduce energy consumption without affecting the energy service to the user, through co-generation, decentralized energy production, peak-shaving of electricity consumption and rebalancing daily and seasonal demands to avoid or defer the construction of the infrastructures required for a safe and high-quality electricity service.

Upstream actions on factors that determine energy demand are also concerned, even including the questioning of some travel modes and the siting of housing, for example by rethinking the spatial distribution of functions in cities that result in unnecessary travel.

The *magnitude of the potential savings* from energy efficiency in the various Mediterranean countries depends, therefore, on the level of their present 'wastage' and their determination to reduce it by adopting new behaviour. Potential savings of the order of 20–25 per cent of total energy demand by 2025 are certainly feasible through using technology that is already available. This figure is at the lower end of the range and could be higher in the SEMC (50 per cent of total energy demand by 2025), and even higher in very voluntary or proactive scenarios (for example through control of urban sprawl, development of public transport). Many energy audits in Mediterranean countries have confirmed these orders of magnitude. Some studies show impressive savings, varying from 10 to 60 per cent in the construction sector. In Morocco, for example, such audits estimate potential savings of 10–25 per cent of energy demand in the industrial sector, or 1Mtoe per year, and scope for saving 15–20 per cent in public administration buildings, or potential savings of about 9.5 million euros per year, corresponding to 20,000toe per year[24] (which, however, have not been implemented because of a lack of start-up funds and inter-institutional coordination). On the French electricity market, the ADEME agency estimates 30TWh per year of possible savings in less than ten years through demand-control measures. Of course these estimates must take the *net* global energy balance into account and include the 'upstream' energy needed to achieve 'downstream' savings (for example, some insulation materials may have a high energy content). Nevertheless, the potential savings remain considerable.

There is a very large potential for saving *traditional energies* in the mostly rural SEMC, through improved household technologies or substitution by more efficient energy sources (such as gas).

The *electricity production and distribution* sector also has a significant potential for savings, with distribution network losses often approaching 30 per cent. These losses, when added to low production efficiencies, can result in very low overall efficiency. Given the magnitude of the projected growth in electricity infrastructures in the SEMC, this sector seems particularly strategic in questions of energy efficiency; the current restructuring of these sectors (opening up, modernization, FDI, role separation) should promote improvements in network performance and efficiency. There is also a need to fight the very common practices of theft of electricity and non-payment of electricity bills by many users. Such behaviour discourages investors, cancels the effects of incentives in pricing policies and prevents rational energy management.

The *industrial* sector, representing more than a third of final commercial energy consumption in the Mediterranean countries, is where gains in efficiency are the easiest to obtain because of the small number of organizations concerned and the more immediate benefits that industrialists can derive from them (Box 2.3). It is also the only one of the three uses (domestic, transport and industry) for which the technological and substitution effects can exceed the negative consequences of growth and the rise in living standards.

> **Box 2.3   Energy savings by clean production technologies**
>
> Introducing 'clean' production processes in the industrial sector allows very significant energy savings, at a lower cost. Many examples can be found in the Mediterranean countries.
>
> In **Morocco**, the implementation of 'clean' production technologies has allowed a fish cannery to save the equivalent of 9 tons of fuel per year, a saving of 2200 euros per year, following an initial investment of only 1740 euros with a return on investment period of only nine months.
>
> In **Turkey**, a textile firm from the Y. Bosna-Istanbul region successfully reduced its energy consumption by 3kWh/day, using clean production technologies and applying stricter environmental regulation. In this case, the technologies put into practice generated total annual savings of 2000 euros at no investment or operational cost.
>
> In **Spain**, a plant for assembling power transmission components reduced its electrical consumption from 465,100kWh/year to 118,200kWh/year. The cost subsequently fell from 50,800 euros/year to only 8,880 euros/year. However, the return on investment period was longer than three years.
>
> Finally, in **Croatia**, a dairy factory near Zagreb offers a good example of energy reduction: thermal energy savings reached about 500,000kWh/year for a total investment of 31,000 euros. The annual savings are ten times higher: 328,000 euros. The return on investment period was only one month.
>
> *Source:* MAP/Regional Activity Centre for Clean Production, Barcelona; www.cema-sa.org

However, in the Mediterranean, the priority sectors are the rapidly growing *residential and tertiary* sectors (nearly 40 per cent of final commercial energy consumption in the SEMC) where considerable potential for savings can be exploited.

In the *construction* sector, savings in lighting, heating, air-conditioning and clean hot-water production are among the easiest to justify economically given the present conditions of the energy market. For example, the European Directive on energy efficiency in buildings (2002) seeks to save up to 22 per cent of energy demand from now to 2010 in the EU by simple measures, such as standards for new (residential and tertiary) buildings and renovation of buildings of more than 1000m^2 (insulation, heating, ventilation, lighting, the use of renewable energies, and the location and orientation of the building). An energy certification system for all buildings and regular control of heating and cooling systems are also required. By 2025, 7 million additional households are projected in the NMC and 33 million in the SEMC, which will result in a large increase in housing demand. The issue of energy saving building technologies is crucial. In France, implementing building heating regulations has resulted in that a new dwelling, built today, consumes 50 per cent less energy than one built in the 1970s.

The Mediterranean region has traditionally developed exceptional expertise in architecture and urban design, showing great ingenuity in adapting to the climate (mild, sunny winters, very high temperature peaks in summer). The choice of layouts, openings, materials, patios and fountains, and narrow, shady streets have produced an urban development and architectural heritage perfectly adapted to the climate. But, more recently, urban expansion and changes in lifestyles have been contributing to the emergence of urban developments and architectures that, by their very design, are too disconnected from the climate, leading to an over-consumption of energy for ventilation, air-conditioning and heating. Glass-encased skyscrapers are increasing in Mediterranean cities despite their being totally unsuited to summer heat. Yet a few simple building rules, satisfying new demands for comfort and adapted to the climate, would enable substantial cumulative savings for the whole life cycle of the buildings.

Box 2.4   Bio-climatic building to save energy

To get back in touch with its climate, Mediterranean architecture could reintroduce a few simple building principles that it has jettisoned rather too rapidly. The principle of bio-climatic architecture is to meet new demands for comfort while reducing, even avoiding, energy expenditure on active heating, ventilation and cooling technologies. Architecture in the region could thus exploit several parameters:

1. Using the exterior of the building to absorb winter sun, reduce the heat input in summer, protect the building from wind and noise, and light the inside by exploiting the particular luminosity of the region. The architecture can reduce heat loss by good organization of space, a compact shape, good wall and ceiling insulation, protection against predominant winds, window sizes adapted to exposure, insulated night-time protection of windows, main fronts oriented to the south, and a high thermic inertia of the building. It can reduce cooling needs by solar protection in windows, walls and roofs, good ventilation, heat retention linked to night-time ventilation, the planning of outside space to reduce outside temperatures, shady areas (trees, green spaces), reflecting colours for walls and outer materials, etc.
2. Using active technologies (heating, cooling). The traditional Mediterranean habitat sometimes had no heating. Nowadays, residents demand comfort (heating, cooling, ventilation), which is not necessarily more expensive, since there have been more energy efficiency improvements in the construction sector than in household appliances, for example. There is considerable scope for reducing heat loss and the costs of ventilation and cooling.

Experience is beginning to accumulate that shows the big advantages of bio-climatic architecture. The Energie-Cités (EC-GD Energy, Altener) project, for example, records initiatives such as the building of several bio-climatic hospital complexes in Murcia in Spain. The design of these complexes, for 300 people, has resulted in nearly a 70 per cent saving in energy costs for an extra cost of 5 per cent compared with a standard construction, and a return time of less than eight years for recovering this initial extra cost. In Lebanon, undergoing reconstruction, a FFEM project to support the design of energy-saving buildings (insulation of terraced roofs, water-proofing of walls, mechanical ventilation, double glazing) combined with the use of collective solar water-heating and compact fluorescent lamps is resulting in 30–60 per cent savings in energy consumption compared with existing buildings.

Source: Arene, ADEME, Projet Energie-Cités; www.energie-cites.org

After heating (for which the saving potential in the EU is an estimated 10–60 per cent, depending on the country), *hot-water production* is the second biggest energy consumer in the residential sector in the EU, with savings potential estimated at between 5 and 50 per cent.[25]

But along with the growth in the number of households, the number of *household appliances* has also increased (which explains the steady growth in electricity demand), the energy potential saving of which is estimated at 20–50 per cent. Among the main consuming items are cooling devices (refrigerators, accounting for one-third of the electricity consumption with a savings potential of 30–50 per cent) and lighting. In the Mediterranean countries of the EU, energy labelling of household appliances has been mandatory since 1995. Standards for minimum performance levels were set in September 1999, to complete a set of measures for domestic cooling devices, which has resulted in a major change in the market for cooling devices in terms of better energy performance. Thus, between 1992 and 2000 the average electricity consumption of refrigerators and freezers in Europe fell by 30 per cent. Given the energy performances of their household appliances, many SEMC are looking closely at the implementation of similar labelling and standardization programmes. The risk is that poorly performing equipment will increasingly appear on the markets of countries that have not yet adopted comparable legislation.

In the *lighting* sector the potential for savings simply by changing behaviour and by using compact fluorescent lamps is enormous (savings of up to 70 per cent with a pay-back period of less than two years). A recent evaluation of the potential for specific electricity uses in 400 dwellings in the EU (including half in Italy and Greece) identified a potential saving from approximately 20 per cent (Greece) to 40 per cent (Italy) based on simple measures (the replacement of cooling equipment by models with better performance, the replacement of incandescent lamps by compact fluorescent lamps and the suppression of stand-by settings).[26]

The very large expansion of *air-conditioning*, which particularly concerns the Mediterranean countries, and is expected to continue with the expected growth in urbanization and possibly global warming, also requires consideration of the savings potential through adapting construction technologies (natural ventilation, insulation, exposure), and using high-performance devices and control equipment.

The *transport* sector (between 24 and 33 per cent of final commercial energy consumption) also offers considerable energy-saving potential, but this is a sector where the trends are the most unfavourable and the inertia is the greatest. The present system based on road and private car transport is extremely costly, as shown in Chapter 3. This question is inseparable from that of *urban-*

ization. Indeed, purely technical solutions (more efficient diesel engines, hybrid or electric engines, fuel cells, substitute fuels) cannot compensate the effects of growth in transport demand. Transport energy consumption can also be reduced by better urban planning and management. Increased urban densities, bringing daily services (schools, shops) nearer to residences, the redevelopment of activity areas to minimize the transport of merchandise, and public transport are some of the vital actions for meeting the challenges of urban transport in a region that will have 100 million additional urban inhabitants between 2000 and 2025 (see Chapter 4).

The potential for renewable energy

The considerable potential for renewable energy in the Mediterranean is mostly underexploited, whether for electricity production or domestic purposes.

The potential in the SEMC is very high: it is one of the world's sunniest places (about 5kWh per m² per day) and the needs, both for thermal and electric applications, are many. There are many sites suitable for wind farms, considerable geothermal resources, for example in Turkey and Italy, significant possibilities for developing small hydroelectricity plants, and the use of biomass is an important energy option in many areas. A recent OME study has shown that the potential electricity production from renewable energies (excluding hydropower) in the SEMC could be about 105TWh per year, or between 10 and 15 per cent of total electricity production in 2020.[27]

Table 2.2 Electricity production potential from renewable energies in the SEMC in 2020

	Potential (MW)	Electricity produced/yr in TWh/yr
Wind energy	10,000	20
Photovoltaic	2500	5
Solar thermal	6000	15
Biomass	8000	48
Geothermal	2900	17
Total	**29,400**	**105**

Source: OME, 2002

Biomass is an important resource. In the SEMC (for example Maghreb and Turkey) firewood often constitutes a vital part of non-commercial supply, particularly for the most deprived. The challenge of improving efficiency of use is crucial, especially since the increase in energy demand risks aggravating an already visible over-exploitation of woodlands.[28] In the NMC a massive switch to biomass (especially the sustainable exploitation of woodland products) could enable a significant increase in the share of renewables in the primary energy balance.

For large *hydropower plants*, the Turkish potential remains considerable, and the very ambitious projects currently being implemented have not exhausted the possible sites. In the other SEMC, useable sites are being exhausted, and a number of dams are silting up, for example in Morocco. In the NMC all the possibilities have been exploited. *Micro-hydroelectricity*, a proven and now mature technology, is ideal for electrifying remote sites. It also provides a back-up for national electricity production, which is especially attractive at times of peak demand. Italy and France remain the leaders in terms of installed power with, respectively, 2230MW and 2020MW. Spain has made the greatest efforts recently. The Balkans, Greece and Turkey still have a large potential. Morocco has a potential of 3630MW of micro-hydroelectricity stations[29] at 200 sites that could be exploited within its borders.[30]

Wind power is being called on to play a significant role in the development of renewable energies in the region, as in the rest of the world. In recent years, technological progress in wind generators has been considerable and they have now become a genuine industrial sector. The most significant example in the Mediterranean is the Spanish market, with a total capacity of 3660MW in 2002. In the other NMC, launching the wind sector seems ongoing with an annual growth rate of more than 20 per cent in 2001. This includes Italy, with an additional 308MW (79 per cent) in 2001. In the SEMC there are wind sites with high potential in Morocco, Egypt and Tunisia, and all these countries have seen strong growth in wind powered electricity generation. Most of the other countries have begun to set up detailed wind power atlases, and projects are being studied for new sites. The question of environmental impact must be carefully monitored by the project designers, especially on the coasts. The question of the price for buying back the electricity produced is crucial; it should be the first to be decided for all national development plans.

Box 2.5 The situation and development perspectives for wind energy in the SEMC

In **Morocco**, installed wind power capacity could reach 200MW by 2004, or 6 per cent of the total electricity production capacity of the country. Moroccan authorities aim to reach 1000MW by 2010, adding 80MW of capacity per year, increasing the renewable energy contribution to total Moroccan primary energy demand from less than 1 per cent in 2000 to 10 per cent by 2010.

In **Tunisia**, current wind capacity is 10MW and *Société Tunisienne d'Electricité et du Gaz* forecasts 250MW for 2010, which means that wind energy would provide 3.4 per cent of annual electricity demand. A 400MW scenario has also been studied by the Tunisian Authorities.

In **Egypt**, there is an installed capacity of 60MW in Zaafarana and another 60MW project is under way. The national programme for wind energy development aims to reach 600MW installed capacity by 2010, or 2 per cent share of wind energy in overall electricity production. The low cost of gas restricts the development and use of the large wind energy potential of this country.

In **Turkey**, current capacity is 20MW; planned projects would increase this to a total capacity of 500MW with a guaranteed buy-back price system. A 2010 scenario developed by the General Directorate of Electrical Power Resources Survey and Development Administration aims at reaching 2000MW of wind power or 2 per cent of total electricity production. With the aim of joining the EU and implementing the European Directives on renewable energy for electricity production, considerable efforts have to be made in order to rely on wind power in a more intensive way. By 2010, with a total electricity production of 300TWh, a wind power share of 4 per cent would require the setting up of a huge 4800MW capacity.

Source: OME from national sources

Note: Wind load factors (assumed by national companies) are 40 per cent in Morocco, 27 per cent in Tunisia and 37 per cent in Egypt. For Turkey, the assumed load factor is 29 per cent (IEA Renewables information, 2004).

Because of the large amount of sunshine, the Mediterranean countries have a very large *solar energy* potential.

Solar photovoltaic (PV) energy has a very large potential in the SEMC, especially in the countries where the electricity network covers only part of the country. This is the case in Morocco where the rate of electrification is low and where nearly 10,000 villages still have to be electrified, or over 300,000 households in rural areas. A project for electrifying 16,000 rural households was begun in 2002.[31] The total potential in Morocco is estimated to be 200,000 PV systems, and in Tunisia 14,000 systems.[32] Turkey is also interested in developing PV, particularly for applications related to water pumping, electric signals and telecommunications. The PV sector has grown enormously (23 per cent of additional capacity installed in the NMC in 2000). But the market remains small and its development is dependent on the implementation of more ambitious national programmes and the creation of a specific industry to manufacture 'solar silicon', which would enable the sector to enter another dimension. The installations still show some vulnerability to meteorological hazards (sand storms, saline deposits), which increase maintenance needs.

There has been large-scale growth in *solar thermal energy* use in the NMC. A total surface area of about 4.5 million m² of solar panels has been installed in the four EU-Med.[33] In France in 2001 the surface area installed increased by 17 per cent, and the national market is beginning to feel the effects of 'Plan Soleil' launched by ADEME in the middle of 1999. In this context, the target set by French industrialists of one million m² installed each year from 2010 appears achievable. However, these developments are less ambitious than those in other European countries, Germany for example, which, although having less sunshine, already has more than 4.2 million m² installed. Solar energy in the SEMC is used mainly for producing clean hot water. Turkey is one of the world's leaders in this field with a total installed capacity of 3.5 million m² of solar panels. At present solar energy contributes 290 kilotonnes of oil equivalent (ktoe) of the country's total energy production, and by 2010 this should reach 600ktoe. The importance of the national energy context is particularly relevant to

Figure 2.12 Map of average solar radiation, April, 1981–1990

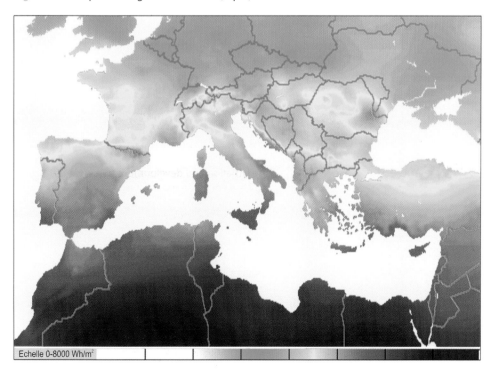

Copyright: this map is issued from The European Solar Radiation Atlas 2000 (vol 1), published by Les Presses de l'Ecole des Mines (www.ensmp.fr/Presses collection sciences de la terre et de l'environnement)

the development of the solar thermal sector. For example, the Palestinian Territories, very dependent on energy imports, had over 1 million installed m² in 2001. Tunisia, with about 90,000 m², saw a 37 per cent growth in 2001. In contrast, energy exporting countries, such as Algeria and Egypt, have a relatively small number of collectors installed (less than 1000m² in Algeria and 2000m² in Egypt).[34] In general, it is necessary to consolidate the emergence of an autonomous and sustainable solar thermal market. This means organizing the sector, training professionals, increasing public awareness and using high-quality materials developed in the framework of south–south as well as north–south exchanges.

Italy and Turkey are the leaders in *geothermal* energy in the Mediterranean. Italy and Greece aim to increase their installed capacity for producing electricity, to 912 electric megawatts (MWe) in Italy and 210MWe in Greece by 2010. As far as low-temperature geothermal energy is concerned, there are only limited data on the number of installations. The main producers are France (326 thermal megawatts (MWth)) and Italy (325MWth). In Turkey the geothermal potential is 31,500MWth, only 3 per cent of which is currently being exploited, and the target for 2020 has been set at 2000 installed MWth.

Renewable energies therefore have enormous potential that is just beginning to be exploited, especially in Morocco, Israel, Spain and Italy. We are seeing a build-up of increasingly structured industries and the emergence of mature technologies. The best examples are wind energy, photovoltaic energy and solar thermal energy. Spain has developed a first-class wind power industry. France is a leader in bio-fuels, and Italy and Turkey in geothermal energy. The diversity of situations and the complementarity of experiences makes the Mediterranean one of the world's regions with very significant potential (natural and technical) for renewable energy. In many cases, the renewables are the most economical option for decentralized rural electrification (for areas beyond the reach of electricity networks).

However, it must be stressed that industrial sectors and the market are much more developed in the north than in the south, while the potential, especially for solar, is found mostly in the south. Apart from wind and solar energy, and excluding a few isolated cases, the development of renewable energies remains dependent on many factors (see the next section): the local energy context, technological progress and a reduction in equipment costs, the implementation of appropriate regulatory institutional frameworks and the establishment of well-adapted and innovative financial systems. But, even more important, it is vital to undertake far-reaching information and dissemination programmes so as to raise the awareness of both decision-makers and users.

Conditions for developing energy efficiency and renewable energy

Despite the large potential and clear-cut benefits of energy efficiency and renewable energy, these have been largely underexploited in the Mediterranean. The conditions for a better realization of these potentials are analysed in the light of experiences in the Mediterranean.

Rethinking energy systems, involving all actors

Realizing the potential for energy efficiency and renewables requires a total change of approach in the thinking, planning and management of energy systems. In outline, this means shifting from the classical approach of increasing energy supply in the conventional energy sectors that are characterized by a fairly centralized and concentrated organization with a small number of actors (a few companies, even a single company and very often state-owned companies) to a *new approach* where a large number of actors are involved in energy supply and consumption. Changing the behaviour of users and suppliers (companies, local authorities, households, and energy producers and distributors) towards practices that optimize the energy resource is particularly complex (a problem that the more classical centralized approaches do not have) and requires the use of several possible measures:

- *Voluntary agreements*: for example the successful industry-wide agreements in Europe in the household appliance field (TVs, washing machines) that have enabled considerable energy efficiency gains quickly and inexpensively.
- *Regulations*: for example in the fields of construction and housing insulation (thermal regulations in Europe).
- *Financial incentives* for investment in energy-saving equipment.
- *Pricing energy* to internalize the externals, that is the impacts on the environment. This means introducing an element of 'truthfulness' when comparing various energy sources (and its corollary, the suppression of energy subsidies, which makes it possible to reduce demand).
- *Professional training and motivation*: the lack of awareness by salespeople has to some extent acted as a brake on the dissemination of efficient equipment in Europe.
- *Awareness-raising for the general public*: this is vital for ensuring acceptance of the need to change behaviour and prices; in this regard, the residents of the Latin countries seem less aware of these questions than those in the north of Europe. With growing urbanization, the general public is losing the link between its lifestyles and the natural resources it consumes, and people are mostly ignorant of the long-term impacts of their behaviour. Awareness-raising should enable this link to be re-established.

European experience in the field of renewable energies shows that it is the combination of all these measures that makes them effective.[35]

Given the number of actors and professional sectors involved in energy consumption and the complexity of their interactions, energy efficiency and renewables call for a significant amount of collaboration, teaching, adaptability and a capacity for innovation in the funding and regulatory systems. Choices may have to be made between the sometimes divergent interests of actors, including between the public and private sectors. Moreover, energy efficiency covers the whole spectre of economic activities. Strategies for energy efficiency and renewables cannot be designed without considerable political will and without being integrated into all economic sectors so that coherent signals are sent to users. Stimulating energy-management programmes with sufficient vigour therefore requires groundwork in awareness-raising and thought-out strategies. *Complementary levels of public action* must also be well articulated by ensuring convergence between economic measures (such as taxes on petrol) and structural measures (such as public transport supply). Failing that, the measures will remain ineffective and/or unfair. Such strategies, therefore, have a 'governance cost', and often a 'political cost'. They face obstacles that are often more of a cultural, institutional or political rather than a technical nature. Requiring new approaches, they also have a high 'learning cost'. This means time, patience and an exploitation of knowledge.

Public initiatives and national strategies

For all these reasons, actions to support energy efficiency or renewable energies, although often successful in the medium and long term, must in the short term overcome resistance and therefore require strong initiatives from public authorities supported if necessary by competent NGOs. Such actions are not triggered spontaneously by market forces alone, which lead, at best, to only a limited dissemination of effective technologies in a few niche areas. Public action remains indispensable for modifying user behaviour that is in the general interest, by including long-term signals in today's decisions by economic agents and ensuring a stable institutional and regulatory framework. This is why a growing number of Mediterranean countries are adopting official national strategies around the two complementary objectives of energy efficiency and the promotion of renewable energies (Boxes 2.6 and 2.7).

The *four EU-Med countries* (Spain, France, Italy and Greece) were among the first to take this path, and vigorous energy management policies followed the first oil crisis in some countries. A degree of relaxation was observed in the 1980s, but we are now seeing a return to interventionism, less from the fear of a new oil crisis than from the climatic consequences of the greenhouse effect.

In these countries, the European Directives on energy efficiency and renewable energies are stimulating the renewal of such strategies. In its White Paper on renewable energy, the EU proposes to its members the ambitious overall objective of doubling the share of renewable sources in primary energy consumption by 2010 (increasing the share from 6 to 12 per cent). As set out for the electricity sector by *Directive 2001/77/EC on promoting renewable energies for electricity production*, this is equivalent to 22 per cent of electricity being produced from renewable sources, including large hydropower, by 2010.

All these objectives are incorporated in national energy plans. However, even in countries with 'official' strategies, the results rarely match up to the stated ambitions, and the savings made are often less than those hoped for. Progress with renewables (although very rapid) is overwhelmed by the growth in total demand. Thus, for example, Greece and France will find it difficult to meet the European objectives for an increase in the share of renewables in total consumption. The only alternative is to increase the share of renewables significantly by combining this with substantial action to control demand.

Box 2.6 Examples of energy efficiency and renewable energy strategies in the NMC

Three countries in the north, France, Italy and Spain, alone account for two-thirds of energy demand in the Mediterranean Basin. An increase in demand of more than than 0.5 per cent per year up to 2025 is projected.

Italy is among the countries most dependent on others for its energy supply. Over the past few years it has implemented a vigorous policy of energy management and the development of renewable energy sources. Improving energy efficiency has been a prime objective of the National Energy Plan since 1988. By setting detailed objectives, quantified in time, for electricity producers, Italy has seen a spectacular increase in renewables: in 1999 nearly 600GWh of electricity were produced from biogas, more than 400GWh from wind and more than 200GWh from biomass, whereas in 1990, none of these sources produced more than 2GWh. Quantified energy savings objectives have even been set for sectors up to 2006 (decree 2001). Italy has the best Mediterranean results in lowering energy intensity, which remains one of the lowest in the world.

France has weaker energy efficiency objectives than the northern European countries. Given the strong continuity of public development of nuclear energy and private investment by the major fossil-fuel producers, French energy efficiency policy has so far been characterized by remaining in line with the evolution of the situation, with public expenditure on energy management quite closely following fluctuations in oil prices. Actions have nevertheless borne fruit. Results in the areas of thermal regulation of buildings and labelling of household appliances have been striking. Energy efficiency efforts made since 1973 have thereby led to savings of more than 30Mtoe compared with a 'laissez-faire' approach. A National Plan for combating global warming was drawn up in 2000, containing some 100 national measures that are still to be implemented. The renewables are seen as serious strategic alternative options. A price structure for electricity from new renewables has been implemented. In the thermal sector, ambitious objectives have been set for 2006 (thermal solar, fuel-wood, biogas and geothermal energy). The budget of the French Agency for the Environment and Energy Resources (ADEME) was increased in 2002[36] but thereafter decreased.

Spain voted for an energy efficiency plan for the 1991–2000 period, which enabled savings of 6.3Mtoe during the period (corresponding to an energy efficiency improvement of 10.4 per cent). Despite this progress, energy intensity during the same period increased.

Source: OME, ADEME, Plan Bleu

The history of energy management in the *other Mediterranean countries* began only in the mid-1980s, with more modest results. With a few exceptions, few Mediterranean countries have ambitious national strategies. This is the first major hurdle in developing energy efficiency and renewable energy activities, which, at best, are reduced to technological and isolated actions that are not enough to create a genuine market that can provide services.

Box 2.7 Examples of energy efficiency and renewable energy policies in the SEMC

Turkey, Egypt and Algeria alone are expected to account for 30 per cent of the primary energy consumed in the Mediterranean by 2025. With forecasted growth rates of more than 3 per cent a year, they will account for three-quarters of the additional demand by the SEMC between now and 2025. However, they have no explicit energy efficiency objectives in their energy strategies.

Turkey has to face up to a large increase in its primary energy demand. Its energy savings potential, although enormous, is not being exploited. Energy demand management is largely lacking in Turkey's energy policies. Energy prices even remain subsidized. Urgent action is therefore needed to promote the energy demand management essential for this demographically and economically growing country. In this context the resources allocated to the NECC (National Energy Conservation Centre) founded in 1992 are inadequate.[37] Turkey has not officially approved the Kyoto Protocol.

Egypt has been interested in energy efficiency for some years, but there has been only limited action. Several projects have been carried out with international support to identify the potential and the barriers to improving energy efficiency, and to strengthen capacity. A study has shown that a policy of rational use of local energy resources should make it possible to achieve savings of about 1 per cent of GDP and more than 10 per cent of annual CO_2 emissions by 2017 and help the country reach its ambitious development objectives in a sustainable way.

Algeria has long preferred to increase its energy supply capacities. However, faced with financial requirements, environmental impacts and growth in internal demand that risk compromising its export commitments by 2020, Algeria passed a law in 1999 and adopted a national strategy for energy demand management in 2003. Many measures have been set up: a national programme, a fund and a strengthening of the Agence pour la Promotion et la Rationnalisation de l'Utilisation de l'Energie. The use of renewables has been encouraged by a National Strategy and by creating an agency for promoting and developing renewable energies, the NEAL (New Energy Algeria) and several projects: a 120MW solar thermal power station, a hybrid wind-photovoltaic-diesel installation in Timimoun, electrification of the south using photovoltaics, the promotion of a local solar water-heating plant, a research programme on solar energy, and the marketing of LPG in the south to counter deforestation. A decree setting a minimum penetration (5–10 per cent) of renewable energy in energy supply is under preparation. Although it is too early to assess the results, the factoring of new energy management objectives into strategic planning documents demonstrates a national awareness of energy efficiency issues that may gradually spread to other producing countries that also have high energy intensities (such as Syria and Libya).

Experience in **Tunisia** also deserves mention; it implemented a national programme for energy demand management based on an institutional and regulatory framework in 1980 with the founding of the Agence Nationale des Énergies Renouvelables (National Agency for Renewable Energies) that later became the Agence nationale de maîtrise de l'énergie. The framework has evolved through a strengthening of its knowledge base, the standardization of incentives and a strategic and institutional realignment of the activity. It has contributed to the drop in energy intensity over the past decade and the reduction of emissions, but, given the development of the country's energy situation, a new energy-control programme has been relaunched for 2001–2010, involving two kinds of action:

Short-term priority actions:
- strengthening awareness-raising and energy management information;
- implementing adequate regulatory frameworks for promoting private investment;
- involving the public sector in exploiting the potential;
- mobilizing the required financial resources;
- strengthening local capacities and programmes in research and development.

A ten-year plan up to 2010, including:
- large-scale dissemination of solar water heaters, optimization of rural electrification by photovoltaic systems and the development of wind farms and biogas for electricity production;
- intensification of energy audits, promotion of clean technology and the development of energy-efficient programmes in all economic sectors.

Source: OME, 2002

Agencies to promote energy efficiency and renewable energy

To carry out these public actions, most Mediterranean countries have set up specific institutions for energy efficiency and the promotion of renewable energies. These public or NGO bodies are assigned actions of awareness-raising, training, making financial aid available and technical consultancy, depending on the situation; they also constitute a force for relevant proposals for tax or regulatory reforms.

These agencies play an essential role in *aid to decision-making*, the first step in energy efficiency strategies, by identifying potential energy savings, advocating renewable energies, demonstrating the economic rationality of actions and prioritizing the actions that are the easiest to implement. Their role may also prove to be fundamental in setting up projects and making use of knowledge of new methods and technologies. In addition to technical expertise, these bodies must also have a capacity for organization, dialogue, rapid intervention and an understanding of the problems and constraints of very varied partners.

However, their institutional situation, especially in the SEMC, remains fragile, and their financial and human resources insufficient. In general, institutional strengthening of these agencies (staff and operating budgets) is an essential factor in launching energy efficiency strategies. Their status and institutional position are also very important, since most of the potential energy savings call for direct intervention through sectoral policies (transport, housing, etc).

Funding energy efficiency and renewable energies

Immediate costs and deferred benefits

One of the main obstacles to disseminating energy efficiency and renewable energy technologies is financial. Being new sectors, energy efficiency and renewable energy development strategies have costs that are related to their complexity, their state of knowledge and, quite simply, the need for equipment. Like the older energy sectors in the past, they call for considerable financial resources for:

- aids to decision-making, to introduce these new options to the decision-makers;
- research and development;
- training and the exploitation of knowledge;
- organization of new sectors (training, organization);
- increasing awareness among users;
- above all, *investment* in infrastructures and equipment.[38]

From a *macro-economic* point of view, the 'profitability' of expenditure in the field of energy-demand management comes from the savings made. Many cost–benefit analyses show very advantageous internal rates of profit or very short return-on-investment times. Saving a unit of electricity costs from three to ten times less than producing it.[39] A study in France shows that the financial savings resulting from a low level of electricity consumption can be considerable, 2.5 thousand million euros per year on average for the electricity system (compared with total public incentives for energy savings of 7 million euros).[40] These savings provide ample room for manoeuvre in funding public policies for energy efficiency since they represent an expenditure of an additional 2.6 cents per kWh on electricity savings measures (or an additional equivalent expenditure of 130 euros on an average refrigerator). In the field of renewables, 'profitability' depends on technical progress being achieved, but also on how the costs of the different energy sources are compared, mainly on whether the costs of the externalities of the 'polluting' energies or, on the other side of the balance, the benefits of the non-polluting energies are entered into the accounts (see Box 2.8 p136).

From a *micro-economic* point of view, initial investment costs are often too high when compared with other traditional or alternative energies, and this continues to dissuade such investment, even though, in the building or industry sectors, it is common to have return-on-investment times of less than five years, even two years, as a result of the savings made (high-performance equipment in household appliances: fridges, compact fluorescent lamps and electronic devices with reduced 'stand-by' consumption).[41] The main obstacle to improving the energy efficiency of household appliances is no longer technical since economical appliances exist; it is mainly consumer buying habits, which *energy labelling* and the *label* itself aim to influence.

In most cases, bearing the cost of these investments is all the more difficult because the benefits that result, in the form of energy savings, reduced gaseous emissions, infrastructures not required and less vulnerability to risks (price variations, geo-political risks), are deferred in time or transferred to other beneficiaries (future generations). This is why public intervention remains essential.

The need for public support

Since private funding alone does not generally cover these actions, many countries have implemented *public funding incentives* for clean energies and energy efficiency. These incentives aim to pay for all or part of the costs (energy analyses, feasibility studies, subsidies for buying efficient equipment) in the form of direct aid, tax deductions, investment subsidies or bank guarantee funds or through funding the staff and operations of energy efficiency agencies. However, these financial incentives are often woefully inadequate and seldom on the same level as the public support given in the past (or even currently) to 'conventional' sources such as nuclear or coal. Particularly in the SEMC, the few public resources that exist are allocated to other priorities. For example, investment in energy efficiency in Morocco between 1994 and 1996 was less than 8 million euros out of a total of about 2650 million euros invested in the energy sector.[42]

Massive participation by public authorities in *research and development* on renewable energies would be one way to speed up the transition to such energies by accelerating cost reductions through technological progress, the strengthening of the field of activity and the economies of scale of an expanding market. Experiments and research on energy efficiency are also needed to promote synergies between the short and mid-term. Here too, however, public budgets for R&D are very small and still oriented mainly to the extraction technologies and the production of fossil and nuclear energy. In 1998 France earmarked only 1.8 per cent of its R&D budget for energy efficiency and renewable energy with 93 per cent going to the nuclear sector.[43] Of the Mediterranean countries making the greatest efforts, Italy in 1998 allocated 35 per cent of its energy R&D budget to energy efficiency and renewable energies (US$73 million) and 44 per cent to nuclear. Spain, devoted 47 per cent of its energy R&D to energy efficiency and renewable energy (US$22 million), Greece 68 per cent (or US$9 million) and Turkey 34.5 per cent (US$1.1 million).[44]

But it is mainly efforts to promote the *dissemination* of already operational technology, rather than technological research, that could play a vital role in lowering costs through the emergence of critical size markets. Figure 2.13 shows the influence of the 'installed power' factor on average investment costs per KW for various energy sources. Public authorities can also stimulate technological change through socio-economic measures such as guaranteed purchase prices for renewables, competitive auctions and the trade in green certificates (imposing renewable energy quotas on suppliers).

Energy efficiency measures often require *adjustments in*

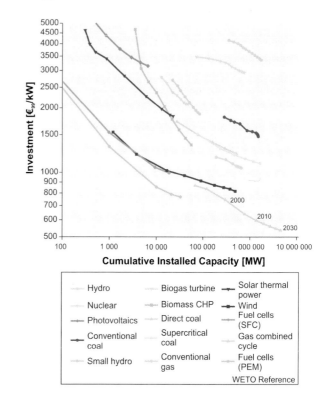

Figure 2.13 Learning curves, the increase in installed capacity strongly reduces costs

Source: EC-DG research, 2003

Note: Points in curves indicate time steps of five years. WETO Reference: scenario made of a coherent set of assumptions used for technological forecasting, based on the analysis of historic trends.

financial mechanisms, in the regulatory framework or the banking system, in order to adapt to their specific characteristics, that is their high initial cost (investment), which may not always be covered by the savings made (operations). Indeed the investor often does not benefit from the savings (as is the case with the owner and the tenant of a dwelling) or cannot recover part of the savings made in order to invest further (as is the case of the French municipalities, which, because of public accountancy rules, cannot use the financial savings made by saving energy in public buildings for further investment in energy efficiency). The small size and wide range of energy efficiency and renewable energy activities also require flexible, specially adapted methods of funding (micro-credits).

Because of limited public funding resources, the 'leverage effect' of public aid is increasingly sought as a way of jointly involving the *private sector*. Over the past few years new financial products have appeared, developed by operators in the public–private sector, which make it possible to finance the additional costs of initial investments through a mechanism for recovering these costs from savings made subsequently, either through an earmarked *investment fund* (for example a mixed public–private investment fund to finance the projects, such as the FIDEME, managed by ADEME in France[45]), or through the intervention of an outside third party (third-party investor or eco-energy service company). Public authorities can limit their involvement to covering the risks related to the novelty of the activity, by using bank guarantee funds to cover the investment-related risks (e.g. FOGIME in France[46]).

Providing clear price signals with energy prices and taxes

For users to change their behaviour and for investment in renewable energies to reach a critical level, 'price' signals sent to users and investors need clear objectives and duration. In simple terms, a price to the user of fossil fuels that remains high encourages energy saving and the search for alternative sources. For example, high prices per kWh in the residential sector is the main reason for the strong development of solar-powered water-heaters in Cyprus, Israel and the Palestinian Territories, while low summer energy prices in France slow exploitation of the solar potential, and, in Egypt, the very low price of natural gas (2.5 cents ($) per kWh in 2000) limits the development of renewable energies.[47]

Public authorities therefore play a major role (consciously or not) through *energy taxation* or the setting of prices, and they can directly influence the behaviour of the various actors through the relative final prices of the various energy forms. In order to introduce such distortions into the relative prices of different energy sources and send a price signal to users (prices that would integrate short- and long-term environmental and social objectives), public authorities may, for example, tax the most-polluting energy sources more heavily. This provides a threefold advantage:

1 it internalizes some of the negative environmental externalities in the price[48] and therefore facilitates awareness of the sustainable development objective in energy choices by giving an advantage to the least-polluting energy sources;
2 it increases the price of energy, thereby encouraging energy savings;
3 it creates possible resources for funding energy efficient actions.

There are currently surprisingly large differences in electricity (particularly residential), heating and fuel prices between countries around the Mediterranean. Although there is an attempt to harmonize energy taxation in the EU,[49] the taxation levels in the other Mediterranean countries vary considerably.

This variation is shown by the *pump price of super-grade petrol per litre* (Figure 2.14), which shows indirectly the low level of petrol taxes in the hydrocarbon-producing countries (Algeria, Libya, Syria and Egypt) and differences of 1–5 between the prices in Libya and Italy, also between Greece and Italy (1–2). Italy has some of the Mediterranean's highest prices and one of the lowest energy intensities in the world.

Figure 2.14 Pump price of super-grade petrol

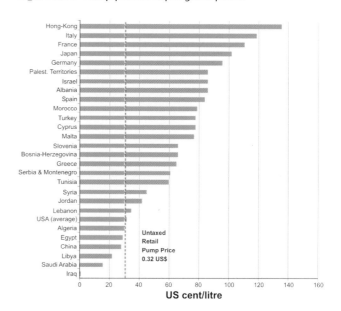

Source: GTZ, 2001

Note: $1 = 1.16 euros

The level of $0.32 per litre is considered as the average hypothetical cost per litre of super-grade petrol (including distribution costs but not taxes). Below this price, the petrol is probably subsidized; above this price, it is certainly taxed. This graph indirectly shows the taxation level of petrol in various countries. The hydrocarbon-producing countries are among those that tax petrol the least, or even subsidize it.

The price of *low-tension electricity per kWh* (especially residential) also varies widely in the countries around the Mediterranean, whereas, basically, there is almost a single price for oil, and the prospect of interconnecting networks and exchanges of energy flows should result in a similar trend for electricity prices. Table A.24 in the Statistical Annex shows the magnitude of these variations in the SEMC. The lowest price is found in Syria (0.006 euros for the first bracket), the highest in some Palestinian and Moroccan villages connected to generators (up to 0.30 euros), some 50 times higher.

However, the general level of *energy taxes* is fairly high and provides a substantial share of public income. The producer countries levy from 30 to 90 per cent of oil income. The consumer countries apply specific taxes to oil products. Taxes on automobile fuels account for more than 50 per cent of the final price in all EU-Med countries and in Turkey.[50] These taxes contribute 10 per cent of total tax income in France, and 7 per cent in Italy. They contribute significantly to national budgets, which might explain a certain internal resistance to reforms in this field. On the other hand *environmental taxes* on energy are not widely used, except in the four EU-Med countries, which increased them gradually between 1985 and 2001 (coal tax, etc.). Moreover, the costs of the impacts of fossil fuels on human health and the environment are often paid by the community or transferred to future generations, but this is far from being included in the price to the user.

Box 2.8 Energy prices and costs

Energy prices currently do not always reflect the full costs to society, because they seldom account for all the impacts of energy production and consumption on human health and the environment. Estimates of these external costs for electricity, for example, are about 1–2 per cent of EU GDP and reflect the dominance of environmentally polluting fossil fuels in energy production.

Energy subsidies between 1990 and 1995 remained focused on the support of fossil fuels and nuclear power, despite the environmental impacts and risks associated with them.

With the exception of diesel and unleaded petrol for transport, energy prices in real terms fell between 1985 and 2001. This reflected trends in international fossil fuel prices and the move towards liberalized gas and electricity markets, which stimulated greater price competition. The reductions occurred despite increases in energy taxation – other than for industrial electricity where energy taxes fell.

Source: EEA, 2002

The introduction of taxes at a sufficiently high level to act as an incentive, without, however, adding excessively to existing mandatory charges or penalizing the poorest, requires *fiscal* reforms,[51] which involve the adoption of fairly sophisticated and complex tax measures to which public authorities are often reluctant to commit themselves. Among the easiest reforms to get accepted, the introduction of tax rebates for labour following the application of an incentive tax on energy may enable a partial transfer of productivity improvements from labour towards energy. Such reforms should, on the other hand, make it possible to guarantee access to energy for the poorest and get the measures accepted. Even in the industrialized countries, a steep rise in energy prices can lead to violent social disturbances, as shown by the scale of the truckers' strikes in Europe in 2000, following too steep a rise in fuel prices. One of the keys to energy taxation that favours sustainable development is therefore to introduce tax reforms and cross-subsidy mechanisms that make *apparently* antagonistic objectives compatible. It may indeed seem contradictory to want both inexpensive energy, accessible to the poorest, and sufficiently expensive energy to encourage savings and to integrate all the externalities related to its use. However, some countries have managed to invent operational pricing systems with socially based brackets that differentiate between users, or cross-subsidy mechanisms for the poorest, that make it possible to reconcile these objectives. Cross-subsidy mechanisms have been implemented between urban and rural areas, between regions and, more generally, between tax-payers and users or even between different categories of users. Morocco and France fund part of the investment needed for rural electrification (nearly 50 per cent in Morocco) by a charge on all customers of the national electricity grid.

Poorly targeted reforms (for example social brackets that are too large, subsidies for coal in some producing countries, tax exemptions for some professional categories or some areas), considerably limit the effectiveness of environmental taxes and obscure the signal for users. Many countries maintain the price of electricity well below its actual production costs for social reasons and through public aid. This is the case in Algeria, Egypt, Lebanon and Syria where most subscribers pay less per kWh than the average cost,[52] estimated at about 0.06–0.07 euros per kWh for low voltage distribution. Financial incentives and tax systems in the energy field in Mediterranean countries often remain extremely complicated and opaque; they overlap with one another. Social objectives or national independence are often put above environmental objectives. There results a clouded price signal that is generally unfavourable to more efficient energy use and 'clean' energy.

Such measures therefore require explicit, vigorous and coherent policies if they are to be implemented in anything other than a purely symbolic way. Renouncing the use of subsidies that encourage the waste of resources, and implementing energy taxes that are well adapted to the objectives of sustainable development are essential elements in the positive evolution of energy use, and may be facilitated by the following conditions:

- Strong determination and good communication efforts. To be accepted by consumers, who are familiar with a high level of taxation, tax reforms need to be explained by raising awareness about what is at stake in the long term, and applied in a fully transparent way.
- A certain consistency in time to avoid some reductions in energy costs, related, for example, to fluctuations in world prices, which should not be passed on to the consumer and thus cloud the message.

- Using the occasion of a falling trend in energy costs resulting from productivity gains in the energy production and distribution sectors (technological progress and the opening of energy markets to competition) to introduce tax reforms that internalize the cost of the externalities and thus restore the environmental objective into energy choices. Currently, except in the transport fuel sector,[53] taxes in EU countries have not compensated for the average reduction in the price of energy to the consumer between 1985 and 2001, resulting from liberalization of the gas and electricity market. They have therefore had little effect on reducing demand[54] (see Chapter 3). Also, given that a car in 2000 uses half the petrol it did 20 years ago, a rise in petrol prices resulting from the introduction of an environmental tax is likely to be more acceptable if it is seen to be partially compensated by this technological improvement.
- Increasing the price elasticity of energy demand in the short term by developing the possibility of substitution. Economic measures (such as energy taxes) can affect company profits or household incomes without necessarily resulting in behavioural changes unless alternatives are immediately available as a result of complementary structural changes (for example town planning that enables improved mobility, better engines in cars and better public transport).

Thus economic measures are invaluable for implementing more sustainable energy policies in some circumstances, but they have their limits and must be integrated into a wider framework that combines all the other measures: regulatory (regulations in the building sector), technological, and awareness-raising, which is sometimes the most effective (for example leading to industry-wide agreements in the household appliances sector).

The need for regional cooperation

Despite national differences, the need to improve energy efficiency and develop renewable energy is common to all countries. Regional cooperation is vital for progress down this path and for helping countries that urgently need to follow less costly development paths in terms of energy and environmental impact.

Regional Mediterranean cooperation is included in the framework of the major conventions and international agreements, particularly the Framework Convention on Climate Change and the Johannesburg Plan of Action (Box 2.9). They recognise the *energy development needs* of the poorest countries and confirm the importance of *energy efficiency strategies* and the *promotion of renewable energy* for more sustainable energy development. These paths will be a priority for international cooperation and this is reinforced by the commitment of the G-8 countries in 2002 to promote renewable energy and the International Energy Charter.

> **Box 2.9** The main recommendations of the Johannesburg Plan of Action in the energy field, 2002
>
> **Access to energy**: improve access to reliable, affordable, economically viable, socially acceptable and environmentally sound energy services, including the target of halving the proportion of poor people by 2015.
>
> **Energy efficiency**: integrate energy considerations, including energy efficiency, into socio-economic programmes conducted with the support of the international community. Accelerate the development, dissemination and deployment of energy conservation technologies related to energy efficiency programmes, including the promotion of research and development.
>
> **Renewable energies**: diversify energy supply and substantially increase the share of renewable energy sources.
>
> **Energy markets**: reduce market distortions by restructuring taxation and phasing out harmful subsidies.

Funding SEMC energy development

Major efforts on access to energy are needed in many SEMC, given current delays in providing the necessary equipment and the very strong growth in demand that is predicted. The funding needs of the energy sector in the SEMC are estimated at 200 thousand million euros to 2010, 55 per cent for the electricity sector and 45 per cent for the hydrocarbon sector.[55]

The role of Official Development Aid (ODA) in this funding is getting smaller and smaller. There is a significant and continuing decrease in bilateral or multilateral public capital likely to be invested in structural projects in the Mediterranean region (for reinforcing institutions or for large infrastructure expenditure). The average amount of ODA for the energy sector in the Mediterranean countries between 1973 and 2001 was US$300 million per year,[56] which is little compared with the needs (Figure 2.15 p139).

Governments have already begun to liberalize the energy sector in order to attract direct foreign investment. The strength of the big international companies has become a major determining factor. The 'return' to Algeria of foreign companies and the intensification of their presence in the other countries is significant. In the electricity sector, for example, to satisfy the very strong growth in demand, several SEMC have changed their laws to enable investors to build, exploit, even own independent electricity production plants (IPP) through different forms of concessions. More than 21,500MW of IPP projects have been awarded to private consortia, and 47,000MW are planned by mid-2004. Large developments are underway in Turkey favouring operations that give foreign or mixed companies responsibility for installation (from building to operations) with the investors being reimbursed by payments from customers.

Clearly the specifications concerning respect for the environment in this kind of operation are vitally important given the risk of 'low cost, low environmental standards' in tenders and the difficulty of imposing environmental rules and managing them.[57] Although liberalizing the sector may in some cases contribute to modernizing networks and reducing distribution losses, the strategies of private operators may conflict with the search for energy efficiency by their client consumers.

Regional cooperation in energy efficiency and renewable energy

Mediterranean regional cooperation has an essential role to play in collective learning and the development of Mediterranean 'know-how', the importance of which has been seen in the implementation of new energy efficiency strategies and renewable energy developments. The most active countries and companies in this cooperation will, in the medium term, benefit from additional growth potential because of the savings achieved and the reductions in environmental impact. They will also have a technological head-start in sectors that are certain to develop in the run-up to the 'post-oil' era.

The original MEDENER initiative, the Mediterranean network of energy efficiency agencies, deserves mention. This network promotes the exchange of experience in the energy efficiency and renewable energy fields. The extension of such a network to the whole of the Mediterranean and the strengthening of its resources could play a very significant role in this collective learning process in a region with large renewable energy potential.

> **Box 2.10   MEDENER, a Mediterranean network for energy efficiency agencies**
>
> The objective of the MEDENER Association, founded in 1997, is to contribute to the development of partnerships among its members by promoting exchanges of experience and the sharing of know-how in the fields of rational energy use and the development of renewable energies, as well as the protection of the environment both on a local and global level in relation to the energy sector. At present, MEDENER gathers 12 energy efficiency agencies from the Mediterranean countries. Within the framework of its mission and since its founding, MEDENER has collaborated with a view to executing joint projects, in particular the current MEDA projects, 'applying solar thermal energy in the Mediterranean Basin' and 'urban environment in the Mediterranean cities'. MEDENER also organizes Mediterranean events on topics where it is active.

Cooperation could also play a fundamental role in reinforcing and supporting energy efficient agencies and the emergence of *energy efficiency* programmes in the Mediterranean countries, especially in the electricity field. Some progress has been made in the field of energy planning in the SEMC, particularly in awareness raising and training of energy teams at the local level. Pilot projects for managing electricity demand for public lighting and co-generation have also been implemented (Box 2.11).

Regional cooperation can also play a role in the harmonization of standards and coordination of several countries' activities. For example, the role of the European Commission was very important for the EU-Med countries in harmonizing methods and standards. The power of a group of 15 countries certainly contributed to voluntary agreements with industrialists (for example on refrigerators). The many European Directives (energy efficiency, labelling on household appliances, the harmonization of energy taxes, thermal regulation, renewable energies, etc.) have certainly played a very important role in the emergence of national strategies. Mediterranean cooperation could also influence this kind of harmonization.

International public aid has a vital role to play in promoting the funding needed to implement energy efficiency actions and develop renewable energy in the southern and eastern countries. Official Development Assistance (ODA) could help the poorest countries to emerge from a vicious circle: 'to develop at the lowest energy cost you must already be rich enough to be able to afford energy-saving technologies which generally involve high initial expenditure'. Given the financial stranglehold in which a number of SEMC find themselves, ODA can trigger the process by paying the immediate 'additional cost' of taking long-term requirements into account (the benefits of such costs being collective and deferred). Such aid could be conditional on the implementation of the national instruments necessary to make such actions lasting. Outside the traditional methods of public aid to development, systems are being developed for exploiting the new financial mechanisms set out by the Convention on Climate Change and to increase incentives for local saving (such as micro-credits).

But cooperation in energy efficiency and renewable energy remains minimal compared with the scale of the challenge. It consists mainly of many isolated projects, with inadequate long-term structural capability (Box 2.11).

In the field of energy, the ODA for the Mediterranean countries has been falling significantly in absolute terms since 1991 (Figure 2.15). Moreover, for the whole of the 1973–2001 period, ODA for energy efficiency and renewables was only about 10 per cent of total ODA received by the Mediterranean countries in the energy

Energy

Table 2.3 The distribution of the ODA intended for the Mediterranean countries in the energy sector, Cumulative, 1973–2001

	Amount (thousand $)	% of total
Biomass	14,346	0
Coal-fired power plants	445,410	5
Electrical transmission/distribution	2,980,980	33
Energy education/training	17,526	0
Energy generation and supply	6596	0
Energy policy and administrative management	723,171	8
Energy research	6143	0
Gas distribution	236,602	3
Gas-fired power plants	607,773	7
Geothermal energy	5873	0
Hydroelectric power plants	1,558,174	17
Nuclear power plants	39,382	0
Oil-fired power plants	92,010	1
Power generation/non-renewable sources	2,004,112	22
Power generation/renewable sources	47,696	1
Solar energy	10,183	0
Wind power	128,087	1
Total	**8,924,063**	**100%**

Source: Blue Plan from OECD-DAC data

Note: in green, actions in energy efficiency and renewables.

sector (US$947 million out of US$8,925 million). Funds remain devoted mainly to the production and distribution of electricity (Table 2.3).

Figure 2.15 ODA total amount intended for the Mediterranean countries in the energy sector, 1973–2001 (in 1,000 US$)

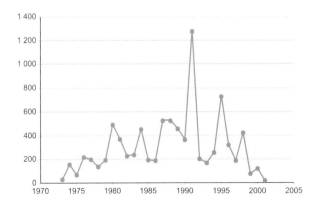

Source: Blue Plan from OECD-DAC data

Note: The cumulative total between 1973 and 2001 comes to roughly 9 thousand million dollars, or an average of 300 million dollars per year. Some of the donor countries in diminishing order: Germany, Japan, the US, Italy and France cover over 80 per cent of the ODA.

If we look at MEDA regional and national projects since 1997, we find a more significant effort on energy efficiency and renewables, about 35 per cent of the total for the energy sector. Nevertheless the amounts remain very small (seven projects totalling about 24 million euros between 1997 and 2003) when compared with those invested elsewhere in the sector. For comparison, the EIB alone made 2 thousand million euros of loans to the energy sector in the SEMC, of which only 1 per cent went to renewable energies.

Box 2.11 Some energy efficiency and renewable energy projects in the Mediterranean region

This is an indicative glimpse of a few significant cooperation projects in energy efficiency and renewable energy in the Mediterranean. Given the budgets involved and the lack of time to assess such recent projects, it is difficult to evaluate their real influence other than as a demonstration of what is possible.

The French, Spanish and German governments support such projects, but their budgets remain small. French cooperation is about support to local authorities and proposing a wide range of technical transport solutions that are adaptable to the specific circumstances of each town (for example high-capacity buses, tramways with rubber-tyred wheels, light or heavy metro systems).

The GEF has also supported projects for promoting energy efficiency in Egypt, Palestine and Syria (US$10 million in all) and developing photovoltaic energy in Morocco (US$5 million), solar heating in Tunisia (US$7.3 million with participation from Belgium Funds) and a solar power station in Egypt (US$20 million).

In Lebanon, projects in 2002 aimed at reinforcing energy efficiency in construction (FFEM, 1 million euros), UNDP/GEF on adopting thermal regulation for buildings (US$500,000), a multisector project for improving energy efficiency (UNDP/GEF, US$5 million) and a more general project for restructuring of the energy sector, particularly in terms of liberalizing the sector and supporting a reduction in prices and improved energy efficiency (EU, 2.85 million euros).

Through various programmes (regional and national MEDA, Save, Alterner, Synergy changed to Coopner, Life-Third-Countries), the European Commission supports some interesting projects but is very limited by a lack of resources:

- a pilot programme for rural electrification with photovoltaic energy in northern Morocco;
- support for national energy-efficiency agencies;
- between 2000 and 2003, one to three projects per year funded in the Mediterranean by the SYNERGY and COOPNER programmes;
- six 'Regional MEDA' projects launched in 2000 and 2001 (for example urban energy and the environment, training in energy policies, photovoltaic pumping of water). Of these, the 'application of solar energy in the Mediterranean Basin' project is an example of an integrated approach. Its aim is to spread to the southern Mediterranean the concept of Guaranteed Solar Results (GSR), used successfully in many solar installations in Europe and making it possible to contractually guarantee the technical and economic performance of solar water heaters. This project should eventually contribute to the development of local small and medium-sized industries and networks of craftsmen, following a process similar to that under way in the northern Mediterranean.

Among the projects put forward in Johannesburg, a regional initiative MEDREP (Mediterranean Renewable Energy Programme) was launched for promoting renewable energy in the Mediterranean region. The programme, bringing together many partners, is supported by the Italian Ministry of the Environment. It aims to develop a renewable energy market system through financial instruments, and strengthen policy frameworks and private sector infrastructures. Within this framework, a Mediterranean Renewable Energy Centre (MEDREC) has been set up in Tunis.

Source: Plan Bleu, European Commission

Energy efficiency projects need to overcome obstacles related to the *scattered nature* of the actions, the size of the projects and the multiplicity of financial sources which are not always well adapted to funding energy efficiency projects. Traditional donors have, till now, seldom supported the kind of project which is justified, at least partially, by factors other than financial profitability. Another problem is related to the generally small size of energy efficiency projects, that are poorly adapted to the usual funding procedures. What is needed is to group such projects together (for example 'low consumption' lamps and small-scale co-generation) so as to obtain an overall project that is likely to interest local or international banks.

Many proposals uphold the idea of creating a *support instrument* for setting up project funding. Such a public-private structure (specially created or included in an existing body) could provide aid for designing and regrouping projects. It could benefit from preferred access (rapid set-up, simplified procedures) to the right to draw from various existing public funds (according to well defined criteria) and private funds that could be implemented within the framework of a renewed partnership. The potential of the flexibility measures set out in the Kyoto Protocol, for example, could be exploited.

The Kyoto Protocol provides an opportunity to build and test new forms of mutually beneficial partnerships on the Mediterranean scale through the *Clean Development Mechanism* (CDM). This mechanism enables the so-called Annex I countries, committed to stabilizing or reducing their GHG emissions, to acquire emission credits by contributing, at least-cost, to reducing the emissions of developing countries that are not committed to the Protocol. Although the detailed methods of implementing clean development mechanisms are complex[58] and not yet finalized, this is an area where cooperation between the Mediterranean countries should be developed to help promote energy efficiency and renewable energies. The benefits are mutual, that is by funding such projects in the SEMC, for example, the European countries could acquire (at least-cost because of their lower energy intensity) emission rights while contributing to the development of their southern shore neighbours and the protection of the Mediterranean eco-region. In return, the SEMC would benefit from technological transfers, relax their immediate financial constraints and reduce their medium-term energy and environmental costs.

Most of the SEMC have identified projects for reducing greenhouse gas emissions that might be implemented within the framework of the CDM. Most of these are renewable energy and energy efficiency projects, and the CDM could cover the additional costs of the projects compared with coal- or gas-burning plants. But given the scattered nature of energy efficiency projects, the CDM is particularly appropriate for the development of renewable energies (more particularly wind power). One study[59] has shown that in terms of the costs of GHG emission reduction, the EU would be well advised to develop CDM-renewable energy projects in the Mediterranean, since it would allow it to reduce its costs significantly by substituting CDM interventions for expensive domestic actions in the Mediterranean. Up to 2030 the savings that might be made through investment in CDM-renewable energy projects in the SEMC would be about 1.8 thousand million euros for a total investment of about 1.2 thousand million euros (1999). The additional installed renewable energy capacity would be about 18,000MW by 2030, corresponding to approximately 11 per cent of the electricity produced from renewable energy sources.

To facilitate the implementation of the CDM potential in the Mediterranean and thus to prevent the CDM from mainly favouring the large carbon-producing countries (China, India and Brazil, which have greater greenhouse gas-saving potential than the SEMC), a *Mediterranean Carbon Fund* (MCF), uniting all CDM initiatives and projects, could be created. This fund could be managed by the above-mentioned regional structure, which would organize and facilitate the bringing-together of initiatives in the SEMC in exchange for carbon emission credits.

However, the CDM mechanism, despite its recognized effectiveness, is likely to resolve only a small part of the energy problems encountered by the Mediterranean countries. It will apply mainly to the development of wind power. It may have no impact on other major areas (heavy transport infrastructures, urban planning and housing), which will govern energy demand in the medium and long term (with the number of dwellings expected to double between now and 2010 and an increase in passenger and freight transport of more than 5 per cent per year in the SEMC). *More international aid targeted on energy efficiency and renewable energy* could catalyse the emergence of more sustainable national energy strategies.

The EU's role in Mediterranean energy cooperation

Within international cooperation, *Euro-Mediterranean cooperation* plays a particular role given the complementarity between the two shores: the significant potential for renewable energy is in the southern-rim countries while the northern-rim countries possess most of the technology. For Europe the security of its supplies also depends on its Mediterranean neighbours (considerable gas and oil reserves in the southern-rim Mediterranean countries, the security of transportation of hydrocarbons from the Gulf, southern Russia and the Caspian Sea). Least-cost environmental and social economic development of its southern shore is also a fundamental challenge for Europe. For the hydrocarbon-producing SEMC, Europe represents a vital outlet (they export 50 per cent of their oil and 90 per cent of their gas to other Mediterranean countries). For the other SEMC, the possibility of exporting renewable energy sources may not be insignificant in market terms ('green' certificates).

It is therefore natural that energy figured among the six priorities of the Euro-Mediterranean Partnership launched in 1995 in Barcelona, which stressed its structural importance (particularly alongside transport, water and the environment). Since then, there have been several Euro-Med forum where 27 energy ministers have met to promote dialogue on energy questions and orient regional cooperation. Although the developmental objective of energy efficiency and renewable energy policies is specifically mentioned in the guidelines for this cooperation, it nevertheless seems that most of the efforts and thinking have focused on *liberalizing* the energy sector in the SEMC (institutional and regulatory framework reforms, restructuring the energy industry, privatization and the introduction of competition), on modernizing infrastructures, and on pursuing interconnectivity. Here again, the process seems focused mainly on energy supply without considering the possible conflicts with developing the alternatives of energy efficiency and renewable energy. There is no doubt that restructuring supplies can contribute indirectly to improving energy efficiency, but it may also produce price distortions unfavourable to managing energy demand.

This new and original Euro-Mediterranean dialogue building on energy is an opportunity to pursue the energy efficiency and renewable energy objectives more vigorously by readjusting budgets in their favour (national and regional MEDA). But to do that, the projects should be undertaken by the countries themselves and included in coherent national strategies (the importance of which was discussed above) that guarantee the sustainability of actions, through the implementation of the regulatory, institutional and financial mechanisms needed for pursuing them, for longer than just the duration of projects. The *Euro-Mediterranean partnership* (particularly the bilateral actions) could well play an even more significant role for the emergence of policies more directed towards energy efficiency and renewable energies. In this way, the EU would be faithful to its Kyoto commitments and its strategies in the eastern European countries would be more clearly directed at improving energy efficiency. From this point of view, the path from the baseline scenario to an alternative proactive scenario as regards energy efficiency and renewable energy, might look something like the following:

- Providing aid to countries to develop national energy efficiency and renewable energy strategies (for example institutional support for an energy efficiency

2 A Sustainable Future for the Mediterranean

Figure 2.16 Map of the Euro-Mediterranean energy trade (gas, oil, electricity), 2000

Oil trade

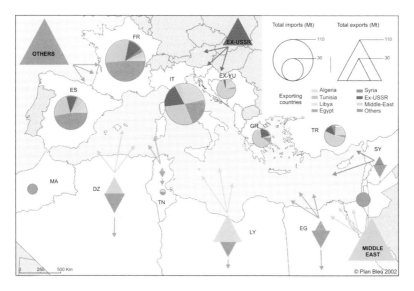

Electric trade

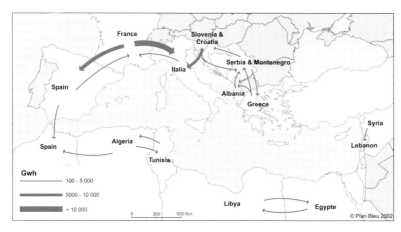

Gas trade

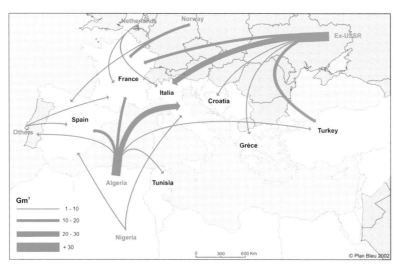

Source: data from OME, 2002

plan) and for using the experiences of other Mediterranean countries (Tunisia, Morocco, Egypt, Italy, France and Spain, etc.).
- Setting energy efficiency objectives in all projects funded by the EU (indicators) and making more use of pre-feasibility studies in the initial stages of energy investment projects, in order to establish whether some of the proposed infrastructures could be avoided or deferred by less expensive measures of energy efficiency or renewable energy.
- Exchanging experience and training in this specific area (agency networking, training courses), particularly with more demonstration projects.
- Setting up financial systems adapted to the characteristics of the projects (small size, many actors); ODA could facilitate the provision of funds for financing the immediate 'additional costs' of energy efficiency and renewable energy projects.
- Setting a minimum quota for funding in the energy efficiency/renewable energy field, which might be a way of guaranteeing that priority is transformed into practice and not systematically put below liberalizing markets and interconnecting projects.
- Exploration of the potential for 'green' exchanges of energy between Mediterranean countries.

The challenge: An alternative scenario to 2025

To sum up, potentially, there is a path to significantly improving the prospects for energy efficiency and renewable energy and, fortunately, this is already being followed by some Mediterranean countries. But it calls for an all-out effort by all actors, in particular public authorities, to:

- develop national energy efficiency and renewable energy strategies that have quantified objectives and include all the sectors and actors concerned;
- reinforce financial resources earmarked for energy efficiency agencies and the funding of investments;
- implement favourable institutional and regulatory frameworks;
- gradually reform energy taxation and implement innovative cross-subsidy mechanisms;
- reinforce regional cooperation (exchange of experiences, technology transfers, funding mechanisms).

The benefits of such an orientation can be illustrated quantitatively by means of an alternative scenario, proposed below, which would differ from the baseline scenario through:

- *exploitation of potential energy savings* of approximately 20–25 per cent (depending on the country) of total energy demand, a potential that it should be possible to exploit over 25 years, given currently available technology;
- *faster development of renewable energy sources*, with renewable energy sources[60] providing 14 per cent of primary energy (11 per cent would be other than large hydropower) and 40 per cent of electricity production (24 per cent if large hydropower is excluded) by 2025, compared with 4 per cent of primary energy and 21 per cent of electricity production (8 per cent if large hydropower is excluded) in the baseline scenario.

These objectives have been differentiated by country, depending on their current energy structure and their energy potential, and quantified by the OME. The findings are shown by country groups in the Statistical Annex (Tables A.19 and A.21). Compared with the baseline scenario reduction of an average of 0.7 per cent per year from now to 2025, *energy intensity* in the Mediterranean countries in the alternative scenario would fall roughly twice as rapidly (–1.3 per cent per year).[61]

The total feasible *energy savings* in this scenario compared with the baseline scenario could reach 208Mtoe per year by 2025 in the entire region, about half of the projected growth in demand between 2000 and 2025. Approximately 60 per cent of this potential is in the SEMC and 40 per cent in the NMC. Figure 2.17 compares the energy supply structure of the two scenarios in 2025. In the alternative scenario, the share of oil is 34 per cent (40 per cent in the baseline scenario), which, with a reduced total demand, would lead to a *stabilization of oil demand* in 2025 at the 2000 level, compared with the baseline scenario projection of a 40 per cent increase in demand (150Mtoe) between 2000 and 2025. It also projects a saving of 92Mtoe in the demand for natural gas, compared with the baseline scenario. This is equivalent to half the present demand and would reduce the need to import and transport hydrocarbons, with its related environmental risks.

The alternative scenario would reduce the average *dependency index* of the Mediterranean countries by 20 percentage points compared with the baseline scenario in 2025. Generally speaking it would go from 34 per cent for the whole basin to 18 per cent between 2000 and 2025, compared with 38 per cent in the baseline scenario in 2025.

It would also enable very substantial *financial savings*. Assuming a linear exploitation of the savings potential over the next 25 years, the cumulative primary energy saving in the alternative scenario over the period would be around 2600Mtoe[62] for all Mediterranean countries, or the equivalent of US$455 thousand million[63] (or US$18 thousand million per year on average over the period). Of course this figure is very approximate, and the costs of implementing energy-demand management policies

2 A Sustainable Future for the Mediterranean

Figure 2.17 Feasible energy savings with the alternative scenario, 2025

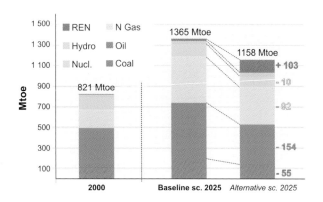

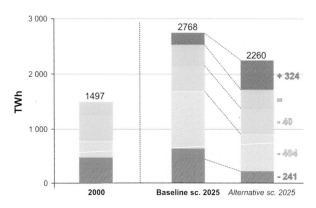

Source: Plan Bleu and OME data

Note: MED – all Mediterranean countries together

would have to be deducted, but these are generally much smaller. By reducing the costs of energy services, this scenario reveals a considerable potential for economic growth.

The impact on *employment* is harder to assess. However, to give an order of magnitude, Morocco estimates the number of jobs likely to be created at 11,500 between now and 2011 as a result of increasing the share of renewable energy in total energy supply from 0.2 to 10 per cent.

The *environmental* impact in such a scenario would also be considerably smaller than in the baseline scenario. Gaseous emissions in 2025 would be 25 per cent lower than in the baseline scenario, both in the north and the south, or 858Mt of CO_2 emissions avoided by 2025. This corresponds to 45 per cent of current emissions. The share of the Mediterranean countries in the world's CO_2 emissions by 2025 would be less than 7 per cent (9 per cent in the baseline scenario) and would enable the four EU-Med countries to get closer to their Kyoto Protocol targets. This would also contribute to an improvement in urban air quality. Many energy supply infrastructures could be avoided or deferred, and their impact and related environmental risks would be reduced accordingly. Thus, in the alternative scenario, which is optimistic but not utopian, 154 power plants[64] could be avoided by 2025 compared with 400 additional power stations to be built in the baseline scenario.

Figure 2.18 Total CO_2 emissions according to the two scenarios

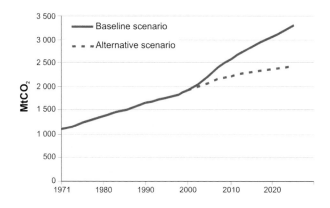

Source: OME, 2002

Obviously, these orders of magnitude have no predictive value. They simply illustrate quantitatively the considerable potential benefits of a possible alternative strategy that also reduces vulnerability to several geo-political, socio-economic and environmental risks. This alternative scenario clearly shows the need to act as early as possible. Given the strong inertia of energy systems and the irreversibility of some infrastructures, today's choices are critical. Other, even more proactive scenarios could be considered. They would question more profoundly some highly energy-consuming development and urbanization patterns of the Mediterranean countries.

* * *

Energy supply in the Mediterranean countries over the next 25 years presents a real *challenge* in terms of environmental, socio-economic and geo-political risks. To meet this challenge and accelerate the energy transition towards more sustainable supply modes, there is room for manoeuvre, which is already being explored by some Mediterranean countries, through strong national public policies that promote a fragile combination between managing demand and a more diversified, less polluting and less expensive energy supply.

Progress with *supply* has been achieved through the progressive development of gas, which is gradually replacing oil. However, the transition towards a larger share of renewable energies could be accelerated by more active public policies. Only the renewable energies can minimize the risks of growing energy dependence and the impacts on global warming without presenting new technological risks. Their potential in the Mediterranean is enormous and underexploited, although it could significantly (and at least-cost in many cases) improve the living conditions of millions of people, particularly in rural areas. The countries most advanced in this transition are investing in the future and will be winners in the mid-term from all points of view, economic, social and environmental, while also benefiting from technological progress.

However, such efforts on the supply side could be in vain if they are not accompanied by a search for energy efficiency on the *demand* side, which, again, will in the medium term be of most benefit to the countries that are most advanced in this field. The potential for savings is considerable. An alternative scenario has enabled us to illustrate some orders of magnitude of the possible savings for the whole Mediterranean Basin: nearly half of the additional energy demand and CO_2 emissions projected for 2000 to 2025 in the baseline scenario could be avoided. These figures are obviously very approximate but indicate the scale of the benefits available, which would open the way to a more sustainable energy future even without exploiting the major technological innovations that will be necessary between now and 2025. However, despite some progress, the considerable benefits of energy efficient strategies at all levels are not yet fully appreciated. Actions do not match the challenges and encounter obstacles that are more institutional and cultural than technical. Such strategies call for perseverance and determination but also an exploitation of knowledge.

In addition to purely national policies, *Mediterranean cooperation* has a major role to play. Faced with such challenges, not all Mediterranean countries are at the same level of awareness and action. Without a complementarity, well understood by both the most advanced and the less advanced countries, the technological and institutional gaps could increase, in the energy field as well as elsewhere, and efforts could become too thinly spread. Cooperation in the Mediterranean in terms of energy efficiency and renewable energy offers real possibilities for synergies with mutual benefits on the way to more sustainable development in the region.

Notes

1. Defined as the ratio between a country's total primary commercial energy consumption and its GDP.
2. IEA, *World Energy Outlook*, 2000.
3. In 1998, the US consumed 7930koe (*World Development Indicators*).
4. Measured directly at the end users.
5. This contribution may be more significant depending on the reality of applying the European Directive for promoting renewable energies.
6. Turkey alone is planning a considerable increase in its coal production: 40Mtoe between 2000 and 2025 for electricity production, mainly low-grade lignite.
7. In the SEMC the index drops from –57 per cent to 10 per cent between 2000 and 2025. In producing countries this index increases.
8. IEA, *World Energy Outlook*, 2001 (before the Iraq war). At the beginning of 2005, IEA prepared a new report that integrated the considerable rise in oil prices that has occurred since 2003.
9. European Commission (2000) *Livre vert*, p38.
10. IEA, *World Energy Outlook*, 2002.
11. From a 2004 survey (data communicated by the Syrian Ministry of Local Administration and Environment).
12. Lahbabi et al, 1996.
13. Max Planck Society, Smog over the Mediterranean, *News Release*, 25 October 2002.
14. Other greenhouse gases (under 20 per cent) emitted by the energy sector consist mainly of CH_4, from gas pipeline leaks or coal mines.
15. Also less NO_x and almost no SO_x.
16. Estimates based on emission coefficient averages used by the International Energy Agency, or 3.9 tonnes of CO_2 emitted per toe of burned coal; 2.8t CO_2/toe of oil and 2.3t CO_2/toe gas, (*World Energy Outlook*: 235; 354–357).
17. Or about 520Mt C.
18. In this regard flexibility tools such as the Clean Development Mechanism (described below) will have an important role to play.
19. In France they represent about half of the surface area occupied by industry and less than 0.13% of the total
20. Estimates on the basis of the power stations' average operating time being 6600 hours per year.
21. Grenon M., *Plan Bleu*, 1993.
22. The electricity intensity, defined as the ratio of electricity consumed to the GDP, has risen even more strongly than the energy intensity in most countries. In the NMC it has gone from 0.19kWh per $GDP95 in 1971 to 0.23kWh per $GDP95 in 2000, and in the SEMC it has nearly tripled over the same period, going from 0.23 to 0.60kWh per $GDP95.
23. EEA, Environmental signals 2001
24. Lahbabi et al, 1996.
25. Save Programme of the European Commission, project SA/263*98/Fr.
26. Enertech, 1998.
27. The alternative scenario, developed below, assumes exploiting about 80% of this potential by 2025.
28. Montgolfier, 2002.
29. Study carried out by the ESHA (European Small Hydraulic Association).
30. National Strategic Plan for the Development of Renewable Energies – CDER – October, 2001.
31. Information communicated by ADEME.
32. MED2010 Project, co-funded by the European Commission, coordinated by the Mediterranean Energy Observatory.
33. Including 2,970,000m² in Greece, 660,000m² in France, 440,000m² in Spain and 400,000m² in Italy – Source: information communicated by ADEME.
34. The 'Applications of Thermal Solar Energy in the Mediterranean Basin' project, the European Commission/ ADEME.
35. EEA, 2003.
36. ADEME's renewable energy budget was first increased to about 45M euros per year in 1999, then to 78M euros in 2002, 15M euros of which went on R&D.
37. IEA, OCDE, 2001, Turkey 2001 Review.
38. For example, the EU quantifies investments required for doubling the share of renewables in the EU's energy consumption between 1997 and 2010 at 165 thousand million euros.
39. Grenon M., *Plan Bleu*, 1993.
40. Charpin J-M., Dessus B., Pellat R., 2000.

41 This is especially true in the countries where the price of energy is high; targeted communications should then be enough for the development of the market for well-functioning machines.
42 Lahbabi et al, 1996.
43 In 1998 the US devoted 33 per cent of their R&D budget to renewables compared with less than 12 per cent to nuclear.
44 IEA, 2002.
45 FIDEME (French Investment Fund for the Environment and Energy Resources) is an own-funds financial action fund. The action is done through the intermediary of a joint venture capital mutual fund, one-third of which is met by ADEME and two-thirds by banks. In 2002 the fund was worth 45M euros, aiming through a leverage effect at triggering investments of around 300M euros in the energy efficiency and renewables sectors (the needs of which between now and 2005 are put at 850M euros).
46 FOGIME (French Guarantee Fund for Energy Resource Investment), created by an initiative of the SME Development Bank and ADEME, later joined by EDF and Charbonnages de France, is intended to promote investments made by small and medium-sized companies for energy efficiency or renewable energies, by backing loans contracted by them from banks up to a 70 per cent ceiling.
47 Cornut B., ADEME, 2001.
48 Or, in the opposite direction, enhance the environmental benefits of certain energy sources (for example a tonne of CO_2 avoided).
49 A European Directive is trying to reduce the spread in excise duties of fuels by setting a minimum rate of taxation.
50 IEA, Energy prices & taxes, 2001a.
51 This is the budgetary neutrality principle that generally goes along with the option of a tax incentive.
52 Including all production, transport and distribution costs.
53 Where, in any case, price elasticity is low, substitution being difficult.
54 EEA, Energy and Environment in the European Union, 2002.
55 The projects concern oil and gas production, oil and gas pipelines, LNG refineries and factories, electricity power stations, electricity transport and interconnections, the distribution of electricity and natural gas and renewable energy projects (OME).
56 i.e. around 1.7 per cent of the total ODA granted during that period.
57 Chatelus M., Plan Bleu, 2000.
58 Because what is required are procedures and institutions capable of observing the savings effectively made in the countries that receive the investment.
59 A research project supported by the European Commission, coordinated by the OME, which analysed the potential of the CDM renewable-energy projects in the SEMC with the POLES model to 2030.
60 Hydro+REs excluding bio-mass (see Statistical Annex, Tables A.19 and A.21).
61 Decrease of 1.7 per cent per year in the NMC and 1.4 per cent per year in the SEMC (Table A.18 in Statistical Annex).
62 208Mtoe × 25:2 = 2600Mtoe.
63 At the average price of US$175 per toe, corresponding to US$25 per barrel of oil.
64 Of 500MW, operating an average of 6600h per year.

References

Charpin, J.-M., Dessus, B., Pellat, R.; Commissariat général du Plan. (2000) Étude économique prospective de la filière électrique nucléaire. Rapport au Premier Ministre. Paris: La Documentation française.

Chatelus, M., Plan Bleu. (2000) Libre échange et environnement dans le contexte euro-méditerranéen: Volet industrie.In Plan Bleu. Free Trade and the Environmentin the Euro-Mediterranean Context. Montpellier/Mèze, 5–8 October 2000, vol II. Regional and International Studies (MAP Technical Reports Series no 137, Athens, 2002

Commission des Communautés Européennes. (2000) Livre vert: vers une stratégie européenne de sécurité d'approvisionnement énergétique. Bruxelles: CCE

Commission Européenne. DG Énergie. (1999) Énergie pour l'avenir: les sources d'énergie renouvelables. Livre blanc établissant une stratégie et un plan d'action communautaires. Bruxelles: Commission Européenne

Commission européenne-DG Energie, Bruxelles; Agence Régionale de l'Énergie Provence-Alpes-Côte d'Azur (ARENE), Marseilles ; Institut Català d'Energia (ICAEN), Barcelona; Punto Energia, Milan. (1999) L'énergie dans la programmation des bâtiments en région méditerranéenne.

Cornut, B.; ADEME. (2001) Les tarifs des énergies dans la région méditerranéenne et leurs impacts sur le développement des énergies renouvelables. Forum pour le développement des énergies renouvelables dans la région méditerranéenne, Marrakech, May 2001

Dessus, B. (1999) Énergie, un défi planétaire. Paris: Belin

EEA. (2001) Environmental Signals. Luxemburg: OPOCE

EEA. (2002) Energy and Environment in the European Union. Luxemburg: OPOCE

EEA. (2003) Europe's Environment: The Third Assessment. Luxemburg: OPOCE

ENEA; CNEL. (2002) Rapporto ENEA sullo stato di attuazione del Patto per l'Energia e l'Ambiente 2001. Roma: ENEA

Enertech. (1998) Le projet Eureco. Etat des lieux et evaluation des gisements d'économie potentiels des usages spécifiques de l'électricité dans 400 logements de la Communauté européenne. Commission Européene, Programme SAVE, Bruxelles

European Commission, DG Research (2003) World Energy, Technology and Climate Policy Outlook 2030 – WETO. Luxemburg: OPOCE,

Grenon, M. et al (1993) Énergie et environnement en Méditerranée. Enjeux et prospective. Les Fascicules du Plan Bleu, no 7, Paris: Economica

GTZ. (2001) Fuel Prices and Vehicle Taxation. Eschborn: GTZ, www.zietlow.com.

IEA. World Energy Outlook. (1999, 2000, 2001, 2002) Paris: IEA

IEA. (2001a) Energy Prices and Taxes. Quarterly statistics. Second quarter 2001. Paris: OCDE; IEA

IEA. (2001b) CO_2 Emissions from Fuel Combustion 1971–1999. Paris: OCDE; IEA

IEA. (2001c) Energy Balances of Non-OECD Countries, Paris: IEA

IEA, OCDE. (2001) Energy Policies of IEA Countries: Turkey 2001 Review. Paris: OCDE.

IEA. (2002a) Bilans énergétiques des pays OCDE et non- OCDE. Paris: IEA

IEA. (2002b) Renewables Information. Paris: OCDE, IEA

IFEN. (2002) L'environnement en France. Paris: La Découverte

Lahbabi, A.; Maroc. Ministère de l'Environnement. Direction de l'Observation, des Études et de la Coordination; PNUD. (1996) Plan d'Action National pour l'Environnement. Énergie et environnement. Note technique. Version provisoire. Rabat

Maîtrise de l'Énergie et développement durable. (2002) Les Cahiers de Global Chance, no 16. Suresnes: Global Chance

MedEnergie, la Revue Méditerranéenne de l'Énergie (2001, 2002) Alger.

Montgolfier, J. de. (2002) Les espaces boisés méditerranéens. Les Fascicules du Plan Bleu, no 12, Paris: Economica

OME. (2002) Energie en Méditerranée. Rapport pour le Plan Bleu, May 2002

Pauwels, J.-P. (1997) Géopolitique de l'approvisionnement énergétique de l'Union européenne au XXIème siècle. vol II. Bruxelles: Bruylant

World Energy Council. (2001) Survey of Energy Resources. 19th ed. London: World Energy Council

World Energy Council, IIASA. (1998) Global Energy Perspectives. London; Luxemburg: World Energy Council; IIASA

Chapter 3

Transport

The Mediterranean has always been an area of trade and high mobility. Although transport can provide extraordinary freedom (through the private car) and economic development,[1] it is also, in addition to the continuing toll of accidents, the cause of ever-increasing pressures on the environment through land being taken up by infrastructures, air quality being degraded by emissions of pollutants into the atmosphere, noise nuisance and discharge of pollutants into the sea.

This fundamental contradiction has a strong resonance because of the region's particular characteristics. Narrow coastal zones are criss-crossed by infrastructures that make them completely artificial; cities, poorly adapted to the car, lose their quality of life; mountain passes and valleys become corridors for lorries, and the environmental integrity and attractiveness to tourists of the sea itself is threatened by accidental or intentional pollution by regional and global maritime traffic. Thus the very identity of the Mediterranean is directly threatened by the development of transport.

While other regions of the world are beginning to take notice of this contradiction (in particular the European Union, the development of transport in the southern and eastern Mediterranean, particularly that resulting from the liberalization of trade, is proceeding with no effective implementation of vital regulatory environmental and social measures.

The *Blue Plan*'s baseline scenario to 2025, developed below, shows a growth in transport by land and sea that is strongly governed by the needs of trade and tourism. Above all, it projects significant impacts on the land and the environment despite substantial technological progress and an embryonic partnership between north and south.

Such developments are far from inevitable. Indeed, the area could develop along the lines of an alternative scenario, with an increase in traffic slightly below that in the baseline scenario, with substantial benefits in terms of environmental and social impacts. To move to such a scenario, people will have to adopt a more integrated transport system that combines:

- a rationalization of transport taxes and subsidies in environmental terms and the promotion of alternatives to road for land transport;
- the effective implementation of international maritime safety conventions and a genuine regional partnership based in particular on the concept of a multi-modal Euro-Mediterranean corridor.

This chapter aims to warn on the risks associated with the baseline scenario, give an order of magnitude of the benefits of, and explore ways of achieving, the alternative scenario.

A boom in transport benefiting the road sector

A first approach might be to examine the issue of transport using traditional analytical indicators of transport development (such as increase in car ownership) and studies of traffic flows. But from the point of view of environmental impacts and economic analysis, the key requirement is to study what modes of transport (plane, road, sea, rail) are being used. Moreover, the logic of modal split depends markedly on the type of traffic considered, for example domestic, international, or intercontinental sea traffic. Indeed, the intensity of traffic flows, in terms of frequency and volumes, has a direct influence on each mode. Above all there are major differences depending on whether one is considering passenger or freight transport. The following paragraphs will therefore be based on a geographic breakdown of both passenger and freight transport. The data for 1970 to 2000 show:

- an increase in transport that is much greater than that in the economy or population, largely dominated by road for land transport and transit traffic for maritime transport;
- a concentration of traffic at some points (road traffic through the Alps and the Pyrenees and transfer hubs, particularly for airplanes and ships).

Passenger transport

Analysis of the development of all modes of passenger transport requires caution, since the data submitted by countries often confuse international and national traffic. Moreover, they are not continuous, particularly for road transport. Passenger transport data for air and rail are available for 1970 to 2000 from all countries. However, road transport data between 1970 and 2000 have had to be estimated by *Plan Bleu* for some countries, for which only the size of the vehicle fleet is known.

Despite these data problems, the link between economic activity and the level of traffic is clear: the richer the country, the more passenger traffic there is. Traffic in the four Mediterranean countries[2] of the EU (France, Italy, Spain and Greece designated 'four EU-Med') in 2000 accounted for 67 per cent of all passenger traffic in the Mediterranean Basin. Moreover, the development of transport reflects the political and economic events that have marked the Basin. Wars resulted in a sharp drop in traffic in the eastern Adriatic and Lebanon. On the other hand, the boom in passenger traffic in Turkey appears to be a consequence of strong economic growth.

Generally speaking, the annual increase in passenger traffic between 1970 and 2000, for all modes except sea, was much larger (4.9 per cent per year) than that of population (1.4 per cent) and the economy (2.9 per cent),

Figure 3.1 Passenger traffic (air, rail and road), 1970–2000

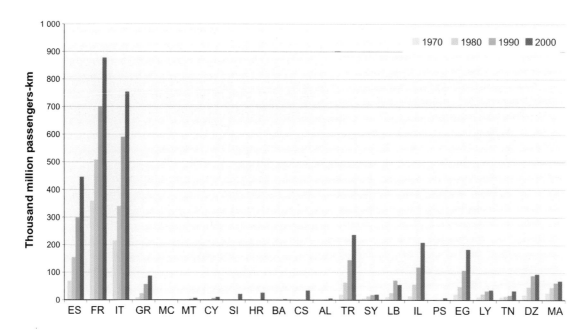

Source: National Statistical Institutes, IRF, 1999; UIC, 2002; OACI, 2001

Figure 3.2 The car fleet in the Mediterranean, 2002

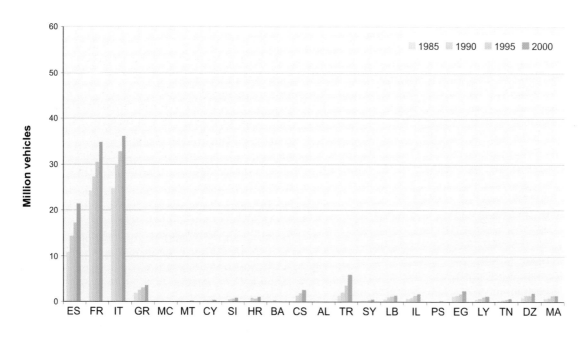

Source: MOTORSAT database 2002 (http://perso.club-internet.fr/motorsat), IRF, 1999

in both the northern Mediterranean countries (NMC) (0.5 per cent for population, 2.7 per cent for the economy and 4.4 per cent for traffic) and the southern and eastern Mediterranean countries (SEMC) (2.4 per cent for population, 4.6 per cent for the economy and 6.6 per cent for traffic) (Figure 3.1).

Although most of the data do not distinguish between domestic and international passenger traffic, it is important to do so since they use different transport modes. To enable this, the relationship between car fleets and the resulting traffic has been calculated for the Mediterranean countries where data are available and extrapolated to countries with similar geography and economies where only car fleet data are known (Serbian data were used for the eastern Adriatic countries, French values for Monaco, Greek for Cyprus, etc.).

Domestic passenger traffic: The dominance of road

The general increase in *car ownership* between 1984 and 2000 (4.1 per cent per year overall, 3.8 per cent in the NMC and 4.5 per cent in the SEMC) has resulted in an increase in car use to the detriment of rail (Figure 3.2). The increase in car ownership was 32 per cent greater than *Plan Bleu's* 1985 projection for the European Union, 18 per cent less for the eastern Adriatic, Turkey, Malta and Cyprus, 172 per cent less for Maghreb and 166 per cent less for the Mashrek.

In countries where the development of *car traffic*[3] is regularly recorded, the data confirm these trends (3.7 per cent per year increase in road passenger traffic between 1984 and 2000, 2.7 per cent per year for rail). This preference for the car is shown by the way the modal split is dominated by road (Figure 3.3).

Figure 3.3 The modal split of domestic passenger traffic in the Mediterranean countries, 1999

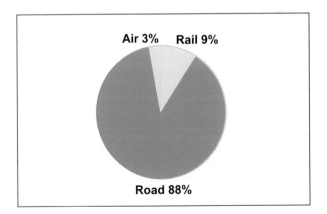

Source: MOTORSAT database 2002 (http://perso.club-internet.fr/motorsat); UIC, 2002; OACI, 2001

Note: In this diagram all rail traffic was considered to be domestic; and 20 per cent of road traffic in Spain, France, Italy and Monaco, and 10 per cent of road traffic for the other countries was international.

Despite the dominance of road, *rail* continues to play a role, mainly in Egypt (47.3 per cent of domestic trips in 1999) and in the eastern Adriatic countries (23 per cent in the same year). *Maritime traffic* also plays a significant role in Italy, Greece and Spain (where more than 50 per cent of passenger trips by sea in 2000 were domestic). *Air traffic*, on the other hand, plays a significant role only in France and Spain (5 per cent in 1999).

There are many reasons for these developments, but essentially they result from:

- demographic and societal changes: household sizes are decreasing, car ownership is increasing, resulting automatically in a huge rise in transport by car;
- urban sprawl: cities are expanding considerably, increasing the distances between daily functions (home, work, leisure, etc.). In most of the big cities, it now takes an average of half an hour to go from home to work (see Chapter 4).[4]

International passenger traffic

There are no up-to-date statistics on origins and destinations for sea and road travel. The data given in this section therefore relate to trips by country and include all origins and destinations.

The driving force behind these trips is tourism and migration (see Part 1). The traffic is therefore mainly by air, with sea in second place.

Air traffic

The increase in air traffic remains high (7.3 per cent per year between 1984 and 2000), although this figure is lower than the comparable increase in the EU.

The link between air travel, tourism and emigration/immigration is clear. In 2001 more than 102 million passengers, arrivals and departures combined, passed through the airports of the non-EU Mediterranean countries (Figure 3.4). This was highly concentrated, with the main airports of only seven countries handling 75 per cent of this traffic. Nearly all flights were for tourism and emigration or immigration.

The liberalization and deregulation of air transport explains the fall in prices and the strong growth in this sector. But in the Mediterranean it has also contributed significantly to the development of mass tourism and the equipping of large airports at tourist destinations (including Palma, Antalya, Malta, Cyprus and Djerba). It also favours a concentration of traffic at 'hub' airports (Palma, Nice and Malta, for example).

This concentration has two essential consequences:

- Long-term planning of airport investment is more difficult, since traffic at a hub is not determined exclusively by local factors. It basically depends on the shape of company networks and often bears no

2 A Sustainable Future for the Mediterranean

Figure 3.4 Air passenger traffic between the SEMC and the EU, main regular connections, 2001

(Map showing air routes between European cities — Amsterdam, London, Paris, Dusseldorf, Francfort, Stuttgart, Munich, Hanover — and SEMC cities — Algiers, Tunis, Athens, Istanbul, Antalya, Larnaca, Tel-Aviv. Legend: Number of passengers (in thousand): >600, 500–600, 400–500, <400. © Plan Bleu 2003)

Source: Airport International Council (2002) *World Report*, May–June

Box 3.1 The hub system

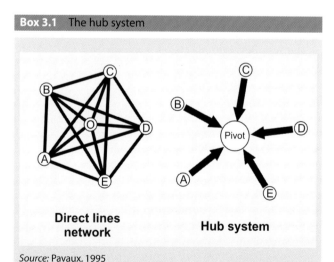

Direct lines network — **Hub system**

Source: Pavaux, 1995

Compared with direct-line networks, control of a hub (or star-like network) considerably reinforces the power of the carrier that has set up there, in the sense that it is very difficult for a potential competitor to penetrate a hub network either in a radial direction or on a 'crossover' route joining two peripheral cities (for example A and C in the above figure).

relationship to the transport generated by the region served by the airport. This results in the building and development of airports with no relation to national planning for the area.
- There is increased congestion at many hubs. The concentration of traffic by one or two carriers at a hub tends to increase traffic at the busiest times, since rush-hour traffic then includes a dominant proportion of transit passengers, with a considerable increase in the frequency and number of flights and the associated noise and pollution.

Maritime traffic

After a phase of relative stagnation in the 1980s, maritime traffic in the Mediterranean countries of the EU (except France) saw a clear increase in the early 1990s. In the SEMC, where there is little travel by sea (slightly under 5 per cent of the traffic in the NMC in 2000), growth has nevertheless been spectacular (5.5 per cent per year between 1985 and 2000 compared with 1.3 per cent per year in the NMC) (Figure 3.5).

Generally, in the entire region, the growth in passenger traffic by sea remains well below that in GDP (1.5 per cent per year compared with 2.7 per cent for GDP).

Figure 3.5 Port passenger traffic, 1985–2000

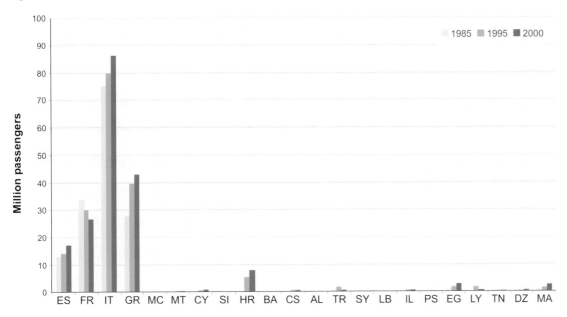

Source: Dobler (EcoMar) (2002) ; Eurostat New Cronos (2002) (http://epp.eurostat.cec.eu.int); Spécial trafic ports du monde (1997)

Note: The data are not expressed in passenger-km since there is no origin-destination matrix for maritime routes. Transit traffic is not taken into account because of lack of data. Moreover, for the sake of comparison with land-traffic data, national statistics are factored in, thus including non-Mediterranean port traffic in France, Morocco, Spain, Turkey and Egypt.

Figure 3.6 Freight traffic (air, road and rail), 1970–2000

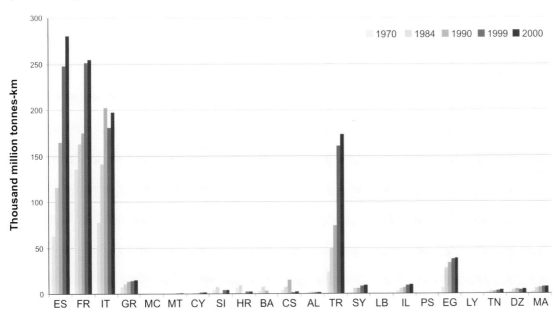

Source: National Statistical Institutes, IRF, 1999; UIC, 2002; OACI, 2001

Note: Road traffic has been assessed on the same bases as passenger road traffic.

Box 3.2 High speed vessels

Data for some Mediterranean countries (including Malta, Croatia, Lebanon, Albania and Israel) show that passenger transport by sea between 1984 and 2000 grew nearly as fast as that by road (by 9.5 per cent per year in the above-mentioned countries with 11.5 per cent for road). A good proportion of this vitality in coastal transport is attributed to the High Speed Vessels (HSV) that have made it possible to halve the time required to reach many destinations on the Dalmatian coast or to cross the Straits of Otrante.

Some see, thereby, an end to the relative isolation of the Mediterranean islands and a modal switch from air to sea for short- and medium-haul links. This does not, however, seem very realistic. It also does not appear particularly desirable from the point of view of the environment and marine safety:

- *Not realistic*, because the operating profit of these vessels is extremely sensitive to the price of fuel, given the power of the engines needed to reach speeds of 40 knots. The very high cost of speed will limit the growth of HSVs for freight transport. Many experts believe that they will remain restricted to tourist services to the major islands for a targeted tourist clientele ready to pay higher prices, wishing to travel with their own car and fearing long crossings on more conventional car ferries.
- *Not desirable*, since HSVs continue to raise as-yet unresolved safety and environmental protection problems, in particular increased collisions between these craft and cetaceans.

Other international passenger transport modes

The lack of a unified *rail system* prevents it from playing a significant role at the international level, even between France and Spain and Italy.

Private cars play an important cross-border role for tourism between Spain and France and to a lesser degree Italy and France, but a lack of data makes it impossible to give details of their share of international passenger traffic.

Freight traffic

Freight traffic by *air*, *road*, *rail* (but not sea), like passenger traffic, increased faster than the economy and the population (3.8 per cent per year between 1970 and 2000 in the whole of the Mediterranean, 3.2 per cent in the NMC and 6.4 per cent in the SEMC) (Figure 3.6). Even more than for passenger traffic, this reflects differences in living standards (France, Spain, Italy and Greece alone account for 71 per cent of the traffic), and political and

Figure 3.7 Maritime freight traffic, 1985–2000

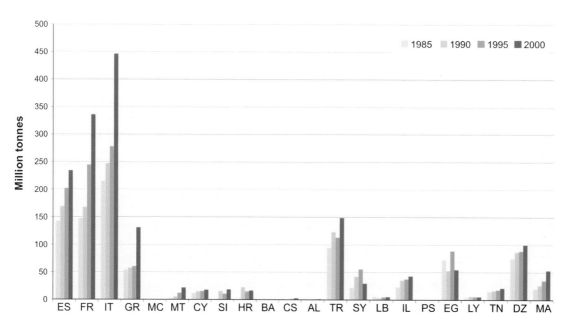

Source: Dobler (EcoMar) (2002) ; Eurostat New Cronos (2002) (http://epp.eurostat.cec.eu.int); Spécial trafic ports du monde (1997), the United Nations Statistical Yearbook (2002)

Note: For the same reasons as for passenger traffic, non-Mediterranean port traffic is factored in. Transit traffic (non-stop Black Sea and Suez–Gibraltar links) has been calculated by assuming that the transit traffic/global traffic relationship observed by Blue Plan 89 has remained constant. There is, in fact, no up-to-date work resulting in the quantification of traffic on the major maritime routes in the Mediterranean.

economic developments (in particular the annual average growth of 6.8 per cent in freight traffic in Turkey between 1970 and 2000).

Like road and air, the growth of freight traffic *by sea* was faster than GDP (4 per cent per year between 1985 and 2000 in all of the Mediterranean, 4.7 per cent in the NMC, but only 3 per cent in the SEMC) (Figure 3.7). As with passenger traffic, it is important to distinguish domestic traffic, which is closely linked to growth in GDP, from international traffic, which is totally dependent on trade flows.

While domestic freight traffic is increasingly dominated by road in all countries (despite resistance from rail in a few areas), international traffic has three distinct components related to trade flows: trade between the four EU-Med, trade between the EU and the other Mediterranean countries, and trade in transit through the Mediterranean.

Domestic freight traffic dominated by road

Road is the dominant mode in the four EU-Med and many SEMC, accounting on average for 82 per cent of all freight traffic (83 per cent in the NMC and 76 per cent in the SEMC) (Figure 3.8).

Figure 3.8 The modal split of domestic freight traffic (excluding maritime) in the Mediterranean countries, 1999

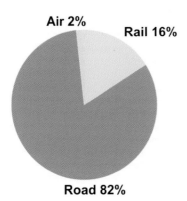

Source: UIC, 2002; OACI 2001, National Statistical Institutes

Note: The external part of the report's road cargo has been deduced (imports + exports)/GDP.

However, *rail* is putting up stiff resistance in countries where there is a tradition of rail freight and the possibility of combined rail/sea transport.

Recent developments in the four EU-Med show a high GDP growth elasticity[5] for domestic rail freight traffic with especially high values in Spain and Italy and to a lesser degree in France where this elasticity is close to one. Recent studies suggest that this elasticity itself is growing with the growth of economic activity. Despite this relative success of rail, road haulage dominates in each country.

Even though rail investments have recently been considerable in Spain, there were unprecedented efforts during the 1980s to build a full motorway system. Although far from finished, road investment in Greece is considerable and expansion of the rail network is limited by physical constraints. In France and Italy the motorway systems are older even if their network is still being completed or their capacity adapted.

The *road situation has become critical* in all these southern European countries, with growth in congestion and pollution, particularly around the Mediterranean coast where road investment has become very difficult. Local freight traffic adds to longer-distance national traffic, and all freight traffic competes with passenger traffic, which is concentrated on the same roads in the same areas.

In the other Mediterranean countries, the approaches are similar: the flexibility of road freight remains the easiest and fastest response to growing economic needs without, however, proper development of rational solutions for the logistical organization of flows. For example, the first two official counts in Syria in 1993 and 2000 showed an increase in road traffic on some sections of the Latakia–Damascus motorway of more than 80 per cent, with around 6000 trucks a day travelling between Latakia and Homs in 2000.[6] Alternatives are limited and exist only where there is a rail line or waterway, such as the Danube or the Sava in the eastern Adriatic countries. The average annual increase in rail freight between 1970 and 2000 in Syria was 9.5 per cent compared with only 3 per cent in Israel, 1.8 per cent in Turkey and 1.3 per cent in Tunisia.

Except in the four EU-Med and the eastern Adriatic countries, where large-scale projects remain to be fully realized, there is no genuine rail network in the Mediterranean worthy of the name (Box 3.3). Railways can only develop gradually, and require a search for national or regional opportunities, between neighbouring countries, that can be combined with sea and road transport. Rail interconnections in the Mediterranean will only work in the very long term, although rail could prove effective in some places, for example crossing the Alps or the Pyrenees.

There remains the possibility of sea transport within national waters for servicing Mediterranean coasts, and, in some cases, alternative, less congested inland routes are being sought. Neither of these has yet led to any significant developments.

International freight traffic

International freight traffic, resulting from international trade, is essentially by sea and road, both of which have seen considerable growth over the past 30 years (4.9 per cent per year for road between 1970 and 2000, 4 per cent for sea) in contrast to rail (a reduction of 0.1 per cent)

2 A Sustainable Future for the Mediterranean

Box 3.3 Railways in the SEMC

The first railway in the Middle East was opened in 1854 between Alexandria and Kafar Zayat in Egypt. Networks have been built in Morocco, Malta, Algeria, Tunisia, Turkey, Syria, Lebanon, Iraq and Libya (only Cyprus has never had railways). The most important route was the Hijaz line linking Damascus with Medina via Amman, built at the beginning of the 20th century (between 1900 and 1905). These lines were designed by the colonizing countries for reasons that were more strategic rather than economic. With the different colonial powers having different technical standards, the rail gauge of the 32,000 kilometres built before the First World War varied from one country to the next: 1m in Tunisia, Algeria and Iraq, 1.05m in Lebanon, Syria and Jordan, and 1.055m for part of the system in Algeria and Libya. Standard gauge was only really adopted from the beginning in Egypt, Morocco and part of the Algerian network.

Today, most of the SEMC have committed to a programme for standardizing their lines in order to reach a standard gauge of 1.435m by 2010–2015. Despite this undeniable progress, most networks remain single track: in 1997, of 33,682km of track, only 2766km were double and 2457km electrified. The different gauges contribute to reducing the speed of commercial shipments. But above all, competition from roads, and to a lesser degree air, have reduced rail to a minor role in these countries. It is true that rail traffic between 1984 and 2000 increased by 4.3 per cent per year, while air, sea and road traffic increased on average by 6.8 per cent, 2.5 per cent and 4.2 per cent per year respectively, but rail traffic volume has been 5–10 times lower than road traffic since the early 1980s.

Dissatisfaction with rail has even been expressed in some countries simply by abandoning the system. For example the line linking La Valette to Rabah has not worked since 1932, and this effectively sealed the fate of the Maltese system. Two important countries have virtually abandoned their rail systems: Libya where not a single train has run since 1965, and Lebanon where the 399km of track, seriously damaged during the civil war, has never been repaired.

Despite these setbacks, governments have announced their desire to see rail playing a greater role in future. With this in mind, the Union of Arab Railroads was founded in 1979. At their 1997 council of ministers meeting, the dotted rail routes shown in the figure were chosen as priorities.

This would require the laying of 20,000km of new track. In 2000, 35,325km of track were in service in the whole SEMC. Not one of these routes is among the projects currently funded by the Euro-Mediterranean Partnership.

Actual and planned rail network in the SEMC

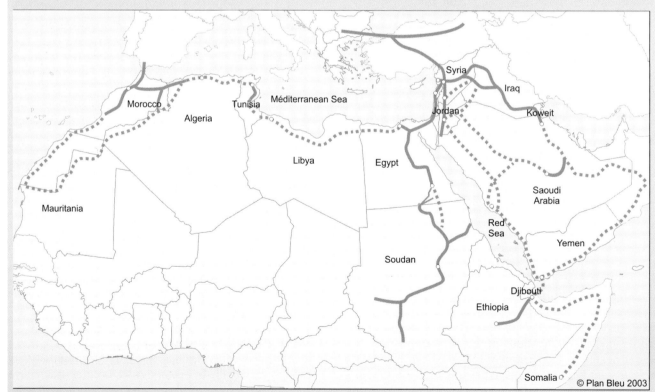

Source: Arab Union of Railways

Note: Existing routes are solid lines and projected ones dotted.

(Figure 3.9). The highest rate of growth during the same period was in air freight (10.6 per cent per year) but this still accounts for only a relatively low fraction of the total.

Figure 3.9 Modal split of international freight traffic (excluding sea) in the Mediterranean countries, in 1999

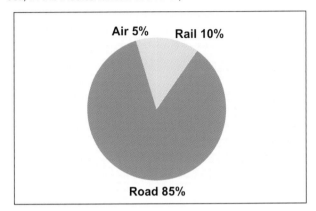

Source: UIC, 2002; OACI, 2001, National Statistical Institutes

Note: To split road cargo between international and national, the same method was applied as for domestic traffic.

Analysis of international freight traffic, discussed below, must differentiate transport between the four EU-Med from that between the EU and the SEMC, both of which demonstrate the weak competitivity of land-based transport. Moreover, particular attention will be given to international trade through the Mediterranean Sea.

Trade within the four EU-Med countries

As the countries that account for most of the value-added produced and traded in the Mediterranean, the associated flows of freight are particularly important and occur essentially between Spain, Italy and France, with a growth in trade between Greece and these three countries emerging during the past few years. Road traffic dominates, with 71 per cent of the trade in 1998 (Figure 3.10).

Figure 3.10 Intra-southern Europe trade in 1998

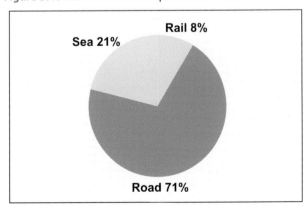

Source: Comext database (http://fd.comext.eurostat.cec.eu.int/xtweb)

Statistics for the EU show a rapid growth in trade that goes mostly by road, although rail traffic between France and Italy accounts for about 25 per cent of the land-based trade between these two countries. Trade by sea is far from insignificant in terms of tonnage, but is limited mainly to some bulk products.

Box 3.4 Traffic across the Alps and the Pyrenees

Between 1990 and 2000 the total volume of freight between Spain and France across the Pyrenees increased by 80 per cent, and that between France and Italy across the Alps by 33 per cent.

Although all transport modes benefited from this increase (road: 140 per cent between France and Spain, 77 per cent between France and Italy; sea: 41 per cent between France and Spain; rail: 32 per cent between Spain and France; and 13.5 per cent between France and Italy), road accounted for 80 per cent of the growth in trans-Pyrenean traffic and 77 per cent for trans-Alpine traffic.

In 2000, the modal split was:

- **trans-Pyrenean trade**: 150.6 million tonnes, 80 million of which was by road, 66 by sea and 4.6 by rail, or 53 per cent by road, 43 per cent by sea and 4 per cent by rail.
- **trans-Alpine trade**: 51 million tonnes, 49.8 million of which was by road and 10.2 by rail, or 79 per cent by road and 21 per cent by rail.

For the residents of the Pyrenean valleys this means a 130 per cent increase in the number of heavy lorries driving over the Perthus and Biriatou passes over the past ten years (a daily average in 2000 of 9500 over the Perthus pass and 7500 over the Biriatou pass) and a 47 per cent increase in heavy lorry traffic for the residents of the Riviera and Alpine valleys, distributed between the Fréjus, Montgenèvre and Ventimiglia tunnels (average daily traffic in 2000: 326 lorries through Montgenèvre, 2900 through Ventimiglia and 4254 through Fréjus).

This continuing growth in road traffic has led the French, Italian and Spanish governments to study possible alternatives:

- **Alpine traffic**: although the Lyons–Turin rail link is going to be brought into service by 2015 and the 45km of single track that still exists in Liguria can be doubled, this will not come close to meeting the growing needs for rail transport. With this in mind, serious discussions about a new high speed rail project have begun at the latest Franco-Italian summit conferences.
- **Pyrenean traffic**: at the Franco–Spanish summit of October, 2001, it was decided to create two working groups, one on short-sea shipping, the other on developing rail freight in the medium and long term. The 20-year scenario envisaged includes plans for nearly doubling the volume of freight, by 100 to 120 million tonnes, with 20 per cent of the additional tonnage going by rail and 25–30 per cent by sea.

> In the light of these analyses, consideration is now being given to modifying and developing the road tunnel from Somport (Franco–Spanish agreement of April, 1991) and a combined high speed/freight line along the Mediterranean.
>
> In both cases, the question of improving existing crossings that are not yet used for freight remains (Pau to Saragossa and Nice to Cuneo), and has not yet been resolved by the authorities involved.
>
> *Source:* Data from Observatoire franco-espagnol des trafics dans les Pyrénées and Office Fédéral du développement territorial de la Confédération Helvétique in 'Trafics de transit circumméditerranéens', Bulletin GIR Maralpin, no 25/26, 2000

The contribution of combined sea/road transport became much more significant for Greece's external trade following the closure of transit roads through the former Yugoslavia during the various wars there. These inter-modal routes appear to have been adopted and maintained after the end of hostilities. This new trend has been confirmed by the spectacular increase in maritime freight traffic in Greece (6.8 per cent per year between 1984 and 2000) and the corresponding collapse in rail traffic in the eastern Adriatic countries (by 17.9 per cent per year between 1984 and 2000 in Albania, and by 12.6 per cent in Serbia and Montenegro, 11.5 per cent in Croatia and 4 per cent in Slovenia between 1990 and 2000).

Trade between the EU and the other Mediterranean countries

Trade between the EU and the other Mediterranean countries, in terms of both volume and value, is much less important than that between the EU Mediterranean countries, especially Spain, France, Italy and Greece. However, it is still growing rapidly (see Part 1) and is dominated by sea transport. This growth may eventually be threatened by the poor speed and efficiency of transport systems in the SEMC, particularly when compared with those in central and eastern Europe.

The situations are highly dependent on and influenced by traditional links between countries. Two groups of countries need to be considered:

- *Trade between the EU and Maghreb*: highly polarized between Spain and Morocco, and Tunisia and Italy, with French trade apparently more equally distributed, although Algeria has a preferred position. This trade is very much influenced by a few bulk products (fertilizers and chemical products for Morocco and Tunisia, oil and gas for Algeria). For general trade, in the west of the region, road traffic from Morocco has risen sharply, crossing the Straits of Gibraltar and often continuing overland north across the Pyrenees to other European countries. Nevertheless this road alternative to sea transport is not yet well developed. The trend is rather for the development of *roll-on roll-off* for freight (loading lorries onto ships).
- *Trade between the EU and the eastern Mediterranean*. Turkey has experienced a strong phase of expansion, which has resulted in an intensification of trade with the EU. Of the other eastern Mediterranean countries, Egypt, Lebanon and Israel also have a high level of trade with Europe, although this is partly counterbalanced by important trade with the Arab countries and Africa and with the US and Asian countries.

Trade with Turkey can be either by land or sea. There have been notable developments of ingenious inter-modal links, using roll-on roll-off sea transport for a large part of the route, as shown by the example of the 'bridge-head' established in Trieste to promote trade and continued transport overland to the heart of the EU. International trade has not yet been fully re-established with the countries of the eastern Adriatic except for Slovenia. Rail and the Danube link play only a limited role.

The lack of competitiveness of land transport systems

The Euro-Mediterranean 'system' of freight transport is therefore characterized by predominantly maritime traffic with international road traffic involving mainly Turkey and the EU and a small contribution from rail. This is currently not very competitive because of the low volume of EU–SEMC trade and the complexity of the administrative and banking operations of the transport systems. It costs as much to send goods from Tunis to Marseilles as it does to Rotterdam, and more to send a container from southern Europe to Maghreb than to Japan or the US.[7] The use of combined transport by developing port areas on the northern and southern shores was put forward as a solution in the early 1990s. However, the eastward EU enlargement turned this type of analysis upside down, and the strategy of strengthening the competitiveness of Mediterranean transport systems currently seems to concentrate on improving the ports of southern Europe. These would be the only ones able to benefit from the huge increase in freight traffic demand in Europe and could, as an indirect consequence, bring back to the Mediterranean traffic (including for the SEMC) that has naturally tended to concentrate in the north-eastern quarter of Europe.

Even this 'modest' strategy is, however, regarded sceptically by many operators, since the SEMC market alone does not and is unlikely to justify large investments in transport infrastructures. For other reasons (essentially improving intra-community north–south relations) investment is planned in the southern European ports, and these may contribute to developing 'short-distance' Mediterranean links. But they will not change the present situation significantly. Apart from improving a few nodal points such as Algesiras–Tangiers, Trapani–Tunis and Igumenitsa–Brindisi, investment is generally limited to improving the profitability of port infrastructures and maritime companies between the two shores. Moreover, although investment in infrastructures is an indispensable condition for the development of inland traffic from ports, it is still insufficient in the light of the benefits of the inter-modal system that it could support. It is proving difficult to develop effective and complete transport systems (providing continuous service quality from one end to the other, complete coverage, adequate speed, integrated prices, etc.) with a well-adapted network and satisfactory connections with more local networks throughout the region.

> **Box 3.5** The logistical difficulties of a German investor in Tunisia, Leoni Tunisie
>
> Leoni Tunisie SA is a subsidiary of a German corporation specializing in electronic components for car manufacturers, in particular Daimler-Chrysler. It was founded in 1977, employs 2400 people and invests more than 3.5 million euros every year in Tunisia.
>
> Demand-pull production obliges the Leoni Tunisie logistics system to achieve first class results. Leoni therefore subcontracts all its logistical requirements to an international company with a subsidiary in Tunisia and has tailored its production cycle in such a way that this company uses its trucks to between 95 and 98 per cent of capacity.
>
> A full production cycle including transport lasts nine days: raw and intermediate materials come from Europe, Asia and the US and are collected at Leoni's head offices in Germany. Seven trucks leave Germany for Tunisia every week. The trailers are checked by German customs in Leoni's factory after which the logistics company takes charge of the shipment. It drives them to Genoa or Marseilles in 2–2.5 days, then loads them without a driver on a roll-on roll-off ferry that delivers them to the port of Radès 20–24 hours later. There the trailers are taken by trucks and delivered 2–3 hours later in Sousse. Within 24 hours, the components are assembled, checked on site by Tunisian customs officials and sent back the next day by the same route. Eight trucks carrying between 320 and 350 tonnes of finished products leave Tunisia each week. Leoni is particularly satisfied with this transport chain since, as a major exporter for Tunisia, it benefits from off-shore status regarding customs dues and special treatment in Tunisian ports and is the envy of many small companies.
>
> However, the demands of 'just-in-time' production are so great that they threaten Tunisia's position. Instead of the current nine-day cycle, customers are now demanding six days. The internal production processes have been rationalized to such an extent (for example assembly in 24 hours) that the only possible time savings are those made possible by transport. In 2002 then, Leoni chose to invest 12 million euros in Romania, creating 1700 jobs, rather than to expand its plant in Sousse. Leoni's directors were motivated not by the salary levels or the political context (from this point of view Tunisia is more competitive) but by the logistical advantage offered by Romania, which enabled a saving of one day in each direction and 1000 euros per trailer. According to Leoni Tunisie's directors, if Tunisia is not in a position to offer air or faster boat links, the relatively cheap manpower is no longer enough to maintain its attractiveness to foreign investors.
>
> *Source:* World Bank, 2002

Transit trade through the Mediterranean

The Mediterranean Sea accounts for only 0.7 per cent of the world's seas but handles 30 per cent of maritime trade traffic and 20–25 per cent of hydrocarbon traffic. Its shipping lanes are of major strategic importance for Europe and America. The redistribution of rapidly increasing world trade plays a growing role, to such an extent that these transit flows account for nearly 40 per cent of the whole Mediterranean traffic (Box 3.6).

Sea transport has been able to adapt and to retain its market share of long-distance trade in the face of competition from road and air, despite transit difficulties through the eastern Adriatic countries for trade between the EU, Turkey and the Middle East and Central Asia.

Gas and oil account for huge amounts of intra-Mediterranean trade (see Chapter 2). The region remains one of the main transit areas in the world for both oil and container traffic. Fixed petro-chemical port facilities governed these flows until recently, although this could change as a result of the Middle East crisis and also with the opening of land routes for conveying products from central Asia through Turkey. The trade in essential chemicals and bulk agricultural freight comes second to gas and oil in volume and also results in specific flow patterns, as does the trade in textiles and perishables from south to north.

This marine traffic is increasingly 'containerized', with more and more goods being packaged in containers and no longer stored individually in holds. *Containerization* became dominant in Europe in the 1970s, and subsequently on the major east–west shipping lanes. It is gradually becoming universal. In 1999 traffic between Europe and Asia accounted for 8025 million TEUs (twenty-foot equivalent units). Containerization is spreading to all products and is gradually replacing conventional transport. The size of ships has increased in consequence. By 1995 the biggest container carrier could ship 3000–3500 TEUs, and we are starting to see the first 8000 TEU (ships 340m long).

Since the early 1990s the Mediterranean has also seen *new international maritime 'hubs'* with the very rapid development of several container ports such as those of Gioao Tauro in southern Italy, Malta and a whole batch of ports

2 A Sustainable Future for the Mediterranean

Box 3.6 Maritime traffic in the Mediterranean

The main maritime routes into the Mediterranean are Gibraltar, Suez and the Bosphorus. The most important intra-Mediterranean shipping lanes are those linking Greece with Egypt and Libya with Italy.

Of the total of 412 million tonnes of oil imported into Europe in 1999, 121 million came by intra-Mediterranean routes, and 278 through Gibraltar, Suez or the Bosphorus (the remaining 13 million tonnes correspond to trans-Atlantic trade).

The same is true for container traffic that amounted to 1560 million TEUs (twenty-foot equivalent units) for the intra-Mediterranean in 1999 compared with 8025 million of Europe–Asia trade through Suez, Gibraltar or the Bosphorus.

The main maritime routes in the Mediterranean

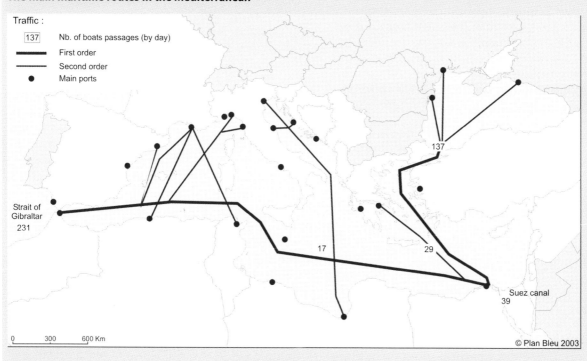

Source: REMPEC (2002) from Lloyd's Register of Shipping (2001), London

Maritime oil traffic – main trade inter-zones (in million metric tonnes)

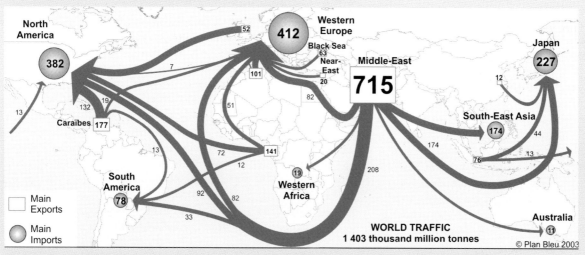

Source: IMO News (2000), no 3

that serve as hubs for the great shipping companies of southern Italy. Such growth rates have never been seen before, following the earlier example of Algesiras, with sites also being sought in the western section of the Mediterranean, nearer to Gibraltar, that offer opportunities not only for redistribution towards northern Europe but also towards Africa. There are also comparable developments at the exit of the Suez canal, to reorganize container carrier distribution networks for the whole of the Mediterranean, and even ports in the Black Sea. Limassol, Port Said, Piraeus and Haifa have profited from these new moves to reorganize maritime traffic on a global basis for trade, the shipping costs of which have been reduced to very low levels. This unusual phenomenon has benefited all Mediterranean ports, even those without strong inland industrial development.

Traffic projections to 2025

The main driving force for this *increase in traffic*, by all modes and to all destinations (with a traffic/GDP elasticity sometimes greater than two, or even higher in some countries such as Turkey) seems to be a global liberalization of trade and specialization of economies as well as changes within industry and logistical structures (particularly the pooling of stock-taking points). This has already been observed within the framework of the North American Free Trade Agreement (NAFTA). The Mediterranean is no exception. A *Plan Bleu*[8] study concludes that including Morocco in the Euro-Mediterranean free trade zone would result in a doubling of the passenger car fleet by 2020 compared with what would have occurred without free trade.

A look at past traffic changes in some Mediterranean countries seems to show that every advance in international free trade was preceded by changes in the productive and logistics systems of the country involved, leading to an increase in traffic and then a levelling off, followed by a new rise due to the acceleration of world economic growth (Figure 3.11). Such a trend seems limitless and not very sensitive to the 'dematerialization' of economies.

Apart from the move towards free trade, two other factors also contribute to the growth and concentration in freight traffic:

- The appropriateness of road transport to new production patterns and the nature of the goods transported (more perishable, fragile and individualized goods such as chemical products and foodstuffs, cars and machines and less coal, iron or steel) demand for door-to-door transport in relatively small quantities, which excludes modes adapted to bulk such as rail and water (rivers, canals). Furthermore, cutting-edge technologies make it possible to decentralize production within a 'just-in-time' framework, with frequent,

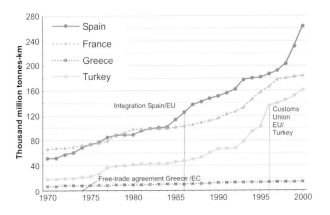

Figure 3.11 Trade opening and road freight traffic in some Mediterranean countries, 1970–2000

Source: World Bank; IRF, 1999 and National Statistical Institutes

small-volume deliveries that reinforce the preference for road transport. On-line buying is increasing these trends with home-delivery of products replacing travel by buyers. Several studies in the EU are confirming these trends.

- The lack of internalization of the environmental and social costs of road traffic, which is resulting in an increasingly favourable price structure for road freight transport and priority being given to road infrastructures by many countries and donors. Between 1987 and 1998, the four EU-Med countries invested 0.2 per cent of their GDP in rail; between 1984 and 2001 they invested 0.67 per cent in roads.

The key factor for changes in *passenger traffic* is the increase in the number of people living in *suburbs*. In French suburbs, for example, between 1980 and 1995, daily average distances travelled increased 3.5 times faster than in city centres.[9] This concentration of people in suburbs is being accompanied almost everywhere by two other changes, which also generate more traffic:

- *the development of hypermarkets*: this type of city and suburb supply system is very cost-effective for the company, and is essentially based on road haulage. But the last stage of delivery is usually ensured by the customers themselves, who need a car to get to the sales point and, of course, pay the associated costs. One of the most significant consequences of this phenomenon is the sharp fall in nearby business activity, destroying urban life. Energy consumption and the disturbances engendered by this kind of organization are very considerable compared with proximity shops;

● *changes in packaging*: shifting to a 'consumer' society has led to big changes in the packaging of goods. New packaging is often characterized by large volume (blister layers, boxes, voluminous protective coverings, etc.). Liquids are more often than not sold in throw-way packages. All this leads to an eruption in the volume of household waste, which results in large transport flows in urban areas. In France, for example, the 29.5 million tonnes of household waste produced in 1997 led to a transport activity of 800 million tonne-km.

An important factor that may help to slow the demand for transport is the 'non-material' content of future growth, in other words the share of services in relation to trade in physical goods. In the Mediterranean a *dematerialization of the economy* is underway; except in Egypt and Algeria, services increased their share of GDP in all countries between 1990 and 1999. This is coupled with a dematerialization of physical production, that is a reduction in the amounts of heavy goods produced and a general reduction in the weight of products (particularly important in the southern and eastern Mediterranean because of the weakness of national industries and the resulting need to import a major share of manufactured goods). It is therefore possible that traffic/GDP elasticities may eventually diminish. The baseline scenario does not, however, take this possibility into account, partly because the SEMC are just opening up and are looking for strong economic growth, and partly because retrospective analysis of traffic/GDP elasticities does not show any such shift in any Mediterranean country. Such shifts may be possible with policies of the type proposed by the EU in its White Paper, but there is no sign of such policies in the southern or eastern Mediterranean.

The consequences of all these changes on the baseline scenario are of two types:

● First of all, trade will increasingly concentrate along a smaller number of routes. The flexibility of road delivery is well matched to company practices that involve managing stocks en route and pooling stock-taking points. The increase of traffic on these routes will be much higher than elsewhere. An example is given by the concentration in north–south corridors in the Alps and the Pyrenees and the major airline hubs.
● Roads will maintain their dominant position in land transport. Except in France, Italy and Spain, which are trying to maintain rail freight traffic, the other countries do not have domestic rail systems that can really compete with roads. A few major international links could probably be re-established, but these isolated projects cannot prevent roads from playing a major role in the growth in land transport. This near-monopoly of roads makes it even more important to search for alternative sea routes.

Taking into account all these general changes, the baseline scenario for traffic to 2025 uses the same 'traffic/GDP' elasticities as those observed for the 1970–2000 period[10] and the same modal split as that observed in 2000. These assumptions lead to the following baseline scenario projections presented by major country groups (the figures on traffic include national and international traffic for each country) (Figures 3.12 and 3.13).

Passenger traffic

Figure 3.12 Passenger traffic (rail+air+road), baseline scenario, 2025

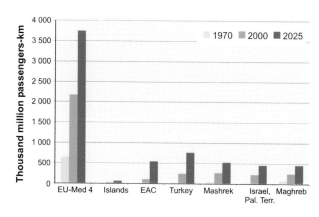

Source: Plan Bleu

Figure 3.13 Passenger port traffic in 2025, baseline scenario

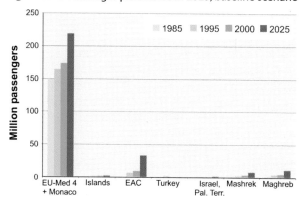

Source: Plan Bleu

Note: This only includes data relating to recorded port traffic. The projection has been made following the same method as for land traffic.

According to the baseline scenario, the general demand for transport continues to increase at a steady rate in both the NMC (4.6 per cent per year between 2000 and 2025) and the SEMC (7.6 per cent per year) and still at a *greater rate* than the *population* (0.1 per cent per year in the NMC and 1.3 per cent in the SEMC) and the *economy* (2.5 per cent per year in the NMC and 4.2 per cent in the SEMC).

From environmental and safety points of view, the projected modal patterns are the least satisfactory way of fulfilling this mobility demand. Thus for the whole of the Mediterranean in 2025 *roads* are projected to account for 85 per cent of the traffic (84 per cent in the NMC and 86 per cent in the SEMC), air 8 per cent (10 per cent in the NMC and 5 per cent in the SEMC), with rail remaining steady at 7 per cent (6 per cent in the NMC and 9 per cent in the SEMC). Maritime traffic is projected to grow more slowly (by 0.9 per cent) than the economy in the EU but at a very steady rate in the eastern Adriatic (especially Croatia) and the SEMC (especially Morocco at 4.7 per cent).

This must, however, be qualified since *rail* will generally keep up with air and road in relative terms (an average annual increase between 2000 and 2025 of 2.2 per cent compared with 2.6 per cent for air and 2.9 per cent for road). The relatively good maintenance of the share of rail is even clearer in the SEMC (annual increase of 3.4 per cent for road, 3.2 per cent for air, mostly as a result of tourist-related international traffic, compared with 3.2 per cent for rail). The 'performances' of the Mediterranean countries vary considerably from one to the other and do not always match with the north shore/south shore distinction. For example, the use of rail in Egypt is projected to increase by an annual average of 3.1 per cent compared with only 1.4 per cent France. Economic and political constraints play a large part in traffic fluctuations, for example in Bosnia-Herzegovina, where the high annual rate of increase in rail and air traffic (11.7 per cent) is because it is starting from very low levels as a result of war. The same applies to Croatia for road transport (6.8 per cent).

Freight traffic

Figure 3.14 Freight traffic (air+rail+road), baseline scenario, 2025

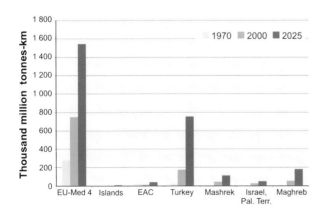

Source: Plan Bleu

Note: The same method as for passengers was used for evaluating road traffic in the countries that do not measure vehicle flows.

Figure 3.15 Maritime freight traffic, baseline scenario, 2025

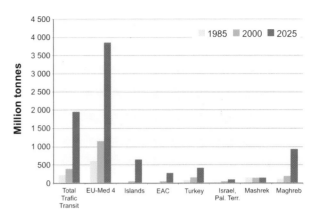

Source: Plan Bleu

Note: For this traffic, the elasticity of international trade (national for the port traffic of each country and global for the transit traffic) was used and reduced according to the same coefficient as the GDP elasticity for land traffic. The 2000–2025 growth rates of trade have been deduced from the GDP elasticities of IFRI's projection paper 'Le commerce mondial au 21ème siècle'. The growth rate of world trade was linked to the growth assumptions for world GDP of the baseline scenario, i.e. 3.8 per cent per year between 2000 and 2025 (see Part 1 of the report).

As with passenger traffic, freight traffic is projected to increase *more rapidly than the economy or the population* (by 3.8 per cent per year between 2000 and 2025 for the whole Mediterranean, 3 per cent in the NMC and 5.4 per cent in the SEMC), with the least 'sustainable' modes continuing to carry most of the freight. The baseline scenario has roads taking 85 per cent of the traffic (84 per cent in the NMC, 89 per cent in the SEMC) and air 3 per cent (4 per cent in the NMC and 2 per cent in the SEMC) compared with only 12 per cent for rail (19 per cent in the NMC and 9 per cent in the SEMC). The rate of increase in total traffic will be comparable to that between 1970 and 2000 (3.2 per cent per year in the NMC and 6.4 per cent in the SEMC) and rail freight will retain its share. However, the NMC will see a clear-cut *recovery of rail* with an annual increase of 2.4 per cent compared with a 0.5 per cent per year reduction between 1970 and 2000.

Stimulated by transit traffic (increasing by 6.6 per cent per year between 2000 and 2025), *sea freight* is projected to continue to grow more rapidly than the economy (5.6 per cent per year). The share of sea transport in Italy and Greece in particular is projected to increase compared with the other modes. The same applies to the eastern Adriatic countries (especially Croatia). This means that the inter-modal road–sea transport links opened because of the Balkan wars will maintain their market share. This contrasts with a stagnation in freight traffic in the SEMC (probably related to the reduction in conventional traffic). There is also a strong projected

2 A Sustainable Future for the Mediterranean

Figure 3.16 The population exposed to road noise over 55 decibels, 1970–2000

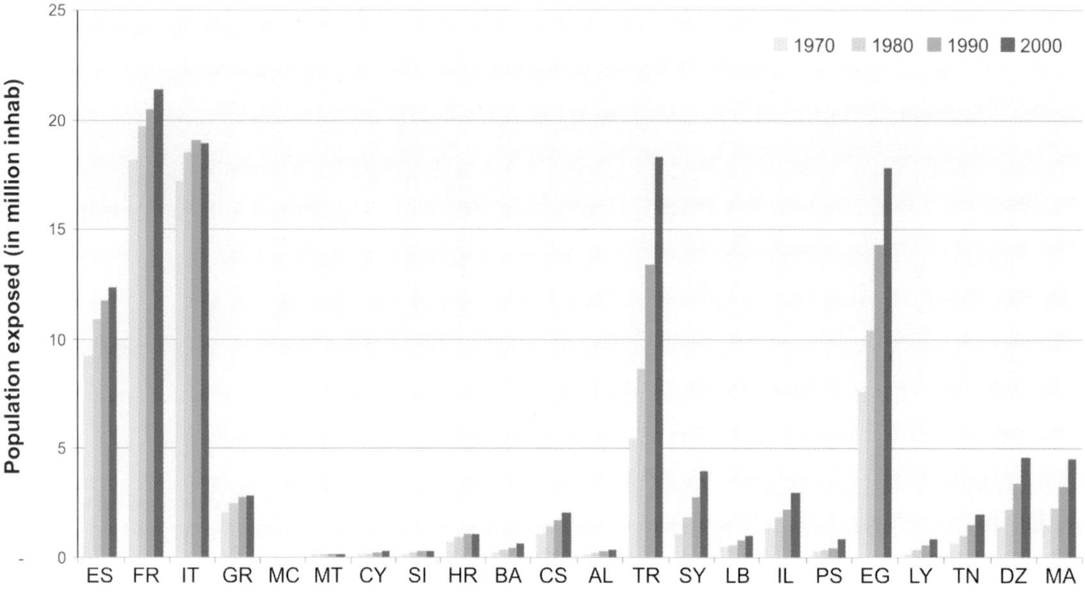

Source: Plan Bleu from European Environmental Agency data

Note: An estimate based on a 1998 evaluation by the EEA of the percentage of the population in the EU exposed to road noise over 55 decibels. Based on the traffic intensity of GDP in each Mediterranean country, the percentage of the population exposed in each has been deduced. Changes are related to the rate of urbanization.

increase (13.5 per cent per year in Malta) in trans-shipment ports, the optimum location of which will be on the major international sea routes.

This analysis of the driving forces and magnitudes of the passenger and freight flows will serve as a framework for evaluating the impacts of the Mediterranean transport system up to 2025.

The large impacts of land and air transport

Transport in the Mediterranean today often results in a waste of natural resources, a reduction in living standards (air, space, noise), considerable loss of human life and high costs for countries, companies (congestion) and individuals (the share of transport in household budgets). Many of these problems could be avoided with policies that will be discussed further below.

Significant environmental and social impacts

Environmental impacts

Noise pollution
Traffic is one of the major sources of noise pollution. Road traffic is the main cause of exposure to noise (Figure 3.16) except for those who live near airports. Noise-related environmental disputes in Europe are increasing. Nearly 32 per cent of the population lives in a noise environment of more than 55 decibels from road traffic (and nearly 12 per cent more than 65 decibels), 9 per cent from rail and 3 per cent from air traffic. In the SEMC, large suburban areas, for example in Cairo and Algiers, are also experiencing a large rise in noise level. This is all the more worrying since noise is increasingly seen to be a serious public health problem.

Some epidemiological studies have shown the extent of the problem in Europe. The same is likely to be the case in the SEMC, but general data are not yet available.

Noise pollution is a real problem. In the EU Mediterranean countries, more than a quarter of the population is affected (including 36 per cent in France and 33 per cent in Italy). Some other countries are already reaching a very high level (45 per cent in Malta, 51 per cent in Israel). This is all the more worrying since the numbers exposed are increasing rapidly (1.8 per cent per year between 1970 and 2000 in the Mediterranean, 0.7 per cent in the NMC and 3.5 per cent in the SEMC).

Changes in this indicator suggest that the problem is likely to get considerably worse in southern and eastern countries (Turkey, Maghreb and the Mashrek) while remaining high in the north. By 2025, 32 per cent of Turks (compared with 27 per cent in 2000), and approximately 35 per cent of French, Italians and Egyptians, will be

Figure 3.17 People exposed to road noise above 55 decibels, baseline scenario, 2025

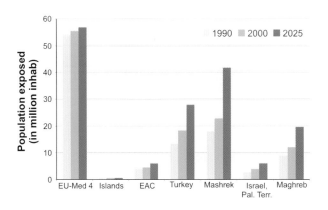

Source: Plan Bleu

Note: The forecast has been made by projecting to 2025 the 1970–2000 elasticity of the variation of exposed populations/urbanization rates.

subjected to road noise of more than 55 decibels (Figure 3.17). The lack of any prospects for technological solutions to these problems by manufacturers does not give much hope of reversing the trend.

Atmospheric pollution

Atmospheric pollution is mainly the result of transport-related *gaseous emissions*. Indeed, transport:

- accounts for 70 per cent of nitrogen oxides (NO_x) emissions in the major European urban areas;
- is entirely responsible for peaks in ozone pollution;
- is the main source of emissions related to hydrocarbon combustion (volatile organic components [VOCs], benzene, etc.);
- is also the source of 30 per cent of CO_2 emissions and about a third of particulate emissions, especially those resulting from diesel.

Gaseous emissions may be classified as primary pollutants, present in the atmosphere in the state in which they are emitted, or secondary pollutants, which result from chemical transformations in the presence of components known as precursors. Because of the availability of data and coefficients of emission per passenger-km and tonne-km, information is provided for emissions of CO_2, VOCs and NO_x (Figures 3.18, 3.19 and 3.20). VOCs and NO_x are suspected of being the cause of the increase in death rates from cardiovascular and respiratory diseases resulting from atmospheric pollution (see Chapter 4) and account for much of the emissions related to transport.

Figures 3.18, 3.19 and 3.20 show time-series of transport-related gaseous emissions.

The increase in CO_2 emissions (4.3 per cent per year between 1970 and 2000, 3.7 per cent in the NMC and 8 per cent in the SEMC) is due to the lack of any significant technological improvements. On the other hand, there has been a significant drop in VOC emissions (0.6 per cent per year between 1970 and 2000, 1.8 per cent in the NMC but an increase of 2.1 per cent in the SEMC) mainly due to improvements in road vehicle engines. And it is because of less progress in improving road vehicles that NO_x emissions have remained relatively constant, increasing by 1.6 per cent between 1970 and 2000 for the whole of the Mediterranean, 0.8 per cent in the NMC and 3.4 per cent in the SEMC.

This 'emission profile' in the Mediterranean results in a combination of two situations:

- in the *northern countries*, concentrations of particulates have generally fallen. NO_2 concentrations have remained stable. However, limit values, above which short-term harmful effects to humans may occur, are rarely exceeded. Thus for nitrogen dioxide, the average concentrations on the French Mediterranean coastline are roughly 50 micrograms per cubic metre while the quality targets set by European directives is put at 135;
- in the *southern and eastern countries*, measuring networks are still embryonic. However, initial results show concentrations similar to those recorded on the northern shore which are also decreasing for particulates and are remaining constant for nitrogen oxides (see Chapter 4).

Despite these relatively low concentrations and the reductions in emissions, a recent epidemiological study,[11] carried out jointly by France, Austria and Switzerland, estimates that 6 per cent of the overall *mortality* in these three countries, or 40,000 deaths per year, may be attributable to atmospheric pollution. Half of these deaths are related directly to emissions from road traffic, or twice the death toll from road accidents in these countries.

In the SEMC, quantification of the impacts is not as complete, but the few isolated surveys available show an alarming situation. In Algeria, diseases related to air pollution were the second most important cause of hospitalization in 1990.[12]

Different transport modes have different impacts; public urban transport and rail are the least-polluting modes in terms of VOCs and NO_x, but they are clearly declining slightly everywhere in the Mediterranean. The introduction of catalytic converters for light vehicles and

Figure 3.18 VOC emissions, 1970–2000

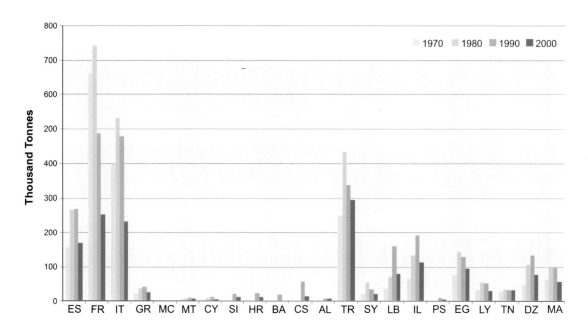

Source: Plan Bleu

Note: VOC and NO$_x$ evaluations used the passenger-km and tonne-km emission coefficients of the Austrian Ministry of the Environment quoted by the European Environmental Agency in its publication, 'Are We Moving in the Right Direction? Term 2000'. To adapt them to the Mediterranean situation, we used them as they were for the four EU-Med countries and Monaco, with a deferred reduction of five years for Malta, Cyprus, Slovenia, Croatia, Bosnia-Herzegovina, Israel and Serbia-Montenegro, and ten years for all the other countries. In the case of Turkey, Tunisia and Lebanon, the results were matched with the official estimates of the countries.

Figure 3.19 NO$_x$ emissions, 1970–2000

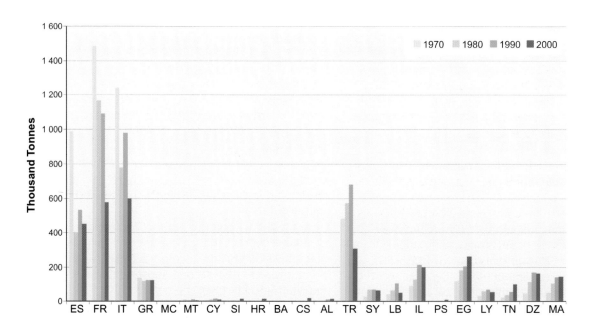

Source: Plan Bleu

Note: Same remark as for calculating VOC emissions.

Figure 3.20 CO_2 emissions, 1971–2000

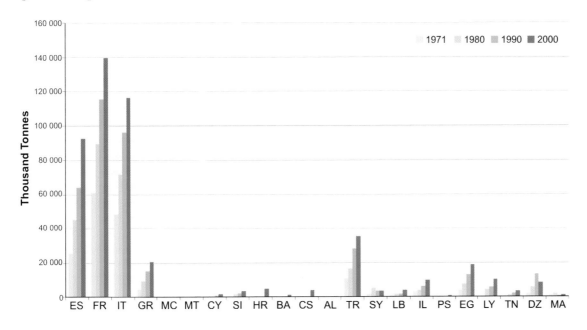

Source: Plan Bleu

Note: Estimates from the IEA database

electrifying the rail network (in Morocco, for example) have resulted in a considerable drop in emissions per passenger since the beginning of the 1990s.

In future, the trend lines for VOC and NO_x emissions are projected to converge, and fall dramatically (by 5.6 per cent per year for VOCs and 4.5 per cent for NO_x) (Figures 3.21 and 3.22). Any continuing difference between VOC and NO_x emissions would be increasingly attributable to air transport, for which the improvement in emission control is expected to be less rapid than for road transport (NO_x coefficients for air transport will be reduced by 3.9 per cent per year between now and 2025 and those for road transport by 8.6 per cent per year) resulting in a clear increase in the share of air transport in NO_x emissions.

But the fall in emissions does not mean the end of atmospheric pollution problems. First, there is no linear relationship between emissions and concentrations, and *synergistic* effects between pollutants, even at low doses, may continue to have significant impacts on health. Secondly, with traffic becoming more concentrated on a few routes and the fact that emission coefficients relate to *average* traffic conditions, it is not impossible that, even if total emissions fall, there may be *local* increases in concentration, particularly in urban areas. Moreover, CO_2 emissions are projected to continue to increase (by 1.9 per cent per year), increasing the Mediterranean share of the world total of CO_2 emissions (Figure 3.23, see also Chapter 2).

Figure 3.21 VOC emissions, baseline scenario, 2025

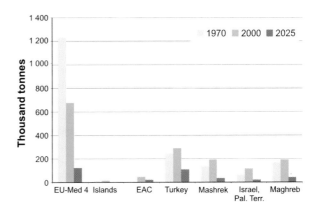

Source: Plan Bleu

Note: The traffic-emissions 2000 coefficients were divided by ten.

Energy consumption

In addition to noise and air pollution, the transport sector is also the biggest consumer of energy in Europe (in 1996 it accounted for 30 per cent of total energy consumption). Consumption is increasing continuously (by 3 per cent per year). Road transport is responsible for nearly 75 per cent of the total energy consumption of the transport sector.

Figure 3.22 NO$_x$ emissions, baseline scenario 2025

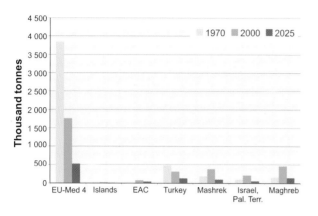

Source: Plan Bleu

Note: The average annual reduction in emission coefficients observed from 1984 to 2000 have been extrapolated to 2025.

Figure 3.23 CO$_2$ emissions, baseline scenario 2025

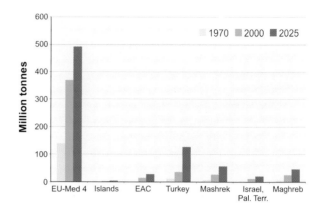

Source: Plan Bleu

Note: The CO$_2$ emissions of each country calculated by the IEA in 2001 were carried over to the total in 2000 CO$_2$ emissions calculated in Chapter 2, Energy. This percentage was then applied in 2025 to the total CO$_2$ emissions estimated in Chapter 2.

Different transport modes have very different energy consumption levels. Energy consumption per passenger in a private car is twice that in public transport. Although fuel consumption per vehicle has dropped, the total energy intensity of passenger transport[13] in Europe has not changed since the early 1980s. This is related to changes in fleet composition (more powerful vehicles, older vehicles) and modes of use (decrease in vehicle occupation rates). In the SEMC the trend is less clear, probably because of the rate of use of vehicles, which is still high. However, it is the attitude of people to their cars that is fundamental.

Individual buying behaviour, partly urged on by manufacturers, tends towards cars that are, most certainly, safer and more reliable but also more powerful (4x4). Although very considerable technical improvements in the energy performance of engines have been made by manufacturers, consumption by vehicles has increased because of greater power and the addition of accessories and safety systems (catalysers, airbags, power steering, etc.), which add weight. An additional hundred kg consumes 0.3–0.8 litres more fuel per 100km, and emits 11–29 grams more CO$_2$ per km. Consumers are also tending towards ever faster models (in Maghreb in 2001, the manufacturers offering the heaviest and most powerful ranges recorded the strongest increases in sales, and the models sold in that market are all 'tropicalized', that is equipped with air-conditioning and components adapted to the hot climate). Speed has an environmental cost. An increase in maximum speed by 10km/hr means a real consumption increase of 0.2–0.3 litres per 100km. Air-conditioning also continues to sell strongly, including in the north, leading to higher consumption, typically 7 per cent of annual average consumption, additional NO$_x$ emissions and emissions of chlorinated and brominated compounds. Converting fleets to diesel also has a high impact (because of particulate emissions). Over 20 years in France, for example, between 1975 and 1995, the share of diesel in the fleet owned by households increased from 2 to 30 per cent. In Algeria in 2000 the share was 27 per cent, and Morocco 49 per cent.[14]

As regards *vehicle use*, studies show that one in every two car trips in Europe is for less than 3km, one in every four is less than 1km and one in every eight is less than 500 metres, including the search for a parking place. These short trips, generally in town with a cold motor, result in high consumption and pollution. Vehicle consumption is also very dependent on acceleration/deceleration phases, so that, on average, when used in urban situations, nearly 60 per cent of the energy produced by the engine is used to overcome the inertial forces due to the vehicle's mass. Traffic regulation is still a theoretical concept in most Mediterranean urban areas, even in the northern countries. Exceeding speed limits also leads to more emissions and energy consumption.

The quality of *vehicle maintenance* is also important. In France, for example, of 30,000 vehicles examined by the French Agency for the Environment and Energy Management (ADEME)[15] between 1991 and 1993, 51 per cent were badly tuned. This means 0.5Mtoe per year of additional consumption. From this point of view, mandatory technical checks that are currently aimed mainly at safety should also include the environment. In France only one of the 52 potential leak points for CO and CO$_2$ emissions is subject to technical control. Only Turkey, Italy and Israel require their vehicles to undergo environmental checks during technical surveillance tests.

As regards *freight transport*, the trend to lower energy intensity has been clear since the early 1970s both in Europe and in the SEMC. The two main causes of this movement have been the penetration of diesel and technological improvements. This has, however, affected different countries in very different ways because of the characteristics of the vehicle fleet in each country, the occupancy rate of vehicles and traffic fluidity. Moreover, in Europe since the early 1980s, energy efficiency gains have levelled off, whereas in the SEMC, energy performance per tonne transported continues to improve (see Chapter 2).

Impacts on land

The building of transport infrastructures leads to an *irreversible loss of land* that is not insignificant. Nearly 30,000 hectares in Europe were covered by roads between 1990 and 1998. In some north European countries, nearly 4 per cent of the national territory is covered by transport infrastructures. Roads take the lion's share and account for 93 per cent of land uptake. Rail uses 3.5 times less land per passenger conveyed.

Although the amount of land used for transport infrastructure is clearly increasing in some countries (Malta, Spain and Egypt in particular) the total increase in land occupied remained modest between 1984 and 2000 (1.6 per cent per year for the Mediterranean, 1.7 per cent in the NMC and 1.4 per cent in the SEMC) (Figure 3.24). In no country except Monaco does land occupation by transport infrastructures exceed 3 per cent.

This increase in land area occupied by infrastructures is therefore of concern mainly from the point of view of indirect impacts.

Land-take can have several consequences:

- more demand for building materials (especially aggregates, which are often extracted directly from river beds);
- very many pollutant emissions, particularly during construction (dust, toxic effluents from the application of coated materials for road construction, waste from bottom-dredging when building ports, etc.);
- irreversible 'artificialization' of some exceptional areas (for example harbours and parking lots on the coast);
- accelerated random urbanization (roads and motorways along seashores).

It is the *sealing* of soils that is the greatest concern. The rainfall regime in the Mediterranean shows wide variations that directly influence the water regime and

Figure 3.24 Areas occupied by land transport infrastructures, 1984–2000

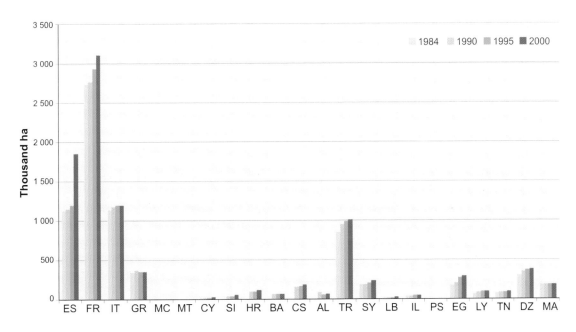

Source: Plan Bleu

Note: These findings have been obtained by multiplying road and rail line information supplied by the International Road Federation and the International Union of Railroads by the coefficients of ground-occupancy used by the EEA ('Are we moving in the right direction?', TERM 2000). Airports were excluded, since their spatial expansion on the scale of a country is not significant and not necessarily related to national traffic. The covered surface of ports is dealt with in Chapter 6, Coastal areas.

turns many river beds into torrents at the end of the summer. By sealing soils, the construction of transport infrastructures, and the urbanization that goes with this, reduces the expansion areas for flooding and increases the flow velocity of floods, particularly on the mountainous coasts throughout the basin. Thus every autumn in northern Italy there are serious flooding problems, and in 1994 the French *département* d'Alpes Maritimes recorded disastrous floods. Such sealing also reduces the infiltration of rainwater into the ground, which results in slowing or even compromising the recharging of water tables. In Israel, for example, nearly 0.4 per cent of the country's surface area is already sealed.[16]

Infrastructure construction also threatens nature conservation and *biodiversity*, through emissions of pollutants from vehicles, noise, fragmentation of land, collision risks and reductions in species habitats.

The increase in land-take is expected to continue (3.3 per cent per year for the 2000–2025 period) (Figure 3.25). Some countries (notably Greece and Israel) or regions (the eastern Adriatic) may well record considerable annual increases (more than 4.5 per cent per year) for the rail and road investments necessitated by reconstruction or steep increases in traffic. Total land-take for infrastructures in the SEMC by 2025, however, is projected to remain below that in the NMC in 1996.

Figure 3.25 Areas occupied by land transport infrastructures, baseline scenario, 2025

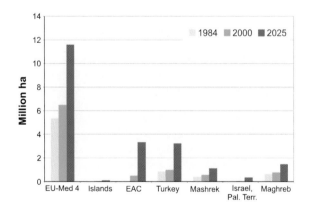

Source: Plan Bleu

Note: Areas have been estimated on the basis of surfaces/traffic elasticity for 1984–2000.

Social impacts

The dominance of the car also has a major impact in terms of the numbers killed in *road accidents*, which account for the largest proportion of transport-related accidents (Figure 3.26). In the EU, for example, there are about 50 times more people killed every year in road than in rail accidents (44,000 for road compared with 936 for rail, or 11.1 deaths per million passenger-km for road compared with 3.4 for rail). The death rate for air accidents is even lower, 916 deaths worldwide per year, or 0.4 deaths per million passenger-km.

The number of deaths per year fell by nearly 50 per cent in the NMC between 1984 and 2000 with most of the improvement attributable to the EU countries. Taking into account the increase in traffic and the number of passenger-km, the number of deaths has dropped by 75 per cent during the same period. These good results are due mainly to the safety measures adopted by the countries (road layout and vehicle design, policing, prevention, etc.). The toll nevertheless remains heavy: 24,000 road deaths in the NMC in 2000. Within the EU, some countries (particularly Greece) have accident rates four times higher than others.

In most of the SEMC (except for Turkey, Syria, Lebanon and Israel), the number of road deaths continues to increase although the number of deaths/1000 vehicles is falling (by 2.7 per cent per year between 1984 and 2000). This relative drop is explained by the very high road accident death toll in these countries in 1984. With a vehicle fleet only one-tenth of that of the NMC, there were 80 per cent more road deaths. Turkey has a large percentage of the SEMC car fleet but succeeded in improving road safety by 10 per cent per year between 1984 and 2000.

The total number of road deaths is expected to continue to increase (by 0.6 per cent per year between 2000 and 2025) with a 4.5 per cent per year reduction in the NMC and 2.8 per cent per year increase in the SEMC (Figure 3.27). Without strengthening road safety policies, the toll could become a veritable slaughter in the SEMC with 50,000 deaths projected for 2025, 19,000 of which would be in Egypt alone. However, the increase in terms of number of road deaths/1000 cars is more moderate (0.4 per cent per year for the whole region, a 1.4 per cent increase in the NMC and a 0.1 per cent reduction in the SEMC).

This road-dominated system leads to a *dispersion of services*: ever-increasing distances between work, school and shops, and homes (see Chapter 4). This can eventually lead to greater social disparity since access to these services becomes more of a problem for the poorest (those unable to afford a car). The problem is becoming all the more acute since access to public transport is also becoming more difficult as investments are concentrated on roads.

This *increasing of distances* between the different functions of daily life are explained by the sociological and demographic factors that govern motoring: demographic development has replaced people who never acquired the habit of driving when they were young with

Figure 3.26 Number of road deaths per 1000 vehicles, 1984–2000

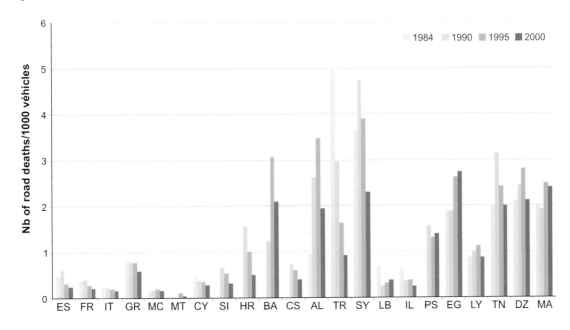

Source: National Statistical Institutes, Eurostat and IRF, 1984–1999.

people who are used to driving and only give it up late in life, the gap in status between men and women is closing, and incomes are increasing. Other factors are more directly related to public policies and contribute to fragmented cities that are more dependent on the car, with dense areas that always carry intense traffic and outlying areas that act as the source of very high traffic flows. Some of the factors that directly and significantly govern transport are regulation of the use of urban spaces, pricing and taxation, and questions of location and urban structures (see Chapter 4).

The main result of this increase in the travel required to satisfy daily needs added to increasing mobility for tourism, leisure, etc. is the *rising share of transport in household budgets* (Figure 3.28).

The share of transport in household budgets varies considerably between countries (from 19 per cent in Israel to 4 per cent in Egypt in 2000). Income per capita is not the only factor, since there are considerable differences between countries with similar standards of living. In 2000, for example, Spanish households spent 11.4 per cent of their income on transport compared with 15.3 per cent in Italy. But the general trend, which is obvious in both the north and the south, is for rising household transport budgets. The average increase for the whole of the Mediterranean between 1984 and 2000 was 3.7 per cent per year (3.9 per cent in the NMC and 3.3 per cent in the SEMC).

In future this increase may well level out in the EU (0.6 per cent per year is projected for 2000–2025) and in the islands (1 per cent). On the other hand, the increase is expected to remain very high in the Balkans (3.4 per cent per year) and to a lesser extent in the eastern Mediterranean (1.7 per cent). However, similar disparities can be expected in 2025 as in 2000 since developments

Figure 3.27 Number of road deaths, baseline scenario, 2025

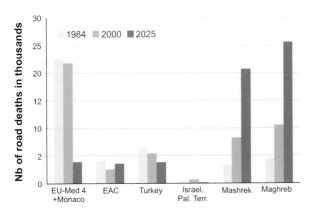

Source: Plan Bleu

Note: The projection was made by applying the annual average growth rate in number of deaths per 1000 vehicles from the period 1984–2000 to that of 2000–2025.

Figure 3.28 Share of transport in household budgets, 1985–2000

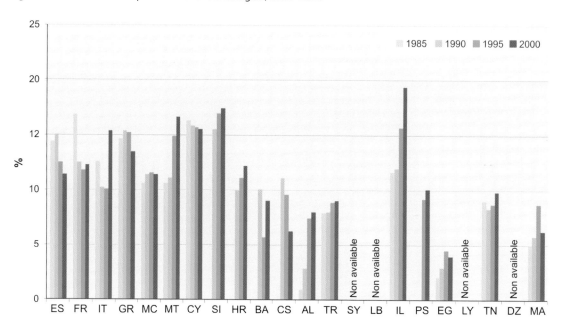

Source: Plan Bleu

Note: Data come from national statistical yearbooks. The Turkish series was reconstituted from household surveys from 1987 and 1994 based on a linear increase. Algeria conducted a survey in 2000 but does not yet have a time series. The same applies to Lebanon for which the World Bank (WDI) only reports 1998.

seem to be related more to the physical and technical characteristics of each country's transport systems than to differences in living standards (Figure 3.29). But throughout the Mediterranean the indicator remains high and increasing, compared with the rest of the world.

Another important consequence of increased road traffic is *road congestion*, which has costs in terms of lost work time, extra pollution, premature wear and tear on vehicles, user stress, accelerated wear of some roads, etc. An INFRAS/IWW[17] study, the findings of which were published in 2000, put these costs at 35 thousand million euros in the EU in 1995.

These costs for the whole of the Mediterranean were estimated to be about US$41 thousand million dollars in 2000 (Figure 3.30). Here again the differences between countries are enormous (US$14 thousand million in France in 2000, US$1.6 thousand million in Turkey and US$0.4 thousand million in Egypt), but the trend is a general and very rapid increase (16 per cent per year between 1995 and 2000). These costs already accounted, in 2000, for between 0.3 per cent of GDP in Bosnia-Herzegovina and 1.7 per cent in Greece.

The general increase in congestion costs is projected to continue, both in absolute terms and as a percentage of GDP (Figure 3.31). By 2025 the problem may well become significant since the costs for the whole Mediterranean are estimated at US$185 thousand million but will mainly concern the EU, Slovenia, the islands, Israel, and Serbia-Montenegro. Only those countries may have

Figure 3.29 Share of transport in household budgets, baseline scenario, 2025

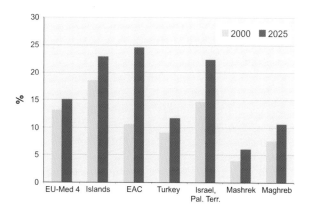

Source: Plan Bleu

Note: 'Percentage of household expenditure on transport/passenger road traffic' elasticities for 1984–2000 were applied to 2000–2025 projections of road passenger traffic.

Transport

Figure 3.30 Costs of congestion, 1995–2000

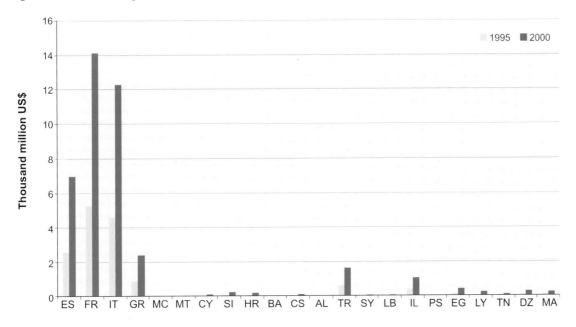

Source: Plan Bleu

Note: Based on the 1995–2000 INFRAS/IWW estimates for the four EU-Med countries, the percentage of GDP absorbed by congestion costs was calculated for these countries and applied directly to estimate the costs for Malta, Cyprus, Slovenia and Israel, increased by 55 per cent for Monaco and reduced by 20 per cent for Croatia, Bosnia-Herzegovina, Serbia-Montenegro and Turkey and by 40 per cent for all other countries.

more than 1.4 per cent of their GDP absorbed by congestion costs. For the other countries, this proportion may well increase between 2000 and 2025 by between 25 per cent (Syria) and 85 per cent.

The sum of all these effects suggests that the explosion in land and air travel will impose *increasing environmental and social costs* on all Mediterranean countries, north and south, which will be harder and harder to bear, especially insofar as this increased transport benefits road and air travel to the detriment of rail. Reversing these trends is all the more difficult in that the transport sector (like all service sectors) has played a significant role in economic growth over the past 15 years (see Part 1).

Inadequate responses

Four main policies are being used in the Mediterranean to try and staunch the rise of the above problems:

1. decoupling economic growth from transport demand through national spatial planning policies and controlling urban travel (discussed in Chapter 4);
2. improving the eco-efficiency of each mode, providing as much or more transport but with less social and environmental damage;
3. changing the modal split by switching all or some traffic to less-polluting modes;
4. improving road safety.

Figure 3.31 Congestion costs, baseline scenario, 2025

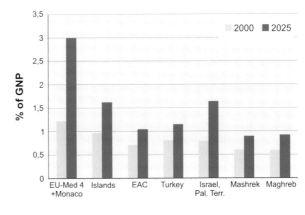

Source: Plan Bleu

Note: The increase projected by INFRAS/IWW for Europe to 2010 was applied to all the Mediterranean countries.

Controlling transport demand by national planning and good urban planning

The most important measure is proper planning of urban passenger and freight transport. Two countries have started using this kind of measure on a large scale: France (with Plans de Déplacements Urbains, PDU) and Italy (with Urban Traffic Plans, UTP).

In both cases, local authorities are required to direct urban passenger and freight transport on the basis of:

- reducing car traffic, with a parallel development of public transport and economical and non-polluting modes, and promotion of car-sharing in companies;
- developing and using the main transport networks in each urban area in a way that optimizes the use of the networks by each transport mode;
- reducing the impact of freight transport and delivery on traffic and the environment.

It is difficult to draw conclusions about these experiences because they are relatively recent and because their impacts are difficult to dissociate from urban management as a whole (see Chapter 4).

Favouring less-polluting modes

The atmospheric pollutant emission coefficients per passenger-km of *rail* are far less than those of road and air. One passenger-km by train emits 0.02g of VOCs and 0.08g of NO_x, compared with 0.5 and 0.6g respectively for cars and 0.07 and 0.7g for planes. The ratios are more or less the same for tonne-km.

Rail also occupies less ground area, for a given network length, than roads (on average a linear rail kilometre uses 3.5 times less ground area than a linear road kilometre). But rail is not the only 'sustainable' substitute for road transport. Coastal shipping and combined rail/road/sea freight transport, are possible ways of diversifying transport by 2025. However, there are only a few very localized examples of the latter in the Mediterranean.

Favouring rail traffic

There are three types of situation in the Mediterranean:

1. Spain, France and Italy, which are vigorously constructing high-speed lines (LGV);
2. the eastern Adriatic countries, which have old, already organized rail networks that could again play a relatively large role with some significant investments, but are, instead, prioritizing roads;
3. the other Mediterranean countries, which, with the exception of Egypt for passenger transport, do not have a strong tradition of rail and whose networks remain very underdeveloped. The special case of Turkey should be noted, however; its network has been vigorously developed over the past ten years.

It is therefore mainly in Europe that there is any chance of a significant modal switch from road to rail during the next 25 years.

The priority objective set by the EU (60 per cent of infrastructure investment by EU countries to 2010 to be earmarked for rail, 30 per cent for roads and 10 per cent for other modes[18]) is to build a European network of LGVs. The north-western Mediterranean countries very quickly adopted the innovative technology of the High Speed Trains (TGV); in 2002, there was 1354km of LGV track under construction and nearly 9000km planned, and industries to produce rail equipment adapted to this technology have been established. Spain, France and Italy are among the dozen countries provided with this equipment (Box 3.7). However, the rail networks of the Mediterranean Basin were long thought to be doomed to marginalization, even to disappear. Uneven topography and difficult construction conditions explain why until around the mid-1980s the search for economies won out over the determination to establish high-speed lines adapted to freight. Even today there are limits to the capacity and speed of trains: 75km per hour between Ventimiglia and Genoa, 85km between Barcelona and Port-Bou, 105km between Madrid and Barcelona, 78km

Table 3.1 Plans for investment in rail infrastructures in the north-western Mediterranean

Country	Amount (thousand million euros)	Period	Reference	Annual investment (thousand million euros)	Rate per capita (euros)	In % of GDP
Italy	62	2001–2010	General transport plan 3/2001	6.25	109	0.59
Spain	38	2000–2006	Infrastructure Plan 2000/2006	5.34	136	1.03
France	18.2	2000–2010	CIADT 18/12/03	1.8	31	0.14

Source: Plan Bleu

between Cannes and Saint-Raphael and 96km between Marseilles and Toulon. Correcting the shortcomings of these networks is difficult and expensive; moreover, building LGVs requires the projects to meet certain imperatives: the existence of large traffic flows between destinations a few hundred kilometres apart; the will to find free slots on saturated lines to develop piggyback transport; and the promotion of a non-polluting but sometimes noisy mode of transport. Despite these difficulties, the TGV is beginning to dominate in France, Italy and Spain.

> **Box 3.7** The development of high-speed rail lines (LGV) in Italy, Spain and France
>
> **Italy** was the world's second country (after Japan) to begin building LGVs in the mid-1980s. A first line was put into service between Rome and Milan and a second between Rome and Florence. An extension of the latter to Naples was to be completed in 2004, then later through the Apennines to Bologna and Milan. Changes in the initial Rome–Florence section should increase speeds. The result of these operations will mean an appreciable increase in speed, although this will remain modest compared with Italy's neighbours. The Italian approach is to combine the exploitation of conventional and new lines to maximize flexibility and capacity, with many connecting lines between the two networks to enable freight trains to use both. This will require expensive additional construction and explains why the Italian LGV network will not be completed before 2015. However, even when completed, this new system will not relieve the roads leading towards the eastern Adriatic countries and Spain. Indeed nothing is planned for the Ventimiglia–Milan route or the Milan–Lecce route (which still takes ten hours). However, the Italian industry, Eurostar Italia (ETR), has delivered high-performance tilting trains, so there is room for manoeuvre for modal shifts which remains to be explored.
>
> In **Spain**, the difference in gauge between the Iberian peninsula and the rest of Europe (1.67m as against 1.43m) has long impeded the development of rail transport. The construction of a new network using European standards based on LGVs, for freight and passengers, seemed essential to give rail a real boost.
>
> The first line, put into service in 1992, preceded this overall programme. It was built between Madrid and Seville on the occasion of the world exhibition. This is the only operating European-standard line. An ambitious programme has been designed to provide as much national coverage as possible. The new network will be focused around Madrid, with a series of routes linking Madrid to the main provincial cities, with many branch lines since, in contrast to France, Germany and Italy, there is no inter-operationality between the old and new networks, which would require the use of variable-gauge rolling stock. Two cross-country lines are planned connecting Grenada with Algesiras and Almeria with Barcelona, joining the Madrid–Barcelona line at Tarragona.
>
> The first project underway, the Madrid–Barcelona–Perpignan line, should come into service by 2005. By then, Spain will have a 1300km LGV line from Seville providing proper access to European rail systems. Later, this line will be completed with international standardization of the gauge of the Pau–Saragossa line, due in 2006, and a twin tunnel reserved for freight through the Pyrenees to be implemented by 2020.
>
> The trains to be used on the Spanish network will have an average commercial speed of 239 kmh.
>
> The **French** situation is paradoxical: although one of the pioneers of the TGV, after bringing the LGV line linking Paris and Marseilles into operation, the public authorities seem to be hesitating about completing the network towards Italy and Spain. Initial studies on extending the LGV between Aix and Fréjus were only begun in 2001. And the Languedoc branch (Manduel–Perpignan) is still awaiting funding.
>
> Thus even in Italy, France and Spain, a modal switch from road to rail using the TGV (depending on the freight/passenger mix) is unlikely to be implemented before 2010/2015. In particular, the concentration of traffic flows on a few lines (essential if rail is to compete with road) requires an institutional environment that enables such networks to emerge, that is one that enables negotiations between rail operators and governments. This modal switch may occur sooner for higher value-added products, particularly using sea containers within the framework of an inter-modal rail–sea system. Support will then be needed for a high-performance 'core network' that would enable a new kind of rail operation relying on the major corridors. Such a scheme again requires a redefinition of the 'rules of the game' for rail transport. The current hesitancy of the French, Italian and Spanish governments towards rail makes it unlikely, for the time being, that such winning strategies will be developed.

In *Turkey*, the re-opening of the rail routes in the eastern Adriatic countries and significant investment in rolling stock should help to maintain the current healthy state of the main operating company. But this will only make sense if the Turkish governments make a break from the 'road-dominant' culture which marked public investment policies during the 1990s, when roads accounted for nearly 25 per cent of capital public expenditure[19] compared with just 4 per cent for rail.

In the *Near East* the situation is in limbo until the Israeli–Palestinian conflict is resolved and the possible re-opening of the Lebanese network (Box 3.8).

> **Box 3.8 The prospects for re-opening the Lebanese rail system**
>
> The civil war dealt a fatal blow to the operations of the Lebanese rail system, which consisted of three main lines: Tripoli to Sidon (130km), Beirut to the Syrian border (60km) and Tripoli to the Syrian border (40km). In 2000, the Lebanese government decided to refurbish the rail link joining Tripoli to the north-eastern border with Syria. Two phases are planned: the operation of a single track along the present route beginning at the end of 2003 and of a second track by 2005. This project is being funded within the context of the national budget (about 20 million euros). The contractor (the Office for Railroads and Public Transport, a public agency under the aegis of the Ministry of Public Works and Transport, which enjoys a 50 per cent state subsidy) is planning to issue a call for tenders for operating the line within the context of a public–private partnership. Although the main motive behind this project is to give the port of Tripoli a rail link that should help to develop inland areas, this first rail operation, after decades of neglect and delay, could produce a 'snowball' effect for Lebanese railways in general.

In the *Maghreb* countries, objectives converge around an improvement of rail transport. The network in Morocco has been greatly modernized and strengthened during the past five years, and the government has announced an extension of the system to the north and south. In Tunisia efforts are focused on greater Tunis and there are plans for public transport with dedicated lanes to complete the existing light metro line. In Algeria, the government invested more than 95 million dinars in rail between 1990 and 2000 without, however, preventing a fall in rail traffic (from 2.7 thousand million tonne-km of freight in 1990 to 2.0 in 2000, and from 3 thousand million passenger-km to 1.1). With the lifting of the embargo in 1999, the Libyan government is once again planning to build a line between Sfax and Tobruk. However, neither the priorities of the national transport policies nor the future projections of the Arab Maghreb Union (UMA) show any sign of a large-scale programme to develop a Maghreb network serving all the countries of the UMA, even in the long term.

In summary, while waiting for the implementation of the LGV network, maintaining the market share of rail, even in Mediterranean Europe, requires *determined policies* that aim to reorganize and significantly improve the performance of rail. Beyond aiming to maintain the status quo, political decisions are needed, aimed at giving freight priority over passenger transport and providing investment to increase the capacity of the rail system. Between 2000 and 2025 rail will barely achieve a market share of more than 4–5 per cent, which, compared with current trends, would constitute an important recovery and maintain a credible alternative to road transport.

Favouring combined rail/road/sea/river freight transport

A modal switch to combined transport is particularly important in the western Mediterranean and even more so in Mediterranean Europe, which generates and whose ports receive most of the area's freight traffic. Outside Mediterranean Europe, maritime traffic is mainly in transit. In the eastern Mediterranean, political crises have postponed structural investments in transport to the distant future, although there are some projects, such as the rail project proposed by Israel and the Palestinian Authority within the framework of the Regional Economic Development Working Group (REDWEG) project, described below.

For Europe, a comparison between northern European ports and those in southern Europe show how far there is to go in the Mediterranean before combined freight can be developed further. For example, the share of road in carrying freight from ports in 1999 in the north (Benelux, UK and German Atlantic coasts and the English Channel in France) was less than 60 per cent compared with 80 per cent in Mediterranean Europe (Mediterranean France, Italy, Greece and Spain). The difference results mainly from the use of combined water/rail transport.

Mediterranean Europe has some big *rivers* (the Rhone and the Po). The Rhone in particular is navigable by large ships. That is why many analysts have long been suggesting the following uses of these rivers:

- *In the Rhone basin*: cancelling the tax levied when passing through the Port of Marseilles; improving the navigable sections to end the near-mandatory use of a pilot for sailing upstream; and recognizing the international nature of the Rhone to enable ship-owners to benefit from tax exemptions for fuel. As far as capacity is concerned, the Rhone offers considerable potential up to Lyons (current traffic is only about 10 per cent of capacity). The Rhone can accommodate river- and sea-going vessels of 2500 tonne dead-weight up to Lyons, which opens significant possibilities for combined sea/road/rail transport with the opening of the multi-modal Saint Exupéry complex in Lyons.
- *In the Po basin*: major initial improvements are needed to make navigation possible up to northern Italy's main economic zone (Milan and the Lombardy region, which generates nearly one-third of Italy's imports and exports). The navigable sections of the Po only accept ships of 1250–1450 tonnes, but international traffic calls for the ability to accept ships of at least 1500 tonnes.

Beyond these two river routes, which offer the only real possibility of large-scale combined rail/road/sea transport in the Mediterranean, there are some examples of successful combined transport (Box 3.9).

> **Box 3.9** Examples of successful combined transport
>
> **Italy** is without doubt the leader in combined transport. Its geography, the distances between north and south, and a population of more than six and half million people living in islands, have resulted in this possibility being exploited. Italy uses combined sea/road/rail transport at both national and international levels.
>
> At the domestic level, it supplies the islands of Sicily and Sardinia and ports along the Adriatic and Ionian Seas (between Venice and Catanzaro). Competition between this form of combined transport and road transport has long been nearly impossible. Only ships able to do 24–28 knots could hope to compete with trucks. But such speeds were impossible for freighters. With the entrance of operators already active in other modes (such as Benetton and the Chargeurs Réunis) combined sea/road transport has become a reality, and completing the LGV network will add rail to the combination in a few years' time.
>
> At the **international** level, the ports of Brindisi, Ancona, Ravenna and Trieste are the mandatory landing points for road traffic from Greece, Turkey and the Middle East. Regular services are also developing on the west coast between Genoa and Barcelona, and the Grimaldi shipyard in Naples is becoming a leading operator as a storage and handling hub between Tunisia and the Mediterranean-to-Rhine corridor. Political turbulence in the former Yugoslavia and the Middle East resulted in road transport being so unsafe that combined sea–road transport took its place.
>
> For this traffic, the **Greek** maritime companies play the dominant role with the Italians usually reduced to the role of maritime agents, handling companies in ports and managing unloading–reloading to rail or road (combined transport companies such as Combinare). Most traffic is concentrated in the ports nearest the Greek coast (Bari, Brindisi), but Ancona and Venice have grown following the introduction of super-fast vessels. Turkish haulage contractors have also entered this market through a road hauliers association (UND), which started by fitting out two ships in 1990 and now owns 12, devoted exclusively to this traffic.

Thus with a few minimal improvements, there is genuine room for manoeuvre for combined road/rail/sea/ river transport in the western Mediterranean. However, there seem to be few prospects in the rest of the Basin although some inter-modal links have occasionally emerged, particularly as a result of the cutting of the land and river routes in the former Yugoslavia.

The effectiveness of coastal shipping

Coastal shipping as an alternative to road freight transport is based on the idea of 'seaside highways', parallel to coasts (similar to what combined transport has already made possible in Italy) or as direct links (Mediterranean or Atlantic). It assumes high-frequency traffic, which depends on fast ships and maximum efficiency at ports (high-grade infrastructures and precise regulations, that do not slow transit speeds too much). It also requires the active participation of road hauliers who must adopt international integration and occasionally new packaging techniques (for example containers and unaccompanied trailers).

The need to satisfy all these requirements probably explains the failure of several attempts to open 'short-sea' routes in recent years (notably between Toulon and Savona and Sète and Palma) (Box 3.10).

> **Box 3.10** The Franco-Italian short-sea shipping project
>
> Marseilles ship-owners had already launched the 'coastal métro' concept in 1995, before the accident in the Mont Blanc tunnel in 1999 gave an even more direct impetus to an alternative link between France and Italy by sea rather than through the Alps. In 2002, based on the transport potential of the inland areas of Marseilles, southern France and the Italian peninsula, five shipping companies (CMA-CGM, CMN, Marfret, SNCM and Sud-Freight) founded the Société des Autoroutes Maritimes du Sud. With 880,000 euros of aid from the European Marco Polo programme, the Marseilles project provides a link between Fos and Savona and aims to capture 2 per cent of the road traffic passing through Ventimiglia. The launch was, however, postponed in early 2004 because of a lack of public funding.

A greater supply of SSS (short-sea shipping) between the western Mediterranean ports could provide a wide range of services:

- regular services between ports (roll-on roll-off, with or without loading lorries onto ships);
- river–sea services between coastal and inland ports (river ports) of the Rhone and Po basins;
- transhipment of local traffic linked to intercontinental deliveries (containers).

Achieving this would require:[20]

- rationalizing and harmonizing administrative and pricing procedures at loading/unloading points (sea and river ports);
- implementing dedicated ticket-offices similar to those being introduced in the rail sector, enabling customers to have a single agent organizing door-to-door transport; this has succeeded for rail and should be transposable to short-sea services;
- providing financial support (as often enjoyed by combined transport in general and particularly com-

bined rail/road in Europe) to enable survival beyond the early years of operation that are financially difficult even when large customers (in terms of volumes transported) use the service from the beginning. Current allocations of EU resources (for example the PACT and Marco Polo programmes) are generally considered to be insufficient;
- research and development to design and try out technical options (at both interface and navigation-equipment levels), that might enable the economic performance of these very specific maritime services to be improved.

Freight transport projects are included in trans-Pyrenees and trans-Alpine traffic scenarios.

Previous developments of transport policy frameworks towards pricing for the use of infrastructures that includes realistic internalization of externalities are required as for other alternative rail and water transport modes.

The scope for an increase in short-sea transport is huge. In 1997, despite an increase of 4 per cent between 1990 and 1993 and 18 per cent between 1993 and 1997, it accounted for only 6 per cent of the tonnage transported between EU countries (compared with 84 per cent by road), and 33 per cent of the tonnage of non-community traffic (45 per cent by road).

Increasing the eco-efficiency of transport modes

The results of the improvements in engine technology declared by car manufacturers can be measured by analysing pollution emission curves. However, this improvement must be interpreted in the light of the very advanced age of the car fleet in the southern countries (83 per cent of the fleet in Morocco and Algeria is more than ten years old) and the slow rate of fleet renewal. Moreover, these positive effects can be offset by the purchasing choices and vehicle-use patterns of consumers on the northern shores and limited purchasing power on the southern shores.

Technological improvements do not apply only to engine performance; they can also involve the use of motor vehicles in urban areas (small vehicles) and traffic organization (for example forecasting traffic peaks). Generally speaking, in Europe, these technological improvements have been swamped by increased traffic.

There are technical limitations to the potential for *clean fuels* (for example LPG and electric vehicles) and also environmental and economic limitations (biofuels). Lead-free fuels and catalytic converters are promoted in most Mediterranean countries (in the EU-Med countries and also in Tunisia, for example, with a law passed in August, 2001). LPG is very effectively promoted in Algeria where, thanks to its attractive price (a third of that of petrol) and a partial subsidy by the state for installing gas tanks, demand increased by 730 per cent between 1995 and 2001. Favouring clean fuels also penalizes the most polluting fuels. In Lebanon, an experiment was conducted in 2002, banning diesel engines for vehicles that carry fewer than 12 passengers (the measure was subsequently repealed).

In many SEMC, for example Turkey, Lebanon and Tunisia, the use of *collective taxis* or private cars for commercial transport results in a significant decrease in emissions per passenger-km, because of the higher occupancy rate of passenger vehicles. Moreover, this informal transport provides a significant source of occasional income. But the attractiveness of this kind of solution must be considered in the light of the need for improvements in conventional urban passenger transport. In the NMC there is no policy for promoting car-sharing, common in the US for example.

Improving road safety

The frequency of road accidents can be reduced by active road safety policies in both the north and the south. However, the cost of some of the policies and the infrastructures that would be needed to implement them means that some can only be considered in the northern-rim countries.

In the SEMC

There appear to be five priorities for improving road safety.

1. *Statistics*. In most SEMC, the recording of accidents is of poor quality. The lack of standard recording and filing databases that enable the storage and analysis of accident data in time and space precludes a proper understanding of the way the situation is developing.
2. *Coordination*. Road safety is at the crossroads of many institutions and social and economic actors: ministries (transport, national planning, education, health), car manufacturers, etc. A common framework must be created if progress is to be made. Some countries are beginning to implement such frameworks, for example in Algeria, where a National Prevention Council for Road Safety was set up in 2000.
3. *The road environment*. Road networks in the SEMC still have many high-risk sections; traffic signals and street marking are still inadequate and there are frequent obstacles on the road (for example animals, plants).
4. *Road traffic regulations*. Regulations in a number of SEMC have been improved since the mid-1990s, for example the mandatory use of seat belts nearly everywhere and speed limits. But their application still remains too theoretical. The lack of policing and above all the small fines imposed is the main

reason for dangerous driving behaviour continuing. The implementation of technical vehicle verification centres (such as those in Algeria and Morocco[21]) will certainly have an effect on the level of accidents, since, in contrast to the north, vehicles in the south are sometimes the cause of accidents (for example 13 per cent in Algeria compared with 1 per cent in France in 1998).

5   *Training and information.* Training in driving schools is very uneven. A reform of qualifications for driving school instructors seems to be a prerequisite of any serious attempt to improve the situation. There are also too few campaigns for raising awareness of violence on the road.

In the EU Mediterranean countries

Accident rates in the four EU-Med countries seem to have changed little during the past decade or so. Traditional policies to reduce law-breaking (making excessive speed a crime in 1999 and installing radar speed trap on roads in France in 2003) are measures with a proven effectiveness However, the emphasis of accident prevention has now been shifted to the interaction between the road and the driver through the 'intelligent road' concept. Several avenues to 2025 have already been identified:

1   *Traffic management.* This is a matter of knowing what is happening in real time on the network in order to be able to react and forecast. There is greater use of systems that provide immediate information about what is happening and what is likely to happen on the major roads (the motorway networks): vehicle speed, traffic flows and 24-hour forecasts.
2   *Information on traffic conditions.* Motorway system managers are increasingly installing systems for broadcasting information to the authorities and to drivers. These use local sensors, radar and video cameras, making it possible to direct traffic around accidents very quickly and prevent chains of accidents.
3   *Navigation information.* On-board computers on which drivers can programme their route with a series of maps.
4   *Driving aids.* The aim is to automate driving by fitting roads with systems that take the place of the driver. Two recent projects illustrate what may be in store: the HAS project in the US (1997) aimed at equipping the San Diego freeway with a system that uses magnetic studs every metre to control driving automatically and double the carrying capacity of the road without incident, and the *Prometheus* project on 55km of a Dutch motorway near Amsterdam, at a cost of 1.22 thousand million euros, with the same objective.

These various approaches give an indication of possible improvements in land transport systems. However, they seem to have resulted in only minimal benefits (maintaining the share of rail at the current level, isolated examples of combined transport and coastal transport, and little progress in road safety). The measures that have been taken appear to have been too sectoral and scattered in time. The alternative scenario described below assumes the integration and simultaneous adoption of these different measures in a coherent way to achieve sustainable transport development and much greater progress.

Reducing pollution and the risks of maritime transport

Without an overall and up-to-date matrix of origin/destination data, it is impossible to give a precise picture of the actual maritime traffic in the Mediterranean. By default then, it is the Mediterranean-flag fleet that will be described in the following paragraphs even though this fleet sails in many seas. This will provide an indirect indicator of the state of the fleet circulating in the Mediterranean and what is being done to make traffic safe.

The Mediterranean merchant fleet has characteristics that are poorly matched with the requirements of sustainable development. As with land traffic, maritime traffic is increasing significantly. This is already leading to major environmental impacts that risk becoming much more severe unless there is a substantial improvement in safety.

An unsafe and poorly controlled Mediterranean merchant fleet

The fleet licensed under Mediterranean flags[22] carries a large proportion of *environmentally dangerous substances*. In 2001 nearly 42 per cent of the Mediterranean merchant fleet carried hydrocarbons and chemical products, with the possibility of pollution from routine operations and accidents (Figures 3.32 and 3.33).

As a second risk factor, the fleet includes many ships with *open-shipping registrations*, which are or may be 'at risk'.

Open-shipping registrations (Malta, Cyprus and Gibraltar), that allow the licensing of shipping companies with almost no formalities or prior checks, generate considerable savings for ship owners in terms of manpower and lack of inspections of their vessels (given the disproportion between the sizes of the countries that hold these registers and those of the fleets flying their flag). Between 1987 and 2001, the deadweight (transport capacity) of the Mediterranean merchant fleet increased by 27 per cent, but the open shipping registrations by 123 per cent. Moreover, it was

Figure 3.32 The Mediterranean-flag merchant fleet, breakdown by freight type, 2001

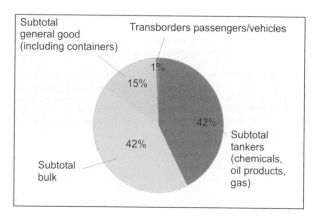

Source: Dobler (Eco-Mar), 2002

Figure 3.33 The Mediterranean-flag merchant fleet, breakdown by country, 2001 (million TDW)

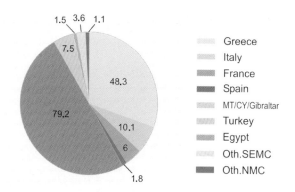

Source: Dobler (Eco-Mar)

Note: TDW – deadweight tonne (the maximum capacity of transport authorized for ships)

tical practices promoted by the European Commission for the eradication of 'sub-standard' ships. Until now such ships have been relatively scarce under the national flags of EU countries. However, with the entry of Malta and Cyprus, the situation may change completely since 80 million dead-weight tonnage (TDW)[23] of open registry will be added to the 67 million of the traditional marine country flags. The EU intends to tackle these deficiencies by improving inspections of ships entering its ports. This response may prove inadequate since, even in Europe, most states monitor less than half of the ships that put into ports and, once the deficiencies are noted, a ship is rarely held for very long at the quay given the financial stakes involved. Moreover, most of the Mediterranean countries do not have a sufficiently large (and qualified) maritime affairs administration to reduce the deficiencies in the supervisory checks required by the flag status.

The *high average age of the fleet* constitutes another risk. It is very worrying to find fleets under Mediterranean flags with an average age of often more than 20 years (19.7 years world average): 23 years in Egypt and Slovenia and also Greece, 24 years in Turkey and Algeria, 27 years in Albania, 29 years in Croatia, 33 years in Lebanon, 34 years in Serbia-Montenegro and 43 years in Syria. An analysis by type of ship shows an average age of 55 years for the four Turkish 'cruise ships' and 68 years for the two Maltese 'passenger-freight ships'. The risks to the environment from these old ships are confirmed by the organization of the Paris Accord Memorandum[24] that classifies all the world's flags every year according to the social and environmental risks involved. The lists published by this organization are drawn up according to the number of detentions for non-conformity with international rules following inspections carried out in the ports of Europe and North America. In 2002 Algeria, Lebanon, Syria, Turkey, Libya and Morocco figured on the black list as posing a very high risk, Egypt was classified as high risk, followed by Malta and Cyprus as medium risk. The grey list included flags considered dubious in the eyes of the Memorandum's inspectors and thus likely to require special attention. In the Mediterranean, this is the case for Gibraltar, Italy and Spain.

Controlling operational and accidental pollution

Operational pollution

Operational pollution from ships by hydrocarbons includes emissions of several kinds of hydrocarbon and hydrocarbon mixtures generated on ships, including oil tankers, during routine operations. The expression encompasses ballast waters[25] (slop), tank washing residues, sludge and bilge waters.

the open registrations of the flag countries least endowed with maritime administrations and with almost no national ship owners that recorded the biggest increase in dead-weight. For example Gibraltar (with limited tonnage) and Cyprus experienced growth rates of about 30 per cent while Malta recorded a huge increase in registrations of 542 per cent in 14 years. Given its population size and resources, it is unlikely that Malta will ever be able to carry out all the technical and social inspections required by the size of the fleet registered under its flag.

The fact that Malta (as well as Cyprus) is now a member of the EU poses a problem concerning the poli-

Transport

In the Mediterranean, in contrast to what happens in most of the world's oceans, the international waters are regulated. The MARPOL 73/78 convention states that the Mediterranean is a 'special area' and as such, any discharge into the sea beyond territorial waters of 12 nautical miles of *hydrocarbons or hydrocarbon blends* from a tanker (or any other ship of a certain size) is forbidden.

Despite this, sludge and residues are still being emptied into the Mediterranean, mainly because of the lack of adequate reception facilities in some ports and terminals. Sea monitoring can be done with remote sensing, and satellites and could be further developed and coupled with monitoring of seaways near the busiest passage points. However, the quantities of residues from de-sludging have been falling steadily since the mid-1970s, to an estimated 100,000–150,000 tonnes of hydrocarbons (Figure 3.35 below),[26] although the WWF reckons that annual emissions are 12 times greater.[27] This is certainly due to developments related to the MARPOL convention, which has required the inclusion of separate ballast tanks in ship construction since 1979. However, diesel and engine-oil residues are continuing to grow.

For *other polluting products* carried or used by ships (chemical tankers in particular), the Mediterranean is not a special area, and, apart from the ban on dumping less than 12 nautical miles from shore, there are no rules.

One of the immediate answers to the problems of operational pollution remains equipping Mediterranean ports with recovery and liquid and solid waste-treatment systems. According to a study carried out by a professional organization of Greek ship owners in 1997, 54 of the Mediterranean's 123 ports studied were equipped with this kind of system.[28] There is still the problem of funding the use of these systems, even where they exist they are often underused, since their cost remains relatively high. EU Directive 2000/59/EC of 27 November, 2000 makes it mandatory for Member States to include the costs of using these systems in port charges. In the Mediterranean, non-members of the EU are waiting to see the results of this initiative before committing to such procedures.

Accidental pollution

Since 1977, under the Mediterranean Action Plan, the Regional Marine Pollution Emergency Response Centre (REMPEC)[29] has methodically recorded *accidental* pollution by compiling reports on accidents that have or could have caused marine pollution by hydrocarbons. Between August 1977 and December 2000, 311 accidents of this type were recorded, 156 of which were followed by an escape of hydrocarbons. Only two accidents recorded in the Mediterranean between 1981 and 2000 led to emissions of more than 10,000 tonnes: 18,000 tonnes from the Cavo Cambaos in 1981 and an unspecified amount from the Haven in 1991. The number of accidental emissions recorded per year increased from 2 in 1982 to 11 in 1991, 1992 and 1993, while the quantity emitted fell from 36,500 tonnes in the 1980s to 21,700 tonnes in the 1990s. But behind this apparent improvement, the *nature of the spilled products* is of increasing concern. Between 1996 and 1999, the proportion of persistent hydrocarbons (heavy diesel and crude oil) was 75 per cent, compared with 65 per cent between 1981 and 1995. Moreover, given the sensitivity of the region for tourism, the costs of a disaster of the Prestige or Erika type in the Mediterranean are certain to be far greater than anything that has occurred in the oceans.

For pollutants other than hydrocarbons, REMPEC has recorded at least 79 accidents between 1988 and 1997 involving ships carrying noxious substances (minerals, sulphur, chemical products, LPG, ether, glycol, etc.). The total discharge of such pollutants is difficult to assess, but REMPEC's accident files record four 'incidents' in 1996 that resulted in the spilling of 1500 tonnes of phosphate, 2703 tonnes of chromium ore and 7600 tonnes of phosphoric acid. That may seem negligible, but with the expected increase in traffic and the size of ships (the development of container-carriers), and the Mediterranean's geophysical and geographic characteristics, ecological disasters are increasingly likely.

Observed trends and scenario to 2025

The general trend is of a spectacular drop in cumulative operational and accidental pollutant emissions (by 17.6 per cent per year between 1985 and 2000). But this should be examined at two levels. First, two important sources (probably increasing[30]) are unknown: accidental emissions of non-hydrocarbons, and losses of containers at sea. Second, the general trend hides increases in two operational pollutants, the volumes of which are increasing: fuel and oils residues (sludge) from all ships and the emptying of ballast water from chemical tankers which is not subject to any specific regulation (Figure 3.34).

The baseline scenario projects an overall increase in emissions (by 5 per cent per year between 2000 and 2025) essentially resulting from an enhancement of past trends (increasing dumping of ballast water from chemical tankers and sludge) and a slight increase in accidental emissions and tanker de-sludging, resulting from the very large projected increase in marine traffic (Figure 3.35).

Improving the safety of maritime transport

The international community has made efforts to remedy the specific situation in the Mediterranean.

Figure 3.34 Accidental and operational polluting emissions into the sea related to maritime traffic, 1985–2000

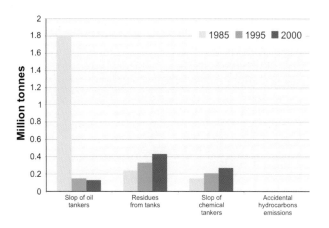

Source: Plan Bleu

Note: Fuel oil consumption by Mediterranean maritime traffic was estimated at 30 per cent of world consumption by an international assessment carried out by the EU in 1997 entitled 'Study on the Feasibility of a Mandatory Discharge System of Ships Waste to Shore Reception Facilities in Ports'. Extracts of de-sludging coefficients were also taken from this study (0.5 per cent of the transported cargoes) and the percentage of cargoes de-sludging in the sea (83 per cent according to the analysis of the Rotterdam case). For ballast water, the coefficient of 0.47 per cent of the used oil fuel finishing up in the form of fuel residue or ballast water was also taken from this study. The respective percentages of hydrocarbon and chemical-product transports were deduced from J. P. Dobler's estimates in 1985 for the Blue Plan. From the same author, a reduction by a factor of 20 between 1985 and 2000 in oil de-sludging emissions was assumed for assessing the development of this kind of pollution. Data on accidental emissions come from REMPEC.

Figure 3.35 Accidental and operational polluting emissions into the sea related to maritime traffic, baseline scenario 2025

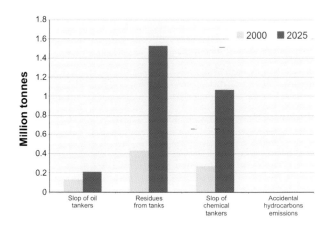

Source: Plan Bleu

Note: Fuel-oil consumption in 2025 is based on the baseline scenario assumptions for energy intensity (0.9 per cent decrease per year in energy intensity from now to 2025). For de-sludging, the 1985–2000 trend was extrapolated, and for accidental emissions, the 2000 ratio of emitted tonnage/traffic was projected to 2025.

The legal foundation for international cooperation in preventing, reducing and abating pollution in the Mediterranean is based on the *Barcelona Convention* for the protection of the marine environment and the coastal region of the Mediterranean (1976), ratified by all the Mediterranean-rim countries and the European Community, and amended in 1995. A Protocol relating to cooperation in combating pollution by oil and other harmful substances in the Mediterranean in cases of emergency, commonly known as *Emergency Protocol* (1976), was revised in 2002 and came into effect in 2004. It covers the prevention, preparedness for combating, and combating marine pollution. The revised text is in accordance with other international legal requirements, particularly the OPRC 90 Convention (a Convention on preparedness, combating and cooperation in combating hydrocarbon pollution) and takes into account the European Community's contribution to implementing international maritime safety standards and preventing pollution from ships.

By committing themselves in such a way, the Contracting Parties to the Barcelona Convention accept that navigation is an international activity that must obey rules adopted on an international scale (the International Maritime Organization – IMO) and that these rules apply in the Mediterranean without the need for additional legal provisions.

This continuing concern for safety is beginning to be expressed in domestic legislation and the inspection levels of ships using ports. Since 1983, the costs of ship inspection measures and intervention units in cases of pollution in Israel have been met by an eco-tax on oil imports (0.01 cent per imported litre) and a one-off duty on all ships entering an Israeli port. The monies collected contribute to a national fund directly managed by the marine environment sector of the Ministry of the Environment. Similarly, in Europe, it is the Mediterranean nations that show the highest ship inspection rates in ports (27 per cent on average in Europe in 2001, 43 per cent in Italy, 28 per cent in Greece and 30 per cent in Spain, but 10 per cent in France).[31]

Following the Prestige and Erika accidents, the EU introduced a certain number of measures[32] and is seeking to harmonize the legislation of the Mediterranean countries with those in force within the EU through the Euro-Mediterranean Transport Forum.[33] It is also trying to get Mediterranean countries interested in the European satellite navigating system (GALILEO), which should be operational in 2008. The MEDA programme is developing projects aimed at training appropriate administrations in the SEMC in the use of this system.

Some problems, however, remain unresolved, in particular that of the penal and civil responsibility of charter fleets, which remains a meaningless word in the light of the limitations under the IOPC provisions (international compensation funds in cases of accidental hydrocarbon spillage).[34]

However, international regulations and structures do not appear sufficiently opposable to operating companies and states. So, even if the International Maritime Organization decides to give the Mediterranean a 'sensitive area' status, the application of its recommendations and standards will remain within the jurisdiction of each state. Some countries have agreed to limit traffic and preserve their biological richness (the Franco–Italian agreement for the Bouches de Bonifacio) or to implement specific procedures in case of accidents (the *Ramogepol* agreement between France, Monaco and Italy for the Ligurian Gulf). A strategy prepared by REMPEC aims to improve safety by training administrators in maritime affairs and exchanging information between countries, currently under discussion on a Basin-wide scale. Some proposals from NGOs such as Greenpeace and the WWF even aim at instituting systematic checks of the largest ships near straits (Gibraltar, Suez and the Bosphorus). It was along these lines that Morocco, in 2002, built an assistance centre in Tangiers for the navigation of vessels passing through the Strait of Gibraltar, at a cost of US$15 million. But these initiatives run up against two principles that make them difficult to implement: the principle of free maritime circulation (thus Turkey finds itself unable to regulate the Black Sea–Mediterranean traffic despite the accidents that have already occurred) and the extra-territorial nature of the high seas beyond the 12-mile coastal limits.

Real improvements in safety requires two types of measure:

1   Improving the *scope of the conventions* from the IMO, particularly MARPOL, to cover sectors that are currently beyond their range (particularly the dumping of pollutants other than hydrocarbons, and ships operating in territorial waters) and to increase regional coordination with a view to their ratification. In 2002, of the 420 ratifications needed to obtain a Mediterranean Sea completely covered by the whole of the IMO apparatus, only 233 were agreed. An adaptation that recognizes the extra-territorial nature of the straits is needed because of the exceptional traffic in the Mediterranean (231 ships passing through Gibraltar every day and 137 through the Bosphorus).
2   The creation of a *regional intervention system* in case of accidents to enable action for those involving spills of all dangerous substances, not only hydrocarbons. This could be done in two stages. The first is to create rescue units, the costs of which would be shared according to the coastline of each rim country and the amounts of hydrocarbons and dangerous chemicals transported by each country. The investment needed for four units of this type (one near the Strait of Gibraltar and the others in the Tyrrhenian, Adriatic and Aegean Seas) has been estimated at 45 million euros, which can be put in perspective by comparing it with the costs of the spillages from the Erika (300 million euros) and the Prestige (one thousand million euros) as evaluated by the French and Spanish administrations. The second would consist of selecting places of refuge at various points along the coast. These would be sheltered from the high seas, thus near coasts, where a damaged vessel could be taken or towed if threatened by sinking, in order to lighten its load or, if it is dangerous or polluting, transfer it entirely.

Towards a more sustainable transport system

In addition to improving eco-efficiency and sectoral policies aimed at changing the modal split of traffic, general measures are vital for decoupling an increase in traffic from economic growth and for a massive switch towards greater use of rail and public transport. In the Mediterranean, as elsewhere, economic measures and international cooperation also have a role to play. But this can function only on two conditions:

1   the problems of infrastructure funding have to be resolved;
2   economic measures and international cooperation have to be part of an integrated vision of transport development that involves all actors. A combination of measures that link incentives with sanctions is likely to be the most effective way of moving towards a more sustainable transport system.

This integrated vision is the basis of the alternative scenario presented at the end of the chapter.

Rationalizing taxes and subsidies

Economic analysis enables transport externalities to be evaluated, even if some improvement in the analysis is still needed. It also indicates optimal price levels for preventing or minimizing the effects of these externalities.

The preferred way of reaching this optimum is *price setting*: through a tax on the harm generating activity being equal to the marginal cost of that harm, the invisible hand that the market makes work automatically in the case of merchant goods is recreated in a conscious way. This is the polluter-pays principle, with pricing at the marginal social costs or even at the cost of congestion in the case of loss of time in traffic (users pay the value of the delays that their presence creates for the other users).

The improvements that are currently being made by integrating the environment into transport policies essentially derive from regulation (for example banning sulphur additives in petrol, speed limits, banning single-hull tankers and MARPOL). In comparison with the many regulatory tools, examples of well-adapted *taxes* are rare. The increase in car taxation in Lebanon from 20 per cent of the price of the vehicle in 1997 to 55 per cent in 2002, was aimed mainly at reducing public deficits, and the proceeds did not go to financing rail or mass transport infrastructures. In France, the purpose of the 'axle tax' (*taxe à l'essieu*) first levied in 1970 was to get the owners of lorries to pay some of their external costs (in particular the increased damage to roads caused by their weight); it has not since been revised. The adjustment of motorway tolls around Paris was a modest experiment, partly abandoned. Congestion taxes are being tried from Asia to London (since January 2003). Hong Kong and Singapore set up an electronic toll system based on congestion. The results of these very recent experiences seem promising (for example, a reduction by 45 per cent of traffic in the areas involved in Singapore and 20 per cent in London). There are similar projects in Italy and Spain. Aeroplane noise is fought through a tax on many airports (Italy, France, Spain, Tunisia, Cyprus, Malta, Greece and Turkey), but this tax has a financial aim (to compensate local residents) whereas it could vary according to congestion. Paying for parking is a form of taxation on urban transport, but its effect is ambiguous, since it tends to increase the number of vehicles in circulation by favouring short journeys.

On the other hand there are many situations of *inconsistent pricing* in almost all the Mediterranean countries. Motorway tolls tax traffic flows, not congestion; in urban areas two motorway routes coexist, one toll-free, and the other with a toll (for example Nice, Athens, Marseilles, Madrid). Fuel price developments between 1991 and 1998 favoured the car: the prices of premium grade petrol and diesel fell (in constant dollar terms) in all Mediterranean countries and in most cases with a significant advantage for diesel. Diesel and petrol are taxed at very different rates with no relationship to their impacts on the environment. In the hydrocarbon producer countries (Algeria, Libya, Syria, Egypt), petrol benefits from a low level of taxation (see Chapter 2). The pricing of rail has become completely anarchic in Europe with prices varying by a factor of ten from one country to another without the disparities being justified by differences in cost or demand. Other measures (permits-to-pollute markets) are little used in transport.

Thus, the shortage of anti-pollution taxes and the many pricing inconsistencies contribute to the pollutant-payer principle not being applied. Why is there this gap between theory and practice?

- A first set of difficulties has to do with the *problem of evaluating externalities*, for example for congestion or road noise (Box 3.11). To overcome these problems, the use of transport is taxed in other ways, the main way being taxes on fuel. But this does not nearly reflect the cost differences between cars. The other measures that affect cars (road tax discs, parking, purchase tax, tolls) are even blinder. All these taxes have contrary effects that limit their impact. Those of payment for parking have already been noted. Fuel tax has different short- and long-term effects; it affects traffic more in the short than in the long term and influences vehicle design and consumption per vehicle. These may well be praiseworthy objectives but they are not related to the fight against congestion. All of these pricing measures affect traffic volumes but not the impacts on the environment. However, trying to reduce these impacts through reducing traffic is not very effective.[35] The fight against environmental damage should focus on old, poorly maintained vehicles with single occupancy and on driving styles (for example, bonuses when scrapped, standards on vehicle use and tolls on single occupancy).

Box 3.11 Examples of difficulties in assessing externalities

Measuring the cost of congestion is often difficult and nearly impossible to calculate with precision.[36] For example, for a single road, the cost of congestion is relatively well correlated with traffic intensity, since one vehicle slowed down by another cannot pass; the average speed reduces almost linearly with increasing traffic intensity, and calculating the cost of congestion is relatively easy. But when the road has a variable morphology, the calculation gets complicated. When several vehicles are slowed down on the same road by local narrowing, a queue forms. Traffic is higher than the capacity of the road, and the cost of congestion is no longer simply correlated with traffic intensity.

For road noise, the main source is friction between the tires and the road. Who is responsible? The motorist who uses his tires or the community that manages the road? And in what proportions? Thus, unlike the situation with most goods and given the present state of technology, making motorists pay directly for the use of their cars is not easy.

- A second set of difficulties derives from the *variability of optimal prices* depending on circumstance. The costs of congestion on roads can vary by a factor of 1000 between quiet rural roads and rush-hours in cities. Pollution from the same kind of vehicle can vary by a factor of one to five depending on whether it is well or poorly maintained. Similarly, if infrastructure costs are considered as externalities, road damage increases as the fourth power of axle weight: a 13-tonne axle should pay four times more than a 10-tonne axle, and the amount to be paid would differ from one road to another. Clearly price differences that depend on the vehicle and the circumstances would be very difficult to implement. It is also easy to see that the equalizations that would be needed would reduce the effectiveness of the pricing system.
- A third set of difficulties has to do with the *multiplicity of decision-makers* with inconsistent objectives. Pricing is not implemented through a single price only but by the combination of many different charges. This would not be too troublesome if the sums were coordinated towards a single objective. But in many countries this is not the case. Taxes on petrol depend on the energy ministries in most of the SEMC and on the economy ministry in the Mediterranean states of the EU. They have macro-economic objectives. In Europe, road tax discs are often set by local authorities and are paid directly into their coffers (a purely financial objective). And the rationale behind road tolls is the financing of infrastructures. Taxes on parking, often non-existent in the south and north, frequently result from local decisions that are not coordinated with neighbouring districts.
- Finally, *transport taxation has considerable redistribution effects* and is seen as a form of general taxation. Thus, congestion taxes, apart from raising large amounts, result in no change in the well-being of the consumers, and it is the public authority that recovers the value of the time lost in traffic jams and uses the resulting community surpluses.

Incentives against pollution can have similar effects. They can be administered either in the form of taxing the polluters or subsidizing those who clean it up, or a combination of the two. The consequences of the three possibilities are similar at the effectiveness level but very different in terms of redistribution. All possible ways of changing the state of things in terms of external effects have consequences in terms of distribution, which sometimes results in rather strong reactions from the groups affected. It is easy to understand why the divergent interests involved use the uncertainties in the calculations to guide them in a direction that is favourable to them. It is also understandable why pricing reforms are very difficult to implement (problems of equity and social acceptability).

However, some northern European countries are tackling the task, and in southeast Asia the pricing of parking and the use of the private car are beginning to be rationalized.

What can be learnt from this for the Mediterranean?

The first recommendation would be to pay more attention to the problems of distribution and social acceptability when introducing eco-taxes. These issues have generally been neglected in the relationships between decision-makers and specialists. Regional forums on the greenhouse gas effect show that this kind of concern is gradually emerging. Attention should also be paid to compensation of the affected groups and ways of limiting the harsh effects of over-rapid change. Tax reforms of fuel prices should develop in a planned way (see Chapter 2).

To prevent special interests from exploiting the uncertainties in the calculations, the values of the basic parameters on which they rely should be set after the widest possible debate, as is happening in several Scandinavian countries. In the Mediterranean countries, such political choices are made by university and administrative experts and thus have little legitimacy.

On both the northern and southern shores of the Mediterranean, there is strong resistance to an ecological rationalization of prices. Of course crises can arise, making large changes possible and unblocking the situation. At the technical level, electronic data transmission offers enormous possibilities that are likely to drastically alter the way in which external effects are internalized. But with the present state of the transport system, it is difficult to see a significant change in the relative prices of the various modes through taxes and prices to the 2010/2020 horizon. Only an integrated approach (used in the alternative scenario described below) can offer the prospect of such reforms. Indeed if the EU, the Mediterranean's biggest client and supplier, redirects its transport systems towards rail and combined transport and decouples its economic growth from increased traffic, the Mediterranean countries would be obliged to adapt their networks to this new situation. The challenge of Europe–Mediterranean interconnection is structural, and also in terms of sustainable development.

Ensuring the funding of transport policies and infrastructures

The coming years are expected to see continuing pressures on public finances, not just for the southern and eastern Mediterranean countries but also for the southern and eastern European countries since the large amount of EU structural funding devoted to the development of the EU southern regions may well be transferred to central and eastern Europe, to the new Member States, which are well behind the southern member states in terms of development.

Moreover, in addition to these quantitative developments, in deciding on transport investments the EU may choose to stop supporting the major road projects of the SEMC to the benefit of a more inter-modal approach involving rail and coastal transport (see the Commission's White Paper), whereas most of the SEMC investment programmes essentially concentrate on roads.

In this changing international scene, the transport policies of the SEMC, and often those of their regional or urban institutions with responsibility for transport in their areas, have also changed, because of a double constraint. First, these countries have been obliged to follow the general trend of opening up transport markets, if only to benefit from new knowledge and possibly promote investments by foreign operators. Second, they have had to limit their expenditure to reduce their budgetary deficits, create a climate of confidence and attract foreign capital, even if the last of these is also attracted by good infrastructures.

This explains the serious difficulty in initiating the ambitious investment policies needed to set up a new transport organization. For cities, these budgetary constraints are often expressed by a limitation of major projects for mass transport equipment, leaving an increase in car use the easier option.

Today, the question of *public investment* is the subject of difficult debates,[37] including within the EU for funding major projects such as the rail links through the Alps and Pyrenees. This is why the idea of cross-financing from resources devoted to roads is making progress (rather in the way chosen by Switzerland); it is raised in the EU's White Paper as a possible way to resolve the problems of funding 'sustainable' infrastructures. Given the huge road investments that most Mediterranean States are prepared to make, such a strategy would guarantee significant development of rail infrastructures. In the relations between the north and the south, the possibility of linking reimbursement of the SEMC external debt to implementing an investment programme based more on rail and rationalizing transport taxes should also be considered.

In addition to this public funding, *private investment* is also a way of financing infrastructures in the SEMC. The weakness of such investment (4.5 thousand million euros for all the SEMC between 1995 and 2000)[38] raises doubts about the viability of such a strategy. Even if liberalization of the transport markets were to attract a significantly greater share of private investment, most specialists agree that a massive input of public money would still be needed. In this context, the European Commission and the EIB have established close cooperation in order to systematically build public–private partnerships for funding EU-Mediterranean interconnection projects. The measure used for achieving this objective is the Facility for Euro-Mediterranean Investment and Partnership (FEMIP), which could result in an increase in EIB funding to the SEMC of 1.4–2 thousand million euros, during 2000–2006. If this were more targeted on non-road infrastructures, it would certainly be good news for sustainable development.

Improving international cooperation to regulate liberalization

At the EU level, the decision to enlarge and to define a common transport policy has created new market framework rules for sustainable development. It is in this context that the liberalization that dominates at the international level must be assessed.

In the SEMC the dominant factor is indeed the *liberalization* of the transport sector. This is particularly important since there is no international agency or authority with a remit for transport in the area. In this situation the national authorities are repositioning themselves: they remain responsible for applying rules and standards and are the main financers of infrastructures, but increasingly look to the private sector to manage the flows of equipment, passengers and freight. This liberalizing movement has encouraged a number of carriers, in particular in the air sector, which has strengthened some services (Tuninter in Tunisia, Royal Air Lines in Morocco, Egyptair in Egypt).

In the maritime sector, the development of strategies by the world's big ship owners, who have selected the Mediterranean to site their hubs, is an example, with the development of whole regions now being shaped by decisions the rationale for which is international and purely economic. For example, Gioao Tauro in southern Italy has become a huge transfer port with no significant relevance for the surrounding inland area. Environmental impacts (noise, air, sea and land use) is considerable in return for minimal local development and employment. Of course, the effects of such installations are not all negative, for example the installation of proper service networks between the various ports has benefited many of them in southern Europe, and even a few in the eastern Mediterranean near the exit from the Suez Canal. The Mediterranean has, therefore, regained a central role in international maritime matters that had long been centred around northern Europe. The entire port complex has been made more dynamic, and port operator policies for opening up are seen almost everywhere in the SEMC.

But this liberalizing movement, together with a very large increase in traffic, has more generally resulted in an aggravation of the environmental and social impacts discussed in this chapter, and at the local level, in effects on the quality of the service provided to customers.

> **Box 3.12** Liberalization of road transport, examples of impact in the city of Batna, Algeria
>
> Since the land-transport market was liberalized in Algeria in 1988 the country's public road transport fleet has increased hugely. According to the Ministry of Transport, between 1988 and 2000 it developed in the following ways:
>
> - **Freight transport**: freight capacity increased from 710,000 to 1,560,000 payload tonnes. The number of vehicles increased from 60,000 to nearly 150,000 (all types), a 150 per cent increase.
> - **Passenger transport** (excluding informal transport): the fleet grew from 12,600 to 44,700, for all types except taxis, an increase of nearly 255 per cent.
>
> Although the objective of liberalization can be considered to have been generally met quantitatively, there remain serious problems of national planning, continuity of public service and, more generally, sustainable development. In the passenger transport sector and in urban areas, the example of the city of Batna (a medium-sized city of 300,000), is typical of the situation found in most Algerian cities.
>
> The mass urban bus transport market in Batna has splintered into a multitude of operators, with 138 local operators in 2000. The companies seem rather like cottage industries (1.63 vehicles per carrier in 1996, 1.41 in 2000).
>
> To survive in such a competitive environment, operators do not operate on the routes allocated to them (theoretically the length of the network is 90km; in reality it is only 34km) but concentrate on the profitable routes. A rota system per day and operator is enforced, concentrated on the formerly profitable routes, confirming the existence of over capacity. However, there are routes that are not serviced, because of weak demand and the poor condition of the roads.
>
> There is practically no system for monitoring the competence of the operators, which explains the lack of professionalism and know-how, resulting in inadequate safety. The lack of requirements for financial capacity (as required elsewhere by the European directives or the LOTI law – Loi d'Orientation sur les Transports Intérieurs – in France) makes it impossible to ensure the continuity of the service and, most importantly, a renewal of the ageing fleet: 56 per cent of the fleet is more than 10 years old and some more than 45 years old, resulting in safety and pollution problems.
>
> The Algerian government seems to have taken the measure of the problem, and is trying to improve public services and the quality of public transport through a new law passed in 2001 dealing with the organization of land transport.
>
> *Source*: Boubakour, 2003

In the general context of the consequences of competition, the regulations covering social and environmental protection provide safeguards that are often not very effective. Thus, in addition to open maritime registry, many road transport companies (including on the northern shore) are among the economic sectors with the longest working hours and the lowest hourly salaries, and the frequency of take-offs and landings at airport hubs is increasing.

The most worrying aspect of liberalization for the future is that it appears to have no limits. Current limits are essentially technical (especially in the Mediterranean): inadequacies in domestic transport systems and door-to-door operations with good connections between long-distance transport and national terminals after crossing borders. But these limits will eventually be lifted under pressure of the need to ensure the economic competitiveness of each country's transport system, with a concomitant worsening in the social and environmental damage described above.

The necessary *regulation* of this liberalization seems inconceivable at the national level: each country is far too dependent on being attractive to foreign investors and the efficiency of its export chains in purely economic terms to consider unilateral modification of the conditions of competition between different transport modes and a reduction in traffic growth.

The essential regulatory factor for this liberalizing movement can only be E*urope* and its opening onto the Mediterranean. In the 1990s, the EU became an important centre of influence for new transport regulation, to be incorporated in the national legislation of its members and to guide transport policy in central Europe and through wider cooperation. The community experience of transport policy serves as a reference for many SEMC. This influence was initially a conveyor of liberalization and the opening-up of markets, reinforcing the dominant trends at the international level. Currently it promotes *integration* and cohesion, and is concerned with questions of harmonization, the construction of networks and impacts on the environment. The Commission's White Paper advocates shifting modes, with inter-modal links involving rail and coastal shipping, and stressing the importance of the need to protect natural areas. This is also seen in the distribution of ODA towards the Mediterranean countries (Figure 3.39 p190).

Without the active participation of the EU in regulating the liberalization of transport in the SEMC, transport does not seem to have a sustainable future.

What form might this international cooperation take, placed under the sign of a liberalization regulated by the community *acquis*? Three 'concentric circles' can be distinguished:

1   The EU, where the *liberalization of road transport* has been a reality since 1993, and the *liberalization of maritime transport* is underway for coastal shipping between European countries (not yet very developed for freight). The opening up of rail networks has been the subject of specific measures since 1991. The *liberalization of rail* is following a model of separating infrastructures from operations. This is taking longer to implement for air transport: the *opening up of air space* was done gradually following a number of predefined stages and lead to a system with several 'degrees of freedom'.

2   The central and eastern European countries of the eastern Adriatic, where the principles adopted by the EU are being transferred fairly rapidly, either formally to satisfy membership requirements, or indirectly when defining new priority transport corridors to match those in the EU. So the priority corridors agreed at the first pan-European conference on transport in Crete in 1993 were added to by defining 'priority transport zones' in eastern Europe and the Mediterranean, agreed at the Helsinki pan-European Conference in 1997. For the eastern Adriatic countries, the Commission recently defined a 'strategic network' with the states, intended to concentrate European aid as part of the stability pact.

3   A still wider circle that includes the rest of the Mediterranean for which several European initiatives have been launched but where transport markets are still very opaque and where operators are still often directly or indirectly monitored by states. These initiatives are dominated by the challenge of opening-up markets but also by the need to develop networks and criteria for evaluating projects.

In concrete terms, this international cooperation could be structured around the funding of networks drawn up according to the *Euro-Mediterranean multi-modal corridor concept*.[39] At present, these corridors include both the eastern and western Mediterranean and the Balkans and serve the EU as reference points within the framework of its thinking about establishing a Euro-Mediterranean transport network.[40]

Figure 3.36 Corridors of International Euro-Mediterranean networks in the eastern Mediterranean

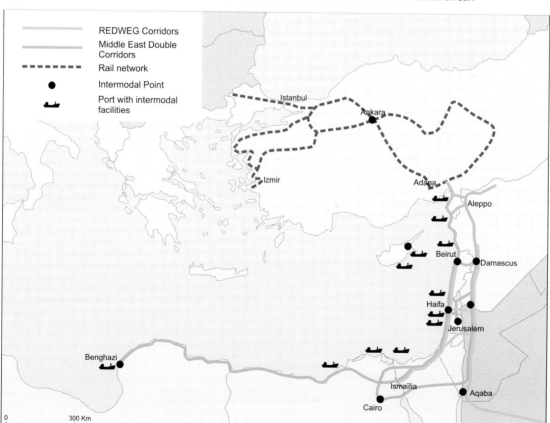

Source: INRETS DEST (1997) in Reynaud (2002)

Figure 3.37 Corridor of International Euro-Mediterranean Maghreb Networks

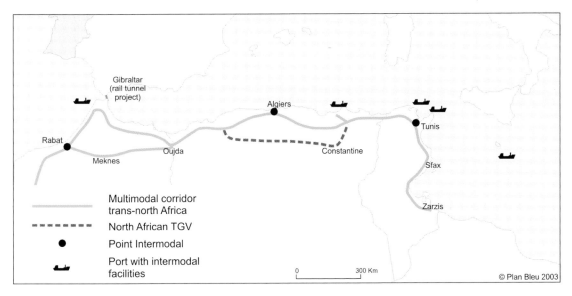

Source: INRETS DEST (1997) in Reynaud (2002)

Eastern Mediterranean

On the heels of the Middle East peace process launched in the late 1990s, economic cooperation was agreed between Syria, Lebanon, Israel, the Palestinian Authority and Egypt. Five committees were set up to consider issues concerning refugees, water, the environment, arms control and regional economic development (REDWEG). This last committee defined 'corridors', routes to be developed as priorities as multi-modal channels by trying to optimize existing infrastructures. Even before REDWEG, initiatives that began in the 1980s at a regional level (the TER and TEM programmes), particularly within the framework of the Arab League, came up with the idea of the double eastern Mediterranean corridor (Figure 3.36). Projects have been assessed for each of these corridors and proposed for community financing. However, they remain in abeyance awaiting an end to the political crisis that is still engulfing the Near East.

Western Mediterranean

Crossing the Alps by rail (Lyons–Turin, operational by 2015), and the trans-Pyrenean central rail tunnel project, which has been adopted as a priority project by the Commission, have already been discussed. As seen above, these projects should relieve the road traffic bottlenecks in southern Europe, even if their funding (especially for the latter) is still problematical.

As far as Maghreb is concerned, international cooperation in transport matters rests mainly on the CORRIMED process (International Euro-Mediterranean Network Corridors), which is similar to that used in the eastern Mediterranean and aims to:

- integrate with Europe and serve the inland areas around ports and airport entry points;
- create connections between countries and regional integration;
- carry out the planning and economic development of the Mediterranean countries.

Since the UMA (Arab Maghreb Union) was established in 1989, the Maghreb countries have made considerable efforts in the transport sector with a view to reaching bilateral and multilateral accords and harmonizing technical and regulatory standards.

The multi-modal trans-Maghreb corridor (Figure 3.37) includes the trans-Maghreb Unit Motorway, the trans-Maghreb Train, coastal shipping and port services and north–south connections. It is a matter of regret from the sustainable development point of view that only road investment has started.

The Balkans

The multi-modal corridor concept in this region is defined through the TIRS network (Transportation Inter Regional Survey).[41] This essentially involves the rehabilitation of road and rail infrastructures in the former Yugoslavia as shown in Figure 3.38.

International cooperation also has a role to play in *air transport*. One of the essential elements in the Partnership would be to enable the emergence of an efficient and reliable Euro-Mediterranean air system, enabling a strengthening of the links around the Mediterranean. Priority efforts should at first be concentrated on monitoring and air traffic control infrastructures. Within this

Figure 3.38 Priority infrastructure network for the Balkans

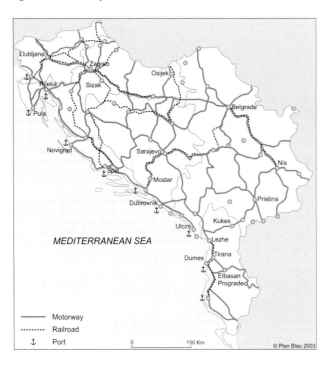

Source: Revue Transports, no 418, March–April 2003

remains ridiculously low. In 25 years, the Mediterranean countries (excluding France, Italy, Spain, Greece and Monaco) received only US$6.251 thousand million of ODA. In contrast, the EU is planning to spend 2.780 thousand million euros in five years in the framework of its trans-European Transport Network strategy, and the motorway investment requirements alone in Maghreb from now to 2010 amount to US$7 thousand million.[42]

Figure 3.39 Official Development Assistance in the transport sector, 1973–2001, towards the Mediterranean countries (US$ millions)

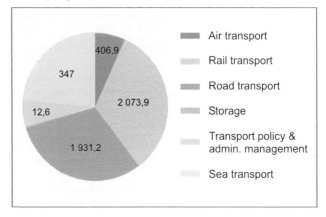

Source: OECD-DAC 2000 database

framework, the development of an SEMC system integrated with the European system is a priority. Setting up infrastructures of reception, land-links, and calculation for satellite data, based on schemes defined by the OACI, as well as harmonization of procedures, will enable a full extension of the European network. This should enable the regulation of traffic peaks and parking on hubs.

Thus international cooperation (especially Euro-Mediterranean) is vital to meeting the objectives both of sustainable development and of practical infrastructure projects. The national policy changes that have so far been pursued have been indispensable (more transparent liberalization, rationalization of taxation and subsidies, modal switches), but it is difficult to see how these changes might actually occur without the new impetus that introducing European policies into the transport sector could provide. In this regard, the distribution of ODA among the different modes is significant (Figure 3.39): between 1973 and 2001 roads took only 31 per cent of the funds compared with 33 per cent for rail, 7 per cent for air and 24 per cent for inland waterways and sea transport.

Only financial resources can truly lead the Mediterranean onto the path to sustainable transport. Unfortunately ODA for transport in the Mediterranean

Even if the imperative of sustainable development is more widely appreciated, notably at the EU strategy level, the thrust of transport development in the Mediterranean seems directed towards ever more traffic and roads. However, starting with current shifts in transport policies, a more sustainable future could be built. This is the subject of the alternative scenario.

An alternative scenario

The paths examined above towards more sustainable transport are brought together here in the form of an alternative scenario, which assumes that the group of measures described above are adopted simultaneously and in a combined way. This integrated approach should lead to a slightly lower increase in road traffic than in the baseline scenario and to an improvement in the environmental performance of transport through changing the modal split and regulatory improvements. The scenario includes:

- *Decoupling economic growth from traffic growth*: this implies adopting *fiscal measures* that increase the cost of road transport (congestion charges, fuel taxes, removal of the tax advantages of diesel, lorry taxes, etc.); *subsidies and tax incentives*, for virtuous practices relating to the

occupation or loading of vehicles (pooling freight, favouring car-sharing, the use of information technology for driving vehicles and traffic management, company programmes aimed at voluntary reductions in transport needs,[43] etc.), *promoting non-motorized modes* (infrastructures for pedestrians and cyclists, car-free tourism).
- A *modal split* that favours rail over road or air by redirecting infrastructure investment and internalizing all the social and environmental costs of road transport that are currently paid by the general public.

What would be the quantitative effects of such developments? Policies aimed at reducing road traffic are rare and have been applied only on a very restricted geographical scale. The results have nevertheless been spectacular. In Singapore, for example, there was a 45 per cent drop in traffic between 1975 and 1991 in the area targeted by the congestion tax. This is why it is not unreasonable to assume a decoupling of economic growth from traffic twice that in the baseline scenario. For the modal split, a 20 per cent traffic share for rail seems realistic since it is the objective of several local scenarios for traffic development (e.g. the trans-Pyrenean scenario to 2020).[44]

Traffic

Based on these assumptions, passenger and freight traffic is projected to evolve as illustrated in Figures 3.40 and 3.41.

Figure 3.40 Passenger traffic, alternative scenario, 2025

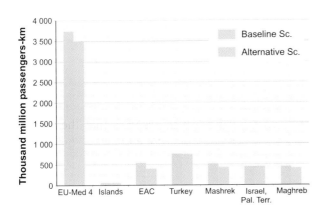

Source: Plan Bleu

Reductions in traffic are most marked for the EU-Mediterranean, eastern Adriatic and Mashrek countries. They account for a modest 8 per cent drop compared with the baseline scenario.

Figure 3.41 Freight traffic, alternative scenario, 2025

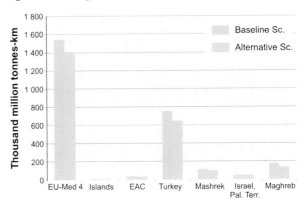

Source: Plan Bleu

The reduction in freight traffic compared with the baseline scenario is also modest (11 per cent).

Impact

However, these small differences in traffic translate into considerable gains in terms of impacts.

Figure 3.42 People exposed to road noise above 55 decibels, alternative scenario, 2025

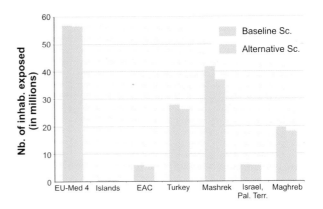

Source: Plan Bleu

Note: A drop in transport/GDP intensity is assumed to result in an equivalent drop in exposure to road noise.

For road noise, the gain appears relatively limited. This is due partly to the projection method used and partly to the fact that there is little scope for technological improvement. Nevertheless in absolute terms, nearly 9 million Mediterranean people are spared a road noise level of more than 55 decibels, compared with the baseline.

The improvement in terms of CO_2 emissions is clearly greater than for noise pollution: 191,000 tonnes of CO_2 avoided and a significant reduction in all sub-regions. The modal-split effect (the transfer of some road traffic to rail) is of great importance for this indicator.

Figure 3.43 CO_2 emissions, alternative scenario, 2025

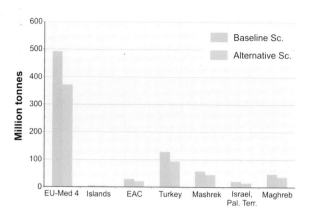

Source: Plan Bleu

Note: The drop in energy intensity in the alternative scenario in Chapter 2 has simply been transposed to transport.

Figure 3.44 VOC emissions, alternative scenario, 2025

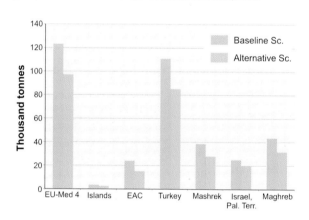

Source: Plan Bleu

For VOCs, gains are significant: 90,000 tonnes, nearly 25 per cent less emission than the baseline scenario.

Figure 3.45 NO_x emissions, alternative scenario, 2025

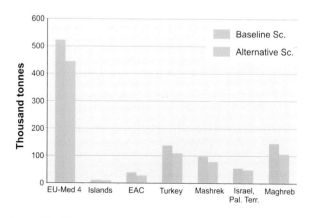

Source: Plan Bleu

For NO_x the gain is considerable: 185,000 tonnes, 18 per cent less emission than the baseline scenario, although this is less than for VOCs.

Figure 3.46 The number of road accident deaths, alternative scenario, 2025

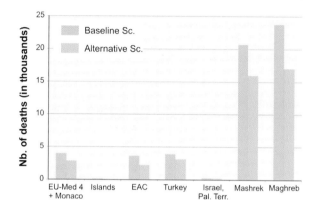

Source: Plan Bleu

For road accident deaths the gain is much greater: 15,000 fewer per year than in the baseline scenario. With a more proactive road safety policy in the SEMC, the improvement would be even greater.

Figure 3.47 Congestion costs, alternative scenario, 2025

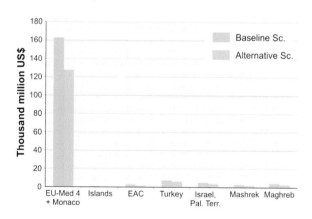

Source: Plan Bleu

The effect of decoupling GDP from traffic is particularly important for congestion costs: the alternative scenario foresees 41 thousand million euros less cost in 2025 than in the baseline scenario, or the equivalent of the total cost of road congestion in the Mediterranean in 2000.

The impact of the alternative scenario on the share of transport in household budgets is minimal, except in the eastern Adriatic countries (25 per cent in the baseline compared with 18 per cent in the alternative scenario). Other policies linked to housing would need to be

Figure 3.48 The share of transport in household budgets, alternative scenario, 2025

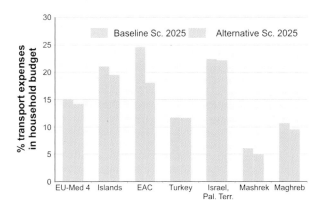

Source: Plan Bleu

combined with the decoupling of GDP from transport and changes in modal split if the share of transport in household budgets is to be really stabilized or even reduced.

For *maritime transport*, the alternative scenario assumes extending MARPOL to chemical products and including the costs of de-sludging and ballast water treatment in port dues. The projected effectiveness of such measures is based on the observed result of the mandatory separation of ballast tanks on emissions from de-sludging (a 30-fold reduction).

Figure 3.49 Polluting emissions into the sea related to maritime traffic, alternative scenario, 2025

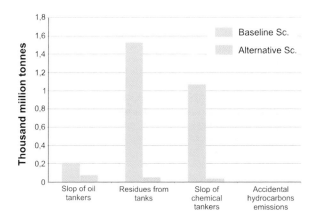

Source: Plan Bleu

The gains are spectacular: 159,000 tonnes of pollutant emissions in 2025 compared with 2.81 million in the baseline scenario, a 94 per cent drop. This is particularly attractive since the measures used in the alternative maritime transport scenario do not require as much change as those that govern the results of the alternative land transport scenario.

Thus a small drop in traffic compared with the baseline scenario would result in very considerable gains at the environmental and social levels. These have been quantified mainly to show the orders of magnitude and give an indication of the challenge; such quantification is always somewhat arbitrary. The effect that such a scenario would have on the economy still has to be determined. This will depend to a large extent on resolving the problem of funding of Euro-Mediterranean cooperation and on the capacity of the Mediterranean people to change their future.

* * *

To conclude, questions need to be asked not only about transport projections but also about the whole transport system.

The baseline scenario, with its very large increase in freight and passenger traffic, is unsustainable: the environmental impacts get considerably worse for some people (especially from emissions of pollutants from vehicles and ships), the social impact becomes increasingly unmanageable (road accident deaths, the impoverishment of suburban households) without guaranteeing the economic competitiveness of the transport systems. In contrast, the strengthening and more effective application of marine safety rules and some transfer from road to rail, albeit minimal, are likely to help both the environment and the economies of the Mediterranean countries.

So the question is how to shift from one scenario to the other.

There is no doubt that the volume of freight transported will depend on economic and population growth. The underlying economic picture is one of specialization and diversification of production on the international scale. Globalization and new technology result in a very considerable growth in all kinds of traffic (local, national and international) and in the economic specialization of regions of the world. The Mediterranean will not escape this, at the risk of marginalizing itself in the general context of global growth.

Nevertheless *Europe* can play a growing role as a centre, or even as an original model, for regulatory development, although it is not yet easy to see how this could be achieved. Trade, economic, social and cultural exchanges and migration give Europe a particular responsibility for the future of the Mediterranean, particularly since so much of the traffic in the sea is in transit between Europe and other continents. Of course, significant traffic flows still depend on local initiatives or bilateral cooperation, but the problems of governance of this traffic are the same, independent of scale, and require integrated policies that closely associate the future of transport with the choice of technology and development paths at local, national and international levels. Such choices can only be made through strategic planning, aimed at taking control of the future.

From this point of view, the alternative scenario assumes some changes already recommended by the European Commission for its own transport policy.

- *The use of new technologies* for vehicles and the organization of transport and communication modes. Vehicles should be adapted to Mediterranean cities and to their future needs, with a particular place for small vehicles in narrow historical centres, two-wheeled as well as non-motorized. Communication tools play a part in controlling traffic and monitoring transport systems in order to ensure the quality of the service, but can also have new roles in distant areas or difficult to get access areas, such as islands, to ensure servicing to remote sites.
- *Spatial planning* to ensure a decoupling between economic growth and the growth of traffic, including tourist traffic. Transport is not just for providing individual satisfaction. People often suffer from it, particularly in areas congested by daily commuting. This presents a double challenge: to plan a living environment that respects the natural environment, history and culture, and to control traffic. This double challenge relates back to the setting of objectives for and by the community and therefore to the problem of decentralizing institutions, and of cooperation between them.
- *Preserving the marine world*. The Mediterranean sea is at the heart of the eco-region, not on its edges, in contrast to many other regions or organizations of regional areas in the world. This is why it is particularly important to control the traffic across it, including planning for the coastal areas and for short and long journeys. A revitalization of coastal shipping is being considered for short journeys. The potential of major hubs is reaffirming the importance of the Mediterranean for international trade flows. These hubs are, as yet, only partially integrated into the organization of Mediterranean transport. But, without them, the region risks being marginalized in the international organization of the economy. The question of their siting in the context of their surroundings and the monitoring of intercontinental ship movements through the Mediterranean are becoming crucial.
- *Partnerships between the Mediterranean peoples*. There are already some European examples of progress towards more sustainable transport (high-speed trains, the timid renaissance of coastal shipping, harmonization of the European skies). These initiatives have an important role to play in developing in-depth contacts between neighbouring peoples for the benefit of better mutual understanding.
- *The strengthening of future oriented analyses*. Prospects relying on objective data and transparent arguments can help to build an understanding of major changes and define the debate about what transport system to aim for.

The crucial points for shifting from the baseline to the *alternative* scenario are at three levels:

1. *Local*: actions aimed particularly at more dense fabrics in cities through planning, aid for housing and parking, and organic linking between urbanization and mass transport.
2. *National*: using laws, conventions and contracts with local authorities to implement sustainable transport at the local scale by undertaking an exhaustive assessment of the external costs of current systems; by implementing a tariff system for transport use (especially for the car) and a system of subsidies that favour protection of the environment and social equity, all within the framework of regional integration, guided by the new directions of European transport policy.
3. *International*: by substantially increasing European funding of Mediterranean transport systems that are not exclusively road-oriented and by ensuring the effectiveness of international maritime safety conventions, particularly by monitoring shipping in transit and adapting the international law of the sea to the situation in the Mediterranean.

The projections in this chapter are, of course, only indicative. The intention is not to foretell the future. There will undoubtedly be many changes in the areas analysed (particularly relating to urban congestion) that may accelerate the advent of an organization of transport quite unlike today's. But simply drawing attention to the 'hidden' costs of current trends may help to predict the adverse effects of such changes and alter the relationship between each Mediterranean citizen and his mobility.

Notes

1. Transport in a number of Mediterranean countries represents a significant proportion of economic activity. Thus in 2000, the transport and communications sector accounted for 10.7% of GDP in Syria, 9.3% in Egypt, 6.7% in Algeria, 6.1% in Morocco and 10.9% in Tunisia.
2. Before the integration in 2004 of Cyprus, Malta and Slovenia into the EU.
3. Spain, France, Italy, Greece, Monaco, Serbia and Montenegro, Albania and Turkey.
4. World Bank, 1998.
5. The ratio between the rate of change in the traffic and the rate of change in GDP for the same period.

6. Lablanche, 2003.
7. EU DG 16, 1991; World Bank, 2002.
8. A. Jorio, 2000.
9. Orfeuil, 1997.
10. The 1970–2000 elasticities have been reduced for the projection to 2025 by 18.2 per cent for cargo and 48.3 per cent for passenger traffic. These figures reflect the reductions in projected traffic/GDP elasticities by the EU in the baseline scenarios of its White Paper and its Climate-change Programme. However, the 1970–2000 elasticities for the countries entering the EU in 2004 (Cyprus, Malta and Slovenia) and for Bosnia-Herzegovina in reconstruction have been kept.
11. Kunzli et al, 2000.
12. Algeria, MATE, *Report on the state and future of the environment*, 2000.
13. The relationship between the quantity of energy consumed and the number of passengers carried multiplied by the distance they covered.
14. INRETS' data for France; Boubakour 2002, Va-t-on vers des transports écologiquement viables au Magreb? for Morocco and Algeria.
15. Orfeuil and Morcheoine, 1998.
16. Israel, Ministry of the Environment, *Towards Sustainable Development*, 1999.
17. INFRAS/IWW 2000.
18. CEC, 2001.
19. The Turkish General Directorate of Highways 2000. Yücel Candemir, Brief Presentation of Transport Policies in Turkey, MEDA TENT 2003.
20. ECMT, *Short Sea Shipping*, 2000.
21. Measures not always well received by people as seen in Algeria in April 2003, when the bill's introduction triggered a transporter strike.
22. To navigate, a boat must have a licence number (like a car) that is issued by a national register. As with car licences, this operation requires that a certain number of obligations contained in the registering country's laws be met. Obligations are theoretically verified through inspections.
23. TDW: deadweight tonne (the maximum capacity of transport authorized for ships).
24. Vende, 1999; Dobler and MARPOL, 2002.
25. When a tanker navigates empty there may be problems of stability, so part of the tanks (ballast) are filled with sea water. Before tankers are again filled with oil, the ship empties its ballast (de-sludging) to make room for the oil. Unfortunately the ballast water has often had time to become impregnated with residual oil that sticks to the walls of the tanks after each unloading. Thus large quantities of hydrocarbons are expelled into the sea on desludging.
26. Studies of the Marine Board of the National Research Council of the United States National Academy of Science.
27. WWF, 2003
28. Helmepa, *Evaluation of the Port Reception Facilities in the Mediterranean*, Athens, 1997. These data are being updated by REMPEC.
29. REMPEC also operates under the aegis of the International Maritime Organisation (IMO).
30. Chemical freighters are not subject to the same regulations as oil tankers, and the spilling of chemical products into the sea is much less visible than hydrocarbons (but often more harmful for the environment). Containers are being stacked higher and higher, even on decks, meeting financial and economic imperatives, a procedure which, at least in the short term, is not being questioned.
31. EU, DG-Transport, 2001.
32. These proposals concern the banning of transporting heavy fuel-oil by single-hull tankers, the introduction of penal sanctions resulting from flagrant negligence, the implementation of a community approval system of certificates of sailor competence delivered outside of the EU, improving the system of reporting by port pilots and improved protective measures for coastal waters, in particular territorial waters and exclusive economic zones.
33. The Euro-Mediterranean Transport Forum was founded in 1998 and is the forum for developing regional cooperation in the transport sector. Organized and presided over by the European Commission, it is composed of high transport civil servants of the EU Member States and the Mediterranean countries.
34. The IOPC (International Oil Pollution Compensation Fund) acts in cases of accidental hydrocarbon spillage, on behalf of the producing and distributing oil companies fitting out ships. The fund is fed by these companies in return for an exoneration from their responsibility beyond a certain financial amount. At present the IOPC has nearly a thousand million euros, which corresponds to the damages assessed in the Prestige accident alone.
35. In urban situations, a 5 per cent reduction in traffic can reduce congestion by half; but the environment will hardly be improved. This has been demonstrated by the experiences on traffic control during pollution peaks.
36. Economists define it as the delay that vehicles impose on one another in terms of the speed/traffic flow relationship in conditions where the use of the transport system approaches the system's capacity (Wyrnbee, 1979, *Journal of Economics*).
37. In Algeria, the government is planning to create a road fund that will be fed from road-user fees and open the funding of rail to the private sector by conceding network operations to it (Chafik, 2003).
38. FEMISE report on the change in trade and investments structure between the EU and its Mediterranean partners, March 2002.
39. Multi-modal corridors are the major traffic routes around which researchers and engineers, funded by the EU since the late 1990s, are trying to structure transport infrastructures in the mid- and long-term that promote combined transport solutions and the development of south–south trade.
40. EC, *Communication sur le développement d'un réseau euro-méditerranéen de transport*. COM(2003) 376 of 24/06/2003.
41. A study financed by the French Development Agency (AFD) carried out in 2001–2002 by the Louis Berger office, with a view to identifying priority transport infrastructures for the Balkans.
42. Chafik, 2003.
43. In some states in the US and in the Scandinavian countries tax deductions or subsidies to companies that promote car sharing or encourage their employees to live near their place of work are granted.
44. In countries for which the baseline projection of the share of rail is over 20 per cent, this has been increased either by 10 per cent (for freight in France), or by 5 per cent (Slovenia, Croatia, Bosnia-Herzegovina and Serbia-Montenegro for freight; Egypt and Morocco for passengers).

References

Baumstark, L. (rapp.), Commissariat Général du Plan. (2001) *Transports: Choix des investissements et coût des nuisances*. Paris: Documentation française

Boubakour, F. (2003) *Vers des transports durables en Algérie?* Paper for the Blue Plan. Sophia Antipolis: Blue Plan

CCE (2003) *Communication de la Commission au Conseil et au Parlement européen sur le développement d'un réseau euro-méditerranéen de transport*. Bruxelles: CCE. www.europa.eu.int/comm/energy_transport/euromed/

CEC (1998) Fair Payment for Infrastructure Use: A *Phased Approach to a Common Transport Infrastructure Charging Framework in the EU*. White Paper presented by the Commission. Brussels: CEC, COM (1998) 466 final

CEC (2001). *White Paper. European Transport Policy to 2010: Time to Decide*. Brussels: CEC, COM (2001) 370 final of 12 September

CEMT (2001) *Le transport maritime à courte distance en Europe*. Paris: OCDE

Chafik, J. (2003) *Evaluation et perspectives de développement des transports dans les pays de l'Union du Maghreb Arabe*. Introductory Report. Table ronde sur les modalités de mise en oeuvre du programme d'action de l'UMA en matière des transport dans le contexte du NEPAD. Rabat,

20–21 October

Conseil Général des Ponts et Chaussées. (2003) *Rapport d'audit sur les grands projets d'infrastructures de transport*. Paris: Documentation Française

Dobler, J. P. (EcoMar) (2002) *Les Flottes Marchandes des Pays Méditerranéens*. Report for the Blue Plan. Sophia Antipolis

Dobler, J.-P., MARPOL. (2002) *Updated transboundary diagnostic analysis of the pollution in the Mediterranean Sea*. Report prepared for the Secretariat to the MARPOL Convention

ECMT. (2001) *Short Sea Shipping in Europe*. Paris: OECD

EEA. (2000) *TERM 2000. Are we moving in the right direction? Indicators on transport and environment integration in the EU*. Environmental issues series no 12, Luxembourg: Office for Official Publications of the European Communities (OPOCE)

EU DG 16. (1991) *Combined Transport: An Analysis of the Profitability of a European System*. Brussels: EC

Eurostat. (2000) *Évolution du transport dans les pays de la Méditerranée*. Luxembourg: OPOCE

Eurostat. (2002a) *Le transport aérien dans les pays MED 1998–2000*. Luxembourg: OPOCE

Eurostat. (2002b) *Transport de marchandises et de passagers 1997–2000*. Luxembourg: OPOCE

Eurostat. (2002c) *EU Intermodal Freight Transport: Key statistical data 1992–1999*. Luxembourg: OPOCE

INFRAS/IWW (2000) *External costs of Transport: Accident, Environmental and Congestion costs in Western Europe*. INFRAS/IWW

IRF. (1999) *World Road Statistics*. Washington DC: IRF

Jorio, A. (2000) *Modes de consommation, environnement et libre échange au Maroc*, Plan Bleu; ENDA Maghreb

Kunzli, N. et al (2000) Public Health Impact of Outdoor and Traffic-related Air Pollution: a European Assessment, *The Lancet*, no 356

Lablanche, F. (2003) Transports et espace syrien, *Ann. Géo*, no 630

Müller-Jentsch, D., World Bank, EC. Programme on Private Participation in Mediterranean Infrastructure. (2002) *Transport Policies for the Euro-Mediterranean Free-Trade Area*. Washington DC: World Bank

OACI. (2001) *Statistiques de l'aviation civile dans le monde*. Montréal: OACI.

OCDE. (2002) *Pour des transports écologiquement viables: les mesures à prendre*. Paris: OCDE.

Orfeuil, J.-P. (1997) *Les coûts externes de la circulation routière*, INRETS

Orfeuil, J.-P, Morcheoine, A. (1998) Transports, énergie, environnement, *Transports*, no 390, July–August

Pavaux, J. (dir.), Institut du Transport Aérien. (1995) *Le transport aérien à l'horizon 2020: éléments de réflexion prospective*. Paris: Institut du transport aérien.

PNUE/PAM/REMPEC. (2002) *Stratégie pour la mise en œuvre du protocole relatif à la coopération en matière de prévention de la pollution par les navires en cas de situation critique, de lutte contre la pollution de la mer Méditerranée*. Gzira: REMPEC

Reynaud, C. et al (1996) *Transports et environnement en Méditerranée*. Les Fascicules du Plan Bleu, no 9 Paris: Economica

Reynaud, C. (2002) *Transport et développment durable en Méditerranée*. Report for the Blue Plan. Sophia Antipolis

Spécial trafics ports du monde. (1997) *Journal de la Marine Marchande et du Transport Multimodal*, no 4071, 26 December 1997

UIC. (2002) *Statistique internationale des chemins de fer 2000*. Paris: UIC

Vende, B. (1999) *Le Memorandum de Paris sur le contrôle des Navires par l'Etat du port*, University of Nantes

World Bank. (1998) *World Development Indicators*. Washington, DC: World Bank

World Bank. (2002) *Transport Policies for the Euro-Mediterranean Free-Trade Area*, Washington, DC: World Bank

WWF (2003) *20 Prestige souilleront la Méditerranée cette année*. WWF

Chapter 4

Urban Areas

The future of the Mediterranean Basin up to 2025, 2050, and even beyond, will depend largely on the urban areas that have always been important in the region. Some 380 million inhabitants, or 80 per cent of the population, are soon expected to be living in and around cities.

For thousands of years, the different civilizations that have shaped the region have relied on the vitality of cities, their economic, political and cultural functions, their exchanges and struggles, on their functional and balanced spatial layouts and on the interlinkages between them. Cities have been tightly integrated into the Mediterranean environment, characterized by its specific geopolitical and bio-climatic situation.

From a historical point of view, the issues of unsustainable urban development in the region have arisen quite recently, only during the past few decades. The acceleration of urbanization in the south together with the resulting boost in demand, uncontrolled urban sprawl in the north, the concentration of people in large agglomerations, the development of inequalities and segregation, all contribute to the transformation of societies, and are challenging what used to be the Mediterranean *city*. The spatial and temporal relationships between cities and their immediate or more distant surroundings have changed profoundly.

Responses to current unsustainability issues do exist, and in some cases have existed for a long time. Examples are policies for physical planning and regional development, land management, and town planning; these are all responses relevant to a frame of action much broader than local development. But no Mediterranean country has yet had the courage or the resources to develop policies, which, if they are to succeed, require national Governments and urban authorities to work together, and people to turn into true citizens again.

The challenges are often serious. Some courses of action that are put forward may lead to a change of scenario. The levels of action are necessarily interlinked. Urban structures, regional development policies, moves towards decentralization or de-concentration as well as mutual support between states, regions and towns all need to be taken into account in each country. More active cooperation between Mediterranean local authorities and an acknowledgement of the specific urban situation in the Euro-Mediterranean process are also crucial.

Widespread urbanization, once in the north, now in the south

At the beginning of the 21st century, Mediterranean peoples and societies are mainly urban (Figure 4.1). The total urban population[1] of the countries bordering the Mediterranean increased from 94 million in 1950 (44 per cent of the population) to 274 million in 2000 (64 per cent).

The *northern Mediterranean countries* (NMC – from Spain to Greece) have experienced strong urban growth throughout the 19th century and up to the 1970s, and their current very moderate growth rates are projected to continue.

The *southern and eastern Mediterranean countries* (SEMC – from Turkey to Morocco) have, for the past two or three decades, seen an accelerated rate of urbanization, 3–5 times faster than the northern countries. Whereas European countries needed a whole century to absorb urban growth, southern and eastern ones take only a few decades. Thus, in spite of a recent slowdown in population growth in the southern countries – along with a spectacular decrease in fertility rates and a relative decrease in rural–urban migration – the strong trend towards urbanization is expected to continue during the next 50 years.

It would certainly be more stimulating to look into the future in more than one way, but, except for immigration policies, the prospects to 2025 more or less follow current trends. In absolute figures, the urban population could reach more than 243 million in the southern and eastern countries (145 million in 2000) and nearly 135 million in European coastal countries (129 million in 2000). Approximately a third of this growth would take place in the *coastal regions* (Figure 4.2).

According to United Nations (UN) estimates, the urbanization rate in the SEMC in 2000 was already slightly higher than in the NMC. By 2030, three-quarters of the population will be urban (Figure 4.3).

Apart from the convergence observed in urbanization rates, urban issues in different countries are seldom comparable.

The southern and eastern Mediterranean countries experienced very clear urban growth (an annual average

Figure 4.1 Urban population in the Mediterranean countries, trends and projections

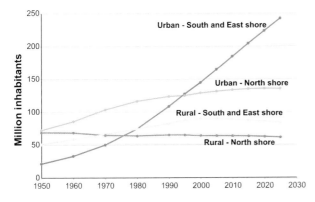

Source: Attané and Courbage; *Plan Bleu*, 2002

2 A Sustainable Future for the Mediterranean

Figure 4.2 Urban population in the countries and their Mediterranean coastal regions, projections 2000–2025

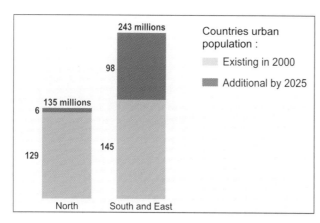
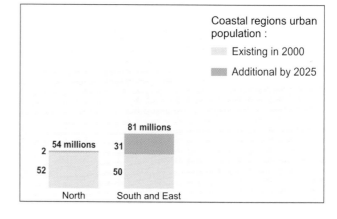

Source: Attané and Courbage; Plan Bleu, 2001

of 4 per cent between 1970 and 1990, 3 per cent in later years). In spite of the decreasing fertility rate in most countries, population growth will only slow after 2030 or even 2040. Till then, the challenge will be to achieve the economic development needed to be able to deal with urban growth.

- The driving force behind this growth is becoming more and more endogenous (natural growth, very young urban populations), fed by internal redistribution,[2] inter-city migration and rural to urban migration, which is either decreasing (Egypt, Tunisia) or being maintained (Turkey, Syria, Morocco).

Figure 4.3 Urbanization rate in the Mediterranean countries, 1950, 2000 and projections to 2030

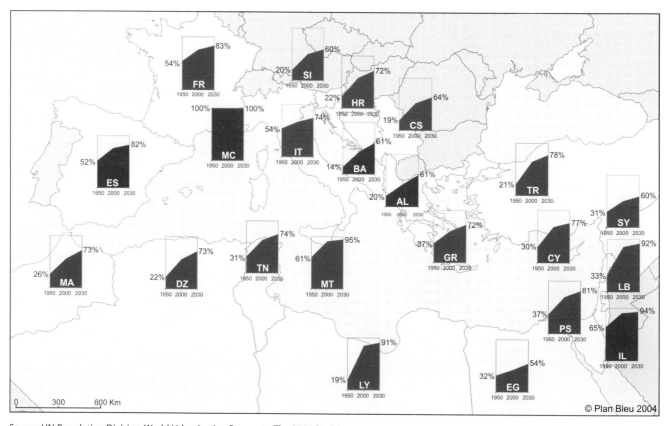

Source: UN Population Division, World Urbanization Prospects. The 2003 Revision

200

Urban Areas

Figure 4.4 Cities with 10,000 or more inhabitants, 1950 and 1995

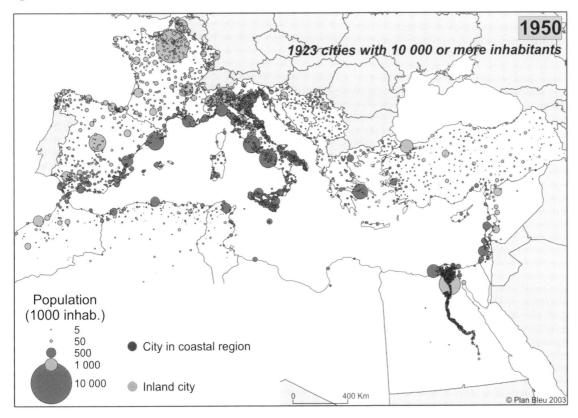

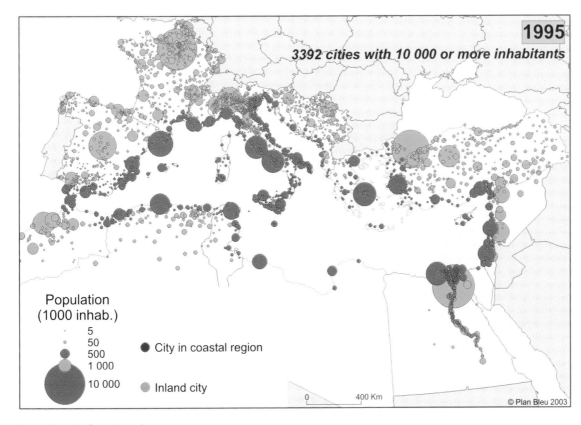

Source: *Plan Bleu* from Géopolis

- There are, and will continue to be, many young people, with the economically active population growing by 3–4 per cent per year. These people need training and employment. According to the forecasts, net entry into the employment market will be 4.2 million per year between 2005 and 2009, falling gradually to around 2.3 million per year by 2025–2029.[3] Jobs and the appeal of working in the city are no longer the major driving force of urban growth (unlike 50 years ago). The unemployment rates in Mashrek and Maghreb countries are among the highest in the world and the young are especially affected, for example in Cairo where 20 per cent of 15–29 year olds are unemployed.
- Urban demographic pressures and the lack of jobs encourage the so-called 'grey' sector: informal jobs are estimated at about 30 per cent of active urban employment in Turkey, 45 per cent in Egypt and Tunisia, about 40 per cent in Algeria and Morocco.
- The poverty of a large part of the urban population is of real concern in Mashrek and Maghreb countries. In Egypt, 42 per cent of the poor live in urban areas. In Beirut, 25 per cent of the population lives below the poverty line. There is certainly more poverty in rural areas, but the number of urban poor is increasing.[4]
- The spread of urban areas significantly complicates city management, since public services cannot keep up with the speed at which new neighbourhoods, often not officially recognized, are created.

With nearly 100 million extra inhabitants between 2000 and 2025, including 23 million in Turkey, 36 million in Egypt, 10 million each in Algeria and Morocco respectively, cities are, and will be, undergoing major economic, social and environmental change (Figure 4.4). Economies of scale certainly need to be taken into account, but the marked concentration of populations creates difficult problems in the fields of employment, infrastructure and services, pollution management, and waste production – where there is an alarming increase in quantity.

In the northern Mediterranean countries (mainly Spain, France, Italy and Greece, but also Malta and Cyprus), as in all European countries, the transition from the rural settlements of a largely agriculture-based economy to the urban settlement patterns characteristic of an industry- and service-based economy has now ended. The urbanization process has reached 'saturation', with the population concentrated in the central areas of agglomerations apparently unable to grow and sometimes even tending to decrease. It is true that around the Mediterranean basin, strictly defined, cities have seldom been transformed by the industrial revolution. However, they do show the demographic stagnation that prevails in other European towns.

But when considering total land area and urban land use, it appears that urbanization is far from over. The urbanized area continues to expand. At the local level, cities reach out towards the outskirts, even if the total population has stopped growing. Urban sprawl started in north-western Europe, spread to countries in the western Mediterranean, and then to the east (Athens, small Greek towns, Nicosia, etc.) Although this sprawl is occurring with time-lags, with limited effects, and in particular ways, it is becoming extensive and widespread, creating heterogeneous peri-urban areas with transport and commercial infrastructures, small businesses and low-density housing. At the same time, the agricultural and rural areas around cities, whether it is good or poor quality, is shrinking and often being converted to urban use. In addition, tourism, common in many cities, is adding to 'urban' growth, in the form of built-up areas that are often only seasonally occupied.

The process of extensive urbanization in the north, and increasingly in the south, has become so widespread that the traditional concepts of town/country, urban/rural, are no longer of great help when considering these new types of area, which are occupied by residents and 'urban' activities, but are not *towns* in the traditional sense. The phenomenon calls for new regional development policies that can limit or forestall mushroom development, wastage of land, and long-term human and environmental impacts such as the greenhouse effect and pollution.

Small towns and large cities: Is it possible to choose?

According to population projections, the urban population could reach more than 135 million by 2025 in the NMC (6.5 million more than in 2000) and nearly 243 million in the SEMC (98 million more). The spatial distribution of this growth, between cities of different sizes and between coastal and inland towns, throws up many questions: can the main metropolises increase their size even more, and by how much? Are we going to witness their hypertrophy or will there be a redistribution in favour of medium-sized towns? What future is there for small and medium-sized towns? Are these open questions? Is there, in fact, a choice?

Changes in agglomerations[5] since 1950 have been measured for 1887 towns in the NMC and 1505 in the SEMC, all having more than 10,000 inhabitants in 1995, and classified according to their size (Figure 4.5). In spite of the extreme variability in growth between different towns, the persistence of urban structures is evident and particularly clear in the NMC. For the SEMC, the graph shows how towns have grown spectacularly during recent decades, and the imbalances affecting the urban framework within countries, with a trend towards a macro-cephalic pattern. For coastal/inland distributions and the attraction of the coast, see Chapter 6.

Urban Areas

Figure 4.5 Population in cities according to town sizes (thousand inhabitants)

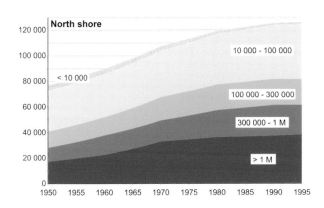
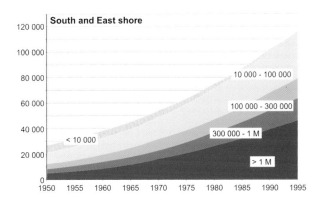

Source: Plan Bleu 2002, from Géopolis

Note: The shaded class corresponds to very small towns of less than 10,000 inhabitants which, during the period under consideration, have grown and passed to the next size.

The number of *very large cities*, with more than a million inhabitants, has grown significantly: 10 cities in 1950, 29 in 1995 (Figure 4.6). This class includes many metropolises located in the coastal regions, like Barcelona, Marseilles, Rome, Naples, Athens, Izmir, Beirut, Tel-Aviv, Alexandria, Tripoli, Tunis and Algiers, but also the three megacities (Istanbul, Cairo and Paris) and other inland metropolises (like Seville, Madrid, Lyon, Milan, Ankara,

Figure 4.6 Metropolises with more than one million inhabitants

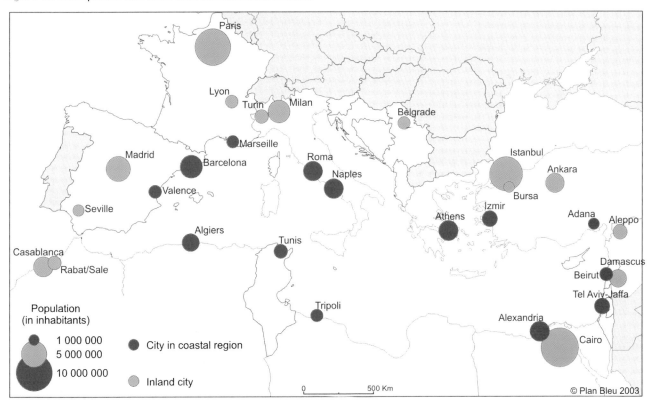

Source: Plan Bleu, from Géopolis

Note: In the NMC, 7 big metropolises contained 22 per cent of the urban population in 1950; in 1995, there were 13 with more than 30 per cent of the urban population. In the SEMC, 3 very large cities contained 15 per cent of the urban population in 1950; in 1995, there were 16 large metropolises with 28 per cent of the urban population.

Aleppo and Damascus). The 'weight' of this class is considerable; the proportion of the urban population gathered therein continues to increase.

Small towns, with between 10,000 and 100,000 inhabitants, continue to multiply: 3062 in 1995 compared with 1776 in 1950. Their relative weight is characteristic of the settlement patterns in the region, even though their proportion of the urban population is tending to decrease (in the NMC, 1369 small towns with 41 per cent of the urban population in 1950, 1705 with 34 per cent in 1995; in the SEMC, 417 small towns with 34 per cent of the urban population in 1950, 1357 with 32 per cent in 1995).

Between these two extremes, the 218 *medium-sized towns* and the 83 *intermediate cities*, which contain about a third of the urban population in the region, are changing in ways that are less easy to interpret.[6]

- Medium-sized towns (from 100,000 to 300,000 inhabitants) have certainly grown in number (in the NMC, 74 in 1950 to 123 in 1995; in the SEMC, 25 in 1950 to 95 in 1995), but their power to attract seems hardly to have changed: 16 per cent of the urban population in the NMC in 1950 and in 1995, 14 per cent and 13 per cent respectively in the SEMC.
- Intermediate cities (from 300,000 to 1 million inhabitants), on the other hand, maintain their attractiveness but in a less marked way than very large cities. In the NMC, 21 intermediate cities contained 14 per cent of the urban population in 1950; in 1995 there were 46 with 18 per cent of the urban population. In the SEMC, 7 towns with 12 per cent in 1950 became 37 with 15 per cent of the urban population in 1995.

Apart from the distribution of urban populations according to town sizes, the very significant *differences in growth rates* (Table 4.1) provide a better indication of urban dynamics on the two shores, even though the average changes hide substantial fluctuations within the same class of town size during each period. The three megacities – Istanbul, Cairo and Paris – are here given a separate classification (more than 5 million inhabitants).

In the NMC, all sizes of town have registered the same type of cyclical change in their growth rate, with a maximum rate during the 1960–70s. Subsequently, urban growth slowed considerably (to 0.26 per cent per year in 1990–1995 compared with 1.22 per cent per year in 1960–70) and the growth rates across the various sizes tended to level out. Is this a new cycle? In any case, urbanization in the northern countries seems to have undergone a large change, resulting in the expansion of urban areas.

In the SEMC, changes appear to fluctuate more between periods, but the salient feature is the strength of their growth. In spite of a gradual slowing in growth rates during the period (from 3.2 per cent per year in 1950–60 to 2.6 per cent per year in 1990–95), growth remains very strong for all town classes. With 2.5 per cent annual growth, populations double in less than 30 years; with 3 per cent, in less than 23 years.

Table 4.1 Growth rates of cities, per town class

Town population (inhab)	No of towns in 1995	Rate of average annual variation (%)					Average 1950–1995
		1950–1960	1960–1970	1970–1980	1980–1990	1990–1995	
Northern Mediterranean countries							
10,000–100,000	1705	1.22	1.40	1.24	0.81	0.38	1.07
100,000–300,000	123	2.09	2.32	1.31	0.62	0.13	1.42
300,000–1 M	46	1.73	2.11	1.30	0.54	0.15	1.27
1 M–5 M	12	2.46	2.72	1.23	0.29	−0.05	1.48
> 5 M	1	1.66	1.34	0.42	0.43	0.41	0.90
Total	1887	1.13	1.22	1.05	0.67	0.26	
Southern and Eastern Mediterranean countries							
10,000–100,000	1357	3.06	2.79	2.82	3.14	2.58	2.90
100,000–300,000	95	5.04	4.20	3.75	3.33	2.95	3.93
300,000–1 M	37	3.85	4.16	3.75	3.49	3.14	3.73
1 M–5 M	14	4.84	4.42	4.27	3.18	2.56	3.99
> 5 M	2	4.48	4.59	3.75	3.50	2.49	3.90
Total	1505	3.23	2.94	2.91	3.15	2.59	

Source: Plan Bleu 2002, from Géopolis 1998

Urban Areas

Behind these average trends, in 1990–95, Cairo grew noticeably more slowly than Istanbul (1.5 per cent and 3.4 per cent per year respectively). More generally, in Turkey, Syria and Morocco, all countries where rural–urban migration remains important, many towns, and especially those of medium size (100,000–300,000 inhabitants) still have a growth rate of between 4 per cent and 5 per cent per year, which means that their population will double in only 15 years. In the exceptional cases of Israel and the Palestinian Territories, several towns show an annual growth higher than 8 per cent; the population could then double in less than 10 years.

Metropolitan and regional development

When extrapolating only current population trends (the baseline scenario), urban patterns vary little or not at all. To examine future urban prospects, it is necessary to refer to the economic dynamics at work over the last decade, where, in the context of a growing interdependence of economies, urban areas and metropolises have become the preferred destination for capital flows and business internationalization.

Metropolitan development marks both a change in scale and a proliferation of relationships and functions. The process refers not only to the growth of large cities, but also to an increasing concentration within them of business, wealth, capital investment in information and communication technologies, and technological developments that affect transport, infrastructure management, construction and public works. The metropolitan functions that currently characterize and differentiate large cities are linked to integration processes in various kinds of networks (transport, research, knowledge, finance) on the basis of which more and more wealth is produced, and the power of metropolises is built.

The concentration of investments in the main metropolitan areas raises the issue of their increasing weight with the resulting costs of urban concentration. Globalization means relocation, increasingly large investments, and investors favouring the places with the most advantages. Given a choice between a medium-sized town and a metropolis with high-quality services (transport and telecommunications infrastructures, business facilities and services, a well-trained workforce), they choose the latter.

In European countries, several observers[7] consider it most likely that this major trend towards metropolitan development will be reinforced in future decades as a result of improved city networking. Along the *European Mediterranean coast*, however, there are relatively few metropolitan areas that have acquired a dominant position at international level. But major urban projects proliferating around the Mediterranean demonstrate the willingness of large cities to affirm their position in European and worldwide networks (see the section entitled 'Urban regeneration and renewal' p220).

However, it seems evident that not all large cities will become metropolises in the globalized world. Some will have a regional metropolitan role, within a country or a group of countries, others will remain medium-sized towns with specific features and needs. Although urbanization is occurring everywhere, differentiation between different types of town remains just as important as globalization, expanding networks and convergence. The trend towards widespread urbanization requires policy-makers and land planners to think about different types of towns.

For the *southern and eastern countries*, prospects are contrasting and uncertain. However, with increasing economic liberalization and within the context of the Euro-Mediterranean free trade area planned for 2010, existing territorial inequalities will be highly accentuated, including those within towns:

- Economies in the SEMC, especially in the Maghreb, are already showing a clear dependence on trade with European Union countries. This south–north orientation of trade flows could intensify the growth of coastal towns even more, at the cost of inland towns.
- The removal of customs barriers and relocations may have profound consequences for spatial organisation, with unequal (positive or negative) effects on urban areas.
- In a situation where public budgets are tight and there is no access to structural funds such as those that exist in the EU, only a few regions in the SEMC will be able to offer attractive conditions for foreign or national businesses that are coping with stiff global competition.
- Under such conditions, the likelihood of an even balance developing between very large urban agglomerations and medium-sized towns are jeopardized. The trend is for too high a concentration of economic, political, administrative and cultural power in a very limited number of metropolises.

Faced with this threat of increasing regional inequality, can Mediterranean decision-makers balance the growth of metropolises with that of medium-sized towns, as India is trying to do?

Such policies are necessary but not easily achieved. State intervention cannot take the route of controlling migration. Spatial planning policies can only be implemented through financial support to housing and facilities, which influence people's choices. And incentives for job creation, creating attractive business areas, directing foreign investments, and creating fiscal advantages for certain priority areas are all effective levers, easier to implement by states or regions, but so far hardly used.

A Sustainable Future for the Mediterranean

Increased expansion and mobility, vulnerable and fragmented towns

Expansion and the fragmentation of urban areas is a current characteristic of urban dynamics at the local level, with major impacts on the quality of life and immediate and more distant surroundings.

More than 25 years ago, the UNESCO Man and Biosphere programme proposed an ecological approach for the 'urban system'. It envisaged a city as an ecosystem with flows of people, materials, energy and information. Today, other approaches aim to assess the *ecological footprint* of urban areas, that is their impacts in terms of consumption of natural resources and damage to the environment, which often go far beyond the local or regional level. Generally, these impacts rise with the increasing wealth of cities. This approach shows a 'predatory' city, but cities are also places of human development and societal organization, capable of housing millions of people, and of limiting their impact on the natural environment. The problem is not so much the number of city-dwellers but urban lifestyles that are too wasteful, and the way cities are managed. Policies aimed at regenerating and renewing ancient and more recent urban fabrics seek to counter the expansion of suburbs and at the same time reduce environmental impacts.

Wasting space

Throughout the region, cities, which used to be compact, are spreading like oil stains, with sprawling towns and ribbon development along the coasts. They invade their surroundings, absorb previously independent small towns and villages, and use up vast amounts of land. They drastically reduce agricultural land, especially on the coastal plains, which are a limited and scarce resource in the region. Many such sprawling cities have become vast, loosely connected conurbations expanding at a far more rapid rate than that of the population in the agglomerations.

The trend towards urban sprawl is very common and occurs in both northern and southern agglomerations. However, the mechanisms and forces at work differ considerably between countries. Responses intended to channel such sprawling are rarely applied and often have no real effect.

In the north, increased mobility and dispersed activities are the main cause

In countries on the northern shore, where natural population growth in towns has now almost ceased, changes are characterized by a decreasing population in city centres, dispersion of people and employment, spreading of suburbs, creating urbanized areas that are further and further away from the centre; in short, resulting in ever-increasing artificial areas. Many factors affect urban sprawl. Determinants are: greater mobility, more use of private cars, households and businesses organizing their travel on the basis of their location, more affordable house prices far from the centre, and the choice of suburban, individual family houses. The result is a self-reinforcing spiral that moves urban functions away from the centre and invalidates urban planning intentions.[8] Analyses of urban sprawl suggest that the process is partly a result of the market operation and of urban lifestyles, but also of several public mechanisms that actually facilitate sprawl (such as housing, transport, and tax systems).

Car traffic has increased at a rate that would have been unthinkable 50 years ago. It has encouraged and accentuated urban sprawl, allowing towns to spread even more loosely. Private car ownership has grown everywhere. In Mediterranean EU countries the average is more than 590 cars per 1000 inhabitants. The length of daily trips has also grown and trips between suburban centres have increased even more (see further below).

Depending on the circumstances and the context, the *housing market* (with substantially rising house prices, while car prices are falling) and grants given by housing organizations have led to households moving to the edges of towns. In France, for example, measures to finance housing, in force for more than 20 years, have encouraged home ownership by households with modest incomes in areas where land prices are low, such as low density built-up areas increasingly distant from the centre.

The demand for out-of-town locations by businesses and shops often accounts for half of urban sprawl. In suburban areas, employment possibilities are growing rather more than in town centres, and this generates *taxation difficulties* in such large urban areas.

Moreover, standard zoning regulations, which recommend separating urban functions (such as housing, work and shopping), have, in the same way as the supply of cheap land, too often encouraged single function estates on the outskirts. The observed trend is characterized by *chaotic patterns of shopping malls* and hypermarkets (developments of some 10,000m^2, sometimes as much as 20,000m^2) far from town centres, and, in some cases, built on what were originally green areas. These commercial areas offer free parking leading to daily polluting emissions by the huge number of motorized customers. They are widespread and common on the outskirts of towns on the northern rim (such as most French and Spanish towns, the major Italian metropolises, Athens, Thessalonica and Split), and are beginning to proliferate elsewhere (in cities such as Istanbul, Ankara, Beirut, Cairo, Alexandria and Tunis). A gradual erosion of the retail trade (especially small shops) and employment in the inner cities is expected in the medium term.

In the south and east, dynamic 'spontaneous' housing development feeds urban sprawl

In most SEMC, change since the 1970s has been characterized by a steady rapid increase in unregulated construction, in large as well as medium-sized cities and smaller towns. This is a new pattern of construction often resulting in vast neighbourhoods that provide access to housing mainly for house-owners in the poor or middle classes. This form of 'spontaneous' and unregulated urbanization (without permits), often anarchic, takes place either on land where construction is forbidden, or in areas designated for construction and building houses that do not comply with housing or building regulations.

This process of spontaneous house construction adds to the stock of precarious housing (shanty towns)[9] that already existed. But instead of land being squatted or rented, land plots are sold by landowners on a parallel real-estate market, intended for those who do not have the means to buy into the official market themselves. Almost all occupants own their houses. Land prices are often higher than in regulated neighbourhoods, but the size of each plot is much smaller than the officially regulated size. The constructions are 'solid' and evolve over time, sometimes with several floors. Construction is not clandestine and occurs in full view of the urban authorities.

The exact extent of spontaneous housing is rarely known. In fact, depending on the country or city, between 30 and 70 per cent of town-dwellers cannot find a building plot and settle unless they use informal channels (Table 4.2).

Unregulated housing development shows the existence and dynamism of unofficial or illegal actors in urban land, that is, a parallel market in building land. At the same time, it reveals weaknesses in central state policies that were intended to guarantee access to housing for all. In fact, restrictions on public finance in the SEMC have prevented programmes from being carried out in full. Only the demand from the middle classes has been met, while the needs of less solvent people have not been covered.

In terms of *impacts*, unregulated housing is responsible for considerable sprawl, leading to a rapid gnawing away of suburban areas (including agricultural land) and a lowering of urban density in central areas. But it also results in other unwanted social developments, such as derelict areas on the outskirts due to speculation, with consequences for access to water and other basic services.

Faced with the rapid increase in spontaneous housing over the past 25 years in the SEMC, *urban policies* for precarious housing and/or unregulated neighbourhoods have changed significantly. Although starting at different times in different countries, policies have followed a similar pattern: 'zero tolerance' and the eradication of shanty towns in the 1970s, followed by 'tolerance' and some 'recognition' since 1980–85. For shanty towns, policies consist either of eradication while transferring inhabitants to new estates, or of rebuilding such neighbourhoods on site. For unregulated estates, priority is given to bringing land tenure in line with regulations, acknowledging the right to self-construction, improving infrastructure and, if possible, facilities.

Table 4.2 Estimates of the extent of unregulated housing (averages unless otherwise indicated)

As a percentage of the total urban population %

Albania		70
Turkey	Large cities	50–70
Syria	Aleppo	40
Lebanon	Beirut	20
Egypt	Average of Egyptian cities	34
	Greater Cairo	58
Algeria		30
Morocco		30

As a percentage of the annual production of new accommodation %

Syria	Damascus	60–80
Egypt		60–80
Tunisia		40
Morocco		40–50

Sources: Plan Bleu from Chaline, 2001; M. Chebbi, personal communication, 2003; UNHSP, 2003

Note: In Tunisia, spontaneous housing has decreased in recent years: 22–25 per cent of the housing stock in 2000 against 34 per cent in 1993.

Gecekondu in Turkey, *tanake* in *Beirut*, *ashwaiyat* in Cairo, *mudun safi* or *brarek* in Morocco, are some of the terms used for informal housing. Cities on the European coast are not immune to illegal construction either 'abusive' in Rome or Naples, 'un-planned' habitation in Athens, and several similar forms in the eastern Adriatic countries.

Everywhere, very strong pressures on land surrounding towns

At first sight, at the national level, the area of artificial land seems to be low (3–10 per cent of total land area, depending on the country), but there are nevertheless strong pressures on agricultural land and on the environment. As artificial areas increase, the surface becomes less permeable and this can increase the risk of flooding, for example. An analysis of land use changes shows that more land becomes artificial every year, but few such areas change back to their original use. For example, in France, the total built-up area increased by 12 per cent between 1992 and 2000, roads and car parks by 10 per cent and non-built artificial land (such as gardens, lawns and construction sites) by 17 per cent. In total, 620km^2 of natural land are lost each year, half of which is for gardens around houses.[10]

At the regional/local level, it is easier to observe the wastage of land resulting from significant land use

2 A Sustainable Future for the Mediterranean

Figure 4.7 Land use changes in Marseilles, Padua-Mestre, Nicosia and Istanbul

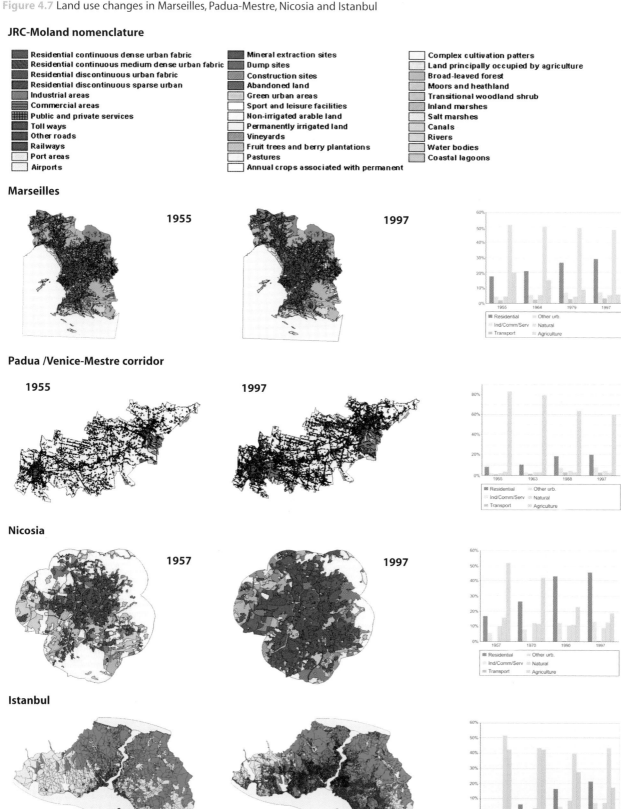

Source: EU-Joint Research Centre, Ispra (MOLAND)

Urban Areas

Figure 4.8 Land use in Great Sfax, Tunisia. Trends and scenarios

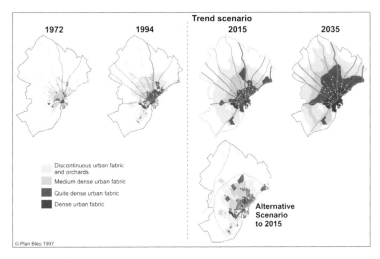

Land use changes in Great Sfax (1972–1994)

Land use	1972 Surface (ha)	(%)	1994 Surface (ha)	(%)
Artificial areas	7285	10.80	16,206	24.02
Agricultural lands	33,823	50.13	24,923	36.94
Forests and semi-natural milieu	30	0.04	30	0.04
Wetlands	1815	2.69	1815	2.69
Water surfaces	24,518	36.34	24,498	36.31

Source: Plan Bleu, 1997

Note: Urban fabric changes in Sfax from 1972 to 1994 show a clear increase in artificial areas, with an urban sprawl along the coastline and along transport infrastructures. Over 22 years, artificial areas have more than doubled, at the cost of market gardens. In the baseline scenario, the average decreasing size of agricultural land (1.4 per cent per year) is extrapolated to 2015 and 2035. An alternative scenario to 2015 is based on the assumption that measures are taken to redirect urban growth and preserve market gardens, mainly the implementation of land management policies, the development of secondary urban centres linked between them, a green belt surrounding the urban area and green 'lungs' within the agglomeration.

changes. In the five examples given in Figures 4.7 and 4.8 (Marseilles, Padua-Mestre, Nicosia, Istanbul and Sfax) the increase in built-up area (housing, commercial areas, transport infrastructure, industrial areas) has led to substantial losses of agricultural land. Within 40 years, 45km² of agricultural land was lost around Marseilles (107 hectares per year), 25km² around Nicosia (62 hectares per year), and 116km² in the Padua/Venice-Mestre corridor (276 hectares per year). Within more than half a century 561km² was lost around Istanbul (1021 hectares per year), and in the Sfax region 89km² of market gardens were lost in 20 years (405 hectares per year).

Apart from quantitative losses, a major problem is the irreversible loss of good quality agricultural land and natural land of high ecological quality.[11] The loss of good agricultural land is estimated at 7500 hectares per year in Turkey and at 12,000 hectares per year in Egypt, especially around Cairo, where highly productive (cotton) irrigated land is found. In Damascus, urban expansion is mainly taking place in the agricultural areas of the Ghouta oasis, jeopardizing a rare resource and polluting the water table that supplies the town. In Greater Algiers, poorly controlled urban sprawl continues to gnaw away fertile land, especially on the Mitidja plain. In Lebanon some coastal plains are already threatened (Box 4.1). Other details are given in Chapter 5.

Box 4.1 Land speculation and pressures on the agricultural plains in Lebanon

In the Lebanon, the war years (1975–1990) had a direct impact on agriculture and uncontrolled urban expansion. Thousands of hectares of agricultural land were abandoned, and rural poverty encouraged a movement away from an agricultural economic system towards land and real estate speculation, transforming lands into 'forests of cement'.

The peripheral areas around Beirut, Saida, Zahle and Tripoli gradually spread over a wider area. The Lebanese coast from Tripoli in the north to Tyre in the south (140km) has become one long stretch of densely populated urban settlements, haphazardly built and lacking services. This sprawl crosses Dahr-el-Baidr and the Beqaa valley, where main roads are lined by low-density urban ribbon development and the scattered urbanization of rural areas.

Although no reliable figures are available on the absorption of agricultural land by planned or speculative estates, it is estimated that, over the last two decades, urbanization around cities and motorways has taken nearly 20,000 hectares or 7 per cent of all cultivated land, and a higher proportion (15 per cent) of irrigated land.

Some coastal agricultural plains are seriously threatened. The plains of Minieh, Nahr Ibrahim and Damour, and the Saida orchards, could loose their agricultural function within less than 10 years. Agriculture could disappear even sooner in the plains of Tripoli, Batroun and Nahr-el-Kalb. In contrast, the most important plains, those of Akkar and Kasmieh, offer good chances of long-term preservation.

Source: Blue Plan 1999. Mediterranean Country Profiles. Lebanon

Is it possible to cope with uncontrolled urbanization?

Regulations intended to protect agricultural areas against urbanization have been formulated in most countries, but with limited, or even minimal, effect. Prohibitions and threats of punishment can only be effective if alternatives to land speculation are made available and town planning is enforced. Town planning measures need to be rethought in order to integrate the real city (official as well as informal areas), the dynamics at work, and the 'malfunctioning' that is the expression of existing interests.

- Drawing up *master plans*, inspired by the industrialized countries, was common in the 1970–80s in the southern Mediterranean cities. Experience has shown that these measures have often been ineffective, especially because of the lack of regulatory and budgetary resources for implementation and, above all, the lack of land tenure control. Although these planning measures are still being applied in most cities, today a balance is being sought between suburban extension and concerns for renewing and regenerating the existing fabric, both the ancient fabrics and those handed down from colonial periods. Town planning becomes less coercive and more strategic (see the section entitled 'Urban regeneration and renewal' p220).
- Sometimes, a solution was sought by creating *new towns* in the outskirts, intended to function relatively autonomous from the inner city. Egypt adopted this

Box 4.2 New towns in Egypt

In Egypt, urbanization takes place at the cost of arable land, since the great majority of towns are located on the fertile land of the Nile delta and valley. Soil improvement measures have been taken on 'new' land won from the desert. However, the quality of this land is lower than that in the valley. Since 1979, the Egyptian government has also made huge efforts to build new towns intended to absorb the Cairo overflow and safeguard agricultural land in the Nile delta and valley, which is under pressure, but the results have been below expectations.

Of the six new towns being built in the desert some 30–40km outside Cairo, two (Fifteenth of May and Tenth of Ramadan) seem promising with about 70,000 inhabitants, but this is small compared with the intended 250,000–500,000. Other attempts for ten new settlements on the Cairo suburban fringes, with a range of building plots without basic facilities, have also not reached their targets. In the 1990s, four new residential towns were started to the east of Greater Cairo. In fact, major difficulties are occurred with the development of the new towns because the projected rail connections are still missing.

Source: El Kadi, 2001

Box 4.3 Development scenarios in the French Riviera conurbation

The French Riviera conurbation has been spreading since the mid-1960s. Its development pattern is determined by secondary housing, car use and road development. This process results in an intense consumption of space. Between 1970 and 2000, the urbanized area was multiplied by 2.4 and the population by 1.4. Suburban houses with large gardens are common. Commerce and industry have relocated; new economic centres emerged, such as Sophia Antipolis, linked to the motorway network. Located in a typical Mediterranean situation (strong topographical constraints and scarcity of available space), real estate is scarce. The result is a substantial rise in real estate prices, reinforced in the case of the French Riviera by very high pressures from tourism.

1960–1990 changes: car logic

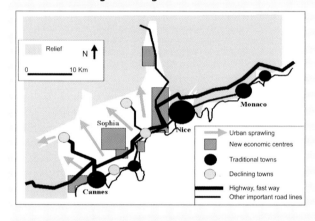

'Rhine' scenario

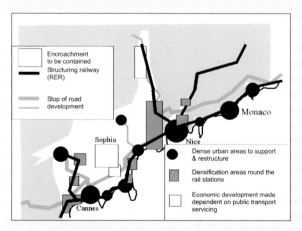

Up to the mid-1960s, the French Riviera conurbation had developed mainly around railway lines. Thirty years later, a sustainable development scenario (a 'Rhine scenario', referring to German and Swiss towns where urbanization remains concentrated around public transport lines) could have channelled and redirected urban growth around the railway

Urban Areas

and trams lines and through rehabilitated town centres, making use of opportunities to increase densities around railway stations and tram lines, and supplementary measures to discourage car use in dense areas and limit the spread of the road network. For such a scenario to succeed it would have been necessary for the public authorities to ensure a strong commitment to implementing land management policies, regulating the real state market, and developing public transport.

However, the option recently chosen by the public authorities through the Directive territoriale d'aménagement (a planning document drawn up by the central State and subjected to a public enquiry in 2002) would take a different direction. It does envisage some dense population poles along the railway line and encourages the rehabilitation of town centres. But it also plans to strengthen the road networks and continue the development of the outskirts. This choice shows how difficult it is in urban planning to integrate local public action and national level strategies. The central state has to respect the autonomy of local authorities and can hardly impose a scenario like the 'Rhine scenario'.

Source: Jourdan, 2003

policy to de-congest Cairo (Box 4.2). France did this in the 1970s by launching a dozen new towns around Paris and regional metropolises. Algeria may follow with its 'Algeria' project, a city to be built in the desert, 200km south of Algiers, for some 400,000 people, intended to become a business centre. These fully planned public operations seem only rarely to attain their desired goals, especially the mixing of functions, the job/residence balance and the urban atmosphere.

- In countries where traffic has exploded, innovative policies are trying to influence mobility at source, by organizing towns in more appropriate ways. In the last few years *joint town and transport planning*, linking population density with dense transport facilities, has been introduced in some industrialized countries.[12] Emphasis is being put on making existing urban fabrics more dense, especially around public transport stations and correspondence points and along transport corridors, in order to prevent excessive car use for urban travel. Given that town planning measures usually intend to control too high densities and almost never too low densities, the innovation lies in defining minimum densities in order to increase population density. The concept of minimum density seems suited to a period of urban expansion and falls within a sustainable development approach (energy saving for transport, less polluting emissions, safeguarding natural areas). In Naples, the combined urban and metropolitan transport systems illustrate this search for measures to integrate town and transport planning (Box 4.8 p219).

- In general, countering the excessive expansion of big cities calls for determined overall (regional) development policies for the entire country, and for putting in place a set of measures, both stimulating and discouraging, to promote the development of medium-sized city *networks*, mainly by improving accessibility.

Social and ecological inequalities in fragmented urban areas

With urban sprawl, fragmentation within a city increases, undermining overall social cohesion. Social differences have always existed, but this old issue is reappearing in new and overwhelming forms. The process of fragmentation is probably most visible in the vast metropolitan areas of both the south and the north, where a mosaic can be seen, where differences are accentuated, where large cities are fragmented. Some observers even speak of a *metropolitan disaster*, which would turn cities into fragmented, centrifugal (scattered) urban areas.[13]

In any case, in large cities, there are considerable differences between neighbourhoods in terms of quality of life and environment. These inequalities are very often added to those of income, access to work, even survival. Apart from extensions, some inner-city neighbourhoods tend towards gentrification with extremely high land prices, luxury dwellings and offices. Other areas decline, where only those inhabitants and businesses that cannot afford to go elsewhere remain. In these less favoured neighbourhoods, socio-economic and environmental problems are accumulating, such as poor access to transport, housing and water, and consequent health risks.

With increasing inequality and urban fragmentation, people sometimes start to feel insecure, giving rise to withdrawal processes within groups. When social conditions harden, fragmentation leads to increasing confinement of both poor and rich neighbourhoods. 'Sometimes, entire neighbourhoods hide behind walls, some to protect themselves against thieves, others for protecting themselves against the police'.[14] The development of new residential neighbourhoods where the middle and upper classes take refuge in gated areas, can be observed in many parts of the world. This 'urban development' pattern has also appeared in the Mediterranean (Box 4.4).

Box 4.4 Gated cities in the Mediterranean

Secure residential neighbourhoods are generally composed of villas and/or buildings surrounded by a protective wall, provided with special services and equipment (such as video-surveillance, shops, green spaces, swimming pools, golf courses). Access to these walled complexes is limited, they are guarded 24 hours a day. Management is handled by the co-owners and falls entirely outside the responsibility of local public authorities.

In **Lebanon**, the first gated residential complexes appeared during the civil war from 1975 to 1990, arising from a demand for protection and for security of service supplies. In the 1990s, the post-war building boom, particularly in luxury accommodation, took place in a context of re-established public administration but without land tenure regulation. Investors developed projects on land located on the urban fringes and the coast, promoting a 'global lifestyle'. By 2000 gated residential complexes accounted for about 2 per cent of all housing in Lebanon.

Gated cities in **Turkey** have grown significantly for the past 15 years, especially around **Istanbul** and on the Mediterranean coast. First, small luxury housing complexes were built on the land of old properties, followed by holiday villages on the Marmara Sea coastline. With the development of the motorway network these villages became permanent fringe cities, resulting in the gated cities of the 1990s. By the end of 2001 there were about 270 gated cities in the Istanbul metropolitan area, the largest being 'Alkent 2000' (720ha) and 'Acarkent' (230ha in a forest). Gated cities are spreading more and more to the outskirts of the urban area, at the shores of the Bosphorus, near woodlands, and near the Marmara Sea and the Black Sea, on land that does not fall under the jurisdiction of the Istanbul metropolitan municipality.

In **Egypt**, exclusive and protected residential areas were built in the 1980s along the beaches of the north-western coast, and later along the Red Sea coastline. In **Cairo** the phenomenon developed in the city suburbs in the 1990s, facilitated by the ring road. 'Mirage City' in New Cairo, 'Mena Garden City', 'Dreamland', 'Golf City' and 'Beverly Hills' are among the main projects carried out or under construction, some of them offering more than a thousand houses per estate.

Walled real-estate complexes have arisen in **Spain** along the Mediterranean coast, both as holiday homes and primary dwellings.

In **France,** protected enclaves for the upper-middle classes are proliferating, with four in Montpellier, three in Avignon and 20 in Toulouse. The French Riviera has several protected residential enclaves including the 'Domaine des Hauts Vaugrenier' (170ha) between Nice and Antibes and, more inland, the 'Domaine de Terre Blanche' (265ha) not far from Cannes.

Source: Plan Bleu from Etudes foncières, several issues 2002–2004.

These drastic changes point towards a *break* with the mixed and compact urban model of the Mediterranean and Europe. Does this mean that cities are in transition towards a fragmented model with ghettos? In any case, the rapid increase in private, secure urban areas shows a purely market-driven development, where urban planning and the law appear to be powerless to realize the ideal town presented in plans inherited from the ancient city.

In the Mediterranean, where towns were traditionally well balanced, the process of developing gated cities seems to have started. Will it lead to fragmentation with 'well-to-do' enclaves and 'neglected' housing or informal towns? The *challenges* for the region are important. Fragmented cities put the very essence of the Mediterranean city at risk, an essence based on the principle of intermixing.

What efforts are being made to avoid segregation and combat exclusion in both the north and the south? Answering this question would require taking a look into the social policies carried out in the different countries, which is outside the scope of this report. However, segregation may also be approached through its spatial and environmental aspects.

In line with the 1992 Rio Earth Summit and the 1996 Istanbul Conference on Human Settlements (Habitat II), the international community defined Millennium Development Goals in 2000 and the Johannesburg Action Plan in 2002, which aim at integrating the social, environmental and economic development dimensions. They encourage looking at the *linkages between social equity and environmental quality* in order to facilitate this integration.

Urban vulnerability to natural and technological hazards

Warning signals are periodically provided by scientists who forecast a significant rise in urban vulnerability to disasters in the 21st century. This would be a result of urban growth (population and/or spatial growth), and climate change, causing more frequent and intense hydro-meteorological disasters at both global and regional levels.

The Mediterranean region is especially vulnerable to extreme weather events and even more to *earthquakes*, the effects of which tend to be concentrated in urban areas. Countries such as Morocco, Algeria, Italy (with an added volcanic hazard), Greece and Turkey are the most vulnerable to earthquakes. *Catastrophic floods*, which are linked to the sometimes violent rainfall of the Mediterranean climate and aggravated by deforestation and building on slopes, are also a major risk for many cities in Spain, France, Italy and Algeria.

In future, inequality of exposure to risks could increase between the northern and the southern shores. Indeed, if no substantial progress is made on the three fronts of hazard prevention (stopping building in disaster-prone areas, reducing makeshift housing, and implementing safety standards in urban dwellings), the vulnerability of the southern and eastern large urban areas will increase, together with the impacts of disasters in these huge urban concentrations. Floods, earthquakes and landslides have common features wherever they occur, but the challenges from the human point of view are not the same. The highly populated urban areas in the SEMC, with their unregulated housing construction, are especially vulnerable, as unfortunately demonstrated by recent fatal disasters (Table 4.3).

Table 4.3 Balance of recent natural disasters in the southern and eastern Mediterranean

		Fatalities	Homeless
Izmit (Turkey)	Earthquake (August 1999)	17,200 killed/ 44,000 injured	600,000
Algiers, Bab el-oued	Floods (November 2001)	920 killed	50,000
Algiers and Boumerdès	Earthquake (May 2003)	2200 killed/ 10,200 injured	120,000
Al Hoceima province (Morocco)	Earthquake (February 2004)	600 killed/920 injured	30,000

For *technological hazards*, lessons can be drawn from the accident that occurred in Toulouse in 2001 (Box 4.5), which, after the Seveso accident in a Milan suburb (1976), reopened the debate on the presence of hazard-prone industrial sites, which have been classified as such, in or near densely populated areas. The problem is even more worrying in Toulouse, where the city expansion towards the south-west occurred under legal circumstances.

Box 4.5 The industrial accident in Toulouse (France)

In September 2001, the explosion of a warehouse in the AZF chemical plant in Toulouse was the largest industrial accident recorded in France in 20 years. This chemical plant, which specializes in manufacturing agricultural fertilisers, falls under the Seveso Directive[15] and is located in the south-western suburb of Toulouse, less than 5km from the town centre. The depot that exploded contained stocks of substances (about 300 tonnes of ammonium nitrate) that did not comply with the required specifications and standards. The huge blast set the plant on fire and shook the ground (equivalent to an earthquake of a magnitude of 3.2 on the Richter scale).

The effect was very serious: 30 people were killed and 2400 were injured by the explosion; 2500 houses, several schools, a hospital, and buildings and residential halls of Toulouse University were destroyed by the blast. The AZF plant was built in the 1920s outside the town, but since the 1960s it had been gradually engulfed by urbanization. Today some 30,000 people live within the direct perimeter of this chemical area. The situation had already been condemned for several years by resident associations and ecologists.

Source: IFEN. L'environnement en France, 2002

What can be done at the national, the local and Mediterranean level? First, one should reject the idea of inevitability since the impacts of many accidents or disasters would be much less serious if timely action were taken during the urban development process. At the closure of the International Decade on Natural Disasters,[16] it was emphasized that disaster prevention and sustainable development approaches are part of two different worlds that totally ignore each other. In a sustainable development approach, the classical view of risk management is reversed: risks are integrated up front in development and spatial planning issues, and not the other way around. This is not usually the way it works in the world. Two proposals for action formulated within this framework are recalled here:

- Move gradually towards a less sectoral natural hazard management, by also considering hazards as economic, social and environmental issues, and integrating them into development or regional planning policies and programmes at all levels. For example, in France, plans for preventing predictable natural hazards, elaborated for more than 2700 municipalities, should be integrated into detailed land use planning and in construction rules and regulations.
- Give priority in sustainable development strategies to policies aimed at reducing vulnerability to the most important hazards, particularly the vulnerability of the most disadvantaged groups, especially those in the large cities of the south. Izmit, Bursa and other cities in the very vulnerable Marmara region in Turkey, provide examples of this approach. Their Local Agendas 21 have, since 2000, been redirected towards prevention and preparedness for natural disasters.

For *industrial hazards*, prevention policies need to integrate hazard reduction at the source (setting up safety management systems), organize specific inspection systems for hazard-prone activities (in the EU Mediterranean countries industrial risk prevention and management is regulated by two 1996 directives: Seveso II and the IPPC directive on integrated pollution prevention and control), and include measures in town planning to deal with technological hazards (for new establishments, granting permission only in areas far from residential areas; for older plants, granting no permission for new residential areas nearby).

Finally, without better guidance being given to urbanization, and if the processes through which a city is actually built are not taken into account from the start, technical responses for managing disasters, emergencies and crises can only involve damage control. Sustained efforts are necessary on two fronts:

- Reducing disaster vulnerability requires real alternatives to land and real-estate speculation (in order to stop urban settlements near hazard-prone industries or in flood-prone areas), to buildings not complying with anti-earthquake standards, and to deforestation and building activities (including infrastructure) on steep slopes.
- Improving disaster/emergency responses and hazard management techniques so as to mitigate the impacts of disasters.

Cooperation at the Mediterranean level has generally been focused on disaster management; in future, hazard prevention needs to be further supported and strengthened.

Ever longer travel in urban areas

Urban transport is an essential component of the urban quality of life and of accessibility, for people living both inside and outside the central urban area. There are two types of concern regarding the environmental impacts of urban transport:

- public health and the local environment, both of which require reduced air pollution and reduced noise from transport.
- greenhouse gas emissions, essentially carbon dioxide and nitrogen oxides (CO_2 and NO_x), the effects of which are global (threats to climate and ecosystems) as well as local, and which require national action plans (see Chapters 2 and 3).

We will here deal with travel modes both in central areas and on the fringes of towns. The current trend is characterized by an increased use of private cars, a mode that is costly to the environment in terms of energy consumption and polluting emissions. On the other hand, urban travel modes are closely related to *urban morphology* (urban structure). International comparisons carried out on 30 large cities[17] have shown that car use in urban areas increases as urban density decreases.

Individuals living in low-density suburban areas cover daily, by car, distances two to three times greater than those living in central areas, without spending more travel time. They are more or less forced to take the car because of the relatively long distances between their point of departure and destination, poor public transport systems, the good road traffic conditions and car parking. The situation is quite the opposite in most densely populated areas, where using the car is very much constrained (scarcity of parking spaces, congestion) and where public transport is attractive.

So, commuting distances vary greatly, depending on location in a given urban area. As a result, fuel consumption and polluting emissions increase by up to five times when comparing central dwellers with those living in the suburbs or in rural villages in urban fringes.[18] This situation prevails in cities of the Mediterranean EU countries, but how does it apply to the southern and eastern Mediterranean cities?

The 'automobile transition' in the south and east

The three Mediterranean shores, as we have seen, are experiencing expanding urban areas and increasing daily mobility. As travel distances continue to get longer, people are forced to use fast motorized transport modes: private cars and/or public transport. The main reasons for travel are work and study, while leisure is becoming more important in the urban areas in developed countries.

In 2003, compared with the northern developed coastal countries (where car ownership averages at more than 590 cars per 1000 inhabitants), car ownership in the SEMC is fairly low (on average 124 vehicles per 1000 inhab) with rates varying between 64 cars per 1000 in Egypt to 520 in the Lebanon (see Table A.25 in the Statistical Annex).[19] These are national average figures, however. Rates are higher in the cities, for instance in Cairo where the motorization rate may be higher than 110 per 1000 inhabitants.

Contrasts between the north and the south remain strong, but for how long? Differences in income levels, in the purchasing power of urban populations, as well as uncertain economic growth forecasts, suggest that widespread car ownership in the SEMC will not happen overnight. But other factors such as decreases in car prices, car attractiveness, particularly among the middle classes, a political will to democratize access to cars through significant reductions in customs duty, and the Euro-Mediterranean free trade area expected for 2010, suggest that the shift towards the car is already in progress in southern and eastern cities.

On a world scale, global prospective modelling of passenger mobility and the modal split to 2050,[20] suggests that the Middle East and North African countries will experience mass car ownership by 2010 (Figure 4.9). These models do not predict, but provide a first indication of ongoing trends: the SEMC seem to have entered the 'automobile transition', something to be examined more closely.

In order to clarify these trends in southern and eastern Mediterranean towns, travel modes have been analysed in six large cities[21] (Istanbul, Beirut, Cairo, Tunis, Algiers and Casablanca). The emphasis has been put on motorized travel in the large urban areas, even though it is known that pedestrian mobility, little researched and generally underestimated in mobility surveys, remains relatively important in small and medium-sized towns,[22] as well as in central areas of large Mediterranean cities.

Urban Areas

Figure 4.9 Passenger travel modes in the Middle East, North Africa and western Europe (eastimates and forecasts (km/person/day))

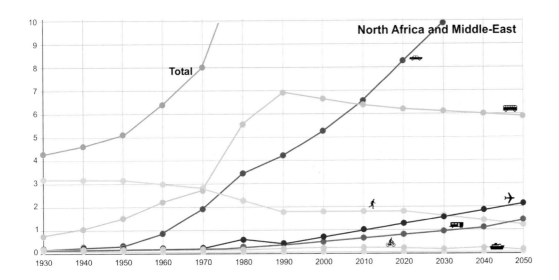

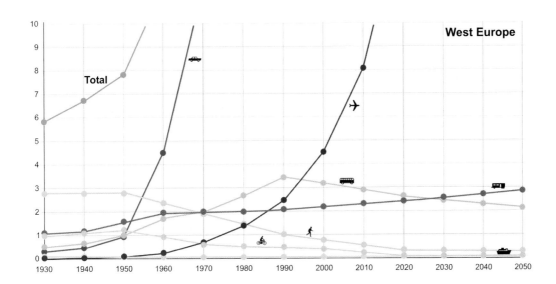

Source: F. Papon (Inrets) for the Blue Plan, from A. Schafer's data for car, bus, train and airplane modes

Note: Forecasts through 2050 are based on historical trends everywhere in the world. Individual mobility increases with revenue, travel time remains relatively constant (1–1.5 hour per day), faster transport means are used to cover ever longer distances within the same time.

In North Africa and Middle Eastern countries, motorized travel, essentially by bus and by car, has risen noticeably in the last 20 years. By 2010–2020, and even more beyond, the car will largely be the dominant travel mode (10km/inhab/day by 2030), while the bus will continue to decline.

In west European countries, the 2050 prospect suggests that the mobility share of high speed transport (airplane, high speed trains, car) will significantly increase to cover longer distances. In 2004, car travel represented 60km/inhab/day (out of scale in the graph).

A Sustainable Future for the Mediterranean

Table 4.4 Motorized travel share in six very large cities of the SEMC (estimates, in per cent)

City	Istanbul 2000	Beirut 1994	Cairo 1998	Tunis 2000	Algiers 1998	Casablanca 2000
Public transport	71	29	74	50	52	45
Private car (+ individual taxi)	29	71	26	50	48	45 (+10%)
Total	100	100	100	100	100	100

Source: Godard, 2003 from recent household surveys and estimates from various sources

Note: Classifying the different modes is not easy. For instance in Casablanca, small taxis (metered taxis operating within the urban boundaries) are classified as an individual mode in this table, but could also be classified as a public transport mode, since they actually serve as both (taxi-sharing is permitted).

Growing use of private cars and decline in organized public transport

Table 4.4 shows the modal share of motorized travel. For each urban area, average figures hide considerable disparities between social categories, particularly between those using private cars and those dependent on public transport.

Public transport still dominates in Cairo, Istanbul and Algiers, but its supremacy seems to be seriously threatened in Casablanca and Tunis. In Beirut private cars account for more than 70 per cent of journeys – an untypical figure for the region;[23] in 2003, this could well be 85 per cent. In Tunis car use increased in the 1990s, in spite of the light rail system in the central area; projects are being studied to improve public transport systems at the metropolitan level, without which the car share of motorized travel could reach 60 per cent by 2010.

In all cities, the private car share seems to be rising rapidly, as is well illustrated by the Cairo example (Table 4.5).

Table 4.5 Changes in transport modal share in Cairo

Motorized modes	Market share (%) 1971	1998
Car and individual taxi	13	26
Underground railway	0	17
Tram	15	2
Bus and minibus	62	19
Shared taxis	0	28
Other (rail, boat)	9	7

Source: Godard 2003, from 1971 and 1998 Household surveys in Cairo

Note: The share of individual modes is significantly in the minority, although it is rising noticeably due to road network development and the rapid increase of flyovers and other heavy infrastructures (bridges over the Nile, urban tunnels, the ring road), most of which are not open to public transport and favour cars. Within public transport, the traditional bus and tram system has been supplanted by microbuses (shared taxis), which are now the most important mode (28 per cent of travel), and by the underground, which has had a remarkable success (17 per cent), even though it so far covers only part of the metropolis transport routes.

So, individual motorization seems to have grown appreciably across the region in the 1990s, perhaps with the exception of Algiers.[24] The progressive liberalization of vehicle imports (new and second-hand vehicles), national programmes encouraging access to cars (Tunisia, Morocco) and the crisis in organized public transport systems, have no doubt contributed to foster car purchases, whether as a first car in a household or as second or third car in wealthier households.

Formal and informal public transport systems

Public transport systems have changed noticeably, and the mid-1980s was a crucial period for these changes.

The poor condition of public *bus and coach* companies offering suburban or intercity connections, and their deteriorating supply compared with growing travel needs, have led most countries (except Tunisia) and large cities to begin opening up to the private sector, although with mixed results (on deregulation in this sector, see Chapter 3). In 2000, bus travel, with either public or private systems, took a minor share of public transport (figures vary between 20 per cent in Casablanca and 1 per cent in Beirut). Only Tunis is an exception, here buses catered for nearly 68 per cent of collective travel in 1998.

Other transport modes, operating on small scales, have taken over from the bus: for example formal and informal *microbuses and shared taxis*, which play an increasingly important role in urban and suburban travel (Table 4.6). Their development is relatively recent, and is both a response to and a sign of the crisis in organized public transport. In Istanbul, the *dolmus*, an informal shared taxi, used to be an important traditional transport mode, but was opposed by the authorities. A new and flexible small-scale form has emerged, known as the 'service', for collecting company employees or schoolchildren by minibus. In Beirut microbuses were rare in 1994, while in 2002 they represented a dynamic transport mode. In Tunis, on the contrary, small-business transport with microbuses (*louages*) only operates for suburban services, while the predominant system is organized and run by larger businesses, the main one being a new public company.

Urban Areas

Table 4.6 Share of informal systems in total public transport (at the beginning of the 2000s)

City	Description	Share
Istanbul	*Dolmus*, services and private minibuses	30–38%
Beirut	Taxi services (shared taxis halting on demand, not fitted with meters, carrying a red plate)	79%
	Microbuses, for 9–15 people, operating like taxi services between Beirut city and the whole metropolitan area	9%
Cairo	Shared taxis (small minibuses with some ten seats), operating on licensed fixed routes allocated by one of the three *Governorships* in Great Cairo	28%
Algiers	Minibuses	50%
Casablanca	Shared taxis and small taxis	23% and 10%

In general, these vehicles are of low-capacity (for 10–15 people), and are sometimes little more than converted vans. Today this public transport mode is expanding rapidly.

Finally, upmarket public transport systems, such as the *underground* or the *tram*, register varying success in the six cities, because of the considerable investments they require (Box 4.6).

Box 4.6 Underground and tramways in six Mediterranean metropolises

Cairo has successfully undertaken an investment programme in the underground. A first regional line opened in 1987, which transports more than 1 million passengers a day, and an urban line opened in 2000, which transports 500,000 passengers per day. The share of the underground has thus become significant in travel in the agglomeration (17 per cent in 1998), with ticket prices that are especially favourable to schoolchildren, students and civil servants. A third underground line is planned. At the same time, the tramway has experienced a decline (2 per cent of public transport travel), because of a lack of investment in modernization.

Istanbul has a variety of rail-based public transport (light-rail metro, tramway, underground, suburban train), with as yet a limited market share (some 6 per cent of all public transport travel). Important investments would be necessary for developing a regional rail service to cover the metropolitan area.

Tunis has made notable progress with its quite extensive light-rail network in the central area, built over a period of 15 years. The success of this flagship investment (25 per cent of all public transport travel) is now counterbalanced by the saturation of the network. A Regional Rail Network project is under consideration; implementing it would re-energize public transport in the city and the metropolitan area.

Algiers has chosen to build an underground and work began nearly 15 years ago. Many setbacks of all kinds (political and economic) have slowed progress, but completion should come about within a few years.

Beirut has not gone down this path, in spite of the 1994 Transport Master Plan proposals (IAURIF).

Casablanca has been discussing its underground project for 20 years but is faced with the financing obstacle, and no start was foreseen in 2003.

A recurrent question on rail and light-rail transport is that of beneficiaries and users: are car users encouraged to switch over? In Tunis the light metro did attract car passengers at the beginning, but this effect gradually diluted. The metro service quality, with overcrowded trains during peak hours, is certainly inadequate to attract and retain car drivers as long as road traffic conditions are not too bad. In the case of Casablanca, with wide income disparities, a major issue relates to whether an underground system would attract both disadvantaged groups and the car-relying middle classes. In an unfavourable financial context, it seems difficult for the underground to satisfy several objectives at the same time: to provide mass transport on radial suburban lines, to maintain a satisfactory cost-recovery rate from revenues, and to attract car users. The underground financing and pricing system both depend to a large extent on the answer to this question.

Improving transport management in urban areas

Southern and northern cities have many difficulties in common:

- for land use planning: continuing urban sprawl with a decline in population densities in central areas;
- for urban transport: increasing car traffic and numbers of cars, decreasing public transport travel, longer trips, especially by car, congestion and traffic jams;
- for local air pollution: increasing ozone pollution, the presence of several other air pollutants (see the section entitled 'Air quality and public health' p223) and high noise levels.

Many of these difficulties are similar, although some are more acute in certain urban areas. Policies will be conditioned by institutional shortcomings, town sizes and their economic specialization. Challenges and room for manoeuvre differ from country to country.

In the southern cities, better linking of public transport systems

In large cities in the south, maybe to a lesser degree in Beirut, minibuses, buses, formal and informal shared taxis provide a significant travel share for urban and suburban areas. The future seems open for the southern cities and countries, although they will need to innovate, since the attractive car-dominated European model will be difficult to reproduce. In addition, the model would lead to the same impasse found in northern cities. Several paths can be followed:

- Making access to the town easier and strengthening public transport before mass car ownership occurs. However, re-conquering the public transport market will have to face the financing issue. In most cities, authorities rarely have the resources (their own revenues through transport-specific local taxes) to maintain cheap fares. State intervention thus plays a decisive role in stopping the downward spiral of reduced passenger numbers, growing deficits and higher fares, and the decline of bus or rail transport.
- Taking advantage of typical informal/small business public transport systems in cities (microbuses, shared taxis and similar modes), which represent an undeniable asset. Their flexibility makes it possible to connect urban areas generally neglected by standard public transport systems, under attractive conditions. Their high occupancy rates make it possible to maintain a relatively low rate of polluting emissions per passenger-kilometre. On the other hand, this advantage is counterbalanced by the age of the vehicle fleet (10 years, sometimes up to 30 years), with serious local air pollution impacts. Major anti-pollution efforts are thus necessary (such as fleet renewal, use of less polluting fuels, and regulation of licences).
- Imagining a town planning system where automobiles gradually adapt to the town rather than the opposite, so that urban motorways or bypass roads do not disfigure cities as it has happened in northern urban areas.

All these issues concern urban management and constitute policy challenges for local and regional authorities as much as for central Governments and international cooperation bodies. Here it can be noted that national and international financing for urban and regional rail networks is allocated mostly to the capitals (political or economic) of countries. Since capitals are of strategic importance, this targeted financing seems logical, but only up to a point, because it occurs at the cost of financing transport needs in towns of secondary and tertiary importance. A better *overall* view of financing issues would contribute to a better balance among the various towns within a country.

In the northern towns, the struggle to reduce pollution

For north Mediterranean urban areas, and particularly those in the four EU countries,[25] the major challenge is to reduce air pollutants as well as their contributions to the greenhouse effect, where the share of urban transport continues to rise.

France and Italy have set up large scale innovative urban transport policies. Urban Travel Plans (PDUs) in France and Urban Traffic Plans (PUTs) in Italy[26] are intended to promote public transport and less-polluting transport modes. Although being old measures, 1982 for the French PDUs and 1986 for the Italian PUTs, they have been relaunched during the 1990s, within the framework of the two countries' commitment to reduce their greenhouse gas emissions (Kyoto Protocol). It is still too early to draw conclusions from these recent experiences, but it is useful to present the main innovations introduced (Box 4.7).

> **Box 4.7   Urban Travel Plans in French cities**
>
> According to the law on air and the rational use of energy, adopted at the end of 1996, all French towns with more than 100,000 inhabitants are legally required to set up urban travel plans (*Plans de déplacements urbains* – PDU) for passengers and freight. The plans have compulsory objectives and content: reducing car traffic, developing public transport and less polluting and economic travel modes, and re-assigning the road network for use by different travel modes. PDUs are developed by the urban transport authorities, within the urban transport perimeters of a town. By mid-2003, out of the 72 cities with more than 100,000 inhabitants, 43 had an approved PDU.
>
> The law stipulates that PDUs must plan to reduce **car traffic**. Only the Ile-de-France PDU is targeting a 3 per cent car travel reduction by 2005; the other cities aim at reducing a few points in car travel, generally by 2010. This will not necessarily lead to car traffic reduction, due to the ever longer distances travelled and increasing car purchases in suburban households. Cities with overall coherent policies over time (particularly Grenoble, Nantes and Strasbourg) have managed to contain the continuous increase in the share of private cars in motorized transport. The Mediterranean cities have not yet achieved such progress.
>
> Travel plans envisage a gradual 2–6 per cent increase in **public transport share** by 2010. Trams are often preferred (Strasbourg, Nantes, Grenoble, Bordeaux, Montpellier, soon to be followed by Nice, Paris, etc.). Trams have technical advan-

tages compared with buses (high flow in good conditions) and economic advantages compared with the underground (cost ten times less per km). Strategies to improve bus networks include bus lanes and priority at traffic lights. Other measures are designed to improve their attractiveness: accurate timetables, frequent services and integrated fares. 'Green' transport (bicycles and cycle lanes) is also encouraged. All these measures mainly concern densely populated areas, while they occasionally also enhance suburban rail systems.

Parking control is an important measure in PDUs to curtail car invasion in town centres. Measures concern overall supply and differentiated pricing, so as to encourage residents to leave their car in the car park, penalize commuters who use the car for going to work, and facilitate short-stay car parks for shopping. In addition, park-and-ride systems are set up along the public transport lines.

As for the **road network**, urban motorways should generally be restricted. Most PDUs are planning to create or complete bypass roads, designed to take transit traffic away from central areas. However, opposite effects may occur: although designed to reduce traffic, town bypassing actually sucks in cars, and new ring roads are shortly afterwards saturated during rush hours without any traffic reduction. To counterbalance new road supply, a third of the PDUs envisage road reclassification (Grenoble, Lyon, Marseille, Nantes, etc.), where roads are shared with alternative transport modes on roads going into the town. Other cities are considering underground tunnels, which would in fact encourage urban traffic.

Finally, the **search for coherence** between transport planning and town planning will require further efforts. According to the 2000 law on solidarity and urban renewal (SRU), town-planning measures should better integrate traffic issues, by defining areas that can only be developed on condition that public transport is provided. All PDUs were supposed to comply with the SRU law by the end of 2003.

Source: Environnement Magazine, October 2002, and CERTU/GART (2002) Bilan des PDU de 1996 à 2001 (www.certu.fr)

At least two lessons can be drawn from the French urban travel plans.

- Although intended to reduce car traffic, the great majority of plans only propose modal shifts. The relative reduction of car travel by such shifts is overwhelmed by increasing car numbers and distances travelled, since overall traffic depends first and foremost on growing numbers of households and car owners. So, the key problem is not modal shifts but car traffic reduction.
- Most measures in favour of public transport apply to dense areas, where modal shifts can be encouraged because public transport in such areas is most profitable. But much traffic comes from the outskirts, where the population is growing. Therefore, it is by taking action on the longer commuting distances at the urban region level[27] that the struggle against pollution and congestion will bear fruit.

Conclusions of more extensive work carried out in the framework of the European Conference of Ministers of Transport (ECMT) indicate the same direction. Currently, measures taken to combat central congestion seem to be relatively effective nearly everywhere. In contrast, traffic outside town centres is getting worse. Building new underground or surface infrastructures, with or without tolls, is not a suitable response to traffic congestion in urban areas. It is counterproductive since the problem of traffic reduction is evaded.[28]

Towards more sustainable urban transport systems in the north and the south

The challenge of more sustainable urban transport systems is thus linked to efforts to contain traffic in the periphery of cities, in suburban and even intercity connections. This will be a big challenge for towns in the years to come, in order to reduce negative impacts on citizens' life and on the planet's health. At the same time, by taking action in areas with poor public transport services, transport can become a tool for preventing segregation and some areas of towns becoming excluded.

In the search for more sustainable transport systems, *four lines of action* are here emphasized:

1. *Conceiving the transport system as a whole*, at the level of the enlarged urban and metropolitan areas (as Naples is trying to do, Box 4.8) is the first path, which raises matters of coherent public action and of defining transport authority.

> **Box 4.8** Integrated urban and metropolitan transport in Naples
>
> In the early 1990s public transport in Naples (1 million inhabitants) provided a very mediocre service. The city centre suffered from weak accessibility due to gaps in the transport infrastructure interconnectivity for serving the entire urban area (3 million people) and to the 'fragmented' public transport management among six operating companies run by the municipality, the province, the region, and the central state (railways).
>
> In Naples, daily travel between the city centre and the suburbs reflects important sprawl: for each person leaving the town, more than six enter. Most journeys are for work purposes (advanced tertiary and universities). Over the years private car travel has become predominant. Moreover, taking advantage of the weak public transport system, a parallel and illegal network has grown, using mini-vans for passenger transport.

This network was controlled by the local mafia, made possible through the lack of a regulating licences system and of effective agreements between local governments.

In 1993 the new municipal team (first municipal direct-voting elections in Italy) set up, among other things, a coordinated and integrated physical and transport plan in order to combat downtown ghetto forming. Consistency between the two plans was ensured in the new *Piano Regolatore Generale* (GRP, similar to a master plan) and in defining a public transport policy at the *municipal level* first, and later at the *metropolitan level* through involvement of the province of Naples (3 million inhabitants) and the Campania region (5.6 million people).

The municipal transport plan emphasized setting up a network structure and enhancing already-existing infrastructures rather than building new unconnected systems. The public transport networking in Naples itself and in part of the metropolitan area took place through institutional innovation, that is by founding the *Napolipass* consortium in 1995, in which the six operating enterprises are associated. The various transport tickets of the different companies involved were integrated into a single ticket (*Giranapoli*) issued and distributed by the consortium. Public transport integration gradually covered the outskirts of the Naples municipality. Revenues from the integrated ticket system clearly grew (50 million euros in 1999 against 27 million in 1995).

Later on, other municipalities drew up agreements with the consortium for special-fare subscriptions, so that 43 municipalities of the metropolitan area were gradually covered. Finally, an experimental single-fare system (the *Unico* ticket) was implemented in 2000 as an initiative of the Campania region for the entire metropolitan area, that is, the province of Naples and parts of the Caserta and Salerno provinces.

For Naples, the integration of the six operating companies into the consortium proved to be a stable system, more than the temporal agreement formula (fairly common in other European cities), which would have added uncertainties to an already weak public transport system. The innovation made it possible to maintain regular customers, attract new ones and combat the illegal parallel mini-van transport system.

Source: Floridea Di Ciommo. In: Jouve, 2003

2   *Integrating transport planning into town planning* in order to facilitate access to the town and to contain suburban developments is a second line of action, implying one or more of the following measures: renovating town centres and inner-city areas to reverse the trend of downtown depopulation, and regenerating old fabrics and port or industrial wasteland sites; limiting or forbidding the development of large peripheral shopping centres, which generate long individual journeys and many polluting emissions, as Barcelona and several French towns have undertaken to do; locating activities that generate travel according to accessibility to public or private transport, which helps to reduce car dependence.

3   *Extending the public transport range*, so as to include inner and outer suburbs, and residential areas far out in the agglomeration. Serious and targeted efforts to provide these areas with good functioning public transport would improve accessibility to the town and reduce pollution due to suburban motorized travel. This touches on the problem of financing mass public transport in unprofitable areas.

4   Finally, *the search for solutions to financing public transport* faces the difficulty of reconciling financing needs with the price the user can pay. Public service obligations, such as offering reduced fares and other welfare advantages, need to be carefully considered so that public transport bodies are not undermined. Although there is no single model to be followed, the ECMT emphasizes the advantage of integrated approaches to urban travel, as they may hold the key to financing public transport. In particular: (i) the allocation of resources from other elements of the system such as road taxes, integrated pricing of public transport, parking fees, environmental protection funds; (ii) the use of incentive and dissuading economic measures, through direct and indirect taxation (for example on fuel and on firms). 'Green' taxation measures on fuel and city tolls have so far been little explored but are interesting ways to encourage the use of public transport and to respond to the crisis of financing public transport (see Chapter 3).

For all these measures to bear fruit, at least two conditions must be met. First, *a flexible national framework for action* needs to be set up that favours the integration of town planning and transport planning, so exploiting synergies between national objectives for transport, environment and health, and objectives at the regional and local levels. Second, *sectoral policies need to be linked*, since local and regional policies intended to stem congestion and urban sprawl risk being ineffective when national policies on housing and real estate, for example, continue to encourage the building of houses or businesses on the outskirts of towns.

Urban regeneration and renewal: An opportunity for Mediterranean cities?

For two decades the globalization of the economy has encouraged metropolises and big cities in the Mediterranean, and elsewhere, to take up positions in international competition and attract growing investment flows, mainly in the tertiary sector. Faced with the

Urban Areas

risks of marginalization in the competition or poorly managed metropolitan development, cities are tending more and more towards strategic planning of their development. Despite highly varied institutional, economic and cultural local contexts, 'urban projects' are being reproduced from city to city, often linked to a major event, such as the Olympic Games in Barcelona or Athens, Seville's Universal Exposition, Columbus Celebrations in Genoa, the Mediterranean Games in Tunis or the Jubilee in Rome. Such events are used to transform the city and change its image. The strategies implemented aim to link the short-lived with the permanent.[29]

Beyond the short-term, urban policies are undergoing real changes in concerns and priorities. In the European Mediterranean countries, town planning oriented to ever more expansion towards the outside is being replaced by the will to slow suburban sprawl and rebuild the city in the city. In the south and east, cities are already facing the dual task of recovering important urban heritage sites likely to host part of the city's growth, thus limiting land and real estate pressures on suburban areas.

Urban regeneration policies have been developed as a response to the effects of economic and social change, to processes of degradation or abandonment affecting the urban fabric, the level of activities and the social functioning of some quarters. However, the objectives are more economic than social;[30] transforming the city's economic bases, introducing new activities (such as urban tourism and 'new economy'), revitalizing declining quarters and areas, creating jobs and providing the municipal budget with new sources of income. Regeneration experiences aimed at reconverting degraded industrial or dockland areas are being multiplied around the Mediterranean (Euroméditerranée in Marseilles, restoring industrial wastelands in Naples, Taparura in Sfax). Other examples of urban innovation projects are the seafront in Genoa, and the Balat and Fener quarters in Istanbul.

The *four examples* presented in Box 4.9 (Barcelona, Naples, Tunis and Aleppo) do not, of course, cover the whole range and complexity of operations in cities, scheduled for decades, but illustrate the different, sometimes combined, issues at the basis of urban metamorphoses. Examples include: regenerating obsolete industrial or dockland areas, renewing neighbourhoods that show various signs of vulnerability, and enhancing the historic and cultural heritage. With varied levels of intervention, the four examples show the opportunities taken or created by cities, the determination of local authorities, and the support or initiatives at higher levels (the state or regions), all in order to shift the development of a city and change its image.

Box 4.9 Urban regeneration and renewal in Barcelona, Naples, Tunis and Aleppo

Since the 1980s, **Barcelona** has undergone constant urban redevelopment, and has seen the renovation of obsolete functions to the benefit of emerging activities, the restoration of public spaces and the creation of new central areas. The founding in 1974 of the Barcelona Metropolitan Corporation (3 million people in 28 municipalities, including Barcelona) and the approval of the Metropolitan Master Plan in 1976 marked the beginning of regeneration and urban policies that are still succeeding one another. The transformations between 1986 and 1992 in preparation for the Olympic Games were used to stimulate a broad urban policy on a metropolis-wide scale, aiming both at long-term urban development and the affirmation of Barcelona as the capital of a European region. The major projects focused on improving and extending infrastructures (rail equipment, drainage and wastewater disposal networks, sub-surface networks, ring roads and traffic distributors), renovating the old town, improving the working-class suburbs, and *rehabilitating the seafront*, probably the city's most symbolic project. A special plan for the seafront made it possible to change the land use of the space occupied by derelict industrial facilities, re-order the infrastructures in the area (such as beaches, railways, the coastal road, sewers), ensure connectivity with the neighbouring areas, and integrate the quarter into the urban system. Mobilizing the financial resources for infrastructures and the Olympic village was handled by a private municipal company, and later by the Spanish State and the Barcelona Municipality together. Overall investment amounted to 503 million euros for the public sector and 613 million euros for the private sector. Three successive Strategic Plans have been implemented since 1988, mobilizing nearly 200 institutions to cope with changing conditions and to guarantee urban project coherence over time.

The continuing redevelopment of Barcelona constitutes a remarkable example of urban, economic and social rehabilitation of historic quarters, regeneration of vast wasteland areas, combining land uses, and channelling investments to new central zones. Barcelona's population has identified itself with the city's image and supported the process.

Urban renewal in **Naples** is mainly on the municipal scale. As a popular city with 1 million people, Naples has long suffered from an especially negative image, due to urban disorder, insecurity, criminality and trafficking. The urban–port regeneration policies carried out since 1993 aim at reversing this image, by operations relating to reviving the city centre, and also to a more global urban project seeking to replace the crisis-hit industrial and docklands with new developments founded on local heritage, cultural and tourism resources of the city and the Campania region. Changes are carried out within a rigorous budgetary context. The 1999 *Piano Regolatore Generale*, diverging from previous plans, envisaged the restructuring of the city by dividing it into sections, so as to act modestly but immediately on each section. Initial actions were launched when preparing the G-7 Summit in 1994 and benefited from European financing. The main achievements in the *historic centre* have been the restoration and re-opening of many public and religious buildings, revitalizing public spaces and facilitating their access through pedestrian-only-areas. Less progress was made in the housing sector, the vast degradation of which would need exceptionally high investment, not affordable by the city in the present context. The *interface between the city and the port* (5km) has been considered within the broader frame of the city/coastline relationship, made possible by a 1994 reform of Italian legislation on harbours, which mandates port authorities to set out a development scheme, in concert with the city authorities. In Naples, this joint effort was made possible in the western dock area where the city and the port authorities were both motivated to re-boost the tourism sector, which had been in sharp decline since the 1980s. Recovering the artistic and cultural potential has enabled Naples to become once again a valued tourist destination (cruises and ferry connections with the islands). In contrast, in the eastern dock area, opening up to the sea would require a relocation of oil industry activities, so that 200 hectares in a very sensitive sector could be freed up (poverty, degraded urban fabric, insecurity and urban ghettos). For the time being proposals are in conflict: the City authorities are willing to put in a 160 hectare park and a high-tech activity zone, while the Port authority would prefer a free commercial zone and a container terminal.

The Naples case illustrates the possible affirmation of a metropolitan function within a framework of budgetary constraint. By promoting a return to normal city life, instead of a spectacular American-style waterfront, Naples has prefered to enhance the existing and the forgotten, as well as providing basic urban facilities.

Tunis is undergoing a complete metamorphosis. The city is surrounded by two lakes and two sebkhas, for which four important development projects have been started, which are at various stages of development. The point of departure for all four projects is to restore a degraded ecosystem before starting any urban development. Located in the heart of the capital, the *lagoon* is undergoing a dual project for developing its banks: the northern lake has been entrusted to a mixed Tunisian-Saudi company, the southern to a government company. Both companies are in charge of cleaning up the lakes and developing the shores (some 2500 hectares of land after completion). In both cases the intervention is on virgin land and seeks a high-quality multifunctional urban fabric as well as efficient public services. Given the investments available for the embankment and for services, the focus is on luxury housing, although some social housing is planned for the banks of the southern lake. Land is state-owned, work began in the 1990s and the banks will be urbanized by 2030–2050. The two *sebkhas* are fragile areas where the problems of disturbed ecosystems and uncontrolled urbanization are accumulating. The Sijoumi sebkha project, at the southwest of the medina, was initiated by UNDP in 1994, with the Tunis Municipality supervising the work. In the Ariana sebkha, located at the north of the Tunis agglomeration, its northern shores have been taken over by tourism and its southern and eastern banks by illegal urbanization. The project is part of a rehabilitation programme by the Agency for coastal protection and development (APAL). A luxury residential area is foreseen around a planned marina, as part of a plan to open the sebkha to the sea. With the financial package for the projects not yet finalized, the sebkhas' future is still open.

The Tunis case shows a case of renewed urban planning, where several concerns are present in the reasoning and the actions of the authorities: a restoration of the link between the city and its lakes, which used to be repellent; the environmental and sustainable development themes that can attract international funds; and 'nature in the city' as an urban marketing argument intended for the well-off classes.

Aleppo, Syria's second biggest city, has a population of about 2.7 million. Its 11th century citadel (400ha and 120,000 residents), a UNESCO World Heritage site, is undergoing rehabilitation of its specific habitat (houses with interior courtyards), which had evolved little during the Ottoman period. The high heritage value of this urban fabric has not always been taken into account. For example, the 1954 Master Plan for the city intended to build roads through the old town and was implemented in some 20 per cent of the area (destroyed habitats, parking-lot construction, displacement of residents, etc.) until 1978, when an important local movement lobbied the national government to protect the site. With UNESCO's intervention in the early 1980s, the Master Plan for the old town was set aside and a project for *preservation and rehabilitation* initiated. Implementation began in 1992 with technical and financial support from the German government and the Arab Fund for Economic and Social Development. A new structure (the Old City Directorate, directly dependent on Aleppo's mayor) is in charge of managing and developing the area, mobilizing local and international institutions, and seeking new donors. The old town's development scheme, approved in 2001, aims to ensure greater harmonization with the whole agglomeration and moving the developmental dynamics towards rehabilitating the old town. Main priority actions are the rehabilitation of the sewage and water supply systems, renovation of the fronts of houses and of buildings with heritage value, a traffic scheme, and an advantageous loan system for facilitating some urgent housing rehabilitation.

In 2004, with a new Master Plan, Aleppo is seeking to create investment-attracting conditions, encourage residents to get involved in stable economic activities, define activities that might be moved out of the old town, and develop new tourism-related opportunities, thus choosing to combine heritage preservation with urban development.

Source: de Forn, 2003; Bertoncello & Girard, 2001; Barthel, 2003; Qudsi, 2003

Despite the specificities unique to each city and the very different political cultures, some common elements of the four examples can be considered. First, as regards the forms of action, partnerships are vital, both public–private and public–public forms of agreement, but the initiatives remain public, that is 'hyperactive' municipalities in Barcelona and Naples, after a change in political majority, a state in partnership in Tunis, local mobilization and support from international cooperation in Aleppo. Scales of collective action become ever more interwoven, from the local, regional and national levels to the European or international levels.

In the four cases considered in Box 4.9, the modes of public action are no longer founded only on regulatory measures. Urban planning and regulation, and their use, has changed to the benefit of *urban projects* guided by a long-term vision of the city. The urban project approach is taking over from the coercive urban planning approach. More generally, the urban project sometimes leads to a new tool, the Strategic Plan, very widely used in Spanish cities for transforming entire urban areas. The formula has not had the same success in France and Italy, although some master plans or regulatory schemes do approach metropolitan development in terms of 'strategies'. As for the contents of the transformations, a certain, not always resolved, tension can be felt between urban projects aimed at accessing the globalized economy on the one hand, and city-dwellers demands for living space, on the other.

From a spatial point of view, regeneration becomes inseparable from *transport network reform* in order to ensure access to the enhanced areas and the spatial coherence of the urban project. This means articulating regeneration operations at the agglomeration scale or, better still, the whole metropolitan area scale, since the framework of regeneration projects is based on transport infrastructures and structuring facilities.

Finally, in Mediterranean cities, *making the economy more dynamic and social revitalization* are closely linked with *enhancing the historic and cultural heritage*. This means that the dynamics of the quarters concerned should be taken into account when combining new functions, bearing two risks in mind. The first is the risk of a 'conservationist' approach, which considers cities as unchanging entities (for example measures to safeguard ageless old towns or medinas); the other is the risk of 'developmental' approaches, which build new urban images from enhanced heritage assets that no longer have any meaning for residents (for example some urban–port tourist developments that follow imported models).[31] In the examples given, these risks seem to have been avoided. However, a systematic resort to the higher tertiary sector, to commercial, leisure and cultural activities, to ephemeral events or to luxury real estate operations, risk producing standardization, a commonplace production of architecture and public spaces, possibly resulting sooner or later in a loss of the particular character of Mediterranean cities.

Urban environment: Air, water, waste

Contrary to some current statements, cities are not the enemies of the environment, especially not in the Mediterranean. Their advantages have always been derived from economies of scale, higher productivity, and rationalization of services. In addition, they have a stock of resources and flows (air, water, soil, waste, etc.), which can a priori be conveniently managed, both quantitatively and qualitatively. But cities also have ever greater 'global' responsibilities, such as those related to the greenhouse effect (suburban and urban transport) and to water resources. Significant efforts are needed to decrease air pollution, improve access to water and reduce wastage, and slow the alarming rise in household waste.

Air quality and public health

Knowledge of the harmful effects of the main air pollutants on health is steadily improving, although there remain sizeable gaps. Outdoor air pollution is characterized by the presence of five main pollutants, all responsible for serious health effects: carbon monoxide (CO), nitrogen oxides (NO_x), fine particles and dusts (PM), sulphur dioxide (SO_2) and volatile organic compounds (VOC_s). In addition, tropospheric ozone (O_3) is formed from the emissions of nitrogen oxides and volatile organic compounds under the influence of light. In the Mediterranean, such ground-level ozone is the main secondary pollutant, likely to create or aggravate respiratory diseases. It irritates the eyes, throat and bronchial tubes, increases inflammatory reactions in the bronchus, causes breathing difficulties for people with respiratory sensitivities, and aggravates asthma.

Topography, climate and the wind regime play a very important role in pollutant dispersion and concentration in the atmosphere. In the Mediterranean, where calm anti-cyclonic weather is frequent, town centres give rise to ascending warm air masses that are reflected back to the ground when they come into contact with colder air masses from above. During summer, such temperature inversions of atmospheric layers generate a quasi-permanent pollution 'dome' in many Mediterranean towns. This, for instance, is the case in Athens and its *nefos*, Cairo, Genoa, Barcelona and Marseille-Berre. In addition, small particles are transported over long distances with the often short, erratic and unpredictable Mediterranean winds (land and sea winds).

Air quality monitoring networks on the *northern shore* (essentially in Spain, France, Italy, Slovenia and Greece) assess air quality changes, pollutant by pollutant, in large cities every year. Regularly published statistics show decreasing concentrations of some pollutants over a decade, especially for SO_2 and NO_2. For example, between 1991 and 1999, sulphur dioxide concentration (SO_2) fell in Marseille from 23μg/m^3 (micro-grams per cubic metre) to 12μg/m^3, in Montpellier from 13 to 6, while in Athens the reduction was nearly 20 per cent between 1988 and 1997. For nitrogen dioxide (NO_2), concentrations remain rather stable: in Marseille 39 μg/m^3 in 1996, 38 in 1999; in Montpellier 39 in 1995, 36 in 1998; in Athens, a 3 per cent reduction between 1988 and 1997 in spite of traffic reduction measures.

More generally, over two decades European countries have gradually set up policies to reduce polluting emissions from different sources (heating, conventional power stations, some industrial processes, etc.), while motorized transport emerged as a major source of urban air pollution. Today, motorized traffic accounts for 27 per cent of the fine particle emissions (PM10) and 52 per cent of ozone precursor gases (in particular NO_x, VOC and CO). Despite a 30 per cent reduction in PM10 emissions between 1990 and 1998, 40 per cent of the European population is exposed to dangerous levels of this pollutant.[32] The danger of small particles to health is in inverse proportion to their size: those with a diameter of less than 10μm (thousandth of a millimetre) deposit in the trachea and bronchus, and may enter the blood.

In the *south and east*, concentrations of pollutants are occasionally measured on a piecemeal basis. They have remained stable for 20 years, sometimes above the recommended norms. For instance, the SO_2 concentration in central Cairo has varied for 20 years between 100 and 300μg/m^3 depending on the season and the measuring points. In Ben Arous (Tunisia) the average concentration was around 100μg/m^3 in the 1998–2001 period (the limit-value recommended by WHO is 150μg/m^3).

Fine particle emissions are becoming a major concern. In Cairo, 2001 results from the air quality monitoring system showed PM10 concentrations in the central business area that, nearly every day, were 5–10 times higher than the standard threshold value. Main identified sources are smoke from diesel buses, open-air incineration of waste, and industry.[33]

Although air quality monitoring networks are only slowly being put in place in southern cities, progress has been made in measurements and data analysis for the existing monitoring stations. However, the information published hardly reflects the linkages between the environment and health, such as the exposure of the most vulnerable groups to pollution. Furthermore, data are not presented in a way that is easily understood by decision-makers, which hampers the adoption of necessary policy actions. Hence the growing interest in *epidemiological studies*, carried out jointly by environmental monitoring institutions and public health institutions (Box 4.10). These studies are still rare in Mediterranean towns.

> **Box 4.10** Air and health, developments in epidemiological approaches
>
> The study of the link between air quality and human health has for a long time been relatively simple. When massive mortality episodes are linked to air pollution in towns, peaks in death curves come a few days after the peaks in air pollution. Until the 1970s, epidemiology restricted itself to considering only the health effects of short-time exposure to very high doses of pollutants. With the adoption of preventive measures, air quality has improved in many towns, and urban populations are these days exposed to lower air pollution levels than before, but over longer periods.
>
> The first long-term study in Europe[34] carried out in the Netherlands, found a correlation between mortality and the distance between the place of residence and a major traffic artery. The risk of dying from cardiopulmonary disease is twice as high for people living within 100 metres of a motorway or 50 metres of a main road. This result shows that the use of average rates of exposure can be misleading, because the health impact of air pollution is then minimized. Approaches have changed towards monitoring the air quality that a population breathes in daily over long periods.

Epidemiological studies on the health impacts of air pollution are now being carried out in 32 European cities under the APHEIS programme for epidemiological monitoring set up in 2000, in which Valencia, Seville, Barcelona, Marseille, Rome, Athens and Tel-Aviv participate. The latest study assessed short- and long-term exposure to PM10 as well as acute and chronic diseases in 19 towns, with PM10 pollution levels between 14 and 73μg/m^3 that entail health risks.[35] The study showed that even very slight reductions in particulate pollution can produce beneficial effects on public health: 5500 annual deaths due to long exposure to PM10s could be avoided by reducing PM10 concentrations by 5μg/m^3 in all the studied towns, even those with low pollution levels. And 15 per cent of these premature deaths (832 deaths) could be avoided by reducing short-term exposure to PM10 by 5μg/m^3. These results amply justify the adoption of preventive measures, even in cities with moderate pollution levels.

Surveys with a comparable scope do not yet exist in the southern and eastern Mediterranean countries. However, the results of the eco-epidemiological study conducted in Casablanca over 18 months in 1998–1999, are worthy of mention (Box 4.11).

> **Box 4.11 Air pollution and health in Casablanca**
>
> Given Casablanca's urban, industrial and harbour context and the high proportion of diesel vehicles, the monitoring station set up for this study assessed the following pollution indicators: sulphur dioxide, nitrogen dioxide and especially air-suspended fine particles (the black smoke type). Pollutants were monitored for a year, to take seasonal variations into account.
>
> A noticeable rise in ill-health events (mortality, consultations for conjunctivitis among children aged over five and adults, consultations for respiratory infections among under five-year-old children) was observed when pollution rose from the base level ($9\mu g/m^3$ of black smoke) to a medium level ($22\mu g/m^3$) and to a high level ($87\mu g/m^3$). The study does, however, note that black smoke is just one air pollution indicator and cannot be considered as solely responsible for the health effects revealed. A second study was then carried out in Mohammedia, an industrial centre also subject to car pollution. Here the accent was placed on a particularly sensitive population: asthmatic children.
>
> *Source:* Morocco. Secrétariat d'Etat chargé de l'environnement & Ministère de la Santé. Étude Casa-AirPol. Rapport final. Rabat, 2000

This type of survey is of undeniable interest. In the Mediterranean region, such epidemiological surveys should be carried out systematically and on a large scale, by increasing the number of monitoring stations in urban regions and in several agglomerations.

Improving air quality and public health

In European cities, improvements to engines, such as catalytic converters for petrol vehicles, and fuel switches (use of natural gas or liquefied petroleum gas – LPG instead of petrol) have, during the past ten years, led to important reductions in motor vehicle emissions. In contrast, emissions of CO_2, the main greenhouse gas, continue to increase (see Chapter 3 for past CO_2 trends and projections to 2025). Emissions of fine dust and nitrogen oxides are not decreasing as quickly as would be necessary to reduce public health risks.

In southern cities, traffic growth, the age of vehicles and the high proportion of diesel engines (notably buses) will continue to degrade air quality in towns. Diesel vehicles release less carbon monoxide and hydrocarbons, but three times more nitrogen oxides (NO_x) than a catalysed petrol vehicle, and 30 times more particles than petrol vehicles.[36]

In future, an explosive growth of motorized traffic, both in the north and in the south, will continue to aggravate air pollution in urban areas *locally*, and increase CO_2 emissions *globally*, in spite of technological innovations introduced by car manufacturers that cannot compensate for the rising traffic. Thus, one major challenge is to continue reducing NO_x and PMs released by light and heavy diesel vehicles. The progressive introduction of high-efficiency particle filters should help to bring down the level of PM emissions, including the smallest particles.

However, since nothing seems to halt the expansion of motorized transport, considerable efforts will also be needed in the next 20–30 years to contain passenger and freight traffic. In addition to the already-mentioned measures on linking urban transport and town planning, and introducing particle filters for diesel engines, efforts will be required in the following fields:

- limiting infrastructure development (if we really want to protect human life); the more roads, motorways and tunnels are built, the more polluting gas emissions are present;
- measures for limiting or forbidding private car use (pedestrian areas, urban toll roads), mainly in central town areas;
- sustained development of non-polluting public transport (light-rail, rail and other clean modes) and improved servicing of the different parts of an urban area or region;
- halting the open-air incineration of waste (see further below).

Does unequal access to water lead to resource wastage?

For this particularly precious resource in the Mediterranean region, cities are not the greatest water consumers: in terms of total consumption, their share is 3.7 per cent, while that of agriculture (with irrigation) is 95 per cent for the whole Mediterranean basin.[37] Nevertheless, taking cities' increasing water demands into account, important efforts are needed in towns, especially in the SEMC. For further details, reference is made to Chapter 1, and some challenges are briefly recalled here:

- Reducing drinking water consumption through awareness and education programmes, pricing policies (beginning with the installation of water meters), sharing water prices, and combating leakages in urban networks that sometimes exceed 30 per cent.
- Responding to urban sprawl (even limiting it) by installing basic infrastructures for drinking water and sewage, and by using primary or low-cost water treatment techniques, suited to the realities and capacities of southern and eastern countries.
- Reducing geographical and social inequalities in water provision, taking poor neighbourhoods and populations into account in pricing policies.

The last point will be underlined here. The issue of *unequal access to water in cities* is linked to the objective of *'reducing by half, by 2015, the proportion of people without access to safe drinking water or not having the means to obtain it'*, adopted in the Millennium Declaration of the United Nations General Assembly (2000), and reaffirmed in the Johannesburg Action Plan in 2002.

According to available statistical data on access to drinking water in the Mediterranean, in 2002 some 30 million (urban and rural) people had no access to safe water, a third of whom were living in Turkey and another third in the Maghreb countries. In the southern and eastern countries, official rates of urban servicing (about 95 per cent of the urban population) do not reflect the connection difficulties of many shelters in disorganized spontaneous urban neighbourhoods. These often illegal urban developments result in habitat types with multiple status, a variation that is also echoed in the access-to-water conditions. Some neighbourhoods are entirely connected to the network and others completely marginalized. Without first resolving the issue of unregulated housing, the demand from many users cannot be satisfied. Many neighbourhoods, both in town centres and on the outskirts, are thus dependent on standpipes or water sold by traders.

Since the 1980s, and especially during the World Decade for Water and Sanitation, important efforts have been made for individually piped water supply, but these policies mainly focused on the 'legal city'. The integration of informal quarters into the official city through drinking water and sewage networks has encountered enormous difficulties. Individual user-paid connections can hardly be realized in informal housing areas and shanty towns, since water distributors are not authorized to provide individual connections unless the requesting households comply with town planning regulations. The result is an accentuated fragmentation of urban areas in access to water, as illustrated by the case of Rabat-Sale (Box 4.12).

Box 4.12 Access to water in different quarters in Rabat-Salé, Morocco

The Rabat-Salé urban area is regularly supplied with water, and is relatively less often faced with water scarcity than other towns situated further south or further inland in Morocco. Within the agglomeration, contrasts exist within the central area (medinas, the old colonial town), between the city centre and the outlying settlements (high-quality or cheaper housing estates, as well as unregulated housing and shanty towns), and between the urban and the rural areas. This spatially heterogeneous and discontinuous urban system is unevenly serviced by basic infrastructures, such as road, water and sewage networks.

In many legal and regularized neighbourhoods, home to a poorer population with very irregular incomes, the practice has developed of sharing a meter among several households in the same building; a practice that is forbidden but hard to control. The bill is shared by the households, however, not in proportion to their individual consumption, but in equal shares. The result is that despite progressive water pricing, the water bill weighs more heavily on the budget of low-income households than on wealthier social classes. After all, the total volume consumed per water meter exceeds the 24m³ of the first consumption block, even though the consumption per household is actually much lower than 24m³.

In the illegal neighbourhoods, with either permanent or temporary constructions, water standpipes play an essential role. These are drinking water fountains situated in the medina, or stand-alone taps. Such forms of collective water provision are common in Salé and in fringe areas. However, despite their importance, standpipes tend to be closed down for economic and financial reasons, thus accentuating the fragmentation between Rabat and Salé, between the centre and the suburbs, and at the very heart of some central quarters where slums persist. Standpipes represent 3 per cent of the total consumption of the Rabat-Salé *wilaya* (in the 1990s), but are considered unhealthy and wasteful, and are held responsible for much of the budget deficit of local authorities.

Source: El Mansouri, 2001

The case of Rabat-Salé is not unique; a similar fragmentation can be found in Istanbul, Izmir, Damascus, Casablanca, and many other big cities. The situation becomes even more difficult in cities with serious overall water scarcity problems, such as Algiers, where competition for access to water is fierce (Box 4.13).

Box 4.13 Competition for access to water in Algiers

As in all large Algerian cities, Algiers suffers from a lack of water. In the 1980s a rationing programme was set up and has since become more drastic: water is distributed in time slots with an average duration of six hours a day; distribution generally occurs at night; and some quarters receive water only once in three days, sometimes even one day in six.

According to field surveys, the Algiers urban area can be split into a western zone where the wealthier people live (continuous water delivery, from 100 to 200–300l/cap/day), poorer quarters in the east where the water situation rapidly deteriorates (rationed distribution, from 80 to 100l/cap/day), and the city's central and northern zones, where the most deprived population lives receiving an extremely low allocation (less than 80l/cap/day). Finally, in the illegal quarters (that cannot be legally supplied) the situation is really critical. Slums are supplied by water trucks hired by the municipalities; households located on the edge of illegal quarters that can afford it, hook up illicitly to the main distribution pipes that provide a permanent water supply to military barracks and hospitals; and other households purchase water from private resellers at prices of sometimes ten times those charged by the official water-distribution company. The geography of water consumption by households thus becomes an indicator of the inequalities in the standard of living.

Source: Chikr Saïdi, 2001

These two examples illustrate how official rates for urban drinking water servicing hardly reflect realities on the ground in the SEMC. Moreover, changes in public water services, including delegating management to the private sector, have not given full attention to the *right to water*. Access conditions have become increasingly dependent on income levels. In fact the access-to-water issue in urban areas is not limited to the supply-and-demand equation, but relates to strategies driven by the key actors, where the objective of people's well-being may conflict with the economic and financial objectives.

Two lessons can be drawn from a sustainable development point of view:

1. There is a discrepancy between water policy and housing policy, whereas water connections should be installed before housing, or at least simultaneously. The two policies are envisaged solely from a sectoral and technical angle, without being steered jointly, thus making urban management very difficult and the action of water institutions much more complicated.

2. Water pricing seems unsuited to the functional and human dimensions of real towns. The question of pricing brings up the question of the role of public authorities and societal choice, since combating inequality requires deliberate action from public bodies, even when the management of the service is delegated. Table 4.7 presents a range of concrete measures for ensuring access to water, some of which are practised in Mediterranean towns and countries.

Household waste multiplied by three?

Since the mid-1970s, urban waste has become a concern of utmost importance in urban management in the Mediterranean countries. It mainly concerns waste produced by households, collected and treated by municipalities or by private operators on their behalf.

A sometimes explosive increase in waste

Volumes of waste have doubled or tripled in 30 years (Figure 4.10). On the northern shore, the very strong growth in waste volumes reflects the excesses of

Table 4.7   Examples of solidarity measures for ensuring access to drinking water

Measures	Examples
Water free of charge	Collective water provision: maintaining drinking fountains and standpipes (Morocco), installing more standpipes or other watering places
	Water piped to dwellings: supply of a minimum free volume per household or per inhabitant, the remaining consumption being paid at full price (in Athens, free water for less favoured households)
Reducing obstacles to access to water	Reduction or suppression of individual connection charges, of secure deposits and of advances on consumption
Affordable water: pricing measures	Spain, France, Italy, Malta, Greece, Turkey, Israel, Egypt, Tunisia, Morocco, etc. – Progressive pricing where the unitary cost per cubic metre in the first consumption block (between 15 and 25m^3) is the lowest (lifeline tariff) – Reduction or suppression of general taxes (VAT) of the fixed part (subscription fee) in water invoicing
Targeted social pricing	Price variation according to the socio-economic characteristics of the consumers: abatement for retired people, families with several children, the poor (Seville, Athens)
	Subsidies ranging between 25 and 85% of the water price, related to a fixed provision with upper limit, for less favoured people (in Seville, abatement of 50% within a fixed threshold)
	Geographical cross subsidies for reducing differences in water prices within a region, among regions (in Turkey, highly different prices according to towns, with targeted measures for the less favoured towns)
Interdiction of cutting-off piped water	Measure targeted for poor households (France). This limit to contract law is sometimes based on court jurisprudence ('excessive constraint liable to disturb public order') and does not seem to have serious economic consequences
Aid for water invoice payment	Creation of a fund for helping those with difficulties to pay water invoices, financed by the consumers or the taxpayers
	Reduction or removal, in case of social services intervention, of penalties on payment delays, of costs of cutting-off/connecting-up; abandoning jurisdictional recourse when the household is living in penury (France)

Source: Plan Bleu from Smets, 2002 (see also Chapter 1)

consumer societies as well as the rapid increase in packaging. On the southern and eastern shores, the challenge for local authorities is to be able to cope with a growth in waste closely coupled to urban growth.

Figure 4.10 Production of household waste (Kg/inhab/year)

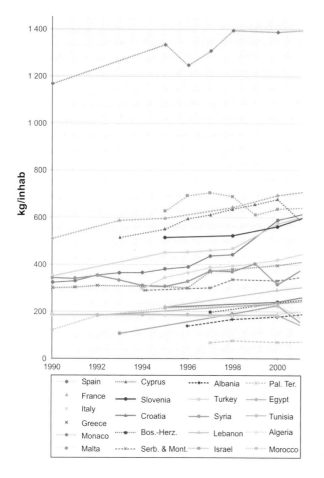

Source: Blue Plan estimates from Eurostat data and national sources

Note: Between 1990 and 2000, the yearly urban waste production per inhabitant has increased from 323 to 588 kg/inhab./year in Spain, 509 to 694 in France, 350 to 573 in Italy, 300 to 397 in Greece, and 513 (1993) to 677 in Cyprus. Monaco is out of scale in the graph, its production being some 1.4 tonne/inhab./year during the decade.

On the eastern Adriatic, the upward trend is also general (1995–2000 period) but with pronounced contrasts: from 515kg/inhab/year to 562 in Slovenia; clearly lower volumes in the other countries (from 218 to 241kg/inhab./year in Croatia, 197 to 237 in Bosnia-Herzegovina, 138 to 178 in Albania).

In the SEMC, growth is as clear, with important volumes in Israel (637kg/inhab in 2000) and lower ones in the other countries: from 297 (1994) to 419kg/inhab/year in Turkey, 290 to 333 in the Lebanon, 106 to 226 in Syria, 183 to 234 in Egypt. In the Maghreb countries, 1990–2000, volumes are 184kg/inhab/year in Tunisia, from 183 to 190 in Algeria, from 120 to 226 in Morocco.

Over the 1980–1998 period, the amount of organic matter has generally decreased: from 38 to 29 per cent in France, 43 to 32 per cent in Italy, 60 to 44 per cent in Spain, 62 to 47 per cent in Greece, and 85 to 64 per cent in Turkey. There has been a more or less parallel increase in packaging, plastic and toxic waste in dispersed quantities, such as solvents, car batteries, paint and pharmaceutical products.

If these trends continue, the future risk of running into a very difficult-to-manage situation is high, particularly as regards treatment facilities.

Predominance of uncontrolled dumpsites in the south and east

First, the collection rate varies from 0 per cent (in rural areas) to 90–100 per cent in the large urban agglomerations and is closely correlated with the income level in the areas under consideration: collection is provided in wealthy neighbourhoods, while disadvantaged neighbourhoods are poorly served.

Uncontrolled dumping is still largely predominant. Since the mid-1990s, countries have begun to remove the main black spots, closing some open dumps and replacing them by controlled landfills, while recovery and recycling policies are being introduced. In 1991 Turkey abandoned incineration and developed a policy of systematic recycling through a legal obligation to annually reclaim a certain tonnage for some economic sectors that are large consumers of packaging. In Syria, Damascus opened its first controlled landfill in 1998 and the government decided to set up a controlled landfill in Aleppo. In the Lebanon, Beirut closed the Bourj Hamoud site in 1997, built two composting/incineration units (but with a tiny share for incineration) and a controlled landfill in Naameh. In 1999 Egypt opened three controlled landfill sites in Giza, Alexandria and Cairo, and has built several composting units since 1990. In 1998, Tunisia granted a concession to a private operator for its first really controlled landfill in Djebel Chekir, meant to serve Greater Tunis. In addition, a national inventory listed 400 sites used for illegal storage of household waste, which would need to be reclaimed.

As regards legislation, ongoing efforts are trying to fill inadequate texts and vague classifications. Syria, Lebanon and Morocco are preparing new laws and in Algeria a new law sets the framework for solid waste management. However, the law is often not enforced. When national plans are defined, implementation difficulties appear at the local level. In addition, the intervention of multiple donors with different approaches to waste clouds the issue through a lack of synergy with the objectives of national plans.

Recycling is making progress in the north

Generally, incineration with or without energy recovery is taking the place of landfilling, and recycling and recovery are gaining ground. In the EU countries, European legislation sets the framework for policy actions:

- The 1975 Framework Directive on Waste, amended in 1991.
- The 1994 Directive on Packaging Waste, amended in February 2004, encouraging prevention, re-use of packaging material, recycling and especially reducing final disposal of this waste. The 2008 recycling objectives are set between 55 and 80 per cent.
- The 1999 Directive on Landfill of Waste lays down a fairly tight framework for landfills and forbids landfilling of non-processed waste from 2002 onwards. Enforcement is not yet fully operational in all countries, but the road has been paved.
- The 2000 Directive on Incineration of Waste strictly regulates this process, making it very costly (gaseous effluents and treatment of residues from purification of smoke from household waste incineration).

Implementing these regulations will be slower than expected essentially because of the inertia of incineration-based systems, and the very complex different processes for urban waste treatment (mainly for toxic waste in dispersed quantities). Other factors also come into play, such as the power of operators who have often acquired quasi-monopolistic positions on the waste collection and processing markets, or the weakness of local authorities.

An alarming situation

Progress observed in both the south and the north, should not, however, hide the fact that, except in France, Italy and Spain, dumping is still widespread (80–90 per cent depending on the country).

Waste dumps have very large impacts on environmental health (such as soil, air and groundwater contamination). Run-off drains toxic matter from household waste (leachate), which then flows into surface and groundwater systems. Open dumpsites often generate emissions of nitrogen oxides and methane from pockets of anaerobic fermentation. Pockets of methane trapped in the subsoil can prove dangerous when coming into contact with an inflammable source. This is why waste dumps are frequently places where fires break out in the Mediterranean, for example the big fire around Marseille in 1997 and in Portugal in the summer of 2003.

Poor management of household waste finally results in negative human health impacts. For instance, open-air incineration of municipal waste, currently practised in Egypt, Syria, the Lebanon and Morocco, has been identified as a major factor in decreasing air quality, especially through dioxin emissions.

The struggle for less waste

With rising urbanization rates in Mediterranean countries and changes in lifestyles and consumption patterns, the challenge for the future lies in reducing volumes and in recycling; recycled material is actually a valuable resource. Only policies designed to decouple waste production from economic growth could curb the increasing waste trend and the deteriorating situation in the Mediterranean region.

Two scenarios for waste production have been worked out (Table 4.8). Past trends are extrapolated and related to GDP growth. The resulting baseline scenario shows a difficult-to-manage waste volume by 2025: the NMC would produce 1036kg per capita per year (that is one tonne) compared with 566kg/cap/year in 2000; for the SEMC the volume would be 587kg/cap/year compared with 282kg/cap/year in 2000.

The alternative scenario is based on the assumption that policies are set up that aim at *reducing waste production at the source*, as well as making *recycling* increasingly common. Significant savings could be realized by all Mediterranean countries. By 2025 the annual average amount of waste produced under the alternative scenario would be 150 million tons less than under the baseline scenario. And the financial savings would be an annual average of some US$3.8 thousand million.

Following the alternative scenario path requires adopting deliberate measures in order to change collective and individual behaviour in consumption and particularly in packaging (Box 4.14).

Box 4.14 Ecological management of household waste in Barcelona

Since 1997, Barcelona has set very ambitious goals for managing household waste within the municipal waste management programme for the metropolitan area (33 municipalities). The programme envisages recycling about 60 per cent of the 1.5 million tonnes produced by the metropolis each year by 2006, and mobilizing about 751 million euros for that period. These funds have already led to the construction of three recovery facilities for collected waste (including one incinerator with energy recovery), linked to a network of drop-off centres where people can leave their waste free-of-charge, and to sorting centres staffed by skilled staff for separating the collected waste (iron, plastics, paper, etc.). For the organic component, a set of Ecoparks has been built to carry out totally integrated treatment, combining mechanization and composting.

The project's expected economic impacts are the creation of 5600 jobs, 3000 of which will be permanent at the end of the project, resulting in an injection of nearly 114 million euros to the regional GDP. From an environmental point of view, the project should lead to a significant reduction in quantities of waste to be treated, and to the closing of the Garraf dumpsite which, to date, is a major source of Barcelona greenhouse gas emissions and a threat to the local hydrological system (untreated leachate that continues to flow in significant quantities).

Source: Barcelona metropolitan area, 2004. Metropolitan Plan for urban waste treatment. www.ema-amb.com

Table 4.8 Two scenarios for waste production through 2025

Baseline scenario

Total (million tonnes)				Per capita (kg/cap/year)			
	2000	2025	AAGR (%/year) 2000–25		2000	2025	AAGR (%/year) 2000–25
Total NMC	109	204	2.5		566	1036	2.4
Total east	37	111	4.5		386	857	3.2
Total south	29	81	4.2		210	409	2.7
Total SEMC	66	192	4.4		282	587	3.0
Total Med	**174**	**396**	**3.3**		**410**	**756**	**2.5**

Alternative scenario

Total (million tonnes)				Per capita (kg/cap/year)			
	2000	2025	AAGR (%/year) 2000–25		2000	2025	AAGR (%/year) 2000–25
Total NMC	109	139	1.0		566	704	0.9
Total east	37	62	2.1		386	480	0.9
Total south	29	49	2.1		210	249	0.7
Total SEMC	66	111	2.1		282	340	0.8
Total Med	**174**	**250**	**1.5**		**410**	**477**	**0.6**

Source: Eurostat, ETC-AEE, national statistics yearbooks for 2000, *Plan Bleu* for the prospects

Feasible savings through the alternative scenario for 2025, compared with the baseline scenario for 2025

	Reduction at source		Recycling		Total savings in 2025			Cumulated savings 2000–2025 million US$
	million tonnes	million US$	million tonnes	million US$	million tonnes	million US$	In % of treatment and collection costs as per baseline scenario %	*period 2000–2025**
	2025	2025	2025	2025	2025	2025		
Total NMC	65	3845	14	860	80	4706	39	58,824
Total east	49	2043	1	53	50	2095	46	26,190
Total south	32	807	1	17	32	824	40	10,302
Total SEMC	80	2850	2	69	83	2919	44	36,491
Total Med	**146**	**6695**	**16**	**930**	**162**	**7625**	**41**	**95,315**

Source: Plan Bleu calculation and estimates; financial data on waste treatment and collection from the MSW Strategy for MENA countries-2010 study, World Bank (1998)

Note: AAGR: Annual average growth rate

* assumption of linear gradual exploitation of savings potential

In the baseline scenario, the quantities of waste produced in the years up to 2025 have been estimated from 1985–2000 elasticity for quantities of municipal waste produced as a percentage of GDP. This elasticity has been smoothed following an analysis of changes in the 'household waste/GDP' link between 1985 and 2000 in a sample of 30 countries (European countries and the US). In the alternative scenario, elasticity has been divided by 3 in line with current trends in the country with the most virtuous reduction at source, the Netherlands (an annual reduction in waste production of 2 per cent per capita in the 1990–2000 period). The 2000 recycling rate has been multiplied by 2 up to 2025, on the basis of progress observed in France and the US between 1990 and 2000.

Commercial outlets, the distribution sector and consumer groups all have a role to play, supported by the law. Along these lines, Italy has limited the use of plastic packaging and France is beginning to do so. Public awareness and people's daily involvement are decisive for implementation. It mobilizes associations at the neighbourhood level but actually begins in school.

The alternative scenario also takes into account the socio-economic aspects of waste collection and recycling. The prevalent technical-economic responses practised in the NMC should not be extended to the SEMC without adaptation, since they could threaten certain sections of the informal economy of waste, which provide far from negligible employment and sources of income (Box 4.15). Forms of collaboration can no doubt be envisaged between, for example, waste treatment companies and the people involved in collection, sorting and recycling.

> **Box 4.15** Household waste in Cairo and Alexandria: the *zabaleens* and foreign companies
>
> Every day, urban areas in Egypt 'produce' some 30,000 tonnes of waste, including 8200 tonnes in Cairo and 2800 tonnes in Alexandria.
>
> According to a study conducted by the Alexandria *Governorate*, waste volumes rise from 2800 tonnes per day in winter to 3400 tonnes in summer. However, the three landfills available to Alexandria, with a total area of 42.4 hectares, have a daily capacity of 2100 tonnes. So in winter, 700 tonnes of waste end up in uncontrolled dumps every day or lie in the streets and, in summer, the volume rises to 1300 tonnes of untreated waste.
>
> In 2001, the Alexandria *Governorate* called upon a foreign company for collecting and processing domestic waste. Onyx, a subsidiary of Vivendi Environnement, signed a US$446 million contract to handle waste processing for 15 years. In 2002, the *Governorates* of Giza and Cairo in turn signed 15-year contracts with Italian and Spanish firms for waste collection and treatment.
>
> In the contracts with these foreign firms in Cairo and Giza, there is no specific clause concerning the *zabaleens*, the tens of thousands of traditional rubbish collectors in all big cities in Egypt who collect about 30 per cent of the waste and recycle about 80 per cent of the collected waste. In Alexandria, Onyx has signed a more or less formal agreement with the *zabaleens* of Alexandria, before starting to implement the contract. In Cairo and Giza there are considerably more *zabaleens*, and a competition appeared between these *zabaleens* and the multinational operators who had become the waste owners.
>
> *Source:* Plan Bleu from El Kadi, 2001 and Cedare, 2004.

What possibilities for action do cities have?

In parallel with the changing urban development, other changes are occurring in the Mediterranean region, related to the organization of public authorities, their effectiveness and level of representation at all levels, the participation of the general public and civil societies in collective choices, and the relationships with the central State.

Weak decentralization and urban governance in SEMC

Gradual changes are taking place in the distinctive features that have characterized public institutions in the Mediterranean countries for the past 40 years: a heavily centralized administrative, economic and political power, and a top-down administration pattern, hierarchic and often with little flexibility.

These forms of organization in the SEMC were set up after national independence, often applying a centralizing logic in order to ensure national unity and stability of the nation-state and its administrative bodies. But for nearly 20 years, due to the effect of widespread economic liberalization and concerns about decentralized participation, these forms of exercising power and even the way of governing are undergoing significant change:

- the boundaries of public action have become less clear and the intervention levels are increasingly numerous and entangled: trans-national, national, sub-national;
- the number of actors involved has multiplied: community groups, non-governmental organizations and the private sector have a growing, and increasingly sought-after, role in the choices and decisions made by the public authorities;
- the way in which these different actors interact has also changed, becoming less hierarchical, more transverse and more flexible than before.

Decentralization is often presented as a possible response for taking account of the actual needs of the population; as a response to declining commitment by central government; as a way to bring decision and implementation levels closer together, and to facilitate the generation of investments by private sources. However, the choice of decentralization for a country is always complex and risky, since the decentralization process presupposes: political and financial devolution, without which local authorities cannot be deemed responsible to the people for their choices and commitments; a simultaneous reform of central state and de-concentrated services, crucial for improving the functioning of local governments; and the existence, at different local levels, of networks of actors (public, private, associations, non-governmental organizations) capable of mobilizing local resources in synergy with resources obtained at other levels.

The Mediterranean appears today as one of the regions of the world where centralization is strong. Centralized systems are part of the historic heritage in the region, and the different empires – Roman, Moorish, Ottoman, Napoleonic – have fashioned, over time, a 'Jacobin' organization of the state. This is generally characterized by uniformity of rules; the will to cover the whole national territory with a central administration, through de-concentrated administration levels; a combination of elections and nominations at the sub-national levels; and the supervision of the centre over all levels. Public expenditure ensured by local governments reveals to some extent this centralization of power (Figure 4.11) and shows the southern Mediterranean as being behind other regions of the world.

Figure 4.11 Local level expenditure as a percentage of total government expenditure, 1997–2000

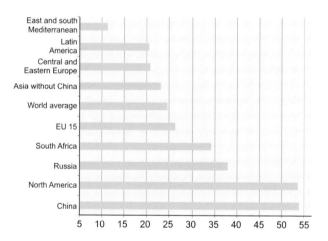

Source: Urdy, 2003, from World Bank Decentralization Database, 2002 and national sources for Mediterranean countries

Note: East and south Mediterranean: Turkey, Israel, Egypt, Tunisia, Morocco; Central and eastern Europe: Albania, Croatia, Slovenia, plus the other 15 countries of this sub-region; World average: sample of 56 countries.

For several observers, the 'participatory' and 'decentralized' processes that international organizations present as essential elements of 'good governance' are still theoretical in the SEMC, because they imply a significant degree of devolution of political and financial power and an equitable distribution of power, while neither of these are tangible realities today in these countries.[38]

Whereas EU-Med countries (Spain, France, Italy) have chosen a three-level sub-national organization of representative public authorities – Municipalities, Provinces or *Départements*, and Regions, the SEMC are still characterized by the absence of an intermediate, decentralized level between central government and the municipal level (Table 4.9). Therefore, de-concentration prevails, being present in the Provinces, W*ilayates*, *Governorates* and M*uhafazates*, where the executives are, in all cases, appointed by the central Government.

Except for the particular case of Israel, inspired by the Anglo-Saxon tradition of self-government,[39] attempts to create an intermediate decentralized level are rare. Only Morocco has set up regional local authorities in 1997.[40]

Almost everywhere the basic administrative unit is the urban municipality. These are in fact relatively ancient institutions in the Mediterranean, introduced by the *Tanzimat* reform in the Ottoman Empire in the 19th century, in 1858 in Tunisia (creation of Tunis' first municipal council), in 1864 in the Lebanon, and in the early years of the 20th century by colonial or mandated powers elsewhere. Today, all municipal councils are elected by direct universal vote and also the Mayors (whatever their local names) are elected, except in Egypt. However, the institutionalization of local municipal powers remains fragile and incomplete.

Administrative decentralization without financial means

Available data concerning decentralization practices in southern and eastern Mediterranean countries are extremely patchy, often inaccurate and difficult to compare. Practises can be analysed through different aspects.

Local authorities' duties

The question of shared responsibility between different levels of local authority has great relevance to the actual degree of decentralization, since the implementation of decisions can suffer from overlapping decision-making powers. The roles allocated to the different levels vary greatly from one country to another, but common features are found in most cases (Table 4.10 p234):

- Generally, the same areas of responsibility are passed on to local levels. In the SEMC, the duties transferred most often concern local services and facilities: land and town planning, social services, water and sewage, household waste, road networks, etc.
- As regards the spread of responsibilities, two typical situations occur on the Eastern and Southern shores. In the Middle East (Turkey and Mashrek countries) the areas of involvement have in general been clearly defined, so that the duties of different territorial levels overlap little or not at all. In the Maghreb (essentially

Table 4.9 Authorities at sub-national levels in south Mediterranean countries

Levels	Basic unit Level 1	Intermediate units Level 2	Level 3
Turkey	3228 Municipalities Mayor, elected by direct universal suffrage Municipal Council elected by direct universal suffrage	Province Executive appointed by the Council of Ministers	
Syria	95 Towns Chairman, elected from among the Town Council Town Council, elected by direct universal suffrage	*Muhafazat* Executive appointed by presidential decree	
Lebanon	708 Municipalities Mayor, elected from among the Municipal Council Municipal Council, elected by direct universal suffrage	*Muhafazat* (administrative level, with no legal identity) *Muhafiz*, civil servant appointed by central power	
Egypt	227 Towns Mayor appointed by the Governor. Council of the People, elected by direct universal suffrage	Governorate Executive appointed by the President of the Republic	
Tunisia	262 *Communes* Chairman, elected from among the Municipal Council Municipal Council, elected by direct universal suffrage	Governorate Executive appointed by presidential Decree	
Algeria	1541 *Communes* Chairman, elected from among the People's Communal Assembly (APC) APC, elected by direct universal suffrage	*Wilaya* Executive appointed by Decree of the Council of Ministers	
Morocco	249 urban *Communes* Chairman, elected from among the municipal Council Municipal Council, elected by direct universal suffrage	Province or Prefecture Executive appointed by royal *Dahir*	Region Executive appointed by royal *Dahir* and elected from among the Regional Council

Source: Plan Bleu

Algeria and Morocco) areas of responsibility overlap to a greater extent. This can lead to local authorities that act merely as managers of short-term operations, while central government keeps hold of both local and national strategic programming.
- The town is seen as the relevant level for the 'day-to-day' management of matters touching the population directly – social services, culture, environmental services and civil registration, while strategy formulation falls on the higher levels.
- Education, highways and economic development present particular cases of shared responsibility between at least two levels.

The degree of financial decentralization

Financial decentralization is a good indicator of the effectiveness (or not) of decentralization: it appears to be very low, as shown by Tables 4.11 to 4.14. From the point of view of the areas of responsibility allocated, the process in the SEMC shows strong resemblances with decentralization in European countries, whereas it appears distinctly more modest when considering the money involved. The share of local public expenditure of GDP is noticeably lower in the southern and eastern countries, and the same applies for public expenditure as a whole. The effort of local authorities in the SEMC appears to be lower than in European countries.

Looking at the *resources* aspect of financial decentralization (revenue from local taxes, tax revenue shared between several levels of local authorities, grants/transfers from central government), the standard practice is a fund that provides the local authorities, and from which they can also borrow.[41] Changes in the proportion of transfers from Central Government between 1990 and 2000 vary greatly from country to country.

2 A Sustainable Future for the Mediterranean

Table 4.10   Responsibilities of local authorities in the Mediterranean

Responsibilities of local authorities in some southern and eastern Mediterranean countries

Countries and levels / Responsibilities	Turkey 1	Turkey 2	Lebanon 1	Lebanon 2	Israel 1	Israel 2	Egypt 1	Egypt 2	Tunisia 1	Tunisia 2	Algeria 1	Algeria 2	Morocco 1	Morocco 2
Land, town planning	■		■		■		▫		▫		▫	▫	▫	▫
Social services	■				■		▫		▫			■		
Education	■				■							■		
Culture	■		■		■		▫							
Water, sewage	■		■		■		▫					▫	▫	
Household waste	■		■		■		▫							
Public road network	■				■		▫							
Economic development	■				■					▫	▫	▫	▫	▫
Registry office	■				■		▫			▫		▫		▫
Law and order	■				■			▫		▫		▫		
Green spaces	■		■		■				▫			▫	▫	
Housing	■		■						▫			▫		
Health	■				■		■			▫	▫	▫		■
Sport and leisure	■				■							▫	▫	▫
Urban transport	■										▫	▫		
Trading activities	▫							▫			▫	▫		
Energy supply	▫										▫	▫		

Source: Urdy 2003 from Dexia, CFCE, ISTED 2002.

Note: Levels: 1 (municipalities, cities); 2 (wilayas, governorates…).

Specific responsibilities in dark colour; Shared responsibilities in lighter colour

Responsibilities of local authorities in some European Union countries

	Spain	France	Italy	Greece	Germany*
Social services	3	1-2-3	1-2-3	1-2	1-3
Education	1-3	1-2-3	1	1-2	1-3
Health	2	–	1-3	2	1-3
Water, waste	1-3	1	1-2	1	1
Public road network	1-2-3	1-2	1-3	1-2	3
Transport	1	1-2	1-2-3	1-2	1
Economic development	3	1-2-3	3	2	3

Source: Urdy, 2003; from Dexia, 2002

Note: Levels: 1: municipality; 3: region; 2: intermediate local authorities (provinces, départements) between the region and the municipality

* Germany is cited simply to show the absence of a level 2 (between the Länder and the municipalities).

Urban Areas

Table 4.11 Share of expenditure by local authorities

Economic weight of local expenditure (% of GDP)

Turkey	4	1999
Israel	6	1999
Tunisia	4	2000
Morocco	2	2000
EU15	11	2001

Share of local expenditure in total public expenditure (%)

Turkey	11	1999
Israel	16	1999
Egypt	18	1995[a]
Tunisia	9	2000
Morocco	7	2000, towns
EU15	24	2001

Source: Urdy, 2003, from Dexia, CFCE, ISTED, 2002 and [a]E.S. Ghanem, 2002

Note: All levels of local authorities except when otherwise stated; the same applies to the next three tables.

Table 4.12 Proportion of transfers to local authorities from the Government budget (%)

	1990	2000
Turkey	18.69 (1997)	14.02
Israel	4.54	6.31
Egypt	1.69	2.10
Tunisia	12.33	6.56
Morocco	0.00	6.02

Source: Urdy, 2003, from World Bank decentralization indicators & IMF-GFS

Table 4.13 Local authority resources

Locally raised resources/total resources (%)

Turkey	9	1999
Egypt	16	1995[a]
Israel	59	1999
Tunisia	63	2000[b]
Algeria	10	1999[c]
Morocco	27	1998[d]

Government grants and transfers/total resources (%)

Israel	41	1999
Egypt	84	1995[a]
Tunisia	37	2000, city[b]
Morocco	45	1998[d]
Spain	70 (municipality), 31 (province)	
France	23	
Italy	62	
Greece	63	

Source: Urdy, 2003 from Dexia 2002 and ; [a]E. S. Ghanem, 2002; [b]UN Pogar; [c]Ceneap, 2002; [d]J. H. Pigey, The Urban Institute 2001

The source of income differs according to the type of local authority: for authorities at the intermediate level (*mohafazats*, provinces, etc.) most income comes by way of government grants, while the municipalities raise the majority of their income themselves, mainly through property taxes (built-up and non-built-up land) and other taxes (on commercial premises, entertainment, motor vehicles, etc.). In the case of property taxes, fiscal systems often show poor efficiency due to the absence of or outdated necessary elements for implementation, such as a land register for taxation, an inventory of individuals or legal entities, or assessments of property values.

Operating *expenditure* absorbs a large part of local budgets (60 per cent in Turkey, 66 per cent in Tunisia and Morocco, 80 per cent in Egypt). It is used mainly to finance the salaries of staff who work in social services, planning, health, etc. On average, the share of capital expenditure (very variable from country to country) appears to be noticeably higher than in European countries.

Actually, it could be that central government releases itself from capital spending on infrastructures so that this charge falls on local authorities. Given the level of revenues, these obligations would far exceed the resources available. This can engender a substantial infrastructure backlog or a growing imbalance in local budgets. A worrying level of municipality debt is observed in some countries, for instance, in Turkey and Algeria, even though these unbalances are finally covered by the public treasury.[42]

Given the level of infrastructures available, the powers granted in terms of investment would require, in fact, a transfer of resources out of any proportion to what currently occurs. From this point of view, the rules governing *borrowing* would constitute a determining

Table 4.14	Capital expenditure by local authorities (% of total expenditure)	
Turkey	40	1999 (local authorities)
Israel	23	1999
Egypt	18	1995
Tunisia	33	1998 (cities)
Morocco	34	2000 (cities as % of local authorities)
Spain	31	
France	21	
Italy	12	
EU15	13.5	2001

Source: Urdy, 2003, from Dexia 2002 and sources of Tables 4.11–4.13

factor: although by law, the general rule is that local authorities have the right to borrow money, only Israel allows local authorities to take out loans from private sources. Could this constitute a lever for change? On the other hand, there would be a risk of 'decentralizing' countries' external debt.

Control and supervision

In itself, monitoring the legality of the actions of local authorities and their spending powers is a fundamental element of the democratization of the life of a nation. In general, this control may take one of two forms: supervision exercised by a central government authority, or monitoring the correctness of income and expenditure, carried out by an authority independent of the executive and legislative powers, such as external auditors from a revenue court. The general rule in the SEMC is supervision. Most often, the central government (Ministry of the Interior or Ministry of Local Governments) supervises the legality of actions and audits the budget. In addition, *ex ante* political control is exercised by central Government, through the appointment of executives at the intermediate level.

Thus, the actual degree of decentralization is weak, particularly for financial resources. In parallel, there has been a real de-concentration of state powers, the state usually being too far away to manage the complex realities on the ground. However, the initiated decentralization appears to be slow in affirming the action of representative authorities to collect taxes and to make expenditures with reduced supervision. Moreover, decentralization is obviously not complete unless the population, companies and associations are allowed to participate from the very beginning, in defining collective choices. This was clearly specified in the 1992 Rio Agenda 21 and its chapter 28, encouraging local governments to adopt Local Agendas 21 through participatory exercises and a search for consensus.

Expansion of inter-municipal cooperation and metropolitan entities

In the SEMC, steady growth in the number of towns is matched by a rapid increase in urban municipalities. In Turkey, the number of municipalities rose from 809 in 1955 to 3200 in 2003. In Morocco, the administrative and territorial reform of 1992 led to an increase in the number of communes, both urban and rural (249 urban communes in 2005 compared with 99 in 1992). The same phenomenon is seen in Algeria, Tunisia and elsewhere. In small municipalities, the response to people's needs is compromised by their small size. Without a real tax base, with a rudimentary or non-existent administrative body, without skilled human resources and in chronic need of staff, the services they offer may be of poor quality and require very high per capita investment. In practice, they have almost no real possibility of organizing and planning urban services.

In the EU-Med countries, which have a large number of municipalities, the majority being of small or very small size, *inter-municipal cooperation* represents one possible response to coping with municipal fragmentation. Italy more than half a century ago, Spain in the 1990s, Greece in 1998 and France today, have turned to functional groups of communes (*comprensori*, *mancomunidades*, *communautés de communes* or *communautés de agglomerations*, *districts*). Even though most of these experienced difficulties when first set up, they do offer a possibility as yet hardly attempted in the SEMC. Inter-municipal collaborations are found only in Mashrek (Union of Northern Metn municipalities in the Lebanon, Joint Services Councils in the Palestinian Territories, Councils of Shared Services in Jordan).

The heterogeneity of municipalities raises another question. It is difficult to consider metropolitan areas with some million people on an equal institutional footing with towns of a few thousand. In Syria, cities that are capitals of *mohafazats* have their own financial resources, but the overwhelming majority of communes have not. In Turkey, the largest municipality has several million inhabitants while the smallest numbers less than 700 inhabitants. In the SEMC, but also in the NMC, applying the same communal status to all municipalities in a country has two major inconveniences: very small communes cannot provide quality services, and large metropolitan areas do not enjoy a status commensurate with their productive capacity and their actual tax base. The rare special statuses that exist are limited exclusively to the capitals, such as Beirut and Greater Cairo.

Some practices with *metropolitan bodies* are beginning to emerge. Turkey created a special status for 16 Metropolitan Municipalities in 1984; Morocco is also exploring this way with 14 Urban Communities by mid-2003. In Algeria and Tunisia, attempts to create metropolitan entities were never carried through. These are all experimental practices, however, for example in Italy, where the 1990 law creating Metropolitan Cities has still not been put into practice because of strong opposition from the existing local authorities.

In any case, institutionalized cooperation between local governments responsible for urban areas that extend beyond the basic municipal boundaries has become a real need, especially clear for large urban areas. Unions, districts and other entities of this kind are being created for linking sovereignties that have less sense today.

Public–private partnerships

Today, cities are the areas that bear the brunt of the effects of the major macroeconomic reforms adopted in the SEMC over more than 20 years. They benefit from the effects of better economies of scale in agglomerations and increased profitability, when these occur. But they also suffer from the principal external diseconomies mainly related to the mismatch between exploding social demand and the capability of towns to act and invest.

Urban dynamics in the SEMC contribute towards this mismatch between needs and capabilities to invest and perform. Each year Istanbul grows by more than 400,000 inhabitants, a metropolis like Casablanca would increase by 60,000 people each year. Municipalities are primarily confronted by their daily functioning problems, whether related to facilities or to human resources. A chronic lack of staff in the past has lead to a significant rise in the number of local government personnel who, without being sufficient, act as a brake on capital investment. In Tunisia, municipal staff increased by 23 per cent between 1991 and 1997; in Morocco staff numbers multiplied by a factor of 5.5 in 25 years, from 26,500 in 1977 to 143,259 in 2001. And the fact that governments relied on local authorities to absorb an excessive workforce supply has only increased the problem.

Progress made in providing services for nearly 20 years cannot, however, disguise the strong imbalance that persists between big and small cities, between city centres and suburbs, or between wealthy neighbourhoods and deprived ones. These inequalities are accentuated by the concentration of activities, financial resources, and wealthier populations in a small number of cities. For example, 80 per cent of private industrial establishments in Syria are concentrated in Aleppo and Damascus, and these two cities are almost alone in benefiting from private investment for the whole country. Tunis would draw about 60 per cent of investment in the country.

The needs thus remain dominant in the SEMC towns. For many public services, the municipality and its elected council are in the front line of responsibility for the quality and quantity of services provided. In managing urban utilities, the most usual methods are either direct management by the local government or through publicly owned enterprises, sometimes owned by the local authority. In the SEMC, national, de-concentrated, technical bodies, have often taken over responsibility for the management, investments and maintenance of water, sewage or electricity systems. These days, there is more and more reliance on contracting private enterprises. Over the last decade, municipalities have been encouraged to call on private companies for financing and managing urban facilities, and an increasing development of public–private partnerships is seen in most Mediterranean countries.

The term 'public–private partnerships' in fact covers many different forms, such as delegated management by contract, concessions and privatization.[43] These partnerships are currently receiving much attention, partly because they are recommended by international donors, and partly because large multinational companies increasingly dominate markets (Figure 4.12). These companies often provide the necessary capital for large investments in exchange for long-term contracts (20–30 years). In the Mediterranean, there have been many examples of the delegation of powers in the last ten years, especially for electricity, telecommunications, waste and transport, but also for drinking water and sewage management.

Figure 4.12 World distribution of public–private partnerships (per cent)

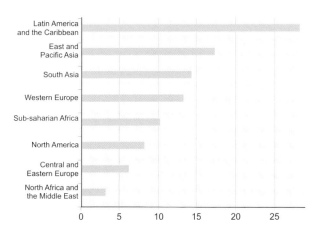

Source: Urdy 2003 from Franceys R., 2000, WEDC-IHE PPP Database (www.silsoe.cranfield.ac.uk/iwe/cws/pppdatabase/)

This process lies within a context of globalization and deregulation that limits the coverage of public utilities, where the concept of public utility or public service is roughly handled, depleted and put into question when the agreements are not rigorous or clear.

Given the growing importance of public–private partnerships for the future of the SEMC, some observations can be made in the light of ongoing experiences in the

Mediterranean and elsewhere. In addition to the specific forms depending on particular economic and political circumstances, two difficulties often arise. First, an imbalance between the partners, in which local authorities that have little room for manoeuvre find themselves in a weak position faced with multinational companies – only very large cities may have the capabilities to delegate authority. Second, there is a lack of regulation of the market for public utilities (water, sewage, waste, transport, etc.).

Public–private relations are being re-examined in some countries for a necessary *public regulation of partnerships*. Delegating management under good conditions should be exercised not only when calling for tenders (and legislation on public markets needs to be reviewed in many countries), but also at the time of implementation and subcontracting. Some countries have introduced interesting practices by making reference to environment and sustainable development criteria.

> **Box 4.16** Public regulation of concessions in Morocco
>
> **Morocco** has recently, but very rapidly, become involved in granting concessions to private actors (currently, according to estimates, 80 per cent of the water distribution, electricity and sewage sectors are managed through private–public partnerships). And, since the late 1990s, there has been lively debate on whether or not a law is needed to regulate these partnerships. There are two possible ways to proceed.
>
> When there is a legal vacuum, these operations are agreed in the conventional way and depend solely on the decisions of the two signatories: the contracting party (the municipality or a group of municipalities) and the contractor (a private company – usually foreign). When a law exists, it can set some clear principles for the delegation of management (transparency, general tendering rules, equal treatment of tender applicants, public registry of tenders), and provide a precautionary framework to avoid misuse and subsequent breaches of contract.
>
> In Morocco, the question is being resolved by law. The draft law on private–public partnerships to manage public utilities is a broad text, flexible enough to facilitate the diversity of cases; it establishes the principles of transparency, competition and equality of access to public utilities and infrastructures. With this new legal framework, local governments retain an important role because there will be no centralization in granting concessions, and the regulatory body is an ad hoc commission supervised by the public authority that initiated the proposed concession (instead of a permanent national commission in charge of the regulation of public–private partnerships).
>
> *Source:* Urdy, 2003

Since agreements are established for long periods, it is very much in the delegating authority's interest to define precisely the performance targets to be met by the private operator, on the basis of quantified indicators that specify the economic, technical, environmental and social requirements. These are, for example: population to be served, environmental quality targets, re-employment of municipal staff and/or job preservation, vigilance on the financial independence of the private operator and pricing, the proposed pricing structure and its social and spatial impacts, and on the procedure to be followed in cases of dispute.

In spite of the highly specific nature of each public–private partnership operation, and in order to reduce the above-mentioned mismatch between the delegating local authority and the private operator, particularly as regards access to information,[44] some consensus is emerging among a number of experts and non-governmental organizations on the need to set up a methodological support facility for delegating authorities in the developing countries. Such a supporting facility could be included in cooperation in the Mediterranean. There are examples such as 'Service public 2000', an association created in France at the initiative of mayors, for providing specialized and independent expertise; and PSIRU (Public Services International Research Unit)[45] database at the University of Greenwich, a worldwide watch facility on contracting-out and privatization in the water sector, which also covers North African and Middle Eastern countries.

Concerted approaches to sustainable urban development

Sustainable development initiatives, although still in their infancy, are emerging throughout the Mediterranean region. Launched by municipalities or urban agglomerations, sometimes even entire regions, local strategies aiming at a more sustainable development take on several shapes and names, such as Local Agenda 21 (focusing mainly on the environment) and Strategic Plan (for economic development). All these initiatives, sometimes covering different thematic areas, sometimes having global objectives, share the same ambition of *changing the fate of a region and its people.*

Sustainable development initiatives open up new perspectives for territorial planning. They can stimulate the involvement of local actors and revive democracy that may be losing momentum or is trying to find its way. Responses that stem from a public debate can be found for the many challenges that confront cities. More than 500 Mediterranean local governments are engaged in this kind of initiative (Box 4.17).

Even though they are not yet numerous, ongoing experiences show a rich variety of approaches. There are many keys to achieving sustainable development, depending on cultural and territorial specifics;[46] they also depend on choices made by urban authorities.

- *The quality of life as a priority.* In Spain and central Italy, the urban environment is seen in the built-up, architectural and aesthetic dimension and gauged by

Urban Areas

> **Box 4.17 Local Agendas 21**
>
> Agendas 21 were born after the 1992 Rio Earth Summit, which defined an Agenda (in Latin, 'what has to be done') for the 21st century on the world scale. Its chapter 28 encouraged the launching of Local Agendas 21. International organizations such as the International Union of Local Authorities (IULA), the Fédération Mondiale des Cités Unies (FMCU) and the International Council of Local Environmental Initiatives (ICLEI) have been supporting the movement by signing the Aalborg Chart for sustainability.
>
> In the Mediterranean, there is no specific survey on Local Agendas 21 processes. At the global level, the ICLEI survey of local authorities and associations in 2002 shows that of 6416 Local Agenda 21 processes in 113 countries, the majority of actions take place in Europe (the total number of Local Agendas 21 in 36 European countries represents nearly 80 per cent of the global results). In the Mediterranean, the number of Local Agendas 21 registered by ICLEI in 2002 breakdown as follows: 900 processes in the four EU-Med countries (of which 360 were in Spain and 430 in Italy), 30 in the eastern Adriatic countries, 50 in Turkey, 15 in Mashrek and 20 in Maghreb.
>
> It is, however, difficult to use these data as if they were statistics, without taking into account the size and the number of municipalities in each country, the diversity of processes and the nature of projects. In certain countries, projects follow a national-level campaign; in other countries, the initiative is taken by local authorities and the state only provides support to a number of selected projects. Some processes relate to a global and integrated strategy that is elaborated with local actors' participation; others are of a sectoral nature and relate to a specific town policy (housing, natural hazards, transport, energy, greenhouse effect, tourism, etc.).
>
> *Source:* Blue Plan from ICLEI. Second Local Agenda 21 Survey 2002 (www.iclei.org)

the open public spaces. The accent is put on hospitality and the quality of the urban living environment. In Barcelona, for example, urban planners place particular importance on the use of public places (promenades, squares and parks).

- *The heritage approach* to renewed cities favours operations of ecological and social rehabilitation, especially for old fabrics and the constant re-use of urban fabrics: 'cities that renew themselves', as in Italy where 'recycling' urban fabrics has long been implemented.
- *Combating suburban sprawl and the greenhouse effect* characterizes initiatives launched in some Mediterranean cities, also in northern European countries (the Rhineland countries) where space is scarce. Many initiatives seek to promote compactness, pedestrian areas and go-slow traffic zones, and to improve the living environment in fairly densely populated neighbourhoods. It is the search for a 'short-distance town' like the ancient Mediterranean cities, such as Ferrara in Italy, winner of the European Campaign for Sustainable Cities prize in 2003.
- *Improving environmental services* is a 'leitmotif' of approaches in several southern and eastern cities, sometimes within the framework of the Urban Management Programme for the Arab Countries. Some ten Tunisian towns are engaged in this way, most of them located in the Medjerda Valley. But large cities, such as Tunis, Sfax and Nabeul-Hammamet, have also started processes to develop city development strategies, with support from the United Nations Human Settlements Programme and the World Bank.
- *Decentralizing local decision-making processes* through people participation (women, young people, the elderly, etc.) and partnerships to raise local investment, is often encouraged in the SEMC, boosted by international development cooperation actors. In Turkey, for example, the programme for promoting Local Agendas 21 was launched in 1997 by the Mediterranean and Middle-East Section of IULA (International Union of Local Authorities) with support from UNDP. First introduced into nine towns, by 2003 more than 50 were included. In the Marmara region, the rehabilitation and rebuilding process after the big earthquake in August 1999 occupies a central place in Local Agendas 21. In Izmit (430,000 inhabitants) the Agenda 21 action plan was completely redirected in this sense in 2000. In Bursa (1.2 million inhabitants) the objective is to strengthen capacities for preventing and preparing for natural disasters with support from the UNDP and Swiss development cooperation.
- The principles of *eco-efficiency innovation* (e.g. recycling waste, promoting renewable energies, adopting environmental management standards such as ISO 14001, applying the High Quality Environmental standard in new buildings) are promoted in many approaches initiated by local authorities in France, Italy and Spain.

Only a few of these initiatives manage to use integrated sustainable development approaches by building bridges between the different sectors, between the local environment and the planet's well-being, and between the short- and the long-term.

Finally, most of these initiatives benefit from involving inhabitants, associations and businesses, and by encouraging consultation processes among the various actors, each with different concerns in community life, with varying status, uneven powers and differing legitimacies. But participation cannot be decreed, it needs to organize itself. Public authorities can only facilitate it by relying on associations of all kinds, particularly since these approaches are still fragile.

The need for national support

States have a major role to play in encouraging the flexible development of concerted approaches at the local level, with local authorities, civil societies and businesses as protagonists on various scales: regions, urban areas, municipalities and neighbourhoods. This implies setting up integrated policies and specific instruments to facilitate local actions.

In the EU-Med countries, national or regional policies acting under specific names but relevant to what is called 'integrated policies for sustainable urban development' have blossomed since the early 1990s.[47] These integrated policies have been supported by the EU since 1994 with the URBAN Community Programme.

In *France* the policy of social development of neighbourhoods goes back to 1982, followed in 1988 by the *Politique de la ville* (Town Policy) and Town contracts in 1992 (230 in 2000), and several other policies in the 1990s. In 2000 the law on Solidarity and Urban Renewal (SRU) placed social and functional mixing at the core of new town planning schemes and introduced the obligation of building 20 per cent of social housing in urban municipalities. The 2003 orientation and programming law on urban renewal relaunched the town policy and focused on the massive restructuring of deprived neighbourhoods (750 sensitive urban areas). In *Italy* the 1990s registered multiple initiatives, which included integrated action programmes, the *Programmi di recupero urbano* for degraded social dwellings and urban regeneration programmes for abandoned industrial estates, followed by the policy of Neighbourhood Contracts in 1997 (Box 4.18). In *Spain*, where there is no integrated urban development policy at the central state level, regions and cities have responsibilities for carrying out such programmes, also by relying on the European URBAN Programme.

Whether or not they are called 'policies for sustainable urban development', all these integrated policies are territory-specific and cross-sectoral policies, carried out through partnerships that seek to go beyond the traditional sector-based logic of public action by implementing various forms of cooperation. Apart from the above-mentioned inter-municipal cooperation, these policies seek to innovate in *vertical* cooperation (between different decision-making levels) and in *horizontal* or cross-sectoral cooperation. In addition, the multiple ongoing experiences show that these policies are not decreed by a domineering power but are formulated together from a given social reality.

Vertical *multi-level cooperation* takes on different forms and names depending on the country: contract, convention, agreement, signed between the different political/administrative levels (central state, region, province, municipality and today, Europe). These contractual forms of cooperation are part of a context of redistributing duties and powers (general for the European countries) with a dual movement of transfer of responsibilities towards supra-state control (the EU) and decentralization towards infra-state government levels. The movement is fed by the *subsidiarity* principle, according to which policies and projects are worked out at the government level that is closest to their implementation; the higher government levels act in a given field of responsibilities only if the lower levels are not in a position to provide a response themselves.

All integrated policies for urban development are part of a multi-level approach that implies reciprocal commitments between partners, and therefore the establishment of a contractual form. Of varying durations (renewable on a yearly basis in Italy within the framework of multi-annual programming; five years, followed by seven years in France), these contracts deal with a wide variety of topics (housing, security, employment, culture, etc.). As for budgets, although the committed sums may appear small, the contracted subsidies are in fact additional resources that serve as *levers* for mobilizing other public or private resources.

This form of cooperation has sometimes led to controversy. Local authorities regard it as a central government mechanism for transferring obligations without transferring the corresponding resources. In any case this is a process of passing on responsibilities by contractual means, enabling gradual decentralization. For the SEMC that have made little progress in financial decentralization, this way can prove very useful for fully supporting town development, especially for medium-sized towns.

Cooperation between the various sectoral policies (housing, urban planning, transport, environment, social and education, etc.) and between different departments, agencies and bodies in charge of their application is the hardest to implement in all countries. It opposes very heterogeneous sectors of skills that inevitably lead to confrontations between approaches, methods, and very different professional cultures. Because of the traditionally sectoral-management approaches, it is essential for *governing the interfaces* to set up specific mechanisms (task forces, working groups and programmes, 'department conferences', etc.) for removing barriers between professional cultures that are traditionally ignorant of one another.

At the central level, ministerial guidance of these integrated urban development policies is far from stable. It is a new field of action implying the re-composition of sectoral policies in order to integrate, at the highest possible level, the complexity of urban issues. And even though integrated urban development policies may sometimes be seen as an overdose of complications, the actors directly involved see this apparent complication as a 'necessary evil', for overcoming the obstacles.

At the local level, local authorities, especially municipalities, experience the same difficulties in integrating administrative action from a traditionally sectoral management. Here too, practices almost everywhere[48] show that these integrated policies succeed mainly when they are in the hands of the mayor or the head of the municipal or inter-municipal executive, the only ones able to make 'an overall city policy'.

The Italian experience is presented in some detail (Box 4.18) to show, apart from particularities, that several ingredients are vital for the effectiveness of integrated urban development policies, in particular: clear objectives pursued by national policy, the elaboration of which has taken into account the limits of previous policies; synergies between the various echelons and stakeholders (the state, the regions, municipalities, the private sector, European support) resulting in cross-financing; the responsibility of municipalities for detailed planning and implementation of the projects, and thus for their success; and the flexibility of the procedure, making it possible to experiment with integrated thematic approaches and concerted actions in cities.

Box 4.18 Integrated urban development policies in Italy

In Italy, the first integrated urban renewal programmes (*Programmi integrati di intervento*) were launched through a 1992 law aimed at promoting the upgrading of the urban fabric and the urban environment by relying on previous local experience (e.g. Turin, Naples). The 1994 *Piano di ricupero urbano* (PRU) identifies urban spaces to be redeveloped (social housing estates) and provides for actions that associate public authorities with the private sector, without, however, implementing complementary economic and social measures. Faced with the deterioration of some neighbourhoods in city centres and surrounding suburbs, the Ministry of Public Works then launched the *Contratti di quartiere* policy (Neighbourhood contracts) through the 1997 finance law. It concerns mainly the areas with public housing built in the 1970s: 12 per cent of these are in town centres, 31 per cent in the first ring of suburbs (consolidated fabric), 48 per cent in the outer suburbs (recent fabric) and 9 per cent in peripheral areas.

The main innovations of *Contratti di quartiere* are the multidimensional approach and the methods for renewing neighbourhoods and diversifying their functions so as to re-integrate them into the city, while rejecting the concept of zoning. Apart from architectural rehabilitation and the organization of urban transport networks for example, an important social/economic field of action is included. Sectoral approaches for the town are integrated during the project's formulation and implementation, and procedures from ministries other than just Public Works, can be mobilized, in particular in the ministries of Economy, Education and Social Affairs.

As for the regulatory framework (a January 1998 decree), the choice of neighbourhoods depends on a call for tender launched by the Ministry of Public Works on two levels: tender documents are sent to their regional government by each interested city, the regional body then assesses each project's feasibility, grants an initial budget and sends a maximum of five dossiers to the Ministry for state funding. In the first series of contracts, 83 municipal proposals were selected at the regional level and 55 were funded by the central government. Implementation of the Neighbourhood Contracts is always the municipality's responsibility.

Proposals concern areas with degraded housing covered by a PRU or by an urban planning tool. Cities can choose one or several experimental topics from four quality objectives:

1. morphological (structural) quality of the urban fabric (re-enhancement of degraded fabrics, conservation and enhancement of historical fabrics, transformation with functional integration);
2. quality of the ecosystem (bio-architecture and urban ecology, resource savings, improvement of environmental quality);
3. urban quality and social actions (accessibility, new ways of using housing, particularly by vulnerable groups, professional inclusion of young people);
4. quality system (management and control of construction quality).

Launching Neighbourhood Contracts was facilitated by the local government reform of 1993 (election of mayors by direct public vote) and by the 1997 Bassanini Law dealing with the reorganization of administrative services, which set up several new measures: a 'single branch', and 'departmental conferences' to facilitate consultation between administrations; joint public–private companies; 'territorial pacts' for mobilizing different strategic actors at the local level; 'social round tables' at the municipal level. At the level of neighbourhoods, specific participatory forums arose such as 'neighbourhood workshops', sometimes true mechanisms for revitalizing neighbourhoods.

Cross-financing is mandatory. The Ministry of Public Works created a special fund for Neighbourhood Contracts, constituted from a withdrawal system equivalent to a 1 per cent building tax (about 300 million euros in 1997). Local governments allocate part of their own funds, which are fairly significant amounts since Italy decentralized. Other ministries act through agreements, with the Ministry of Public Works or with the municipalities concerned. Private companies also enter into the contractual procedures. This policy is complemented by the European URBAN programme: 16 cities selected in 1994 and 10 more for 2000–2006.

Source: Les contrats de quartier en Italie. Les exemples de Padoue, Rome, Turin, Parme et Venise. Rapport final pour la Délégation interministérielle à la ville. Paris, 2001 (www.ville.gouv.fr)

Mediterranean and Euro-Mediterranean cooperation: Beyond the challenges

In spite of Mediterranean diversity, improving the urban living environment concerns all countries in the region. Regional cooperation is one way of making progress and helping southern and eastern countries to engage in urban development that is equitable and respects human health and the environment.

Introducing the urban dimension in international cooperation

It is hard to gauge the flows of official development assistance (ODA) for the urban sector in the Mediterranean: the region is often split into various areas in which international donors operate. In addition, there is a specific drawback due to the fact that many funds concern actions that contribute indirectly to urban development, without allocating funds to projects with urban development as their main goal. Euro-Mediterranean cooperation is still seldom targeted at urban development.

Some (partial) trends can be drawn from the OECD's Development Assistance Committee (DAC) database. Between 1973 and 2000 ODA specifically concerning urban development represented only 0.55 *per cent of the total* ODA earmarked for the Mediterranean countries. In absolute figures, the cumulative total for the period was US$683 million (an average of US$26 million per year). The 1990s saw clear progress in aid for the urban sector (US$59 million per year compared with US$7.4 million per year in the 1980s). The aid falls under donations (US$526 million, 77 per cent of the total aid for the urban sector) and loans (US$158 million, 23 per cent). The aid covers urban planning and development projects, the rehabilitation of underprivileged neighbourhoods, and, more and more, activities to reinforce the management capabilities of local governments.

Aid from *multilateral agencies*[49] represents only 7 per cent of total urban ODA. In *bilateral cooperation* (93 per cent of the cumulative total), the US is the main donor for urban development (US$469 million committed, or 68 per cent of the total for urban-sector aid), followed by France (15 per cent of total aid or US$109 million, including US$101 million earmarked for Tunisia), Germany (4 per cent) and Spain (1 per cent).

During 1990–2000, Egypt was the main beneficiary of ODA in the urban sector in the SEMC (57 per cent, US$335 million), followed by Tunisia (17 per cent) and the Palestinian Territories (3 per cent). For the eastern Adriatic countries, rebuilding after the war, Bosnia-Herzegovina benefited most from aid to the urban sector (13 per cent of the commitments) and Albania (5 per cent).

As for financing by *European institutions*, to date the European Investment Bank (EIB) has financed only a very limited number of urban development projects outside the EU,[50] and only two urban projects for the SEMC can be identified: a 300 million euro facility earmarked for rebuilding Izmit in Turkey after the 1999 earthquake, and 165 million euros for Algeria after the Algiers floods in 2001. However, EIB loans for infrastructure developments in cities (water and sewerage, urban transport) are significantly higher: about 1.4 thousand million euros for water distribution and sewerage projects in cities (or 16 per cent of total loans made to the SEMC during this period); 374 million euros for urban transport, particularly developing public transport such as the light metro in Tunis (30 million euros in 2000) and extending the Cairo underground (50 million euros in 2002). In total, nearly 2 thousand million euros (an average of 266 million euros per annum) would have gone to projects favouring cities between 1995 and 2002, a quarter of EIB loans for all sectors to the SEMC.

Finally, in the *Euro-Mediterranean* process, the urban dimension of development has been absent from MEDA's predominantly sectoral programmes (energy, transport, the environment), as shown by the small proportion allocated to urban development projects between 1995 and 2002, with only two projects: the renovation of the Balat and Fener neighbourhoods in Istanbul, classified as UNESCO World Heritage sites (7 million euros in 1998), and the improvement of insalubrious neighbourhoods in Tangiers, Morocco (7 million euros in 1999). On the other hand, among the sectoral sections of MEDA, several projects benefit cities: water distribution and large-scale sewerage (214 million euros committed), energy and urban environment (1.9 million euros in 2000), pilot projects in the programme for the environment (SMAP) concerning urban solid waste management or air quality improvement. In addition, after the suspension of the MED Urbs programme in 1996, the MEDA programme supports decentralized cooperation between cities and local governments through projects coordinated by Bordeaux, Rome, Genoa and Barcelona (5 million euros in 2005).

In summary, the urban sector share appears weak in the development aid flows. The share is more significant if the funding of various sectoral infrastructure projects that benefit cities is included, but it is still not possible to get a clear image of the real scale and even less the impact of this funding for the southern and eastern Mediterranean cities. In any case, the specificity and complexity of urban development issues are not taken into account in the prevailing sectoral approach of international financing.

In future, the EU's growing role in the financing of development in the Mediterranean, and the new Neighbourhood Policy that will enter into force in 2007, call for extending the European structural support programmes (URBAN, Interreg) with specific funding for the urban areas of the southern and eastern Mediterranean countries that are in crisis.

Within the EU, where the urban dimension is not part of the Community competences because of the subsidiarity principle, urban policies are nevertheless at the heart of EU efforts to build a strong and competitive Europe while maintaining social cohesion. The *Framework for action for sustainable urban development*, adopted by the Commission in 1998, recognizes the importance of the urban dimension in sectoral policies and stresses in particular the possibilities offered by the regional development programmes co-financed by the Structural Funds. Alongside the structural funds earmarked for European regions that are slow in their development or in decline,[51] the URBAN community initiative (Box 4.19) is exemplary.

> **Box 4.19** The URBAN programme: Giving a dynamic boost to European urban areas in crisis
>
> URBAN is an instrument of European policies for cohesion aimed at revitalizing urban space and neighbourhoods in difficulty. Launched in 1994, the initiative comes under the European Regional Development Fund (ERDF).
>
> In 2000–2006, URBAN II is supporting the formulation and implementation of innovative economic and social regeneration strategies in a limited number of small and medium-sized towns or neighbourhoods in big cities. These are 70 urban areas (a total of 2.2 million people throughout Europe) facing serious difficulties such as unemployment and criminality rates twice as high as the community average, and twice as many immigrants as in the average urban community. The 70 target areas include Granada in Spain, Bastia in France, Genoa, Carrara, Pescara and Taranto in Italy and Heraklion in Greece. The ERDF contribution is set at more than 700 million euros, which may turn into a total investment of 1.6 thousand million euros, when including local and national co-financing.
>
> Two other programmes contribute to improving the targeting and effectiveness of URBAN programmes. Since 2000 the Urban Audit has made it possible to assess the quality of life in some 200 European cities of varying sizes through a series of urban indicators of socio-economics, the environment, education and training, participation in the life of the city, and culture and leisure issues. URBACT (a European network for exchanging experiences) highlights good practices and makes it possible to draw lessons from the strengths and weaknesses identified in the URBAN programmes.

Finally, a thematic strategy for the urban environment is to be proposed in 2005 for improving the living environment and reducing the impacts of EU-25 cities on the local and global environment (some 500 cities of more than 100,000 inhabitants).[52]

Insufficient city-to-city cooperation

Exchange and cooperation between cities within true urban networks have typified the history of this region since ancient times and until the emergence of nation states. Great geographers have highlighted the importance of such networks, and emphasized that cities and urban networks have strengthened the Mediterranean region. Today these are no more than a weak web. In the 1950s they were almost non-existent, the international influence of cities being minimal and often contrary to national laws. Over the past 30 years inter-city networks have re-emerged and now constitute an embryonic pattern that may contribute to rebuilding the future.

The web being woven today is part of global networks of local authorities. Since 2004 Barcelona has hosted the headquarters of United Cities and Local Governments, a new international town organization, born by merging FMCU (Fédération mondiale des Cités Unies), IULA (International Union for Local Authorities) and Metropolis.

International agencies also encourage town networking. For example, since 1988 the WHO Healthy Cities network has been bringing together 100 towns at the pan-European level, but also including Turkey and Israel, aiming to fully integrate health concerns into urban development. Since 1977 UNESCO has been encouraging the networking of small historical coastal towns in the Mediterranean and the Adriatic. And in 2004 Marseilles and the World Bank (Middle East and North Africa section) promote partnerships between some 20 European and Mediterranean cities aimed at identifying urban management projects that could lead to funding. Europe has given an important boost to twinning since 1945, and to north–south interaction supported by EU programmes such as MED Urbs and MED Campus (1992–1995), and, currently, MED Act. The Council of Europe has played an active role, but with limited funds. South–south interactions, however, currently represent only a very small portion (about 5 per cent) of inter-city exchanges.

Intra-Mediterranean exchanges have been developed, through thematic networks, for example on towns (the Urbama research and knowledge network on cities of the Arabic World), energy saving (Medener, between national energy-management agencies) and natural hazards (Stop Disasters, which aims to contribute in a more incisive manner to preventing natural disasters and helping affected Mediterranean towns). Other networks, such as Medcities, are more 'generalist' (Box 4.20).

> **Box 4.20** The Medcities network
>
> The Medcities network of Mediterranean cities was founded in 1991 on an initiative by Metap (Mediterranean environmental technical assistance programme of the World Bank, the EIB, the

European Commission and the UNDP) to strengthen the environmental management capabilities of local administrations.

Preference was initially given to coastal cities other than capitals. In 2004, Medcities includes more than 25 cities: Barcelona (Spain), Marseilles (France), Monaco, Rome and Palermo (Italy), Koper (Slovenia), Dubrovnik (Croatia), Tirana (Albania), Thessalonica (Greece), Limassol and Larnaka (Cyprus), Gozo (Malta), Izmir and Silifke (Turkey), Latakia and Aleppo (Syria), Al-Mina and Tripoli (Lebanon), Haifa and Ashod (Israel), Alexandria (Egypt), Benghazi (Libya), Sousse and Sfax (Tunisia), and Tangiers and Tetouan (Morocco). Az Zarqa (Jordan) is also in the network.

The areas of cooperation have gradually been expanded, from carrying out environmental audits and plans (on urban solid waste, air quality, etc.) to projects on urban mobility, upgrading marginal quarters and preparing sustainable urban development strategies. Projects are financed mainly by the EU, but also by the cities themselves, Spanish development cooperation and the World Bank.

Sustainable development is a new cooperation area that could be developed between the 300 to 500 Mediterranean municipalities already engaged in an Agenda 21-type process. But this would imply going beyond 'sustainable development campaigns' or signing short-lived charts, and working in areas such as:

- training of municipal and regional personnel on sustainable development issues;
- promoting interfaces between activity sectors through joint projects, such as transport and urban planning, energy/urban transport and the greenhouse effect, risk prevention and urban planning/housing policies, water/sewerage and housing policies, poverty and the environment, and the environment and public health;
- mobilizing and better targeting of internal and external financing, access of cities to private funding under strict budgetary control;
- promoting multi-player actions (state, local governments, companies, associations) and forms of citizen participation;
- assessing and monitoring strategies and action plans (urban indicators).

Within the Mediterranean Action Plan, the Mediterranean Commission on Sustainable Development formulated a first set of orientations in 2001, stressing the initiatives taken by the Mediterranean countries for sustainable development with the new Barcelona Convention (1995), to which the Euro-Mediterranean process makes reference, and the need to reinforce cooperation networks among cities.

***

With nearly 80 per cent of the population projected to be living in urban areas by 2030, the future of the Mediterranean region will depend largely on cities and the lifestyles of urban populations. The challenge is particularly great for the southern and eastern countries, with 100 million additional urban dwellers projected between 2000 and 2025. The challenges in the northern countries are somewhat different, with urban sprawl continuing to predominate, despite very moderate population growth.

Throughout the region, the major objectives for curbing urban trends and moving towards a more sustainable future are to *guide urban development*, *improve the living environment*, and *promote less wasteful and polluting lifestyles*. The necessary actions and policies, described throughout this chapter, relate to these three objectives; they are already being implemented in many Mediterranean cities and some countries.

The future will certainly rely on policies being implemented at the very *local level*: the entire city, the heart of the city, the neighbourhoods. Municipal and inter-municipal authorities can do much to induce changes, through more strategic urban planning, public participation and by taking into account the overall human, economic, environmental and cultural dimensions of urban issues; results may take 20–30 years to be seen. But city sizes, economic specialization, institutional capacities or weaknesses, regional and national contexts, will all govern the policies that need to be designed; room for manoeuvre differs between different cities and countries.

Faced with the immense challenges of sustainable development, however, cities will not be able to do everything alone. A change of scenario requires simultaneous efforts by cities, provinces or regions, and states, with Euro-Mediterranean support, if it is to create the necessary synergy and strive in the same direction.

At the *national level*, without regional development policies that favour medium-sized and inland cities, concentration and congestion in big metropolises and unbalanced urban structures will be accentuated. Without substantial efforts at the national/regional/local level aimed at improving the quality of life in cities, a certain 'urbanization in poverty' could take place with serious effects on societal balance, insecurity and increased environmental degradation.

At the *Mediterranean and Euro-Mediterranean level*, there remains a need for regional cooperation to cope with urban challenges. Without such cooperation, city networks and decentralized cooperation projects will continue to experience difficulties with longer term projects, and positive effects for cities and for countries will remain limited. Neither can decentralized cooperation alone cope with the challenges. The region would benefit from a Euro-Mediterranean cooperation programme that could mobilize national and EU support for medium-sized towns and urban areas in crisis, and strive more effectively towards more sustainable development.

Notes

1. Population living in urban areas of more than 10,000 inhabitants.
2. In Tunis for example, 70 per cent of the households living in suburbs come from the town centre.
3. Attané and Courbage, 2001; *Plan Bleu*, 2001.
4. UNHSP, 2003.
5. International comparisons of urban statistics have always been very difficult, because of the variety and flexibility of definitions of 'urban' (in addition definitions change between censuses). By applying a standard definition worldwide, the *Géopolis* database, created by F. Moriconi-Ebrard, enables the problem to be surmounted and comparisons to be made between urban areas with more than 10,000 inhabitants. This is the base used in this chapter so that the population and growth of towns at the Mediterranean level can be compared.
6. It should not be forgotten that all classifications of this kind made to enable gross comparisons, are always somewhat arbitrary, and that a town that is classified as medium-sized in one country may be considered very large in another.
7. Cattan et al, 1999; Carrière, 2002.
8. Wiel, 1999.
9. Shanty towns in Arab countries began to host rural migrants in the inter-war period, and consist of waste-material dwellings, built on small plots and squatted land.
10. IFEN, 2003.
11. Huybrechts, 2001.
12. Netherlands, Great Britain, Norway, etc.
13. Dematteis and Bonavero, 1997.
14. Haeringer, 1999.
15. The European directive on major hazards from certain industrial activities, which since 1996 also applies to the storage of dangerous substances (the 'Seveso II' Directive).
16. United Nations Conference on the 'Prevention of natural disasters, land improvement and sustainable development', Paris, 17–19 June 1999.
17. Newman and Kenworthy, 1989.
18. Orfeuil, 2000.
19. World Road Statistics, 2003 (www.irfnet.org/wrs.asp).
20. A. Schafer, MIT, 1997.
21. Based on household surveys on mobility and other studies on transportation in these urban areas.
22. Data on Oran in Algeria, late 2000, indicate a very substantial proportion of walking journeys (63 per cent) whereas motorized transport amounted to 37 per cent of the total, which can be explained by poor access to cars in the context of an economic crisis, and by the crisis in the supply of public transport.
23. By way of comparison, the car market share (motorized travel done by cars) is 66 per cent in Paris, 75 per cent in Grenoble and Lyon, and 90 per cent in less dense cities such as Montpellier (Certu, 2000 www.certu.fr).
24. In Algiers, the latest household survey on mobility dates from 1990, but recent household surveys in Annaba and Constantine show that between 1987 and 2000, all-modes mobility is stagnant or regressing, while modal transfers take place from walking or private cars in favour of public transport, the supply of which has noticeably improved in this period. These changes in Algerian cities seem to run against the growing car use trend.
25. In 2000, Spain, France, Italy and Greece were responsible for 70 per cent of all CO_2 emissions in the region (see Chapters 2 & 3).
26. Italian municipalities are also setting up interesting time management strategies, in order to break up traffic jams in the rush hour.
27. This is the case for the Paris region (11 million inhabitants), with many administrative layers having responsibility for transport. Here central Government is responsible for the PDU.
28. ECMT, 2002.
29. Carrière, 2002.
30. Chaline, 2003.
31. Unesco, 2004.
32. EEA. *Environmental signals*, 2002
33. Siversten et al, 2001.
34. Hoek et al, 2002.
35. APHEIS, 2001.
36. Special dossier on air quality. *La Revue Durable*, January–February 2003.
37. Net consumption relates to quantities used, not rejected by the users. From Margat, 2004.
38. ESCWA, 2001; UNDP, 2002.
39. In Israel, the notion of 'level' is all relative. The responsibilities of municipalities (30,000 inhabitants), and of local and regional councils, vary according to the population represented.
40. The 16 Regions of Morocco are run by regional councils elected by indirect suffrage; however, very few financial resources have been granted to the Regions, which do not have their own administration either.
41. Bank of Provinces in Turkey, *Caisse Autonome des Municipalités* in the Lebanon, *Caisse des Prêts et de Soutien aux Collectivités Locales* in Tunisia; *Fonds commun des collectivités locales* in Algeria; *Fonds d'Équipement Communal* in Morocco.
42. In Turkey, the high rate of indebtedness of municipal authorities has led government to consider reforming the conditions under which local authorities have access to loans. In Algeria, the level of debt in *communes* was estimated at 22 billion dinars in 1999; in an attempt to bring this down to 2 billion dinars, the Algerian authorities decided at the end of 2001 to allocate more than 15 per cent of a budget that was initially intended to relaunch the national economy (Urdy 2003 from Ceneap 2002).
43. Delegated franchises and management are part of the 'French-type' system in force since the 19th century in the water sector; this system is widespread The privatization or management by the private sector is the predominant model in Great Britain, less widespread as yet.
44. Multinational operators are in regular touch with the major companies of business lawyers; local authorities do not have similar legal advice.
45. University of Greenwich. Public services international research unit. www.psiru.org
46. C. Emelianoff, in *La Revue durable*, no 5, May-June, 2003 Fribourg (Suisse).
47. De Decker and Vranken et al., 2003; Jacquier, 2003.
48. Cf. *Villes du XXIe siècle. Quelles villes voulons-nous? Quelles villes aurons-nous?* Actes du colloque de La Rochelle, Certu, 2001.
49. International Development Association, UNDP, etc. In the CAD-OECD database, the file which makes it possible to make a sectoral breakdown of the projects, loans by the World Bank are not included.
50. From 1988 to 2001 the total number of urban development projects funded by the EIB within the EU amounted to 29.5 thousand million euros (10 per cent of the EIB's total loans, 2 thousand million per annum). Cf. Rapport d'évaluation. Projets d'aménagement urbain finances par la BEI à l'intérieur de l'UE. July 2003.
51. 70 per cent of the European programmes coming under goals 1 and 2 (providing 15 thousand million euros for 2000–2006, or more than 2 thousand million per annum) deal with urban problems.
52. EC. *Towards a Thematic Strategy on Urban Environment*. COM (2004) 60 final.

References

APHEIS (Air Pollution and Health: A European Information System). (2002) *Health Impact Assessment of Air Pollution in 26 European Cities. Second-year Report, 2000–2001.* Institut de Veille Sanitaire: Saint-Maurice (www.apheis.net)

Attané, I., Courbage, Y. (2001) *La démographie en Méditerranée. Situation et projections.* Les Fascicules du Plan Bleu, 11. Paris: Economica

Barthel, P.A. (2003) Les lacs de Tunis en projets, reflets d'un nouveau gouvernement urbain. In *Annales de Géographie*, no 633

Bertoncello, B., Girard, N. (2001) Les politiques de centre-ville à Naples et à Marseille: quel renouvellement urbain? In *Méditerranée*, no 1.2

Carrière, J.P. (dir.). (2002) *Villes et projets urbains en Méditerranée*. Maison des Sciences de l'Homme, Université de Tours

Cattan, N. et al (1999) *Le système de villes européennes*. 2nd ed. Paris: Anthropos

CE. DG Politique Régionale. (2000) *L'Audit urbain. Vers un référentiel pour mesurer la qualité de la vie dans 58 villes européennes*. http://europa.eu.int/comm/regional_policy/urban2/urban/audit/

Chaline, C. (2001) *L'urbanisation et la gestion des villes dans les pays méditerranéens. Évaluation et perspectives d'un développement urbain durable*. Report for the Blue Plan

Chaline, C. Blue Plan. (2003) *Urban regeneration: a need or a chance for cities' development in the Mediterranean region*. Workshop on Urban Regeneration in the Mediterranean Region. MAP/PAP, Split, July

Chikr Saïdi. F. (2001) Alger: des inégalités dans l'accès à l'eau. In *Revue Tiers Monde*, April–June

De Decker, P., Vranken, J. et al (2003) *On the Origins of Urban Development Programmes in Nine European Countries* (UGIS collection 2). Antwerpen-Apeldoorn: Garant.

De Forn, M. *Le processus de réaménagement urbain de Barcelone (1979–2004)*. MAP/PAP Workshop, Split, 21–22 July 2003

Dematteis, G., Bonavero, P. (dir.) (1997) *Il sistema urbano italiano nello spazio unificato europeo*. Bologne: Il Mulino

Dexia (2002) Finances locales dans l'Union européenne - Tendances 1996/2001, *Note de conjoncture*, October.

Dexia, Isted, CFCE. (2002) *Autorités locales du monde*. www.almwla.org

EC-Joint Research Centre; EEA. (2002) *Towards an Urban Atlas. Assessment of spatial data on 25 European cities and urban areas*. Luxemburg: OPOCE

ECMT. (2002) *Implementing Sustainable Urban Travel Policies*. Final report. Paris: OECD.

El Kadi, G. (2001) *L'urbanisation et la gestion des villes dans les pays méditerranéens. Égypte*. Report for the Blue Plan

El Mansouri, B. (2001) *L'eau et la ville au Maroc. Rabat-Salé et sa périphérie*. Paris: L'Harmattan

ESCWA. (2001) *Decentralization and the Emerging Role of Municipalities in the ESCWA Region*. New York: United Nations

Gafsi, H., Ministère de l'Environnement et de l'Aménagement du territoire. (2002) *Rapport introductif sur les villes durables en Tunisie*. Tunis: Oted; GTZ

Ghanem, El Sayed, A. M. A. (2002) 'Fiscal decentralization in Egypt'. Communication presented to the Fourth Mediterranean Development Forum (MDF4), Amman, 9–10 June

Godard. X. (2003) *Synthèse sur la motorisation, la mobilité et les systèmes de transport dans les villes du sud et de l'est de la Méditerranée*. Report for the Blue Plan

Haeringer, Ph. (1999) L'économie invertie. Mégapolisation, pauvreté majoritaire et nouvelle économie urbaine. Essais 1988–1998. In *2001 Plus*, no 50

Hoek, G. et al (2002) Association between mortality and indicators of traffic-related air pollution in The Netherlands: a cohort study. *The Lancet*, no 9341

Huybrechts, E. (2001) *L'urbanisation et la gestion des villes dans les pays méditerranéens. Liban, Syrie, Turquie*. Report for the Blue Plan

IFEN. (2003) L'artificialisation s'étend sur tout le territoire. *Les données de l'environnement*, January–February

Jacquier, C. (2003) *Politiques intégrées de développement urbain durable et gouvernance urbaine en Europe*. Paris: Délégation interministérielle à la ville, November

Jourdan G. 2003. *Transports, planification et gouvernance urbaine: étude comparée de l'aire toulousaine et de la conurbation Nice Côte d'Azur*. Paris: L'Harmattan

Jouve, B. (ed.) (2003) *Les politiques de déplacements urbains en Europe. L'innovation en question dans cinq villes européennes*. Paris: L'Harmattan

Keles, R. 1999. *Urban development and sustainable management for the Mediterranean towns. Turkey*. Paper for the Meeting of the MCSD Working Group. Split, 26–27 April

Le Galés, P. (2003) *Le retour des villes européennes*. Paris: Presses de Sciences Po

Margat, J., Plan Bleu. (2004) *L'eau des méditerranéens: situation et perspectives*. MAP Technical Reports Series, no 158.

Newman, P., Kenworthy, J. (1989) *Cities and Car Dependence*. Aldershot: Gower

Orfeuil, J.P. (2000) *L'évolution de la mobilité quotidienne. Comprendre les dynamiques, éclairer les controverses*. Paris: Synthèse Inrets

Pigey, J. H. (2001) 'Morocco – Municiple management and decentralization', The Urban Institute, draft version

Plan Bleu; CEDARE. (2000) *Policy and Institutional Assessment of Solid Waste Management in Five Countries: Cyprus, Egypt, Lebanon, Syria, Tunisia*. Sophia Antipolis: Plan Bleu (www.planbleu.org)

Programme de Gestion Urbaine (PGU). Région des États Arabes. (2001) *Options de politiques municipales pour une meilleure gouvernance*. Le Caire: Cedare, PGU, Habitat.

Prud'homme, R. (1996) Managing megacities. In *Le courrier du CNRS*, no 82.

Qudsi, A. (2003) *Urban Regeneration in the Old City of Aleppo*. MAP/PAP workshop, Split, 21–22 July

Rodrigues Malta, R., (2001). Naples, Marseille: waterfront attitude. In *Méditerranée*, no 1.2

Schafer A., Victor D. G. (1997) *The Future Mobility of the World Population*. Discussion Paper 97-6-4, Cambridge, MA: Massachusetts Institute of Technology, Center for Technology, Policy, and Industrial Development.

Signoles, P. et al (1999) *L'urbain dans le monde arabe. Politiques, instruments et acteurs*. Paris: CNRS Editions

Siversten, B. et al, EEAA. (2001) *Air pollution in Egypt*, August. Cairo: EEAA

Smets, H. (2002) *Le droit à l'eau*. Académie de l'eau. Nanterre

Theys, J. (2000) Développement durable, villes et territoires. Innover et décloisonner pour anticiper les ruptures. *Notes du CPVS*, no 13

UN (2002) 'Programme on governance in the Arab region – POGAR', www.pogar.org

UNCHS. (2001) *The State of the World's Cities*. Nairobi: UNCHS

UNDP. (2002) *The Arab Human Development Report*. New York: UNDP

UNEP/MAP/MCSD. (2001) *Urban Management and Sustainable Development in the Mediterranean. Recommendations formulated by the Mediterranean Commission on Sustainable Development and adopted by the Contracting Parties to the Barcelona Convention*. Monaco, 14–17 November 2001

UNEP/MAP/PAP. (2003) *Workshop on Urban Regeneration in the Mediterranean region. 21–22 July 2003*. Split: PAP

Unesco. (2004) *D'Istanbul 1996 à Venise 2002: la revitalisation socialement durable des quartiers historiques/From Istanbul 1996 to Venice 2002. Socially sustainable revitalization of historical districts*. Paris: UNESCO

UNHSP. (2003) *The Challenge of Slums. Global report on human settlements*. Nairobi: UN-Habitat

Urdy L. (2003) *La décentralisation fiscale en Méditerranée? Investir dans les formes urbaines*. Report for the Blue Plan

Wiel, M. (1999) *La transition urbaine ou le passage de la ville pédestre à la ville motorisée*. Liège: Mardaga

World Bank (2003) *Decentralization indicators database*. www.worldbank.org

Chapter 5

Rural Areas

Over the past 10,000 years, mankind has transformed the landscapes around the Mediterranean into a large variety of characteristic rural regions ('*terroirs*' in French), resulting in a complex mosaic of gardens, arable fields, vineyards, olive groves and other fruit trees, pastureland, semi-natural woodlands, *macchia* and *garrigue* and dry grasslands. This transformation has been slow and gradual, rural areas typically being areas where it takes a very long time to change things.

However, Mediterranean rural areas have experienced radical changes over the past few decades:

- the old equilibrium, established over time between man and nature, has largely been disrupted, both in the northern and in the southern and eastern Mediterranean countries;
- today, what perhaps differentiates the EU Mediterranean countries most from the southern and eastern countries is the differences in standards of living in rural areas.

The main symptoms of the crisis in rural areas are the widening regional gap between the marginalized hinterland and the coveted agricultural plains; rural poverty in many southern, eastern and east Adriatic countries; and degradation of the environment and landscapes. For a superficial observer, this evidence is usually hidden behind the facades of new constructions and coastal tourist resorts. But the dust carried by the wind from areas undergoing desertification, uprooted rural populations that end up in the suburbs of big cities in the south and east and who are often likely candidates for emigration to Europe, forest fires, floods and growing tensions over water resources are there to remind us of another reality.

This serious crisis affects the value of a priceless heritage and is a source of increased risks and instability. As well as degrading vital ecological systems and aggravating poverty with its related risks, it threatens biodiversity and landscapes that are exceptional on the global scale, and the very identity of the Mediterranean region and its wide range of valuable know-how (landscapes, traditions and products). It also has serious economic repercussions, since environmental degradation, poverty, rural migration and marginalization of many areas weigh heavily on national economies.

With globalization and the liberalization of agricultural trade presently under discussion, there is a risk of an increase in these difficulties for many fragile rural areas in the south and east, which could have repercussions well beyond these areas.

However, Mediterranean rural areas are also areas of the future. The undeniable, although late, 'rural revival' that characterizes many rural areas in the developed countries of the northern shore and some recent agricultural advances, both in the north and in the south and east, bear witness to this. This revival depends on the mobilization of local and professional actors and the implementation of adapted policies. It is stimulated by the establishment of new relationships with cities in response to new demands (recreational, residential, quality products). This development leads to a growing distinction between 'urban-influenced' rural areas, those governed by 'rural market towns', and 'remote' rural areas. These new dynamics are not necessarily positive. Increasing use of private cars and the demand for land in the countryside lead to significant urban encroachment into agricultural lands and lifestyles that are highly taxing on the rural environment.

The whole issue of sustainable rural development is therefore posed to decision-makers involved in achieving Euro-Mediterranean stability and prosperity. Action is urgently needed since the break points are sometimes very close, especially in ecological terms. There is a need for more awareness raising, more stimulation of public opinion, and for redirecting or improving management policies.

In meeting these challenges, the two shores of the Mediterranean are unified by an ecological and socio-economic interdependence and the need to succeed in reconciling environment and development. Despite what separates them, on the global scale they form an entity that is held together by multiple bonds. This entity will only become part of the globalization process, and have enough weight in it, if its *identity* and its *geographical unity* are reinforced through sustainable partnerships and effective solidarity in development aid.

The originality, fragility and diversity of Mediterranean rural areas

Rural areas, living areas where vegetation cover continues to evolve and retains a certain reversibility, provide basic functions that range from *production* (agriculture, grazing and forestry) to *protection* of vital ecological functions (water cycle, soils and biodiversity).

The rural areas in the region owe their uniqueness to the specific Mediterranean climate and vegetation. Because of summer droughts, which become more pronounced the further south one goes, and the high variability in precipitation, the issue of water supply for cultivated and natural vegetation (pasture and woodlands) is always crucial. Dry farming, rarely giving high yields, is only possible when there is sufficient rainfall.

Another essential characteristic is the diversity of soils and the large contrasts between vast mountain ranges, hills and the high steppe-like plains, and less extensive but highly prized coastal and fluvial plains. Due to the mountainous nature of the relief (except in

Egypt and Libya) and the scarcity of large plains, a sharp contrast exists in local climatic conditions, which leads to great diversity in local situations.

Such diversity is no doubt a source of wealth, but environmental constraints (low availability of fertile soil and water, summer droughts and irregularities in the water regime, mountainous relief) weigh heavily on agriculture and rural dwellers. They explain the efforts that have always been made to enhance soils and control water.

Rural landscapes built over time

Farming was born in the eastern Mediterranean in the 8th millennium BC. By the 4th millennium BC, controlling flood waters, and later irrigation, made it possible to found great agrarian empires around the large rivers in Egypt and Mesopotamia (the Nile, Tigris and Euphrates). Outside these privileged regions, in mountains, hills, dry plains and plateaux, it is the transition from cultivation through 'slash and burn' to more permanent systems 'with light animal traction' that has, since antiquity, shaped the rural landscapes, a heritage that has not always survived to the present day. During this long period, Mediterranean rural land has been organized into four major, complementary components: the *hortus* (intensively cultivated home gardens for fruit, vegetables and various other household plants), the *ager* (grain fields planted in two-year crop rotation with fallow periods), the *silva* (forests providing fuelwood and grazing land), and the *saltus* (more or less wooded pasture land, where cultivation of annual crops is sometimes possible). Where small plains existed as isolated 'islands', agricultural development has contributed in a decisive manner to the development of the ancient city-states (compact towns depending on agriculture and maritime trade), and later to those of the Middle Ages. The art of irrigation progressed greatly with Arab civilization and expansion.

Agricultural progress in the north European Middle Ages, shifting to cultivation with heavy animal traction and three-year crop rotations, did not benefit the Mediterranean, but contributed significantly to the economic and political supremacy of northern Europe. Intensification in the Mediterranean was limited to the development of *terraced cultivation* on slopes, *irrigation* and *arboriculture*. The development of the unhealthy plains, which required high capital investment, was long and difficult, holding back progress in the southern and eastern Mediterranean, for example in the Mitidja near Algiers (early 20th century), the Salonika plain (around 1925), and the Ebro delta and the Pontine marshes (just prior to the Second World War). This development, together with other events that have marked recent history (wars, colonization, decolonization, collective farming, liberalization, globalization) have had profound consequences on land tenure systems, water management, production systems and migration. European colonization of the Maghreb plains led to the start of modernization in these areas, but also pushed indigenous peoples towards the steppes and mountains. These peoples did not always regain their former rights after decolonization.

Since the 1960s the development of *irrigated lands* and *specialized crops* such as olive orchards, vineyards, citrus and other fruit, vegetables and flowers, has become a major trend in all Mediterranean countries. In the developed northern countries this change has gone hand in hand with increasing *abandonment* of less productive and thus less competitive areas. In the developing southern and eastern countries with high population growth, however, stress on the marginal lands has never been so high, despite low and often shrinking yields because of desertification.

Biodiversity and rural landscapes of global value

The rural Mediterranean areas are rich in biodiversity and landscapes of high international value. The climates in alternating geologic periods, and the great geographical diversity in environmental conditions (soils and climates) have favoured an exceptional floral wealth that has adapted to the diversity of 'ecological niches'. With 25,000 species of seed-plants, the region contains 10 per cent of the world's plant species on only 1.6 per cent of the land surface. The number of endemic species (13,000) is particularly high on the islands and in the mountains. This wealth, and the stresses associated with it, makes the region a *global biodiversity hot spot*. In contrast to other hot spots, where vegetation is still mainly 'natural', the Mediterranean has been 'inhabited' for thousands of years. Areas with real natural vegetation (in line with climax conditions) account for a maximum of 50,000km^2, or 2.1 per cent of the total area of the eco-region. With the exception of these relict areas, the region's landscapes are thus a fairly remarkable example of *co-evolution* between man and nature. To be conserved, both the landscapes and the rich biodiversity of these semi-natural environments require specific pastoral, agricultural or forest management, which is relatively exceptional on the worldwide scale.

The eco-region is also one of the eight most important centres in the world from which cultivated plants originate. It is the *cradle of plants* that provide a decisive share of the world's food production (wheat, barley, rye, oats, lentils, cabbage, vineyards, alfalfa, leek, almonds, etc.). It also possesses a large number of cultivars (selected varieties of cultivated plants) and domestic animal breeds.

Rural Mediterranean landscapes, with their originally high ecological diversity, are the result of action by societies,

Rural Areas

which themselves also have a rich cultural diversity. All this diversity results in the typical natural and rural landscapes of the region. This 'unity in diversity', both natural and cultural, is a very valuable heritage. In particular the rural Mediterranean contains some of the world's most beautiful 'cultural' landscapes (for example the landscapes of Tuscany, Provence, Puglia, the Cap Bon Peninsula, Cyrenaica, the Iberian dehesas, the mountains and valleys of France's Cévennes, Morocco and Lebanon, the Syrian and Anatolian foothills, the 'sea' of olive trees of Delphi, the Greek and Croatian islands, and the oases in North Africa). These landscapes embody the region's long history and its cultural and natural diversity. Their beauty, which is a considerable *tourist and cultural asset*, and the need for their preservation, are still insufficiently perceived by communities or taken into consideration by decision-makers.

Fragile areas

Rural Mediterranean areas are fragile landscapes, vulnerable to many pressures. These are linked to climate (variability, droughts) and soils (restricted amount of fertile, easily tilled lands), but also have an economic origin.

Natural fragility originates from specific, serious *risks* that are characteristic for these environments. Risks include: *water erosion* of soils that are often shallow and occur on steep slopes; *wind erosion* in semi-arid or arid areas subject to desiccating winds; serious *droughts* even in usually well-watered areas because of the climate's large inter-annual variability; *flooding*, even in usually water-poor areas because of torrential rains; *salinization* or alkalinization of some irrigated soils when poorly drained or infiltrated by saline groundwater; and *forest fires* due to drought and wind. To adapt to these constraints, Mediterranean societies have managed to invent and develop local technological and governance solutions (terracing of slopes, small water works, procedures for collective management of limited resources, property and customary rights, specific agricultural and forestry applications such as arboriculture and floral and perfume-flower cultivation). However, these solutions are based on cheap labour and could only develop in economies that had no competition from other regions not experiencing such constraints. *Economic vulnerability* has been the result. It appeared as soon as these regions were affected by population, political, legal, and technological developments and when societies and economies opened up; all trends that characterized the 20th century. One of the most unsustainable consequences was the destabilization of many rural societies and the imbalance in ecosystems.

Figure 5.1 **The Mediterranean's main agricultural and natural systems**

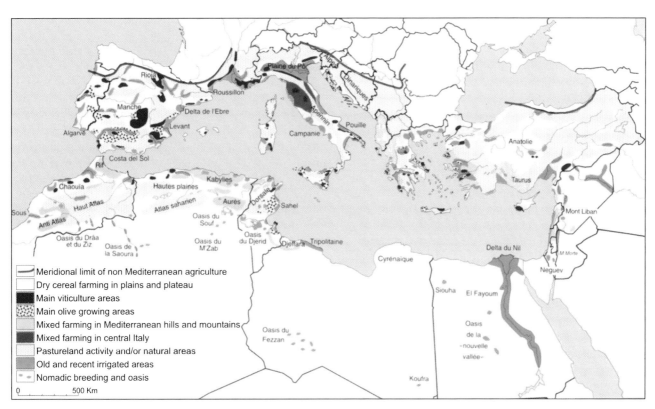

Source: Méditerranée, vol 97, no 3–4, 2001

Abandoning good cultivation practices and land maintenance, consequences of either agricultural abandonment (in the north) or overexploitation (in the south and east), often had very negative impacts on rural areas: increased erosion of non-maintained terraces, salinization due to poor irrigation practices, degradation of plant cover and soils due to over-grazing or improper tillage, loss of biological and landscape diversity, and large forest fires. However, the abandonment of agricultural practices has also had positive 'biological revivals' when very degraded areas, too marginal to sustain agriculture, are restored, thus favouring a return of forests with very extensive management.

Diversity in rural areas and in issues

The specific Mediterranean character explains the large regional diversity of Mediterranean rural areas. Figure 5.1 shows this inheritance in a schematic way. It gives a good picture of the current distribution of the main agricultural and 'natural' systems in the region. The map covers most, if not all, the rural areas of the region (as well as some countries not bordering the Mediterranean such as Portugal). Only a small part of the agricultural and rural areas of France and the north-eastern Adriatic countries are Mediterranean. To avoid too large statistical distortions, only the southern part of France will be considered as far as possible in the rest of this chapter, that is only the three regions bordering the Mediterranean: Provence-Alpes-Côte d'Azur, Languedoc-Roussillon and Corsica, together accounting for 12 per cent of France's land area.

Using farming intensity as a differentiating criterion, the major *Mediterranean agro-systems* can roughly be divided into two main groups:

1 The first group corresponds to the fertile areas: coastal and fluvial plains, hills and oases. These include large irrigation schemes, large rain-fed agricultural systems in areas with more than 400mm of rain, suburban agricultural areas and the specific oasis systems. These areas are generally highly populated and intensively farmed. The land is owned mainly by farmers and residents. The main issues in terms of sustainability are related to sustainable water management and irrigation (risks of exhaustion and pollution of water resources, risks of salinization, and insufficient value given to rain and irrigation water), and to the gradual encroachment of rural areas by urban sprawl and infrastructure expansion. This encroachment threatens the last 'natural' areas of the plains if not protected (forests and wetlands), but even farmland itself and, more generally, the rural character and quality of the surroundings of suburban areas. Apart from safeguarding the last natural environments, a sustainable management of these fertile areas will depend on the individual discipline of landowners in water management and in maintaining the agricultural aptitude of land, provided that political mobilization and appropriate policies with regulations and incentives are put in place.

2 The second group corresponds mainly to areas with permanent handicaps, *semi-arid areas* and *mountains*. Semi-arid zones are typical of the southern and eastern Mediterranean and include plains, plateaux, dry mountainous areas and steppes. Marginal farming interferes with pastoral economies that predominate in the steppe regions. The Mediterranean mountains are generally areas with an agro-sylvo-pastoral civilization. Both in the north and the south, they play an important role as 'water towers' for the entire population.

The main problems in these inland areas (mountainous and semi-arid zones) in terms of sustainability are social, economic and cultural marginalization, soil degradation (erosion, desertification), deforestation and loss of biodiversity in the south and east, and abandonment of cultivated and grazing land in the north. These problems trigger others, such as increasing 'natural' hazards (landslides, floods, forest fires), or the silting-up of downstream reservoirs, and the more general problem, relevant in all rural areas, of the population distribution in the countries (urban/rural population balance).

In the north these inland areas are mostly privately owned and are being increasingly abandoned by their owners, the land returning to its natural, vegetal state. In such situations, sustainable management of the area cannot rely on individual initiatives, since they lack both the means and the incentives for managing the territorial heritage. The solution can therefore only come from a collective approach to the problem, making the landowners accountable within the framework of long-term management schemes, while ensuring an appropriate mobilization of development funds. In the southern and eastern countries, the areas corresponding to the second group of agro-systems are still largely *used collectively*, including wooded areas. Where farming is possible, this collective use is challenged by the strong trend towards land privatization. However, ecological constraints in these vast areas offer no choice other than collective use, the only viable option. Several innovative experiences in participatory management of catchment areas, forests and natural pastureland are given below to illustrate this point.

This *diversity of territories* potential and issues call for different approaches and responses. These require a proper identification of the issues through a quantification of past, present and future socio-economic and ecological developments for each of the major agro-systems. However, despite the enormous number of studies, maps and statistics available, this is still an exercise in its infancy. The paradox comes from the fact that the immense quantity of data about ecosystems is generally not combined with socio-economic information. Each approach develops outputs while using incompatible principles with the other approach; one uses bio-geographic zones and the other administrative boundaries. Consequently, even simple questions about socio-economic issues in either of the agro-ecological zones cannot be answered. What, for example, is the value of mountains in the entire Mediterranean Basin in terms of population, farm sizes or surface? Data are often available to formulate the answers, but experience shows that, with rare exceptions, quantifying studies are not made, because the problem of incompatibility has not been solved.

Within this context, the analysis of major indicators (both retrospective and future developments towards 2025) dealing with rural and agricultural population (see the next section), degradation in rural areas (the third section), and agricultural and food perspectives (the fourth section), can only be generic. Through further analyses and examples, an attempt will be made to highlight a few major 'territory'-specific challenges (the fifth section). The focus will be mainly on the second group components: the semi-arid and mountainous areas that are the most extensive and most threatened rural areas. The 'sustainability' of suburban rural areas will also be looked at briefly, while a full chapter of this report tackles the issue of water management. A final section will highlight the main goals and conditions of an alternative scenario for sustainable rural development in the region.

Rural and agricultural population: Divergent and unsustainable developments

In 2000 the region[1] had 133 million rural dwellers and 83 million farmers. The criteria and definitions used in each Mediterranean country to distinguish 'rural' from 'urban' differ widely, for example numbers of people not included in urban areas classified as 'cities' or 'municipalities', the number of residents of municipalities, the size of gathered populations, the population density, the level of service infrastructure, the agricultural population as a percentage of the active workforce, and the population thresholds (ranging from 400 residents in Albania to 10,000 in Italy). This makes international comparisons difficult. To enable comparisons, we will use the *Géopolis* database (the 10,000 resident threshold)

or, when possible, the population density criteria. The OECD considers an area to be 'rural' when there are fewer than 150 residents per km^2; this has the advantage of illustrating the typical rural characteristic of low building density. It should be noted that the OECD statistics differ considerably from the FAO data from countries. FAO data are used here for data relating to agricultural populations.

Rural and agricultural populations have long been considered as one and the same. This historical situation changed considerably in Mediterranean Europe during the 20th century. The proportion of farmers as a percentage of the rural population is now no more than 28 per cent (the agricultural population is 12 million, the rural 43 million). In the southern and eastern countries, however, the two populations are still widely blended with a proportion of 79 per cent (71 million farmers and 88 million rural residents), although agriculture is no longer the only source of income. Rural poverty is such that it must be countered through job mobility and the search for income outside the agricultural sector, including emigration.

In the north, a collapse of agricultural manpower and the start of a 'rural revival'

The collapse of agricultural manpower

The total agricultural population in the northern Mediterranean countries (NMC) fell from 46 million in 1960 to 12 million in 2000, 74 per cent in 40 years. The drop has been particularly high in the EU-Med countries (Spain, Mediterranean France, Italy, Greece, Malta, Cyprus and Slovenia), but has also been significant in the eastern Adriatic (Figure 5.2). The percentage of farmers in the total active population fell from 35.7 per cent to only 7.4 per cent in 2000 (3.3 per cent in France as a whole and 5.3 per cent in Italy, despite the economic importance of agriculture in both countries).

Changes often occurred long ago and may be profound, depending on the country. Mediterranean France was the most and earliest-affected region (starting at the end of the 19th century), followed by Italy (since 1950), Spain, Greece, Slovenia and Croatia (since 1960) and Cyprus (since 1970). The drop has only just begun in Albania.

Despite this rapid reduction, the agricultural share of total employment in the EU-Med regions (especially those on the fringes, i.e. Greece, the Mezzogiorno, Sardinia and Andalusia) is still two to three times higher than the EU average because of the relative importance of multi-activity and the remarkably strong attachment that many Mediterranean people (especially in Greece) continue to have with their native villages. However, analysis of age groups shows a sharp *ageing* of farmers

2 A Sustainable Future for the Mediterranean

Figure 5.2 Total agricultural population in the Mediterranean countries, trends and projections to 2025

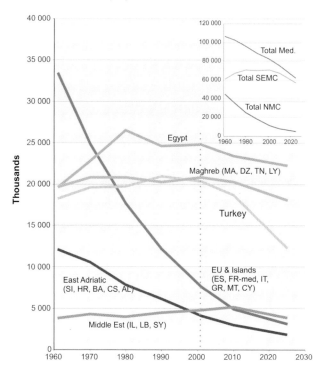

Source: FAO data, *Plan Bleu* projections

in Greece and 78 per cent in Italy (compared with only 27 per cent in France).

Such a structuring of the farming sector is largely contrary to northern Europe's actual directions (including non-Mediterranean France) which are mainly very capital intensive and large-scale. On the other hand, it is in line with options based on intensification and enhanced quality.

From rural migration to rural revival

The collapse of northern Mediterranean agricultural populations has also resulted in a fall in other rural populations, which has strengthened *urbanization and 'coastal overdevelopment'* (concentration of population and activities along the coasts). However, after a very sharp drop in the agricultural population, there has been some stabilization of rural populations (even an increase in France). This shows some success in rural economic diversification efforts.

A comparative analysis of population changes in the Languedoc-Roussillon of France, the Marche of southeast Italy and Andalusia in Spain over the past century reveals the scale of rural migration, the time lags in countries, and the relative 'rural revival' characteristic of recent times (Figures 5.3 and 5.4).

In Languedoc, the rural population began to shrink at the end of the 19th century and led to a profound devitalization of rural areas, especially in the mountains. The 'rural revival' apparent in the past 20 years bears witness to the new 'attractiveness' of rural areas in an urbanized France that is seeking a better quality of life. In the Italian region of Marche, the drop only began in 1950 but was steep (much more so than in Tuscany) and lasted 20 years. Although less numerous than a century ago, the rural population has since stabilized, despite a continuing fall in the agricultural population. The same trends have been seen in Andalusia, although with a ten-year time-lag.

On a countrywide scale, this change in the rural world is confirmed, although with differences in trends: there has been a clear growth in the rural population during the past 20 years in Mediterranean France; stabilized

(the percentage of individual farmers aged 55 and over is 38 per cent in France, 53 per cent in Spain, 56 per cent in Greece and 62 per cent in Italy). This means that the large reduction in the number of farm workers will continue over the next two decades, including Malta and Cyprus (where the number of active farmers fell from 98,000 in 1961 to 33,000 in 2000 and is shrinking by 3 per cent per year).

The agricultural population of the northern countries is characterized by high economic stratification. Farm size distribution shows a great disparity and a high percentage of *small farms* (Table 5.1). The percentage of farms of 1–5 hectares is 55 per cent in Spain, 75 per cent

Table 5.1 Farm size distribution in EU Mediterranean countries

Size (ha)	Spain		France (total)		Greece		Italy	
	% UAA	% farms	% UAA	% farms	% UAA	% farms	% UAA	% farms
1–5	6	55	1	27	31	75	20	78
5–20	14	28	6	22	23	22	25	16
20–50	14	9	21	24	38	3	22	4
50–100	14	4	32	17	8	1	12	1
> 100	52	44	40	10	< 1	< 1	21	1

Note: UAA: Usable Agricultural Area.

Rural Areas

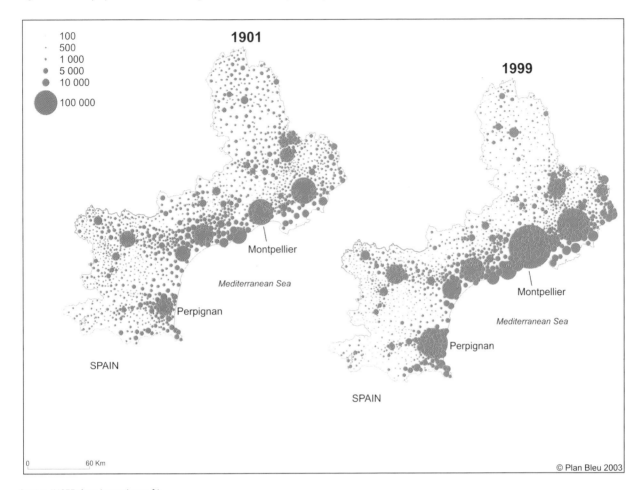

Figure 5.3 The population of the Languedoc-Roussillon, France, in 1901 and in 1999

Source: INSEE data (www.insee.fr)

populations in Italy and Greece for 30 years and in Cyprus for 10 years. Spain continues to lose its rural population (a decrease of 2.3 million people since 1970) and this has only just begun to show signs of stabilization. Changes are equally striking in the eastern Adriatic, with a sharp decrease in Croatia (2.1 million now compared with 2.6 million 20 years ago) and in Bosnia-Herzegovina (2.1 million compared with 2.9 million). A fall is beginning in Albania and Serbia-Montenegro, but the rural population in Slovenia remains the same.

These overall figures hide important internal differences: the rural population has again become relatively dense in rural areas near cities and on agricultural plains (increase in residential functions), whereas the density has remained low or very low further inland, particularly in remote mountain areas.

Nevertheless, in the more developed countries, the population and economic revival is affecting even the remotest rural areas.

Towards a new social structure in rural areas

Farmers in northern developed countries are only a small and ever decreasing proportion of the rural population. Their farms continue to grow, and farmers are less and less concerned about the remaining, more marginal land that has been abandoned (left fallow or returned to woodland). They are often becoming '*multifunctional entrepreneurs*', with varying parts of their income derived from activities other than farming. Income supplements come mainly from food processing and agro-tourism. In Cyprus, for example, only 25 per cent of farmers make their living exclusively from agricultural work.

The activities of non-agricultural rural residents are becoming more and more diversified. With the collapse of the agricultural labour force, *labourers* have often become the largest professional category (important in the building sector, increasing in the food-processing industry, and decreasing in traditional activities such as textiles, leather and furniture). Moreover, there is a large

Figure 5.4 Urban, rural and 'remote' rural populations in three EU Mediterranean regions

Population of Languedoc-Roussillon (France) municipalities since 1881

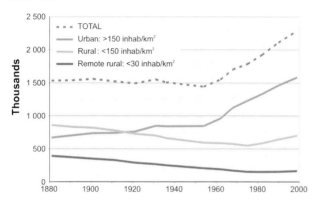

Source: INSEE

Population of Marche (Italy) municipalities since 1881

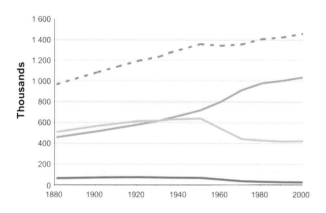

Source: ISTAT

Population of Andalusia (Spain) municipalities since 1900

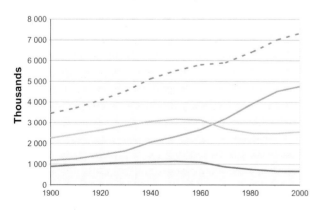

Source: INE

number of *retired households, holiday homes*, and, in urban-dominated rural areas, many middle-class young couples with children.

In recent years there has been a remarkable growth in 'executive and upper intellectual professions', especially in France, where this group represented 20 per cent of the rural working population in 1990, and is still growing. *Tourist facilities in rural environments* are also increasing: 415,000 B&B rooms and furnished tourist apartments in 1988 in Mediterranean France (with a spectacular increase of 631 per cent in eight years) and 342,000 camp sites, including small ones in nature areas, compared with only 146,000 hotel rooms.[2] Tourist stays are less and less focused on 'topical' activities (the main ones being hiking and visits to cultural sites). What matters is the living environment, reception on a human scale, friendliness, a variety of activities and the quality of products.

The intense relationships established with cities and the convergence between rural and urban lifestyles have not prevented the continuation of marked differences. *Differences between urban and rural dwellers* remain steady and even increasing in the field of public services: reduction in public transport, closing of primary schools, cinemas, small businesses, local bank branches, post offices, etc. Although household incomes remain lower than in cities, rural dwellers benefit from lower unemployment rates and cheaper housing and food (a significant part of expendable income). In suburban areas, households demonstrate their difference from towns by willingly buying directly from the farm.

Hunting remains a very important social and cultural activity, but often sets 'traditional' rural dwellers up against 'neo-rural' and town dwellers. Friction also exists between hunters, a growing percentage of whom no longer have their primary residence where they hunt, and farmers and foresters who are victims of the increasing damage caused by larger fauna. Also general opinions about managing forests and nature differ.

The fragmentation of the rural population in the north into categories with different concerns (of a professional, environmental, aesthetic and cultural nature) creates specific problems when formulating local development strategies, which should be based on a common, shared interest and vision for the future. However, the new social diversity is also an opportunity for innovation and convergence between the many actors now involved in rural areas. The dynamics of associations, the emergence of new generations of elected officials representing this diversity, and innovative businesses bear witness to an undeniable renewal in many rural regions, even remote ones. Added to that are the effects of a certain cultural renaissance that tends to create regional identities, sometimes exclusive ones, but fortunately, often being open and creating attractive environments for tourism and local activities.

Rural Areas

A sharp decrease in agricultural employment by 2025, with the rural population remaining stable

In the baseline scenario, the current trend of a large decrease in the agricultural labour force continues. For the four EU-Mediterranean countries[3] this is projected to decrease from 3.6 million in 2000 to 2.25 million by 2010 and 1.43 million by 2025. The percentage of the working farm population in 2025 is projected to be less than 1.5 per cent in France and 2 per cent in Italy. Increasing farm sizes and abandoned farmland will have many negative consequences in terms of sustainability. The geographically increasingly concentrated farming systems will be decreasingly able to maintain the quality of landscapes. The decrease in the agricultural labour force will reduce professional community solidarity and the ability of the agricultural world to resist urban sprawl, especially on the coastal plains. Mountain herders, even less numerous, will have to fall back to difficult, risky techniques such as large-scale pastureland fires. To compensate, society will increasingly have to assume land management responsibilities, using public funds (for river, footpath and hedgerow maintenance, forest management, fire-fighting and the regulation of large fauna).

On the other hand, the baseline scenario confirms a stabilization of the rural population and economic diversification (Figure 5.5).

Despite the drop in the agricultural population, the total rural population of the four EU-Mediterranean countries is projected to remain fairly stable (Box 5.1). An increase is not excluded if the French example (with a new attractiveness of rural areas) were to be widely followed by other Northern countries. The impact of population changes will differ between areas: positive in intermediate and 'remote' areas, which will succeed in their economic and cultural revival; negative in other areas (particularly in some remote mountain areas), which will see little in the way of a revival, but rather a stronger decline; and negative for the suburban rural areas, where encroachment of farmland and loss of 'rural character' will increase sharply.

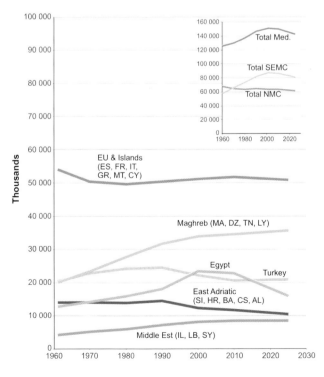

Figure 5.5 The rural population in Mediterranean countries: trends and projections to 2025[4]

Source: Attané and Courbage, Plan Bleu 2001 (see Statistical Annex)

Box 5.1 Four possible scenarios for the rural areas in the developed Mediterranean countries

With the constant decrease in the number of farmers and the ever increasing demand for countryside, the future of rural areas in developed countries is called into question everywhere. A recent prospective exercise for France, but widely applicable in other northern countries, describes four possible scenarios to 2020. The trend scenario is a liberal one with a reduction of EU Common Agricultural Policy (CAP) funding and farming, and an evolution towards a *'generalized residential countryside'*. It meets the short-term demands of the middle classes, but is costly in environmental terms (an all-car scenario with increasing areas of built-up and other artificial land). It also comes with growing social inequality (dreams of a life in the countryside remain inaccessible to the poorer classes) and social risks (growing urban insecurity). The second scenario, *'sustainable cities and rural agribusiness'*, is one of sudden change. It would see residential development being jeopardized by restrictive laws, and a conversion towards eco-friendly farming and agricultural policies, which remain or become strong again. Governments, big cities and a 're-shaped' farming profession would be the driving actors. Agriculture would integrate environmental concerns, combining scientific progress and the 'green lung' function for major urban areas. Served by public transport, edge cities would become denser. The third scenario, *'sustainable cities and nature marketing'*, is a variation on the second, where government intervention would recede, farming would not succeed in its environmental transition, and environmental markets would become more widespread. The fourth scenario, *'industrious and competitive countryside'*, shows a rural life with local initiatives, entrepreneurs and lively territories, while urban areas are in crisis. This scenario would see weakened central power (reduced state influence) and strong local powers. A collective economic identity would grow around rural centres, small towns and rural areas thanks to active and diversified entrepreneurship. The risks of this dynamic scenario are the creation of inequalities between rural regions, and the undermining of salaried middle classes, which have until now been relatively protected.

Source: DATAR, 2003

In the south and east, large and vulnerable populations: Agricultural duality, poverty and rural migration

Very large agricultural populations and major internal disparities

In the southern and eastern Mediterranean countries (SEMC), agricultural populations have not fallen as in the north. On the contrary, they have and continue to increase in absolute terms. The total agricultural labour force in the SEMC was 32.4 million people in 2000 (compared with 25.4 million in 1960), while the total agricultural population reached 71 million (compared with 61 million in 1960).

Changes, however, vary between countries: both the agricultural labour force and the total agricultural population in Syria, Egypt and Algeria are growing. The agricultural labour force in Turkey, Morocco and Tunisia is still growing, while the overall agricultural population has now stabilized. Lebanon and Libya have recorded a spectacular drop. Israel has recorded a smaller decline.

Despite this increase in absolute terms, the *relative* percentage of the agricultural labour force has decreased considerably, because of strong urban growth. In 2001 it nevertheless remained very high in Turkey (46 per cent), Morocco (36 per cent) and Egypt (33 per cent), high in Syria (28 per cent), Tunisia (24 per cent) and Algeria (24 per cent), but was very low in Israel (2.7 per cent) and Lebanon (3.7 per cent).[5]

These overall figures hide extremely large internal differences as shown by the data on farm structures (Table 5.2).

The developing countries are characterized by an *agricultural duality* where modern, competitive farming co-exists with subsistence farming, and duality still tends to increase. Land and capital concentration on one hand, and the fragmentation of small-holder farms on the other, can be observed. The large farms (>20ha), which represent less than 2 per cent of farms in Lebanon, 5 per cent in Turkey and 12 per cent in Tunisia, contain 20 per cent, 35 per cent and 61 per cent, respectively, of the usable agricultural area. At the other extreme, small-holders with 0–5ha are in the great majority (97 per cent of all farms in Lebanon, 68 per cent in Turkey and 53 per cent in Tunisia), but possess only a small part of the usable agricultural area (55 per cent, 24 per cent and 9 per cent respectively).

In a context of natural resource scarcity (land, water) and economic insecurity, *agrarian structures* have been disrupted, suffering from increasing fragmentation. By comparing census data for 1962 and 1995 in Tunisia, it is apparent that the number of small farms (less than 5ha) increased in a spectacular way (89 per cent) during that period. In Morocco the total number of farms remained stable between 1974 and 1996 (1.43 million); the number of large farms increased slightly (110,000 in 1996 compared with 102,000 in 1974); the number of pastoral farms dropped sharply through a settling process (from 307,000 to 65,000 between 1974 and 1996). A large percentage of farms are still subsistence farms. Particularly poor farmers with land under the viability threshold (less than 3ha for drylands or 1ha for irrigated land) have no survival strategy other than to seek other income through temporary migratory work. Farms of less than 3ha (41 per cent of farms) cover only 5 per cent of total irrigated land.

The 'duality' of agrarian structures and territories has been reinforced by the lack of equitable land policies and excessive polarization of public investments. Concentration in landownership and capital is occurring everywhere and tends to be amplified, including in the steppe regions with large-scale cattle raising. Although irrigated areas have become islands of relative prosperity (agricultural intensification, collective equipment), rain-fed farming areas are suffering from marginalization, which also has a cumulative effect (lack of economic dynamics, lack of infrastructures). An aggravation of social and regional inequalities can thus be observed.

Table 5.2 Agrarian structures in selected southern and eastern Mediterranean countries

Size (ha)	Turkey %UAA	Turkey % farms	Tunisia %UAA	Tunisia %farms	Lebanon % UAA	Lebanon % farms
0–5	24	68	9	53	55	97
5–10	20	18	15	20	14	5
10–20	21	9	15	15	11	<1
20–50	20	4	24	9	11	<1
>50	15	1	37	3	9	<1

Source: Plan Bleu from CIHEAM; Ministère de l'agriculture du Liban; FAO

Rural Areas

The scale of rural poverty in the SEMC

In the SEMC, the *rural population* is still mostly identified with the agricultural population (for example 78 per cent of the rural households in Morocco[6]). Despite much rural migration, in Turkey and Morocco in particular, the rural population has increased in absolute terms. This increase was much less than that of the urban population, despite fertility rates that are significantly higher in the country than in towns, thus confirming the scale of rural migration. However, there are big differences between countries. Egypt, Syria, the Palestinian Territories and Algeria have recorded a high and continuing growth in rural population, with a slower growth in Israel and Libya. On the other hand, after very large increases over the past few decades, the rural population in Turkey has begun a sharp decline in the past decade (being a country where the scale of rural migration and its related impacts are worrying, especially the extremely fast growth of the Istanbul urban area). In Tunisia, Morocco and Lebanon the rural population has stabilized.

Egypt seems to be a special case, since rural migration seems almost to have stopped. This could be explained by several factors: farming is all irrigated (thus suffering less from drought, while yields are much higher), the 'usable' national territory is densely populated, has a good transport system and operates like a vast 'conurbation', and urban poverty is now as severe as rural poverty.

The major issue in most of the developing Mediterranean countries (and in some eastern Adriatic countries) is *rural poverty*, which manifests itself in monetary poverty, educational poverty (in particular illiteracy), and basic services poverty (Box 5.2). Almost all indicators show a very great inequality with urban dwellers, even though the rise of urban poverty is tempering this situation.

Box 5.2 Rural poverty in the Mediterranean developing countries, the magnitude of the disparities with urban dwellers

- Percentage of the population living under the poverty line: 30 per cent of rural Algerians in 1995 against 15 per cent of urban dwellers, 27 per cent against 12 per cent in Morocco in 1998–1999, 13.9 per cent against 3.6 per cent in Tunisia in 1995, 29.6 per cent against 20.1 per cent in Albania in 2002, 19.9 per cent against 13.8 per cent in Bosnia-Herzegovina in 2001–2002.
- Access to drinking water in 2000: 56 per cent of rural dwellers in Morocco against 98 per cent of urban dwellers, 58 per cent against 92 per cent in Tunisia, 82 per cent against 94 per cent in Algeria.
- Access to sewage in 2000: 42 per cent of rural dwellers in Morocco against 100 per cent of city dwellers, 47 per cent against 90 per cent in Algeria, 81 per cent against 98 per cent in Syria, 70 per cent against 98 per cent in Turkey, and 87 per cent against 100 per cent in Lebanon.
- Access to electricity: 26 per cent of rural dwellers in Morocco in 2000 against 91 per cent of urban dwellers, 77 per cent against 98 per cent in the Middle East.
- Access to health services (1990–1996): 50 per cent of rural Moroccans against 100 per cent of urban dwellers, 84 per cent against 96 per cent in Syria, 85 per cent against 98 per cent in Lebanon, and 95 per cent against 100 per cent in Algeria.
- Schooling rate in primary education: 77 per cent of rural children in Morocco against 93 per cent in towns.
- Schooling rate in secondary education: 8 per cent in rural Morocco against 49 per cent in towns, and 50 per cent against 78 per cent in Turkey. In Albania, only 3 out of 10 children attend secondary school in rural areas, compared to 7 in Tirana.
- Rate of adult illiteracy: 81 per cent of farmers and 79 per cent of rural Moroccans in 1994 against 41 per cent of city dwellers, 71 per cent in rural Algeria against 43 per cent in 1987, 60 per cent against 32 per cent in Tunisia in 1989, 17 per cent against 12 per cent in Lebanon, 63 per cent of illiterate women and 36 per cent of illiterate men in rural Egypt against 34 per cent and 20 per cent respectively in towns.

Source: UNSD Millennium Indicators; Statistics Global Report on Human Settlements 2001; World Resources 1998–1999; World Development Indicators 2004; UNESCO yearbooks; national statistics institutes, UNDP

Such gaps are perhaps demonstrative of a lack of consideration for rural populations by the dominant urban elites. Important efforts have recently been started, however, to better equip countrysides in countries where such deficits have accumulated. A greater effort is underway towards equity between rural and urban areas, and a better understanding of challenges for the national economies. For example in Morocco, a plan for grouped supply of drinking water was launched for the rural population in 1995 (PAGER), funded by many donors, charging higher water prices to consumers. This plan aims at filling a major gap in access to drinking water: from 14 per cent access in 1990 to 97 per cent by 2007. In Tunisia, in 2003, the access rate to drinking water was 86 per cent in rural areas (against 100 per cent in urban areas).[7]

The continuing degradation of natural resources (desertification) and the rise in the number of droughts over the past decade have increased poverty and affect most national economies in the region. The fluctuations in wheat yields since 1990 show the magnitude of the reduction, excluding Egypt (where farming is less vulnerable because it is entirely irrigated).

Lost crops mean lost income. Moreover, in subsistence farming, crop loss can force farmers to borrow in order to make ends meet. This leads to household debts and an increased vulnerability of smallholders, who are sometimes even forced to sell their land. These losses weigh very heavily on all national economies, as shown by the comparative development of agricultural and overall GDP in Morocco between 1992 and 1999 (Figure 5.6).

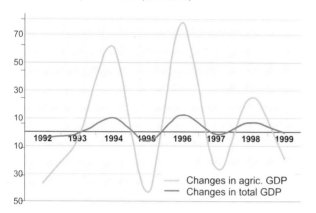

Figure 5.6 Drought-related fluctuations in agricultural and total GDP in Morocco, 1992–1999 (annual %)

Source: Direction de la statistique, 2000 in Reiffers, 2000

Increasing poverty through free trade?

The liberalization of agricultural trade, under negotiation between countries at various levels of development, could lead to an increase in rural poverty. Liberalization has been on the agenda since the Marrakech Uruguay Round farm agreements (WTO in 1994), the implementation that same year of the North-American Free Trade Agreement (NAFTA), expanded to include agriculture, and the signing of agreements between the EU and countries such as Mexico and Chile, and between the US and Jordan or Morocco. Although agriculture is still not included in the Euro-Mediterranean agreements, negotiations are underway for more effective free trade with the southern and eastern partners. Free trade, such as that between the US and Mexico, could open currently protected cereal-crop farming and cattle-raising to competition. These systems have very different land tenure structures and set-up conditions in terms of natural, human, technological and institutional capital.

In several developing Mediterranean countries, import custom rates on cereals in 2003 averaged between 10 and 77 per cent.[8] Such protective measures are in addition to internal support for the consumption of these products, a strategic measure to maintain social stability. High grain prices provide an annual income for middle-class farmers on good agricultural lands and affect countries' manufacturing competitiveness, but also enable the economic survival of millions of poor farmers. These protective measures are, however, not comparable with those of the US or the EU, which provide average subsidies per producer per year of 21,000 and 17,000 euros, respectively (average 2000–2002, OECD).

Developments in Mexico after NAFTA came into force in 1994, and observed by the Commission for Environmental Cooperation (a tripartite commission comprising the US, Canada and Mexico), confirm the potential risks in terms of sustainability for fragile rural areas in developing countries entering north–south free trade zones and competing in essential agricultural production (cereals and livestock) (Box 5.3). The loss of income of small Mexican corn producers, due to lower prices after the gradual de-protection of tariffs, has led to growing demands for re-examining the agreement or coupling free trade with structural funds. National policies (direct aid to farmers and rural development) were then implemented to support the necessary transition.

Box 5.3 The impact of NAFTA on Mexican agriculture and the environment

Impacts include:

- a strong growth in trade, including agricultural trade. The export of Canadian tomatoes to the US has increased by 3000 per cent, and the sales of American beef and veal to Mexico by 400 per cent, followed by grains, animal fodder and vegetables;
- vertical integration of the agricultural sector is especially rapid, particularly in Mexico where it is stimulated by American investment;
- the gradual adoption of modern agricultural production methods and the resulting specialization, even on small farms, has lead to a simplification and mechanization of production methods. Local resources (manpower, traditional seed, subsistence farming, natural pesticides and fertilizers, and locally produced fodder) are gradually substituted by commercial products (machines, hybrid cultivars, chemical products, and industrial animal feed) that are in general more harmful to the environment;
- a growing dependence on genetically modified crops, which is considered a risk to the long-term stability of agricultural production, as well as a threat of epidemics caused by plant diseases or parasites;
- agricultural production is less and less diversified, removing the stabilizing effects of such diversity (increased dependence of farmers on social assistance programmes);
- a drop in income in absolute terms (GDP per capita) in the poor southern regions of Mexico since 1994. The growth of rural poverty is considered a major cause of environmental degradation in general, of land degradation and loss of

biodiversity (both habitats and species) in particular. An increased loss of biodiversity will be all the more worrying since it concerns a region containing 10 per cent of the world's biodiversity, and which is losing 600,000ha of forest land per year;
- the macroeconomic effects of production concentration, market integration and specialization are otherwise hard to measure. Some think that the predicted benefits have not necessarily materialized, since the savings realized through lower prices are benefiting the food processors and distributors and not the consumers.

Source: Commission for Environmental Cooperation of North America (CEC), 2002

Recent *Plan Bleu* studies confirm that a drop in cereal prices not accompanied by compensation for farmers would lead to a vicious circle of *impoverishment* for subsistence and semi-subsistence farms (the majority) without alternatives (Box 5.4).

Box 5.4 The impact of free trade on rain-fed agriculture in Morocco

A study carried out by *Plan Bleu* shows that an immediate alignment of Morocco's cereal prices with world prices would probably result in:

- a 20 per cent drop in usable agricultural area for cereals;
- a drop in the area actually used for soft wheat, durum wheat and sweet corn;
- an actual increase in the area for growing barley.

Subsistence and semi-subsistence farms of less than 5ha (67 per cent of the total number of farms) would be the most threatened, given the area devoted to soft wheat (33 per cent of the UAA, low margin per quintal and their dependence on the market).

Source: Jorio in PNUE-PAM-*Plan Bleu*, 2002

It is therefore legitimate to fear that *non-regulated* liberalization of international trade in agricultural products may have very negative effects on fragile rural regions in some southern and eastern Mediterranean countries, in particular rain-fed farming countries that have a large population of poor farmers, so in fact on the whole region. The risks of such a scenario (increasing poverty and instability) call for consideration of an alternative scenario.

Given the magnitude of the social and food security challenges, and the risks of instability in several rural southern and eastern countries, it is likely that their shift to agricultural free trade will be gradual or at least partial. The *transitional forms set out by the* WTO make this possible: by accepting the principle of taking aspects other than the commercial agricultural ones into account, and by a special, differentiated treatment of developing countries through a more gradual abolition of protective tariffs and internal support. Such measures might serve as a basis for vulnerable countries (even if they are not WTO members) to negotiate their transition to agricultural free trade, knowing that no obligation other than a contractual one would be required of sovereign countries. A concerted Euro-Mediterranean approach, founded on a mutual recognition of the legitimate interests of the partners and the common interest of the region, would be an important progress in bringing together the viewpoints of rich and developing countries, and influence negotiations where, until now, the Mediterranean appears to be divided.

Between now and 2025: A stabilized rural population and the start of a reduction in the agricultural labour force

FAO projections point to a still limited reduction in the total *agricultural population* in the southern and eastern countries (from 71 to 67.5 million by 2010) despite an increase in the agricultural labour force (from 32.4 to 34.2 million). By 2025, the Blue Plan baseline scenario assumes a sharper decrease in both the agricultural population and the labour force, which may reach 57.1 and 28.3 million, respectively, figures that still remain pretty high. Changes will vary according to the countries, in some of which (Algeria, Syria, etc.) total agricultural population is still projected to grow in absolute terms until 2010 or 2015, whereas the decline observed in Turkey is expected to be confirmed (and should speed up with the scheduled integration into the EU).

Despite the moderate projected decrease in agriculture population, all various available projections predict a rather stable, even increasing rural population in absolute terms. Egypt is again an exception, where the projected decrease in the rural population will not result from rural migration but from a growth of rural villages into 'towns' with over 10,000 residents, while still maintaining a not particularly urban character.

The baseline scenario in the developing countries is therefore one of *large stable rural 'agricultural' populations* (and therefore stress on natural resources), in particular up to 2015. Unless there are major improvements in agriculture and above all in economic diversification, rural poverty (which undermines the development of national economies and results in migration to cities and emigration) is projected to continue despite current efforts for better equipping the countryside.

Land dynamics and the extent of environmental degradation

The region is characterized by the speed at which changes in land use are occurring and by the scale of the

A Sustainable Future for the Mediterranean

observed environmental degradation. Some forms of degradation are common to both shores, others are more sub-region-specific depending on differences in socio-economic, population and environmental conditions.

Cultivated lands, pastureland and forest: Pressures in the south and east, abandonment in the north

Cultivated land, pastureland and forests occupy similar land cover areas in the eco-region. Their relative percentages are, however, unequal, depending on the country (Figure 5.7).

Arable land (wheat, etc.) and permanent crops (vines and fruit trees) cover limited areas. They represent only 102 million ha (41 million ha in the NMC, or 32 per cent of the total land area, and 61 million ha in the SEMC, or 9 per cent).

Pasture land, on the other hand, is quite extensive with some 118 million ha of permanent meadows and pasture lands (27 million in the NMC and 91 million in the SEMC).

Forests occupy 91 million ha, including some 34.2 million ha of 'other woodlands', that is mainly areas covered with low-grade wood and shrubland (including *macchia* and *garrigues*). They are very unevenly distributed in the eco-region with 42 per cent forest cover in the NMC (from Spain to Turkey) compared with only 4.7 per cent in the SEMC (7 per cent in Lebanon and 1.6 per cent in Algeria).

Cultivated land: The importance of permanent crops

Most arable land is used for *grain crops* (43.5 million ha, including 28.6 million ha of durum wheat). *Permanent crops* (18 million ha) are increasing rapidly and include 7.7 million ha of olive trees and 3.6 million ha of vineyards. They are major components of the Mediterranean landscape.

With population growth in the south and east, the amount of arable land per capita (Figure 5.8) has become extremely low in several countries (less than 0.06ha/cap in Israel, Lebanon, Egypt, Malta, and the Palestinian Territories). The fall is rapid in all SEMC. Also in the NMC, arable land per capita is decreasing due to abandonment and urban sprawl (0.14ha in 2000 in Italy compared with 0.25ha in 1961).

In the *south and east*, the total amount of cultivated land (arable land plus permanent crops) increased from 55 million ha in 1960 to 61.2 million in 2000. Land hunger has led to a strong growth in cultivated land, in particular in the five North African countries. The increase over the past 40 years is considerable: Morocco (40 per cent), Egypt (28 per cent), Tunisia (18 per cent), Algeria (16 per cent) and Libya (9 per cent). The crop rotation rhythm has diminished over the past few years since the possibilities of bringing new land into cultivation are finite. However, new farmland includes pastoral areas, forests and deserts. This new land is of much poorer quality than arable land lost to

Figure 5.7 The relative percentage of forest, cultivated land and pasturelands

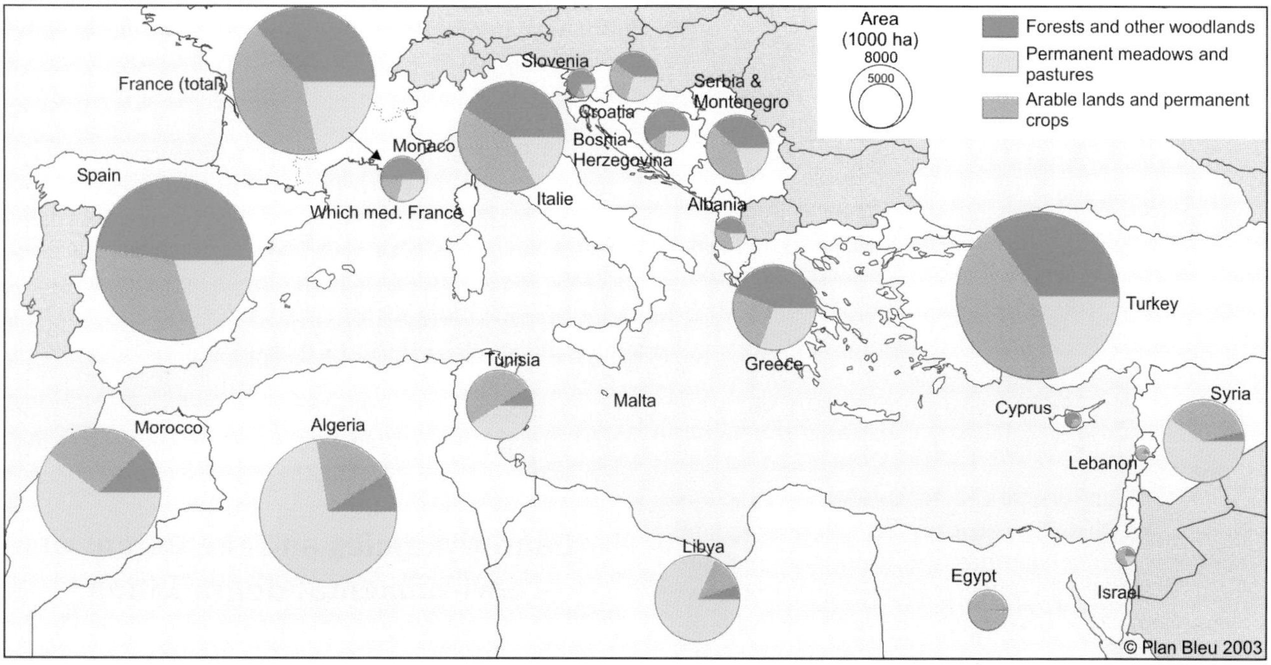

Source: FAO

Rural Areas

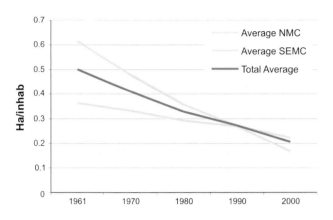

Figure 5.8 **Arable land per capita**

Source: World Bank

urbanization. In several southern and eastern countries, the pressure will remain very high at least up to 2015, which will lead to increased degradation of land resources (deforestation and desertification).

In the *North*, on the other hand, the area under permanent crops in the four EU-Mediterranean countries is stable or slightly increasing, while arable land is continuing to drop sharply (from 27.8 million ha in 1990 to 24.6 in 2000). The decrease is particularly sharp in Italy (38 per cent in 40 years). It is not insignificant in Spain either, where it has accelerated over the past few years (13 per cent in ten years), and in Mediterranean France. It has also fallen in Bosnia-Herzegovina, although for other reasons. Some countries have recorded significant drops after a continuous rise: Turkey (an increase of 13 per cent from 1960 to 1980 followed by a decrease by 6.3 per cent from 1980 to 2000) and Greece (an increase by 5.5 per cent from 1960 to 1980 and a decrease by 2.5 per cent from 1980 to 2000).

Pastureland, a major component of the Mediterranean landscape

Pastureland corresponds to the old *saltus*, and is still an important component of the Mediterranean eco-region, whereas it has almost completely vanished from the agricultural regions of northern Europe. Vast grazing lands (plateaux and steppes) are found mostly in Maghreb: Algeria with 31.5 million ha of 'permanent meadows and pastures', Morocco with 21 million, Libya with 13.3 million, Turkey with 12.4 million, Spain with 11.5 million and Syria with 8.4 million.

The natural grazing lands in the NMC were largely vanished from the rural landscape when the extensive, migratory herding of the old days disappeared. Considerable reductions in grazing land (2 million ha) also occurred between 1980 and 2000. A main consequence is that the abandoned land is gained by dense shrubs, which, in the absence of repeated fires, may evolve towards forests. In areas where extensive herding has continued, population density becomes too low for the herders to be able to properly manage the lands as they used to. Although there is a lively debate on this issue, the extensive herders of Corsica and the Pyrenees have, for example, been accused of causing many forest fires, since burning pastures is a necessary but hard-to-control practice in these conditions.

Grazing land in the SEMC on the other hand, has maintained its importance as a basic resource with

Figure 5.9 **Wood cover in the Mediterranean region**

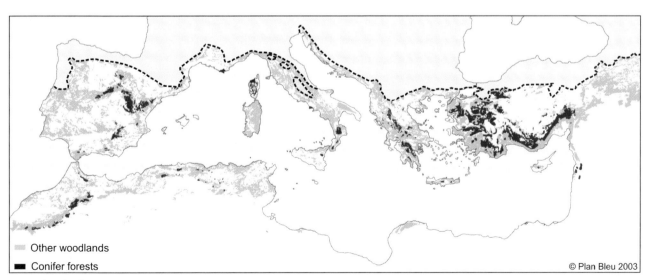

Source: *Plan Bleu*, from UNEP-WCMC (www.unep-wcmc.org)

A Sustainable Future for the Mediterranean

considerable and *extensive grazing by sheep and goats*. This component of the rural areas is, like the others, subject to population pressures, manifested in the overexploitation of grassland resources by livestock densities that are too high (in the Maghreb countries the number of sheep and goats is estimated at 50 million). Traditional practices for managing pastoral lands that ensured the maintenance of main ecological balances are no longer applied. Social frameworks have mostly disintegrated and the collective grazing lands tend to be increasingly privatized. Some experiences (see the section entitled 'Territorial challenges in the Mediterranean', p282) show that approaches exist that may lead to the restoration and sustainable management of these resources.

Woodlands: Degraded in the south, risks of forest fires in the north

In the southern and eastern countries: are woodlands increasing or declining?

In the southern and eastern Mediterranean, a first look at data on woodland areas shows that, with the exception of Lebanon, the forested area is stabilizing in Morocco, Tunisia, Syria and Turkey (and even increasing in Algeria, by 1.3 per cent per year) after a long period of decline (Figure 5.9, Box 5.5). This interpretation, however, requires some caution, because 'forest land' covers widely varying situations. Stabilization or increases in North Africa and the Near East result mostly from *reforestation*. This also explains the large annual increases recorded over the past few years in forest-poor countries such as Israel (4.9 per cent), Libya (1.4 per cent) and Egypt (3.3 per cent). Therefore, the need to be cautious with these figures, since forest statistics include replanted areas, even if the survival rate of planted trees is very low.

For *natural forests*, in contrast, the picture is much less encouraging. In general, the naturally wooded area is declining, while the composition of the wooded land is degrading. Many forest areas with tall trees are changing into shrublands such as *macchia* or *garrigue*, or habitats are being fragmented by increasingly large clearings, which nevertheless continue to be counted in the statistical category of 'forest land' (Box 5.6).

Box 5.5 Changes in forest land and irrigated areas in Tunisia over a century

The graph below shows a two-stage change in Tunisia. The decrease in forest area, especially significant between 1900 and 1970 (more than 500,000ha was cleared) is explained mainly by strong population growth not combined with sufficient economic diversification, and by mechanization of the best lands in the plains. This led to marginalization of a large part of the farming community, forcing many to fall back on clearing and exploiting marginal lands. Since 1970 the change (which started as early as the 1950s) has been characterized mostly by agricultural intensification through irrigation. Pressures on marginal lands remain very high, however, and the loss of natural forest is partly compensated by considerable reforestation.

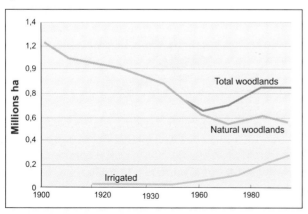

Source: Blue Plan from Auclair, 1997

Box 5.6 Shrub encroachment

Shrublands are areas composed of low-grade woody plants, often very dense (including *macchia* in acid soil, *garrigue* in calcareous soils and other types of shrubs). Shrub encroachment is a process in which taller trees gradually disappear while shrubs increase. Once degradation is well advanced or when the natural conditions are very hard, the shrub stage is often overtaken or even skipped, turning into steppe land, which can lead to bare soil or to the formation of a mosaic of grass or brush tufts separated by bare soil, a condition that is hard to reverse. Shrublands originate from old overexploited forests because of rural poverty (practices such as the cultivation of marginal lands that are susceptible to erosion, over-grazing by goats, and overexploitation for fuel wood) or from degraded forest because of (too) frequent forest fires; they can also occur on former agricultural or pastoral land that has been abandoned after rural migration (or simply abandoned) and naturally recovered by vegetation. A large percentage (34.5 million ha) of the 'other wooded land' class (FAO classification) in the region corresponds to shrubland, which is found all around the Mediterranean. These areas are therefore an expression of a particular part of Mediterranean history: forest degradation and/or the regression of agricultural and pastoral land.

There is thus a discrepancy between the impression (based on statistics) that the forest situation is generally improving in the SEMC, and the sometimes very alarming findings of many specialists who estimate that forest decline could reach 2–4 per cent per year in a number of these countries. Given the 'land hunger', which is projected to continue in the south (and to a lesser degree in the east) until at least 2015, it is likely that the pressure on forest land will remain very strong and result in *continuous* and sometimes irreversible *degradation*, unless appropriate policies are developed.

In the North, increasing wooded areas and risks of forest fires

In the NMC, recent changes have seen a large or very large increase in forest cover: 0.9 per cent per year in Greece, 0.6 per cent per year in Spain, more than 10 per cent in 12 years in most of Mediterranean France (sometimes even 28 per cent), and 3.7 per cent per year in Cyprus between 1990 and 2000. The exceptions are Albania with a 0.8 per cent decrease per year between 1990 and 2000 and Serbia-Montenegro (with a 0.1 per cent decrease). Here abandonment of farmland, as seen in the other countries, is still a very recent process.

This rapid increase in forest cover has partly been the result of continuous reforestation policies, very often carried out or subsidized by the state to re-occupy abandoned private land with marketable woody vegetation (mainly Austrian pine). However, a large part of the re-vegetation has occurred spontaneously, when *dense shrubs* and *macchia* – with little commercial value – encroach on abandoned agricultural land.

Given the projected fall in the number of farmers between now and 2025, and the change in farming practices, this very rapid growth of forests and shrublands is expected to continue at a similar rate. This 'repossession' of rural areas by a wooded biomass with positive economic and ecological effects is contrasted by negative effects such as the formation of 'closed' landscapes or under-forest layers with low differentiation. But above all, the resulting *increased risks of major forest fires* have and will continue to have the most important ecological, economic and human impacts at both social and political levels. The gravity of these risks is confirmed by recent data showing a rise in the number of fires. Between 1990 and 2000, 40,000 fires were recorded per year on average (including over 17,500 fires in Spain and 11,000 in Italy) compared with 31,000 between 1981 and 1988, or 30 per cent more in ten years. On the other hand, although the number of fires has increased, the burned surface (about 400,000ha per year), which had continued to rise for several decades, is on the decline. This may be due to improved fire-fighting efficiency. Despite this progress, the baseline scenario, which assumes a decrease in cultivation and grazing, and also global warming, shows a growing risk of forest fires, including major, very hard-to-contain fires. The disastrous fires in the summer of 2003 in France (in the Var) and Portugal remind us of the likelihood of such a scenario. Searches for a shift towards a lower-risk scenario would imply a certain reorientation of forest management by, where possible, supporting grass land, sylvo-pastoralism and choosing less inflammable tree species (like Iberian *dehesas*). A forest with some grazing and holm oak or cork oak, and well-managed plantations of almond trees, olive trees or vineyards within peri-urban forests, is less inflammable than a pine forest or shrubland. Experiments in the frame of contracts with local authorities and livestock farmers are being carried out in several countries.

Another consequence of reforestation and increasing shrubland areas in the north is increasing water consumption (more competition between forests, cities and irrigated agriculture), and the very rapid and uncontrolled growth in the number of *large ungulates* (wild boar, deer). The latter are a source of increased pressure on some rare plant species (bulbous) and cause increasing damage to agriculture, forestry and road users (collisions).

Irrigated land, growth and salinization

The total irrigated area is large (20.5 million ha, 8.5 million in the north and 12 million in the south), but only represents 20 per cent of total arable land and permanent crops. The areas are unevenly distributed:

- E*gypt*, with 100 per cent of its cultivated land irrigated, remains unique. The issue of rural development is therefore different from that in the other Mediterranean countries where dryland farming and pastureland generally predominate.
- Apart from Albania, the percentage of irrigated land in the *eastern* Adriatic countries, Slovenia, Serbia-Montenegro, Croatia and Bosnia-Herzegovina, is very low (less than 1.5 per cent).
- Irrigation occupies a relatively important place in most of the *other Mediterranean countries* (Figure 5.10): the Gaza Strip (63 per cent of arable land and permanent crops), Albania (49 per cent), Israel (46 per cent), Greece (38 per cent), Lebanon (31 per cent), Mediterranean France (30 per cent), Cyprus (27 per cent), Italy (25 per cent), Syria (23 per cent), Malta (22 per cent), Libya (22 per cent), Spain (20 per cent), Turkey (17 per cent), Morocco (13 per cent), Tunisia (8 per cent) and Algeria (7 per cent).

The irrigated area has about *doubled* in 40 years, accounting for 20.5 million ha in 2000, compared with 11 million in 1961. The biggest increases in absolute terms were in Turkey (3.2 million ha), Spain (1.7 million ha), Mediterranean France, Greece and Syria. Growth was also very strong in Maghreb (1.39 million ha, including 0.43 million in Morocco and 0.35 million in Libya).

This development is important for both its direct and indirect economic effects (the creation of areas of relative prosperity), and is projected to continue in those coun-

A Sustainable Future for the Mediterranean

Figure 5.10 **Percentage of irrigated lands in the Mediterranean in 1995**

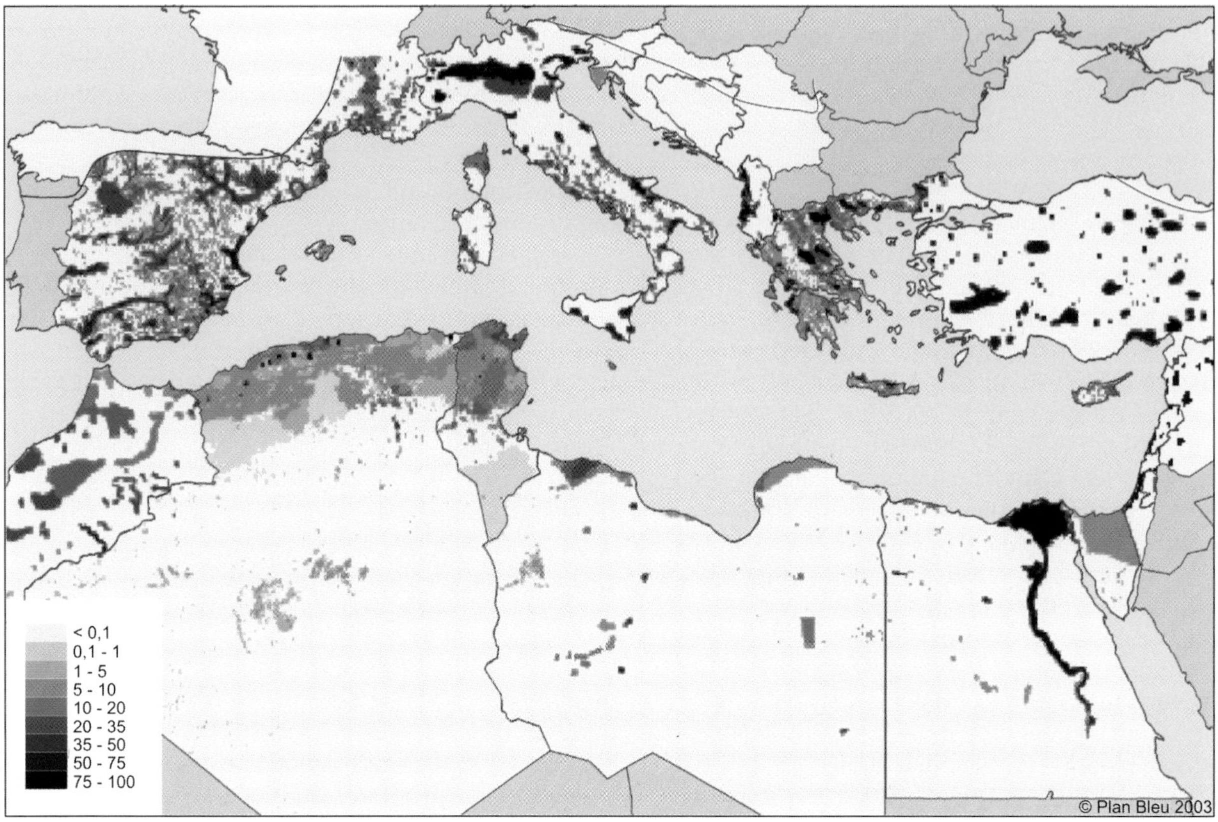

Source: FAO

tries that still have potential for new exploitation and continue to devote a significant percentage of their public resources to major water works. Although Tunisia has already reached its limits and Syria is making little headway, since its potential was quickly used up, the spread of new irrigated areas is projected to increase from 10 per cent (Algeria and Lebanon) to 20 per cent or more (Morocco, Egypt and Libya). The large projects in the eastern Sahara, such as the 'Great Artificial River' in Libya (inaugurated in 1991 and 1997) and the 'New Valley' in Egypt, contribute to this but have not yet led to the expected agricultural development. Farming areas under development are limited to those in Syrte (10,000ha) and Suluq to the south of Benghazi (32,000ha) in Libya, and in the Peace Canal (170,000ha to the east of Suez) and the Tushka depression (200,000ha) in Egypt.

Turkey deserves closer inspection, since its irrigated area could increase by nearly 1.5 million ha without using up its potential (a one-third increase involving lands in the Mediterranean bio-climatic area but located outside the catchment area). More than half of the expansion of irrigated areas in the south and east would be due to Turkey which is already becoming the biggest regional agricultural producer, both in terms of area and potential, able to energize an original strategy. The expensive mobilization of its water resources should not, however, be dissociated from Syria's and Iraq's needs, these two countries being located downstream of the Turkish basins of the Tigris and Euphrates Rivers. Growth in irrigated area is also expected to remain strong in the EU-Med countries. Table 5.3 summarizes projections to 2025.

However, it is clear that this growth in major irrigation can only increase the pressure on water resources and ecosystems that are already highly degraded. It will thus make water demand management measures indispensable.

It will also increase the risks of *soil salinization*, the main form of degradation of irrigated land. The main aggravating factors to be considered are: the invasion of coastal fresh water aquifers by sea water; irrigation of land with water with too high a concentration of salt; water-logging due to a rise in the ground water table; and the rise of salt-water tables because of poor drainage. The figures, often differing by source, show high levels of degradation, especially in Turkey (1.5 million ha) and Egypt (1.2 million ha) (Box 5.7).

Table 5.3 Land and water used by agriculture, 2000–2025

Country	Total UAA 2000/2025 % per year	Total UAA % potential used 2025	Irrigated area (ha) 2000	Irrigated area (ha) 2010	Irrigated area (ha) 2025	% potential in 2025
EU Med +islands	−0.2	—	10,482	11,000	13,000	—
EAC	0.4	—	369	450	500	—
Turkey	0.3	81	4500	4900	5800	70
Syria	0.5	74	1211	1205	1250	100
Lebanon	0.3	98	104	120	150	80
Egypt	1.0	97	3291	3700	4200	98
Mashrek	0.3	—	4606	5025	5600	98
Libya	0.4	36	470	520	580	80
Tunisia	0.4	96	380	402	450	100
Algeria	0.5	62	560	590	630	88
Morocco	0.3	75	1305	1390	1540	95
Maghreb	0.4	—	2715	2902	3200	91

Source: Blue Plan; Labonne, FAO

Box 5.7 Some data on the aggravation of soil salinization

Turkey: 1.5 million ha have become unsuitable for farming because of salinization,[9] that is 33 per cent of the total irrigated area in 2000.
Egypt: 35 per cent of the cultivated land suffers from salinity.[10] FAO indicates that 1,210,000ha have become saline, or 37 per cent of the total irrigated area in 2000.[11]
Syria: 60,000ha have become saline (FAO[12]); national sources indicate that 125,000ha are affected (5 and 10 per cent, respectively, of the total irrigated area in 2000).
Algeria: *sebkhas* and *chotts* have gained several thousand ha.
EU-Med: 1 million ha have been affected by salinization. In France, salinity and alkalinity affect about 100,000ha,[13] 4.5 per cent of total irrigated land in 2000. In Italy salt crusts during droughts would affect some 800,000ha;[14] according to FAO, 300,000ha are affected by salinization (30 and 11 per cent, respectively, of the total irrigated area in 2000). In Greece, 150,000ha of soils in the plains have very high salinity, 10 per cent of the total irrigated area in 2000. In Spain, saline soils are said to cover 190,000ha (FAO: 240,000ha) and salinization would affect 3.5 per cent of irrigated lands.[15]
Israel: irrigation without sufficient drainage has salinated about 300ha, and the intrusion of salt sea water in fresh water aquifers is endangering soils in the coastal areas.[16] According to FAO, 100,000ha, or 25 per cent of the country's cultivated area (52 per cent of the total irrigated land in 2000), are affected by salinity.

The lack of comparable data, both in area and in time, makes a quantitative projection for 2025 difficult, but the risks of worsening salinization are real and will force countries and irrigating organizations to make much greater efforts to achieve sustainable management of irrigated land (*drainage and irrigation water control*).

Box 5.8 Tunisia's 1000 hillside-lakes programme

The '1000 hillside-lakes' programme (which is planning to build 2000 groundwater-recharge sites and 2000 water-spreading works) was launched with EU support within the framework of Tunisia's 1991/2000 strategy. Hillside water retention combines surface water harvesting and protection from water erosion. These constructions are being built as priorities in semi-arid areas for the benefit of small farms. Because of their medium size they may be a good vector in the decentralized management of rural development. A mid-term evaluation (1995) showed that the programme was progressing satisfactorily, although a significant problem was that of the little use of water by the beneficiaries, probably due to their low level of participation in the design and management of the works. More recently, greater attention has been given to awareness-raising and participation, since when the water began to be better used.

Source: Labonne, 2003

By continuing to concentrate most resources (water, capital and technology) on a limited part of a country's area, the development of major water works also risks accentuating *internal dualities* between the high potential irrigated areas and the rain-fed farming areas, drylands and mountains. Some countries have fortunately begun to introduce more balanced policies by investing in small and medium-sized water works or improving the *agricultural management of run-off water*, as in Tunisia with its 1000 hillside-lakes programme (Box 5.8). A shift towards an alternative scenario would require this kind of approach to become more common.

Agricultural pollution

France, Israel and Egypt are among the biggest fertilizer users in the world, with more than 260kg per ha per year. The *overall use of fertilizers* in the northern countries has stabilized and even decreased since 1990 after a phase of strong growth between 1960 and 1990. The southern and eastern countries have experienced some slowdown in the growth in fertilizer consumption since 1990, particularly following structural adjustment programmes, which have led to a suppression of aid (Figure 5.11). Nonetheless, apart from the Maghreb countries, there is continuing growth in overall fertilizer use (an increase of 4 per cent per year in Syria, 2 per cent per year in Turkey and Israel, and 1.3 per cent in Egypt between 1985 and 1997).

Figure 5.11 Fertilizer consumption

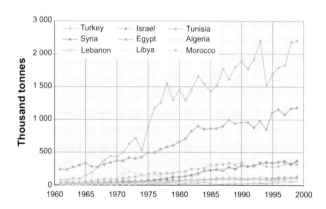

Source: World Bank, WDI 2002

Pesticides were long overshadowed by nitrogen fertilizers, but because of their high persistence they are becoming a growing concern for health institutes and consumers. Data on their use and presence in the environment are very scarce. Since 1960, 620,000 tonnes of 200 different kinds of pesticides have been used in Egypt.[17] In France, pesticide use reached 85,000 tonnes in 2002. It seems to have stabilized over the past ten years, despite a recent drop. The costs incurred are very high, since 20 per cent of households receive water that no longer complies with drinking water standards. Surveys have also shown that the entire coastline is contaminated. Lastly, it is estimated that more than 6.5 million ha of vineyards, orchards, corn and industrial crop areas could experience worrying releases of heavy metals if the pH were to increase.

With the reforms of the Common Agricultural Policy (CAP), the promotion of an 'agriculture respectful of the environment' and progress in organic farming (Box 5.12 p277), fertilizer and pesticide use in Mediterranean Europe should be stabilizing, if not reducing. In the southern and eastern countries, where fertilizer and pesticide use is still relatively low, use will probably increase with agricultural intensification to meet increased food demand. Fertilizer use could increase by 70 per cent in the east up to 2025 (especially in Turkey) and 50 per cent in the south.

The development of specialized cultivation and irrigation (flowers, fruit and vegetables) has been one of the factors in the increase of agriculture based pollution on all Mediterranean shores. Food processing, such as olive oil production, can also cause serious pollution (Box 5.9).

Box 5.9 The development of olive-growing and the environment in the Mediterranean

Olive groves are a major component of the identity, landscape and biological diversity of the Mediterranean region, and olive-growing has contributed much to the region's history (including technological). Olive culture also has a strong social component since labour costs represent about 80 per cent of the oil's production costs. In the EU, this sector is responsible for 2.5 million jobs and providing a livelihood for nearly 8 million people, spread over 5.4 million hectares.

In 2003, 95 per cent of world olive-oil production (more than 2.5 million tonnes) was in the Mediterranean. Over the past 40 years the world market for olive oil has undergone a sharp rise in production and consumption, which has practically doubled, while trade has increased by a factor of five. Syria is a notable example, with production that has increased 33 times in the coastal region and 10 times in the whole country in 30 years (1970–2000). Olive oil's reputation as a healthy product has become a determining factor in the increasing consumption. For the past 15 years new markets have opened up where consumption has increased by 8–10 per cent per year, for example in the US, France, Canada, Australia and Japan. These non-traditional markets now represent 14 per cent of the world market compared with 8 per cent in 1990.

The continuous increase in production (encouraged in the EU by the CAP) can have serious impacts on the environment: soil erosion in olive orchards, over-consumption of water in irrigated olive groves, the spreading of fertilizers and pesticides (which will alter the quality of soils and water), and deterioration of river water quality (due to largely non-biodegradable 'black liquor' wastewater from the oil production process). The CAP reforms now underway (decoupling aid from production, eco-conditionality) should make it possible to reduce the negative environmental impacts and strengthen the positive effects of olive groves (biodiversity and landscapes). Starting in 2005, EU product-support, about 1300 euros per tonne, will be converted into income-support (a single payment per farm) and olive-groves support (per hectare or per tree).

Main olive-oil producing countries in 2003

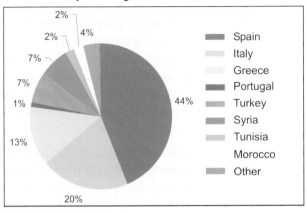

Source: from FAO data (http://faostat.fao.org/)

Promoting technology for black wastewater treatment is needed to reduce its very negative impact on rivers and downstream users. Lagooning made it possible to considerably improve the water quality of the very polluted Sebou wadi (Fes, Morocco). Research projects are underway in several countries to perfect new techniques (biological treatment, electro-coagulation and bio-methane processing, electro-chemical treatment of wastewater, etc.) that will make it possible to recover biogas (methane), and upgrade sludge to compost or cattle fodder. Research and promotion of two-step extraction processes (instead of three) is to be encouraged. This would make it possible to reduce the production of black wastewater by half and significantly limit the amount of water used.

Source: Benyahia, N. and Zein, K. (2003) Analyse des problèmes de l'induxtrie de l'huile d'olive et solutions récemment développées. Contributuin spéciale de Sustainable Business Associates (Suisee) á SESEC II (www.sba-int.ch/)

Soil degradation by erosion and desertification

The term *desertification* should be understood as the *'degradation of lands in arid, semi-arid and dry sub-humid areas following on various factors, including climatic variations and human activity'*. It is thus not an 'advance of the desert' but a process of gradual loss of soil productivity and reduction of plant cover attributable mainly to *human* activities in dry areas – where the aridity index or the precipitation/evapo-transpiration ratio (P/ETP) is less than 0.65. When less protected by plant cover, soil becomes more susceptible to *water and wind erosion*, which leads to its gradual degradation. The main consequences are a loss of fertility and a degraded soil moisture regime, which in turn will have a negative effect on plant life and production. A spiral of degradation is created, which can lead, without intervention, to irreversible desertification.

Estimates of the magnitude of desertification and its degree of irreversibility are often contradictory and highly disputed. Nonetheless, the gravity of the phenomenon in the region is now widely accepted. According to estimates of the early 1990s, 80 per cent of arid and semi-arid areas in the SEMC are affected. In these areas, pastureland (84 per cent) and rain-fed arable land (74 per cent) are the most affected, but also irrigated lands, especially through salinization. Desertification also damages 63 per cent of the Mediterranean Europe arid land in Spain, Greece and Italy (Table 5.4).

Of 245 million ha likely to be affected by desertification in the Mediterranean, more than 188 million ha (77 per cent) were already more or less degraded in 1990.

National data on the extent and nature of soil degradation are still very incomplete and relatively unreliable, but confirm the gravity and magnitude of the phenomenon (Box 5.10), in particular of degradation due to water and wind erosion.

Table 5.4 Arid lands affected by desertification

	SEMC		NMC	
	Arid zones (million ha)	% affected by desertification	Arid zones (million ha)	% affected by desertification
Pastureland	150.5	84	25.6	72
Rain-fed crops	42.6	74	12.6	71
Irrigated lands	6.9	20	6.9	17
Total	**200.0**	**80**	**45.1**	**63**

Source: H.E. Dregne (1984) quoted in Plan Bleu 2003b

Box 5.10 The magnitude of soil degradation in various countries

Cyprus: 112,000ha of cultivated land was affected in 1995–1997 (56 per cent of the total cultivated area); 42 per cent of meadows and pastureland and 79 per cent of arable land may be affected.

Algeria: 21 per cent of cultivated land, or 8.4 million ha affected in 1985, 19 per cent of meadows and pastureland (6 million ha) and 79 per cent of forests and other woodlands (2.9 million ha).

Morocco: 89 per cent of cultivated land, or 8.2 million ha and 13.4 million ha of meadows and pastureland affected (1995–1999).

Syria: 3.2 million ha in Syria are affected (1995–1997), including 1.1 million ha by water erosion, 1.6 million ha by wind erosion, 408,000ha by shifting dunes and 125,000ha by salinization.

Tunisia: 3.5 million ha are affected, or 21 per cent of the total land area (1995–1997). According to a recent study,[18] annual soil losses through various degradation processes are estimated at 37,000ha, of which 13,000ha would be irreversibly lost.

Turkey: Erosion is considered to be one of the most serious problems for the rural environment. Water erosion affects 57.1 million ha, wind erosion 466,000ha. In all, 73 per cent of cultivated land would be affected, and 1000 million tonnes of soil would be lost each year. Although countermeasure programmes began 25 years ago, they have so far treated only 2.2 million ha.

Source: national sources, Eurostat, *Plan Bleu*

The main *causes* of degradation are deforestation, overgrazing, cultivation, overexploitation of biomass, industrial activities and public works. Available data and analyses do not indicate any reversal of trends. The baseline scenario assumes that, with continued human pressure and inappropriate agricultural practices, desertification in arid areas in developing countries will continue to spread.

The adoption in 1996 of the United Nations Convention to Combat Desertification and Drought (UNCDD) is intended to be an initial response on an international scale, but the mid- and long-term effects remain uncertain. An estimate of the technical *costs* of prevention, improvement and restoration of lands degraded by desertification (pasturelands, cultivated land, irrigated land) in North Africa and Mediterranean Europe has been assessed by *Plan Bleu* at about US$1.2 thousand million per year. This only indicative figure is rarely compared with the losses of production induced by degradation and the indirect costs. The fundamental issue remains the need to combat poverty through sustainable rural development, so as to have an impact on desertification. This battle cannot be fought through technical measures to prevent soil degradation and restore degraded land alone, it has to be part of an integrated and holistic process. During the World Summit in Johannesburg in 2002, the Global Environment Facility (GEF) agreed to contribute to financing the implementation of the Convention. Its implementation at the national level will not be without problems. The recently developed national action plans reveal a certain dispersion of content and methods and insufficient diagnoses and evaluation.

Loss of agricultural land and rural character resulting from urbanization and new infrastructures

The irreversible loss of a significant part of the best agricultural land to *artificial land cover* is a serious and worrying trend in the south and the east, but particularly in the north (Box 5.11).

Box 5.11 Loss of high quality agricultural land and rural character in the Mediterranean region

Between 1960 and 1990 **Egypt** lost 315,000ha of high quality land (10,000ha per year), especially in the Cairo region, losses that have continued since 1990 at about 12,500ha per year. In 42 years these losses represent nearly 14 per cent of agricultural lands (recorded in 1994). In **Malta** the percentage is 37 per cent in 90 years. In **Lebanon**, 7 per cent of cultivated land and 15 per cent of irrigated land has been lost over the past two decades. In **Turkey**, 150,000ha of high quality land was lost between 1978 and 1998 (or 0.54 per cent of the land). In **Algeria**, 140,000ha have been lost around greater Algiers. In **Cyprus**, 3200ha around Nicosia in 16 years (between 1985 and 2001). In **Tunisia** the growth of Sfax since 1992 has taken 9000ha of vegetable gardens. In **France** half of market gardening and horticulture is located in 'urban centres' (urban units of 5000 jobs or more). These centres have lost 200,000ha of suburban agricultural land in 12 years (between 1988 and 2000), or a 12 per cent UAA loss (six times more than losses in intermediate or isolated rural areas). In rural areas in **Israel,** all the more precious because they are limited, construction activities drastically increased in the 1990s following town-dweller demands for country houses, the demand of farm workers who had to abandon farming due to increased labour productivity, and, above all, the demand created by the large number of immigrants from the former USSR (nearly 600,000 arrivals between 1989 and 1995).[19]

The loss of cultivated land is especially rapid in **coastal areas.** On the Marbella–Malaga Andalusian coast, for instance, the percentage fell from 26 to 9 per cent (a loss of 64 per cent) in 15 years (1975–1990) (see Chapter 6). On France's Côte d'Azur (a 2-km strip), it fell from 12 to 8 per cent of total cultivated land in the same period (a loss of one-third). In contrast, forest areas, being more efficiently protected, have resisted better.

Source: Plan Bleu, from various national sources

Such losses are projected to continue, especially on coastal plains, resulting in a *nearly irreversible loss of about 1.7 million hectares* of high-quality agricultural land between 2000 and 2025 (a N1 country-level figure, to be compared with 20.5 million irrigated ha), as well as highly valuable natural areas. Apart from the loss of agricultural potential (and job possibilities in the long term), this trend has considerable consequences for ecosystems, landscapes and the living environment. By gradually destroying rural areas that surround cities, both separating and linking them, it reduces the infiltration capacity of soils (reduced permeability), thus contributing to increased risks of flooding. It also fragments ecosystems and reduces the area still covered by natural habitats, with heavy consequences for biodiversity.

Biodiversity and landscapes: Loss of an invaluable heritage

Despite their present and future ecological and economic value, Mediterranean landscapes and biodiversity are now considered to be largely degraded or *under threat*.

Many *species* have become either vulnerable or scarce through the ever-increasing degradation (quality) and reduction (quantity) of their habitat. This is the case for birds, mammals (for example the *dorcas* gazelle in Morocco, wild goats, Anatolian mountain sheep, and the ibex of Spain and the Pyrenees), reptiles, amphibians and invertebrates, and flora. Nearly 3000 of the 4777 plant species endemic to a single Mediterranean country (not including Syria, Lebanon and Turkey) are today considered rare, 180 endangered, and 344 vulnerable.[20] Vulnerability is especially high and worrying on the *islands*. Island stocks, typically having a clearly higher proportion of endemic species, group small populations. The percentage of threatened species on the islands is high (28 per cent of the flora in Malta, 13 per cent in Crete, 12 per cent in the Balearics and Corsica, and 8 per cent in Sardinia). On Crete, for example, of 1820 plant species, 150 of which are endemic, 238 are threatened, of which about 100 are endemic.

Such degradation is largely the result of the serious trends previously noted

In the SEMC, where pasturelands and forests are still overexploited by humans, analyses of plant-life show that a number of very rare groups of plants have completely disappeared or occur only in a residual form, especially in cork-oak forests, evergreen oak forests and cedar forests. Some forested landscapes of the eastern Mediterranean and North Africa, admired since antiquity, have regressed in a terrible way (the cedars of Lebanon, Syria and Cyprus, the *Babor* oak groves of Algeria, the indigenous fir trees of Anatolia and Little Kabylia, etc.).

In the NMC, in contrast, the main threat to biodiversity (especially landscapes) is the gradual disappearance of open rural environments and traditional agricultural practices. The abandonment of agricultural and pastoral land at first allowed for the re-growth of forests (some of which have become very beautiful), and thus favoured the return of some rare animal species (wolves and large prey birds) and plant species dependent on shade and moist soil (orchids and various bulbous plants). However, it also led to a noteworthy and increasing *loss* of plant *species diversity* through overgrowing and suffocation of shrubbery and fields where the most remarkable species are often found. However, this loss, related to the 'mosaic' effect, is less serious than the loss of species in the south. The gradual invasion of open spaces, orchards and cultivated terraces by shrubs is also a cause of *landscape degradation*. Territory-specific quality, so important for the Mediterranean future, is also changed by advertising and encroachment on agricultural land in rural suburban areas, due to urbanization and the development of infrastructures.

In the north, south and east, a rapid degradation in *agricultural biodiversity* is also to be regretted. The continuous decline since the 1930s and the near-extinction of native varieties of wheat cultivated in Greece is a good example. In Italy, the alarm bell has been sounded in the past few years by the disappearance of a very large numbers of fruit cultivars, wiped out by the few big varieties demanded by the big sales outlets. The situation is just as serious for domestic animal breeds: of the 263 bovine breeds listed in the Mediterranean 78 have disappeared and 73 are endangered.

The response to this degradation generally consists of creating protected areas for conservation (national and regional parks, biosphere reserves, and reserves and sites of ecological interest). Several countries have committed themselves to such programmes over the past few decades (albeit unequally, see Figure 5.12). The part of the eco-region covered with protected areas amounted to 5.3 per cent in 2004, whereas the fourth World Parks Congress aimed for a 10 per cent coverage by 2003, and 12.7 per cent of the total area is protected on a world scale.[21]

A number of these areas fall under the World Conservation Union IV and V categories, which well expresses the Mediterranean region's specificity (the importance of semi-natural areas resulting from co-evolution of man and nature). However, few of these protected areas have the authority or the funding and other means to ensure the conservation of biodiversity and landscapes, and associate local communities, enabling them to effectively derive benefits from protection and thus become genuine 'laboratories' for sustainable development, as is sometimes the case in the north.

A Sustainable Future for the Mediterranean

In the EU region, where the protection of nature and development are closely linked, the adoption of directives (on birds, habitats, etc.) has not been without its difficulties. The latest decisions about the CAP, providing for agri-environmental funding, should facilitate the implementation of Natura 2000 (the European network of nature reserves, which aims to preserve the biodiversity of rural areas and forests), an important programme, since 80 per cent of plant communities considered by the Habitats Directive are present in the EU-Med countries.

The costs of degradation

Poor management of rural areas induces considerable costs, which remain largely under-evaluated.

First there are *direct costs* to the rural population involved. The drop in production capacity due to soil degradation (desertification) represents an average annual cost of about 3000 million euros for the Mediterranean countries (Table 5.5), based on an estimate that only takes part of the direct costs into account.

Deforestation and degradation of pastureland and soils also lead to *indirect costs* that are much larger than the direct costs. One of the major consequences of desertification for the southern countries is the accelerated silting-up of reservoirs, considerably shortening their life-expectancy (see Chapter 1). In Morocco, the annual cost of degradation of watersheds (erosion, silting-up of reservoirs and increases in management costs) has been evaluated at 10,000 million Dirhams (925 million euros).

Table 5.5 Annual losses in agricultural production on degraded lands (million US$)

	Irrigated lands	Rain-fed cropland	Pasturelands	Total MED
SEMC	356	1201	888	2445
NMC	293	338	129	759
Total	649	1538	1017	3204

Source: Plan Bleu, based on UNEP

Note: Estimate based on the average values adopted by UNEP in 1992 for moderately affected land (annual loss per hectare is evaluated at $250, $38 and $7 respectively for irrigated land, rain-fed farmland and pastureland). As far as the NMC are concerned, a recent report for the European Commission reinforces this, giving an estimate of 748 million euros for the four EU-Med countries.

Management deficits for watersheds, sealing of soils by urban sprawl and infrastructures, channelling and other water works in rivers and a reduction of natural areas for water spreading, all explain the growth in hydro-meteorological risks (floods and landslides), which are particularly serious in the region. In Italy, for example, it has been estimated that floods and landslides in the past 20 years have impacted on more than 70,000 people and caused damage estimated at a minimum of 11,000 million euros.

Forest and scrub fires create considerable costs for society: material destruction, fatalities and fire-fighting costs. In countries that have developed modern fire-fighting systems (land-based engines and powerful airplanes), the costs have been estimated at about 150 euros per hectare of forest per year, or total annual costs of about 1000 million euros for the whole region. In the SEMC the risks of fires and related costs are fortunately lower, because of a larger human presence, maintenance of grazing land in forests and the regular collection of firewood by local people, but even here the risks are not negligible.

The uncontrolled development of large fauna in the developed countries involves other significant costs. In France, where the wild boar population has increased more than tenfold in a few years (Mediterranean France being the most affected region with a third of the national total), more than 15,000 collisions between cars and big mammals occurred in 1994,

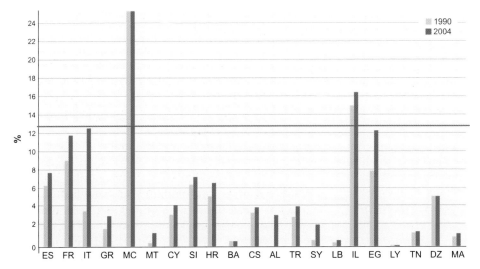

Figure 5.12 Growth in protected area coverage in the Mediterranean (according to the UN list of protected areas)

Source: IUCN, 2004.

Note: The red line indicates the percentage of protected area on a worldwide scale in 2004.

three times more than in 1986, and insurance payouts for damaged crops rose to 18.3 million euros in 1999.

The costs of degradation and the related economic consequences (water regime deregulation, silting-up of reservoirs, rural impoverishment and increased regional imbalances, forest fire protection, loss of long-term potential for development of biotechnologies, etc.) remain *insufficiently evaluated*. A recent estimate by the World Bank/METAP dealing only with degradation of soils, forests and biodiversity and part of the incurred costs showed an annual loss of 1.36 per cent of GDP in Algeria, 1.21 per cent in Egypt, 0.3 per cent in Morocco and 0.32 per cent in Tunisia.

In addition, the accumulated effects and synergies of all these impacts on the environment should be considered. The different degradations of rural areas are indeed often interrelated within the ecosystem (soils, water, biodiversity and landscapes). A number of these impacts are *irreversible* or nearly so: the loss of arable land due to an increase in artificial land cover forms (especially damaging when it concerns high quality land), the disappearance of species and cultivars, desertification in arid areas of the south and east, the exhaustion of non-renewable water resources. The degradation of traditional rural landscapes, and the loss of know-how related to them, also have or are about to reach a point of no return.

Food and agriculture outlooks

The primary function of rural areas is still to produce food. In the Mediterranean this is all the more important because of its virtuous dietary regimes with special energy profiles and health qualities. At the same time, changes in food demand (quantity and composition) partly condition agricultural development and consequently rural area development.

When considering the future, one important question relates to the possible impact of an evolution towards free regional, even global, agricultural trade. This complex question cannot be dealt with without highlighting two major characteristics of the region: the growing disparity in productivity between the northern, and the southern and eastern shores, and the limits imposed on production by the scarcity of natural resources (soil and water). It also leads to the question of the evolution of agricultural performance and society's demands on agriculture. All these different factors will contribute to the future of agriculture in the region.

Growing gaps between north/south productivity and agricultural trade deficits

Everywhere in the Mediterranean Basin agriculture has been and remains a well-organized sector, either under direct administration by states or, as in the EU, through a synergy between states and professional organizations. Since the 1980s, there has been a general convergence towards liberalization with a gradual and peaceful opening up of markets. This is all the more remarkable because agriculture carries strong cultural, social and political values.

In the *EU-Med countries*, the EU enlargement and the multilateral negotiations within the WTO framework have led to a greater consideration of world markets and an ever greater effort of decoupling price support from income support. The relationships between agriculture and processing activities have been gradually freed from state intervention; industrial coordination is taking on sophisticated forms, and calls for the organization of professionals following complex regulations.

In the *southern and eastern countries*, liberalization has been favoured by internal and external factors, in particular by agricultural structural-adjustment programmes, which have often posed serious social problems. Liberalization has promoted the emergence of a new entrepreneurial spirit. In several countries, the private sector has begun to play an ever-more active role in the promising markets for strong value-added agriculture, including organic farming.

The situation is much more difficult in the poor or bruised *eastern Adriatic* countries, where war and the abrupt collapse of the socialist system have ruined the regulatory structures of the economy, while the market has so far not been able to put anything in its place. In Bosnia-Herzegovina, for example, the land currently in production is only two-thirds of the area used in 1991, but the new-found stability and the new mid-term strategy adopted for agriculture should make it possible for the sector to recover gradually.

There is reason to believe that this general transition towards a market economy will be confirmed and amplified over the next two decades.

Despite this convergence towards a market economy, Mediterranean agricultural developments over the past two decades have been marked by a growing gap in productivity between the northern and southern and eastern shores, and by deficits in the agricultural trade balance.

An analysis of *agricultural productivity* (value-added per worker in 2001, in 1995 US$) shows large disparities, varying from 1 to 45 (US$1300 in Egypt and US$60,000 in France) (Figure 5.13). These disparities illustrate the persistence of traditional farming essentially oriented towards food crops and home consumption on the one hand, and the dominant commercial agricultural sector with high technological input in the EU-Med countries, on the other. Productivity per worker is particularly low (US$2000 and even less) in Egypt, Morocco, Turkey, Albania and Algeria. Although better in Syria (US$2700)

A Sustainable Future for the Mediterranean

and Tunisia (US$3100), it is still far from what is observed in Spain (US$23,000), Italy (US$28,000) and Slovenia (US$39,000). Bosnia-Herzegovina (US$7700) and Croatia (US$10000) are in between. Another cause for concern is the increase in disparities. Productivity has nearly doubled in France in only 11 years (between 1990 and 2001), and disparities have become even greater with most countries (those with Turkey have nearly doubled), including Greece.

Figure 5.13 Agriculture value-added per worker

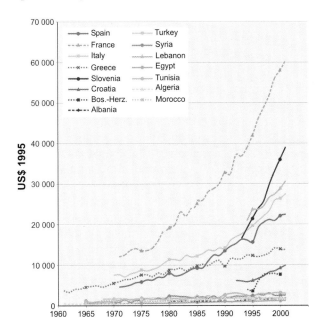

Source: World Bank, World Development Indicators, 2003

Furthermore, soil and water resource scarcity in a context of strong population growth, explains the *growing deficit in the agricultural trade balance* in practically all SEMC (Figure 5.14) and the importance of cereal imports (US$4000 million in 1996–2000, or 10.7 per cent of the world's imports). The overall balance for these countries went from plus US$500 million in 1970 to minus US$9.7 thousand million in 2001. The deficit is especially high in Algeria (US$2.7 thousand million in 2001), Egypt (US$2.8 thousand million), Libya, Israel, Lebanon and Morocco. Only Turkey, more endowed in water and land, shows a positive balance (US$800 million in 2001), as does Syria occasionally. Between 1996 and 2000, 34 per cent of the EU cereal exports went to the SEMC. However, the EU share of this market remains relatively small (21 per cent on average over those 5 years), because of competition between the large worldwide producers, where bilateral aid to importing countries plays a role.

On the northern shore, the balance sheet would also be very negative if France were removed from the calculation, since agriculture in France occurs only marginally in its Mediterranean regions. Only Spain has a positive balance and is progressing. Greece is on the decline, Italy has a very negative balance and the East Adriatic countries are all in deficit.

Figure 5.14 Agricultural trade balance in the Mediterranean countries: 1970–2001

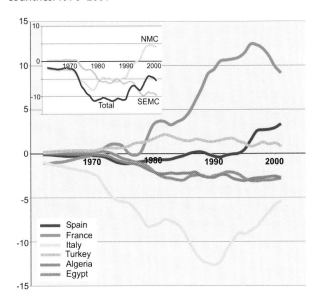

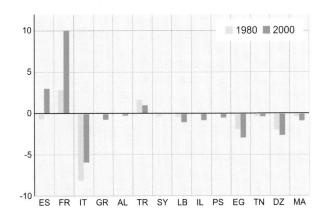

Source: Plan Bleu from FAOSTAT

Benefits and decline of the Mediterranean diet

The Mediterranean food regime

Food is one of the major components of the Mediterranean identity, while at the same time reflecting its biodiversity. The extreme blending of cultures over the centuries did not lead to more uniformity, but to increased food and culinary variation. This diversity is due to clear differences in dietary regimes between countries, but also to increasing differences in income, and, consequently, to differences in the share of household budgets devoted to food.

Given the benefits of the Mediterranean diet for health, now confirmed by epidemiological studies, some common characteristics in this diversity deserve to be highlighted. The Mediterranean model is attracting new and noteworthy interest at international level. In the 1970s and 1980s, the American diet was considered the best example to follow, throwing the Mediterranean diet into turmoil and, indirectly, its agricultural production and rural areas. But this tendency has started to turn around on a global scale during the past ten years. Stimulated by the medical community, Anglo-Saxon countries and the WHO have stepped up information campaigns to promote the Mediterranean model and its specific products in order to limit the growing social costs of major endemic diseases (cardioascular diseases, some cancers). Their recommendations are based on the diet of the 1960s in Crete and southern Italy; a time when the benefits of the Mediterranean diet were greater, since these societies combined a low calorie input with sustained physical activity.

Traditionally frugal, the Mediterranean diet is characterized by *typical products*: a high consumption of highly varied plant products, that is cereals, leguminous plants, vegetables and fruit (fresh and dried); low consumption of milk, meat (especially bovine meat) and animal fat, but fresh cheese, yoghurt, fermented milk, fish, olive oil and the meat of goats and sheep; extensive use of aromatic herbs, spices, lemon and vinegar; and moderate consumption of alcohol, consumed during meals. A convivial atmosphere during meals (many doctors are increasingly stressing conviviality and pleasure as determining factors for good health), as well as an adaptation to natural conditions by following the seasons, are benefits to be added. This adaptation to nature, developed under the influence of the Hippocratic medical system, was widely disseminated and popularized throughout the region by Medieval Jewish and Moslem doctors. It remains present in popular culture today. In contrast to what is happening in the northern world, a large percentage of women in the Mediterranean generally cook every day and actively pass on of their knowledge to future generations.

The percentage of *animal calories* in the Mediterranean diet is low (12 per cent in the south and east, 25 per cent in the eastern Adriatic and 35 per cent in the EU-Med countries, compared with an average of 45 per cent in OECD countries in 2000). This results in much lower pressure on the land, since about seven plant calories are required to produce one animal calorie. Thus, the average intake for a consumer in an OECD country (3365 calories) translates into 12,450 cal in plant-equivalent, whereas the consumption of a southern Mediterranean person (3210 cal) translates into only 5155 cal in plant-equivalent (less than half), and the consumption of a EU-Med person (3564 cal) to 9550 calories plant-equivalent (two-thirds). In other words, one and a half to two times more agricultural land would be required to feed the Mediterranean people if their diet were to imitate that of the present 'north-western' model.

Regional disparities and decline of the health model

In the *developed countries* of the *northern shore*, the healthy diet model that was developed over centuries, has been replaced by a model of excess, in calories, with animal products (22–38 per cent of energy intake in 1995 compared with 13–30 per cent in 1960), and fat (32–40 per cent in 1998 compared with 20–30 per cent in 1960). Despite growing international interest in the Mediterranean diet, the northern shore continues to imitate the northern European and American model (Figure 5.15). The shift from a 'biological rationale' to an 'economic rationale' in agricultural production has also encouraged the introduction of food and practices foreign to the Mediterranean culture. At the same time lifestyles have changed remarkably with accelerated urbanization. This explains why the difference between the originally lower mortality rate from cardiovascular diseases and cancers and the longer life expectancy of the residents of France's Languedoc-Roussillon region, and the higher figures in the rest of France, is gradually disappearing. Obesity rates are now approaching those of other French regions.

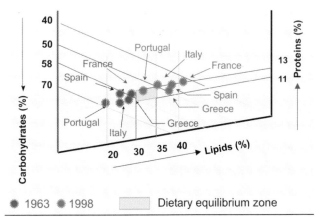

Figure 5.15 The evolution of food intake in the northern Mediterranean countries

Source: Padilla, 2002; *Plan Bleu*

In the *developing countries* of the *south and east*, improvements, supported by active health and food policies, have made it possible to fill some of the dietary deficits of the 1960s. Yet, despite progress in animal production, food intake remains far from the equilibrium zone as defined by international dietary standards. Despite the absence of recent surveys, a deterioration of the dietary situation cannot be excluded after the structural adjustments and privatizations of the 1990s, which led to some disengagement by states and a marginalization of part of the population. As on the northern shore, lifestyles and consumption patterns that are foreign to the region are to some extent imitated. This is expressed in Tunisia, for example, by a sharp decrease in durum wheat (to the benefit of soft wheat, the consumption of which doubled between 1968 and 1995) and a large increase in the number of overweight people.

The *eastern Adriatic* shows strong disparities that are increasing in several countries when comparing 2000 with 1963. Poverty-related food insecurity is still a basic problem in several regions. In Albania, the average food intake is 2521 kcal per inhabitant per day in 2000.[22] The rural poor spend the greatest part of their household budget on food (66 per cent) and less for non-food products (21.2 per cent) as compared to the shares respectively of 48 per cent and 25 per cent (of the relatively wealthy) in Tirana.[23]

A paradox baseline scenario

In the baseline scenario, the developed countries in the Mediterranean, which have already reached food satiety, are projected to maintain consumption at the same level (3565 calories) with a slight rise in the relative percentage of cereals and a slight fall in that of meat. The SEMC would reach a *food intake* of about 3350 calories with a sharp drop in cereals, which would represent only 1500–1600 calories per day. The eastern Adriatic transition countries would reach 3200 calories. These projections would lead to an increase in food demand between 2005 and 2025, as a cereal-equivalent, of 36.5 per cent in the southern and eastern Mediterranean, 9.5 per cent in the eastern Adriatic and 8 per cent in the northern Mediterranean. This would lead to increased pressure on land and/or trade deficits.

In this scenario, the *decline* of the Mediterranean health model would be confirmed, while paradoxically also confirming the international interest in the Mediterranean diet, and a revival of interest in the region in typical quality products like the 'Tupi', a Catalonian cheese once considered a rustic food of primary necessity, which has now become a valued product.

In this baseline scenario, one can fear, however, that the distribution of Mediterranean products (typified by their freshness and authenticity) may be restricted by new *world standards* for food security, under discussion at present in the large food distribution sector. The new 'Global Food Safety Initiative', which comprises 65 per cent of global distribution, aims to market only products adhering to a certified reference (ISO and HACCP standards, Codex alimentarius). This would mean that products not guaranteed by such standards would run the risks of scandal and law suits for distributors in cases of food poisoning. For this reason, some major distribution chains may be led to revise their present commitments in promoting local products in Mediterranean countries and the regions where they set up shop.

Europe: Questioning the high productivity agricultural model and the transition towards a 'rational' agriculture

The rural world of Mediterranean Europe has changed more in 40 years than in several centuries. This is due first to the change from farming based on a 'biological' rationale (the choice of crop production and systems depending on the environmental conditions) to farming based on a new 'delocalized' economic rationale (adapting to broader markets). The new dominant agricultural model is built on the basis of chemical inputs, mechanization, specialization, and concentration and integration with upstream and downstream activities. This high productivity agricultural model, which characterized the 1989 *Plan Bleu* trend scenarios, was encouraged by the implementation of a common market and the Common Agricultural Policy (CAP). Although the CAP has reached and even surpassed its initial objectives (increased yields and production to meet interior food demand with both quantity and low prices), this new dominant agricultural model has also generated serious perverse long-term effects:

- the excessive abandonment of many rural areas, which, no longer maintained for farming, have seen their economies collapse and their environment and landscapes degraded;
- the very strong and rapid fall in the agricultural labour force with all its direct and indirect effects (uprooted populations, broken social bonds, regional imbalances and unemployment);
- the problem of future generations on farms (ever more difficult);
- serious degradation of water and soil induced by pollution, mechanization and the separation of arable farming from animal husbandry (soil compression and loss of fertility, resumption of erosion);
- the loss of biodiversity; increase in transport and the resulting environmental costs; and increases in some long-term risks.

The magnitude of the negative environmental, social and territorial impacts shows the market's difficulty in *internalizing the costs* of the positive and negative externalities of agriculture. The prices of products and inputs do not reflect the scarcity of the natural resources used (especially water), or the services provided by agriculture that have no direct commercial value but receive an important social and environmental demand (for example the demand for quality agrarian landscapes is growing as illustrated in Europe by the development of flourishing rural tourism), or the cost of environmental degradation, some of which is irreversible (loss of biodiversity). And agricultural policies, instead of trying to reconcile the market and sustainability, have all too often amplified these costs by encouraging unsustainable practices.

This model is thus no longer considered acceptable by a part of European society, who have ceased to accept the direction given to very high public subsidies (47 thousand million euros) that favour big industrial farms at the expense of smallholders. This criticism was revived sharply in the mid 1990s with the objective or subjective questions of food insecurity.

Growing awareness of the negative impacts of the dominant agricultural high-productivity model has recently had several major repercussions in Europe, which have led in particular to:

- A nascent consensus on the need to factor-in the *multifunctional dimension of agriculture*.
- The development of an *alternative agriculture* (organic, sustainable, with solidarity, small farms) where the objectives are not maximum yield but optimal production depending on the environment's conditions but still through modern processes (mechanization, marketing). The producer is remunerated by the consumer with a price that corresponds to the health and ecological quality of the products, while also rewarding the long-term maintenance of the soil's health and fertility and that of the ecosystem, in particular by a balanced management of organic matter. This implies a continuation of developments in consumption of organic products, which today represent a significant and growing share of the market (in France today 16.5 per cent of customers regularly buy organic products, and 28 per cent occasionally). In response to this growing demand, many EU-Med countries, but also southern and eastern countries are promoting the development of organic agriculture (Box 5.12).

Box 5.12 Organic agriculture in Europe and the Mediterranean. An example of success in Egypt: The Sekem Group

Stimulated by strong European demand (a 10 thousand million euro market and growing strongly), organic agriculture has recorded strong growth mainly in the Mediterranean, in Scandinavia and in some central and eastern European countries. Germany, a large importer, has set a goal of 20 per cent of its total agricultural land for organic production by 2010. With 1.23 million ha (or 11.4 per cent of cultivated land), Italy is the leader both in Europe and in the Mediterranean region. Growth is very strong in Spain (485,000ha, 1.7 per cent), in Mediterranean France (32,000ha, 3.2 per cent compared with 2.1 per cent at the national level) and in Slovenia (5000ha, 2.5 per cent); development has started in Greece (31,000ha, or 0.8 per cent in 2001 compared with 1500ha in 1997). The increase in organic farming in the EU has been supervised and stimulated through rules (directive 2092/91) and transition subsidies.

Significant progress has also been observed in the south and east, particularly in Israel (70,000ha, 1.7 per cent), Egypt (15,000ha, 0.5 per cent), Tunisia (18,000, 0.4 per cent) Turkey (57,000ha, 0.2 per cent), Morocco (12,000ha, 0.1 per cent) and Lebanon (300ha, 0.1 per cent).

Thanks to the Sekem group, Egypt offers a quite remarkable success story. Specializing in bio-dynamic agriculture (drugs derived from plants, fruit and vegetables, cotton), the Sekem Group was founded in 1977 on a 70-hectare desert area near Cairo. Sekem is an example of a social and environmental project based on north/south knowledge and partnerships (investment, certification and fair trade). Today it is a network of strong companies with 2000 employees, active in, among other things, agriculture, production, craftsmanship and pharmaceuticals. It has its own nurseries, schools and apprentice workshops. A pioneer in producing organic cotton (since 1990), Sekem is behind the founding of the Egyptian Bio-dynamic Association that promotes the development of organic agriculture on nearly 4000 hectares throughout the country on over 400 small and medium-sized farms.

All EU countries, Slovenia, Turkey, Israel, Lebanon and Tunisia have fully operational regulations; in Croatia and Egypt regulations exist but are not yet fully implemented.

Source: IFOAM, 2003 (www.ifoam.org).

- The promotion of *'agriculture respectful of the environment'*. Proposed by agronomic research and supported by the agri-business sector, it aims to limit the present model's environmental impact by a controlled management of inputs (efficient irrigation, reduced use of mineral fertilizers and pesticides).
- *Gradual reform of the* CAP so that agricultural subsidies lead to more compatibility between the market and sustainability, especially by factoring in multifunctionality, that is the positive and negative externalities of agriculture. Subsidies thus spent will find support from the public and better ensure their legitimacy.

Along these lines, the European Council of Ministers of agriculture, in June 2003, took a number of decisions towards reforming the CAP. These moves rely on two principles: *'decoupling'* and *'eco-conditionality'*. Decoupling financial assistance will lead to a gradual decrease in various subsidies for production to the benefit of a single subsidy set up according to farm area and the

A Sustainable Future for the Mediterranean

amount of former subsidies. In addition, subsidies allotted to large farms will be reduced and the resulting released funds will be used for a more dynamic rural development policy. Moreover, a systematic application of an 'eco-conditionality' approach is made compulsory. This means that access to such a single subsidy will in future depend on the respect of 18 health, environmental and animal safety rules, as well as effective land management to avoid abandonment. However, the ambitious nature of this reform remains constrained by the budgetary discipline that has been adopted. In the context of the EU enlargement to 25 countries, a reduction is predicted in the actual allocation of subsidies.

Systematically applying the eco-conditionality principle to an agriculture that depends heavily on subsidies, together with the efforts currently being made in training and extension to help farmers improve their practices, should lead the present dominant model towards an *efficient agriculture* that is *more respectful of the environment* (lower water, fertilizer and pesticide use per quantity produced and less aggressive cultivation methods for the soil).

However, respecting pollution standards and rational water use alone will not be enough. With the continued fall in the number of farmers, and the growing problems of transmitting capital (the continuity of production tools), abandonment of farms and geographic concentration, urban sprawl and the separation of arable farming and animal husbandry will probably continue, which will not be without consequences. The problem of long-term *sustainability* of European agriculture and the evolution of the social contract between consumers, citizens, tax-payers and farmers will persist.

Pressures from civil society, strengthened decentralization and local initiatives, and 'territory-specific' projects and policies, supported by the CAP second pillar, should promote *innovation* and the necessary *plurality* of approaches. Moreover, regions, towns, rural communities and protected areas are increasingly engaged in the defence and promotion of sustainable agriculture by means of a whole range of tools, including land tenure (up to full control through land purchase). Agri-environmental aid will positively contribute to this development by better remunerating the production of amenities (the management of landscapes and biodiversity), which are important for Mediterranean development (agriculture/tourism synergy).

The south and east: An unfavourable trend despite some progress

Production growth, despite droughts and environmental degradation

In the south and east, despite the quantities imported, *cereals* remain the main group of crops in terms of cultivated areas, number of farmers and as staple food. For most of the countries, cereal production is a vital national issue in terms of food security, not to mention jobs and the present lack of credible alternatives for most areas and producers. The expansion of cultivated areas, consequent on population growth, is, however, a source of harmful pressure for soils.

Animal production is the second basic production pillar. This is a combination of an extensive traditional red-meat production system (sheep, goats and cattle) complemented by concentrated food supplements, and an intensive, indoor white meat production system, known as a 'corn-soya' system imported from developed countries. These two animal production forms result in over-grazing and erosion problems, as well as local pollution. Dairy production, mainly by small family herds, has made good progress. Some countries have become self-sufficient (Tunisia, Syria and Egypt) by solving problems of milk collection and protecting their borders against powdered milk imports.

As well as these basic production forms, *vegetable* and *orchard crops* have continued to progress (olive trees and market-garden products such as tomatoes and potatoes play a growing role). They represent nearly 50 per cent of the production value (43.9 per cent in Algeria, 53.4 per cent in Tunisia, 53.4 per cent in Libya, 47.7 per cent in Egypt, 75.6 per cent in Lebanon, 37.6 per cent in Syria, and 46.0 per cent in Turkey). Their development has been bolstered by developing irrigation and through reforms, in particular land reforms. Today they use a large percentage of fertile lands and irrigation water. In some oases, growth has been very strong (Box 5.13). Unlike staple products, these do not benefit from direct agricultural subsidies. They have the advantage of offering much work, but the drawback of inducing a growing use of water and chemical inputs (fertilizers and pesticides). Almost all the marketing is in national markets, while exports only involve a few countries and products (citrus, tomatoes, flowers and olive oil).

Box 5.13 Oasis: Laboratories for land reform and agricultural renewal

After a long period of decline, oasis farming, considered to have no future 40 years ago, has undergone an amazing development over the past three decades (date palms, vegetables, either in open fields or under plastic tunnels). The opening up to the market economy, the remarkable progress in transport infrastructure and the discovery of very large deep-lying aquifers have contributed to this development.

But this renewal also owes much to land reform. Initial development with a social objective strictly organized by governments in the 1970s and 1980s led to overly restricted land allotments (1ha plots). It was followed by the creation of large or very large private properties (20–100ha plots in

Algeria and more than 200ha in Egypt), which corresponded to a new phase of reduced state intervention, often with disappointing results. Then small and middle-sized farms emerged in some countries, large enough (about 5ha) to be very efficient. These farms, which have managed to combine tradition and innovation, peasant wisdom and openness to modern ideas, have been largely responsible for the remarkable development of oasis farming in Algeria. This is illustrated by the date-palm development (12 million trees in 2000 compared with just 5.5 million in 1959, with yields increasing to 50–80kg per tree compared with 30kg in 1959), and market-garden development (for example, El Ghrouss municipality has become one of the busiest markets in Algeria for fresh vegetables, with 800 vehicles per day).[24]

These land reforms have made oasis farming today a sort of technical, agricultural and social laboratory. However, this form of farming accounts for only a very modest percentage of national economies, and the sustainability of this development is not guaranteed. Current urban, agricultural and tourism development is threatening the environmental integrity of many oases. The decrease in water tables (observed everywhere), salinization of soils and the increased amount of waste are worrying signs that should inspire a reorientation of development with a better long-term understanding.

Generally speaking, agricultural production, despite degradation of the natural capital, has grown significantly over the past 20 years because of the mobilization of natural and human resources and technological progress. On the whole, productivity (average production per farm worker) has increased (Table 5.6) despite the disturbing effects of drought (very severe in Turkey and Morocco in 2000) and, for the first time in many years, agricultural growth rates are keeping up with those of population.

Table 5.6 The evolution of production indices per capita in a selection of SEMC

	Turkey	Syria	Egypt	Tunisia	Algeria	Morocco
1980	97	138	96	93	93	71
1990	102	105	101	97	93	94
2000	92	117	125	108	100	76

Source: FAO

The 'technological trend' (a proxi-indicator of agricultural progress) that made this growth possible was about 2 per cent in 2000. It is projected to remain high at least up to 2010, since the scope for progress is large given the expected growth of irrigation and the agronomic improvements that are still possible. After 2010 it is expected to fall as the waterworks potential will by then be realized, leaving less scope for further progress. This would allow many SEMC to increase their agricultural production by half by 2025, thus helping them to maintain *self-sufficiency* without major changes.

Most of the growth in agricultural production will go to feeding the population, leaving only a small percentage for export (Table 5.7). Exports will have to be targeted and concentrate on higher added-value items to bring some relief to an agricultural trade balance that will continue to degrade because of improving diets (Turkey as an exporter is a special case).

Table 5.7 Technological trend and agricultural production growth in SEMC

Technological trend (all production in per cent per year)

Country	1990–2000	1960–2000	2000–2010	2010–2025
Turkey	1.6	2.1	1.1	0.8
Israel and Pal. Aut.	3.3	2.2	0.9	0.7
Syria	5.4	1.5	1.7	1.2
Lebanon	1.0	3.0	1.8	1.1
Egypt	2.6	1.8	1.1	0.8
Mashrek	—	—	*1.3*	*1.0*
Libya	3.5	4.6	1.9	1.2
Tunisia	1.3	1.9	1.3	1.1
Algeria	5.0	1.3	1.5	1.2
Morocco	0.7	1.7	1.6	1.2
Maghreb	—	—	*1.5*	*1.2*

Agricultural production growth (%)

Country	1990–2000	1960–2000	2000–2010	2010–2025
Turkey	1.3	2.8	1.4	1.2
Israel and Pal. Aut.	1.0	2.0	1.4	1.1
Syria	4.0	3.3	2.2	1.8
Lebanon	3.0	2.2	1.9	1.7
Egypt	4.1	2.6	1.8	1.7
Mashrek	—	—	*1.9*	*1.7*
Libya	3.8	4.7	2.4	1.5
Tunisia	0.6	3.0	1.8	1.6
Algeria	2.7	1.1	2.4	1.6
Morocco	0.5	2.7	1.9	1.5
Maghreb	—	—	*2.1*	*1.6*

Source: FAO, *Plan Bleu*, Labonne, 2003

What opportunities will be there with agricultural free trade?

Fruit and vegetables, olive oil, flowers and cotton are the only possible sources of 'agricultural' currency for the south and east. The SEMC would like to benefit from the liberalization of agricultural trade that is presently under discussion, to give value to their *comparative advantages* with these products, and to compensate somewhat for the short-term negative impact of the Euro-Mediterranean free-trade zone on their industrial fabric. The EU is one of the major economic powers most open to importing agricultural products from developing countries. However, the EU still imposes quotas and high tariffs at the entrance to its markets (between 40 and 85 per cent for the SEMC's most sensitive agricultural products, especially high value-added products such as fruit and vegetables and their processing). Their reduction or suppression would make greater specialization possible.

To make such a hope and specialization come true, several constraints and challenges have to be taken into account:

- *real export potential* will remain modest because of the growth in internal demand and the reduced availability of land and water (except for Turkey);
- access to markets and payment for export products require high capacity in *logistical* and marketing *organization* (packaging, transport). These are very developed in the EU and Israel, but remain limited in the developing and transition countries. This weakness reduces competitiveness, especially in exports. An example is the date sector, the export price of which is US$5.45 per kg from Israel and only US$2.25 from Tunisia and US$1.6 from Algeria. These price differences are not due to intrinsic quality variations but to management capacities (organization, management control), marketing (eco-labelling, advertising) and quality control (certification). It is clear that quality and safety are rewarded and that originality of the offer (high quality varieties of organic dates) can open up interesting niches;
- *comparative advantages* are relative. The cost price 'delivered to the European market' of Almeria tomatoes (Andalusia) and tomatoes from Souss Massa (near Agadir) are similar (Morocco's advantage of cheaper manpower is compensated by Spain's advantage in transport costs);
- the hoped-for benefits of free trade (in terms of increasing exports, incomes and a better allocation of resources through specialization) have not occurred everywhere. Countries involved in free trade that have succeeded in their transition to a higher development level (Spain, Portugal, Greece as well as the South-East Asian countries) have generally been able to do so only through much *aid or protective tariffs*. Such a transition has also had significant impacts in terms of sustainability (Box 5.14);

> **Box 5.14** The entry of Spain and Greece into the common agricultural market, EU aid and sustainability
>
> Spain's entry into the European Community meant a radical change in competitive conditions for the other Mediterranean countries. The CAP (Common Agricultural Policy) promoted a clear strengthening of Spanish agricultural specialization (Spain's share in the import of fruit and vegetables in the rest of the EU increased from 4 per cent in 1970 to nearly 14 per cent in 2000; that of the third Mediterranean countries fell from 6 per cent to 4 per cent).
>
> EU aid is intended for agricultural modernization and structural purposes for countries or regions whose development is lagging behind or that are handicapped by natural conditions (mainly fragile rural regions). With 8.8 and 1.8 thousand million euros of structural funds and 5.9 and 2.64 thousand million euros of agricultural funds received in 2002, respectively, Spain and Greece are among the prime beneficiaries. In total, Greece recorded a net budgetary balance of 3.4 thousand million euros with the EU in 2002 (2.4 per cent of national GDP), and agricultural subsidies amounted to 39 per cent of farm income.
>
> This support has created distortions in competitiveness with third countries. Despite the structural funds and a successful transition through much aid, entry into the EU has not been without negative effects in terms of sustainability: a worsening trade balance and a loss of agricultural jobs not compensated by new jobs in other sectors (in Greece); the creation of strong regional disparities between the plains of intensive agriculture and inland areas, a strong intensification of water use and chemicals (in Spain); and loss of biodiversity, abandonment of environment-friendly traditional agrarian systems and strong specialization in particularly polluting fields such as intensive pig breeding and poultry raising.
>
> *Source:* EC. Allocation of 2002 EU operating expenditure by Member State (http://europa.eu.int/comm/budget)

- the EU's already significant technological and organization progress will increase even more with the continuing development of *precision agriculture* in southern Europe. This technology relies on the biological resources of production, involving sharp work on phenomena such as plant nutrition and assimilation (including the role played by light), and genetic improvement of varieties (for example seedless red watermelons). Techniques for controlling environmental conditions (such as crop protection, integrated pest control, the timing of water and nutrient inputs, neutral substrates in hydroponics and aeroponics) have been tested and extended. All this is extremely costly in terms of research, training, extension, operational implementation and sub-sector organization. The province of Andalusia alone could, technically speaking, have over 30,000ha under cultivation in the next ten years using plastic tunnels for producing several million tonnes of fruit, vegetables and flowers, benefiting from the necessary certificates for marketing across Europe. Funding possibilities exist in the country; research, extension and training are already very active, and organizing the entire sub-sectors (*filières*), from the beginning to the end of the production line, has had several restructuring rounds in a number of years.

It is therefore doubtful that the legitimate SEMC hopes of an easy entrance into European markets for their fruit and vegetables, due to market liberalization and the stagnation, if not the drop in yields resulting from protection of the European environment, will be fulfilled. Limited in number, joint ventures between the two shores would be more complementary rather than replacements, and only take place in a low-risk context (security for the northern investors).

Through exports, the SEMC may be able to exploit a few niche markets for *specific products* (some winter fruits and vegetables, organic produce, typical produce), provided that nearby markets able to pay for quality (i.e. in the EU) are reasonably open, that these products are identified and/or protected by authenticated source labels, and that adapted development policies are implemented, as well as promotional and effective marketing networks (there is an important potential role here for joint ventures and for the promotion of 'Mediterranean' food). Countries that already have established export positions (Morocco and Turkey) and, above all, those that have a strong logistical capacity (Israel) should be able to benefit more than the others, who will have problems getting a foot in the door. The example in Egypt (Box 5.12, p277) indicates, however, that some progress is possible in all countries.

The risks of an unsustainable baseline scenario
The baseline scenario for agriculture in the southern and eastern Mediterranean countries is therefore one of production growth stimulated by internal liberalization, agronomic progress and the development of domestic markets and irrigated lands, despite environmental degradation and recurrent and highly punitive droughts. In this scenario, various bilateral trade agreements are assumed to be negotiated (between southern countries, with the EU and the US, etc.) applying the principle of reciprocity and (almost entirely) excluding sustainable development objectives. Development issues and the multiple roles of agriculture (rural employment, the environment, food security) would not be properly taken into account in these agreements. For example, to adhere to the quotas imposed on exports to the EU, farmers are led to growing out-of-season crops, not adapted to the region, which leads to extremely high consumption of water and chemicals. At the same time, a number of seasonal fruits and vegetables, and typical and competitive products of the region (figs, pomegranates, cactus plants, etc.) cannot be exported since they are not included in the agreements. Moreover, for lack of adapted policies (lack of information, technology, organization, capital or preferential access to the European market), the desirable development of organic farming is greatly impeded.

Even if this de-protection is gradual, the risk of the baseline scenario is an *increase* in the already observed *negative trends*, in particular an accentuation of the *internal dualities* and *pressures on resources* that are characteristic of the region:

- risk of impoverishment and social vulnerability of the mass of small farmers, growing marginalization of the least productive areas;
- intensification in the most fertile and best infrastructure-equipped areas and a concentration of means on the minority of farms that have diverse incomes;
- increased pressure on natural resources: degradation of marginal lands (accentuation of survival strategies) as well as some irrigated lands (pollution, salinization), increased water consumption and reduced biodiversity.

These risks would increase even more with a lack of, or insufficiently active rural development and income-support policies and support for upgrading farms. Indeed, only a resolute implementation of such policies would enable the risks to be reduced and the opportunities enhanced.

With the anticipated profits from agricultural exports being harder to realize because of technical and financial difficulties, and with hopes of sufficient qualified manpower to massively adopt precision agriculture, there will be a strong temptation in the south and east to accept *genetically modified organisms* (GMOs) for the major crops, in particular to reinforce their resistance to water stress. The EU's position in defending the multifunctionality of agriculture and on GMOs may therefore risk losing support from the south and east in international bodies. In any case, an assessment of the possible risks of GMOs spreading in the region is necessary (Box 5.15).

Box 5.15 A necessary assessment of the possible impacts of GMOs in the Mediterranean

Since their marketing in 1996, the area cultivated with genetically modified crops has increased very rapidly on a global scale to reach approximately 58.7 million ha in 2002. Nearly all (99 per cent) of these transgenic crops are produced by only four countries: the US (39 million ha or 66 per cent of the total area cultivated with GMOs), Argentina (13.5 million ha, 23 per cent), Canada (3.5 million ha, 6 per cent) and China (2.1 million ha, 4 per cent). Sixteen products and GMOs are currently on the market at the EU level where no significant area is cultivated with GMOs, except in Spain.

Considering the population projections for 2025 for the Mediterranean Basin, it is likely that import or cultivation of GMOs could develop in the region. What would the impacts be on the Mediterranean ecosystem if genetically modified crops were cultivated? Would modified plants that, for instance, have been given good water stress resistance behave like invasive species, threatening the very existence of local species?

It would be necessary that each Mediterranean country acquires expertise in order to evaluate the release of GMOs into the environment, and that regional systems are put in place for an exchange of information and experience, in view of preserving the extremely valuable and fragile biodiversity of the region.

Territorial challenges in the Mediterranean

Southern and eastern Mediterranean arid regions: The battle to save lands and emerge from poverty

Aridity is a dominant constraint in a large part of the Mediterranean region, although with varying intensity. In the north its impact is limited to Spain (and a small part of Greece), while most of the southern and eastern countries are affected, except the wet mountains and the plains and hills that receive more than 400mm of rain per year.

Aridity poses different types of problems that policies have tried to counter, but the results are still not conclusive. The first problem is erosion and land degradation, which has become the dominant theme in combating desertification. A second is climate hazards and recurring droughts. A third is human pressures with increased impacts on the environment due to social changes and their effects on land tenure. New approaches to these problems are currently orienting development policies towards options that make the actual users more responsible, but they run up against the inheritance of 'interventionist' development policies that have prevailed for so long.

Land degradation: Inappropriate technical responses

The *drylands* of the southern and eastern Mediterranean, pastoral or forested areas and rain-deprived agricultural areas, are economically and politically marginalized and ecologically degraded. At the end of the 1990s the situation in the developing countries was discouraging: an increasingly degraded environment (desertification), people getting poorer, and in general terms, a failure of policies for protecting and restoring land and vegetation. Degradation is generalized and has got worse with droughts and structural adjustment programmes that have further reduced the meagre government credits devoted to rural areas, considered as 'not very profitable'.

Yet *soil conservation policies* have a long history in the Mediterranean. They began in the Mediterranean part of France in the 19th century with the *restoration of mountain lands*, the main objective being to protect towns and land located downstream of disastrous floods. After state-approved expropriation, exemplary and large-scale works were carried out (dams, micro-water catchments, reforestation). Problems of poor livestock raisers were resolved through migration to industrial towns and the colonies. In the southern and eastern Mediterranean drylands, *soil protection and restoration*, first implemented by the foresters in Algeria, then throughout the region, consisted of building bench terraces parallel to contour in order to regenerate agriculture and reduce the serious problems of sedimentation of water reservoirs downstream.

Soil and water conservation approaches, implemented since then, combine public works (gully correction and stabilization, stone ridges, banks of consolidated earth and hillside water retention works) with advice to farmers. At the same time, agronomists and pastoralists have tried to promote technical solutions (better pasture management, control of pasture carrying capacity, dry farming, introduction of new fodder crops, etc.).

The results of these efforts have generally been disappointing, and many projects have not achieved their objectives. A survey carried out in Algeria showed, for example, that of the 350,000ha of land with conservation works, more than 20 per cent vanished and more than 60 per cent were damaged, because they were never maintained. Similarly, attempts to reduce the number of animals or to sell grass in an authoritarian manner generally failed. Another example, the introduction of *atriplex*, on which many pastoralists based or are still basing disproportionate hopes, has never truly broken through, despite the food qualities of this fodder crop. For example, in Syria only 3000 of the 35,000ha planted ten years ago, are still holding out.

The causes for poor effectiveness or outright failure in combating desertification in the 1990s can be summarized as follows:

- the problem of desertification was not considered in the overall context of socio-economic development, and actions were not integrated into rural development programmes;

- the approach was sometimes not appropriate to the problems (there was a lack of understanding of processes and insufficient diagnoses);
- actions were developed with little reference to people's needs, priorities or know-how;
- poor coordination between agencies, and insufficient de-concentration/decentralization;
- a preference for investment in more fertile areas, unless for strategic or political reasons.

Because the prescribed solutions did not take the various actors into account, experts and public authorities ran up against difficulties in transferring the technical rationale into social reality. In woodlands, serious disputes developed between local people and forestry services.

Responses to climate hazards and droughts

For the past decade drought has become a structural phenomenon that is hard to cope with in economic and social terms. Governments have had to implement *drought prevention plans*, especially to safeguard livestock. The most common policy consists of subsidies for rehabilitating herds and *feed supplementation*, this one being a normal practice in marginal areas. Nevertheless, such state aid also comes with some perverse effects: it reduces efforts by livestock farmers to cope with climate hazards and reintroduces an animal pressure before pastures have been able to recover. Moreover, it promotes the use of often far-away markets (for inputs and livestock sales), thus increasing the dependency of local economies. Reduction of herd mobility and motorization, which are resistance mechanisms to climate hazards, mainly benefit the big livestock farmers.

Exploiting groundwater (by digging wells with motorized pumps) is another response to droughts. However, its development has generally been poorly regulated, and has led to *overexploitation of groundwater aquifers* over recent years. For example, in some regions of central and southern Tunisia, where new administrative procedures have considerably accelerated the official privatization of collective lands, more than 7000 wells have been dug in 15 years. There has also been considerable expansion of orchards into the steppe, as a response to climate hazards. This phenomenon, of an unforeseen scale, is due to land purchases by local families with external resources (emigration, trade) or by city dwellers.

Faced with the increased recurrence of droughts, some governments have turned to more *sustainable policies of climate hazard management* by developing forecasting instruments.

Concerning livestock-raising, policies are turning to early culls and storing feed supplements to save the reproductive capital. Some projects (as in western Morocco) have succeeded, with the participation of local users, in obtaining crop rotation and opening pastureland depending on its forage potential, according to conditions during the year. Longer-term measures are being studied to introduce cereal supplement irrigation (an already widely practised measure in Spain and Tunisia), and better rainwater management. Ad hoc measures are also planned to encourage cereal irrigation when the risk of shortage becomes clear. Barley crops are also being encouraged in semi-arid regions, even though this crop has been much neglected by research. Agricultural extension is an active part of this strategy with experience showing that a good timing of sowing and agricultural practices could significantly reduce the loss of productivity due to insufficient rain.

Many technical innovations over the past few years have increased hopes for better management of the environment and a more confident adaptation to climate hazards. More effective *rainwater* management occupies an essential place in these options, although its techniques have not been sufficiently explored. Very often all that is needed is to revitalize traditional techniques, which are widely known for their effectiveness. Such techniques include *magden* (ponds), *tabia* (embankments), *jessours* (weirs) and *matfias* (tanks).

Another technique relates to repeated disk ploughing, which is a major cause of farmland degradation. Unfortunately commonly used in the southern countries, such mechanization leads to soil destabilization and erosion. Direct seeding (cultivation with no tillage, or, its even more innovative form, sowing under existing plant cover) has been subject to encouraging professional experiments in Tunisia since 1999 and could prove to be an alternative to disk ploughing.

Spectacular applications of cactus in several Maghreb countries have shown many advantages, for example field closure, protection of soil against erosion, production of export fruits, and livestock fodder. The expansion of rain-fed olive trees in low-rainfall areas should, however, not be seen as a viable alternative. These trees, planted in rows, in broad 24 × 24 metre patterns, only yield fruit every 6–10 years. Moreover, they increase the soil's sensitivity to wind erosion when it is laid bare by repeated disk ploughing. Excluding some marginal areas where olive trees could be planted, this practice should not be allowed when it encroaches on pastureland. The only 'justification' for these futureless investments is the privatization of a public good (and to benefit from unjustified state subsidies). In south-eastern Tunisia, for example, this has had harmful impacts.

Participatory management of natural resources in arid areas

The importance of natural areas in the southern and eastern countries, traditionally collective-use pasturelands, puts a particular emphasis on changes that have occurred in the relationships between communities and land tenure.

In contrast to the intensive production sectors, tribal customs have maintained an important role in pastureland and forested areas. Here, *'belonging to a group'*, which is not necessarily respected by administrative subdivisions, still commands the right of access to pastures or cultivated land, particularly in Maghreb. Forest privatization has not changed user customs; they continue to consider forest grazing land as their own.

However, cultivation and livestock-raising systems in these areas have experienced profound changes over the past few decades. Customary organization used to enable long-term management of resources (with community protection from grazing). Up to the 1950s and 1960s, the areas concerned preserved a relative balance between production and consumption as they did between the population and the environment. Families ensured a sufficient agricultural production while practising what appeared to be sustainable natural resource management. But even then some pastureland was already degraded. Distortions started growing from the 1960s onwards, with the effects of the very large population increase and the regression of customary organization. One significant resulting trend was the *monopolization of pasturelands*, whether for cultivation or 'privatization' (made possible by mechanization or digging of wells). Big farms based on extensive livestock-raising or marginal cultivation have thus been created, opposing their economic and political power to the mass of landholders, who were pushed aside on what remained of collective lands. This development began very early in the Middle East and spread to Maghreb, although with unequal effects.

In an attempt to control this development, several countries have committed themselves to management policies of collective land, mainly through *privatization*. Results have been uneven, particularly because policies have usually been applied without taking the pastoral specificity of most of the collective land into account. In Algeria the law dealing with access to farmland property has led to private investment (largely by urban dwellers) in the steppe-lands of the Upper Plateaux, once enjoying collective land status (*arch*), later privately owned. The results have often been disappointing, the privatized lands not being suitable for agricultural activities. In Morocco, the privatization of collective land has been limited to irrigated land. But other collective land with rain-fed farming has been enhanced by investors who pay a small rent to the local community services that act on behalf of the holders. In Syria, collective pasturelands have been organized into large cooperatives, again with unequal results. In Tunisia, the privatization of collective lands has undeniably boosted farming in the centre of the country and, through secure land tenure and access to credit, enabled major intensification in the form of fruit-tree plantations and cactus reserves. However, the inconsiderate spread of this policy to the arid southern regions has caused growing distortions in pastoral areas (despite legislation prohibiting agricultural activity in these regions).

These observations on the perverse effects of privatizing collective lands in pastoral areas, whether the result of a de facto monopolization or an application of the law, have led decision makers to seek other formulae for managing collective lands of pastoral nature, without necessarily turning to privatization. For example, *managing pasturelands through negotiations with users* has been practised in eastern Morocco in an innovative project based on an organizational formula rooted in the social inheritance of custom rights. The negotiated restoration of the richest but most degraded pasturelands was founded on 'ethno-lineage cooperatives'. In 1996, after three years, 300,000ha of mugwort (*Artemisia vulgaris*) were restored and used for pasture following strict rules accepted by the land users (the area increased to 450,000ha in 2000). In south-eastern Tunisia a new development project has followed a similar approach by promoting an *integrated, territory-specific and participatory approach* (Box 5.16).

Box 5.16 The agro-pastoral development programme and local initiatives in Tunisia

This programme (US$44 million funded by IFAD, OPEC and Tunisia) claims to be a local development process founded on an integrated, territory-specific and participatory approach. It involves 66,000 rural dwellers, many of whom have abandoned their perched villages and moved to new 'water-and-electricity-equipped' urban centres. Livestock raising has remained their main activity despite a long tradition of emigration that began to counter poverty and the development of not very sustainable irrigated agriculture.

The programme includes a series of complementary actions: infrastructure investment (roads, pastoral drilling for water), improvement in natural resource and small infrastructure management (transferred to population groups effectively having rights on pastoral lands, which is an innovation), and the promotion of micro-enterprises and other economic initiatives mainly for the benefit of women and young people. An initial experience highlighted the population's profound knowledge of its natural resources and the solutions for their sustainable management. It also showed their ability to organize themselves and design a diversified development programme.

These new approaches bring hope, but the difficulties should not be underestimated. Their implementation is still being impeded everywhere by a lack of de-concentration and decentralization, by compartmentalization between administrations and the rigidity of administrative rules (lack of flexibility in implementing public support, and managing land tenure and uses). Nevertheless, such a return to traditional wisdom relying on today's technologies and strong voluntary policies would be the only way to escape a very sombre future. This is one possible way forward (and history shows the incredible resilience of human societies), but it also shows that societies can die as a result of their inefficiency.

Mountains: Degradation or conservation and restoration of the 'Mediterranean garden'?

The importance and fragility of Mediterranean mountain systems

A large percentage of Mediterranean peoples (more than 60 million, of whom 23 million are in Turkey alone) live in mountainous areas. The percentage is especially high in Bosnia-Herzegovina (50 per cent of the population), Albania (39 per cent), Lebanon (37 per cent), Turkey (37 per cent), and in Algeria, Morocco, Serbia and Montenegro and Slovenia. It is not insignificant in Syria and the four EU-Med countries.[25] Many come from communities with strong identities (Berbers, Kabyles, Lebanese Maronites and Druze, Kurds, Kosovars, Montenegrins, Corsicans, Sicilians, etc.). These long-inhabited areas are also 'built' areas, with local communities always having sought to develop slopes and control water. The closeness of the *relationships* between *people* and a *natural environment changed* to maximize resources and reduce risks, constitute the main characteristic and real unifying element of the Mediterranean mountains.

Despite their particularities and an isolation that still remains very real in the north, but even more so in the east and south, the Mediterranean mountains have always been closely attached to the cities, coasts and the sea that surround them. For a long time, the mountains, a place of refuge, were suppliers of manpower for cities, pastureland for herds of the plain (transmigration) and timber for ships. They play an essential role as Mediterranean 'water towers' and main shelters for biodiversity. Apart from their coolness, they offer their authenticity and landscapes to all visitors from near and far, amenities that can only grow with globalization.

Despite these varied and essential roles, mountains have been, and often still are, the 'forgotten ones' of development. Since production has a hard time coping with competition from other more privileged regions and life in the mountains is by nature hard, the Mediterranean mountains have seen *massive* and *abrupt migration* throughout the 20th century, the reasons often being cultural and economic (Box 5.17). Abandonment is the sign of a loss of general confidence in the future of their area, by individuals and local communities and can occur even when tourism development offers new horizons to development. This kind of process is underway in several countries, for example Turkey, even though increasing numbers of urban dwellers are visiting the mountains. One trend intersects with the other without really gauging what both are losing.

Box 5.17 Massive migration from mountain areas, Albania as an example

Migration from the mountains, often massive and sudden, can result in a genuine 'desertion' of entire villages, which, once abandoned, are left to fall into ruin. This was the case several decades ago in France, Italy and Greece, and is now the case in Albania. Since the freedom of movement and the right to reside anywhere in the country was affirmed by law in 1991, some mountain villages have lost between 60 and 80 per cent of their population, which indicates total desertion in future and a concentration of migrants on the plains of the western part of the country and large cities such as Korça, Tirana, Dürres and Shkoder. According to surveys findings, 80 per cent of the migrants claim that they migrated for economic reasons. The same surveys show that more than half do not want their descendants to remain in the village and in farming. All the young women wish to marry in towns, but only a third manage to actually do so, with a third marrying on the plains and a third finally accepting to get married in their villages after having given up all hope of a better solution. Only 16 per cent of the people interviewed, all over 50, stated they would accept the idea of returning to their villages if all the living and working conditions were met.

Source: Sabri Laçi, communication to *Plan Bleu* workshop on 'Montagnes méditerranéennes', Sophia Antipolis, 1–2 April 2003

Once abandoned, the 'built' agricultural systems in the mountains deteriorate and can no longer produce all their amenities. When people leave, the land abandoned and the generational links broken, a whole historical culture disappears, a loss in the sense of collective responsibility related to the land and a land 'escheat' becomes general. Latin law, which does not formally link the property right with a (legal) duty to maintain the land, has promoted this development in Mediterranean Europe.[26] The land changes and returns to its 'wild' state, becoming less hospitable. Not enough men and women remain to maintain diversity and quality landscapes, basic services, and receive possible visitors. Another consequence of abandonment is the disturbance of the mountain ecosystem: loss of biodiversity, deregulation of the water regime, increased erosion risks, and in some cases renewed erosion.

Despite some rural revival, the baseline scenario for the northern countries remains one of *gradual abandonment*, which will probably continue, if not accelerate, with the projected fall in agricultural manpower.

A Sustainable Future for the Mediterranean

Abandonment, which is increasingly frequent in the northern countries, is not occurring in most of the southern and eastern mountains. Instead, huge rural poverty and the resulting overexploitation of resources are the predominant trends. The degradation is even more serious than in the north: disappearance of plant cover, sometimes disastrous erosion, and loss of biodiversity (middle Atlas and Tell Atlas). The baseline scenario in the south and east thus risks being even less favourable than in the north. Indeed, it is likely that poverty will get worse and *stress and degradation increase*. However, abandonment would also occur, at least locally and mainly after 2015, and might enable some 'biological revival' given the strong resilience (a return to equilibrium after serious disturbances) of Mediterranean ecosystems.

Reconciling development and conservation

Faced with a fairly unfavourable baseline scenario, especially in the south and east, there are many examples showing that shifts towards a more desirable future for mountainous areas are possible and that their generalized application would enable the necessary changes.

In the *north*, new dynamics are emerging (Box 5.18). Active protection of valuable natural and semi-natural environments and diversification of activities, with typical products of the Mediterranean region inserted in quality processes, is no longer a utopian dream but a reality. This reality is highly dependent on the level of commitment by local actors and decision-makers, and many examples show that this commitment is possible and can be very effective. Such an approach enables the 'Mediterranean garden' with its biodiversity to be reconstituted, at least locally, while meeting the growing expectations of people to discover quality products, visit the mosaic of natural, well-kept landscapes and through these landscapes, rediscover elements of culture, nature, history and authenticity. Thus, an original and complex but *specific* 'Mediterranean mountain model' could develop. This calls for another perception and other responses at the political level different from the 'Alpine agricultural model' (cattle raising) that predominates in Europe, which is today the only well-recognized and supported model, especially by the EU authorities.

One condition is that support be oriented towards more sustainable systems. In the Mediterranean mountains, EU and national aid programmes have had perverse effects such as unjustified reforestation and opening up of forest roads, and aid to livestock-raising based on ill-perceived realities that have aggravated the shift towards practices harmful to the environment (extensive grazing without sustainable management of pastures, an increase in artificial land cover).

> **Box 5.18** Mountain revival in Cyprus, the Iberian dehesas and France's Cévennes
>
> Until 1992 water and irrigation works absorbed 60 per cent of the rural areas budget in **Cyprus**. With the exception of an integrated project in the Pitsilia area (between 1978 and 1982), rural areas and mountains (16 per cent of the country, 160 villages), where farming revenues are the lowest, barely managed to survive. The rapid economic development of the island went hand in hand with massive rural migration. After 1992, policies became more balanced. The development of communication networks and urban stress favoured a stabilization of the rural population. The rural development plan for 2004–2006 is based on sustainable rural development. In addition to conventional aid to agriculture, it aims to support projects linked to the conservation of local breeds, forest development of abandoned lands, maintaining rural landscapes, local heritage and traditional know-how, support for organic and typical quality products, private forestry and small food processing units. Integrating local activities and actors, multiple activities and spatial planning are seen as indispensable principles for these new approaches.
>
> The *dehesa*, a complex agro-sylvo-pastoral system located mostly in the **south-western quarter of the Iberian peninsula**, is a remarkable example of adaptation to the natural environment. On very large farms it combines activities like cork production, extensive husbandry (exploitation of the herbaceous layer under cork and holm oak forests), pig breeding (enhancing the pastoral value of masts) and feed supplementation if needed. The production of cereals and charcoal, once important, has been practically abandoned. This system requires a constant fight against encroaching shrubs and maintenance of trees. It entered a period of crisis in the 1960s after Spain's and Portugal's economies opened up, which led to a fall in the prices of animal husbandry products (competition with new intensive livestock-raising) and a rise in salaries. Part of the area was degraded by shrub development and part was reforested with eucalyptus trees (for the paper industry). A revival of the *dehesa* started some 15 years ago. This has been possible because society started to recognize and enhance the multifunctional role of these areas: interest in quality production favourable to the environment, value of landscapes and ecosystems rich in biodiversity, enhanced value of products (for example *Jabugo* ham), eco-tourism and safeguarding the Mediterranean forest. EC aid (for extensive animal husbandry in disadvantaged mountain areas, structural funds, etc.) brought support, and the *Estremadura* region even voted for a '*dehesa* law' that sets the rules for its sustainable management.
>
> In France's **Cévennes** mountains, tourism and agriculture have been revitalized by developing the production of sweet onions on terraces, apple juice from old apple varieties, goat cheese with the '*pélardon*' label of origin and new chestnut-based products, and by the national park activities (actions in highlands, protection of valuable natural areas, contributing to

preserve pastoralism through land management and support for herders and local endangered breeds, development of agro-tourism and eco-tourism, architectural regulations and aid to the restoration of the built heritage). All this has made it possible, at least at the local level, to conserve and restore the 'Mediterranean garden'.

A promising approach to managing mountain areas is the one adopted by pastoral land associations (several of which are operating in France). It works on the principle of pooling cultivated or pasture lands into a single fund, which is managed by a livestock-raising company, while land taxes are partly paid with the farm revenues.

Source: Plan Bleu, various national sources

In the *southern and eastern countries*, the 'Mediterranean garden' still exists in the mountainous regions, although in a degraded form because of poverty/overexploitation and deforestation. The main challenge is to continue producing a lot while successfully *decoupling production from degradation*. For the future, the long-term objective would be to strive for rural revival through quality products and services while avoiding the excessive abandonment that has occurred in mountains in the northern countries. A scenario *moving towards modernization without breaking with tradition* would make it possible to avoid all the problems encountered today in the northern countries where authenticity has to be reconstituted from residual fragments.

In halting or reducing degradation, responses such as forest protection, reforestation and creation of protected areas, have only had partial success. Their application has too often led to serious disputes between forest administrations and local communities. In reaction to these approaches *integrated and participatory approaches* for the sustainable management of natural resources have begun to be promoted. They seem to be a promising path as shown by the example of the GEFRIF project in Morocco (Box 5.19). This integrated approach is now being factored-in when implementing *catchment-area-development* programmes (sustainable management of natural vegetated areas, reforestation and land restoration) for which complementary actions between actors and private and public areas are needed. However, these programmes seem to face difficulties in gathering the required funds. The Moroccan national development plan for catchment areas, implemented under the water and forest administration, includes strong participation by local communities. However, in the 1990s implementation was only 13,000ha per year, instead of the 75,000 envisaged, and the downstream positive impacts are still hard to gauge. The allocated and planned budgets (about 30 million Dirham) are still far from the 150 million said to be necessary, and the principle adopted in 1988 of a financing system including downstream beneficiaries is still to be realized.

Box 5.19 The GEFRIF project in Morocco

The Rif has a large rural population (100 people per km^2 in the Chefchaouen Province) exerting much pressure on natural resources (30 per cent loss of forest cover in 30 years, up to 65 per cent in some areas, and soil erosion). The main crops are low-yield cereals (6–8 quintals per ha), forest goat raising, arboriculture and cannabis. The cannabis production has developed significantly but is currently in crisis due to policing efforts and a reduction of profits through market saturation.

Changes in the development model (agricultural diversification and sustainable resource management) are underway thanks to Morocco's strong political will, EU financial support and the mobilization of local communities. Negotiations with local people are leading to targeted investment choices in line with needs actually felt to be priorities (for example schools, road access to enable a tenfold reduction in transport costs), and a review of administrative procedures for land tenure and natural resource management.

Within the framework of the GEFRIF project (participatory management of the Rif forest ecosystems), the forestry administration has organized a new land delimitation after negotiations with the communities. Forest stands of quality or protection stands have remained classified as publicly owned, whereas those that were overexploited have de facto been allocated through individual plots. Offences have been significantly reduced, and a new natural forest regeneration dynamic has been observed. Developing a new generation of locally produced bread ovens (metal) should lead to a 50 per cent saving in fuel-wood consumption while lightening the women's work load (wood gathering) and pressures on forests.

This example highlights the changed role required in administrations for shifting from a 'command and control' culture to one of local and participatory governance. Involvement is facilitated by a de-concentrated cross-sectoral organization (the *Agence du Nord*), which pools national and European financing, and benefits from simplified administrative procedures that facilitate implementation.

The spatial distribution of the population could evolve along one of two scenarios. There could be a concentration in towns and villages located on the plains and along the communication lines, with the mountain villages being abandoned or reduced to a role of temporary residence. Or there could be economic diversification of villages, implying a better insertion in the modern world. The coastal road currently under construction could well lead to an evolution towards the first scenario. However, the form given to tourist development in the region may favour a more balanced development.

Source: Montgolfier, *Plan Bleu*, 2002

In the north, as in the south and east, the shift to an alternative scenario implies launching processes and dynamics that seek to *reconcile development and protection*. This means recognizing that in this part of the world,

2 A Sustainable Future for the Mediterranean

perhaps more than anywhere else, the biodiversity to be protected is both (and often simultaneously) 'natural' and 'cultural', and that it is difficult, if not impossible, to discuss development without discussing the protection of nature and vice versa. This would argue for a specific *Mediterranean approach to the Convention on Biodiversity* combining nature and culture. And local development should be added to the two basic missions of national parks (ecosystems protection and welcoming the public) since, in this region, the third mission is often a condition for success of the other two. This means ensuring strict protection through regulation (possibly by land acquisition) of the last 'relict' natural environments, which are all extremely precious for their biodiversity and their landscapes; favouring the ageing of some forests; protecting and promoting the traditional practices and architectural heritage that have shaped today's heritage with its great value (landscape and biodiversity); and making local people benefit from the protection efforts (to which they are called to actively contribute) through the enhanced value of products and controlled tourism development.

Several national parks are fortunately involved in this pathway, the Italian national parks in particular, as exemplified by the *Monte Sibillini* park, which makes its information huts available to young people for developing new sustainable tourism activities. The French experience of *regional natural parks* is also interesting for combining protection and development. These parks generally operate as associations of municipalities, have technical advisory teams and ten-year territory-specific projects involving various partners (through a charter) for ensuring a development process with environmental protection. The 'charter' is reviewed by the national council for nature protection and approved by the state, owner of the 'regional park' label, which can be withdrawn if need be.

A formula particularly well adapted to the Mediterranean is that of *'biosphere reserves'*, first established in 1976 under the aegis of UNESCO (Figure 5.16). Their aim is to reconcile the imperatives of protecting biodiversity and ecosystems with the legitimate aspirations of local populations with a view to satisfying their economic, social and cultural needs. The different functions of each reserve are facilitated by continuous monitoring (scientific follow-up and indicators), and zoning: strictly protected central zones are surrounded by a buffer zone where activities required for conservation can be promoted (active protection) and others regulated; a transition zone is devoted to cooperation and local development in the broadest sense. However, many countries do not yet have any biosphere reserves (Turkey, Syria, Lebanon, Cyprus, Libya, Albania, Bosnia-Herzegovina) or only a few (Greece, Croatia, Algeria and Morocco).

Figure 5.16 **Mediterranean biosphere reserves**

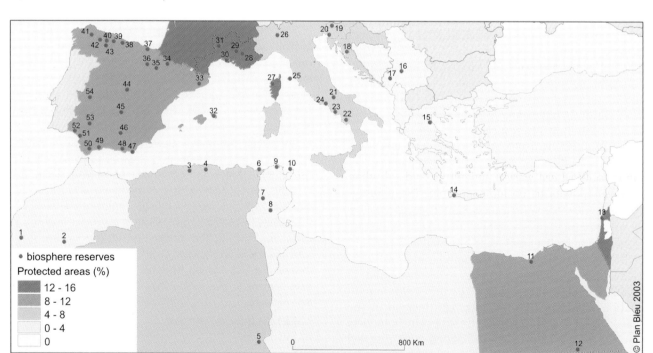

Source: UNESCO

Note: A list of the reserves is provided in the Statistical Annex (Table A.36).

Donors have an important role to play in promoting such processes in countries where they are not very developed. Several integrated projects aimed at reconciling conservation with development are underway with their financial support, such as the Karst Ecosystem Conservation Project (KEC), in Croatia, supported by the World Bank. This programme supports cooperation between the local population, national park authorities, entrepreneurs and NGOs (awareness raising, training, sustainable use of resources, rural revitalization). The KEC project has been implemented since 2004 and its completion is expected in 2007, with funds amounting to US$5.07 million.

Will we know how to conserve farmland and suburban rural areas?

Fertile coastal and fluvial plains cover about 5 per cent of the land in the Mediterranean. Through considerable enhancement efforts (draining, mosquito elimination, irrigation and soil reclamation) carried out by successive generations, rural areas in these fertile plains fully contribute to economic development. These regions and their agricultural and natural areas (forests, wetlands) are nevertheless threatened by the development of infrastructures and by very poorly controlled urban expansion in general.

A shift towards an alternative scenario would require an awareness that could lead society to consider valuable (heritage and economic) and threatened farmland as *areas that need protection*, just as much as some forests and coastal areas. In addition, it would be necessary to develop voluntary policies with combined land-planning tools.

Several countries already have specific laws for protecting their agricultural lands (for example Turkey and Tunisia). However, enforcing the law and ensuring the long-term protection of these lands are made harder because of the considerable difference in value between farmland and constructible land and the enormous subsequent pressure from landowners and investors. Organizing a stronger collective, professional and social dynamic to *defend and promote* agriculture and suburban rural areas appears to be indispensable. The different evolution of two regions with very similar natural conditions and development levels such as the Italian Liguria Riviera (which has managed to keep most of its agricultural heritage) and the French Riviera (which has not) reveals that loss of agricultural lands is not inescapable when major long-term issues in terms of identity and the living environment are taken into account.

Apart from urban planning tools (see Chapter 4), the development of *Local Agenda* 21 processes should make progress possible through a higher level of awareness of the heritage dimension of this agriculture.

New practices need to be developed. The objective would be to make *regional development* a tool for helping in decision making with a rationale of decentralization, participation and sustainable development. To this end, *master plans* negotiated with the various decision-makers and actors involved for the benefit of a shared vision of a desired common future, could spatially identify the agricultural and natural areas of rural regions to be conserved in the long term for the common interest. To make such development possible, innovative measures will be necessary in the coastal regions that are subject to very high pressures, even going so far as making lands legally inalienable (not transferable to another) when their long-term conservation as farmland is judged to be a priority. Enforcing such a land easement would require legal backup, and, if need be, compensation, but would cost less than purchasing farmland by public bodies (with leases to farmers). The latter solution is beginning to be suggested in some coastal regions (for example in *Prospective du Languedoc Roussillon à l'horizon* 2015, INSEE, 2003). Funding could come, for example, from taxes assessed on the land's added value.

Some countries (like Egypt) have developed *new town* policies to protect their best land by reducing population pressures (with relative success), others are planning to do so (as in Algeria) (see Chapter 4). The general question raised here is one of redistribution of people and their activities in the Mediterranean region.

Alternatives for rural areas

The current marginalization and dysfunction of many Mediterranean rural areas is generally a direct consequence of the fact that the economic and technological centres are drifting towards northern latitudes where development models have been worked out over several centuries.

However, the preceding analyses show that there are credible options for shifting from the baseline scenario by choosing a more optimistic vision of the future of the ecoregion up to 2025. Nevertheless, the challenges should not be underestimated. Indeed, committing to a path of a more sustainable process implies *in-depth* changes in:

- the 'vision' of the roles of rural areas, of their possible and desirable futures, and the objectives of an alternative scenario for sustainable rural development;
- the overall development and environmental management policies of countries;
- the patterns of regional cooperation and the factoring-in of sustainable rural development by the Euro-Mediterranean Partnership.

Although the nature of these changes is not identical in the north and the south, for the future the prime necessity for solidarity and interactions that bind the two shores should be stressed. The destinies of the two

2 A Sustainable Future for the Mediterranean

shores have been closely linked throughout the history and continue to be so despite what separates them.

Towards a Mediterranean 'vision' of sustainable rural development

The first change, and probably the most promising for shaping the future, concerns the way Mediterranean societies see their environment and rural areas. Everything mentioned above shows that Mediterranean rural areas, perhaps even more than elsewhere, are ones where ecological and socio-economic dysfunctions generate risks (that can endanger the development of countries and regional stability), but that they also offer considerable potential. Increased awareness of the costs and risks of poor management and of the multiple roles of rural areas (and their wealth) is a basic condition for changing scenario.

Awareness of the many roles of Mediterranean agriculture and rural areas

The above-mentioned analyses show that in addition to agriculture and its traditional economic food-security functions, rural areas fulfil many other environmental and social functions which are essential for all Mediterranean peoples:

- On the *ecological level*, rural areas have a decisive role not only for the quality of the rural environment but also for suburban and urban areas, and beyond that, downstream, for the coastal and marine ecological balance of the Mediterranean Sea itself. This role is played in particular through water regulation, water conservation, soil and biodiversity, water, soil and biomass production, and sedimentary deposits in the coastal ecosystems. Finally, what must be stressed is the importance of farms in maintaining the quality of the countryside and acting as 'green lungs' between cities in a heavily populated region with strong aptitude for tourism.
- Rural areas are a major element of the *rural–urban* population *equilibrium*. If not controlled, too rapid concentration of people in cities and coastal areas could lead to social instability and the high costs that come with congestion and pollution. Moreover, excessive rural migration as currently occurring in some southern and eastern countries, and under-population in rural areas, as is being seen in the north, will lead to costs that can become prohibitive for states and local authorities: public service maintenance costs for sparsely populated regions, public management costs of abandoned areas, and costs linked to induced risks (forest fires, flooding).
- *Rural employment* (agricultural, semi-agricultural and non-agricultural) becomes a decisive factor in the search for a viable social and spatial balance. It is also of prime importance in containing the social costs related to rural–urban migration and between the northern and southern shores. This issue is even more crucial in the Mediterranean because of the importance of regional contrasts and the dualities that result from them, particularly between highlands and lowlands and between populations with different cultures. Rural employment has a vital regulatory social function, conditioning its viability but also the resilience of societies in times of crisis.
- Rural areas ensure a growing *function of accommodation* (residential and recreational), the economic impact of which goes well beyond the limits of the eco-region and which will only gain in importance with globalization.

The observations that can be made based on existing trends and predominant policies show that these multiple functions of rural areas, the importance of which should be obvious for decision-making, are in fact very *insufficiently recognized*. Such an undervaluation appears all the more harmful since a lasting satisfaction of social demands concerning these multiple functions implies management costs that have a clearer and clearer economic reality, a reality which markets are not yet taking properly into account (Box 5.20).

It is not, however, from lack of trying to demonstrate this fact. Environmental economists, for example, try to quantify the total economic and ecological values of forest areas. Agricultural economists, for their part, try to identify and quantify the multiple (and non-commercial) roles of agriculture so as to facilitate their inclusion in policies. These attempts contribute to the vital recognition of the need for sustainable development of the *rural sector's social and environmental functions*, and, within it, the agricultural sector, as encouraged by the UN Commission for Sustainable Development. Such observations demonstrate beyond a shadow of doubt the importance of these issues.

> **Box 5.20 The value of the multiple roles of Mediterranean agriculture and forests**
>
> Recent assessments of the total economic value of **Italian forests** have shown that the production of timber accounts for only 9 per cent of the total value of forest use, compared with 58 per cent for the water cycle, 11 per cent for 'hobbies, mushrooms and hunting', 10 per cent for firewood, 3 per cent for grazing, 3 per cent for sequestering carbon and 5 per cent for other positive functions. In Morocco this value is broken down as follows: 30 per cent for firewood, 23 per cent for grazing and 18 per cent for protecting the water cycle. On the other hand it is negatively influenced by erosion (–11 per cent) and by deforestation (–7 per cent).[27]
>
> What is true of the total economic value of forests is also true of pastureland and cultivated land. Well-managed

Rural Areas

pastures and cultivated terraces are, for example, excellent retainers of water and soils. The obvious importance of extensive livestock-raising in the southern countries, can be added to these poorly perceived effects. Synergies and complementarities between the functions of the different components of rural areas add even more to their overall economic value. For example, it is recognized that the biodiversity and quality of Mediterranean landscapes relies heavily on maintaining a certain mosaic, and thus on a balance between 'natural' areas and forests, and pastoral and cultivated land.

Recent work by FAO on the multiple roles played by **agriculture in Morocco** shows that economic growth of the agricultural sector through pro-agricultural policies should make it possible to significantly reduce the incidence of poverty at the national level. Evaluating the sector's environmental (and social) externalities is seen as essential, so that political decisions take into account the non-commercial services rendered by agriculture to society as a whole. For example, in addition to its contribution to employment and GDP, agriculture generates important social benefits in terms of the rural–urban population balance and furnishes high externalities profitable to other sectors (for example agro-tourism in the High Atlas). Actually, it is also the basis for an important, informal social insurance system by providing functions of social connection and reception for emigrants (the money they send back to their families still in villages could be seen as income from an insurance contract).

Unfortunately, national accounts give a distorted and underestimated picture of the real economic weight of the agriculture sector. Undervaluation of the roles of agriculture concerns not only its non-commercial functions, but also its direct economic and commercial ones. Indeed, the accounts separate strictly agricultural activities from agri-business activities and from agriculture-related upstream and downstream services (transport, trade, insurance and others). Furthermore, they neglect the most important multiplier effects of growth that agricultural growth has for the benefit of other sectors.

These observations emphasize that the value of agriculture, forests and rural areas in the Mediterranean region generally exceeds their contribution to GDP because of their multiple functions and the considerable non-commercial contributions (not reflected in market mechanisms) that these sub-sectors (and rural areas as a whole) make to society. The wealth of heritage of rural areas and the diversity of their functions are therefore important assets and opportunities for the future if politicians had the will to take them into account and if efforts were made to adapt to changes and render explicit the underlying challenges.

Objectives of an alternative scenario

An alternative scenario of sustainable rural development would aim to satisfy this diversity of functions. It would result in *continuous* and *cumulative* progress in three complementary fields: coupling/decoupling agriculture and sustainability, with farmers becoming true agents of sustainable development; improving living conditions, through social, economic and cultural development of fragile rural areas; and enhancing the Mediterranean quality and diversity.

Succeeding in 'coupling/decoupling' agriculture and the environment

The alternative scenario is one of successfully coupling and decoupling agriculture and the environment. Decoupling aims to reduce the negative impacts on the environment and coupling aims to maximize the positive impacts (amenities).

Agricultural development in this scenario takes place on the basis of an agriculture that is *respectful of the environment*, the primary aim of which is better enhancement of *water resources*. The objectives of the alternative scenario mentioned in Chapter 1 would be reached with 34km^3 of water saved in the catchment area by the sector, compared with the baseline scenario. The search for efficiency also affects rainwater, which is better conserved and valued. To achieve these objectives, agricultural policies make sustainable water management one of their primary objectives and implement appropriate tools (aid conditions, pricing, targeted subsidies for developing economical equipment, adapted crops and technical practices, awareness-raising and the organization of producers, etc.).

Developing agriculture that respects the environment also makes it possible to reduce the use of *fertilizers and pesticides* compared with the baseline scenario. Their consumption would thereby drop in the north, and only increase by 25 per cent in the south and 35 per cent in the east, compared with 50 and 70 per cent in the baseline scenario. An even greater reduction would be possible by developing organic farming (which could be facilitated in some southern and eastern countries where farming remains more 'natural'). Organic farming would use 15 per cent of total arable land by 2025; the Italian example would be widely followed in the region and markets in the north would be more open to exports from the south and east.

The alternative decoupling scenario would see a *halt to desertification* resulting from poor management before 2025, by upgrading the living conditions of rural dwellers and improving regulation of agricultural, pastoral and forestry activities in the areas concerned (participatory management, clarification of rights and rules for access to resources).

The alternative scenario also assumes a successful coupling of agriculture and the environment.

The sustainable management of agricultural and naturally vegetated areas (mountains, steppes and 'natural' areas in plains, hills and plateaux) takes their *multifunctionality* into account: biomass production for various types of production (ligneous, plants, pasture, energy), water regulation and erosion control, response to landscape and recreational demand, safeguarding biodiversity, residential demand, etc.). The fundamental role of *grass* (not only of forests) is recognized with the consequent support and supervision of pastoral, agricultural and forestry activities. These are also supported and reoriented with a view to *reducing natural risks* (forest fires and flooding). The major fires projected under the baseline scenario would be avoided.

In this coupling scenario, the positive role played by agriculture in maintaining the great 'green lungs' and rural character around urban areas is at last recognized. The developed countries would thus end the current trend of urban sprawl into suburban rural areas and engage in a scenario of 'sustainable cities and rural agri-business'. The southern and eastern countries that are still in a position to do so would succeed in avoiding the urban sprawl that would result from the expected increase in car use. In all, the region succeeds in reducing the loss of quality agricultural land projected in the baseline scenario (1.7 million ha) by at least a third.

Improving living conditions and developing fragile rural areas

The development of fragile rural areas is another major objective of the alternative scenario. Decisive progress is achieved in three fields:

1. *increase in agricultural productivity*, in particular of rain-fed agriculture;
2. *equipping rural areas with basic public services* (roads, drinking water, electricity, health, education, etc) and *structuring rural centres* (market towns, small towns) so facilitating economic diversification, promoting the attractiveness of the territory and reducing disparities with cities in these poor and under-equipped rural areas;
3. *diversification of economic activities* in mountainous or difficult/marginal areas (particularly the semi-arid regions of the southern countries), that are either underpopulated or, in contrast, likely to maintain a large rural population for a long time: promoting the development of the service sector; diversifying local activities; promoting multi-activity on the basis of labour mobility; linking up with intermediary urbanization; promoting rural and mountain tourism, etc.

There are considerable challenges in the southern and eastern developing countries and in some eastern Adriatic transition countries: putting an end to illiteracy and poverty, re-igniting hope in especially impoverished populations, reducing the risks of instability (urban, emigration), avoiding excessive concentrations of people in crowded coastlines, reducing disparities in domestic agricultural productivity and with the developed northern countries, reducing the vulnerability of societies to worldwide competition, promoting rural revival prior to abandonment in order to restrict the long-term costs induced by excessive desertion of difficult regions (mountains, steppe plateaus), and, finally, developing domestic markets for the benefit of national economies in the region. Success of the alternative scenario could be appraised by measuring the percentage of non-agricultural employment in these areas (which would grow considerably) and the reduction in disparities with cities (basic services, poverty) that would almost be achieved by 2025.

Enhancing the typical character and quality of the Mediterranean (food, products, land)

According to this scenario, the Mediterranean people would value the Mediterranean cultural diversity and identity as well as their knowledge and know-how, in relation to north-western societies.

Regarding *food*, people regain the connection between pleasure and health (avoiding over-regulation of hygiene and enhancing the senses, taste and smell). Well-off consumers quickly return to a more Mediterranean consumption pattern that is more economical in primary energy (in particular with a sharp decline in meat consumption, and thus reduced pressure on land). This scenario of a return to a reasonable and efficient diet implies a strong mobilization of public authorities and consumer associations. Measures could be economic (subsidies, taxation) or information-related (recommendations, pressure on lobby groups, as is already the case in Spain).

This reorientation relies on a desire for 'renaissance' and 'regeneration'. A certain return to authenticity would occur, 'turning the old into new', moving beyond the present trend of demanding standardized and sterilized products. The Mediterranean diet would be revalued through a recognition by the elite of popular ancestral knowledge. As a result, there would be a strong growth in demand for *regional authentic, 'village land'*,[28] *organic products*.

In this scenario, the EU, which has contributed to perverting the Mediterranean model, would become one of its main instigators. The *Mediterranean diet* is widely spread throughout the world. On this assumption, a huge market could open up for the region's products, which could, however, weigh on the environment (increasing pressures on resources and ecosystems following increased production of vegetables, fruits, olive oil and fish). However, there is no evidence that such a development would effectively benefit the Mediterranean regions. The agro-food multinationals might use the Mediterranean image to create markets and appropriate typical recipes of these regions while relocating production outside the region (olive growing is increasingly spreading in the US, Argentina and Chile). The rest of the world could therefore very quickly become serious competitors unless the producers, industrialists and craftsmen of the agro-food industry in the Mediterranean Basin, together with the countries, manage to organize themselves to protect their specific recipes and products by patents and labels of authenticity.

To stimulate the shift towards the alternative scenario, *policies of 'guaranteed designation of origin'* would play a major role. Such policies make it possible to recognize, defend and promote products and areas in their full heritage dimension. The highly positive image of Mediterranean produce relies more on intangible rather than tangible characteristics (evoking sun, village lands, a past, conviviality). Products are rooted in a place and a society, enveloped by know-how transmitted down the generations. A *village land* is more a human area, characterized by a common history that describes the locality, rather than an area with soil and climate homogeneity. The characterization of an original locality's identity is to be found much more in human practices than in its agronomic unity. Local and regional actors (producers, consumers and institutions) are linked through a relationship of confidence; they perceive their product as an attribute of their cultural and social identity as well as a vector of local development. They can mobilize themselves to upgrade their produce and promote it as a cultural good.

This search for quality is not limited only to food and products. It also concerns *territories* as whole units, as in Tuscany where added value is maximized by the quality of products and services, the *synergy of activities* (in particular between tourism, industry and agriculture), economic enhancement, and the dynamic conservation of Mediterranean landscapes. Under the alternative scenario, the greatest possible number of Mediterranean rural areas apply a 'Tuscan-like approach'.

Mountainous regions would find new economic perspectives. Their 'Mediterranean garden' role would be acknowledged, funded and enhanced.

Such a scenario has already begun to develop on the northern shore, but it is also of interest to the southern and eastern shores, where it seems important to look ahead to encourage healthier diets, protect and promote diversity, and create added value in relation to the development of tourism and domestic and export markets.

From top-down and sectoral development to accountable and integrated development

Agricultural and rural development will be implemented mainly by *professional and local actors*. Only a resolute commitment (individual and collective) on their part, within the framework of territory-specific dynamics, will make it possible to initiate a process of accumulating capital, serving the different rural territories and domestic economies. A basic component of the alternative scenario is the shift from a development that remains very much administered from the top (too often benefiting a minority that is more concerned about raking in money than creating true wealth) to a more accountable development.

Experience shows, however, that if rural development is mainly the responsibility of professionals and local actors, its implementation and sustainability do not come automatically. It implies upgrading facilities, ensuring an enabling institutional framework, and the promotion of *integrated* and *participatory* processes that take the long term into account.

Integration must take place at several levels, including vertical integration, integration of environmental and social aspects into economic practices and policies, integration of activities in specific territories (facilities, training, promotion) so as to have sufficient scale effects, and integration of activities with the local heritage in order to exploit synergies and add value. The extreme diversity of local situations that characterizes the region requires integrated management solutions and specific options for progress, which are themselves extremely diversified.

Participation is necessary to enable all actors concerned with rural areas to become as many vectors as possible of sustainable development. This implies that they value their know-how, that they share, as far as possible, a common mid- and long-term *vision* for their territory (which does not necessarily mean uniform views on ways and means), that they take part in the formulation and implementation of projects, and adhere to a collective discipline designed to ensure sustainable management of resources and preservation of territorial quality.

Recent examples throughout the region and in other parts of the world show that such a vision is far from unrealistic. The condition for success is mainly to establish a successful link between a *'filière* approach' for sectors (vertical integration) and a 'territory-specific' approach (horizontal integration). These messages are valid for both shores of the Mediterranean even if the contexts for applying them are different.

The integrated 'filière' approach[29]

This first integration approach relates to well-identified activity sectors, strongly characterized by their vertical and horizontal interaction; in particular agriculture and agri-business, forestry, and tourism.

Agriculture and the agri-food industry

The consolidated approach of this first activity sector is characterized by the constraints that weigh on the Mediterranean areas and in particular on the poorest regions. In most southern and eastern countries, the focus should mainly be on improving the operational conditions of *domestic markets* while production for export is relatively less important. Domestic changes would also extend to a more efficient combination of the market economy and state intervention, and states try to equitably integrate the largest possible number of farmers. To achieve this, priority should be given to *improving land tenure structures* (promoting family farms of a viable size and ensuring secure land tenure), developing small and middle-sized water works, improving water and rain-fed agricultural management, access to credit, training and extension, and improving transport.

Because export possibilities are particularly affected by high tariff barriers imposed on processed products, the added value of which does not usually reach the producing countries, new policies to open-up trade and voluntary actions for 'co-development' need to be designed (particularly in matters such as training, information, certification and joint ventures). Alongside typical and organic fresh seasonal products, the *processed products* sector could also be enhanced as a priority.

Forestry and pastoralism

Such integration should also be applied to forestry, which requires a change in mentality and strategies so that this resource can be managed in a more sustainable way. In the northern countries, where the Mediterranean forest has largely been rehabilitated, people and decision-makers do not yet appear sufficiently aware of the potential of its economic development and wealth. The example of Calabria (Box 5.21) demonstrates this and could be followed elsewhere. Support at the regional level is vital to assist in professional training, identifying and transforming products, economic organization, marketing and enhancing non-commercial products.

Box 5.21 Developing forestry in Calabria

The region of Calabria, in the south of Italy, extends its 15,000km² into the heart of the Mediterranean on a very mountainous peninsula with an average altitude of 550m. In 1950 only 3500 of the 7200km² of *Selva Brutia* forest originating from before the Roman period remained. Recurrent floods in the 1950s and 1960s led the Italian government to promulgate special laws for Calabria, which were the starting point of a huge reforestation effort (167,000ha). Today, 5770km² is covered with forest (38.3 per cent of the region) including 39 per cent plantations (the Calabrian pine in particular). Today the Calabria region considers forestry as an essential component of its economic development. It has adopted a regional forestry plan for the development of new sylviculture for production purposes and the development of human resources through the creation of specialized technical and educational centres. The plan attracts private investors.

In the southern and eastern countries (also to an extent in the north) forest products other than wood can represent a real economic value (Box 5.22).

Box 5.22 Forest products other than wood in North Africa

Egypt exports 11,250 tonnes of medicinal plants worth US$12.35 million per year; Morocco exports 6850 tonnes worth US$12.85 million.

Aromatic plants such as *Thymus* spp, *Rosmarinus officinalis*, *Acacia farniesa* and *Eucalyptus* spp play an important role in **Tunisia** (230 tonnes exported per year with a value of US$3.2 million), **Egypt** and **Morocco**.

A third of cork-oak forests are located in **Maghreb** (21 per cent of which are in **Algeria**). However, this region contributes only 9 per cent (Algeria 2 per cent) of world cork production (350,000 tonnes), which is produced mainly in **Portugal** and **Spain**.

Forests in **Morocco** play an essential role in livestock-raising (Marmora forest). Woody vegetation supply 17 per cent of livestock feed, and it has been observed that animal mortality during droughts is reduced by two-thirds compared with that in unforested areas. In addition, the extraction of edible oil from the argan tree is an important source of income for women.

Source: FAO

At present, limited processing activities make it difficult to enhance these resources. Most of it is exported raw, which limits the benefits in terms of employment and income generation at the local level. Training and investment would be necessary to develop processing technologies and, in particular, marketing in the context of partnerships between the private sector, governments and local communities.

Generally speaking, forestry approaches in the whole region originate from Central Europe, where timber production is the main objective. This vision has dominated Mediterranean forestry and has not been adapted to the main challenges of the region. It would benefit from a redirection that factors-in the diversity of the ecological and economic functions, especially the development of all products, commercial or not, including those often wrongly considered 'secondary' (honey, cork, aromatic and medicinal plants, small fruits, mushrooms, forage resources, recreation, firewood).

These considerations on forests relate in a natural way to pasturelands and extensive livestock raising, which remain of high importance in the south and east. Natural forests and pasturelands together supply a basis for livestock husbandry, mainly sheep and goats, the potential of which could be considerably enhanced. New resource management approaches could be widely introduced, including feed supplements in order to cope in a reasoned way with the consequences of climate variation. Experiences underway in eastern Morocco and south-eastern Tunisia have laid the foundation for such approaches, and the lessons, which could benefit other southern countries, should be closely followed.

The IAMF and *Silva mediterranea* (Box 5.26 p300) are valuable regional cooperation tools to ensure an understanding of the specificity of Mediterranean forests (and other natural vegetated areas), to develop a new shared vision, and promote adapted resource engineering for the benefit of the sustainable development of Mediterranean societies.

Tourism

Tourism is certainly a part of future solutions in rural regions, provided that it is desired, controlled and developed by and for local people. Its development calls for a highly innovative and integrated approach. The objective would be to make it a real vector of sustainable development (in synergy with agriculture, forestry and handicrafts). There are many successful local examples that demonstrate its potential, as well as harmful examples that highlight the risks of wrong approaches. Success, especially in the mountains, implies the capability of localities to develop projects that equitably stimulate the entire local society and enable communities to resist the tensions that a new activity (always disturbing) inevitably induces. Despite the strong increase in demand for rural and nature tourism (bed and breakfast, rest sites, farm inns, furnished flats, open-air campsites, family tourist resorts, rural hotels) development strategies have remained centred mainly on coasts and the hotel industry, especially in the southern and eastern countries. A reorientation, with adapted policy tools, seems to be necessary (development facilitators and training for local actors, investment aid, organized marketing, enhancement of the local heritage). These approaches have clearly proved their worth in several countries and poor rural regions, where tourism has played an important role in moving away from underdevelopment and opening up to the world. To relieve the coastal zone and contribute to rural development, the alternative scenario assumes moving tourists towards inland areas: at least a third of the projected increase in tourist flows to coastal regions would be redirected towards the hinterland by 2025 (see Chapter 6).

Industry, residential development and new technologies

Agriculture (including fish-farming), forestry and tourism could certainly be improved, and this should be encouraged, but the most successful examples of development show that the existence of an industrial activity, even on a limited scale, is vital to provide sustainable solutions for employment and services (this is also true in the mountains). The most promising approach is often to *enhance the value of local products* (mineral resources and products derived from farming and forests) with the help of small units, often operating initially on the basis of the family economy. In this regard, experience shows that rural development has frequently succeeded in moving from single-activity farming through *agro-food processing* and that adapted policies (training, incentives and promotion) facilitate such positive developments. In Turkey, for example, the GAP project (in south-eastern Anatolia) aims to stimulate new activities through management training (university/companies) and support to project carriers (for example the marble sector, basalt tiles, concentrated pomegranate juice).

Residential development is another opportunity, the development of which calls for some control, as with tourism.

Although setting up service centres with *new communication technologies* (teleworking) may appear premature, they can contribute to an escape from isolation and facilitate the arrival of new inhabitants to the Mediterranean mountains of the EU. *Solar energy* could be a significant development for equipping remote rural villages and provide vital elements of comfort at a much lower cost than connecting to the electricity grid – and at extremely low operating costs (2 euro cents per hour of light). This could open a window to the world (radio, TV and Internet links) and limit the rural exodus by reducing the gap between rural life and the 'modernity' of cities.

The 'territory-specific' approach: Giving priority to local development

The basic paradigms of all strategies for sustainable rural development – *integration*, *'territory-specificity'*, *subsidiarity*, *participation* and *partnership* – form the framework for an approach that corresponds to the political concept of 'local development'. This concept integrates rural areas in specific territories with small towns and rural villages. Many local experiences confirm the relevance, effectiveness, need and convergence of these action principles. The issues at stake are: the recognition and respect of local diversities and the value of local heritage (particularly natural and 'cultural' diversity, including resources, know-how, landscapes and architecture); the transformation of production systems and lifestyles towards a greater awareness of the fragility of resources (establishing or recreating a close and more sustainable relationship between people and the milieu that characterizes these regions); re-establishing the link between protection and development; the creation of 'territory-specific goods' by exploiting what localities have to enhance their added value; and the adoption of planned, targeted and iterative processes that are monitored through indicators.

These processes only make sense if they rely on and are implemented by local communities. But experience also shows that outside *professional support* and the *promotion of platforms and bridging facilities* are almost always indispensable to formulating territory-specific and community projects, identify heritage assets and facilitate the necessary links between sectoral approaches, territorial organization and conflict mediation. This external support can, for example, take the form of facilitators, coordinators, local development agents, or mediators. The 'facilitating/coordinating' function with local communities is one fundamental key for success.

Such approaches have been greatly stimulated in recent few years in Mediterranean Europe within the framework of *natural parks* (regional natural parks in France) and through the European LEADER *programme* (Box 5.23). Local development initiatives in the style of 'Agenda 21' are also taking shape in the southern countries, particularly with support from NGOs.

> **Box 5.23 The European LEADER programme**
>
> The European LEADER programme, now in its third phase (2001–2006), is an instrument founded on a bottom-up, partnership, cooperative and integrated approach with the objectives of diversifying the economy, making public policies consistent at the local level, and facilitating sustainable development. It is implemented by 'local action groups' (LAGs) that must be formed mainly by local civil society. Under this programme, many Mediterranean mountainous regions have been able to organize partnerships and implement exemplary action programmes. For example, the LAG in Kalampaka-Kili in the Pindus Mountains in Greece have developed alternative forms of tourism in the famous Meteor area that benefit the entire region, particularly the remote mountain villages.

Decentralization, regional development planning and local governance

The issues of decentralization and regional development (spatial planning) are at the heart of the local development concept. Spain has opened up new paths in terms of local autonomy, and Italy has innovated by instituting 'mountain communities' to which municipalities and regions can delegate very extensive responsibilities. However, low population and economic weakness in rural regions (especially in the mountains), lack of initiative, insufficient participation, and the difficulty of effectively involving local actors in projects, sometimes make it hard to fully exploit the opportunities offered by these forms of organization.

Territory-specific approaches are especially necessary to ensure the sustainable management of natural resources. Master plans and management plans make it possible to decide on planning directions and define rules and limits to resource exploitation, for ensuring a long-term ecosystem balance, which can only be achieved if the actors concerned take part in the formulation of plans and fully adhere to the disciplines deriving from this common reference. Adapting ecological (for example deferred grazing) and economic measures to the specific situation in territories is imperative in the southern countries, because it is the foundation of a strategy for moving rural populations out of poverty and preventing environmental degradation. In the north, for example, seeking a compromise between the presence of wolves and pastoralism needs a territory-specific process as well as a change in pastoral practices.

Territory-specific development approaches require a new way of thinking on two essential problems: the scales and the institutions of local governance. Experience shows that several *territorial levels* correspond to the specificity of local problems. In many situations, the locality level that still exist in a large number of regions (especially in the southern and eastern countries) is the scale where responsible actors can be identified, with whom the long-term development and natural resource management plans will be conceived. But community services will be better accounted for on broader scales, as the French experience of grouping municipalities together has shown. And aspects such as the vital master plans that exploit vertical and horizontal synergies, the relationships between cities and the countryside, and primary polarization of rural villages, can only be properly considered on an even wider regional scale. The issue of *'what is the basic region'* is currently at the heart of all thinking about decentralization and de-concentration, and all local development programmes are facing it. The concept of small rural regions is being highlighted: for example *pays* in France; *comarcas* in Spain, 'project areas' attached to local contracts (such as the LEADER territories and natural parks); in Italy, the *comunità montana* constitute to date the most completed example of 'basic region' organized into a community of rural municipalities endowed with extensive competences that are guaranteed by law over the Italian mountain areas. Areas of a certain size that correspond to present or potential tourist destinations, such as natural parks and biosphere reserves, also appear relevant. This is without doubt one of the new frontiers that local development policies will have to explore.

One issue will be that of *'subsidiarity'* for identifying territory-specific levels of responsibility along the lines of bottom-up decision-making. Another is that of the necessary *inter-linkages between levels* to encourage sustainability. A formula to be encouraged is the ambitious and sufficiently long-term 'territory-specific project contracts' (about ten years), which are defined locally, supported by higher levels (regions, states, donors), and monitored through sustainable development indicators (for example the French regional natural parks). Sound inter-linkage between levels will result from a balanced junction between top-down approaches, demonstrating the determination of the central and regional powers to achieve decentralization while searching for the appropriate level of subsidiarity, and bottom-up approaches from initiatives by local actors, whether they be public or private ones. The role of the regional and local powers is therefore essential in the implementation of such inter-linkage, which is actually one of the key issues for successful local development.

The second problem, that of *local governance*, calls for political structuring of the decision-making process. The present decentralized systems show that territory-specific levels are inadequately represented in the light of their real role in the decision-making chain, while other levels have too large a role in the administrative approval process. In addition, many participatory projects, although showing undeniable successes, only operate with ad hoc structures, often not recognized by the administrative systems (this is, for example, the case for all projects that operate at the village level in Maghreb). Finally, the problem of local governance raises the question of the role played by NGOs and networks of associations in the management structures of local development.

Towards policies for sustainable rural development

Many arguments plead in favour of *rural-focused policies* for rural areas. Today agricultural policies are obliged to integrate environmental and social aspects, and the success of environmental policies in rural areas depends largely on their links with local development. Environment and development are indissolubly linked in the Mediterranean. Equipping rural areas also demands an overall approach centred on priority needs in each given area, rather than sectoral and fragmented approaches.

Although the concept of integrated and coordinated rural development is accepted by all governments around the Mediterranean, its implementation nevertheless requires changes in the skills and relationships between operators, which are sometimes difficult to implement in practical terms.

Box 5.24 Rural development in Morocco, Turkey and Lebanon

Rural development is today considered to be an essential priority, if not **the** development priority, in **Morocco**. It is accepted that progress in other sectors of the economy will inevitably be limited as long as rural poverty continues to block the development of an internal market. Without such a market, no sustainable growth will occur, and there will be no decisive progress in social and political fields as long as such a large section of Moroccan society (14 million people) remains socially and politically marginalized. To meet these challenges, the recently formulated '2020 Strategy for Rural Development in Morocco' sets two basic priorities: (1) to improve agricultural productivity (through growth based on high intensity manpower and responsible management of natural resources by changing land tenure management and water management); (2) to improve the countryside by considerably developing rural towns to stimulate the diversification of economic activities, create new opportunities for multiple activities and develop social and cultural services to better integrate rural dwellers into local political life. The Strategy states that development should be based on actors' initiatives, and assigns new roles to the administration in advisory matters with local institutions and professional organizations (focusing on the importance of support structures close to the populations). The administration must abandon the interventionist approach that has prevailed until now. The greatest obstacle to its implementation comes from the resistance to change by government and administrative institutions. The Strategy is therefore inseparable from decentralization and spatial planning policies.

In **Turkey**, the centralized and sectoral organization of administrative decision-making has often limited the success of integrated rural development projects. The need to remedy this has led to experiments with coordinated operations to structure rural towns. For example the 'Koykent' project launched in July 2000 included nine villages in a pilot phase, covering 212km^2 in Cavdar and its surroundings in the province of Ordu. It upgraded or created basic services (roads and public transport, water and sanitation, land registration, telephones and electricity, the primary school, health facilities, a mobile library, a police station, sports fields, a farming cooperative, the restoration of the mosque, and improvement of village squares), and supported economic activities (farming and forestry). The project made it possible to end rural migration to cities, in fact people began to return to villages, and the population in the region has grown by 30 per cent. The project cost US$8 million; it was carried out under the direction of the Prime Minister and mobilized 32 agencies. Ways of duplicating this experience in other rural regions are under discussion. The authorities are now developing an approach that involves the de-concentrated levels of central administrations, but also local administrations, cooperatives, unions and NGOs. This new approach was tested in the projects of Yosgat at the end of 2001 (irrigation, drinking water connections for 70 villages, rural tracks, reforestation, plantations for firewood and health facilities), and of Ordu-Giresun (training, herd management, forestry, irrigation, drinking water and rural tracks). However, administrative coordination takes time and impedes the expansion of rural development to the desired level. The need to restructure existing organizations remains.

In **Lebanon,** rural development has long been considered a priority. An autonomous public agency, the Green Plan, was set up in 1963 to execute and subsidise infrastructures for collective or individual use. Between 1995 and 1999, the Green

Plan continued actions with 10,000 farmers spread over nearly 2000 villages. It completed 130km of tracks, developed 36,000ha of land and built hillside dams containing 760,000m^3 of water. However, as laid down in its statute, the Green Plan can only act on requests by individuals able to pay the non-subsidised part, and local authorities are excluded from intervening. This does not facilitate rural development coordination or integrated and structured operations that are very difficult to develop.

Source: Plan Bleu; Labonne, 2003; national sources

The main difficulty lies in changing institutional frameworks and administrative practices. Indeed, the objectives of sustainable rural development can only be reached if development programmes are part of an *enabling environment* related to *decentralization and deconcentration policies* (setting up multi-purpose and high-level professional field administrations, in the framework of inter-ministerial missions if necessary), as well as legislative frameworks, fiscal policies and cross-subsidy systems between richer areas and poorer, marginalized areas.

In particular this enabling environment must factor-in risks and changes related to the consequences of climate change and water stress (consider structural droughts, measures for promoting water savings, and full implementation of conventions on climate change, desertification and biodiversity).

In this regard, much innovation is required, in particular in the *financial field*. The development and sustainable management of rural areas implies important means of action (to equip villages, manage catchment areas or support and facilitate local projects); resources that are often out of reach of communities weakened by rural migration in the north or poverty in the south. In the Mediterranean, money is concentrated in cities and coastal zones, which are the first to benefit from the products and amenities produced by rural areas (by mountainous regions in particular, which provide their water and pleasant landscapes at prices that bear no relationship to the management costs). But it is also these cities and coastal zones that suffer from ecological and economic malfunctions that develop upstream (increased risks of flooding, fires and the silting-up of reservoirs). This situation of unequal exchange consequently calls for the implementation of new contracts, in particular between cities and mountainous areas and in the well understood interest of both rural and city dwellers.

Sustainable rural development also requires very flexible mechanisms for action, given the diversity in types of activities and the small size of projects. The creation of inter-ministerial or regional funds for rural planning and development, that are de-concentrated and flexible to use, could, for example, be considered.

Finally, an enabling environment would also require a system of *research and development*, training and communications that is better adapted to the challenges of Mediterranean rural areas and therefore would be much more effective.

Regional cooperation that is up to the challenges

Liberalizing trade between the EU and the southern and eastern Mediterranean countries cannot be an end in itself. Indeed, it is hard to see how it would resolve the problems of farmers and rural areas in the south and east, and put an end to the steady degradation of natural resources and the risks of instability that result from it. This is why many experts think that the nature and importance of liberalizing agricultural trade should be defined in a *cooperation agreement* between the EU and the SEMC dealing with the means of action for *sustainable rural development*.

From this point of view several alternatives to the baseline scenario could be considered.

The first would be the gradual dissemination of the European model to the southern and eastern Mediterranean to support the shock of free trade by ambitious regional back-up policies. In fact *European integration* itself is an example of a possible response to the establishment of a widened common agricultural market between countries with different development levels: from its inception to its enlargement to include countries such as Greece and Portugal and today Poland, the EU has provided solutions to its own tensions through an imperfect, yet real strategy of *economic and social cohesion*. Solidarity is embodied in agricultural funds for upgrading farms and ensuring income for farmers, and through structural funds for equipping countries or regions lagging behind in development (in particular their fragile rural areas). However, the EU is going to be directed, at least until 2011, to devoting most of its energy and means to successfully integrating the countries of central and eastern Europe. A situation of uncertainty and dissatisfaction by the Mediterranean world will result from this, which will concern both the 'third' Mediterranean countries and the relevant EU southern regions. With only one or two exceptions, the latter regions will see nearly all of the EU structural funds being transferred after 2006 to the new Member States. The advent of this period will be even more decisive for the future of the Mediterranean world because the cooperation scheme applied to the central and eastern European countries could, in the end, be adopted by the Euro-Mediterranean Partnership, at least in part. Even if this scheme does not formally lead to 'membership', it should go further than the present MEDA programme and result in a formula of the 'more than association and less than integration' type.

In the face of future difficulties, a second possible alternative would be to seek a more *transitional and differentiated partnership equation* in the interest of all partners. This alternative, centred initially on factoring in the differences in situations and challenges of sustainable development, would lead to the development of a shared vision of the multiple roles of rural areas (development and rural employment, conservation of water, soil and biodiversity), to agree on the modalities of a special and differentiated treatment to be applied between the EU and the Mediterranean rural countries in situations of ecological and economic fragility, and to ensure vigorous programmes of sustainable rural development. The EU could, in this spirit, grant a certain *Euro-Mediterranean preference* to agri-food products from the south without total reciprocity. This would make it possible to husband the transitions in the south, avoid needless destabilization, and, in the long term, realize a model of agricultural development, different from and more 'sustainable' than the northern one (it is in the south and east's interests not to follow a European-style developmental model where more and more is produced using fewer and fewer people, while at the same time doing ever greater harm to the environment). In this second alternative, as a priority, the EU would contribute, along with the southern and eastern countries and the different donors, to gradually *upgrading the rural regions* and *decoupling production from resource degradation*; the developing countries, for their part, would commit themselves to conducting the reforms needed to succeed.

In any case, a *political strengthening* of the Euro-Mediterranean Partnership (and European integration to the northern shore), is among the 'turning points' that would make it possible to direct the Mediterranean rural world towards a more sustainable future. The EU and its partners should, in this transition period and as a response to the increasing questions from the Mediterranean world, give strong political signs of their commitment.

One of the first of these signs would be a meaningful *redirection of the Partnership* around the objectives of sustainable rural development, an area that is presently insufficiently taken into consideration by international donors and regional cooperation actions (Box 5.25). This could lead to the affirmation of common values, a shared view of a desirable long-term future for agriculture and the Mediterranean rural areas (notably of their multifunctional nature), and the establishment of real cooperative projects that ensure both a sufficient *pan-Mediterranean dimension* and a *close relationship* with the realities on the ground.

A major challenge will be a concerted development of approaches to succeed jointly in turning production systems towards having greater respect for the environment. Another major challenge will be the successful support of population movements. These are played out in two directions, or 'at both ends': from where emigrants depart, and from where tourists depart. The vital challenge here is to integrate these movements, making them tools in the service of an interconnected and more sustainable development, tools for dialogue, for the sharing of wealth, and for the transfer of know-how.

> **Box 5.25 The weakness of official development assistance in targeting the Mediterranean rural sector**
>
> Official development assistance given to the rural sector in the Mediterranean amounted to only US$3.7 thousand million for 1990–2000, or 5.2 per cent of the total assistance, compared with 6.12 per cent in the preceding decade. This amount includes grants (40 per cent of the total or US$1.5 thousand million) and loans (2.1 thousand million or 60 per cent). The main beneficiaries were Egypt (1.5 thousand million), Morocco (1.3 thousand million) and Tunisia (1.1 thousand million). The eastern Adriatic countries received US$281 million for the same period. These funds were allocated mainly to irrigation (1.1 thousand million). Rural development assistance increased, reaching 462 million (13 per cent of the total) for the 1990–2000 period compared with 250 million (9 per cent) in the preceding decade. Other important items were agricultural policies and administration (US$451 million, or 12 per cent) and agricultural services (US$281 million).
>
> European funding also remained modest and was broken down as follows:
>
> - The MEDA financial tool of the Euro-Mediterranean Partnership has given only 322 million euros to the rural sector between 1995 and 2004, or 4.5 per cent of all expenses, 28 million euros of which are for regional projects. The main beneficiaries have been Morocco (168 million euros) followed by Tunisia and Algeria, two countries where the total contribution went to a single 50 million euro project.
> - The CARDS programme has provided 72 million euros of support for the rural sector of the Balkan countries (4.6 per cent of the total).
> - The EIB awarded 90 million euros in loans to the rural sector of the SEMC, or just 2.25 per cent of all expenses for 1995–1999. An increase has been recorded since, with 140 million euros at present. These loans cover diverse fields: irrigation, improved accessibility, fertilizer production, etc.

To reinforce the common values and make it a matter not just for governments but also for citizens, the Partnership will need *large decentralized cooperation programmes* adapted to the realities on the ground. Extending to the Mediterranean the community initiative programmes that have proved to be effective (LEADER and INTERREG), or establishing comparable programmes, is one avenue to take. These programmes already foresee the possibility of including partners from third countries, but they do not back this possibility with a parallel mobilization of MEDA funds. Without major financial input, but with a concerted administrative effort, the EU and its Mediterranean partners should be able to rapidly establish a framework with real possibilities for decentralized cooperation in rural areas. The impact of this kind of cooperation on the ground could be very important.

It will also be in the Partnership's interests to develop *'resource centres'* for rural areas and rely on existing regional cooperation frameworks (Box 5.26). These are already numerous in the rural and environmental fields and should help countries to capitalize even more on this experience and facilitate the redirection of policies to make them more effective, more 'Mediterranean', and thus much more in line with the objectives of sustainable development.

Box 5.26 Networks on the Mediterranean rural areas

The eco-region's specificities have led to a gradual implementation of Mediterranean networks of intergovernmental cooperation in the areas of forest management, nature protection, agriculture and rural development.

Silva Mediterranea was launched back in 1911 as an instrument for intergovernmental forestry cooperation. In 1948 it became the Committee on Mediterranean Forestry, a FAO statutory body. In 2002 an agreement was reached with MAP/Blue Plan to contribute to relaunching this programme along the lines of the objectives and strategies of sustainable development.

For the agricultural sector, ICAMAS (International Centre for Advanced Mediterranean Agronomic Studies) was set up in Paris in 1962; currently it is composed of 13 Member States. It relies on four Mediterranean agronomical institutes, which provide training programmes and run networks according to their specializations (natural resource management, agriculture and food policies, etc.) in close cooperation with the institutions of member countries. Regular meetings of agricultural ministers offer orientations for the region in the agricultural and rural sectors.

OSS (Sahara and Sahel Observatory), with its headquarters in Tunis, and ICARDA (International Centre for Agricultural Research in the Dry Areas), established in Syria in 1977, are two other international resource centres more specifically dedicated to the problems of drylands and arid areas.

IUCN (the World Conservation Union) opened its Centre for Mediterranean Co-operation in 2001 in Malaga (Spain). Among the topics developed by the IUCN Mediterranean Programme are the conservation of biodiversity, the use of natural resources (forests, fisheries, etc.), hydrological resources and desertification.

Many NGOs develop Mediterranean-wide actions and cooperation in these sectors, in particular AIFM (International Association of Mediterranean Forests) and the WWF Mediterranean Programme.

Finally, the Partnership would gain by using 'the Mediterranean character' as a handle for regional association and sustainable development. Identifying values common to the European and Mediterranean worlds would essentially mean jointly rediscovering and recreating the Mediterranean identity. And it is only by redefining its identity that the rural Mediterranean world will regain the cohesion vital to successful cooperation at the internal level and advantageous positioning outside the region in the framework of globalization.

Notes

1. Understood to be all coastal countries excluding the non-Mediterranean part of France.
2. INSEE-INRA, 1998.
3. Excluding non-Mediterranean France. The figures for 2010 are from FAO. Those for 2025 are from the Blue Plan.
4. Excluding non-Mediterranean France.
5. The ratios between 'agricultural population' and 'agricultural labour force' vary significantly from one country to another, thus requiring some caution when comparing relative data on the agricultural labour force.
6. Tunisia shows a clear difference in rural employment profile, where in 1997 agriculture accounted for 43 per cent of rural employment (about 417,000 jobs), followed by the construction and public works sectors (303,000 jobs, 31.5 per cent). The 241,000 remaining jobs were divided between services and other sectors (national sources).
7. National source.
8. MFN applied rate on spices, cereal and other dish preparation, in Israel, Jordan, Morocco, Tunisia and Turkey (WTO, 2004)
9. OECD, 1999.
10. EEAA, 1997.
11. FAO Aquastat information system, 2003 (www.fao.org).
12. FAO Aquastat information system, 2003.
13. IFEN, 1998.
14. Conacher et al, 1998.
15. Montserrat Soliva Torrento, DG Environment, April 2002.
16. Gradus and Lipshitz, 1996.
17. *Plan Bleu*; METAP, 2000.
18. *Plan Bleu* from Mhiri et al, 1998.
19. Sofer, Feitelson and Amiran in Gradus and Lipshitz, 1996.
20. Ramade, 1997.
21. IUCN, 2004.
22. Albanian Ministry of Agriculture, 2002.
23. UNDP, 2005 (http://hdr.undp.org/).
24. Côte and Khiari, in Courtot 2001.
25. FAO, 2002.
26. In contrast to the case of Norway where a farm or forestry estate returns to the public domain if exploitation is stopped; this leads families to find solutions and is remarkably effective in maintaining the rural character at the national level.
27. MEDFOREX project (www.medforex.net/).
28. *Terroir* in French.
29. The *filière* approach consists of the integration of all stakeholders (agriculture, transformation, distribution, industry services and government) within the same production sector or sub-sector in order to enhance performance on domestic and foreign markets. Some attempts to translate the expression into English have been: 'producer-consumer chain', 'production and trade networks', 'commodity chain'.

References

Akesbi, N. (2003) *Le nouveau cycle de négociations agricoles à l'OMC et les perspectives pour les pays de l'Est et du Sud de la Méditerranée*. Communication au séminaire CIHEAM/EAAE

dans le cadre du 10ème Congrès de l'Association européenne des économistes agricoles
Allaya, M. et al (2003) *Oléiculture et développement durable en Méditerranée*. Note for the Blue Plan
Bessaoud, O. et al (2001) *Problématique de développement rural des zones sèches dans la région Moyen-Orient et Afrique du Nord*. Contribution à la mise à jour du rapport 'From vision to action for rural development' élaboré par la Banque Mondiale, Washington, DC
Blouet, A. et al (2003) L'agriculture raisonnée: limites et alternatives du modèle agricole dominant. *Futuribles*, no 283, February 2003
CIHEAM. (2002) *Medagri. Annuaire des économies agricoles et alimentaires des pays méditerranéens et arabes*. Montpellier: IAMM.
CIHEAM, Plan Bleu. (2003) *Libre-échange, agriculture et environnement. L'Euro-Méditerranée et le développement durable: état des lieux et perspectives*. Actes du Forum des 30 et 31 mai 2002, Montpellier (France). Montpellier: CIHEAM. (Options Méditerranéennes, Série A, no 52)
CIHEAM. (2004) *Agri.med: agriculture, pêche, alimentation et développement rural durable dans la région méditerranéenne*. Rapport annuel. Montpellier: CIHEAM
Commission for Environmental Cooperation of North America (CEC). *Free trade and the environment: the picture becomes clear*. Montreal: CEC, 2002 (www.cec.org)
Conacher, A. J., Sala, M. (eds) (1998) *Land degradation in Mediterranean environments of the world: Nature and extent, causes and solutions*. New York: Wiley
Courtot, R. (dir.). (2001) 40 ans de géographie méditerranéenne. *Méditerranée*, vol 97, no 3–4
DATAR. (2003) *Quelle France rurale pour 2020? Contribution à une nouvelle politique de développement rural durable*. www.datar.gouv.fr
EEAA (Egyptian Environmental Affairs Agency) (1997) *The Environment in Egypt 1996*. Cairo: EEAA
FAO. (2002) *Agriculture mondiale: horizon 2015–2030*. Rome: FAO
FAO. (2003) *Étude prospective du secteur forestier en Afrique. Rapport sous-régional Afrique du Nord*. Rome: FAO
Fe d'Ostiani, L. (2003) *Watershed management, a key component of rural development in the Mediterranean region*. Report for the Blue Plan
FEMISE. (2003) *La question de la libéralisation agricole dans le partenariat euro-méditerranéen*. Rapport FEMISE, Marseille, November
Gradus, Y., Lipshitz, G. (eds) (1996) *The mosaic of Israeli geography*. Beer-Sheva: Ben-Gurion University of the Negev Press
IFEN (1998) *L'environnement en France*. Paris, Orléans: La Découverte; IFEN
INRA; INSEE (1998). *Les campagnes et leurs villes*. Paris: INSEE (Coll. Contours et Caractères)
IUCN Centre for Mediterranean Cooperation (2003). *A Regional Situation Analysis*. Malaga, May (www.iucn.org)
Labonne, M. (2003) *Agriculture méditerranéenne: situation et perspectives*. Report for the Blue Plan
Lazarev, G. (2003) *Notes sur la stratégie 2020 de développement rural du Maroc et sur le programme de développement des aires pastorales du sud-est de la Tunisie*. Note for the Blue Plan
Mhiri, A., Bousnina, H. (1998) 'Diagnostic agri-environmental de l'état des terres cultivées dans les divers systèms de production en Tunisie'. *Revue de l'INAT, numéro spécial "Centenaire de l'INAT"*, Actes du séminaire international sur les resources naturelles, 10–12 November 1997, INAT, Tunisia, pp209–228
Montgolfier, J. de et al (2002) *Les espaces boisés méditerranéens. Situation et perspectives*. Les Fascicules du Plan Bleu no 12, Paris: Economica
OECD. (1999) *Environmental performance reviews. Turkey*, Paris: OECD
Padilla, M. (2002) *Évolution des modes d'alimentation et enjeux de développement durable en Méditerranée*. Report for the Blue Plan
Plan Bleu. (2003a) *Quel avenir pour les montagnes méditerranéennes? Pour une prospective territoriale*. Résultats de l'atelier 'Montagnes méditerranéennes', Sophia Antipolis, France, 1–2 April 2003. *Montagnes méditerranéennes*, no 17, July
Plan Bleu. (2003b) *Les menaces sur les sols dans les pays méditerranéens. Etude bibliographique/Threats to soils in Mediterranean Countries. Document review*. Les Cahiers du Plan Bleu/Plan Bleu Papers, no 2
Plan Bleu; METAP. (2000) *Environmental performance indicators. Synthesis report. 3 national tests*. Plan Bleu
PNUE-PAM-Plan Bleu. (2002) *Libre-échange et environnement dans le contexte Euro-Méditerranéen*, Montpellier–Mèze, France, 5–8 octobre 2000. 4 vol. MAP Technical Report Series no 137. Athènes: PNUE-PAM.
Pujol, J.-L., Dron, D. (1999) *Agriculture, monde rural et environnement: qualité oblige*. Paris: Documentation française
Ramade, F. et al (1997) *Conservation des écosystèmes méditerranéens: enjeux et prospective*. 2nd ed. Les Fascicules du Plan Bleu no 3. Paris: Economica
Reiffers, J.-L. (ed), (2000) Institut de la Méditerranée. 2000. *Méditerranée: vingt ans pour réussir*. Paris: Economica
World Bank. (1997) *Rural development: from vision to action*. Washington DC: World Bank
World Trade Organization (2004) *Report on World Trade 2004*, WTO
Yussefi, M., Willer, H. (eds.) (2003) *The World of Organic Agriculture: Statistics and Prospects*. Tholey-Theley (DEU): IFOAM (International Federation for Organic Agricultural Movements) (www.ifoam.org)

Chapter 6

Coastal Areas

The coastal zone is more than just a narrow strip of land and sea coveted for development. Its ecosystems, with great natural value, and the importance of its social functions and cultural heritage make it a precious resource.

Coastal areas are subjected to a convergence of pressures, resulting in a general trend of increasing urbanization. Under the baseline scenario, some 20 million additional urban dwellers and 130 million additional tourists are expected by 2025, the high concentration of roads, ports, airports, industrial and power facilities will intensify, half of the area will have some form of artificial land cover, and coastal pollution will continue to have serious impacts. With more artificial land and pollution, overfishing, erosion, salt water intrusion, invasive species and global warming, degradation of coastal ecosystems and loss of terrestrial and marine biodiversity will continue and lead to a proliferation of costs and risks. The trend in some coastal zones towards 'tourism-development-only' also carries the risk of loss of identity and economic resilience.

The responses to ensure a planned and stable development of coastal areas have so far been inadequate. Awareness of these trends calls for drastic changes in practices to *stop and urgently reverse the continuous degradation of coastal areas*. An alternative scenario can be envisaged that would lead to a future with a stronger integration of the environment and development, strengthened policies for coastal areas, a redirection of tourism strategies and policies, implementation of new economic instruments, and an overall enabling framework. The Contracting Parties to the Barcelona Convention would send a strong signal of encouragement if they adopted a regional framework protocol for sustainable management of the Mediterranean coastal region. This protocol would focus on legal and operational reinforcement, requiring the creation of coastal laws and agencies, the mobilization of civil society, promotion of 'integrated costal area management' and implementation of territorial planning and region-specific development.

The coastal zone: A description

The coastal zone, a rather narrow strip

Countries are increasingly lead to drawing the boundaries of their coastal zones to ensure sustainable management (Box 6.1). The Blue Plan *defines* the coastal zone as the ensemble of areas and specific territories that are influenced physically, economically and socially by strong interaction between land and sea. Mediterranean ecosystems, societies and economies, which are subjected in a predominant way to the sea, can also be defined as coastal.

The extent of a coastal zone (environmental, cultural, social and economic) depends on the nature of coasts and the history of societies. It varies from a few hundred metres to some kilometres. The coastal zone is thus not limited to the coastline that separates land and sea.

On the seaward side, the coastal zone corresponds with the 'shallow waters' where nutritional input from the land and light favour a high natural production. The limit is set by the 50m depth contour; beyond this depth, photosynthesis is reduced. As a public maritime domain, this marine fringe is administered by the maritime authorities of each country. On the landward side, the coastal zone includes a mosaic of 'natural' (deltas, dunes, forests), agricultural and urban areas that are part of the public and especially private domain. Their management is framed by rules put forward by land-based authorities.

Deltas and lagoons, tourist destinations such as Djerba, the Riviera, Antalya and Benidorm, the large port cities such as Alexandria, Barcelona, Marseilles, Algiers and Tel Aviv all belong to the coastal zone.[1]

Box 6.1 The coastal zone: Official definitions

Official coastal zone boundaries in riparian countries are often lacking or imprecise. On land the boundaries are often physical distances (coastal strips of a few kilometres or just a few hundred metres) that do not necessarily coincide with the territory inhabited by coastal societies. When boundaries exist in the sea, they usually include all the territorial waters, which extend beyond the coastal zone as such (predominantly 12 nautical miles, but 6 miles in Greece and Turkey).

The inadequacies or difficulties in delimiting the coastal zone explain to a large extent the lack of spatial data that would make it possible to assess, compare and monitor socio-economic and environmental developments over time. Comparisons are made difficult by the different delimitation methods applied in riparian countries.

In **Spain** the law on coastal areas uses the term 19 times without ever defining it, except in an indirect way (public domain, neighbouring areas).

In **Tunisia**, the coastal zone is understood to be the shore with a possible inland extension with varying limits depending on the degree of climatic, natural and human interaction with the sea.

In **France** the legal definition includes seaside municipalities (and in some cases estuaries and deltas located outside the salty limit of the water).

In **Algeria**, the new law defines the coastal zone as including the ensemble of all islands and isles, the continental shelf and a strip of land along the coastline with a minimum width of 800 metres.

Other countries have definitions made outside the law:

- **Egypt** defines coastal zones as an interface between the

sea and the land, including the territorial waters and the hinterland that interact with the marine environment; the zones are at least 30km wide in the desert regions, except when geographical obstacles exist.
- In **Israel** the limit on the landward side varies between 1 and 2km from the shoreline and does not include the sea.

Finally, the recent development of the concepts of 'integrated watershed management' and 'integrated coastal area management' has led to delimitation proposals that go well beyond the coastal zone without coinciding with those of administrative coastal regions. This extension can be relevant to activities related to water management but does not mean that the shortcomings of delimiting coastal zones can be ignored.

Lacking sufficient statistical information about coastal zones, the present chapter will, in some cases, refer to data gathered for *administrative coastal regions* (the NUTS 3 level or equivalent: *wilaya, province, etc.*). Such data may include inland plains and desert or mountain areas that do not really belong to coastal areas. Also, some case studies will be presented that go beyond the coastal zone.

Invaluable coastal ecosystems

The Mediterranean *coastline* is approximately 46,000km long and is more or less equally divided between rocky (54 per cent) and sedimentary coasts (45 per cent). The jagged northern coast, including its many islands, accounts for 73 per cent of this length with a total of 33,480km. The eastern and southern shores account for 6615 and 5735km respectively (14 per cent and 13 per cent), including 3000km of desert coastline. The *islands* alone account for nearly 19,000km of coastline, or 42 per cent. Their insularity results in many specificities (Box 6.2).

Box 6.2 Islands: Assets and drawbacks of insularity

The Mediterranean has 162 islands larger than 10km^2, including two that are 'very big' (Sicily and Sardinia), three 'big' ones (Cyprus, Corsica and Crete) and two island states (Cyprus and Malta). Together they house 11 million people.

Insularity is even stronger for smaller islands that are economically distant from the mainland. This insularity results in considerable specificities: exceptional biodiversity and endemic conditions, relative importance of the coastline, scarcity of freshwater resources (low and irregular precipitation, small catchment areas and rivers with low water yields, problems of water transfer, and scarcity of storage sites), scarcity of soil and energy resources, limited domestic markets, dependence on maritime transport, importance of the sense of identity, diversity of know-how and lifestyles adapted to local conditions (for example the Cretan diet, water management and fishing methods).

Insularity is generally a drawback at the social (health and education services) and economic level. The costs of transport (but also loading and stock interruptions), price tensions (shortages, preventive buying, excessive offers, overstocking), ceilings on sales and weak development outlook are all structural obstacles to economic activities and result in an over-cost on goods and services. GDP levels are generally below the national average, and unemployment rates are higher.

Figure 6.1 Sedimentary coasts and their ecosystems

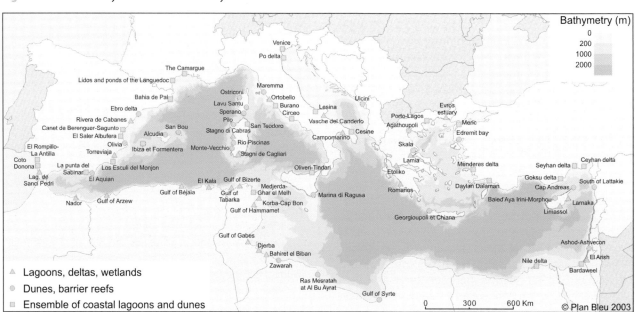

Source: Ramade et al, 1997

Coastal Areas

Financial dependence on the states to which they belong also exposes islands to ups and downs in public spending.

With the development of tourism, pressures on islands, particularly along their coastlines, have grown considerably. Such development has enabled some of them to compensate for their natural drawbacks, but it has also generated considerable environmental impacts.

Biotopes on cliffs on the *rocky coasts* shelter typical animal and plant communities, many of which are endemic. The steep nature of the land extends into the sea where the coastal fringe is not very extensive.

The *sedimentary coasts* are typified mainly by the importance, fragility and environmental value of their *coastal ecosystems*: sandy and pebbly beaches, barrier reefs and coastal dunes, swamps, lagoons, estuaries and deltas (Figure 6.1).

The four big deltas (the Nile, Rhone, Po and Ebro) as well as other less extensive ones, are unstable areas, sensitive to variations in sea level and rapid switches between erosion and sedimentation, as can be constantly seen by the advancing and receding coastlines. Lagoons are mainly present along the coasts of the Languedoc in France, the Tyrrhenian Sea, some parts of the Aegean Sea around Greece and Turkey, Akrotiki in Cyprus, Tunisia (Lake Ichkeul, lagoons in the Gulf of Gabès) and Algeria (El Kala). Egypt alone contains 25 per cent of the total area of the Mediterranean wetlands (Sebkha el Bardaweel, Manzala Lagoon, etc.).

What is typical of these *paralic ecosystems* (deltas, mud flats, lagoons, ponds and coastal marshes) is the simultaneous presence of freshwater and marine species, and species strictly linked to these environments. They are major components of the Mediterranean coastal system. Their main characteristics are derived from the variable salinity and the relative confinement. Despite serious reductions (half of the Mediterranean wetlands are said to have been lost in the past century), the remaining wetlands still extend over 800,000 to 1 million ha, 650,000 ha of which correspond to paralic environments.

The rather rare Mediterranean *dunes* are precious (endemic plant associations such as the *Artemisia-Armerietum* occur in the inland dunes of south-west Spain). Their location near beaches makes them vulnerable to foot traffic and construction activities.

The *marine fringe* (the sublittoral zone) only covers about 5 per cent of the total water surface, but it has a high ecological and fish-related value. Marine seagrass beds, especially the Posidonia meadows (*Posidonia oceanica*), an endemic species, are of exceptional interest. These beds trap sediment, form barrier reefs that stabilize coasts, guarantee the perpetuity of coastlines, are the location of considerable primary production (assessed at some 21 tonnes of dry matter/ha/yr), and provide shelter for the reproduction of many marine species.

The Mediterranean Sea is a rich store of endemic *flora and fauna*; it contains 7 per cent of the marine species known worldwide, while covering only 0.8 per cent of the ocean surface. A total of 694 species of marine vertebrates have been recorded, including 580 fish, 21 marine mammals, 48 sharks, 36 rays and 5 turtles. As far as marine plant life is concerned, 1289 taxa have been recorded. The marine biodiversity is concentrated in the limited areas with shallow water that contain 38 per cent of the invertebrates, 75 per cent of the fish and nearly all seaweeds. Some species, in particular the sea monk, *Monachus monachus*, and two species of marine turtles, *Chelonia mydas* and *Caretta caretta*, have high symbolic value and are under threat.

The natural heritage with its biodiversity and its vital role in the food chain, in purifying water, and hosting the public, has an *ecological and social value* that requires its preservation for future generations. Some attempts aim at justifying this protection from an economic point of view. A survey carried out in Europe has shown the especially high economic value of the environmental benefits supplied by the paralic environments, which could reach some 2.4 million euros per km^2 per year, far more than rivers and lakes, forests or meadows. This exceptional value of coastal wetlands is explained by the multiplicity of services rendered: natural purifying capacity, an environment that is both receptive to and propitious for dozens of fish and waterfowl species and millions of migratory birds to reproduce, climate and water cycle regulation, erosion prevention, biological control, food and raw material production, fisheries, aquaculture and leisure activities, genetic capital, and knowledge, landscape and cultural heritage.

A coveted place for development

For Mediterranean societies and economies, the coast has long been a place to concentrate development. Several capitals (Algiers, Tunis, Tripoli and Beirut) or major economic centres (Alexandria, Tel-Aviv, Naples, Marseilles, Barcelona and Valencia) are located on the coast. Activities such as fisheries, port industries, salt production and agriculture have been established there for a long time and are developing rapidly. Other sectors, such as intensive fish farming and tourism, are more recent but are also expanding rapidly. *Fisheries, aquaculture* and *tourism* depend directly on the environmental quality of the coast.

The nature and development of these varied activities differ with the type of coast. *Rocky coasts* have the advantage of being stable and healthy, offering good port venues; many traditional maritime provinces have developed here historically. These coastal shores, where tourism started to develop more than a century ago (the Rivieras of the Provence, the Côte d'Azur and Liguria), have recently become preferred locations for developing

A Sustainable Future for the Mediterranean

intensive fish farming in cages. Other activities are more typical of *sedimentary coasts*. The main Mediterranean fisheries are found here; the lagoons alone account for 10–30 per cent of the total fish production in the region. Original forms of aquaculture production (brackish water fish farming in the Italian lagoons, shellfish farming), agricultural production (rice in the Ebro Delta, cotton in the Nile delta, bulls and horses in France's Camargue) and salt production have long been established and developed here. The development of the traditionally insalubrious coastal plains, the advent of mass tourism, and the relative availability of accessible land make sedimentary coasts today a prime place for regional development.

A centre of cultural heritage

The long history of the encounter between the Mediterranean people and their coasts has produced a unique cultural heritage. Being aware of this exceptional importance, the Mediterranean countries initiated a cooperative venture in 1987 between '100 historical coastal sites of Mediterranean-wide interest' (Figure 6.2), of which 48 are UNESCO world heritage sites.

This exceptional heritage ensemble contains archaeological sites and historical centres of coastal cities of all sizes. The shores also offer a great diversity of cultural and natural landscapes (islands and isles, lagoons and steep rocky landscapes).

The know-how of local people, coupled with traditional practices for exploiting natural resources, is part of this remarkable heritage (Box 6.3).

Box 6.3 Traditional techniques for exploitation of the Mediterranean coastal waters

Passed on, adapted and perfected from generation to generation, most traditional techniques of exploitation have proved their sustainability. Several are still being used, for example the 'charfias', permanent fishery beds built from palm leaves, one of the main fishing techniques used in the shallow waters of the Kerkennah, Chebba and Djerba Islands of Tunisia. Other examples are sea salt extraction in the Slovenian area of Socelje or *felucca* fishing under sail in the Burullus lagoon in Egypt. But these practices are in danger of vanishing because of competition from new, more profitable techniques and the disappearance of this knowledge. In the same way, the tuna nets, permanent fisheries built on the migration routes of the red tuna, were the main technique for catching the red tuna until about 30 years ago. Nowadays only a few tuna nets still exist in Spain, Tunisia and Italy. Faced with competition from the large tuna-boat fleet, maintaining these tuna nets has more to do with concern for loosing traditional practices than with economic profitability.

Figure 6.2 The 100 Mediterranean historical coastal sites

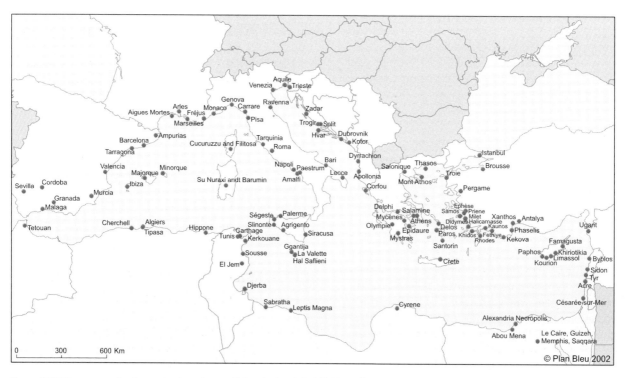

Source: MAP Programme for the protection of coastal historic sites (100 HS), Marseille

Figure 6.3 Pressures on the coastal zone

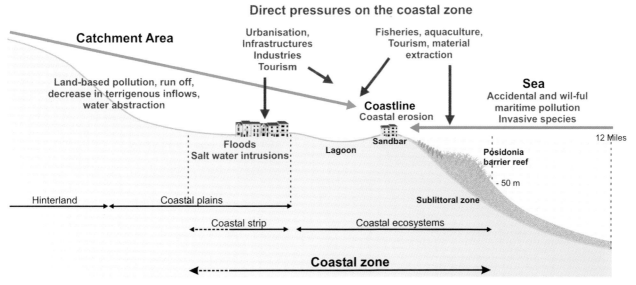

Source: Plan Bleu

A zone subject to convergent pressures

Sometimes chosen because there is no other choice or because it is the preferred area for many economic activities, transport infrastructures and human settlements, and a receptor for land-based and maritime pollution, the Mediterranean coast is subjected to an exceptional *convergence* of direct and indirect pressure (Figure 6.3).

The scale and dynamics of these pressures in the region are illustrated by the Algiers coastal area.

Box 6.4 A convergence of pressures in the coastal zone of Algiers

The Algiers coastal zone, measuring 115km in length and 38km in width from Cap Djinet to Chénoua, was home to 4.3 million people in 1998 (950 per km²); projections show that there will be 5.7 million people by 2010 (1280 per km²). Nearly 40 per cent of the entire coastal road network is concentrated in this zone, as well as the largest port and biggest airport of the country, not to mention two-thirds of the industry with 1000 units recorded in 2001 (metalworking, general chemicals, construction, petrochemicals, pharmaceuticals). Agriculture is present on more than 226,000ha of 'usable agricultural area', nearly 28,000ha of which are irrigated. There are eight fishing harbours with a total catch of 26,500 tonnes. Seaside tourism will undergo major development with plans for an 880ha tourist expansion area (TEA). Illegal extraction of sand has been estimated at 5 million m³ between 1972 and 1995.

These activities exert large pressures on the environment and natural resources: wastewater discharges (225 million m³ per year, with 174,680 tonnes of suspended matter, 14,700 tonnes of nitrogen and 4100 tonnes of phosphates), production of solid waste (2200 tonnes per day) and gaseous emissions (28,300 tonnes NO_x per year, 209,500 tonnes CO and 663 tonnes SO_2).

Since 1960, 13,700ha of high-quality agricultural land has been lost. Urbanization (which is projected to grow by more than 4000ha between now and 2010) and erosion will only exacerbate this loss. The amount of land covered by vegetation is low (24 per cent). The withdrawal of sandy coastlines is occurring along 80 per cent of the coastal zone. The pelagic species of fish are being overexploited (nearly 5000 tonnes of overfishing). Biodiversity is being reduced in sensitive places.

Facing these pressures and multiple impacts, the responses are inadequate (without prejudging the effects of the very recent coastal zone law): weak existing land- and urban-management tools, low rates of urban and industrial water purification (18 per cent), a lack of strategies for sand substitution on beaches and in *wadis*, and a lack of composting units and solid waste treatment plants.

Source: M. Larid, PAC 'Zone côtière algéroise', 2003 (www.planbleu.org)

An increasingly built-up coast: Megapolization?

The coastal zone is home to a large population of permanent residents and tourists as well as many transport infrastructures and industrial sites. In all, the Blue Plan recorded nearly 2300 large coastal settlements in 2000,

Figure 6.4 Coastal city population, 1950–2025

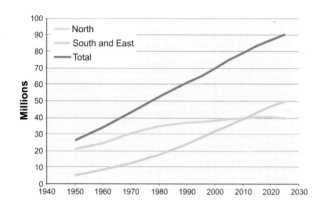

Source: Géopolis, 1998, Attané and Courbage, 2001

Figure 6.5 Number of coastal cities with population more than 10,000, 1950–1995

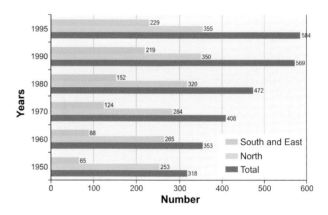

Source: Géopolis, 1998, Attané and Courbage, 2001

an average of one every 20km, including 584 coastal urban areas, 750 'marinas' (yacht harbours), 286 commercial ports, 13 gas complexes, 55 refineries, 180 thermal power stations, 112 airports, 238 desalinization plants and a number of big industrial centres (cement factories, steel mills, etc).The amount of coastline which is built on is therefore quite considerable.

Twenty million more urban dwellers and 130 million more tourists by 2025

The permanent population of the Mediterranean *coastal regions* (N3) is projected to continue to grow to 108 million in 2025 in the south and east (1.4 per cent growth per year), stabilizing at about 68 million in the north (see Statistical Annex).

These growth rates may be higher near the coastline itself, because of its highly attractive nature and the rural-to-urban migration that risks being reactivated when agriculture is liberalized. Population statistics are not available at the coastal district level for the whole Basin. However, population data for *cities located on the coastline* give a good idea of the probable development of coastal populations. The population of *coastal cities*[2] could increase 1 per cent per year between now and 2025 and reach 90 million by 2025 compared with 70 million in 2000 (Figure 6.4), in other words 20 million additional urban dwellers on the coasts in 25 years. This growth will occur mainly on the southern and eastern shores, which will add 18.4 million coastal urban dwellers and will reach 50 million by 2025. On the northern shore, coastal cities may start losing people in 2020 and would have only 1.7 million additional city dwellers between 2000 and 2025. This coastal urban population will increasingly be concentrated in the very big cities. In 1995 41 per cent of the coastal urban population lived in cities of more than 1 million, compared with only 30 per cent in 1970. The number of coastal cities has nearly doubled since 1950, from 318 in 1950 to 584 by 1995 (Figure 6.5). The number of small cities has grown steadily in the south and the east (Libya, Egypt and Turkey) while it has stabilized in the north. There are 196 cities on the Italian coastline alone, nearly a third of the total Mediterranean number.

Tourists have to be added to these resident or permanent populations and can double the numbers during peak periods. The tourist consumption patterns and the seasonal aspect result in an overextension of facilities and services that are costly in terms of space, investment and operations.

The international tourist consumes on average more water and energy and produces more waste than the average domestic tourist and resident. For example, on the Balearic Islands, it is estimated that a tourist produces 50 per cent more solid waste than a resident, and water is consumed to the tune of 90,000m³ per day in winter, but 130,000m³ per day during the high season. Majorca was so short of water that it had to import it from the continent, and later turned to desalinization of sea water.

The construction of holiday homes, many of which eventually tend to become permanent residences, is one of the important causes of land 'consumption'. In some regions, holiday homes can account for more than 30 per cent of the housing stock (Majorca and Corsica). In France the three Mediterranean regions alone hold 30 per cent of the holiday housing stock in the country. Italy is said to have 4 million holiday homes. In Egypt holiday homes are numerous in Alexandria, Port-Said and on the coast towards Marsa Matruh.

The most important impact, however, comes from the construction of tourist facilities (hotels, yacht harbours, etc.) on the coastline itself. This construction contributes to coastal erosion and the degradation of coastal ecosystems.

The projected growth in national and international *tourist flows* between now and 2025 on a country level (N1) and a coastal region level (N3) is very strong, as seen in Part 1 of this report. In total, the coastal regions are expected to receive 312 million tourists by 2025 compared with 175 million in 2000, or 137 million additional tourists in 25 years, some 74 million in the north and some 62 million in the south and east. The high average annual growth rate (2.3 per cent) will be stronger in the south and east (4.2 per cent) than in the north (1.7 per cent). Despite this difference, the coastal regions in the north will still receive 69 per cent of the total flow in 2025.

It is difficult to estimate what proportion of these flows will choose the coast. Given the predominance of the seaside resort model (sea and beach), it will probably be very high, in fact so much so that the figures on a regional level (N3) can be taken as representative of the tourist pressure on the coasts. This is especially true in countries like Tunisia where 91 per cent of night-stays are motivated by seaside tourism or Turkey, where the Antalya province (N3 level) contains 23 per cent of the tourist establishments in the country and 35 per cent of the tourist beds, and concentrates 98 per cent of its tourist establishments and 97 per cent of its beds in a narrow coastal fringe. In the baseline scenario, tourism in the coastal regions will continue to develop, principally near the coasts; this despite demand for diversity, but given the current strategies being used (see the section entitled 'Directing tourism towards sustainable development', p341).

In the coastal regions (N3) resident and tourist populations accumulate during the high season. The indicator of *resident and tourist population density during the peak month* shows strong growth; it is projected to rise from 158 inhabitants per km^2 in 2000 to 209 inhabitants per km^2 by 2025 (Figure 6.6 and Statistical Annex).

When calculated for the coastal zone only, this density is much higher since the coastal regions often include inland desert or mountainous areas that are thinly populated. When reporting the figures for the coastline itself, a *linear density* indicator can be obtained that better illustrates the scale of the pressures on coastal zones. The population of coastal cities reported per kilometre of coastline increased from an average of 580 people per km of coast in 1950 to 1530 people per km in 2000 and could reach 1970 people per km by 2025. The average linear density would thus have more than tripled in 75 years. Its growth would be more than 2 per cent per year until 2025 in Syria, Egypt, Albania, Morocco and the Palestinian Territories.

The tourist linear density, also sharply increasing, could triple on the southern and eastern shores in 25 years, while in the north it would probably increase by only 52 per cent (see Statistical Annex). By 2025 the strongest tourist pressure on coastlines would occur in the east (Israel and Lebanon) and the north (Monaco, Slovenia and Spain).

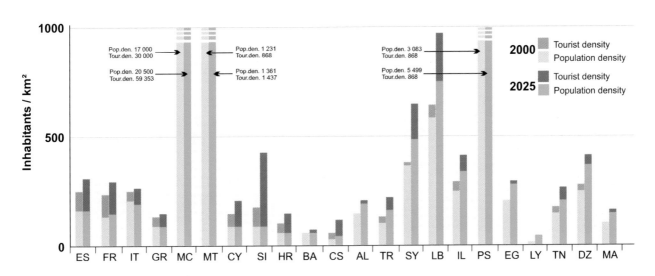

Figure 6.6 Cumulative density of resident and tourist populations in the peak month in coastal regions, 2000–2025 N3 level (inhabitants per km^2)

Source: Plan Bleu, 2003

Note: The cumulative urban and tourist linear densities give an idea of the enormous pressure on the Mediterranean coast in peak periods. The total could reach 3330 people per km of coast by 2025 compared with 2300 in 2000. Pressure would be especially high in the Levant (from Syria to Egypt), Spain, Slovenia and Algeria.

A concentration of roads, ports, airports and industrial areas

Population and socio-economic developments will result in strong growth in coastal infrastructures and facilities, especially for transport and tourism facilities. Population growth in the south and east will also lead to increasing needs in the energy and industry sectors, not to mention desalinization plants.

Coastal areas have substantial sea, air and land - transport infrastructures (see Statistical Annex). Intensively-used *roads* now run along a large part of the Mediterranean coast at no more than a kilometre from the shoreline (Figure 6.7), particularly in Algeria, Lebanon and Israel, but also in Croatia, Italy, Spain and France. The islands are better preserved. Roads, generally built too close to the shores, disturb the physical interactions between land and sea, and generate a ribbon urbanization along the coast. Roads also consume a great deal of space. According to available data, the highest road densities (the relationship between road length and the dimensions of the area under consideration) are found in Algeria (4.1 km/km² in the *wilaya* of Algiers, 1.4 in the *wilaya* of Tizi Ouzou), Syria (1.6 in Tartus Mohafazat) and Cyprus (1.4 in the Limassol district).

Many *airports*, among the largest in the region in terms of traffic, serve coastal cities or cities near the coast. There are a total of 112 airports on the Mediterranean coast (Figure 6.7), some of which have been built on coastal wetlands (Corfu, Larnaka, Marseilles, Rome, Thessalonika and Tunis). Tourism development explains the amount of traffic recorded to some destinations and the observed traffic increase. Opening up small islands increases the need for airport infrastructures.

In 2003, the Mediterranean had 286 *commercial ports* (Figure 6.8), of which only 46 were fitted with receiving equipment for ship wastes. They are often surrounded by industrial areas, sometimes with free zones.

Projections for transport to 2025 show *considerable growth in traffic* (see Chapter 3). The coastal zone will be affected by several major infrastructure projects. In the framework of the trans-European transport network in the enlarged Europe, for example, priority projects include a rail–road bridge at the Straits of Messina (2015), a high-speed railway linking Barcelona, Figueras and Perpignan (2008), Perpignan and Montpellier (2015) and Montpellier and Nîmes (2010), and the Venice–Trieste–Koper–Divaca railway (2015). Plans for new road construction also exist in the south and east, for example the coastal road in Morocco.

Air-passenger traffic is projected to grow between 2000 and 2025 by some 2.6 per year throughout the Mediterranean. This growth would be stronger in the south and east (3.2 per cent per year) because of increased flows of international tourists. It is expected that the expansion of existing capacity and the construction of facilities along the coast will be carried out in coastal areas experiencing a tourist boom (Croatia, Turkey, Egypt, Libya and Tunisia). In Tunisia, for example, according to various growth

Figure 6.7 Road and airport infrastructure along coasts

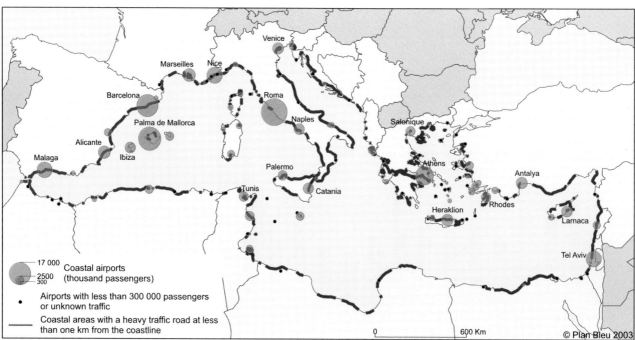

Source: diverse national statistical *sources*

Figure 6.8 Mediterranean commercial ports

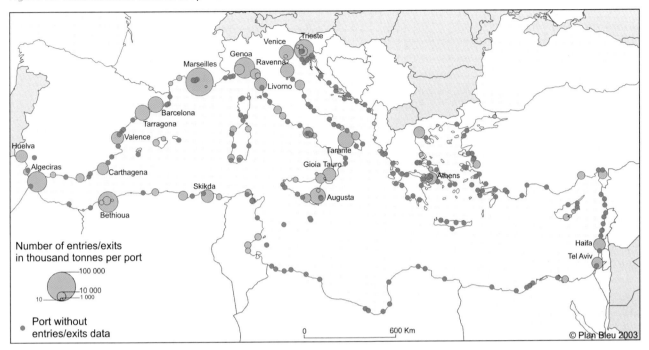

Source: Lloyd's List: Ports of the World 2003; national statistical yearbooks

assumptions,[3] airports with a capacity of 10.5 million passengers per year in 2002 will be receiving between 15 million and 24 million passengers per year by 2020. In Egypt the government[4] has approved a plan for expanding the airports of Alexandria, Port-Said and Marsa Matruh, and building a new airport in El Alamein.

Port infrastructures will probably develop more by expanding existing capacities and/or by changing their specialization than by new construction, with occasional exceptions such as the new Tangiers-Mediterranean port in Morocco. Some examples of planned expansions are the 'seaside highways' in south-eastern and south-western Europe by the European Union (2010) and the development of the port of Dubrovnik in Croatia for passenger traffic.

In addition to the commercial ports, there are 750 *yacht harbours*. New yacht harbours have serious negative impacts on the environment (consumption of land and degradation of surrounding shallow waters, disturbance of the dynamics of coastal currents, and chemical pollution). In countries already well-equipped (Spain, France, Italy, Monaco, Slovenia and Israel, see Figure 6.9 and Statistical Annex), the construction of new marinas will face many difficulties, given the saturation level of coastal land and the degree of coastal protection. To meet an ever-growing demand (projected at 1.5–2.6 per cent per year for France over the coming years[5]), an emphasis will be put mainly on developing, expanding and rehabilitating existing capacity and on developing dry ports, which will create new sorts of pressure on the landward side of coastal zones. For the other countries, the baseline scenario assumes that by 2025 the projected ports (43 in all) will be built in Greece and Turkey. In Croatia, port capacity could be developed mainly by optimizing and converting existing locations and by lighter facilities (anchorage sites). Countries will be put under severe pressure by investors wishing to build marina complexes.

Figure 6.9 Marinas in Italy

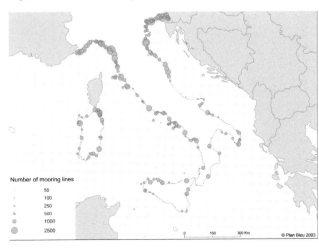

Source: Pagine Azzurre. Il portolano dei mari d'Italia, 1999 edition

2 A Sustainable Future for the Mediterranean

Cruises have experienced strong growth over the past 15 years. 20 per cent of the world supply is met in the Mediterranean with 104 ships and 72,000 beds. The number of passengers is projected to grow considerably by 2025. In Marseilles, the number of passengers increased from 11,600 in 1993 to 260,000 in 2002 and might reach 600,000 by 2025. Tunisia is projecting 380,000 cruise passengers by 2010 compared with 185,000 in 2000. In Cyprus the number increased from 58,000 in 1985 to 135,000 in 2000. This growth involves more specific developments in existing ports rather than the construction of new ones. However, it should be noted that the new Tangiers-Mediterranean commercial port was planned so that the old port of Tangiers could be given over to the cruise business; and an off-shore platform has recently been created in Monaco.

The number of operational *energy plants* (refineries and thermal power plants) located on the coast numbered 112 in 1987 with 43 planned. In 2004 there were 288 in operation (gas plants, refineries and thermal power plants) with 226 planned, excluding the off-shore plants (Figure 6.10). Apart from the pollution generated, energy installations also occupy appreciable areas of land. A refinery can occupy from 1.5 to 35 hectares per million tonnes of processed crude, and a 1000Mw thermal power plant occupies from 6ha (natural gas) to 18ha (coal).

Given the growing energy demands and the fact that on average 40 per cent of the power plants are located near the coastline (maritime access and cooling water), it is projected that by 2025 there will be 360 power plants in the Mediterranean coastal zone, including 160 new ones.

The projected doubling of the intra-Mediterranean gas trade will require new infrastructure to meet demand. At present there are 12 operational gas refineries, one under construction and 17 planned, excluding the new refineries (18 being expanded or planned) that would bring the total to 73.

Industry in coastal areas (chemicals, petrochemicals, metallurgy, food processing, waste treatment plants), at present mostly in Spain, France and Italy, will also increase in other countries. In 1989 the Blue Plan had already estimated the number of major industrial plants on the coast from Morocco to Turkey at 157, plus 67 planned. Strong industrial growth is projected for the south and east to meet increasing demands due to population growth and rising living standards. Stable in the north, *steel* production in the south and east increased from 8.5 million tonnes in 1985 to 21 million by 2002 and could reach more than 50 million by 2025.[6] *Cement* production will probably decline in the north and increase by more than 150 per cent in the south and east. This growth will particularly affect the coastal zone.

Figure 6.10 Mediterranean coastal energy infrastructure

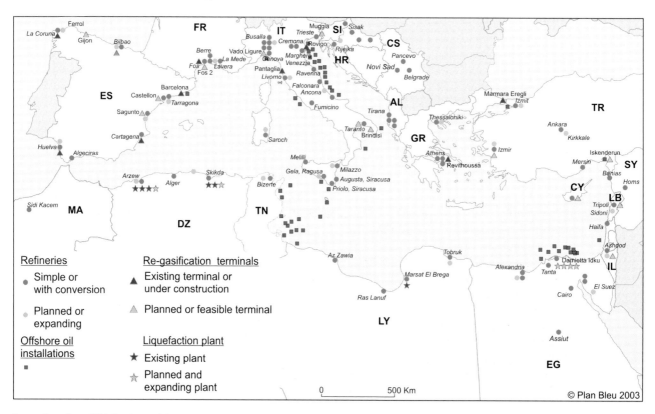

Source: Data from Oil & Gas Journal; OME, UKHO, SHOM Questionnaires

Coastal Areas

To meet freshwater shortages and the growing tourism and urban demand, some countries have turned to desalinization. In all there were 238 *desalinization plants* in 2000 (with an average production capacity of more than 500m³ per day) scattered along the coast.[7] According to national projections for water supply by desalinization in the medium-term,[8] it is estimated that the additional production needed will be more than 1 million m³ per day, which implies 175 new plants with a 6000m³ per day capacity (or 35 with a production capacity of 30,000 m³ per day).

Half the coastal zone under artificial cover by 2025?

The first consequence of continued growth in population, infrastructure and facilities is an increase in built-up and other artificial land cover in the coastal zone.

The *linear nature of coastal urbanization* and the *speed* of the phenomenon is significant (Figure 6.11). The balance between the urban fabric, agricultural land and forests that existed in 1975 on the 66km of coastline in the Malaga region (Spanish Andalusia) was upset in 1990. On this 66km long strip that goes 10km inland, artificial land cover increased by 76km² (3.5 per cent per year) at the expense of agricultural land (which lost 55km²) and forests and semi-natural environments (20km²). This increase in artificial land cover was especially high (50km²) on the coastal strip from 0 to 2km; the coast is now almost completely built-up along its entire length.

On the basis of the European LACOAST programme, which supplies satellite images, changes in the Mediterranean coastal zones of Spain, France and Italy can be assessed (Figure 6.12). The spectacular growth of construction along the Andalusian coast (55 per cent in 15 years on the 0–2km strip) brings it close to the average rate of artificial land cover in Mediterranean France and on the rest of the Spanish Mediterranean coast. The percentage of artificial land cover on the Italian coasts remains slightly lower but is increasing strongly (by 22 per cent in 17 years in the 0–2km strip).

Such comparisons are not available for the Balkans and the southern and eastern shores. Many studies and

Figure 6.11 **Coastal construction from 1975 to 1990 in Malaga province (Spain)**

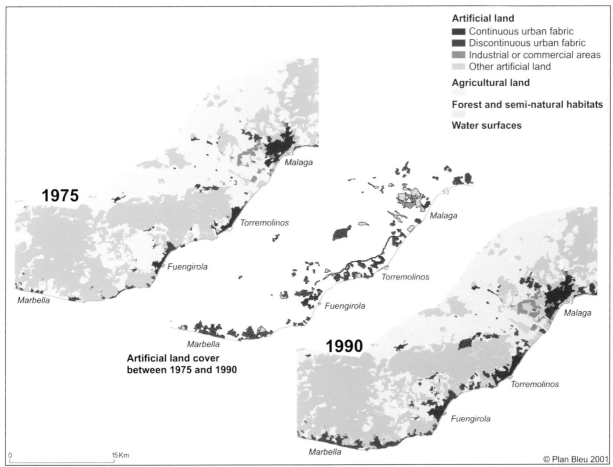

Source: EU-Joint Research Centre, Ispra (LACOAST)

2 A Sustainable Future for the Mediterranean

Figure 6.12 The increase in artificial land covering a few Mediterranean coastal regions (%)

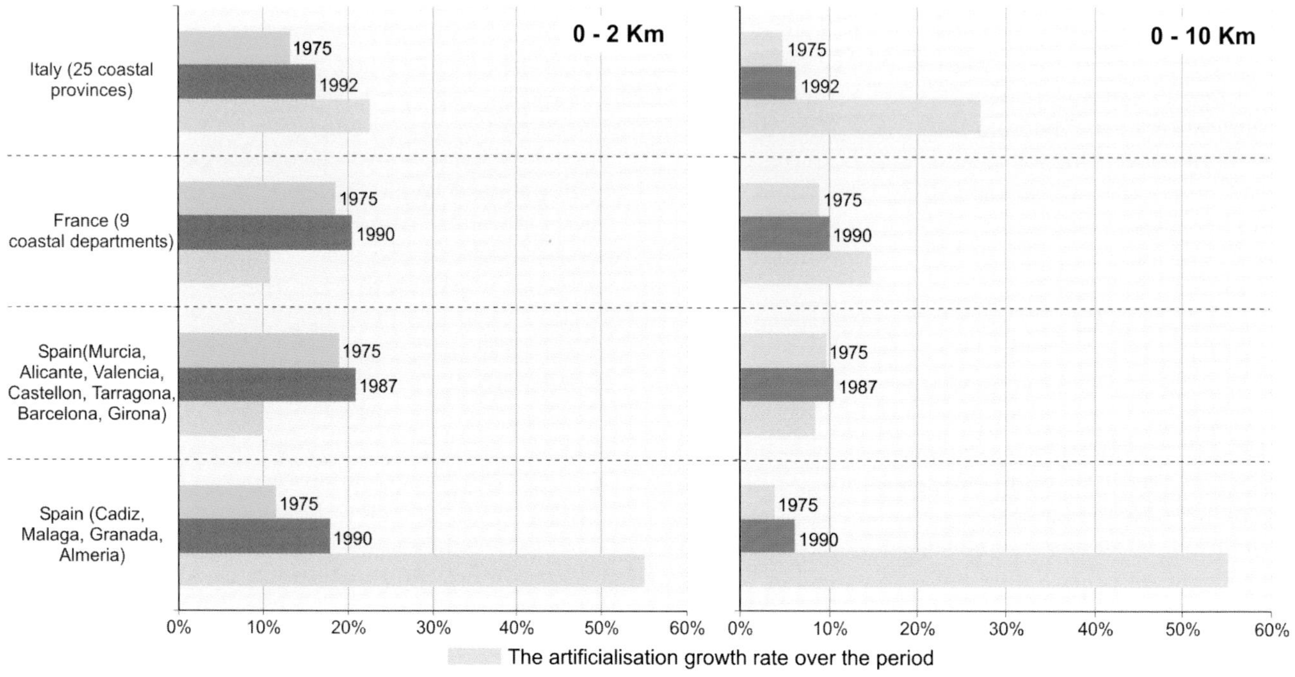

Source: Plan Bleu from JRC data

diverse information, however, show the same trend in linear coastal urbanization.[9] Natural conditions sometimes impose this linearity, as in Alexandria, Egypt. When the land surface makes it possible and urbanization is not mainly the result of tourism, urbanization can develop in the form of a half moon, as in Sfax, Tunisia, where artificial land cover more than doubled between 1972 and 1994 at the expense of agricultural land (see Chapter 4).

Other sources and methods could be used to assess the percentage of artificial land cover of the Mediterranean coastline.

The *Géopolis* database gives the surface area of coastal cities, which can be compared with the surface of the 10km wide coastal strip. These data should be analysed with care, since non built-up areas can be included in the city surface figures[10] and leave margins for densification. The data show that the urbanized proportion of the 10km coastal strip was very high in 1995 in Spain and Lebanon (>50 per cent, Table 6.1). According to *Blue Plan* projections, which assume that city-dwellers will in 2025 consume 1.2 per cent more space than in 1995 (smaller households, increased incomes, increased infrastructures), the process would increase in Spain and Egypt (>70 per cent by 2025) and even more in Lebanon (to saturation point).

Data from night-time light radiation surveys (see Figure 17, p27) show the part of luminous coasts, that is coasts with high density transport infrastructure or built-up areas: about 40 per cent of the total coastal zone seems under some form of artificial land cover. There are large differences between countries, ranging from 7 per cent in Albania to practically 100 per cent in several other countries (Lebanon, Israel, the Palestinian Territories, Malta, Monaco and Slovenia). Except for Albania, Greece and Cyprus, the percentages are in general much higher on the northern and eastern shores (more than 60 or 70 per cent) than on the southern shore (between 20 and 45 per cent). These data must be handled with care, however, because the level of artificial land cover along roads and in villages is underestimated.

Based on these different methods, it is estimated that 40 per cent of the coasts, or 18,000km, were already built up by 2000. This proportion is projected to increase between now and 2025 with population and tourism growth, and with the creation of approximately 500 new major facilities (including 43 yacht harbours, 200 energy plants and a large number of desalinization plants and industrial sites). The development will vary between regions. In coastal zones that are already urbanized and countries where the coastal land is now protected, urban overflow will take place mainly on a secondary line.

Coastal Areas

Table 6.1 Urbanization of the coastal strip in 1995 and 2025 in a number of selected countries

Country	Mediterranean coastline (km)	Coastal fringe surface – 0–10km (km²)	Total surface of coastal cities (km²)		Ratio of total surface of coastal cities and coastal strip surface (%)	
			1995	2025	1995	2025
Spain	2580	25,800	14,182	18,886	55	73
France	1703	17,030	4042	5738	24	34
Italy	7375	73,750	28,320	33,366	38	45
Greece	15,021	150,210	3041	4072	2	3
Lebanon	225	2250	1287	2286	57	102
Egypt	955	9550	3116	7468	33	78

Source: Géopolis 1998; Attané and Courbage, 2001; *Plan Bleu*, 2001 and 2002.

Elsewhere, urbanization may increase the proportion of built-up coasts, especially in regions with a strongly developing tourism industry.

Under the baseline scenario assumptions, an additional 200km of the coastline will, on average, eventually be under some form of artificial land cover each year, or about 5000 additional kilometres in 25 years. By 2025, 50 *per cent of the coast could thus be irreversibly built-on*. In some countries, vast coastal conurbations or urban cordons may extend over tens if not hundreds of kilometres.

Box 6.5 Increase in risks and costs of flooding due to coastal urbanization in France's Languedoc-Roussillon region

Building on coastal zones increases vulnerability to flooding. In the Languedoc-Roussillon region in France, where the population in the coastal districts has doubled in 40 years, 300,000 people live in flood-prone coastal areas. The area dedicated to dwellings, activities and infrastructures (18 per cent more in 2000 than in 1990) has grown faster than the population. The average built-up area in the region increased from 260m² per capita to 420m² per capita in 22 years and could reach 500m² by 2015. The same is true for the flood-prone part of the plains (which will increase by 15 per cent over the same period). Urbanization in flood prone areas reduces the area of natural floodplains, thus increasing vulnerability to flooding. Eighty per cent of the buildings in flood-prone areas have been constructed during the past 40 years and bear witness to the failure to enforce regulations.

The costs of these floods in human and material terms are becoming less and less bearable. In 1988 Nîmes suffered damage valued at 610 million euros. The floods of September 2002 killed 35 people in the French *département* of the Gard and caused damage estimated at one thousand million euros.

Source: IFEN, 2003

This linear urban sprawl results in many impacts along the coasts, including pollution, traffic congestion, degraded landscapes and coastal ecosystems, and increased coastal erosion.

One of the consequences of coastal urbanization is an increased vulnerability to *flooding*, which can be extremely serious along some coastlines (Box 6.5).

Impacts on *ecosystems and landscapes* are virtually irreversible. Although they will probably increase considerably between now and 2025, the degree of seriousness will depend on the capacity to implement appropriate responses. Possible responses to this challenge, such as the ability of countries to develop sustainable management policies for their coastal areas, are analysed in the following sections. Box 6.6 gives an example of a high-quality coastal zone that is expected to undergo strong tourism-related urbanization, while responses are developed to try to control this process.

Box 6.6 Strong tourism-related urbanization on the Croatian coast and measures developed to reconcile environment and development

Croatia has a high tourism potential due to its high-value natural, cultural and historical heritage. These elements are at the core of the national tourism strategy, which aims to increase the competitiveness of the sector and create a clearly identifiable image in international markets.

The coastline of Croatia and its many islands constitute its main tourist attraction. Since 1999 the Adriatic coast has enjoyed the status of a Priority Tourism Zone. This coast, which contains 95 per cent of the national accommodation capacity, has been urbanized along some 850km (15 per cent of the coast). Taking into account existing pressures and local projects, the authorities estimate that an additional 800km will be urbanized in the near future, a length 1650km of coastline (65,500ha of coastal zone), of which 15,300ha will be occupied by tourism along 600km.

Since 1997, the Croatian government has published various documents aimed at reconciling tourism, national planning and environmental protection: the 'land use strategy' (1997)

A Sustainable Future for the Mediterranean

and the 'land use programme' (1999) recommend the privatization of tourism and the implementation of impact assessments. The areas with high ecological value are considered as strategic tourism resources. The Environment strategy and the National environmental action plan for 2002 assess the actual situation and the obstacles, and define the objectives to be achieved. The Report on the state of the environment in Croatia (2003) recommends the creation of a new generation of land-use plans that take the objectives defined in the earlier documents and frameworks into account. The protected environment on the islands is considered a good opportunity for the development of sustainable tourism. Finally, the *2010 Strategy for the development of tourism* that is about to be adopted formulates ten objectives, two of which specifically integrate the environmental dimension (land-use plans that require natural resource preservation, environmental awareness and preservation of ecological wealth).

The challenges are considerable, since the very aspects that attract tourists (preserved natural resources, attractive landscapes and cultural wealth) could very easily be jeopardized by tourism and urban development, which would not sufficiently integrate the demands of territorial quality.

Source: Plan Bleu, based on personal communication from M. Mastrovic, Ministry of Environmental Protection and Physical Planning, Croatia, 2004

Pursuing the fight against coastal pollution

Mediterranean coastal waters are affected by wilful or accidental maritime pollution and land-based pollution.[11] The land-based sources are responsible for 80 per cent of the total pollution affecting the Mediterranean, and responses have been developed over the past few decades to reduce and prevent this. Pollution originating from activities at sea account for only 20 per cent. Hydrocarbon pollution has decreased in recent years, but an increase in pollution risks is retained in the baseline scenario, given the rise in maritime traffic (see Chapter 3).

High-impact land-based pollution, improvements remain insufficient

It is difficult to estimate the quality of coastal waters, and even more its trends, on such a vast scale. The MAP/MEDPOL inventories concentrate on the 101 *priority coastal hot spots* (Figure 6.13) for which a new 2001 assessment showed uneven changes (improvements as well as degradation, depending on the location).

Coastal water pollution affects ecosystems, human health and the economy in many ways (increases in

Figure 6.13 One hundred and one priority pollution hotspots

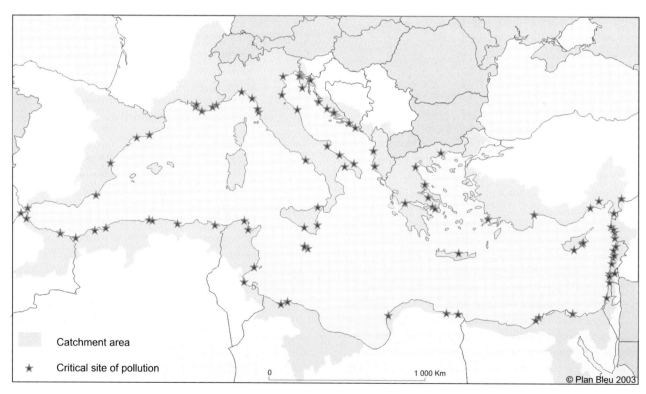

Source: MAP/MEDPOL; Margat, 2004

Note: In 1999 the 101 hot spots contained a population of 35 million, half of whom lived in 11 coastal cities with a population of more than 1 million.

public health costs, reduction in seafood consumption and the related loss of jobs in fisheries and aquaculture, negative effects on tourism). All economic activities in an area can be affected. Izmir Bay in Turkey in the 1950s, for example, enjoyed a high recreational value. Since then, heavy pollution (domestic, industrial and agricultural) resulted in a drop in tourist arrivals of between 1.55 and 3 million people (a loss of 9–18 million overnight stays between 1950 and 1990), a drop in fishing activities (loss of shellfish and fish), a loss of salt production and a surcharge for dredging navigation channels. Bathing is no longer possible.

The most important forms of pollution in the Mediterranean are eutrophication, chemical contamination, including persistent toxic substances (PTS), pollution from organic and pathogenic micro-organisms and hazardous solid waste.

Eutrophication

Eutrophication is a process by which waters enriched with nutrients (nitrogen and phosphorus) stimulate primary aquatic production. It induces an increase in phytoplankton biomass, including seaweed, 'red tides', seaweed scum, the growth of benthic algae and sometimes a massive growth of immersed and floating macrophytes. It has become a chronic problem in shallow waters near deltas (Rhone, Ebro, Po, Nile) and major urban areas, for example near Sfax with its phosphates. However, the Mediterranean is far from reaching the serious situation in the Black Sea or in the Marmara Sea. Between 1965 and 1995 nitrate concentrations in the western Mediterranean increased by 0.5 per cent per year while phosphate concentrations were reduced by 1 per cent, so changing the N/P ratio and thus promoting eutrophication. The northern shore appears to be more affected, with repercussions for biodiversity and tourism (Box 6.7), but it is also the most closely monitored shore.

Box 6.7 Eutrophication and its impacts on the northern Adriatic coastline

Eutrophication has been affecting the northern Adriatic for 20 years and is attributable to discharges (particularly via the Po river) of nutrient-rich substances in quantities that surpass the natural capacity of assimilation. Italy alone contributes nearly 270,000 tonnes of nitrogen per year and 24,000 tonnes of phosphorus. 12,600 tonnes of nitrogen and 600 tonnes of phosphorus are added annually from the Istria region. The first examples were recorded in 1969. In 1975 a massive increase of flagellates led to a general anoxia (lack of oxygen) in deep water accompanied by the disappearance of bathyal fauna and the death of large quantities of fish (7000 tonnes for the district of Cesenatico alone). Similar events continued to occur in the following summers.

The phenomenon has repercussions on tourist visits and biodiversity. The appearance of gelatinous organisms (*mucillagine*), linked to a 'cloudy sea' in many parts of the Adriatic in the summers of 1988 and 1989, led to a 20–30 per cent drop in the number of tourists. The total number of overnight stays in Rimini province fell from 17.8 million (1988) to 11.7 million in 1989. Moreover, recurrent anoxia of deep waters leads to important changes of the benthic ecosystem. The repetition of these dystrophies[12] has led to the disappearance of 15 species of molluscs and three species of crustaceans.

Source: UNEP-MAP, 1996a

Three-quarters of this problem, which many experts consider to be of great importance for the Mediterranean coastal waters, is caused by various *diffuse agricultural discharges*. Farming discards more than 1.6 million tonnes of nitrogen (N) per year and nearly 1 million tonnes of phosphorus. Eutrophication could spread to the southern and eastern shores with agricultural intensification and urban growth. Agricultural projections (see Chapter 5) indicate that the use of fertilizers could increase between 2000 and 2025 by as much as 70 per cent in the east (mainly in Turkey), 50 per cent in the south and 5 per cent in the north. Rational agriculture, making more efficient use of fertilizers, could limit the risks of excessive soil additives, and thus the diffuse discharges of agricultural origin.

Rivers that drain major catchment areas with intensive agriculture (the Arno, Po, Ebro, Pinios and Nile) are the main vectors for transmitting these pollutants to the sea. The input of nitrates by the 80 main rivers flowing into the Mediterranean doubled between 1975 and 1995. Phosphorus has stabilized after its use in washing detergents was restricted (Table 6.2).

Table 6.2 The flow of nutrients carried by river water into the Mediterranean Sea

	$N-NO_3$ flow (Kt N/year)	$P-PO_4$ flow (Kt P/year)	TP flow (Kt P/yr)	$N-NO_3$/ $P-PO_4$ ratio
<1975	333	14	36	23.4
1985–1990	469	38	94	12.5
>1995	605	14	36	42.2

Source: Blue Plan from MAP/MEDPOL data

The discharge of non-treated wastewater from *industry* and *cities* contributes 10 per cent of the total input of phosphorus and 20 per cent of nitrogen. This can aggravate eutrophication locally. Organic discharge lowers the amount of oxygen in water even more, as in the Ebro delta, where it is close to zero.

Intensive aquaculture is responsible for only a small part of the nitrogen input, although is it increasing rapidly

(25 times higher in 2000 than 1990) and can have serious local impacts.

Chemical contamination of coastal waters, impacts on human health and the environment

Of the chemical substances discarded into coastal waters, *persistent toxic substances* (PTSs) are especially worrying because of their persistence and their toxic effect on animal and plant life if concentrations exceed certain thresholds. The flows of industrial *heavy metals*, although low compared with those generated by natural processes, such as mercury, increased by 300 per cent between 1950 and 1990, and this trend has only recently been reversed.[13] Most of the input comes from run-off and the atmosphere, the major contributors being Spain, France, Italy and Greece, with 70 per cent of the lead input and 60 per cent of cadmium. *Organo-chlorides* are the largest group of *persistent organic pollutants* (POPs). They have a high resistance to degradation and accumulate in the fatty tissues of marine organisms. Available information does not make it possible to highlight overall trends, but local levels of PTS can reach toxic thresholds for plants and animals near industrial areas and estuaries (see Statistical Annex). *Hydro-carbons*, particularly polycyclic aromatic hydrocarbons (PAHs) are persistent and originate from land-based sources (incomplete combustion processes and industrial effluents). They can be much larger in volume than the operational and accidental discharges of maritime origin.[14]

Studies measuring the *impact of* PTS *on human health* in the Mediterranean are rare, especially for the southern and eastern shores. Food, especially contaminated sea food, is the source of 70–95 per cent of infections. PTS are difficult to detect in food, but their effects can be very serious in the long run. One-off studies report levels of PTS ingestion that can go well beyond the standards set by the WHO.[15] In Italy, for example, daily intake of PAHs through food has been estimated locally at 3 micrograms per day per capita (1.4 micrograms of carcinogenic PAH per day), or 100 times more than the tolerated daily intake.[16]

The effects of accumulating persistent substances in the food chain and fish oils are particularly important for human health, but also have serious and long-term *economic consequences*. An example: an EU regulation that set limits on the concentration of dioxin and PCBs in fish meant for human consumption, hit exports of herring fished in the Baltic Sea, threatening to ruin an entire economic sector in the riparian countries. These countries will have to pay the price of industrial developments in the past that were not respectful of the environment (paper mills using chloride, waste incineration) since they contributed to high concentrations of dioxin and PCBs in herring (average age between 12 and 18 years). Considering the lifespan of these substances, the countries will have to pay for a long time without any hope of improvement. If the Mediterranean region does not take timely action, it will not be protected against such an 'ecological and economical time-bomb'.

Industrial discharges are the largest source of chemical pollutants, entering coastal waters either directly or via rivers. Pollution risks are expected to increase, given industrial development in the SEMC, with major discharges into the sea and rivers and a low level of pollution clean-up. In the northern countries, there is already clear evidence of contamination by PCBs, PAH and solvents reaching the Mediterranean via the Po, Ebro, Rhone and other rivers (see Statistical Annex).

Armed conflicts are also polluting factors with long-lasting effects. With the destruction of industrial and military sites during the Balkan war, several thousand electric transformers containing PCB oil were damaged and tonnes of persistent hazardous substances were discharged, which accumulated in groundwaters and ended up in coastal waters and finally in the food chain, in particular of fishing families.

Organic and microbiological pollution will increase in the south and east

Particulate *organic pollutants* discharged into coastal waters around outlets of insufficiently treated wastewater lead to an excess of dissolved oxygen consumption, resulting in high environmental degradation of coastal waters.

Microbiological pollution of coastal waters and substances produced by pathogenic micro-organisms in the sea can cause diseases in humans and marine animals. Although considered an important problem in the Mediterranean, this type of contamination remains little known. Monitoring the pathogenic agents is limited to measuring faecal, coliform and streptococcal contaminants, usually by measuring the Biochemical Oxygen Demand (BOD).[17] The main *impact on human health* concerns gastric-intestinal infections from seafood, including salmonella, gastro-enteritis and hepatitis. Some marine animal diseases have been reported, related to infection by a mobilivirus (fish and sponge diseases, the deaths of dolphins in the 1990s). Using biomarkers should make it possible to improve the very incomplete monitoring and assessment systems.

The primary cause of contamination is the *discharge of untreated domestic and industrial wastewater*. The resulting BOD load in the Mediterranean 101 coastal hot spots was estimated at 805,000 tonnes per year in 1999. Discharges from coastal cities participate in the trophic imbalances of coastal waters. Five cities (Alexandria, Naples, Izmir, Barcelona and Beirut) produce a quarter of the total BOD load in the Mediterranean. Total industrial discharge for the 101 coastal hot spots was estimated in 2000 at 410,400 tonnes of BOD load per year. With growing urban and tourist populations, estimated at 2.2 per cent per year in the south and 2.8 per cent in the east, domestic

Coastal Areas

and industrial discharges may see a large increase. In the north they may fall by 0.4 per cent per year because of the decreasing population (see Chapter 1). With tourism developing, especially yachting, discharges may increase considerably if regulations are not kept up to date.

This form of pollution directly affects the quality of bathing water, creating a public health problem. Following Directive 76/160/EEC, the quality of coastal bathing waters in the EU has improved steadily. In the Mediterranean countries of the EU, compliance with standards reached nearly 100 per cent in 2002 in Greece, more than 98 per cent in Spain, 96 per cent in Italy, and just 88 per cent in France where 9 per cent of the bathing areas are insufficiently tested. In the other countries there has been an increase in the number of sampling stations (82 in Tunisia[18]).

Solid waste and hazardous waste: Alarming tonnage by 2025

Coastal districts produce some 30–40 million tonnes of solid waste per year. Plastics alone account for 75 per cent of the waste found either on the sea surface or floor. This waste originates from households (17 per cent), tourism (15 per cent) and dump sites (14 per cent).

Accumulated waste is a source of environmental pollution: marine animal deaths, chemical contaminants, an aesthetic impact and repercussions on fisheries. In the Gulf of Lion plastics make up 70 per cent of large waste and fishing equipment 3 per cent. Waste density is highest in the north-western sector near the large urban agglomerations of the French coast.

In the baseline scenario, the production of solid waste is assumed to increase in all countries between 2000 and 2025, by 2.4 per cent per year in the north and 3 per cent in the south and ast (see Chapter 4). Given the present gaps in information-gathering and analysis this does not promise much improvement. Coastal urban areas are projected to produce about 71 million tonnes of domestic waste by 2025, 40 million tonnes more than in 2000.

Hazardous waste generated in addition to domestic waste has been estimated at about 20 million tonnes per year; 76 per cent originates from five sectors: metallurgy, mining, refineries, the organic chemical industry, and the processing and storage of hazardous waste.

Unlike coastal artificial land cover, these environmental impacts are not irreversible (except for PTS). Solutions exist that can minimize and process the high volumes of pollutants and solid waste dumped into coastal waters.

Mobilizing actors and funds for the Strategic action plan to combat land-based pollution

Over the past few decades we have seen a wealth of *legally binding international agreements*, which are increasingly shaping national regulations. The most important are listed at the end of this chapter.

As early as 1975 the *Mediterranean Action Plan* (MAP) set up a cooperation mechanism for the Mediterranean. The coastal countries and the European Community[19] adopted a framework Convention for the Protection of the Marine Environment of the Mediterranean (the Barcelona Convention) that came into effect in 1978, with Protocols containing technical appendices. In 1995 the 'Barcelona system' – the Convention and its Protocols – was modified to take the developments of international law into account and broaden its scope both in space (adding coastal zones and water catchments, the source of most of the land-based pollution) and content (expanding it to the broader concept of sustainable development).

To combat land-based pollution, MAP has an extremely important protocol, the Land-Based Sources (LBS) Protocol. According to article 5:

The Parties undertake to eliminate pollution deriving from land-based sources and activities, in particular to phase out inputs of the substances that are toxic, persistent and liable to bio-accumulate. To this end, they shall elaborate and implement, individually or jointly, as appropriate, national and regional action plans and programmes, containing measures and timetables for their implementation.

Simultaneously with developments at MAP, the European Union (also a Contracting Party to the Barcelona Convention) added ambitious objectives to its legal apparatus with directives for water protection, including directives on wastewater, IPPC, nitrate and maritime security, and the EU Strategy on waste (see the legal references at the end of this chapter), and the Framework Directive on Water that aims to reach a good state of continental and coastal waters by 2015. These documents, compatible with multilateral agreements on the environment, are gradually framing the environmental regulation of EU Mediterranean member countries and 'accession' countries. The EU budget finances part of the costs of upgrading infrastructures and contributes to the exchange of experiences and strengthening the capacity of its member countries. For the 2000–2006 period, about 3 thousand million euros are to be devoted to the environment each year from structural funds for its member countries (a major portion is for wastewater collection and treatment) and about 300 million euros for wastewater treatment in the ten new EU countries (an average subsidy of 64 per cent of the total costs).

MAP plays a basic role to avoid the gap that would result between the 'European' shore, endowed with regulations and resources to limit pollution, and a 'non-European' shore, less equipped with instruments for implementing environmental policies, but experiencing a considerable increase in pollution over the next 25

2 A Sustainable Future for the Mediterranean

years. Such a division would compromise efforts of the more advanced countries to control their pollution. MAP can help to avoid this division by contributing to the process of defining common priority objectives, harmonizing monitoring methods and regulations, and helping the least-endowed countries to implement them.

After a legal consolidation phase in conformity with European and international provisions, MAP is now moving towards a more operational phase, intended to reinforce the countries' resources for implementation. This is illustrated by the Strategic Action Plan (SAP) formulated to enforce the LBS protocol. Adopted in 1997, SAP sets objectives to reduce land-based pollution by 2025 (for example treating the wastewater of all cities in conformity with the LBS protocol,[20] suppressing the discharge of many persistent toxic substances, zinc, copper and chromium, pesticides and hazardous waste). It also aims to halve cumulative industrial discharges of BOD in *all* countries by 2010. These objectives are defined in *National Action Plans*, which MAP is helping to formulate, and subsequently to implement by lending support in mobilizing the required funds.

With this impetus, regulations are being set up at a national level, foreseeing discharge-permission regimes. Some countries, however, do not yet have legislation for classified installations (Morocco, Turkey and Lebanon).

Major difficulties persist in *implementation*. According to existing inventories,[21] less than half of the liquid industrial waste is purified and less than a third is treated before discharge into the sea or a river. Of domestic wastewater, about 60 per cent of urban waters are discharged into the Mediterranean without previous treatment.[22] It is estimated that 48 per cent of the large coastal cities (with populations of more than 100,000) do not have treatment plants, 10 per cent have primary treatment,[23] 38 per cent have secondary and only 4 per cent have tertiary treatment prior to discharge into the marine environment.[24] Even among the EU member countries, the Directives are only gradually being implemented. While cities with populations of more than 150,000 were supposed to install at least secondary wastewater treatment before 31 December 2000, the EC reported many delays in 2001 in large Mediterranean cities (Figure 6.14).

In 2000 (Figure 6.15), some countries had no wastewater treatment system (Albania and Syria) or only a small proportion of the population was linked up (Italy, Turkey, Egypt and Tunisia). The average yield of treatment plants is low – less than 70 per cent – which corresponds with the spread of secondary and primary

Figure 6.14 Urban areas with more than 150,000 population equivalent in compliance with the EU Waste Water Directive (in the 4 EU-Med countries, 2001)

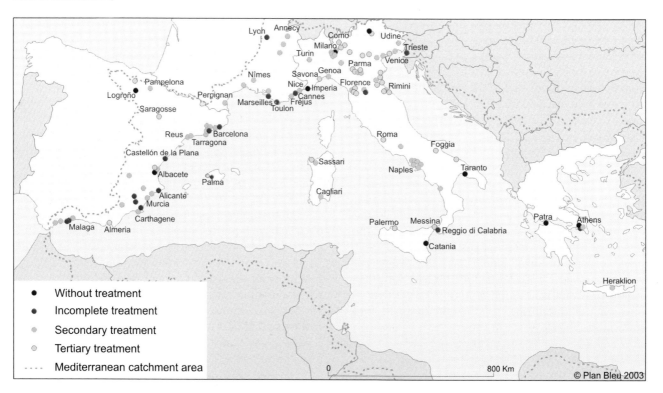

Source: EC Second Forum on Implementation and Enforcement of Community Environmental Law. Intensifying our efforts to clean urban waste water. Brussels, 19 March 2001 http://europa.eu.int/comm/ environment/nsf/city_sevage.htm

treatment plants. Yields are lowest in Slovenia, Lebanon, Turkey and Croatia where plant only provide primary treatment. Average yield in Greece is lower than in other EU countries because of the high proportion of primary treatment plants; but progress has been made since these statistics were gathered, with a secondary treatment plant in Patras completed in 2001 (another one for the sewage of Athens was under construction in 2003). Several countries retain at best some 10 per cent of the BOD_5 (Croatia, Lebanon, Morocco, Slovenia, Turkey and Egypt).[25]

Considerable effort will be required to install treatment plants where they do not exist, to complement primary treatment by at least secondary treatment, and to equip coastal cities that will have to absorb 18 million additional people between now and 2025. This effort mainly concerns the southern and eastern countries.

Likewise, half of *industrial* process discharges and less than 10 per cent of domestic industrial discharges are treated before they enter the Mediterranean countryside (and less than a third before being discharged into the sea or rivers, see Statistical Annex). Thus there is much room for improvement in this field. The use of filters or primary sedimentation of industrial effluents could reduce discharges by 70 per cent and secondary treatment of effluents (activated sludge, for example) would give a reduction factor of 10–20. Better management of techniques presently available for treating industrial effluents, sludge and residues and emissions into the air could reduce pollutant inputs to a tenth of their present levels, thus reducing discharges to coasts. However, given the quantities of pollutant input from rivers, the atmosphere and the Black Sea, efforts to reduce inputs must also occur via comparable measures taken upstream.

The main obstacle lies in the lack of institutional, financial and human *resources* dedicated to pollution monitoring, control and treatment. Funding requirements to remedy the situation in the south and east and to meet future demands are very high. Funds are needed for *investment*, *management* and *maintenance* of infrastructures. For example, the costs required to meet the SAP objective of halving *industrial* BOD discharges before 2010 for all 101 hot spots have been estimated at more than US$600 million for investment and about US$40 million for annual operational cost.

For domestic wastewater alone, the Blue Plan has roughly estimated the cost of upgrading treatment systems[26] for the 32 million people in coastal cities with more than 10,000 inhabitants in the SEMC, and the new infrastructures needed for treatment of wastewater of 18 million additional coastal dwellers expected by 2025 (in cities with a population of more than 10,000) at about 10 thousand million euros[27] between 2000 and 2025.[28] This

Figure 6.15 Wastewater treatment by coastal cities with more than 10,000 inhabitants, 2002

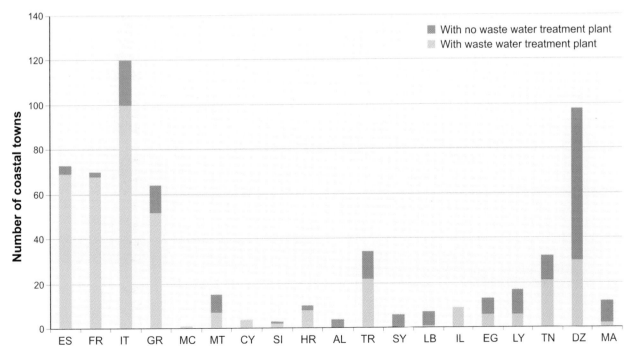

Source: MEDPOL, 2003

equals 2–3 per cent of the GDP of these countries. Operating costs could reach 300–600 million euros per year. Of course these costs should be seen against the cumulative medium-term benefits (impacts on the environment and human health, possibility of re-using wastewater, methane production, see Box 6.8), but this does not make short-term financing any easier.

> **Box 6.8 Factoring in environmental benefits in calculating the treatment tax in Morocco**
>
> The *Régie Autonome Multi-Services d'Agadir* (Agadir autonomous multi-services authority) has launched a sewage treatment programme for greater Agadir requiring an investment of 1570 million DH (150 million euros) spread over the period 2003–2015. As part of a preliminary pricing study applying various cost-recovery assumptions about users, simulations have shown that factoring environmental benefits into the calculation would make it possible to limit the tax increases on users. Examples of environmental benefits of wastewater treatment are: reduced pollutant loads (reduction in human health risks, improvement of bathing waters), wastewater re-use, biogas recovery, mitigation of greenhouse gas emissions (methane) and energy savings. If the Authority were to take into account such benefits, which can be recovered in future, the tax to be paid by users could be reduced by 20–33 per cent. However, the tax was finally set without factoring in these benefits.
>
> *Source*: Jorio, 2003

Faced with such orders of magnitude, it is reasonable to question the objectives of SAP up to 2025 for the SEMC, which will need a massive input to finance and maintain these infrastructures. Of course, autonomous funding mechanisms are preferable, but they need to be put in place gradually and be acceptable in relation to the income of the users. Faced with these financial constraints, the search for *simple and cheap technology* to treat urban wastewater in the SEMC is a challenge. It is also essential to target those actions that are the most advantageous from the cost–benefit point of view. Rather than having very sophisticated facilities in a few isolated locations, more basic standards or technology could be envisaged. The MAP/MEDPOL approach of first addressing priority hot-spots and attacking a few persistent substances is derived from this concern to optimize the available financial resources.

Such step-by-step prioritization would be facilitated by improved *knowledge* of the scale of pollution and its long-term impacts on the coastal environment and human health. Information about the quality of the milieu remains scarce and not easily comparable, even though progress may be expected in the future following current efforts to harmonize methods at the regional level. The use of biomarkers is a promising approach. More research is also required to enable a better assessment of the health risks of PTS (impacts on mortality, reproduction, the cognitive abilities of small children, etc.).[29]

Another way of reducing the costs of *curative* treatment would be to promote *preventive solutions* such as *water demand management* (see Chapter 1) or *cleaner production techniques* in the industrial sector that make it possible to minimize in advance discharge volumes (and thus volumes to be treated). This may prove to be especially relevant for developing countries. Savings made by such waste reduction at the source enable a quick return on investments (Box 6.9).

> **Box 6.9 Examples of savings made by cleaner production techniques in the industrial sector**
>
> Many examples in the Mediterranean show initiatives that introduce clean processes into industrial production, reducing the volume of polluting waste or emissions or the consumption of water and energy. These projects sometimes require no investment, and with the savings they realize they have a very quick return on investment.
>
> In **Malta**, a company producing health care articles reduced the quantity of cardboard used for packaging by 24 per cent just by improving the shape of the cardboard boxes. The savings for the company amounted to US$30,000 per year. And the flow of packaging waste to public dumps was reduced by 36 tonnes per year.
>
> In a potato chips factory in **Lebanon**, a new production process resulted in water savings of 18m^3 per day (consumption and discharges), while the investment was written off within two years.
>
> In **Tunisia** a manufacturer of car batteries discovered 19 ways of preventing identified contamination and pollution (acids, lead scoria and wastewater) and saving lead and energy. The costs of the new measures were US$522,500, but the savings were US$1.5 million per year.
>
> In Alexandria in **Egypt** a textile factory saved 30 per cent on water consumption, 27 per cent on steam and 19 per cent on electricity.
>
> In **Spain** a surface treatment plant saved about 30 per cent on water, 4 per cent on salts for the baths, 60 per cent on chemical products for water treatment, and 22 per cent on treatment tax. Annual savings amounted to US$20,500 while the investment was written off in four months.
>
> *Source*: MAP/Regional Activity Centre for Cleaner Production, Barcelona

Reinforcing ways to control pollutant discharges on the coast and into the sea also need to be considered. Mechanisms for monitoring and control are largely inadequate (the administrations in charge of control have only weak and dispersed financial and institutional resources), especially in many SEMC where the number of inspectors is pathetic compared with the number of installations to be checked. Monitoring and sanctions for offences are rarely strong enough to be dissuasive.

Under the baseline scenario, the objectives advocated in the SAP for the southern and eastern countries would only be partly achieved, unless the mobilization and capacity of all actors involved in combating and preventing pollution (states, cities and industries) are strengthened, regional cooperation based on north–south financial solidarity improves, knowledge is transferred, and adapted technologies are introduced (cleaner production processes). Without such measures, the region could see a growing gap between north and south, with progress on the northern shore while improvements in coastal water quality on the southern and eastern shores are too slow. Development of coastal tourism with its related environmental requirements could, however, play a fundamental role in mobilizing for an alternative scenario.

Protecting coastal ecosystems and their biodiversity

The Mediterranean coasts have been drastically altered over the past few decades, both biologically and physically. Changing water regimes, land- and maritime-based pollution, artificial land cover in coastal areas (infrastructures, construction and development of activities such as aquaculture and tourism) have all resulted in impacts on coastal ecosystems. Coastal erosion has increased as a result of human activities, saline intrusion has increased as a result of excessive extraction from coastal aquifers, and invasive species have become new sources of coastal degradation.

Conventional policies to protect nature, promoted at both international and Mediterranean levels, are being implemented in coastal areas by all bordering countries. However, they are insufficient to ensure the required redirection of current trends.

Altered coastal ecosystems, increasing costs and risks

Coastal erosion and salt water intrusion

Coastal erosion is affecting a significant proportion of the coastline. Sandy coasts are most affected, including beaches and deltas. In Mediterranean France 35 per cent of the beach front has been affected by receding coast. In Italy, Spain and Greece 40 per cent, 35 per cent and 25 per cent of recorded beaches, respectively, are affected. Between 1959 and 1997 the average annual coastal recession in Algeria varied between 0.3m and 10.4m. Similar situations are seen in other countries.

Coastal erosion has increased over the past few decades as a result of extraction and developments in the coastal zones and neighbouring catchment areas. Regulating nearly all the rivers beds in the entire Basin has, over the past 50 years, led to a 90 per cent *decrease in sediment* reaching the sea.

Input from the Ebro in Spain has been reduced from 4Mt per year to 0.4Mt following the construction of dams. Before building the Aswan Dam in Egypt, the Nile deposited 57 million tonnes of sediment per year in Cairo. Today it is only 2 million tonnes per year, resulting in serious erosion of the coasts in the delta, all the way to the Sinai, the Gaza Strip and Israel. Sediment captured over the past 30 years (1962–1991) by the main dams in Algeria[30] amounts to about 264 million m^3, and the dams in the wadis in Morocco that drain into the Mediterranean trap 93–96 per cent of the sediment.

Moreover, *extracting sand* from beaches reduces the sediment load and increases the energy of rising water, which in turn accelerates erosion. In Israel it is estimated that more than 10 million m^3 of sand has disappeared because of sand extractions on beaches.[31] And sediment deposited by coastal currents is far too little to fill this gap. *Construction on the coastline* itself (dikes, groins and water breakers) disturbs currents, leading to erosion or accumulation, locally amplified by degradation of seagrass. A loss of 25–35m of beach has been observed in El Kantaoui in Tunisia since the end of the 1980s following the construction of the yacht harbour. Sometimes these coastal works accelerate accumulation and even trap sediment.

Erosion can have considerable *economic consequences*. The near disappearance of the beach in Tangiers in Morocco in the 1990s after the construction of the fishing harbour and commercial port, resulted in a reduction in tourism. The destination lost 53 per cent of its international tourist night-stays, causing a substantial loss of income from tourism (US$20 million per year), tourist transport (40 per cent) and loss of income for craftsmen (25 per cent).[32] In Italy, 1.9 million m^3 of sand was artificially deposited between 1983 and 1993 on 13.8km of beaches in Emilia-Romana.[33]

According to the baseline scenario, an increase in coastal erosion can be expected, providing a good reason to think about the *shape* of the coastline in 100 years. The main contributors will be the projected growth in infrastructures, construction and extraction of materials along the coast, the construction of new dams on rivers, and the degradation of marine seagrass beds. To counter the impacts of this erosion, protection works will be needed, making the coastline even more artificial.

Extracting freshwater from coastal aquifers and islands, linked mainly to irrigation and growing coastal cities, is on the rise and increases the risk of overexploitation and *saline intrusion*. This is already happening in many coastal plains (Figure 6.16) and is expected to increase over the next two decades.

2 A Sustainable Future for the Mediterranean

Figure 6.16 Sites with groundwater overexploitation and saline intrusion

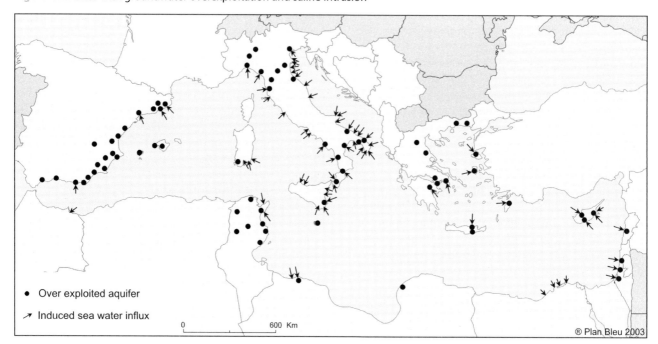

Source: RIVM RIZA, 1991; national sources compiled by J. Margat for the Blue Plan

The growing risks of high-impact biological invasions

On a global scale, *invading species* are the second most important cause of biodiversity loss, after physical destruction of habitats. Introducing species from afar is a particular risk because of the absence of natural predators and competitors. The Mediterranean is particularly affected with its nearly 500 non-indigenous marine species.[34] The introduction of foreign species has increased since the opening of the Suez Canal in 1869, more trade, transport and commerce, the development of aquaculture (Figure 6.17), and a rise in water temperature.

Non-native species arrive with dirt on ship hulls and in ballast water and sediment. In the Mediterranean the species introduced via ships are mostly macrophytes, molluscs and crustaceans.[35] As maritime traffic is projected to increase fourfold by 2025, the introduction of these species via ships will continue to increase. Of land-based animals, rodents (rats, brown rats and mice) are the most invasive in the basin. Of land-based plants, 'ice plant' (*Carpobrotus edulis*), a strong soil-covering species that chokes indigenous plants, is very difficult to eradicate and has spread over many coastal fringes.

The biological invasion receiving most press coverage in the Mediterranean is the alga *Caulerpa taxifolia*, which is threatening the invaded ecosystems.

Introduced accidentally and first sighted in 1984, it now affects six western Mediterranean and Adriatic countries, and concerns 13,000ha of sea floor along more than 180km of coastline (Figure 6.17). It lives at a depth of 0–50m and invades seagrass meadows. The IUCN classifies it as one of the 100 most dangerous invading species. Another alga, *Caulerpa racemosa*, first sighted in 1990, seems even more dangerous because it spreads extremely rapidly and already concerns 50,000ha. This alga, containing toxins and contributing to the depletion of fish stocks, is very difficult to eradicate. Manually removing it is only possible over very small areas. In an optimistic scenario, the Mediterranean would generate its own reaction and limit the expansion. With a more pessimistic view, strong expansion is not excluded, and might affect the biodiversity and economy of the coastal zone.

With increasing transport and the rise in temperature that favours species of tropical origin at the expense of native species, biological invasions can only increase. Major degradation of the ecosystem, comparable to what has been observed in other seas (for example, the Black Sea) cannot be completely excluded.

Degradation of coastal habitats and species

The Mediterranean is considered to be one of the world's most threatened seas. MAP has recorded 104 *endangered*

Figure 6.17 Main ways non-indigenous marine species are introduced and spread of Caulerpa

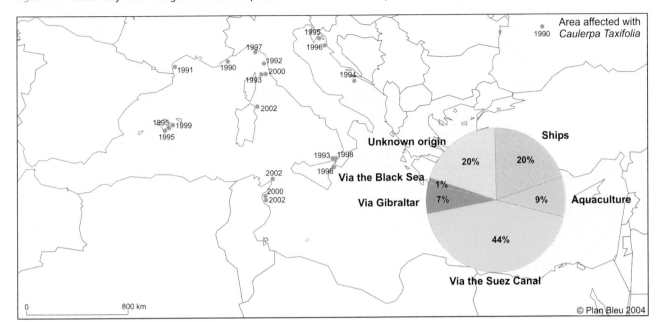

Source: EEA, 2003; Université Nice-Sophia Antipolis, Laboratoire Environnement Marin Littoral, 2002

species, from plants to cetaceans, in particular the monk seal, sea turtles and posidonia.

The reduction in turtle populations has reached the point where two of the three resident species are now considered vulnerable to, if not in danger of, extinction. These are the green (*Chelonia mydas*) and the loggerhead turtle (*Caretta caretta*) that only reproduce in the Mediterranean. The main causes of this reduction are the disappearance of spawning sites (due to sand extraction from beaches) and capture (10,000 turtles are killed every year).

The number of monk seals has decreased steadily over the last half century. Classified by the IUCN as one of the ten most endangered species in the world, the monk seal has, since the 1970s, vanished from the coasts between Syria and Egypt and has practically disappeared from the north-western Mediterranean and the coastal waters of Maghreb. Most of the population can now be found along the coasts of Greece, Turkey and the islands of the Aegean Sea. Causes for its disappearance are pollution, the destruction of its habitat and reproduction sites, and hunting. Given the competition with fishing, overfishing of some areas may also be a cause for the fall in numbers.

The most degraded and endangered *coastal habitats* are the dune ecosystems (as a result of construction and pedestrian pressure), shallow waters (subject to physical changes, pollution and overfishing), wetlands (see below), and some beaches that are used as spawning sites by turtles.

The reduction of *wetlands* has resulted mainly from infrastructures and urbanization (building of airports, roads and marinas), upstream water works mainly for irrigation (reduction of water input), and changes in agricultural practices (drying up of marshes). Levees for flood control also reduce flood plains and can upset ecosystems (subject to periods of drying and salinization or, in contrast, to sudden desalinization after a break in a levee). This happened when the Camargue region in France was flooded in September 1996: the entire *Zoostera noltii* grass beds of the Vacarès vanished following a sudden drop in salinity.

Tourism in particular contributes to the degradation of natural habitats and other biodiversity. Approximately 10–20 per cent of the *Posidonia* meadows of the Côte d'Azur are said to have vanished with the constructions on the seaside and three-quarters of the sand dunes from Spain to Sicily have been destroyed by urbanization for tourism. Open anchorages have also been a source of degradation. In August 2002, along the coast of the Porquerolles Island of France, 1350 anchored boats (outside harbours) were counted at one time, which can mean a floating population of 4000–5000 people without a sanitation system. In addition, the anchoring causes degradation of the sea floor and the dissemination of undesirable species.

Chemical pollution, as well as physical aggression, contribute to the irreversible disappearance of coastal habitats. In 50 years the area of *Posidonia oceanica* meadows has been seriously reduced all around the Mediterranean, in particular near large cities and industrial ports such as Athens, Naples, Genoa, Marseilles, Barcelona, Algiers and the Gulf of Gabes.[36] The example of the Gulf of Marseilles (Box 6.10) shows that the observed degradation is a result of *converging pressures*.

> **Box 6.10   Factors in the degradation of Posidonia meadows**
>
> The main causes identified for the regression of Posidonia meadows in the **Gulf of Marseilles** are:
>
> - Coastal developments that irreversibly destroy the shallow waters. They affect 15 per cent of the sea floor between 0–10m and 10 per cent of the floor between 0–20m in the Provence-Côte d'Azur region.
> - Pollution. The grass beds that grew between 4 and 35 metres in 1947 now only grow between 10 and 25m. The rise of the lower limit is due to a reduction in light, resulting from increased suspended matter and algal blooms. The lowering of the upper limit is due to poisoning by highly concentrated pollutants at that level.
> - Dragging, sand and gravel extraction, anchoring and the repetitive action of boat-based fishing.
> - The presence, since 1984, of an invasive algae of tropical origin, *Caulerpa taxifolia*.
>
> In the **Gulf of Genoa** there is an additional factor, namely the warming of surface water by 2°C in ten years; here some forms of seagrass meadows have quasi disappeared.
>
> *Source:* GIS Posidonie (www.com.univ-mrs.fr/gisposi/)

Although insufficiently assessed, the *present cost of coastal degradation* has been estimated for some southern countries. In 1999 these were estimated at 0.33 per cent of GDP in Egypt, 1 per cent in Tunisia (0.9 per cent as a result of loss of tourism income related to degradation of the environment and 0.1 per cent as a result of loss of fishing), and 0.6 per cent in Algeria (0.59 per cent tourism-related and 0.01 per cent as a result of impacts on health).[37] These estimates do not take account of environmental and social costs or increasing vulnerability to risks such as maritime pollution, flooding and biological invasions.

Over the past few decades, major disasters have been avoided in the Mediterranean, but this does not mean that the sea is protected against pollution such as that caused by the P*restige* in the Atlantic in 2002 and 2003, the cost of which has been estimated at more than 8 thousand million euros.

Finally, *global warming* could be a long-term aggravating factor, in particular for lagoons and deltas where major human and heritage issues are at stake. If in the far future the *sea level* were to *rise* 1 metre, 12–15 per cent of the Nile Delta would be affected. In addition to rising sea levels, climate change could also increase the frequency of exceptional events such as torrential rains leading to increased risks of *flooding*, and could modify the competitive balance between species that are very vulnerable to higher temperatures. This would also increase the impact of invading species.

Mobilization to protect coastal biodiversity

The degradation observed in coastal biodiversity has led to a growing mobilization of civil society, states and the international community. Legal instruments have been adopted over the past 30 years (international treaties, protocols, regional agreements, national laws), some of them specific to the Mediterranean.

Several relevant *international treaties*, not specific to the Mediterranean (see Legal references at the end of the chapter) are important for the region. Under the RAMSAR Convention on wetlands (Ramsar, 1971) it has been possible to classify 81 RAMSAR sites in the Mediterranean Basin (Figure 6.19), to structure the MEDWET network, and encourage specific regional action programmes. The United Nations Convention on the Law of the Sea (UNCLOS), which entered into force in 1994, is an important framework that commits signatory States to protect and preserve their marine environment and cooperate at the international and regional level.

MAP (1975) and its Barcelona Convention (1976, amended in 1995) are an example of regional cooperation and a framework for regulating and acting on the protection of marine and coastal biodiversity. The Protocol Concerning Specially Protected Areas (SPA) came into force in 1986. It was replaced by a new protocol (Barcelona, 1995) relating to both protected areas and biological diversity. It is now applicable to all marine waters of the Mediterranean whatever their legal status, as well as to the sea floor, its underground areas and terrestrial coastal zones. This extension beyond territorial waters was necessary to protect migratory species such as marine mammals. The Protocol stipulates in particular the establishment of a list of 'specially protected areas of Mediterranean importance' (SPAMI). Once an area is included in the list, the Parties shall not permit or undertake activities that could go against the objectives that motivated their creation. It is remarkable and very innovative that the SPAMI could be applied to the high sea.

The Mediterranean is also covered by *sub-regional agreements* such as the RAMOGE Agreement (Monaco, 1976) between France, Monaco and Italy on the protection of coastal waters, and agreements dealing with some endangered species such as the ACCOBAMS Agreement (Monaco, 1996) on the conservation of cetaceans in the Black Sea, the Mediterranean Sea and the contiguous Atlantic area.

The adoption of laws in all coastal countries to protect nature, mobilizing public opinion and NGOs, as well as

Coastal Areas

Figure 6.18 Protected coastal areas in the Mediterranean, 1950–1995 (thousand hectares)

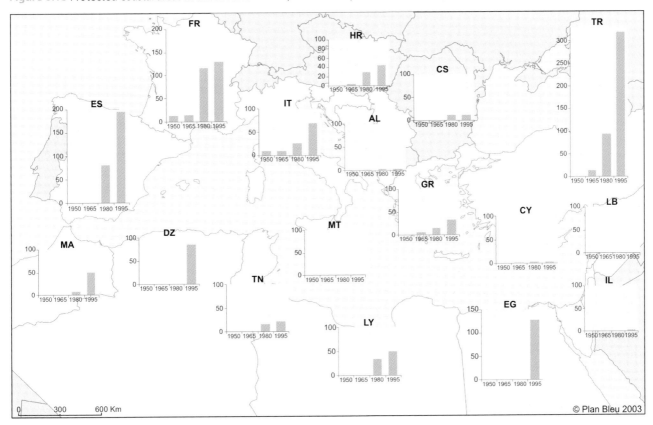

Source: Ramade, 1997; World Conservation Monitoring Centre (www.unep-wcmc.org)

Figure 6.19 Specially Protected Areas of Mediterranean Interest (SPAMI) and wetlands classified through the RAMSAR Convention

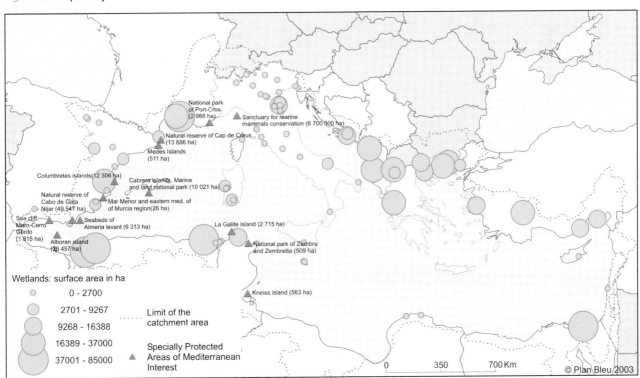

Source: *Blue Plan* from Medwet, 2003 and MAP Regional Activity Centre for Specially Protected Areas (RAC-SPA), Tunis (www.rac-spa.org)

A Sustainable Future for the Mediterranean

the arsenal of international documents have contributed to an increasing number of *protected marine and terrestrial areas*. The total of coastal protected areas in the Mediterranean, including all IUCN categories, reached 1.15 million ha in 1995 (Figure 6.18)[38]. It increased sixfold in 25 years. In 2003 there were 152 SPAs, including 47 marine areas. Fourteen of the 152 SPAs are already included in the list of SPAMI (Figure 6.19), including the 'Franco-Italian-Monegasque Sanctuary' for cetaceans, covering high-sea areas.

However, this progress is relative. In fact, there is a *gap* between the legal status of protected areas and the effective application of measures required for their conservation. If data were available on *budgets for protected areas*, they would show large differences between countries and the meagre resources allocated in most countries. Actually, considerable resources are necessary to develop and comply with regulations, ensure the protection of natural and cultural heritage, organize public recreational access under conditions that are compatible with conservation objectives, and ensure that people living in or around protected areas adhere to the objectives from which they may derive benefits. Many parks may envy the budget of the Port Cros National Park in France (3.9 million euros in 2002 for 3741ha, 1800 of which are marine areas). Still, it is modest compared with the budgets of small cities.

Furthermore, the *relative portion of protected coastal areas (terrestrial and marine)* remains lower than observed in continental regions. This illustrates the difficulty of classifying areas that suffer strong anthropic pressures as protected without directly controlling land property. Most protected coastal areas are protected because of their specific ownership status (state property, local authorities property, or state donations). Many NGOs believe that the proportion of protected land in coastal areas remains much too low, given the projected increases in pressures and impacts. In 2003 the WWF drew up a list of coastal areas that deserve to be classified over the coming years.

Implementing the SPA and biodiversity-related Protocol should not be limited to classifying protected areas. At the same time *action plans* need to be developed to contribute to the Protocol implementation. This is, among others, why a Strategic Action Programme for biodiversity was adopted by MAP in 2003 (Box 6.11).

Box 6.11 The Strategic Action Programme on Biodiversity (SAP-BIO)

In November 2003, in Catania, the Contracting Parties to the Barcelona Convention adopted the Strategic Action Programme for the Conservation of Marine and Coastal Biological Diversity in the Mediterranean (SAP-BIO). The programme was prepared by the Tunis Regional Activity Centre on Specially Protected Areas with financial support of GEF and driven in 2001 and 2002 by bordering countries. It is one result of an assessment and cooperation process involving most international and regional organizations concerned (nearly 120 national and international experts).

SAP-BIO calls for 30 priority actions to reconcile the socio-economic development of the coastal areas with biodiversity conservation in the region. Within the framework for implementing the Convention on Biological Diversity and the expanded SPA Protocol of the Barcelona Convention, it aims to improve governance systems for the sustainable use and conservation of biodiversity.

The funds required for implementing the SAP-BIO (planned to start in 2005) are estimated at US$137 million, 97 million for national level and 40 million for more regional level activities. These funds would be divided as follows:

- 52 per cent for conservation activities;
- 37 per cent for data gathering and research on biodiversity;
- 11 per cent for information programmes and public awareness-raising.

Source: MAP/SPA-RAC

All these protection efforts are, however, not enough to turn around the general trend towards degradation of the Mediterranean coastal biodiversity. A shift to an alternative scenario would require *major changes* in the modalities of development and planning, in particular:

- A more general use and reinforcement of *impact assessments* for projects and programmes, and a development of *sectoral strategies and policies* (tourism, fisheries, aquaculture, transport, energy, water, agriculture) to *integrate* the objective of sustainability upfront.
- The implementation of *specific coastal zone policies* to expand protection instruments beyond conventional nature-conservation approaches and develop an *integrated approach* to coastal zone management and development. In the alternative scenario, new generations of protected areas would emerge. These areas, designed as genuine instruments for sustainable development, would be managed within the framework of various types of partnerships, including management by states, local authorities or communities, NGOs and economic actors.

Ensuring sustainable management of fisheries and aquaculture

The maritime part of the coastal zone produces living resources that people have exploited for thousands of years. Current trends offer new outlooks but could also have serious impacts. The exponential development of intensive fish farming (aquaculture) will not be without problems, while degradation of the environment and stocks endangers the fisheries sector.

Avoiding a decline in fishery

The specifics of Mediterranean fishery

Fishing is an important issue for the Mediterranean. Although it puts only a relatively small quantity of produce on the market compared with the demand, it is an important component of the Mediterranean identity and employment. It accounts for 420,000 jobs, 280,000 of which are fishermen, and the average prices of landed produce are much higher than world prices.

The sustainability of fish resources (and consequently fishing) is favoured by the wide diversity in depth and by the presence of many refuge zones for spawning, two important factors for resilience. The exceptionally high proportion of *small scale operators in professional fishing* is also favourable in terms of sustainable development. Small inshore fishing indeed produces high-commercial-value fish, a source of many jobs compared with the quantities landed, and is much more selective in its catch than industrial fishing (trawl nets in particular).

Small-scale fishing concerns more than 85 per cent of the boats (71,800 out of a total of 84,100). Boats are sometimes not motorized (4000 of the 13,700 boats in Tunisia), and many fishermen have several jobs (80 per cent in Malta, 92 per cent in Syria). The percentage of inshore fishing of the total catch varies between countries (Syria 87 per cent, Cyprus 58 per cent, Greece 56 per cent, Tunisia 44 per cent, Italy 41 per cent, Israel 39 per cent and Slovenia 10 per cent). The industrial fleet is concentrated mainly in the EU-Med countries (57 per cent of the total). Sport fishing accounts for 10 per cent of the total catch, which is a lot.

Unsustainable development

Despite these favourable characteristics, several indicators show that the fishery sector is presently undergoing worrying changes. In addition to the degradation of coastal ecosystems, fishing is suffering from domestic competition. F*ishing activity* has increased in general, for example engines with more power, about 20 per cent more boats between 1980 and 1992, and an increase in the number of trawl nets (137 and 170 per cent respectively in Algeria and Morocco between 1980 and 1992). Industrial fishing practices have also changed. They now exploit all fish resources, fishing up to 800 metres in depth.

Increased catches[39] (about 1 million tonnes in 2001, Figure 6.20 and Statistical Annex) are accompanied with a *drop in yield*, a sign of the start of degradation of stocks. This is clearly the case in the most productive areas where industrial fishing occurs, such as in the Adriatic and around Sardinia. For some species, the overall catch per fishing unit is 60 per cent less today compared with about 20 years ago. The total catch has fallen in several countries, particularly in Italy, the leading producer in the Mediterranean. Catches have effectively dropped from 358,000 tonnes in 1980 to 294,000 tonnes in 2001 (0.9 per cent per year).

Figure 6.20 Fishing: Catches (thousand tonnes)

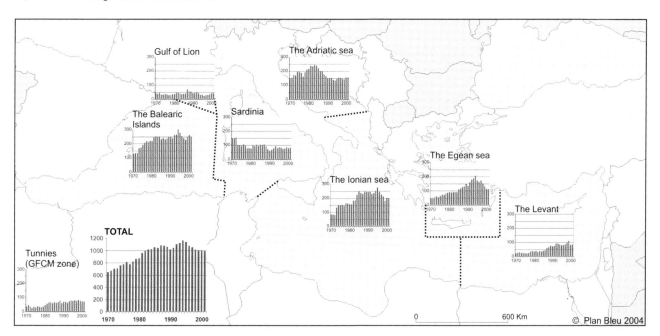

Source: FAO, Fishtat, GFCM capture production 1970–2001

The state of several fish stocks and spawning of commercially valuable species confirms the *degradation in fish resources*. This is the case, for example, with the sea bream in the Alboran Sea, hake in the Gulf of Lion and Atlantic red tuna for which catches in the Mediterranean (all flags included) amounted to 10,000 tonnes in 1980, peaked at 39,000 tonnes in 1994 (a 10 per cent increase per year), and then fall to 23,700 tonnes in 2001. Shrimp catches increased from 10,000 to 25,000 tonnes between 1980 and 1990 (a 9.1 per cent increase per year) to fall back to 14,000 tonnes by 2001 (a 5 per cent annual drop).

Without extensive changes in fisheries management and strengthening of coastal protection, current trends imply a risk of increasing loss of fish resources and corresponding *employment* (several tens of thousands). In 2000, 8000 Italian fishermen lost their jobs, 16 per cent of the total number of jobs in the sector.

More sustainable fishing calls for effective changes

To face up to the risks of unsustainable fishing, changes in operating modes are essential.

In 1995 FAO formulated a 'code of conduct for responsible fisheries', which is, however, non-binding. FAO provides the secretariat of the General Fisheries Council for the Mediterranean (GFCM), which brings together the rim countries, the EU and Japan, organizes scientific fish monitoring, develops regional cooperation and formulates recommendations. The International Commission for the Conservation of Atlantic Tunas (ICCAT) does the same, especially for Mediterranean fishing of Atlantic red tuna.

In 2003 the European Commission proposed its own action plan. A community regulation draft was drawn up in order to ensure the sustainable exploitation of resources in the Mediterranean area. The measures proposed include increasing the mesh-size of trawler nets, banning trawl and drag nets less than 1.5 nautical miles from coastlines and above sea meadows, creating EU and national level 'protected zones' where certain fishing activities would be banned, the implementation of minimum sizes to be landed for more than 20 species, improvement in compliance control, and sharing of management responsibilities between the EU and Member States. Given the multi-specific nature of Mediterranean fisheries and interaction between fishing practices, it also suggested the formulation of management plans before the end of 2004. The second objective of the EU action plan consists of launching a debate on the possible extension of Member State jurisdiction in fishing matters, and where applicable within the framework of a multilateral initiative involving the other rim countries.

As early as 1990 the Blue Plan[40] was warning about the risk of exacerbated competition between small- and large-scale fishing if the observed trends were to continue. It stressed the importance of *strengthened concerted action* between scientists, professionals and decision-makers, an improved knowledge base (for example maps of nursery areas), the reinforcement of regulations (banning trawl nets within the 3-mile limit) and regional cooperation. Promoting *participatory* and national approaches to the *management* of living marine resources as part of local development was also emphasized, keeping the possibility of sub-regional fishing zones in mind, as well as entrusting responsibility to local fishing communities. This kind of approach may make it possible to better adapt modes of exploitation to the environment. One of the most interesting early experiences occurred in Cyprus where trawler fishing slowly declined, leading to a critical situation as early as 1981. The administration imposed a general additional fishing ban in the reproduction period despite opposition from trawler owners but with the support of small-scale fishermen. This reduction made it possible in the subsequent years to increase catches by 40 per cent.

Local development approaches can lead to the development of *artificial reefs* (provided impact assessments are made). Reefs provide shelter and nursing zones for high commercial value species and have proved a deterrent to trawler-net fishing. They promote local fishing as well as sport fishing and eco-tourism. Their recent development in several rim countries owes much to the involvement of local governments and the mobilization of associations and professionals. Spain, with more than 60 locations listed in 2000,[41] Italy and France (each with 25) are the most advanced countries in artificial reef development. Significant projects are underway in Greece, Cyprus, Turkey (Izmir bay) and France (40,000m^3 in the bay of Marseilles in 2006). In addition, spreading reefs around marine windfarm sites is being considered (a project being studied in the Languedoc-Roussillon region). In 2000 the volume of immersed artificial reefs was about 170,000m^3 (92,000 in Spain, 45,000 in Italy and 33,000 in France), which remains quite modest compared with countries like Japan.

At the *sub-regional* level, inter-governmental and professional initiatives can promote the sustainable management of shared stocks. In 2002, for example, the fisheries associations of the Adriatic agreed to establish a regional committee for responsible fishing that promotes a cooperative approach with scientists, the authorities in charge of fisheries and local associations.

These regional, sub-regional and local initiatives with increased awareness illustrate that alternatives are possible. The challenge is to shift from fishing based on ferocious competition between fishermen to *responsible fishing focused on managing ecosystems*. Strengthening regional cooperation and mobilizing actors to develop this kind of process on different geographic scales will be decisive. For non-shared stocks the local level should be preferred, for taking the social context and knowledge to organize fishing into account. At the local level, collective discipline can most easily be implemented to ensure resource conservation.

These approaches imply a continuous adaptation in management and practices. This means reaching agreement on areas reserved for small-scale fishing, banning periods, and the number, place, nature and desired extent of *protected zones*. Close ties are necessary between fisheries and aquaculture, whose development should not add to the problems of fisheries and, on the contrary, contribute to collective progress.

Seeking sustainability in the Mediterranean fishing industry means going beyond the single objective of the long-term conservation of fish resources. In this region, the objective is less related to quantities caught and more to production of the best possible values. What is required is a better recognition and enhancement of the merits of a *'Mediterranean' fishing model*, made up of small-scale fishing units with a strong socio-cultural dimension.

Regulating the rapid development of fish farming

Tripling the production of bass and sea bream between now and 2025?

With the wealth and jobs it creates, aquaculture plays a significant role in the economic development of many coastal areas, especially in some economically hard-hit islands. Such development also contributes to avoiding 'tourism-only' specialization and redeployment of fishermen who have lost their jobs. Most extensive and semi-intense production methods are also beneficial for the environment. In the long term, aquaculture contributes to maintaining and enhancing wetlands, and it strengthens moves to safeguard water quality, the essential condition for its survival.

The strong growth in Mediterranean coastal aquaculture *production* (marine and brackish waters) has been one of the most remarkable facts of the past few decades. Total production (molluscs, fish and crustacean) amounted to 359,000 tonnes in 2001 compared with 149,000 tonnes in 1990 (see Statistical Annex), a 140 per cent increase in 12 years. However, this growth rate of 8.3 per cent per year remains lower than global growth (12.4 per cent per year).

Production and its growth are unevenly distributed. The leading producer is Italy (with about 170,000 tonnes in 2001, 150,000 of which was shellfish) followed by Greece, Egypt, Turkey and France. Growth is very strong (between 25 and 35 per cent per year) in Egypt, Turkey and Greece. A second group of countries with annual production of about 2000–5000 tonnes and very strong annual growth, includes Croatia (26.5 per cent per year), Israel and Cyprus (38 per cent) and especially Malta (72.9 per cent). The fairly low production in Morocco and Tunisia, on the other hand, seems to have levelled off.

Extensive and semi-intensive production, based wholly or in part on the natural productivity of the environment, is considerable (220,000 tonnes of shellfish) and is still increasing. Mollusc production in Greece, 354 farms in 2000 compared with only 5 in 1990, increased from 6000 tonnes in 1990 to 28,000 tonnes in 2000. In addition to Italy's ancient brackish-water fish farming (100 farms along the Adriatic coast produce 5000 tonnes per year of grey mullet, bass, sea bream and eels from 63,000ha), there is a strong growth in mullet breeding in the coastal lagoons and ponds on the Egyptian coastline (raising fish from juveniles fished at sea).

The intensive production of sea perch, sea bass and sea bream has seen even higher growth and today represents by far the largest part of production (in value) (Figure 6.21).

Figure 6.21 Bass and sea bream, aquaculture production in the Mediterranean and the Black Sea, 1990–2000

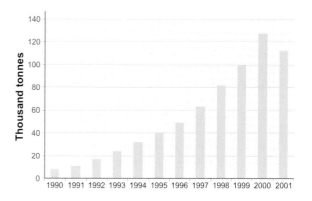

Source: Fishstat, FAO (www.fao.org/fi/statist/FISOFT/FISHPLUS.asp)

Note: Intensive marine aquaculture of bass and sea bream has experienced spectacular growth since 1990. These two species account for a value of 24 per cent and 33 per cent, respectively, of a total aquaculture production, valued at US$883 million in 2001 (compared with US$311 million in 1990). This development was made possible by strong market demands, highly developed artificial insemination, technology transfer, favourable temperatures in the Mediterranean Sea and the availability of appropriate sites. Greece has now become the leading producer with 66,000 tonnes in 2001 compared with 3500 in 1990. Turkey is next, its production started in the early 1990s and reached 28,000 tonnes by 2001. Intensive aquaculture production of red tuna began in 1997 with 670 tonnes, and had grown to 7750 tonnes by 2001. Underway in several countries (Croatia, Italy, Malta, Morocco, Spain and Turkey), this production is exported to the Japanese market. Juveniles are captured in their natural environment and bred for a period ranging from a few months to two years. Fish-raising could develop towards a full cycle (reproduction in hatcheries).

A Sustainable Future for the Mediterranean

In terms of *prospects*, several elements suggest that growth in production would continue at a sustained pace. Regional demand for sea produce in the Mediterranean countries far outstrips supply and is projected to continue to grow with population growth in the south and east, increase in tourist flows and a fall in the price of aquaculture products. Increases are expected in the SEMC where consumption of aquatic products is low compared with the developed countries of the northern rim (5.8kg/cap/year in Morocco, 7.5kg in Turkey and 10kg in Egypt compared with 52kg in Spain, 29kg in France, 26kg in Italy and 24kg in Greece). In Egypt, the national development strategy is to increase consumption and reach 14kg per capita per year by 2017. Strategies of major producers appear to favour policies of growing markets through lower prices. The demand for fish fillets with guaranteed quality and traceability observed in most European markets is expected to lead to the emergence of farming of new fast growing marine fish species such as the croaker and the wreckfish. New offshore production techniques (already under development in Cyprus and Malta) will probably be broadly mobilized to compensate for the limited availability of coastal sites and increasing conflicts between activities.

The growth in marine-fish production is projected to continue, and to more than double by 2025. Should the bass/sea bream thrust succeed on the European and Mediterranean markets, greater growth in demand and production are very likely, as has been the case over the past few decades with salmon in northern Europe.

Risks and regulatory measures

This growth, however, may well run up against difficulties. The production growth in a competitive context risks occurring in a sawtooth manner with crises of overproduction and marketing (falling prices, degradation of the image of products, market saturation, making commonplace). This could raise questions about strategies, especially of the quantitative kind. Local pollution is expected to increase as well as conflicts with other coastal activities. Further difficulties lie in the risks of degrading stocks of species whose juveniles are caught in natural surroundings (red tuna and mullet) if this additional pressure is not accompanied by a greater effort in stock management. Furthermore, loss from cages and contagious diseases should be feared. In the long run, genetic modifications are also possible after interaction between escaped fish-farm species and natural species. Introducing non-indigenous species can also threaten biodiversity.

Several recent *initiatives* aim to reduce and anticipate these risks. The GFCM and ICCAT are formulating directives for sustainable tuna-raising practices in the Mediterranean, and the aquaculture committee of GFCM is planning to launch a regional project on 'sustainable development of Mediterranean aquaculture'. In 2002 the European Commission formulated a strategy for the sustainable development of European aquaculture, which points out four main challenges: promoting economic viability; guaranteeing food security, human health and the well-being of animals; tackling environmental problems; and stimulating research.

The challenge for the Mediterranean is to support and orient aquaculture development through an integrated approach that aims to enhance the diversity of its potential, create employment, put quality products on the market, and reduce environmental impacts. The following would contribute to these objectives:

- *Promoting extensive and semi-intensive aquaculture* that retains the possibilities of development in several countries in both marine and brackish waters.
- Assessing the impacts of aquaculture on ecosystems and *strengthening regulations* (concession or authorization systems) for reducing such impacts and foreseeing risks. Drawing up documents on 'best available techniques' could also contribute. There is also the question of the density of fish-raising in relation to the absorption capacities of sites. Malta has drawn up guidelines, and Cyprus imposes a minimum distance of 3km between two farms. One suggested solution would be to *rotate sites* by granting two sites for a single production unit.
- *Planned development* as part of integrated coastal area development in all rim countries through technology transfer benefiting the developing countries and regions in economic hardship.
- Implementing *eco-labels* and certificates for stimulating the emergence of quality products with less environmental impact and thus enhancing market niches.
- Searching for a *positive synergy with the fisheries sector* and local economies, including support of the aquaculture sector through appropriate instruments (taxes on products) to conserving commercially viable natural species (in particular those caught for raising on farms), reclassifying fishmen, and developing aquaculture methods for the benefit of fisheries (artificial reefs and sea ranching).

To contribute to this necessary development towards an alternative Mediterranean aquaculture and fisheries sector, the strengthening of *regional cooperation* needs to be organized. Regional scientific instruments (GFCM and ICCAT) need to be consolidated along with their synergies with other regional cooperation mechanisms (especially MAP) and the European Commission. The EU also has a fundamental role to play in reducing Mediterranean north/south imbalances through programmes for strengthening capacity and co-development.

Mobilizing for an integrated coastal area management

To tackle the unsustainable heavy trends occurring in the coastal areas, the Blue Plan, in 1989, stressed the imperative of adding *broader responses* to the sectoral responses that have been discussed, integrating development with the environment, promoting inland-area development to relieve pressure on the coastal zones and developing *national policies for coasts*.

Also stressed was the need to strengthen regulatory and legal instruments and accompany them by structural progress:

- establishing administrative entities that are original and specific to the coastal areas;
- prospective studies and integrated planning;
- protecting significant parts of the coastal land and waters;
- resolute interventions in economic mechanisms (pricing) and landownership issues.

These orientations were taken up by the MCSD and the proposals relating to sustainable management of coastal areas were approved by the MAP ministerial meeting in Tunis, in 1997.

Although implementation of the recommendations adopted in 1997 still need to be assesed in detail, available information indicates a growing awareness and mobilization of civil society and the emergence or strengthening of national and local policies for the protection and integrated management of the coastal areas.

Box 6.12 EU mobilization for integrated management of coastal areas

Following a demonstration programme implemented in 1996 by the European Commission, the European Parliament and Council, in May 2002, approved a recommendation on the implementation of a *strategy on integrated coastal zone management*. This EU strategy would deal in particular with the affirmation of principles, setting up national inventories, formulating national strategies, and cooperation with non-EU member countries bordering the same regional sea. The Environment Programme of the Euro-Mediterranean Partnership also included sustainable management of coastal areas as one of its five action priorities. The EU Water Framework Directive (October, 2000, modified in November, 2001) also covers coastal areas since it refers to integrated management at the river basin level and sets out several environmental measures up to 2012 (reducing risks, monitoring and controlling discharges, preserving protected areas, public participation, pricing policies).

At the regional level, MAP and the EU are promoting the national implementation of such integrated management policies and programmes (Box 6.12). The extension of the Barcelona Convention to coastal areas,[42] as agreed in 1995, could lead to strengthened legal and operational frameworks. However, no protocol or regional strategy has yet been formulated that would enable a sharing of efforts to protect the common good, and give the rim countries the necessary impetus for strengthened, coordinated political action in the region.

Creating laws and coastal institutions

In all countries there is administrative separation, even rivalry, between land and sea. The juxtaposition of maritime laws and general environmental and development policies has become insufficient to protect and manage coastal areas in a sustainable way. Specific instruments and institutions are increasingly considered necessary for attaining this objective.

To date, only five Mediterranean countries have a *framework law for coastal zones* (Algeria, 2002; Spain, 1988; France, 1986; Lebanon, 1966; Greece, 1940), three of which apply a broad vision of coastal areas (Spain, France and Algeria). Morocco and Israel are preparing a framework law for the protection of the coastal environment. Three countries have also been innovative in creating *coastal agencies* (France, Tunisia and Algeria). Other specific management mechanisms are either emerging or are planned in several countries (Box 6.13).

Box 6.13 Coastal agencies and action plans in selected countries

France was the first Mediterranean country to pass a specific management instrument (1975), the 'Conservatoire du littoral', to acquire natural coastal land. By 2003 this conservation centre was protecting about 36,000ha and 180km of Mediterranean coastline (11 per cent of its total Mediterranean coastline, 21 per cent in Corsica). Without its activities (together with other measures to gain control over land for protection purposes, including national parks, forests, religious and military properties) virtually no natural space would remain on the Provence seafront.

However, the conservation centre was unable to prevent urban abuse of the coast during the 1980s (housing developments, marinas and other golf/building complexes), facilitated by a lack of spatial planning and the predominance of short-term speculative, political and landownership approaches (property-based electioneering and seeking maximum tax returns on property). These abuses led to an increase in the number of associations, the mobilization of landowners worried about protection, and the coastal area law of 1986 that banned construction on valuable natural areas in coastal districts. State departments and judicial authorities have recently applied a strict reading of this law and its implementation, thus rendering most of the major construction projects inoperable and requiring a revision of district land-use plans. Urbanization pressure has been pushed towards coastal plains and inland areas. This casts doubt on the 'quantitative' model of urban and tourism development in coastal zones, and favours a more 'qualitative' model that aims to optimize the existing built-up fabric.

A recent report[43] warns against an image of the coastal zone that is exclusively tourism-oriented. It describes the loss of identity of a coast that is unstructured, hardly productive and transformed into a linear metropolis, where traditional activities are disappearing to benefit leisure time activities. Agricultural land tends to vanish, with the consequent costs of flooding and fires. New responses are being planned with three objectives: (1) broadening the delimitation of the coastal zone to include the functional interdependence of sea and land; (2) broadening the definition of coastal zones to a more overall heritage context (human, social, cultural and economic); (3) promotion of 'local territory projects'. Management of landownership, spatial planning and the promotion of productive management of landscapes and nature (agriculture, fisheries and aquaculture) are advocated, as well as a Conseil National du Littoral (National Coastal Zone Council) to encourage and disseminate good practices, evaluate policies and move away from an overly administrative view of costal zones.

Tunisia is the only other Mediterranean country with an institution dedicated to coastal zones (since 1995): the 'Agence de protection et d'aménagement du littoral', APAL (Protection and Planning Agency for the Coastal Zones). APAL is in charge of implementing the action plan formulated in 1992 by the Ministry of the Environment and Spatial Planning for the protection and enhancement of the natural and cultural heritage of the coastal zone and its integration into the rest of the country. The action plan requires: identification, protection and management of sensitive natural areas; creation of protected areas; delimitation and definition of a management mode for the maritime public domain; protection and management of beaches; monitoring the coastal zone, and combating coastal erosion. Regulations have been strengthened to control urbanization, develop the coastal zone, protect the marine environment from accidental pollution, and combat attacks on the environment. APAL manages over 700 temporary jobs along the coastal zone that focus on prevention and dialogue. It has set up a 'coastal zone observatory'. In 2000, 17 natural areas had the status of 'sensitive zone', and benefited from a management scheme that makes it possible to protect and enhance about 9500ha along 190km of coastline. Master plans for developing sensitive zones have been drawn up for the whole coastal region, relying on a detailed and dynamic assessment.

Other specific management mechanisms have been set up in other Mediterranean countries:

In **Algeria**, the Commissariat national du littoral, established by law in 2002, is in charge of inventorying coastal zones (human settlements, natural areas, islands regions) and implementing the national coastal zone policy. In **Malta**, a Coastal Resources Advisory Board (CRAB) was proposed in 2001 as part of the Coastal Planning Programme to ensure a sustainable use of the coastal zone in cooperation with local actors, and to create a research centre. In **Slovenia**, the Environmental Agency (Ministry of the Environment, Planning and Energy) established in 2001, ensures the management of coastal zones (ten districts) and catchment areas. It deals with nature conservation, protection of soil and air quality, environmental impact assessments and granting concessions for water use. In **Lebanon**, a draft coastal law, being formulated with assistance from UNDP and the World Bank, provides for the creation of a Coastal Management Council.

In highly decentralized countries, coastal regions are developing often innovative coastal policies. In **Spain**, the Valencia Autonomous Community launched a Strategy for Integrated Coastal Zone Management in 2002, aimed at rehabilitating degraded coastal areas, improving infrastructures and reinforcing conservation and protection measures. In addition, a Permanent Coastal Observatory is planned. In **Italy**, the Tuscany region is working on an Integrated Coastal Zone Management Plan and an investment programme for emergency operations to preserve and return equilibrium along the coast.

Source: Blue Plan from diverse national sources

A recent comparative analysis of national laws[44] makes it possible to present an initial overview.

One of the main problems remains the inadequacy of *institutional coordination* in this area where various rules and administrations in charge of maritime and land-based issues confront each other. According to diagnosis by legal experts there is an institutional disorder in the breakdown of responsibilities, a lack of legal coastal unity and insufficient delimitation and coordination. Specialized consultative committees on coastal zones are still rare (national committees in Egypt, Turkey, Greece, Slovenia, and more informally in France). Coordination is organized mainly around sectoral fields (water, waste, transport or tourism).

In Bosnia-Herzegovina, Croatia, Slovenia, Malta, Tunisia and Israel, coastal areas are in general covered by *land-use plans* along with *impact assessments*. These plans seldom integrate land and sea, and even less so all related activities. Egypt and Spain (and partly Croatia, Lebanon and Turkey) are trying to articulate an *overall scheme* or master plan.

Despite the lack of specific laws, half of the Mediterranean states have at least made provisions to *regulate construction in coastal zones* (Box 6.14).

The principle of *free access to shores* and beaches is accepted everywhere, but has no legal basis yet in Malta, Monaco, Libya and Israel. In Lebanon access is often made impossible by a continuous line of private properties, which in fact leads to the privatization of coastlines. Private beaches also prevent access in Egypt, Lebanon and Malta.

Box 6.14 Regulation to limit construction and roads in coastal zones

Five countries (France, Italy, Lebanon, Tunisia and Turkey) have laws specifically dealing with coastal zone planning and urbanization. Turkey has a 1990 law delimiting the coastal zone, land use and construction on a 100m coastal fringe. In Spain, a 'zone-of-influence' exists where special urban restrictions can be enforced. In France, the width of the coastal fringe is not defined by law, but legal rulings have sometimes allowed it to reach up to 2km from the shore. The width of the zone with a building ban varies considerably between countries, from 6m from the shore in Morocco to 200m in Egypt, 30m in Greece, 50m in Turkey, and a preference for 100m in Bosnia-Herzegovina, France, Tunisia, Libya and Israel. Absolute construction bans exist only in Croatia, Greece, Malta and Slovenia. Easement is applied to all construction except roads in Israel, tourism in Libya and camping sites in Spain and Italy.

Construction of new coastal roads is banned only in Algeria, France, Slovenia and Spain although it is the main cause of the artificial land cover of coasts.

All countries have some form of *permit system* or preliminary declaration for setting up industrial or commercial activities in coastal zones, and impact assessments are widespread (except in Lebanon, Monaco and Morocco). The reach of these instruments can, however, be reduced by the existence of thresholds or the non-implementation of laws.

In general, information about coastal zones remains incomplete. Monitoring the quality of the coastal environment is still centralized, which makes permanent monitoring an illusion. De-concentration efforts (for example in Tunisia with the coastal observatory attached to APAL) or decentralization appear essential to achieve better effectiveness. Existing or planned permanent observatories are rare (Algeria, the EU, Slovenia and Tunisia). The European LACOAST project ensured monitoring of coastal land-use changes between 1975 and 1990, but remains insufficient in detail and coverage (Greece, the islands, etc.) and has no equivalent in the non-EU countries. The coastal dimension is seldom or never taken into account by statistical agencies. In national environmental reports, only six countries (Algeria, Croatia, France, Malta, Morocco and Tunisia) deal specifically with coastal areas, while other countries mention coastal areas only through related themes, including tourism (Greece and Italy), pollution (Egypt), planning (Israel and Greece) or the marine environment (Italy and Turkey).

The fields least covered by law (or its application) are coastal urban planning (importance of *illegal constructions* with frequent laws for regularization[45] in many countries), damaging extraction of sand, coastal industrial activities, waste and sanitation, marinas, monitoring and access to beaches. Obstacles to the law being implemented are the lack of surveillance staff, a sanctions regime that is either non-existent or not enforced for local political reasons, and insufficient financial resources, information and public awareness.

In general, current legal tools seldom contribute to integrated coastal area management. They provide hardly any ways to reduce overlaps or lack of political and administrative responsibilities on the coasts, nor has the issue of using part of the tax and property revenues to finance the protection and sustainable management of coastal zones been considered in law making.

Mobilizing civil society

Without the commitment of often visionary local, national and international personalities, land-owners and associations concerned about the future of coastal areas, many precious coastal sites would have been irreversibly degraded by private or public investment, a number of governance reforms would not have come about, and pollution and risks would have been much more serious. Many protected cultural and natural sites and areas owe their existence to such commitments. The appeals by Commander Cousteau and Mrs Mann Borghese did not go unnoticed when MAP was created. Box 6.15 gives some examples of increasing initiatives.

Box 6.15 Examples of civil society initiatives for the Mediterranean

Mediterranean civil society is increasingly motivated by coastal protection and management. Many local conservationist associations have contributed to protecting some coastal sites from generalized uniformity. Others, such as ENDA, Greenpeace, WWF, Friends of the Earth, MIO-ECSDE or MedForum have acquired international support, and communicate increasingly with local communities and global authorities (Johannesburg Summit). Partnerships are beginning to be formed with local authorities (regions, municipalities) and the business community to achieve sustainable coastal area management.

The **Medcoast** network, which includes 15 institutes from various Mediterranean countries, works to improve understanding and concerted development of Mediterranean research. Since its founding in 1993, its secretariat (Middle East Technical University of Ankara) has organized eight major specialist conferences and ten training programmes (on integrated management of coastal areas and beaches).

Motivated and committed individuals with **local projects** are a decisive factor in mobilizing actors. The **Tyr Coast natural reserve** (MedWetCoast project in Lebanon), the last sandy beach in Lebanon and a Ramsar site, has benefited since 2002 from remarkable initiatives, including assessment studies of the site by NGOs and scientists, protecting sensitive areas, careful development of eco-tourism (organized visits to marine turtle sites, with an increased number of egg-laying sites in 2003), the reduction of tourism pressures on beaches (shops have been pushed back 50m from the shoreline and their numbers reduced), the promotion of organic agriculture with the implementation of new packaging and marketing procedures, grazing control, a ban on hunting, micro-credits and training for developing guest rooms, plans for sustainable fisheries and studies for rehabilitating archaeological heritage. The recruitment in early 2003 of a site manager, the formation of a management committee, many meetings with farmers, awareness-raising of children and citizens, many organized visits and the commitment of local professionals have all contributed to this success.

Helmepa (the Hellenic Marine Environment Protection Association) is an example of a **private business** initiative. Founded in 1982 by Greek ship owners and sailors, it aims to increase environmental awareness in the industrial maritime sector and apply pressure for strengthened protective legislation. Helmepa's 2002 budget was 862,000 euros, of which 80 per cent went into environmental training of professionals and 20 per cent was used for educating young people. Other comparable initiatives have since started in Turkey and Cyprus. **Cymepa**, founded in 1992 in Cyprus with the same objectives in mind, brings together not only professionals of the maritime sector but also many companies with land-based activities (in particular hotel keepers). Cymepa stresses the importance of ethics as a basic value in the move towards more sustainable development. Turkey's **Turmepa** was founded in 1995, has 6600 members, and aims to prevent maritime and coastal pollution by raising awareness (schoolchildren and professionals) through waste-recovery operations and telephone reception intended to register all warnings of pollution.

Towards 'integrated coastal area management'

Integrated coastal area management (ICAM)[46] is a dynamic *process* that brings together various public and private actors in a coastal area to optimize the long-term choices by highlighting the need for a careful use of coastal resources.

ICAM activities are underway in many countries and are encouraged through regional cooperation. MAP has contributed to the implementation of coastal area management programmes (CAMP) around the Mediterranean, through the activities of its Priority Actions Programme (PAP) Centre. About ten CAMPs have been undertaken since 1989 with support from various MAP centres (Box 6.16).

In addition, a dozen projects (out of 21) of the EU Integrated Coastal Zone Management (ICZM) demonstration programme launched in 1996 were located in member countries bordering the Mediterranean.

Several lessons can be drawn from the various experiments. First, implementation of the ICAM approach makes it possible to *bring partners closer together*, contribute to awareness, increase the common vision of priority objectives, and improve decision-making. At the beginning of the process, a prospective study using sustainable development indicators,[47] facilitates this convergence and helps to clarify the challenges to be met depending on the dynamics observed and the economic, social and environmental issues at stake. Analysing the results of CAMPs (Box 6.16) shows improvements in several fields, including *strengthening of institutional capacities* (training specialists and decision-makers in new instruments and techniques), implementation of national information systems, and a better integration of environmental issues into coastal area planning.

Experience shows, however, that real results remain limited to a few, well-defined fields such as the fight against marine pollution, and landscape improvement. Wanting to tackle everything through an ICAM process seems to be a pipe dream. Although 'integrated' in the assessment stage and involving different actors, the process would gain if priority objectives were formulated, for concrete progress along a few limited and well-defined directions that could evolve over time in line with local realities.

Several 'ingredients' are required to give the approach the best possible chance of success. First of all, *political will* makes it possible to legitimize the collective process. This is a basic necessity and may be expressed through inter-ministerial decrees, and may contribute to mobilizing national and local private and public actors. *Local initiatives* are a vital complement to political will. The *time frame* is the third component of success. It sometimes takes five years, even more, to finalize a territorial contract for integrated management. And finally, mobilizing *financial resources* appropriate to the issues at stake is essential. The Bay of Toulon (France) contract, for example, which led to a five-year commitment by many partners in 2002, required ten years of effort to find internal and external financial support to fund the post of the lead person and the total mobilization of nearly 103 million euros for the 2002–2007 period.

The real effect of ICAM approaches is therefore relative. Many serious obstacles explain the implementation problems, in particular:

- the swiftness of changes as well as the power of short-term interests and speculative behaviour that are in contradiction with the long-term objectives of integrated management;
- the multitude of actors and strategies that have objectives that are generally contradictory and can engender conflicts and political barriers;

Coastal Areas

Box 6.16 MAP Coastal Area Management Programmes: Location and some results

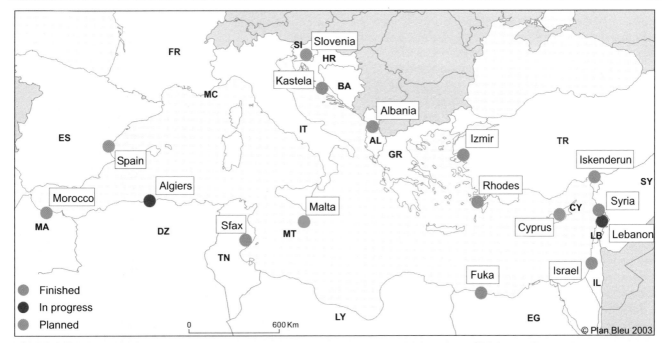

- **Kastela Bay (Croatia):** technical solutions that served as a starting point for the 1999 construction of a wastewater processing system, funded by the World Bank and METAP.
- **Izmir Bay (Turkey):** training of local and national specialists in ICAM, setting up a geographic information system laboratory and a socio-economic and environmental database.
- **The island of Rhodes (Greece):** a database and information system for local development, identification of priority activities (water and waste management), re-organization of administrative facilities and technical services, and improvement of local decision-making.
- **Coastal regions in Syria:** strengthening of institutional capacities and training a multi-technical team of specialists in all sectors.
- **Fuka-Matruh (Egypt):** raising awareness of civil servants and decision-makers about environmental problems in the region, which has led to the adoption of a decree banning all construction in the north-western region until an integrated plan has been approved by all ministries concerned; departure point for a new project ('economic and social planning for spatial development').
- **Albania:** strengthening of institutional capacities and training of local and national experts in ICAM.
- **Sfax (Tunisia):** a database, a geographic information system and a series of multidisciplinary studies on the state of the city's environment and its possible future, and training of local managers and specialists.
- **Israel:** input for changing policies relating to the management of sand and retreating cliffs, reducing pollution in the Nahal Kishon hot spot, a new base for information management, the formulation of land-use plans using remote sensing, and the participation of NGOs and local experts and actors.
- **Malta:** the introduction and dissemination of participatory methods and sustainability indicators; strengthening the cohesion between teams and actors; proposals to set up a 'coastal resources consultation committee', and adopting a 'Coastal area statement for the Maltese islands'.

Outside the CAMP framework, the Blue Plan was requested by the Turkish Environment Ministry and Ankara University to contribute to a forward-looking study on the environmental management of **Iskenderun** bay. This study made it possible to draw the attention of local authorities to the serious pressures threatening coastal areas with high ecological value, since then listed as nature reserves.

Source: MAP Priority Actions Programme, Split (PAP/RAC www.pap-thecoastcentre.org)

- the diversity and complexity of the questions to be considered;
- the fragmentation and overlap of political and administrative responsibilities in the coastal areas as well as the lack of a common culture, communication and coordination between marine specialists and maritime authorities, and spatial planners and land-based authorities;
- the complexity of legal aspects and a certain bureaucratic inertia.

Given these obstacles and the resulting imbalance to the benefit of speculative or sectoral interests, integrated coastal area management appears to be an illusion if not accompanied by *strengthened* legal, regulatory, economic, property and technical *instruments for protecting* and sustainably managing coastal areas not yet occupied artificially.

An *enabling environment* is required to allow the development of ICAM processes. Mobilizing actors and integrating policies at local and regional levels are in fact only conceivable if the higher echelons of government and administration create an integrated legal and institutional framework and promote action at the most appropriate local level.

With *decentralization* and the recent development of *inter-municipal cooperation* in several northern Mediterranean countries, one can see a certain appropriation of the question of sustainable management of coastal areas by local decision-makers, and an increase in processes linking elected officials, scientists, state departments and civil society. Indeed, according to the Spanish experience,[48] it appears that in highly decentralized countries, the *regional scale* (NUTS 2 equivalent) could become an adequate level for promoting integrated management, provided there are mechanisms in place to ensure the factoring-in of long-term issues.

Regional development policies to reduce predictable pressures

The scale of pressures observed on coastal zones is often driven by the marginalization of mountain-region economies in the north as well as in the south. The vulnerability of rural societies and inland cities still cause a movement towards the coast, and free agricultural trade could strongly increase this trend in the developing countries (see Chapter 5).

To relieve the projected additional pressures on coastal areas, some countries are implementing voluntary policies to revitalize inland areas. For example, an approach to create infrastructures and new towns has recently been developed in Algeria (Box 6.17).

However, these voluntary approaches, structured around major projects, do not always have the desired results (for example as seen in the partially disappointing results of the new cities policy in Egypt). Less spectacular and more de-concentrated approaches for planning regional and local development need to be promoted in societies that are turning to a market economy and adopting principles of democratic governance where the population can choose where to settle down. To strengthen or highlight the attractiveness of inland rural areas on the market, policies to *enhance inland cities* and *revitalize rural regions* have the potential to be explored. The EU also supported such efforts in southern Europe, relying on a project approach, local governments and active professional groups. An alternative development would be characterized in particular by the implementation of much more ambitious sustainable rural development strategies and policies (major programmes with small projects), an approach currently being tried in some countries, although not one without problems (see Chapter 5).

> **Box 6.17 Countering the intolerable increase in urbanization along the coast of Algeria**
>
> What characterizes the Algerian coastal zone is the high population density (1300 people per km^2 in the Algiers metropolitan area), which is only going to intensify with population growth. By 2020, 16.8 million of the total of 42 million Algerians are expected to be living on the coast. This will lead to an overall urbanization of the coastal plains, high-quality agricultural land will disappear (an additional 50,000ha of land will be lost in the *Mitidja*) and environmental degradation will increase. At the same time, inland areas of Algeria (the High Plateaux and the south) are in danger of remaining marginal and continuing to suffer from poverty and desertification.
>
> To counter these trends – which are considered unacceptable – the report *Aménager l'Algérie de 2020* (Managing Algeria in 2020) offers a strategy that led to the adoption of two laws in 2001–2002 relating to planning and development of the country's territory, and to the conditions for creating new cities. A national spatial scheme has been drawn up, aiming to reverse migration by voluntarily redeploying 3 million people from the coastal fringes towards the interior. Apart from measures to protect the coastal zone itself, the following measures are planned to protect high quality farmland in the coastal zones: strengthening inland infrastructures (roads, public services), voluntary relocation with financial support from the 'national fund for the development of Algeria's territory', and the creation of new cities with a strong economic performance at the foothills of the Algerian Sahel. The new city of Sidi Abdallah (1998) is oriented towards cutting-edge technologies (cyber-park, medical research). Other possibilities are being considered such as the El Affroun project (agricultural sector) or the Bouînan project (sports and leisure activities).
>
> The objectives are threefold: (1) to rebalance the population distribution in the country and enhance the inland territories, (2) to protect and enhance natural resources by restoring economic vitality and the attractiveness of the inland regions, and (3) to manage urbanization in the coastal zone.
>
> Without prejudging the results of such an ambitious strategy, its orientation in itself signifies a will to reverse a particularly worrying process.
>
> *Source*: Association nationale de développement de l'économie de marché (ADEM). Algérie : perspectives 2010. Alger, 2003

In these regional and local development strategies, reorienting policies and the tourism sector to benefit the hinterland may represent a prime direction for revival.

Directing tourism towards sustainable development

The Mediterranean is the world's leading tourism region. Given the multiple impacts of tourism at economic, socio-cultural, environmental and territorial levels, the development in this sector will be decisive for the future of coastal zones and the whole eco-region in general. The relationship between tourism and sustainable development therefore deserves much attention.

A strategic sector dominated by competition

Tourism has without exception become an *essential economic sector* for all Mediterranean countries because of the size of tourist flows, the amount of money spent, the contribution of the sector to GDP, the number of jobs created (Part 1 and Box 6.18) and the expected growth of domestic and international markets.

> **Box 6.18 Socio-economic imperatives of tourism, some figures**
>
> The tourism sector is of particular socio-economic importance in the Mediterranean since it strongly contributes to the creation of jobs and added value.
>
> In terms of direct and indirect *employment*, the sector represents 10 per cent of the total number of jobs in Greece and 30 per cent Malta. In Italy more than 2 million jobs were generated in 1999. In Tunisia tourism provided work for 355,500 people in 2000, nearly 10 per cent of the total workforce. In Croatia tourism generated 283,000 jobs in 2001, would have risen in 2003 to 294,000 (28 per cent of the total workforce) and is projected to reach 427,000 (40 per cent of the total workforce) by 2013. In Turkey, there were 1.4 million jobs in 2000, projected to reach about 2.12 million by 2010 (8 per cent of the workforce). Such growth represents the creation of 66,000 new jobs per year, about 6 per cent of the net entries per year into the working population for the decade 2000–2009. Tourism represents on average near 12 per cent of Mediterranean *exports* of goods and services (see Part 1 p17). *Expenditures* per international tourist increased from US$96 in 1970 to US$650 in 1995 and US$576 in 2000. Tourism is especially important on the islands, where it provides an alternative to the social and economic difficulties inherent to insularity and where it can lead to radical changes. In the Balearics, for example, per capita GDP reached 18,249 euros, or 3000 euros more than the national average (15,261 euros) whereas the region was among the poorest in Spain before the development of tourism.
>
> *Source:* Blue Plan from Becheri, 2001; Chapoutot, 2002 in UNEP-MAP-Plan Bleu 2005; WTTC, (2002); WTTC, (2001) and World Tourism Organization (www.world-tourism.org)

To benefit from this dynamic, most Mediterranean countries, except up to now Algeria and Libya, have developed proactive growth policies, mainly centred around a *seaside model*. Operational management plans have been formulated for the coastal fringes to create completely new tourist resorts (for example in France-Languedoc Roussillon region, in Italy-Sardinia, Spain, Tunisia and Turkey) or are underway (Morocco). As part of public–private partnerships the states try to manage property, build infrastructures and define operational rules. As long ago as 1973 Tunisia created its *Agence foncière de développement touristique* (land development agency for tourism activities), and a similar process has started in Morocco, which aims to achieve an average annual growth of 15 per cent in tourist flows and 10 million international tourists by 2010 (compared with 4.2 million in 2001). Apart from these resorts, development occurs through individual or collective initiatives by local actors and outside investors. This less state-initiated dynamic does not imply a lack of support from public authorities. Virtually all Mediterranean countries announce considerable growth objectives every now and then and take steps to promote investment (Box 6.19).

In addition to direct measures for tourism as a whole, indirect investments are made that contribute to increased accessibility and attractiveness of specific territories. Investments in transport infrastructures or cultural facilities can be on a large scale, and the operational costs must still be added (for example, promoting destinations on external markets and the costs of ensuring security at destinations).

> **Box 6.19 Planned investments in the tourism sector in the Mediterranean**
>
> An analysis of planning documents shows the scale of planned investments in tourism over the coming years at the national level, with considerable repercussions for the Mediterranean coastal zones. Various formulae are being implemented to encourage private investment. The most matured destinations are beginning to commit to rehabilitation programmes to restore degraded areas and structures and spread tourist flows more evenly throughout the year.
>
> In **Croatia** the objectives are to reinvest US$100 million of income from tourism per year, attract US$50 million in capital per year, and triple the number of low-season tourists. A reform of investment procedures (laws, grants, a one-stop office) has been undertaken with assistance from the World Bank.
>
> In **Serbia-Montenegro**, tourism has been made one of the development priorities for the next 20 years. A 2001 *Master Plan for Tourism* identifies five major regions for tourism activities, mostly on the coast. It is planning a 50,000-bed capacity by 2010 and 100,000 beds by 2020.
>
> In **Albania** the sector is still weak (395,000 overnight stays, 77 per cent of which are Albanians) and is developing in an

unregulated manner, mostly in the coastal zones. 'Priority zones for tourism' have been identified. An assistance and stimulation programme is intended to attract foreign investment in the tourism sector (exemption from customs duty, tax profit and low rents).

The hotel capacity in **Turkey** should reach 1 million beds of an international standard by 2010. The 8th Development Plan (2001–2005) is striving for a more even spread of activities throughout the year by creating new zones for tourism.

In 1997 **Lebanon** set up the Investment Development Authority of Lebanon (IDAL) to promote and facilitate foreign investment, notably in tourism.

The objective of the **Egyptian** authorities is to increase tourist flows by 10 per cent per year in the whole country (from 4 million visitors in 1997 to 27 million by 2017). The necessary bed capacity is estimated at about 315,000 additional rooms by 2012. And private investments are encouraged through laws that exonerate tourism projects from certain taxes.

Plans exist in **Libya** to build several international hotels with a budget of about US$61.3 million, with the overall cost of the planned projects coming to about US$77 million. The Tourism Development Bank was founded in 2003 and is 80 per cent owned by private interests to finance tourism development projects.

The objectives in **Algeria** are to reach 1.2 million foreign visitors by 2010. The whole country has been divided into 174 *Tourist Sites and Expansion Zones* of which 140 are located in the coastal zone (law No 03-03 of 17 February 2003 on TSEZ), and stimulation measures have been put in place to attract investment.

In **Morocco** the government and the private sector have signed a framework agreement intended to triple hotel capacity by creating 160,000 new beds by 2010. It also sets out to strengthen financing, infrastructures (renovation, upgrading) and train 72,000 professionals.

Greece hopes to attract US$321 million of foreign tourism investment over the next six years. A company was set up in 2000 to act as a liaison between the public and private sectors and to develop investment projects.

A *Plan Integral de Calidad del Turismo Español* (PICTE 2000–2006) is being implemented in **Spain** by the autonomous communities. These action plans bring the private sector and political authorities (national and regional) together around several qualitative programmes, including the sustainable rehabilitation of obsolete seaside facilities; enhancement and development of emerging destinations; certification of tourism establishments meeting quality criteria; and alternative tourism (cultural, urban, sporting and rural). The amount of investments realized under the umbrella of the PICTE programme comes to 17.1 million euros in the Valencia Community alone.

In Mediterranean **France**, after decisive action had been taken in 1965, a plan for sustainable development, representing an investment of 306 million euros, is planned for the 2003–2006 period to rehabilitate natural areas, land management and renovate tourism facilities on the coastal zone of the Languedoc-Roussillon region. To coordinate this plan, an inter-ministerial mission for the development of the coastal zone was established in July of 2001 with an accompanying charter for many private and public partners.

In **Italy** investments are being made to upgrade existing facilities and for sustainable tourism. Several *Programma Operativo Regionale*, with an initial period of six years from 2000 to 2006, have objectives related to sustainable tourism, including upgrading saturated infrastructures, taking the environment into account when developing facilities, developing products for tourism that combine natural and cultural heritage, and a better spread of tourist flows, both in space and in time.

Source: Blue Plan from diverse national sources

Mediterranean tourism is also characterized by changes that are minimally or not controlled. Experience has shown that after the launch phase of a destination, the increase in supply and demand can 'explode', preventing any control of development. Such an 'explosion' can be caused by the short-sighted behaviour of individuals as well as by the speculative or commercial strategies of powerful actors.

In*vestors* primarily covet the most beautiful coastal sites, and the outcome of their investment almost always leads to a loss in quality of such a territory. An increase in the number of holiday homes can contribute to revitalizing rural areas but also to fragmentation of the landscape, disappearance of agricultural activities, and a misappropriation of local land property. The tour operator business appears to especially favour competition between destinations and lowering prices. The influence of these professionals is considerable in some of the countries where tourists originate (for example Germany and the UK), in tourist destinations accessible by plane, especially some of the islands (Cyprus, Malta, the Balearics), and in countries where tourism is developing rapidly such as Tunisia, Turkey and Egypt. This influence is not quite as strong in Italy or Mediterranean France where destinations are more easily accessible to European customers coming by train or car, and where the proportion of individual tourism and the diversity of what is on offer are high.

Faced with strong economic powers, development strategies are still characterized by a spirit of *competition* rather than *cooperation*. Given this context, authorities can have a hard time resisting arguments of short-term private interests. In Kemer, Turkey, for example, investors won over the initial development plan of the region (the South Antalya Project) that took the natural and cultural wealth into account. By the end of the 1970s, the increase in tourist-bed capacity and tour-operator policies of low prices shifted this area towards a 'tourism-only' zone, reaching a 65,000-bed capacity, which is far beyond the initially planned 25,000 beds.

This difficulty in managing tourism development is characteristic of many attractive destinations around the Mediterranean. And it makes the question of the relationship between tourism and sustainable development a real and current issue.

Better evaluate the benefits of tourism

Without underestimating the major economic and socio-cultural importance of tourism for countries and specific destinations, it is, however, necessary to put the benefits into context.

Public expenditure allocated by national governments and local authorities to develop tourism is indeed high (infrastructure and operational costs, such as tourist police and environmental management of the coastal zone). *Foreign currency balance sheets* would probably show that in several countries, particularly in the south and east, the proportion of international tourist spending going into national and local economies remains small. The average cost of stays in the large-scale tourist destinations of these countries, which are in competition with one another, is not very high; much of the proceeds go to airlines, tour operators, hotel chains and producers operating outside the countries or destinations visited.

Insufficient control of tourism development generates negative impacts (environmental and socio-cultural) and risks that are considered important by the local people, confirming the *ambiguous effects* of the sector on coastal societies. For example, the main advantages perceived by the people living in Rimini (Italy) and Estartit (Spain)[49] are job creation, increased income, improvement in the standards of living, and the development of services and infrastructures. But almost 60 per cent of those interviewed complained of badly controlled urban growth, a rise in disorder and degradation of landscapes. Nearly half of those interviewed also accused tourism of making economic dependence likely, and approximately a quarter blamed the sector for generating social tension and posing a threat to culture and tradition.

As mentioned several times, tourism has an important responsibility when it comes to *degrading the coastal environment* (coastal urbanization, waste, pollution, noise, ecosystem degradation, and wastage of landscapes). The islands are especially concerned (Box 6.20). In some cases, however, the development of tourism has made it possible to avoid the installation of more polluting heavy industry or inspire the creation of remarkable new heritage (such as villas, gardens).

The risk of *economic unsustainability* should be stressed. The risks of economic dependence and exposure are high in many Mediterranean coastal zones where traditional activities have not been preserved, tourist supply has not diversified and activities other than tourism have not

---

Box 6.20 Tourism puts considerable pressure on island environments

Tourism has become an essential factor in the development of islands, compensating for a decline in the primary sector. It is nevertheless a risk for islands, their identity as well as their natural and cultural resources.

The summer tourism-related *seasonal over population* (even in the least-visited islands the population regularly doubles in summer) affects all fields, social, environmental, cultural and economic: ring-shaped occupation of the territory, oversizing of facilities, degradation and pollution, noise, social tension, a low level of professional training, seasonal saturation of means of transport disturbing the mobility of island dwellers, and risks of land disenfranchisement.

The *tensions generated over limited water resources* are especially tight. To meet summer demands that continue to increase, there is more and more recourse to importing costly water by boat (in Greece, Italy and the Balearics) or desalinization (70 desalinization plants have been built on islands to produce drinking water: Cyprus, the Italian islands, Malta and the Balearics). These trends have growing impacts on supply costs and the environment, and open the way to urban growth, which can make things even worse.

Tourism also generates considerable development of seasonal *transport* and its side effects. Today islands host 46 of the 112 Mediterranean coastal airports. Traffic growth can be phenomenal (Palma de Majorca, for example, saw 19.4 million travellers arrive in 2000 compared with 11.3 million in 1990 and 7.4 million in 1980). Without strong regulatory measures, the near doubling of tourist flows projected between 2000 and 2025 will concern the islands and further increase the pressures on their territory. The development of new fast maritime means of transport (high-speed boats) is especially worrying because of their possible impacts.

---

been developed. In several countries, nominal spending by tourists has decreased considerably, which could lead to a questioning of the quantitative-type approaches (see Box 6.24). Some destinations have evolved towards a real 'tourism-only' attraction, Side in Turkey for example (50km East of Antalya), where, in 2002, 85 per cent of the local population lived on tourism, and Belek (15km East of Antalya), where 90 per cent of the income of the population was derived from tourism.[50] However, a certain obsoleteness must be feared of the seaside model and of ageing coastal resorts that are not well integrated into the local societies and have been dragged down by the business logic of low-pricing. Problems of decline have been observed in buildings constructed in the 1960s in the Languedoc-Roussillon region of France. To meet these problems, rehabilitation programmes are being initiated with public funding. The risks of derelict tourism sites between 2000 and 2025 are not insignificant on several Mediterranean coasts.

Environmental and *geopolitical risks* must be added to the economic ones. They are high for coastal economies that depend too much on tourism, while they are located in a region of the world with a high concentration of maritime traffic in hydrocarbons, a large gap between rich and poor, and frequent conflicts. An oil spill on a par with that of the P*restige* affecting Mediterranean coastal destinations would have disastrous consequences. Recent history has shown to what degree tourism and tourist destinations are often the first victims of conflicts if not even the primary targets of terrorists. At the first warning, tour operators sell other destinations and turn the flows away from the Mediterranean.

What form of governance for sustainable tourism?

To redirect tourism developments towards making a contribution to sustainable development, work by the MCSD/Blue Plan has stressed three complementary objectives: (1) managing territorial and environmental impacts of tourism; (2) promoting tourism as a factor of sustainable social, cultural and economic development; and (3) strengthening Mediterranean cooperation.[51] Although implementation of the recommendations adopted in 1999 by MAP still has to be assessed, it appears that the responses and progress made in some environmental sectors (including resource consumption and pollution) remain insufficient in terms of managing territorial impacts, redirecting strategies and regional cooperation.

Instruments for reducing pollution and resource consumption

Progress has been made on some *environmental issues* such as the consumption of water resources, the fight against pollution (except in the case of transport, which is expected to become an ever-greater problem) and waste management, where the tourism sector may be a driver for improvement. Some countries, such as Tunisia, have implemented differentiated pricing systems for water and sanitation, which contribute to improved performances (Box 6.21).

Increasing the number of labels and certifications contributes to better management. The Blue Flag ecological label awarded to beaches and marinas under good environmental management was created in 1987 by the Foundation for Environmental Education in Europe (FEEE). In 2003 it was in place in 24 countries, 8 of which were Mediterranean, with a total of 1450 beaches and 264 marinas. Malta is just beginning to use the label, while Morocco and Serbia-Montenegro are in the process of joining (Table 6.3).

Box 6.21 The contribution of tourism to the water sector in Tunisia

Tourism can contribute to good water-demand management, as this Tunisian example shows. In this system, pricing is the preferred instrument. The price of water consumed by tourism establishments is invoiced by Sonede, the distributor, at 0.650 dinars per m^3, which is higher than the cost price (0.410 dinars per m^3). Drinking water consumption per tourist bed and per day dropped between 1997 (526 litres per bed per day) and 1999 (473 litres per bed per day), while at the same time overnight stays increased significantly in the four- and five-star hotels. In 1999, drinking water consumption by the tourism sector accounted for only 6 per cent of the total water consumption, whereas hotel keepers paid for about 11 per cent of Sonede income. Similarly in the sanitation sector, tourist users, who accounted for 0.24 per cent of network subscribers and 9.8 per cent of the total treated wastewater volume in 1994, paid 23 per cent of the total bill. Fifty-three treatment plants have been built in the whole country, 35 of which are on the coast, and the eight existing golf courses are watered with treated water.

Source: Chapoutot, J. (2002) in UNEP-MAP-*Plan Bleu* 2005

Tourism-specific environmental regulations remain, however, insufficient, especially for yachting. The Contracting Parties to the Barcelona Convention have asked for a draft regional legal instrument for this high-impact activity. Without strong commitment and forward-looking regulations, uncontrolled development of new activities that disturb coastal zones, such as motorized nautical and land-based sports, are to be feared.

Various charters and voluntary initiatives undertaken by professionals, NGOs and international institutions bear witness to individual or sectoral commitments to 'sustainable development of tourism' or 'sustainable tourism', depending on the ambiguous definition of the subject. Examples are the Tour Operators' Initiative (with support from UNEP, UNESCO and the World Tourism Organization), the International Hotel Environment Initiative (IHEI), which brings together major tour-operators and hotel chains for a better understanding of the environment, and the Green Globe programme, which calls for environmental labelling of companies and destinations on a global scale. Furthermore, the EU has recently implemented a European-wide eco-label for tourism called 'the Flower', meant for tourist lodging facilities that respect the environment. This certification programme concerns, among others, water, energy, training, waste and awareness-raising among visitors. A hotel in Kallithea (Rhodes) is the first candidate for this European eco-label. There are also international certification systems for tourism. In Turkey, the tour operator VASCO in Antalya is the first company of the sector to receive the ISO 14001 environmental certification.

Table 6.3 The Blue Flag as applied in Mediterranean-rim countries

		1990–1998*	2000	2003
Mediterranean regions of Spain	Beaches	170	273	318
	Marinas	42	65	80
Mediterranean regions of France	Beaches	109	121	118
	Marinas	20	39	41
Italy	Beaches	27	145	159
	Marinas	29	41	47
Greece	Beaches	85	319	363
	Marinas	6	8	5
Cyprus	Beaches	11	29	39
Slovenia	Beaches		5	7
	Marinas	1	3	3
Croatia	Beaches	1	10	58
	Marinas		12	16
Mediterranean regions of Turkey	Beaches	12	61	88
	Marinas	4	7	6

Source: www.blueflag.org

* Last year available between 1990 and 1998

Insufficient instruments to reduce impacts on coastal zones

However, progress in limiting the negative impacts of developments, urbanization and tourism infrastructures, is still very insufficient. On this topic the Blue Plan and MCSD have stressed five types of responses to be developed in the Mediterranean: (1) timely impact assessments of tourism-related programmes and projects; (2) definition of and respect for accommodation capacities at destinations; (3) strengthening of legislative, regulatory and landownership management instruments to control urbanization from tourism and protect natural sites; (4) rehabilitation of the environment in mature destinations; and (5) the development of instruments to ensure that tourism contributes financially to the protection and management of natural and cultural sites (the 'the user pays' principle).

Assessments of *accommodation capacities* in destinations are being experimented with in order to limit the supply to a maximum level deemed acceptable. In Malta, the Maltese Tourism Authority intends to plan tourism development while using the conclusions of the accommodation capacity assessment carried out in 2001. It remains to be seen if such an instrument will move beyond the experimental stage, and if the intended limits will be effectively applied over time. It would be desirable to follow the example of Portofino (Italy), which, better than Saint Tropez (France), has maintained its character by banning all new construction and limiting the number of visitors (mandatory parking with limited capacity).

Efforts to *locate buildings away from the coastline* remain too rare. The tourist area of Belek in Turkey (near Antalya) is an example. Here hotels are located at significant distances from the sea – a vital measure for a coast that is a marine turtle nesting site – but have been built on dunes. An integrated coastal area management programme started by Albania in 2002 focuses on the fight against ad hoc tourism development and illegal construction on the coast. A study of the potential accommodation capacity recommends that tourism development projects in some sensitive areas (for example Lazelit Bay) be located at least 300 metres from the coastline, but these recommendations have so far had little effect (many projects are located at 100–150 metres from the coast).

Economic tools, based in particular on the 'the user pays' principle at the level of coastal territories, which is difficult to implement, are also insufficiently developed. They are, however, justified by the fact that foreign tourists, mainly northern Europeans, who enjoy the Mediterranean environment, could contribute to its maintenance and restoration. In addition to conventional instruments (taxes, paid parking garages, fees for accessing sites) new instruments have to be designed to ensure that financial contributions from tourism are in line with the requirements. Examples to be advocated, improved and applied more generally are the tourist eco-tax in the Balearics (even though abandoned these days) and the tax levied on maritime transport going to islands in France that have a protected status (Box 6.22). A reorientation towards an alternative future would depend heavily on such tools.

Box 6.22 Examples of a tourist eco-tax in Spain and France

The contribution of tourism to sustainable development can be facilitated through eco-taxes.

In 2001 the **Balearic Islands** instituted such a tax on stays in tourist lodgings. The revenue supplies a fund which was intended for improving the tourism sector and conserving the environment. This fund could be used to finance projects for:

- rehabilitating tourist areas (quality, water and energy savings, use of renewable energy);
- buying, protecting and managing natural areas and resources in a sustainable manner;
- protecting and restoring the historical and cultural heritage in tourist areas;
- revitalising agriculture by promoting the re-use of treated wastewater;
- managing natural areas to conserve their biodiversity.

The tax (a maximum of 2 euros per night-stay) was charged for every night spent in an accommodation, adjusted by category (hotel, apartment hotel, tourist apartment, holiday village, country hotel or camping ground), and was collected from customers of these various establishments. Children under 12 and stays subsidized by social programmes were excluded. The law entered into force in 2002, and in its first year of implementation income amounted to 17 million euros, 52 per cent of which went to protecting natural areas and parks, 25 per cent to restoring historical and cultural heritage, 20 per cent to rehabilitating tourist sites and 3 per cent to promoting agriculture (cultivation and raising livestock).

However, the law ran up against opposition from professional accommodation-providers who protested that private holiday homes and non-registered accommodations (estimated at 24 per cent of the total offer) did not pay the tax. The new government that came to power after the 2003 election removed the tax, claiming that the number of foreign tourists had declined (a drop of 8 per cent in 2002), for which the tax was certainly not the main explanation.

The 1995 *Barnier law* in **France** imposes an eco-tax on maritime transport companies, based on the number of passengers embarking to visit a classified natural site, a national park, a nature reserve or a site belonging to the *Conservatoire du Littoral* (Coastal Area Conservation Centre). The tax is added to the ticket price charged to passengers and is intended to finance the management of the visited natural areas. Enforced since 1995, it provides 120,000 euros per year to the national park on the island of Port-Cros.

Source: Segui Llinas, 2003; Parc National de Port-Cros, 2004 (www.portcrosparcnational.fr)

Succeeding in changing current trends

Changing current trends would require a reorientation of most of the *national and local strategies for tourism development*. Choices in this sector are often made without prior consultation or sufficient investigation of the possible options and their respective costs and advantages. However, experience shows the importance of looking ahead and successful reorientation, and, conversely, of the high costs of not looking ahead and making corrections.

Regulation of the tourism system should take place at the level of local destinations where the tight interdependence between tourism and the environment calls for the implementation of integration strategies. Such integration implies the emergence of common interests in *projects for specific territories* in order to overcome short-term interests and get the various actors involved in defining a common discipline that they will respect over time (Figure 6.22). Attempts are being made along these lines, notably in Spain and Italy as part of Local Agendas 21 (Box 6.23). Another example is given by the charts of French regional natural parks in which medium-term (ten years) and high quality territory-specific projects have to be formulated.

Figure 6.22 Tourism system in a specific territory: actors and operational levels

Source: Ph. Moisset, Blue Plan, 1998

Making these participatory approaches work requires a commitment and an ability of *local authorities* to bring actors together and develop collective procedures. Monitoring through indicators makes it possible to assess progress over time and decide on re-orientations when necessary. The commitment of major actors (states, regions, international cooperation and professionals in the tourism sector) to support this type of procedure through time remains rare. Such a commitment by tour operators and subsequent changes of business plans would require another level of investment in sustainable development than is currently the case. In current charts the long-term effectiveness at destinations remains uncertain.

> **Box 6.23   The Calvia Local Agenda 21 (Balearics, Spain)**
>
> The Calvia district, the leading tourist destination of the Balearic Islands, adopted its own Local Agenda 21 in 1998. The district has a population of 50,000, receives 1.6 million tourists a year and offers 150,000 hotel beds, five marinas and four golf courses on less than 143km². It started developing tourism in the 1970s with a short-term view that led to overall degradation in the 1990s, related to the destruction of typical island landscapes and the environment, congestion in tourist areas, which threatened tourism attractiveness and local development.
>
> Starting in 1997, an action plan is being implemented along ten strategic lines. Forty initiatives are focusing efforts on the preservation of the natural heritage, modernizing the tourism sector, rehabilitating the architectural heritage and enriching human resources. A revision of the Overall Urban Development Plan made it possible to safeguard more than 1500 hectares from urbanization (89 per cent of the land that was destined for urban use in the old plan). More than 80 per cent of the district area is now protected. Tearing down obsolete hotels made it possible to rehabilitate key areas, and permission to construct new accommodation is only given on the condition that an at least equal number of run-down rooms are demolished in return. The plan enabled the demolition of 12 hotels in five years and the renovation of several others. A mediator is appointed to facilitate the resolution of conflicts between tourists, residents and local governments. A Local Agenda-21 Observatory is in charge of monitoring the sustainability of developments and the quality of life in Calvia.
>
> *Source*: Ajuntament de Calvia. Calvià Local Agenda 21. Observatory on sustainability and quality of life: 1997–2000, www.calvia.com

Shifting to an alternative scenario would also require a reorientation of tourism strategies on a wider regional and national level with the particular objectives of an improved *contribution by the sector* to the overall socio-economic development of countries and to less coastal overdevelopment.

Mechanisms need to be developed for tourism's financial contribution to improving the *living conditions of impoverished peoples*. The Tourism For Development (TFD) initiative deserves to be highlighted. It was started by Egyptian professionals after the terrorist attacks at Luxor in 1998, based on the idea that economic development is the best weapon against terrorism. Voluntary TFD-member hotel keepers commit to paying one dollar per tourist night-stay to a fund used to develop Egyptian villages with a view to equitably re-distributing money derived from tourism. Today the TFD is certifying travel agencies and tour operators (13 in all) who commit to paying 1–2 per cent of the price paid by tourists to financing micro-development projects in the destination country. In the Mediterranean region, only Morocco and Turkey (other than Egypt) have similar initiatives.

Another path relates to encouraging the *diversification of tourism* by promoting different accommodation formulae and better integrated lodgings (existing rural and urban environments can be upgraded with contributions from tourism or in continuity and harmony with tourism). Such diversification would particularly benefit inland areas that often possess unexploited potential. Creating *synergies* with other economic activities (agriculture, fisheries, local services, crafts, industry) is also very important for a balanced development. Such a *re-orientation* seems all the more necessary given the major challenges of urban and rural development (see Chapters 4 and 5) and the role that tourism can play as a driver of economic, cultural and social recovery. It would also make it possible to relieve the pressures on the coastal zones, ensure development with a smaller ecological footprint and enable a better encounter between visitors, residents and the heritage. Strategies and other policy instruments should be put in place to help the emergence of territory-specific projects and encourage project sponsors (training, guiding local development, subsidies, heritage enhancement and tourist information in rural revitalization areas). Many examples of integrated rural and urban development policies show their possibilities and advantages. A detailed assessment of measures implemented in the Mediterranean region would show vast differences between countries but also the insufficiencies in most of them. Some reorientations are underway or suggested, for example in Greece (Box 6.24).

> **Box 6.24   Towards a reorientation of Greek policies for tourism development**
>
> The Greek tourism sector has been growing over the past 20 years, fed by *public-sector investments*. Between 1982 and 1995, 2180 projects were supported by the Hellenic Tourist Office with subsidies covering on average 40 per cent of the total investment. Accommodation capacity is now more than 130,000 beds, benefiting regions that already have a developed tourism infrastructure (such as the South Aegean and Crete).
>
> But this development model is also experiencing problems. The hotel occupancy rate is relatively low (69 per cent from May to September and 34 per cent the rest of the year), and, with growing competition from neighbouring countries, the profit margins continue to fall. Besides, there is a significant *drop in nominal* tourist spending (26 per cent between 1985 and 1998). The developing tourism sector has increased the environmental and socio-cultural pressures and *impacts* on

the coastal zone: ad hoc construction, especially on public land (facilitated by the absence of a land registry), degradation of coastal ecosystems, compaction of beach and dune sand due to the use of motorized vehicles, abandoned litter on the shores, noise pollution (vehicles, discotheques), disturbance of marine turtles by motorboats (20 per cent of the turtles arriving to nest in 1991 on Zante Island were wounded) and by artificial light (tourist facilities and airports), excessive extraction of groundwater (in the high season, drinking water is transported by tankers to 14 islands in the Aegean Sea at an annual cost of 1.5 million euros), a decline in traditional farming, and pleasure-boat discharges of solid and liquid waste.

These trends made it necessary to *reconsider the conventional approach* relying on mass tourism and competitive pricing, looking more for orientations aimed at a better use of existing capacities, better evaluation and improvement of environmental performance, and more diverse products (eco-tourism, conferences, spas) and profiles (ecological, cultural and rural). New criteria for allocating subsidies and tax breaks have been defined. A national land registry is being set up in 2005. And remarkable local initiatives are emerging, including women's agro-tourism cooperatives, special limited-access zones (for example a pedestrian district in Athens and a natural park on Zante Island), and networks of footpaths along the Epirus coast.

The OECD questions the justification of maintaining a policy of subsidies (luxury hotels, marinas, golf courses) by the Ministry of Economy and by the EU (93 million euros under the 'tourism' sub-programme of the European support framework). It is calling for the development of a *strategic action plan* for the sustainable development of tourism in consultation with participating parties. This plan could re-direct part of the funds towards enhancing the natural and cultural heritage, local strategies of integrated development that rely on externalities and positive synergies between tourism, agriculture, fisheries and the environment (for the benefit in particular of the poorer regions), and the promotion of micro-projects. The plan would be assessed using indicators.

Source: OECD, 2000

Recent developments in supply and demand in European destinations call for a reorientation of strategies since the seaside model would be 'running out of steam'. Italy offers an example of a well-articulated development of tourism. The strongest growth is being registered in art and historical cities, and rural tourism is experiencing a large increase. The creation of natural parks is contributing to this. In Tuscany, where non-hotel offers (agro-tourism, self-catering cottages and furnished flats) account for 19 per cent of the accommodation capacity, agro-tourism has registered a 16 per cent growth in one year, while the hotel share fell during the same period. France also has examples of developing rural tourism (see Chapter 5).

Tourism strategies and development would gain much from a more structured *revaluation of the heritage of territories*, whether along the coast or further inland. Heritage inventories (nature, culture and landscapes) and the creation of systems for experiencing heritage (coastal footpaths, special routes in urban, rural and natural areas, museums and centres of the old or modern Mediterranean[52]) would be a major component of an alternate scenario. According to such a scenario, the Mediterranean would derive the greatest benefit from the diversity of its social and economic development. Projects and achievements have recently been put in place (coastal restoration in old and degraded destinations, eco-museums), but they remain insufficient in number and support. This scenario would, among other things, see destinations emerging while looking ahead, avoiding the degradation and restoration costs that mature degraded destinations are facing today. Heritage appropriation and strategies will also be strongly stimulated by the *names* given to coasts (for example the 'Carthage Coast' in Tunisia).

Faced with these multiple challenges, it appears vital to create more *awareness* and *train* more professionals, local government officials and the general public in both originating and destination countries. Several NGOs are already doing this, but a more resolute commitment by the community of countries appears necessary, that is by the European Commission and professionals from originating and receiving countries.

A sustainable coastal management scenario

Implementing good governance in the Mediterranean coastal areas is essential to avoid further unsustainable development, and to shift towards a more satisfactory alternative future. As a necessary complement measure to the liberalization of economies, such governance needs to be organized mainly at the national and local level, although it may be stimulated by strengthening regional cooperation.

The risks of the baseline scenario

The prospects for the Mediterranean coastal areas raises questions and concerns. How will this geographically limited region and its societies be able to sustain the projected increase in the coastal urban population (20 million additional urban dwellers between 2000 and 2025, a 1 per cent annual increase), the near doubling of tourist flows (an additional 137 million in the coastal regions, a 2.3 per cent increase per year), and the even greater increases in transport (passenger transport by 2.8 per cent per year, and freight by 3.8 per cent)? The baseline scenario paints a fairly undesirable picture of the future, stressing coastal overdevelopment, diminished quality and integrity of coastal areas, and increased natural and social risks.

Under present conditions of governance, such trends would result in a continuing *loss* or *degradation* of the natural and landscape heritage of the coastal areas. Many examples have been quoted in the chapter (5000 additional km of urbanized coastal fringe between now and 2025, coastal water polluted by persistent substances, loss of biodiversity), all resulting in impacts on human health and the economy (tourism and fisheries).

With such losses, the *risks* that weigh on coastal societies and economies increase under the baseline scenario. The coastal societies that have managed to eradicate the risks that confronted them in the past[53] will have to face new risks that are enhanced because of increased *contingencies* (global warming, biological invasions, accidental pollution related to increased maritime traffic, terrorism, the versatility of tourists and tour operators in the choice of their destinations), and a *loss of resilience*[54] of their coastal areas due mainly to a *loss of diversity*, as projected under the baseline scenario in various fields. For the *environment* the loss of ecosystem diversity and physical degradation (for example coastal erosion) deprive the Mediterranean coastal zones of their potential to react to future contingencies (bioclimatic and economic). E*conomically* speaking, the development towards 'tourism-only' makes coastal territories more vulnerable to environmental, geopolitical, terrorism and economic risks. Concentrating establishments in restricted risk-prone areas increases the impacts of natural hazards (floods, earthquakes). In *governance* terms, overly centralized approaches discourage diversity in strategies for coastal territories.

Increased losses and risks (and the financial costs they induce) are also the result of a *failure of governance* in managing coastal zones, and, more generally, in regional and local development and integrating the environment and development. This failure concerns all countries, including those most advanced in implementing specific policy instruments. P*ublic policies* have certainly contributed considerably to modern changes in the Mediterranean coastal zones, without taking due account of their possible consequences for natural areas, with their complex physical and biological functions and the overall territorial balance. Too many roads have been built along the coastline, promoting ribbon urbanization. Too many buildings and public facilities have been constructed along the coast or on flood plains, dunes or wetlands, all resulting in serious coastal zone degradation. Too many tourism strategies have led to artificial development models with serious impact and increased urbanization in the coastal zones, rather than promoting a solid and balanced development. Too many subsidies have supported investment projects, such as in the industrial fisheries sector, which are only now showing their negative impacts. In some countries *decentralization* has made it possible to increase mobilization for a sustainable management of coastal zones but has also added negative changes in some cases, because the long-term perspective was not taken into account. Under pressure from particular short-term interests, too many local authorities have allowed their regions to develop towards an unstructured land use, while wasting space.

Without radical changes in policies and behaviour, one cannot exclude the possibility that residents and tourists themselves will start questioning such developments in tourism and transport.

The objectives of an alternative scenario

An alternative scenario for Mediterranean development would focus on *urgently stopping and reversing the continuing degradation of coastal areas*[55] and *reducing risks*, while involving coastal areas in the globalization process and ensuring harmony with their hinterland, all in order to contribute to a balanced economic development. Such an objective mainly implies protecting and enhancing the coastal heritage (including the productive sectors that contribute to the Mediterranean identity), making its transport system more efficient[56] and responsible,[57] integrating tourism and sustainable development more effectively, avoiding development towards 'tourism-only' and 'urban-only' outcomes, and relieving coastal zones as much as possible of pressures resulting from the projected growth.

These objectives can vary according to the coastal region. On coasts already badly affected by urban sprawl, reparation and restoration will predominate (ecosystems, landscapes and buildings). On coasts not yet too badly affected, more innovative anticipatory strategies will make it possible to develop in ways that are less costly in the long run. Some heavily industrialized and international business-oriented coastal regions will favour improvements in their transport network. Other regions will opt for creating more valuable landscapes by playing the quality card and enhancing the character of their territory, and the synergy between tourism and other economic sectors (even if this would mean reducing accessibility and urbanization). In this respect, some islands might well become sustainable development laboratories (Box 6.25).

Depending on its location, history, heritage, developmental needs and particular identity, each country, region, area and island should define its own objectives and find the proper balance between protection and development and between coastal and inland areas.

Despite a wide variety of situations, some of the *alternative paths* described in this report are especially relevant for coastal areas. They show possible significant shifts

> **Box 6.25 Islands as new sustainable development laboratories?**

Changes in the Mediterranean are exacerbated on the islands (the vulnerability of traditional economies, developments towards 'tourism-only'). The importance of social pressures exercised by tourism on a limited space, the scarcity of natural resources (water, soils, energy) and the generally strong awareness of the close relationship between people and nature, are such that the dead-ends of the baseline scenario are more quickly perceived than on the mainland. The isolation of islands, their singularity and originality, and old and new problems can be sources of innovation. Rather than adopting ready-packaged and extremely expensive solutions (for example drinking water transport and building desalinization plants which are often the cause of urbanization serving tourism), they can also benefit from their specificity and innovativeness. There are many possible fields, including the integrated and economic management of resources, the creation of local value-added by enhancing a specific image and short round trips, the introduction of new technologies (in the field of information, aquaculture, renewable energy and food-processing), and implementing new governance instruments (such as Local Agendas 21, eco-taxes on tourist transport, defining limited accommodation capacities, landownership management, transport regulation and specific regulations for fisheries).

Changes along such lines have already started on some islands. For example, Minorca tried to avoid the mass tourism of Majorca and succeeded in evading a too-strong mono-activity. It opted for 'Biosphere Reserve' management that tries to reconcile conservation and development. The Island of Elba is also trying to implement such an approach. Porquerolle Island in France has been able to take advantage of the protection provided by Port Cros Park and to innovate in water management (biological treatment in ponds keeping wastewater discharge from entering the sea and enabling water re-use, thereby benefiting quality wine production, a botanic conservation centre and market garden production). Malta is also cited regularly for its excellent measures for managing water demand. Crete is developing major activities in the field of research and new technologies. 'Virtual campuses' and distance education issue accredited degrees. An optical fibre network connects Crete to several countries, and the use of modern means of communication (data transfer, satellite connections) and medical imaging make it possible to realize tele-consultations and establish 'diagnoses-from-a-distance'. The universities in the six major Mediterranean islands and the Aegean Sea are networked.

Under the alternative scenario, subsidiary principles and contractual agreements (with the upper levels of government) would be confirmed to encourage this role of a laboratory for sustainable development. Strengthening the role of decentralized political institutions and some adaptations in national and community rules could prove to be necessary. This would allow for the emergence of new ways of regulating the fisheries, agriculture, transport, business and tourism sectors to serve sustainable development.

and progress that may be derived from better governance of coastal zones, regional development, and sectoral and economic policies. By integrating the territorial and environmental aspects of sustainable development, it should be possible to considerably reduce the pressures that are projected in the baseline scenario.

Strengthening *policies for coastal zones* would make it possible to reduce land-based pollution in compliance with the Strategic Action Plan of MAP (reducing industrial and urban waste), reduce coastal artificial cover and ensure a more sustainable management of marine resources. An alternative scenario could, for example, between now and 2025, aim to restore the ecology and landscape of 4000km of urbanized coast (in tourist destinations and coastal cities) and conserve/ sustainably manage an additional 4000km of coast that is not yet built-up. This would, in particular, put a brake on the linear urbanization of coastal zones while ensuring the maintenance of a productive agricultural sector and green belts between coastal cities. Thus, by 2025, about one-sixth of the Mediterranean coast, or 8000km, would be improved, compared with the baseline scenario, and projected new urbanization (5000km) could be reduced (less concentrated along shores) and be pushed back by at least half to a sufficient distance from coastlines. In parallel, a shift to fisheries through eco-management would contribute to the protection of the aquatic environment and enable the replenishment of stocks and higher catch levels, complemented by aquaculture, a sector which would grow strongly while complying with sustainability objectives.

The *general integration of the environment with development* would make it possible to reduce the projected pressures. Although particularly important for the tourism and industry sectors, given their direct impacts on coastal zones (pollution and physical degradation), this integration concerns all sectors. As already indicated, a shift towards integrated management of resources and water demand would enable a possible overall saving by 2025 of 54km^3 at the catchment area level. This would avoid the construction of many dams on increasingly problematic sites. These savings would reduce the risk of saline intrusions and coastal erosion, the drying-up of springs, regression of coastal wetlands and the need for new infrastructures and facilities on the coasts (desalinization plants, wastewater treatment plants). More efficient energy management would make the construction of about 80 thermal plants along the coast between now and 2025 unnecessary, so reducing by half the additional pollution and land occupancy compared with the baseline scenario. Successfully 'decoupling' transport growth from environmental degradation, orienting it towards maritime or rail transport and strengthening the fight against maritime-based pollution would also contribute to reduced pressures and risks. A shift to agriculture that respects the environment (well thought-out and organic agriculture) would result in a drop in nitrate and pesticide discharges in the developed countries and a reduction by at least half in their projected increase in the developing countries. Reducing urban waste production at the source could contribute to savings of 28 million tonnes of waste produced in coastal cities by 2025.

Pressure on the coastal zones could also be greatly reduced by *regional development limiting urbanisation along the coasts*, through solid and balanced urban and rural development. A better *distribution of tourist flows in space* and time, particularly by diversifying the forms of tourism (urban, rural and cultural tourism) would enable a shift from the coast towards the hinterland of at least a third of the additional tourists projected for the coastal regions between now and 2025. This would also contribute to reducing rural depopulation, particularly in mountainous regions.

All this cumulative progress could lead to a clear slowing of the continuing degradation of the coastal fringe and marine and land *diversity*, enable the preservation of coastal ecosystems and the health of sea-produce, consumers and bathers, and reduce *risks*.

Conditions for better governance of coastal zones

The alternative scenario assumes, as a primary condition, a strengthened *commitment* by Mediterranean people to sustainable development of their coastal territory. This commitment would be derived both from an ethic dimension, which could be based on the philosophical heritage of the region, and from a better understanding of key issues for the future. In this scenario, coastal societies and national and local government authorities will learn from the past and give themselves mid- and long-term progress objectives. The assessment of diverse experiences and the implementation of collective prospective exercises will highlight the contradictory debate and clarify the possible choices included in *national and local strategies* that commit partners in the medium-term. These forward-looking procedures will be carried out at different relevant geographic scales, not only at the coastal zone but also the broader levels of coastal regions and catchment areas that influence coastal zones.

These objectives for progress will be integrated into different *sectoral policies* (tourism, urban development, transport, water, agriculture, fisheries) that interact directly or indirectly with the coast and will be subject to regular public evaluation by means of indicators. *Coastal policies*, also subject to regular evaluation, will be strengthened through law implementation and dedicated agencies.

The *heritage dimension* will be systematically monitored, as will the threats and causes of degradation. Ecosystems, landscapes and remarkable elements of cultural heritage, traditional production activities and local expertise will all be recorded. This heritage dimension will inspire the formulation of *projects for specific territories* that will enhance such areas economically. Activities of sustainable management in the coastal zone will thus be carried out around a harbour, a bay, an island, a delta or a lagoon complex. They will bring together decision-makers, terrestrial and maritime administrations, donors, scientists and local and professional actors in a spirit of participation. Better targeted, they will aim to reduce pollution, develop specific productive activities, protect, rehabilitate or enhance the coastal heritage (for example eco-tourism), and develop a tourism model that is compatible with the objectives of sustainability. They will gain in effectiveness and will be the subject of innovative procedures such as charters. New public/private/local community partnerships will be experimented with to develop a form of tourism that respects coastal zones and is linked to the development of the hinterland. The management of coastal sites will take highly diversified forms with a growing mobilization of civil society. Protected areas will be developed in the spirit of UNESCO's Biosphere Reserve programme. Their role as instruments of management and sustainable development will be acknowledged, financed (also by the market) and enhanced.

New *economic instruments* for the sustainable management of coastal zones will be implemented. 'Visitor pays back' formulae, levelling out property and tax values, and systems of taxing and pricing inspired by 'the polluter-pays' principle will become more common. Protective districts and farmers and foresters will be paid for their multifunctional role. Their function as 'agents of sustainable development' will thus be recognized and rewarded through regulatory, economic and property-tax measures. In the hinterland, these mechanisms will contribute to the sustainable management of catchment areas, the diversification of tourism and rural revival. Tourists, tourism professionals and coastal cities will thus become major actors in the sustainable development of the Mediterranean.

Coastal policies will be consigned to *master plans* (sometimes requiring the creation of land registries) that organize the spatial distribution of facilities and activities, taking environmental and heritage imperatives into account and defining the areas to be protected or rehabilitated. Special attention will be paid to coasts that are still free of coastal roads. New roads will be built at sufficient distances from the sea with access roads perpendicular to the coastline. Master plans will also organize the protection of marine ecosystems and fish resources and the modalities for development of new activities at sea.

The coastal dimension will be fully recognized and integrated into master plans formulated on the *broader scale of coastal regions*. An infrastructural scheme will be planned to promote a more balanced and fraternal development between coastal and inland areas. Urbanization pressures will be directed to areas that are adequately connected by public transport. Construction rules, adapted to the risks and landscapes and highlighting local expertise, will be defined. Conservation will not be limited to natural areas but will be extended to agricultural land and will incite action from yet-to-be created 'land agencies'.

This alternative scenario assumes *radical changes*, which will not be self-evident because they will run up against political and administrative inertia and dominant, short-term interests. They will not be obtained without establishing, or strengthening, an *enabling environment* for managing coastal zones.

To encourage this shift in the coastal countries, the *regional Mediterranean level* needs to be considered. In line with the orientation given to the new Barcelona Convention, MAP could play an important role by promoting the adoption of an overall *strategy* up to 2025 or 2050, already planned by the Contracting Parties to the Barcelona Convention, and a *regional protocol* for the sustainable management of coastal zones. Such a protocol would make it possible to recognize the Mediterranean coastal areas as having regional public interest and requiring cooperation for their sustainable development. It could propose a delimitation, define the basic principles for its management and monitoring, and serve as a framework for national and local strategies and policies. The *strategy* could set overall progress objectives, invite each country to set its own objectives and develop approaches for integrated management at appropriate spatial levels, specify the desired development of the role of MAP, and encourage the mobilization of other regional partners. Its implementation would be monitored over time using indicators and stimulated by a regular exchange of experiences.

In the strategic tourism sector, implementation of a *regional cooperation mechanism*[58] would make it possible to promote the development of demand towards a tourism that is more respectful of the principles of sustainability (by enhancing the role of the market in this direction), define and promote operational instruments to help the various actors to control developments better, and strengthen Euro-Mediterranean awareness. There are many possible fields for cooperation, including raising public awareness, exchanging experiences, training programmes, technical assistance, strategic consultations, monitoring and forward-looking activities, promotion and marketing, and Mediterranean eco-labels.

The shift towards an alternative scenario, would be facilitated by a greater *synergy* of action between MAP, the Mediterranean countries, the European Commission and donors (and the GFCM for fisheries and aquaculture). This strengthening of regional cooperation would benefit from the adoption of a *common and differentiated approach*, taking situational, particularly institutional, differences between countries into account. It would require increased *financial transfers* towards the southern- and eastern-rim countries, without which the objectives that have been set may well be compromised. Programmes to abate land-based pollution and protect coastal biodiversity would, in this framework, be regularly evaluated and, if necessary, intensified or reoriented. Broader programmes to strengthen coastal policy capacities would be conducted over time. Examples would be: setting up land registries; organizing statistical data, monitoring and assessing the state of the environment; supporting the installation of coastal agencies, national and local strategies, demonstration projects, exchanging experience, and research and promotion of inexpensive technologies for pollution abatement. Networked, structured mechanisms for regional cooperation would be created or strengthened (tourism, coastal cities, fisheries, protected areas, environmental information, islands and coastal agencies) with the participation of professional actors.

Notes

1. But cities such as Rome or even Athens, although near the coast, are not included in the coastal zone.
2. Definition: an urban area with more than 10,000 people, located on the coastline.
3. Saidi, 2000.
4. Egyptian Tourist Authority.
5. AFIT, 2003 (www.afit-tourisme.fr).
6. Updated from Giri, 1991.
7. UNEP/MAP/MEDPOL, 2001.
8. J. Margat, *Plan Bleu*, 2004.
9. New analyses are being developed by EEA and the EU Joint Research Centre in 2004, in addition to a Coastwatch project. These developments will bring new data and methods for monitoring coastal zones.
10. The spatial limits of urban areas are defined by the continuity of built-up areas (a maximum of 200m between buildings).
11. Information contained in this section relies mainly on an analysis of transboundary pollution (*Transboundary Diagnostic Analysis for the Mediterranean Sea* – TDA) by UNEP/MAP/MEDPOL, 2003
12. A condition of water when it is too acid and poor in oxygen to support life, resulting from excessive organic matter content.
13. UNEP/MAP/MEDPOL, 2003.
14. UNEP/MAP, 1996a. The PAH volumes dumped by the refineries of 12 Mediterranean countries investigated in this report were, in 1996, estimated to be more than 780,000 tonnes per year, compared with an estimated 100,000 tonnes from de-ballasting at sea in 2000.
15. UNEP Chemicals Programme. *Mediterranean regional report*, 2002
16. UNEP/MAP/MEDPOL, 2003.
17. Micro-organisms consume oxygen for the oxidation of polluted water. Measuring oxygen demand gives an indirect estimate of the faecal concentration.
18. CITET *news* no 7 2002, Tunis.
19. Under the auspices of the United Nations Environment Programme (UNEP).
20. Cities with more than 100,000 Population Equivalent should be in

Coastal Areas

21 UNEP/MAP/MEDPOL, 2002: 99.
22 UNEP/MAP/MEDPOL, 2003: 70.
23 Primary treatment reduces BOD by only 20 per cent, and secondary treatment reduces it by between 70 and 90 per cent.
24 UNEP/MAP/PAP, 2001.
25 UNEP/MAP/MEDPOL, 2002.
26 Collection and secondary treatment.
27 Based on an average cost of 100 euros per population equivalent for upgrading, and 400 euros per population equivalent for the new infrastructures. (See the EEA Study on countries acceding to the EU, Chapter 1).
28 Compared with 60 billion for all SEMC countries, cf. Chapter 1.
29 UNEP/MAP/MEDPOL, 2003: 110.
30 Larid, 2002.
31 A. Golik. *Dynamics and management of sand along the Israeli coastline*. In Briand and Maldonado, 1997.
32 Snoussi and Long, 2002.
33 Briand and Maldonado, 1997.
34 EEA, 2003.
35 CIESM, 2002.
36 UNEP/MAP/MEDPOL, 2003: 18.
37 Algeria. Ministère de l'Aménagement du Territoire et de l'Environnement. 2002. *Plan National d'Actions pour l'Environnement et le Développement Durable* (PNAE-DD).
38 Since 1995, 23 coastal protected areas have been created in France (5155ha), 2 in Croatia (24ha) and 2 in Albania (6100ha).
39 Data relating to fishing and fish stocks must be handled with care because of the high level of uncertainty.
40 Charbonnier et al 1990.
41 Including 25 in the Valencia region, 13 in Andalusia and 12 in Catalonia.
42 The Barcelona Convention as amended in 1995 and recorded as the 'Convention for the Protection of the Marine Environment and the Coastal Region of the Mediterranean'. It has not attempted to define 'coastal region', leaving this up to the Contracting Parties.
43 France. Conseil national d'aménagement et de développement du territoire, Commission du littoral. 2003. *Communiqué du 8 juillet 2003: Le littoral français. Pour un nouveau contrat social.* (DATAR, www.datar.gouv.fr).
44 This section relies in particular on a comparative analysis carried out by Prieur and Ghezali for MAP in 2000, which has been updated and complemented.
45 In 2003 Italy intended to regularize some 360,000 illegal constructions throughout the country (*Courrier international*, no 674, October 2003).
46 Even though it has been largely adopted as common terminology, the concept of *coastal zones*, traditionally used by oceanographers, is in fact ambiguous because it may not include the land part, which is obviously essential to coastal areas.
47 Such as the latest 'systemic and prospective analyses of sustainability' developed by the Blue Plan as part of the coastal area management programmes.
48 The Andalusian plan for coastal areas 1995–2000; the Balearics management plan for coastal areas.
49 Commission of the European Communities, Tourism Unit, 1993.
50 Larini F., Université de Genève. Mémoire de DESS. 2003. *Tourisme et développement durable en Méditerranée, le cas d'Antalya (Turquie)*.
51 Recommendations adopted by the Contracting Parties to the Barcelona Convention, Malta, 1999.
52 Such as the future Musée des Civilisations de l'Europe et de la Méditerranée planned for Marseilles in 2008.
53 As, for example, the insecurity of sailing or malaria.
54 Capacity to react and adapt to contingencies.
55 As stated by the Contracting Parties to the Barcelona Convention (Catania Meeting, November 2003).
56 By lowering excessive prices that penalize exporting companies.
57 By internalizing its environmental impact in its costs.
58 Recommended by MAP, inspired by what already exits in south-east Asia and the Caribbean.

References

Ajuntament de Calvia. (2001) *Calvià Local Agenda 21. Observatory on sustainability and quality of life: 1997–2000*. Ajuntament de Calvia

Algeria. Ministère de l'Aménagement du Territoire et de l'Environnement. (2002) *Plan National d'Actions pour l'Environnement et le Développement Durable (PNAE-DD)*. Alger

Attané, I., Courbage, Y. (2001). *La démographie en Méditerranée. Situation et projections*. Les Fascicules du Plan Bleu, no 11. Paris: Economica

Basurco, B. (2003) *Mediterranean aquaculture: Marine fish farming development*. Zaragoza: CIHEAM, IAMZ

Bayle Sempere, J.T. (2003) Importancia de las praderas de fanerogamas marinas en la gestion del litoral. *Conferencia sobre Gestion integral del litoral en zonas turisticas del Mediterraneo*, Calvià, Mallorca, Espana, 24–25 de enero de 2003

Bernard, N. (2000) *Les ports de plaisance, équipements structurants de l'espace littoral*. Paris: L'Harmattan

Briand, F., Maldonado, A. (eds) (1997) *Transformation and evolution of the Mediterranean coastline*. Monaco: Musée Océanographique; CIESM (Bulletin de l'Institut Océanographique no spécial 18, CIESM Science Series no 3)

Charbonnier, D. et al (1990) *Pêche et Aquaculture en Méditerranée. État Actuel et Perspectives*. Les Fascicules du Plan Bleu, no 1. Economica, Paris

CIESM. (2002) *Alien marine organisms introduced by ships in the Mediterranean and Black seas*. CIESM Workshop Monographs no 20. Monaco: CIESM

Commission des Communautés Européennes. (2002). *Communication de la Commission au Conseil et au Parlement européen établissant un plan d'action communautaire pour la conservation et l'exploitation durables des ressources halieutiques en Méditerranée*. Bruxelles: CCE. COM(2002) 535 final

Daligaux, J. (2003) Urbanisation et environnement sur les littoraux: une analyse spatiale. *Rives nord-méditerranéennes*, no 15

Di Castri, F., Balaji, V. (eds) (2002) *Tourism, biodiversity and information*. Leiden: Backhuys Publishers

EEA. (2000) *State and pressure of the marine and coastal Mediterranean environment*. Environmental assessment report no 5. Luxembourg: Office for Official Publications of the European Communities

EEA. (2003) *Europe's water: an indicator-based assessment*. Luxembourg: Office for Official Publications of the European Communities

EEA. Peronaci, M. (2001) *Marine and coastal environment*. Annual topic update 2000. EEA topic report no 11. Copenhagen: EEA

Eurostat. (2001) *La pêche en Méditerranée*. Statistiques en bref, Agriculture et pêche, Thème 5–21. Luxembourg: Eurostat

FAO. (2002) *FAO Fisheries Report*, no 684. Rome: FAO.

FAO, General Fisheries Commission for the Mediterranean. (2002) *Committee on Aquaculture*, Third Session, Zaragoza (Spain), 25–27 September

Fiorentini, L., Caddy, J.F., de Leiva, J.I. (1997). *Long- and short-term trends of Mediterranean fishery resources*, Studies and Reviews. General Fisheries Council for the Mediterranean no 69. Rome: FAO

Firn Crichton Roberts LTD, Graduate School of Environmental Studies, University of Strathclyde. (2000) *An assessment of the socio-economic costs and benefits of integrated coastal zone management. Final report to the European Commission.* Strathclyde: University of Strathclyde

Fuchs, J. (coord.) (2002) *Aquaculture et pêche dans les pays du Sud:*

analyse prospective 2025 de la demande en recherche. Ifremer. (Bilans et prospective)
Giri, J. et al (1991) *Industrie et environnement en Méditerranée. Évolution et perspectives.* Les Fascicules du Plan Bleu, no 4, Economica: Paris
IFEN (2003) *L'environnement en Languedoc-Roussillon.* Les cahiers régionaux de l'environnement. Orléans: IFEN
INSULEUR (réseau des Chambres de Commerce et d'Industrie Insulaires de l'Union Européenne), Délégation Régionale Corse. (2002) *Les PME face aux handicaps insulaires.* October (www.insuleur.net)
Jorio, A. (2003) *Étude sur le recouvrement des coûts d'assainissement.* Agadir: Régie Autonome Multiservices d'Agadir
Larid, M. (2002) Le recul des plages en Algérie: problèmes et perspectives. In CIESM. *Érosion littorale en Méditerranée occidentale: dynamique, diagnostic et remèdes.* CIESM Workshop Series, no 18. Monaco: CIESM
Larid, M., Yaker, F., Plan Bleu, Ministère algérien de l'Aménagement du territoire et de l'environnement (2003). *Analyse de durabilité dans le cadre du PAC « Zone côtière algéroise (Algérie).* Rapport du 1er Atelier, Boumerdès, 9–10 février 2003. Sophia Antipolis: Plan Bleu
Le Neindre, M., Université de Corse. Faculté des Sciences et Techniques. (2002) Stage de DESS Ecosystèmes méditerranéens littoraux. *Les espèces introduites et envahissantes dans les îles en Méditerranée: état des lieux et propositions d'action.* UICN, Groupe Méditerranée
Margat, J., Plan Bleu. (2004) *L'eau des Méditerranéens: situation et perspectives.* MAP Technical Report Series no 158). Athènes: PAM
NACA, FAO. (2001) *Aquaculture in the Third Millennium. Technical Proceedings of the Conference on Aquaculture in the Third Millenium, Bangkok, Thailand.* 20–25 February. Bangkok; Rome: NACA; FAO
OECD. (2000) *Environmental performance reviews. Greece.* Paris: OECD
OMT. (2003) *L'activité des croisières dans le monde.* Madrid: OMT
Paquotte, P., Mariojouls, C., Young, J. (eds) (2002). *Seafood market studies for the introduction of new aquaculture products.* Cahiers Options Méditerranéennes, no 59. Zaragoza: CIHEAM-IAMZ
Paskoff, R. (dir.) (2001) *Le changement climatique et les espaces côtiers. L'élévation du niveau de la mer: risques et réponses.* Actes du colloque d'Arles, France, 12–13 October. Paris: Documentation française
Prieur, M., Ghezali, M. UNEP-MAP-PAP. (2000) *National legislations and proposals for the guidelines relating to integrated planning and management of the Mediterranean coastal zones.* Split: PAP
Ramade, F. et al (1997) *Conservation des écosystèmes méditerranéens: enjeux et prospective.* 2nd ed. Les Fascicules du Plan Bleu no 3. Paris: Economica
RIVM/RIZA. (1991) *Sustainable use of ground water. Problems and threats in the European Community.* National Institute of Public Health and Environmental Protection/National Institute for Inland Water Management and Waste Water Treatment. Bilthoven/Lelystad. The Netherlands
Sacchi, J. (2001) *Impacts of fishing technology in the Mediterranean Sea.* FAO: Rome
Segui Llinas, M. (2003) *Les Baléares: un laboratoire touristique méditerranéen par excellence.* Report for the Blue Plan (www.planbleu.org)
Snoussi M., Long, B. (2002) *Historique de l'évolution de la baie de Tanger et tentatives de réhabilitation.* Monaco: CIESM
UNEP. (2002) *Report submitted to the First Meeting of the ad-hoc Technical Committee to select pollution hot spots for the preparation of pre-investment studies within the GEF Project,* Athens, 28–29 January. UNEP(DEC)/MED/GEF/198/3
UNEP, Chemicals Programme. (2002) *Regionally Based Assessment of Persistent Toxic Substances: Mediterranean regional report.* Geneva: UNEP Chemicals
UNEP-MAP. (1996a) *State of the marine and coastal environment in the Mediterranean region.* MAP Technical Reports Series No. 100. Athens: UNEP
UNEP-MAP. (1996b) *Survey of pollutants from land-based sources in the Mediterranean.* MAP Technical Reports Series No 109. Athens: UNEP.
UNEP-MAP. (1996c) *Workshop on policies for sustainable development of Mediterranean coastal areas, Santorini Island, Greece, 26–27 April. Papers by a group of experts.* MAP Technical Reports Series No 114. Athens: UNEP
UNEP-MAP. (1999) *Identification of Priority Pollution Hot Spots and Sensitive Areas in the Mediterranean.* MAP Technical Reports Series No 124 Athens: UNEP
UNEP-MAP. (2000) *Les stations d'épuration des eaux usées municipales dans les villes côtières de la Méditerranée.* MAP Technical Reports Series No 128. Athens: UNEP
UNEP-MAP. (2002) *Revision of pollution hot spots in the Mediterranean. Country Reports.* Athens: UNEP
UNEP-MAP-MEDPOL. (2001) *Seawater desalinisation in Mediterranean countries: assessment of environmental impacts and proposed guidelines for the management of brine.* Athens: MAP
UNEP-MAP-MEDPOL. (2002) *Mediterranean mariculture. Draft report.* Athens: UNEP
UNEP-MAP-MEDPOL. (2003) *Transboundary Diagnostic Analysis for the Mediterranean Sea (TDA).* Athens: UNEP
UNEP-MAP-PAP. (2001) *White paper: coastal zone management in the Mediterranean.* Split: PAP
UNEP-MAP-PAP. (2003) *Feasibility Study for a Legal Instrument on Integrated Coastal Area Management in the Mediterranean.* Split: PAP
UNEP-MAP-PAP, METAP. (2002) *Coastal area management programmes: improving the implementation.* Split: PAP
UNEP-MAP-Plan Bleu. (1999) *Report of the workshop on tourism and sustainable development in the Mediterranean, Antalya, Turkey, 17–19 September 1998.* MAP Technical Report Series No 126. Athens: MAP
UNEP-MAP-Plan Bleu (2005) *Dossier sur le tourisme et le développement durable en Méditerranée = Dossier on tourism and sustainable development* (MAP Technical Reports Series no 159). Athens: MAP
UNESCO-COI, MAB, PICG, PHI, IFREMER. (2001) *Des outils et des hommes pour une gestion intégrée des zones côtières. Guide méthodologique,* vol II. Paris; La Seyne-sur-Mer: UNESCO; IFREMER
Vallega, A. (1999) *Fundamentals of integrated coastal management.* Dordrecht: Kluwer Academics Publishers
WTTC. (2002) *Croatia: the impact of travel & tourism on jobs and the economy.* London: WTTC
WTTC. (2001) *Turkey: the impact of travel & tourism on jobs and the economy.* London: WTTC

Legal references

Coastal water pollution

Multilateral relevant agreements
The United Nations Convention of the Law of the Sea (UNCLOS, 1982)
The Rotterdam Convention on the Prior Informed Consent Procedure for Certain Hazardous Chemicals and

Coastal Areas

Pesticides in International Trade (1998)
Basel Convention on the Control of Transboundary Movements of Hazardous Wastes and their Disposal (Basel, 1989)
International Convention for the Prevention of Pollution from Ships (1973) as modified by the Protocol of 1978 (MARPOL 73/78)
Stockholm Convention on Persistent Organic Pollutants (Stockholm, 2001)
International Convention on the Control of Harmful Anti-fouling Systems on Ships (London, 2001)

Barcelona Convention and its associated protocols

Convention for the Protection of the Marine Environment and the Coastal Region of the Mediterranean, Barcelona 1976, amended in 1995, entered into force in 2004
Dumping Protocol: Protocol for the prevention and elimination of pollution in the Mediterranean Sea by dumping from ships and aircraft or incineration at sea, 1976, in force since 1978, amended in 1995.
Emergency Protocol: Protocol concerning cooperation in combating pollution of the Mediterranean Sea by oil and other harmful substances in cases of emergency, 1976, in force since 1978, amended in 2002, entered into force in 2004
Land-Based Sources Protocol: Protocol for the protection of the Mediterranean Sea against pollution from land-based sources, 1980, in force since 1983, amended in 1996
SPA and Biodiversity Protocol: Protocol concerning Specially Protected Areas and biological diversity in the Mediterranean, 1995 (replacing the 1982 SPA Protocol), entered into force in 1999
Offshore protocol: Protocol for the protection of the Mediterranean Sea against pollution resulting from exploration and exploitation of the continental shelf and the seabed and its subsoil, 1994
Hazardous Wastes Protocol: Protocol on the Prevention of Pollution of the Mediterranean Sea by transboundary movements of hazardous wastes and their disposal, 1996

Other regional treaties

Convention on long-range transboundary air pollution (Geneva, 1979) and its related protocols
Convention on the conservation of European wildlife and natural habitats (Bern 1979, entered into force in 1982)

European Union Directives

Directive 2000/60/EC of the European Parliament and of the Council of 23 October 2000 establishing a Community framework for action in the field of water policy, for the protection of the ecological state of water from local and diffuse pollutions, devised for the integration of the many directives related to water pollution into a single text
Council Directive 96/61/EC of 24 September 1996 concerning integrated pollution prevention and control (IPPC Directive) submitting the industrial plant discharges to a prior permit (provided for by Annex I)
Council Directive 91/676/EEC of 12 December 1991 concerning the protection of waters against pollution caused by nitrates from agricultural sources
Council Directive 91/271/EEC of 21 May 1991 concerning urban wastewater treatment; amended by Directive 98/15/EEC, according to which by end of 2005, all the discharge from cities of more than 2000 inhabitants equivalent should be preceded by a secondary treatment
Council Directive 76/160/EEC of 8 December 1975 concerning the quality of bathing water, under revision
Council Directive 76/464/EEC of 4 May 1976 on pollution caused by certain dangerous substances discharged into the aquatic environment of the Community
Directive 2001/106/EC of the European Parliament and of the Council of 19 December 2001 amending Council Directive 95/21/EC concerning the enforcement, in respect of shipping using Community ports and sailing in the waters under the jurisdiction of the Member States, of international standards for maritime safety, pollution prevention and shipboard living and working conditions (port state control)
Directive 2002/6/EC of the European Parliament and of the Council of 18 February 2002 on reporting formalities for ships arriving in and/or departing from ports of the Member States of the Community
Regulation (EC) No 417/2002 of the European Parliament and of the Council of 18 February 2002 on the accelerated phasing-in of double hull or equivalent design requirements for single hull oil tankers and repealing Council Regulation (EC) No 2978/94

Marine and coastal biodiversity

Multilateral relevant agreements

Convention on Wetlands of International Importance especially as Waterfowl Habitat (Ramsar, 1971)
Convention for the protection of the world cultural and natural heritage (Paris, 1972)
Convention on international trade in endangered species of wild fauna and flora (CITES, Washington, 1973)
Convention on the conservation of European wildlife and natural habitats (Bern, 1979)
United Nations Convention of the Law of the Sea (Montego Bay, 1982)
Convention on the conservation of migratory species of wild animals (CMS, Bonn, 1979)
Convention on biological diversity (Rio de Janeiro, 1992)
Agreement on the conservation of African–Eurasian migratory waterbirds (AEWA, The Hague, 1995)
Agreement on the Conservation of Cetaceans of the Black Sea, Mediterranean Sea and Contiguous Atlantic Area (ACCOBAMS, Monaco, 1996)

Directives and other relevant legal instruments of the EU

Council Directive 79/409/EEC of 2 April 1979 on the conservation of wild birds
Council Regulation (EEC) No 348/81 of 20 January 1981 on common rules for imports of whales or other cetacean products
Council Directive 92/43/EEC of 21 May 1992 on the conservation of natural habitats and of wild fauna and flora
Council Regulation (EC) No 338/97 of 9 December 1996 on the protection of species of wild fauna and flora by regulating trade therein
Commission Decision 97/266/EC of 18 December 1996 concerning a site information format for proposed Natura 2000 sites
Commission Decision 2002/11/EC of 28 December 2001

adopting the list of sites of Community importance for the Macaronesian biogeographical region, pursuant to Council Directive 92/43/EEC

Commission Regulation (EC) No 1808/2001 of 30 August 2001 laying down detailed rules concerning the implementation of Council Regulation (EC) No 338/97 on the protection of species of wild fauna and flora by regulating trade therein

Commission Regulation (EC) No 349/2003 of 25 February 2003 suspending the introduction into the Community of specimens of certain species of wild fauna and flora.

Fisheries

Agreement for the establishment of a general fisheries council for the Mediterranean (GFCM, FAO, Rome, 1949)

International convention for the conservation of Atlantic tunas (Rio de Janeiro, 1966)

Part 3

Summary and Call for Action

Summary and Call for Action

The Mediterranean in motion

The Blue Plan scenarios, published in 1989 and based on 1985 data, highlighted possible futures for 2000 and 2025 for a region with one of the richest histories in the world, not to mention its diversity and fragility. Three years before the Rio Summit, the report called for an increased commitment to sustainable development, a concept that was then still in its infancy. The present report, requested and funded by all Mediterranean-rim countries and the European Commission, makes it possible to arrive at a new diagnosis.

The diagnosis confirms the gloomy trends projected 15 years ago. The drastic demographic developments projected have been realized and trends continue. The practically stable population of the northern Mediterranean countries from Spain to Greece (193 million in 2000) has been overtaken by that in the southern and eastern Mediterranean countries from Turkey to Morocco, which in 2000 totalled 234 million people, an increase of 65 million in 15 years. Urbanization, especially in the coastal zones, has been even more extensive than projected. The growth in tourism and the amount of irrigated land have also been confirmed, and unsustainable consumption and production patterns are continuing to spread in the developed countries.

However, the recent fall in fertility rates in the south and east has been of an unexpected scale and swiftness. It is a factor that draws the two shores closer together and gives these countries opportunities to develop under favourable population ratios: a high percentage of working-age adults compared with retired and young people. It should also result in a more moderate increase in pressures on the environment and natural resources.

Despite this more favourable demographic context, economic growth remains insufficient in the south and east. It has fallen in most countries, also in the north, and north–south imbalances continue to exist. Only the three Mediterranean countries that joined the European Union in 2004 (Cyprus, Malta and Slovenia) come close to the average of the other EU Mediterranean countries (Spain, France, Italy and Greece) and Israel. Standards of living (per capita and in purchasing power parity) are still 3–5 times lower in the developing and transition countries, while east Asia has recorded exceptional GDP growth rates.

Several countries (Croatia, Bosnia-Herzegovina, Serbia-Montenegro, Albania and Algeria) have experienced very difficult times related to international changes (the fall of the Berlin wall and the East-bloc system) or to internal problems. The conflict in the Near East has not yet been resolved and north–south, and especially south–south cooperation, has remained quite limited.

In the environmental field, the risks projected in the Blue Plan scenarios have also been confirmed, especially the greater stresses on water, soils and coastal zones and degradation of natural resources and ecosystems. The social and economic repercussions are becoming increasingly obvious.

Another major feature of the past 15 years has been globalization, leading to a considerable change in the relative prices of products and services and a strengthening of international and regional trade. The trade of Maghreb countries and Turkey has increasingly and strongly focused on Europe, the main regional centre of gravity. The Mashrek and Egypt have also continued to strengthen their links with Europe, but have at the same time maintained a significant amount of trade

3 A Sustainable Future for the Mediterranean

with the rest of the Middle East. A strong American presence can be seen in a few strategic countries.

The local, regional and international impacts of globalization have so far been insufficiently assessed, but they appear to be both positive and negative. The increase in global risks (climate change, loss of biodiversity, continued or worsening inequalities) and certain Mediterranean weaknesses in terms of free trade are real topics of concern.

In contrast, the 1995 initiation of north/south cooperation (the Euro-Mediterranean Partnership) and environmental policies in all countries, called for in the 1989 alternative scenarios, have been positive developments.

However, these policies are not enough. The first observation to be made is the growing disparity in resources and prospects between the northern shore, which benefits from European integration (with, among other things, rigorous directives on the environment and funds for strengthening capacity, infrastructures and agriculture) and the southern and eastern shores, which do not.

By 1985 Europe, already mature but still divided, had little presence in the southern part of the region. Has it now become sufficiently aware of the Mediterranean interdependencies and challenges, risks and strong points for the future of the EU? Has it understood that its future depends to a large extent on the Mediterranean south and east?

Fortunately the international and Mediterranean communities have now become aware of the value of their common heritage, the profound transitions underway and the risks that might result from them.

At the global level, several international summit conferences have warned about the unsustainable nature of developments in the world and called for the implementation of action programmes to remedy the situation. Following the Rio Summit in 1992, which, together with Agenda 21, highlighted the need for environmentally sustainable development, the Millennium Development Goals approved by 189 countries in September 2000 and the direction taken at the Johannesburg Summit in 2002, all reiterate the multi-dimensional character of development, including social questions, problems of access to drinking water, energy, education and health, in short all the conditions required for sustainable human development. The issues of global inequalities and poverty, which had been overshadowed in the 1990s, have returned to the international agenda. A strong emphasis has also been put on the need to change our unsustainable production and consumption patterns.

Dynamic analysis of some indicators of the Millennium Development Goals shows the strengths and weaknesses in the Mediterranean region by comparison with other regions in the world: low rates of infant mortality and extreme poverty, but considerable relative poverty that is not decreasing; good levels of primary education, but still high rates of illiteracy; a substantial delay in gender equality, although it is slowly being overcome; and a rural world that is seriously falling behind. The scale of unemployment among young people is especially worrying. High in the developed Mediterranean countries, unemployment rates in the south and east are approaching record levels. Another worrying trend is that development remains very wasteful of natural resources. And finally, the level of official development assistance in the Mediterranean developed countries remains well below the objectives set at the international level.

The Mediterranean people have become more aware of the risks that threaten their environment. And they have strengthened their common mechanisms for protecting the sea and coastal zones: in 1995 the Barcelona Convention was revised and expanded to include the coastal zones (which entered into force in 2004); a framework and regional strategy for curbing pollution and the protection of biodiversity in coastal zones has been adopted; a Mediterranean Commission on Sustainable Development has been created, and is open to a civil society that is much more active than 15 years ago.

Although the heavy trends may not have been reversed, substantive policies and actions are being implemented in all countries. This has limited degradation and shows the path to possible progress towards sustainable development.

Today the Mediterranean can no longer measure its state and reflect on its future while only considering the enclosed sea and the limited 46,000 kilometres of coastline where its active strengths are mainly concentrated. As required by the Johannesburg Summit, the Mediterranean has to position itself within the sphere of its worldwide responsibilities. Its future needs to be thought about in relation to the organization of major regional poles that tend to be strengthened by globalization, especially the Euro-Mediterranean region, a fundamental key for its future.

The present Blue Plan analysis has endeavoured to measure these developments. It leads to a further warning about the risks of ongoing trends for which the assumptions have been explained in the baseline scenario.

However, this report goes beyond assessment, diagnosis and warning. It identifies possible win–win alternatives (environment and development, north and south), and highlights the obstacles that need to be overcome as well as possible tools for response. It shows that alternative policies and measures are beginning to be put in place and should definitely be spread more

Summary and Call for Action

widely. It stresses some of the actions that could cause a trigger effect capable of changing paths that are anything but inevitable. Finally, it ends with a determined call for action to build a Mediterranean region that is responsible and united.

The risks in the baseline scenario

According to the assumptions of the baseline scenario, by 2025 the Mediterranean region will not have reduced the disparities in internal incomes or those between the two shores. Its relative weight in the world will be reduced, and its natural heritage and lifestyles will be even more degraded, often irreversibly. Its vulnerability to risks will increase.

North/south gap

The main risk for the Mediterranean in 2025 is the prospect of increased social, economic and environmental disparity between the two shores.

In the baseline scenario, five Mediterranean countries (Croatia, Bosnia-Herzegovina, Serbia-Montenegro, Albania and Turkey) are seen to be likely to join the *European Union* which, in the long run, will include 12 Mediterranean coastal countries. These new member countries will consolidate peace, democracy and a social market economy as well as a certain economic and environmental convergence with the rest of the Union.

This convergence will be based on the powerful cultural and institutional pillars of EU values and advantages. Moreover, the EU will provide stimulation funds to facilitate the convergence and consistency of infrastructure investments (environment and transport) and development investment in new member countries. This is how Spain and Greece, in previous decades, were able to benefit from cohesion funds, which, added to other structural and agricultural funds and environmental directives, made it possible to achieve parity with the other EU countries. The recent new members (Malta, Slovenia and Cyprus) are currently benefiting in their turn.

Despite this transitional support, the effects of the large market and the rising standards of living in the member countries will nevertheless increase their environmental and territorial impacts: a rapid reduction in the number of farmers; accelerated urbanization along the coasts; a rapid increase in volumes of transported goods, the private car fleet and the production of household waste; urbanization from tourism in the coastal zones; and rising per capita energy consumption and greenhouse gas emissions.

The *southern and eastern shores* will not benefit from comparable regional dynamics, although cooperation between the EU and the Southern and Eastern Mediterranean Countries (SEMC) has been organized since 1995 within the Euro-Mediterranean Partnership (more recently strengthened with the new European Neighbourhood Policy). This cooperation includes a Euro-Mediterranean free-trade zone (with a target date of 2010), implementation of topic-based cooperation, support of a MEDA fund, and investment facilities. Unfortunately this partnership is presently lacking sufficient resources and political commitment; it is undersized in relation to the huge challenges the region is facing. In addition, it has until recently not been much inspired by the principles of sustainable development. The free-trade zone itself was originally proposed without being preceded or accompanied by an impact assessment study.[1] Nevertheless, progress is being made following decisions by the European Council relating to integrating sustainable development issues into external policies and cooperation.

Under the baseline scenario, it is assumed that the SEMC will have difficulties in engaging in the necessary reforms, and south–south cooperation will remain limited. Europe will probably not be able to provide to its southern neighbours the required resources that would be justified by the importance of its interdependencies and clearly understood interests. Therefore the region will, in general, remain penalized by a lack of dynamism and economic innovation, continued conflict situations and rivalries, and inadequate supporting measures (structural and agricultural funds, environmental policies). The harmful environmental consequences of this scenario would be considerable.

Clear development differences may, however, appear between countries, depending on differences in political, institutional and socio-economic pathways taken and in the degree of success in engaging in reforms. The new EU neighbourhood policy may not alleviate these differences by giving preferential treatment to the countries that are most committed to and advanced in reforms (the differentiation principle). But this individual progress will remain limited because of the lack of perspectives and progress in the region as a whole. Disparities with the northern shore will grow wider.

In this context, the projections for economic growth (4 per cent per year in the Maghreb countries and 4.7 per cent in the Mashrek countries), although improving, will be insufficient to meet the demands of a strongly growing population (91 million more people between 2000 and 2025 for all SEMC, including Turkey) and for jobs (34 million new jobs to be created in 20 years just to maintain present-day employment levels).[2] The disparities in incomes between the two shores will remain wide up to 2025 (between one and three, even up to five times in terms of purchasing power parity).

Thus a whole generation of young people will be partially deprived of jobs and prospects. There will be strong social dissatisfaction and much pressure for emigration. Another consequence will be the relative weakening of the economic power of Europe, which, with its 'neighbours', will represent no more than a fifth of world GDP by 2050 instead of the current one-third (giving way to Asia, with North America maintaining its position).[3]

These trends could result in growing risks of instability, a source of potential conflict. The Mediterranean and Europe as a whole (and the world in general) will suffer the consequences. An evolution towards even darker development paths (such as under a 'fortress scenario') cannot be excluded.

Internal disparities

The risk of domestic disparities is no less worrying. In all countries, urbanization along the coasts and social and territorial disparities will increase, especially in transition and developing countries.

A rural revival will remain limited in the developed countries of the northern shore, and a *marginalization of inland areas* will be confirmed and strengthened in transition and developing countries. Serious signs of decay and rural-to-urban migration will appear in the still highly populated mountainous areas of the northern shore (for example in Albania). Problems of desertification and rural poverty will be much more serious (and may increase with changes caused by free trade and desertification) in some fragile regions of the south and east (mountains, arid plateaux). This will result in precarious social situations and increased recourse to emigration. Lacking prospects in cities or abroad, rural populations in these countries will, however, remain large (little or no reduction in absolute terms) and socially marginalized. Differences in standards of living between cities and rural areas, and the very strong agricultural duality that is characteristic of several countries (the contrast between subsistence farming and modern agriculture) will be maintained or even strengthened.

Even with developments of agricultural policies, irrigated agriculture (an increase of 6 million hectares projected between 2000 and 2025) will continue to benefit from a polarization in investments that will exacerbate pressures on limited and already sorely-tested water resources.

The growth in SEMC agricultural production projected under the baseline scenario (50 per cent) will not, however, result in reducing a trade deficit in agricultural produce.

But more than anything else, it is *urban growth* that will put a mark on the next 25 years. The southern and eastern shores will have to absorb 98 million more people between 2000 and 2025, a good third of which will be in the urban areas of coastal regions, with 20 million along the coastline itself. The vigour of this urban expansion will continue to pose the question of the economic development required to meet it, especially in middle-sized cities, not only in large urban areas. Urban planning will be severely stretched with the continuation of unregulated housing. Local authorities, lacking any real resources for action, will have difficulty in responding to the increasing demand for basic services and infrastructures. The risk of developing new urban poverty is high.

The number of Mediterranean people (both rural and urban) earning less than US$2 per day and without access to water (30 million people in 2000), sanitation (29 million) and electricity (16 million) may thus remain high in many countries.

Urban issues will evolve differently in the northern shore countries with very moderate population growth, where a process of suburbanization and dispersal of population and employment will continue to predominate. In such a situation, urban planning will not manage to compensate for the powerful mechanisms that encourage urban sprawl (the real estate and housing market, transport). The expansion of motorized travel will significantly increase fuel consumption, polluting emissions (CO, NO_x and CO_2) and the proportion of household budgets devoted to transport.

Despite resource stress, this consumption model and urban sprawl, with its high energy and environmental costs, will also develop in the south. Indeed, before 2025 we will see mass motorization (as exemplified by what is happening in Lebanon), boosted by a fall in car prices.

Another large growth sector is *tourism*, with drastic assumptions under the baseline scenario (273 million more tourists in 25 years, including 137 million in the coastal zones). The destinations in full growth in the eastern Adriatic and the southern and eastern regions will be particularly affected (Turkey, Egypt, Croatia, Morocco and Tunisia). But the development will also spread to other coasts (Libya, Algeria, Syria, Albania and Serbia-Montenegro). These countries will see a large increase in domestic tourism. A rise in tourist flows is also projected for the 'mature' destinations in the more developed countries of the north-western quadrant.

Although tourism will continue to be an engine for economic growth, there may be insufficient contributions from tourism to local economies and sustainable development in the region, so threatening a *risk of crisis*. An excessive focus on seaside tourism, the power of large tour operators and competition between countries is tending to result in a standardization of products. This could result in a relative drop in tourist expenditures, of which only a small share gets back to local inhabitants. The repercussions may prove high at the socio-cultural level (loss of identity and a large number of poorly qualified, temporary jobs), the economic level (a fall in income and deterioration of buildings) and the environmental level. Concentrating pressures on the summer season and coastlines will result in over-construction of infrastructures. And with investors seeking out the best venues, this concentration will contribute to increasingly artificial coastal fringes. Land consumption will be higher in destinations where the development of holiday-home tourism, rather than stimulating tourist accommodation companies, has been fostered. Some of the most important impacts from tourism will be pollution from increased transport, which will get worse as holidays become shorter but more frequent. Over-dependence on tourism (moving towards a single activity) will also be a risk in some regions (weak economic resilience, crises in the event of geopolitical instability, loss of territorial identity and socio-cultural effects).

Expanded *transport* will become a major issue in the coming quarter century. In the context of accelerated free trade, maritime traffic (the Mediterranean already handles 30 per cent of traffic in the world) may grow by 270 per cent over 25 years, and freight traffic (road, rail and air) by 150 per cent. This rise, greater than economic growth, will particularly affect road transport with serious territorial and environmental impacts. With urban sprawl and increased tourism, the demand for individual transport will grow in parallel. Passenger traffic could well double and air traffic increase by 90 per cent.

To meet their demands, the SEMC are projected to experience significant growth in their *industrial sector*. Cement and steel production could increase by as much as 150 per cent and 138 per cent respectively. The number of sea-water desalinization plants will grow significantly in all resource-poor countries, including Spain. More than 220 power plants may also be built (including 160 new thermal-power plants). A large part of these industrial structures will be concentrated along the coasts.

Projected developments in cities, industries and transport combined will trigger a strong rise in *energy* demand. Between now and 2030 this could nearly triple in the SEMC. Fossil fuels will provide 85 per cent of this demand. In non-producing countries, it will be accompanied by increased energy dependence, and all countries will see a significant rise in greenhouse gas emissions.

In both north and south development patterns appear to be unsustainable today, and this will become worse.

A neglected environment

The developments under the baseline scenario would have serious consequences for the Mediterranean environment.

Soils would be especially affected, both by urban sprawl and by the crisis in rural areas. Degradation of soils and vegetation cover by desertification (salinization, wind and water erosion, deforestation and the degradation of grazing and cultivated land) would continue at the present pace and could even increase in some regions.

This desertification, which already affects 80 per cent of the arid or dry zones of the southern and eastern Mediterranean, would result in new, irreversible degradation in these areas. This could mean more than 1 million hectares being more affected between now and 2025 (it is difficult to give a precise figure because comparable data and time series are lacking). Accelerated degradation of soils in upstream catchment areas would result in the gradual filling-up of reservoirs; the 21st century would be characterized as the post-dam era.

The change of agricultural land into urban land and infrastructures would be as irreversible and problematic, if not more so, because of the high productive value of agricultural land in all areas around the Mediterranean. More than 1.7 million ha of prime agricultural land (on coastal and fluvial plains) could be lost by 2025.

Up to 2025 *water stress* would continue to increase markedly in the southern and eastern Mediterranean, with 63 million people having less than 500m^3 per capita per year (a threshold defined as a 'shortage'). Because of the increased demand for agriculture and urban water and the scarcity of freshwater resources, one country in three would have to draw more than 50 per cent of its annual volume from renewable natural resources. The portion of non-sustainable water supplies from fossil or overexploited sources would increase (up to 30 per cent in Malta and Libya). Under these conditions some fossil resources would be rapidly used up, and coastal aquifers would continue to be destroyed by salt sea-water intrusions. The quality of water and aquatic systems would also be affected in the SEMC by an increase in industrial and urban waste and the reduction in supply resulting from increased withdrawals. The minimum flows needed to maintain biodiversity in aquatic systems (lakes, rivers) is difficult to estimate and is rarely considered to be an important issue. The retreat of *wetlands* is likely to continue under these conditions; they would also suffer from the construction of infrastructures, real estate and tourist complexes along coastlines.

The effects of *sudden natural hazards* (floods, mudslides, earthquakes, volcanic eruptions and forest fires) would have increasingly serious consequences. The already high costs would probably increase further under the combined effect of increased hazards and vulnerability. Global warming would, among other things, increase climate hazards and risks (intensifying the problems of drought, torrential rains and floods) as well as the water stress typical of the eco-region. Rising sea levels would threaten low-lying coastal areas (deltas) in the long term, some of which are highly populated; and warming may well disturb marine ecosystems (intensifying biological invasions and the 'tropicalization' of the sea). But global warming is far from being the only cause of increasing natural risks. The decline in agriculture with uncontrolled growth in the north of highly combustible biomass, increasing artificial use of soils, the reduction of natural flood plains and continuing construction in high-risk zones or construction poorly adapted to risks, are among the many factors that considerably increase vulnerability. The baseline scenario projects increased risks and costs due to major fires (in the north) and flooding.

The uncontrolled development of urbanization and insufficient public transport systems would contribute to degrade the *urban environment*, except in some ancient centres where programmes of renovation, development of public transport with reserved lanes, and restriction of car traffic will be confirmed. Congestion, having already high annual costs (estimated at some US$41 thousand million) especially in the north (US$14 thousand million in France alone, compared with US$1.6 thousand million in Turkey and US$400 million in Egypt), will increase with the growth of urban areas and the expected mass motorization. It would be accompanied by reductions in air quality and an increase in the number of people exposed to noise (32 per cent of Turkish people could be affected by 2025). W*aste* would also be an ever more serious problem with waste generation more than doubling in the northern Mediterranean Countries (NMC) and tripling in the SEMC. Given the poor supervision of dumping sites in the SEMC, this increase in waste generation will have a detrimental effect on the pollution level of soils, water and air. Impacts on health would be noticeable.

Overall Mediterranean development would contribute to an exceptional concentration of pressures and impacts on the coastal plains, particularly near the *coastlines*. The permanent population density and tourists (during the peak months) per coastal kilometre could reach record numbers (3300 people per coastal km by 2025 compared with 2300 in 2000). The simultaneous growth in urbanization, tourism and infrastructures would confront societies and governments with major problems, answers for which would be difficult and costly.

By 2025 more than 5000 *km of coastline* may be built up, bringing the percentage of artificial coasts to about 50 per cent of the total (only a rough estimate due to a lack of detailed available data). Coastal waters would also see biological invasions, which would only be exacerbated by warming seawater and expanded maritime transport. *Caulerpa taxifolia* seaweed could significantly change the Mediterranean marine ecosystem. The quality of ecosystems and coastal landscapes (both land and sea) would be greatly affected: there would be more coastal erosion, loss of biodiversity and fisheries production. Fisheries would fall victim to degradation of the marine environment and overexploitation of fish stocks (already occurring) with prospects of higher job losses.

With a nearly tripled maritime traffic on such a limited sea, the baseline scenario projects increased risks of *marine pollution* by ships. Although the vast majority of this pollution is of an operational nature, serious pollution from accidents has to be feared because of the large proportion of ships carrying dangerous products (nearly 42 per cent contain chemicals and hydrocarbons), the present state of the fleet (the average age of some national fleets is well beyond 20 or 30 years), and the number of recorded accidents (311 in the past 23 years, including 156 that resulted in hydrocarbon dumping). The coastal economy of several countries would be seriously affected, especially that of coastal regions that are over-dependent on tourism.

The *accumulation* of multiple *environmental pressures* and impacts would inevitably result in degradation of very valuable common goods on a global scale, including the loss of terrestrial (including agriculture) and coastal biodiversity, the degradation of Mediterranean landscapes and historical and cultural heritage (sites and monuments).

Apart from the loss of an irreplaceable potential for future economic development (bio-technologies), the loss of biodiversity would diminish the ability of ecosystems to adapt to future change (loss of resilience). Several symbolic Mediterranean species (monk seals and turtles) and many species related to forest, pastoral and wetland environments could disappear forever.

The Mediterranean region would see its portion of worldwide greenhouse gas emissions increase from 7 per cent in 2000 to 9 per cent by 2025. With an additional 1.4 thousand million tonnes of CO_2 emitted by 2025, the region would thus increase its contribution to global climate change, while at the same time suffering from the effects mentioned above.

The *annual global costs* of these multiple negative impacts, which already amount to several percentage points of GDP, would rise.

Facing sustainable development challenges

Prospective work makes it possible to make warnings about future risks and problems, but it can also help in building up other future outlooks. The future is never pre-ordained, and the ability of societies to adapt to new challenges or anticipate events in order to remain masters of their fate should not be underestimated.

Summary and Call for Action

Principles and objectives of an alternative scenario

This report envisages a win–win scenario for the Mediterranean that is characterized both by stronger economic growth in the SEMC (and thus a gradual convergence of incomes between the two shores, requiring annual GDP growth rates of about 7 per cent) and a better integration of the environment and development.

The report has not delved into great detail on the general question of SEMC *economic growth*, since many other specialized publications tackle this.[4] In general, they stress the efforts required in terms of investment and incorporating technological progress aimed at improving productivity: better labour organization, upgrading production systems and infrastructures, encouraging innovation, improving the quality of products, and, last but not least, training scientists and technicians (who can be encouraged to remain in their countries[5] by financing university programmes and local research centres). They emphasize the profound reforms needed to institute or strengthen the principles of a rule-based state, redefine the steering role of government, improve governance and make that the economic and social actors, companies, administrations, local authorities and households evolve towards new forms of behaviour adapted to a modern market economy (see in particular the UNDP reports on human development in 22 Arab countries).

The *alternative scenario* is based on the assumption that, by 2025, the Mediterranean economy will largely be based on the quality of its environment. It assumes that the ambitious responses of integration and anticipation will have materialized so as to avoid the dead ends projected under the baseline scenario and give more value to the natural resources and territories in the region. A shift to such an alternative development path would enable considerable economic benefits. Even though some objectives would be harder to attain than others, the many concrete examples given in the various chapters of this report show that such a shift is not impossible.

The alternative scenario inverts the rationale that prevailed in the baseline scenario. This relied on end-of-the-pipe approaches, where it is assumed that economic growth can provide the financial means required to repair environmental and social damage that it caused in the first place. In contrast, the alternative scenario relies on a more *integrated and forward-looking approach*, based on development patterns that use resources carefully, integrate and give value to the specific strong points of the Mediterranean eco-region, and contribute to reducing the risks of internal disparities, thus resulting in greater *cohesion*.

Anticipation is especially important for the transition and developing countries, because these countries generally have limited natural resources, and will continue to experience major demographic changes and economic transitions. They will be particularly interested in integrating environment with development from the very beginning and in planning ahead in order to avoid the degradation and costs observed today in the developed countries of the north. These will be constrained to correct their consumption and production practices and consent to costly repairs for restoring lost quality.

The strategic importance of environmental, regional and worldwide public commons (the sea, climate, biodiversity) and the situational difference between the two shores call for the implementation of the principle of both *common and differentiated responsibility*. Responsibility is *common* because the degradation of these common goods, no matter where it happens, affects all countries. It is *differentiated* because the most-developed countries are generally the main polluters (for example the emission of greenhouse gases). However, these countries have the resources and technology to reduce pollution. It is in their interest to invest in the developing countries where the margins for possible progress at lower costs are much greater. Applying this principle will enable a large, rapid and least-cost collective gain. Moreover, by strengthening support for the developing countries, transitions towards sustainable development can be accelerated. The clean development mechanism set out in the Kyoto Protocol and the Global Environment Facility are a first manifestation. The alternative scenario assumes a general application of this principle in several sectors.

To unblock the deadlocks of current trends, the alternative scenario focuses on sustainable cities and towns, revitalized inland areas, protection and enhancement of natural and cultural heritage, and changes in behaviour. The objectives for curbing trends have been specified in the six thematic chapters in Part 2 devoted to water, energy, transport issues, and urban, rural and coastal areas (the cross-sectoral issue of tourism is considered in the last two chapters). The orientations proposed can be grouped around *four major objectives*, which are both complementary and interdependent:

- protecting the natural and cultural heritage and preventing natural risks;
- decoupling economic growth from pressures on the environment;
- reducing internal disparities and promoting in-depth development of countries;
- creating added value by enhancing specific Mediterranean assets, in particular the wealth and diversity of its heritage, and exploiting synergies between activities.

Protecting the natural and cultural heritage and preventing risks

No development can be sustainable if irreversible processes that affect the fundamental common good are not discontinued. To succeed on an alternative development path in the region, priority issues include the protection of the sea, climate, soils, biodiversity and the cultural and landscape heritage, and reducing natural risk-related vulnerability.

Protecting the sea and coastal zones

The coastal zone, terrestrial as well as maritime, is the most valuable, most coveted and most threatened of all Mediterranean areas. Its social (freely accessible public spaces), cultural and economic values are considerable. It is a unique heritage: original highly valued ecosystems with multiple functions (water purification, biodiversity, a basis for fisheries and aquaculture), sites, monuments and landscapes of exceptional quality. What is particularly serious is its degradation by pollution and artificial land cover. B*alanced development* of coastal zones and *ending their continuing degradation* are major objectives of the alternative scenario.

However, sustainable coastal zone management is made difficult by the large number of actors, sectoral policies and short-term economic interests involved, resulting in situations that are often incompatible with public interest. The absence of political and administrative institutions in charge of coastal zones does not facilitate appropriate handling of the problems. However, several responses have been initiated, which show that progress is possible.

The Barcelona Convention with its protocols and the Mediterranean Action Plan (MAP) have been strengthened, and strategies and action plans have been or are being adopted (combating land-based pollution, preventing maritime pollution, the biodiversity programme and the 100 Historic Sites Programme). One condition for their success will reside in the ability to mobilize actors and donors at the right level so that a widening disparity between present and future member countries of the EU (who benefit from directives and cohesion funds) and the other Mediterranean-rim countries can be avoided. Synergies between MAP, the Euro-Mediterranean Partnership (EIB and MEDA) and international funds for the environment need to be strengthened in order to help the developing Mediterranean countries to upgrade their capacities (monitoring systems) and environmental protection infrastructures (pollution treatment and fitting ports with waste-reception equipment).

Although the proportion of protected coastal areas is growing (as seen recently in Egypt), it remains small. Other tools (coastal zone laws and agencies, economic instruments and integrated management programmes) have been initiated in some countries, resulting in noticeable progress. In France, for example, implementing a coastal area law and a coastal conservation centre, and mobilizing public opinion, land-owners and associations, some political decision-makers and judges have made it possible to limit urban sprawl along the coasts, push back second-line urban pressures, and protect and enhance a number of exceptional natural and cultural locations. Regional strategies and approaches for integrated coastal area management have been successfully initiated in the very de-centralized countries such as Italy and Spain. Local authorities, companies and NGOs have initiated exemplary projects in all countries.

However, this progress remains insufficient on the scale of the entire Mediterranean perimeter. To ensure sustainable management of coastal zones, new, more encompassing and ambitious initiatives need to be designed and implemented without delay, keeping in mind that for some regions it is more a matter of restoring artificial and already very degraded coastal land, whereas elsewhere the issue is to integrate the fragility of the natural environment into development choices. One goal suggested in Chapter 6 is to achieve a sustainable conservation and management of additional *4000km of coastline* between now and 2025.

The implementation of a coastal areas protocol under the Barcelona Convention would go a long way in helping countries provide themselves with the tools for protection and integrated management. It would constitute a legal innovation at the international level.

The sustainable management of coastal areas would also benefit from strengthening other action frameworks that contribute to this objective, such as the RAMSAR Convention on wetlands, the General Fisheries Council for the Mediterranean (GFCM), and through improved synergy with MAP. The alternative scenario assumes in particular that the objective of protecting wetlands is taken into account in national water policies (quantification of water demands for ecosystems), as well as a development of fisheries that is based on an ecosystem approach. Fisheries regulations could become a powerful tool for protecting and managing the seaward aspect of coastal zones in a sustainable way. To achieve this, a much more resolute collective commitment within the GFCM will be necessary by the Mediterranean-rim countries and the EU.

Conserving water, soil, terrestrial biodiversity and landscapes

A major component of the alternative scenario concerns strengthening awareness and the effectiveness of policies to protect agricultural land and conserve water, soil, forests, biodiversity and rural landscapes. Such an

Summary and Call for Action

alternative development path will have to be typified by progress in three complementary directions.

The first priority is to significantly reduce *desertification* in the mountains and arid plateaux of the southern and eastern developing countries. Very rare examples show the possibilities and effectiveness of integrated and participatory processes for sustainable management of pastureland and forests, especially in Morocco (see Chapter 5). Cultivation techniques are also being implemented to reduce soil erosion. These practices should become more widespread.

The second priority is to increase the relative percentage of protected areas (currently 7 per cent) and above all to make the role they play more important by favouring formulae such as *natural regional parks* or *biosphere reserves*, which favour the involvement of local communities and make them benefit from the protection and enhancement of their natural and cultural heritage. Several examples in the developed countries of the northern rim demonstrate the effectiveness of this approach. Achieving success requires resources for action in local development. A rapid development of these formulae in the entire Mediterranean perimeter is especially desirable, particularly in mountainous areas.

The third priority should result in reducing of at least one-third of the projected loss of *suburban agricultural land* between now and 2025.

The implementation of sustainable rural development policies and management plans through law enforcement would enable progress in these directions. A more resolute application of conventions on desertification and biodiversity should be encouraged, as well as development along these lines of policies for agriculture, forests, water and soil conservation and nature protection. Relaunching regional cooperation mechanisms (especially the FAO *Silva Mediterranea* network between forestry administrations) could contribute to progress in this area.

Reducing natural risks

The alternative development path calls on countries to resolutely commit to *preventive measures* to reduce an increasing vulnerability to natural risks and the resulting economic and human costs that have become unacceptable. Such measures would complement the very costly corrective responses that are currently being developed. In particular, what is needed is to develop instruments aimed at effectively banning construction in certain high-risk zones (flooding, mudslides and fires), promoting well adapted (present and future) risk-resistant construction and avoid excessive artificial land cover along shorelines and in upstream catchment areas. To reduce the risk of forest fires, a preventive approach would result in improved forest and agricultural policies in sensitive areas (reintroducing forest grazing, planting of more fire-resistant trees such as olive and almond trees, and agricultural fields). A regional structure for cooperation and risk-sharing could be implemented with the support of local authorities.

The alternative development path also requires a real mobilization of countries to contribute to mitigating global warming, through an exemplary implementation of the *Kyoto Protocol* mechanisms (see below). And, finally, a better assessment of the possible risks of the development of genetically modified organisms (GMOs) for the eco-region should be undertaken.

Decoupling economic growth from pressures on the environment

One of the major components of the alternative scenario is the decoupling of economic growth from pressures on the environment (pollution and resource consumption). It aims to reverse the trend that makes economic growth dependent on a parallel, if not stronger, exploitation of the environment, leading to an ever greater irreversible degradation of natural resources and ecosystems. By creating a high added value per unit of natural resource consumed, both a *reduction of negative externalities of the economy* on the environment (levies on resources and pollution), which are in fact paid for by all of society (especially its poorest members), and *large financial savings* are possible. It is also a way to attain the objective of protecting environmental common goods, since it contributes to ensuring the sustainable management of renewable resources, delaying the exhaustion of non-renewable resources, and reducing pollution.

Decoupling would enable the region to commit to a double *win–win development* (the environment and development, north and south).

Recent progress made in the *industrial sector*, through the adoption of cleaner production technologies, demonstrates the possibilities and the economic and environmental benefits of decoupling, both for the business community and for society as a whole (several examples are given in Chapter 6). The challenge for the next 20 years is to accelerate this development (it is a condition for successfully carrying out the MAP programme of reduction of land-based pollution), and, above all, to succeed in expanding it to other sectors of the economy, such as energy, construction, transport and agriculture.

Such an alternative development will not occur by itself. It calls for a break with the current, mostly very demanding, economic, financial and social behaviour that characterizes the region. The accumulation of such behaviour has resulted in enormous wastage and degradation of valuable resources (water, energy, soil, coastlines). Actors should be made more accountable by avoiding waste-encouraging services and subsidies. Strategies and policies need to give an essential place to *demand management* as well as to developing *new services* with lower environmental impacts than traditional ones (public transport, renewable energy, desalinization of sea water). These strategies rely on technological innovations and gradual internalization of the external costs of environmental degradation caused during production or operations.

The required acceleration would be facilitated by studying win–win approaches between countries of the northern and the southern and eastern shores. Investors, who still too often finance infrastructure projects that contribute to wastage, have an important role to play in encouraging the implementation of measures and projects that would achieve a successful decoupling.

Decoupling would also be strengthened by re-orienting the Mediterranean economy, which remains too dependent on, and a consumer of, natural resources (for example agriculture/water, tourism/coastal zones, and residential economy/agricultural land) towards new industrial specializations. New technologies and the information economy, which enable a certain dematerialization of the economy, provide new opportunities that are favourable to sustainable development.

Enhancing water supply by managing demand

It is estimated that a quarter of the current water demand could be saved in the Mediterranean Basin without jeopardizing user needs. The largest potential for savings in terms of volume comes from *irrigated agriculture*: this accounts for nearly two-thirds of the total water-saving potential identified in the region (modernizing equipment, organizing and operating irrigation zones). U*rban drinking water supply* accounts for about 15 per cent of the potential savings (with leaks exceeding 50 per cent in some cities). Given its much higher cost, such savings represent a considerable financial challenge. With the dissemination of cleaner technologies the industrial sector also provides water-saving possibilities. All together, it is assumed under the alternative scenario that by 2025 nearly 54km^3 (from a projected total demand of 210km^3) would be saved, thus stabilizing water demand on a Basin-wide scale at its 2000 level.

This orientation would make a large part of the many constructions planned for water supply redundant (dams, networks, transfers, desalinization plants), and their costs, environmental impacts and energy consumption for producing water would be avoided. The potential financial savings have been estimated at about 10 thousand million euros per year over the next 25 years, which represents 30 times the amount of official development assistance currently received per year by the SEMC for this sector. The expenses required for improving management will be markedly lower.

Compared with the conventional approach that focuses on supply, these win–win solutions limit environmental impacts, risks of conflicts, and cost of access to water, and make way for the potential economic growth and stability in the region. For the NMC, where demand is stabilizing, the challenges relate mainly to controlling water-supply costs, which are increasingly passed on to the user, and environmental impacts. Less well-endowed in water and financial resources and faced by an increasing demand, the stakes for the SEMC are considerable and coupled with social and geopolitical challenges. Social, because through the financial savings and the preservation of water quality, it would become possible to approach the Millennium Development Goal and the Johannesburg action plan: halving the number of people without access to clean water or sanitation between now and 2015. Geopolitically, because the dependence of countries would be reduced without new water transfers or recourse to desalinization (which usually requires imported energy) or imports of virtual water via food imports. Some SEMC have understood this and are far advanced along these paths (for example the confirmed policy of demand-control in Tunisia and Israel).

Desalinization is expected to progress in the Mediterranean with technological advances, a fall in costs (already less than US$0.5 per m^3) and determined policies as recently implemented in several countries (Israel, Algeria and Spain). Quantities are relative though: this resource represented 0.2 per cent of the total water used in the entire region in 2000 and could rise to 0.7 per cent by 2025 (compared with 8.1 per cent for re-used water). Although locally essential for meeting urban and industrial demand (or agriculture) in areas with very few water resources, this solution would only be *complementary* to demand-management policies.

Energy efficiency: A strategic priority

Energy poses a similar problem to that of water. It is a crucial resource for development but a major part of the energy resources (at least 20 per cent) is in fact wasted or poorly managed, whereas it could be saved without changing responses to user needs (tackling low efficiency in energy production and distribution systems, energy waste by users).

If the important increase in oil prices seen in recent years, which has not been integrated in the baseline scenario, were confirmed on a longer term, it will accentuate the need for energy saving policies and renewable energy development.

It is, moreover, not possible to envisage that the rising trends between 1990 and 2003 in greenhouse gas emissions by some northern-rim Mediterranean countries (Spain, Greece) will continue, or that the developing and transition countries will continue to be exempted from commitments to the Convention on

Summary and Call for Action

Climate Change. It is high time to start preparing for post-2012, which can only entail obligations for each country in a spirit of common and differentiated responsibility.

The alternative scenario favours the win–win search for a *more efficient use of energy* combined with more intensive use of the potential for *renewable energy* (solar, wind), still insufficiently explored. On such an alternative development path it would be possible to reduce dependence on energy supplies and improve the eco-efficiency of supply, while creating jobs in the sectors that need to be developed. The use of cleaner technologies in the industrial sector, especially in the *household sector* (building techniques, household appliance standards), represent opportunities for large energy savings, especially in the SEMC with population, urban and industrial growth in full swing. The *transport sector* has considerable savings potential, but these will be harder to realize.

By following these orientations, nearly 208Mtoe (out of 1365Mtoe) could be saved by 2025 compared with the baseline scenario, or nearly half of the growth in demand projected for the region between now and 2025. Along with these savings, nearly 860Mt of CO_2 emissions could be avoided,[6] as well as the risks of pollution related to hydrocarbon transport. The number of facilities required for energy production and distribution would be reduced (nearly 150 power plants avoided between now and 2025) as well as their environmental impacts and their costs.

A rough estimate of the financial savings at stake by such orientations is about 18 *thousand million euros per year* over the next 25 years on a Mediterranean-wide scale, or 18 times the amount of official development assistance currently received per year by the SEMC in this sector.

In the perspective of the inevitable 'post-oil' phase that the 21st century will experience, technological advance is also an issue, as are millions of jobs that could be created in the fields related to this new knowledge. For the NMC, the premier strategic challenge is to reduce energy dependence and global responsibility for climate change. For the SEMC the social challenge is added to this: access to energy for a greater number of people, facilitated by the financial savings that would result.

Transport: Breaking the current vicious circle

The alternative scenario is based on two priority objectives: developing sustainable transport and promoting more efficient modes of transport, both in the socio-economic and environmental fields.

As with the water and energy sectors, *organizing sustainable transport* implies satisfying transport demands while reducing wastage or superfluous travel. One goal could be a rate of decoupling of economic and transport growth twice that in the baseline scenario[7] (3.3 per cent average annual growth of freight traffic compared with 3.8 per cent in the baseline scenario and 2.5 per cent compared with 2.8 per cent for passenger traffic). Dematerializing economies can play a role in this decoupling, with new technologies offering new opportunities. Spatial planning and urban planning (densifying the urban fabric to reduce distances between home and work, choice of distribution patterns, industrial development patterns) as well as tax schemes would contribute to such faster decoupling. The goals would depend on reorganizing company logistics (for example regrouping distribution points and practices for large stock management), loading of vehicles (particularly higher occupancy rates) and promoting non-motorized modes.

Promoting *more efficient transport* means using public and semi-public transport (the shared taxis that exist in many Mediterranean countries are a solution that deserves greater support) and abandoning the 'all-road' model, which is especially costly at the environmental level compared with other means of transport (gaseous emissions, noise) and at the social level (accidents, high costs of travel, congestion). Opting as much as possible for rail or maritime transport would benefit everyone (both are less polluting and less risky than road travel). The potential exists (especially in the west) for maintaining railroads for combined rail/road/sea transport or for coastal navigation. The rail tradition in the Balkans and Egypt are strong points worth enhancing.

This could translate into a modal shift of about 20 per cent between road and rail (the trans-Pyrenees scenario by 2020) or 5 per cent between road and sea (the Marseilles–Barcelona seaside motorway by 2010, for example).

The construction of infrastructures needs to be rethought by taking their potentially negative impacts into account (urban sprawl, the urbanization and destruction of coastal zones and the degradation of landscapes and biodiversity). It is important to build roads some distance from the coast whenever possible. Unfortunately, in some countries, roads are being built along the shoreline itself, sometimes with support from major donors. Transport safety is also a priority to limit the risks of pollution and accidents, which remain very high compared with the mileage travelled. The alternative development path would also see an immediate transfer to the SEMC (with support from the EU) of technological advances recorded for cleaner vehicles and new pollution standards, including diesel engines equipped with particulate filters, the Euro 4 standard for trucks made operational starting in 2006.

Increased economic effectiveness of transport links between the two shores would make it possible to enhance the advantage of their geographic proximity to

Europe in globalization and to develop domestic markets in the SEMC (cost reductions, inter-modal transfers, services to all regions).

However, these shifts will not be realized without a strong input from the EU, the only one that, in agreement with countries, can regulate regional transport, which is largely predetermined by globalization. Seeking fiscal instruments for sustainable transport and putting inter-modal facilities in place that incorporate the objective of sustainability will require Euro-Mediterranean concerted action.

Such shifts are simulated in Chapter 3 (20 per cent of traffic by rail, a faster decoupling of economic/transport growth than in the baseline scenario, progress in enforcing regulations in maritime accords). They would make it possible to avoid many socio-economic costs by 2025: reduced road, rail and air traffic by between 8 and 10 per cent compared with the baseline scenario; 9 million Mediterranean dwellers spared from road noise; 15,000 road deaths avoided; 41 thousand million euros per year of congestion costs saved; stabilized household expenses on transport (18 per cent of total expenditure compared with 25 per cent). Many negative environmental impacts could be avoided by 2025: 90,000 tonnes less VOCs, 185,000 tonnes less NO_x, 191,000 tonnes less CO_2 and 2.6 million tonnes less polluting waste into the sea.

Cities: *Prime candidates for decoupling*

Cities are prime candidates for promoting decoupling through lifestyles that are more 'eco-efficient', less wasteful of space, energy and water, and better equipped to reduce pollution at source.

Potential savings are particularly large in the *energy sector* through construction standards for housing, public buildings and street lighting.

Policies for reducing *domestic waste* at source and recycling could, by 2025, result in savings of more than 145 million tonnes of waste per year throughout the Mediterranean Basin (75 million tonnes per year less than under the baseline scenario) with average annual financial savings of about US$3.8 thousand million.

In the alternative scenario, *upgrading* and increasing the density of the urban fabric would make it possible to reduce suburban sprawl and the excessive use of cars by half, improve the living environment and social cohesion and achieve even greater cumulative social, economic and environmental gains, but over a longer term.

This implies strong anticipation in the transition and developing countries in order to avoid the negative and largely irreversible developments of urban sprawl, which accompanied urban growth in the northern countries. Measures would be enhanced if they were put in place in the early stages of development, for example by protecting agricultural land or siting large shopping centres within the existing urban fabric rather than at the city periphery.

For both sides of the Mediterranean, such an alternative development path would be achieved mainly through *closely linked transport and urban planning*. In this scenario, public transport is developed in the outlying areas, including the outskirts and the suburban areas of agglomerations, while the density of the urban fabric is increased, particularly along public transport lines. The objectives of protecting agricultural land and reducing vulnerability to natural risks are also taken into account in planning. Some cities (especially in Italy and Spain) started to implement such orientations as early as 2000 and thus demonstrate that such a scenario is certainly possible.

Agriculture: *Better water management and soil conservation*

In the alternative scenario agriculture is decoupled from desertification (the degradation of soil and vegetation cover) through new techniques that avoid soil erosion, by better management of irrigated lands to avoid salinization, and through sustainable and participatory management of pastureland and fragile forests.

The alternative scenario also includes a better valuation of *water resources*. The route has already been laid by remarkable examples like the Tunisian strategy for water savings in irrigated agriculture. In Tunisia, success has been ensured by implementing a range of appropriate tools (irrigation associations, targeted subsidies and increased prices) and by a gradual and continuous mobilization of actors that makes them more accountable over time. Putting incentives in place (subsidies for water-saving irrigation equipment amounting to as much as 60 per cent of the cost), and associated improvements in pricing and agricultural performance have been decisive. But no progress is possible in the agricultural world without improvement in incomes. This example shows the need to give an inter-ministerial dimension to water policies. Sustainable management objectives need to be included in each sectoral policy, defined at the national or catchment-area level. The Tunisian example also shows the importance of *subsidies* for successful environmental policies. In the alternative scenario this type of policy would expand rapidly in countries that are not yet very involved in them. The scenario also assumes much progress in *rainwater management*, a sector all too often forgotten although of interest to a majority of farmers.

Shifting to an efficient agriculture would also make it possible to rationalize, and thus reduce, the use of mineral fertilizers and pesticides. The alternative scenario assumes a growth in consumption only half that projected in the baseline scenario. This is often in the interest of farmers, and awareness-raising or cancelling subsidies for fertilizers could be enough for progress to be made. An increasing demand for organic products would also contribute to less fertilizer and pesticide consumption. There are possibilities for Mediterranean agriculture in this sector, as shown by the Italian example (world leaders) and some recent developments in the south and east. The overall goal for the Mediterranean could be 15 per cent of arable land cultivated under organic agriculture by 2025.

The alternative scenario also envisages a coupling between agriculture and sustainable development, recognizing the positive externalities of agriculture (creating amenities) and consequently the role of *farmers as agents of sustainable development*. This recognition would make it possible to reduce vulnerability to forest fires, maintain *green zones* between urban areas, contribute to rural revival, expand sustainable tourism by safeguarding jobs and quality landscapes, and promote healthy diets.

Tourism and the environment: A necessary reconciliation

Tourism plays an important role in the strategy of decoupling economic growth from pressures on the environment in the Mediterranean. It is absolutely in the *interest* of (or *necessary* for) the sector to stop contributing to environmental degradation and work on its improvement (one of the conditions for its development). Tourism could be an *innovative field*, capable of driving collective progress at destinations, as shown by many examples in the region. In the alternative scenario it is therefore assumed that significant progress will be made in decoupling the sector in several directions:

- implementing programmes to *upgrade* mature and degraded destinations;
- *preventive action* in growing and potential destinations to avoid irreversible degradation of coastal zones as observed at mature destinations. This requires building roads and housing at a distance from coasts, promoting local transport with low environmental impacts (hiking, cycling, buses using natural gas or electricity), accommodating commercial ports for yachting rather than building completely new infrastructures, and developing business tourism rather than building private holiday homes so avoiding excessive consumption of space and land seizure from local populations;
- *minimizing consumption*, emissions and pollution (for example management of water and energy demand and waste). This can be achieved through the promotion of *new technologies* (solar, wind, water management, bio-climatic construction and the internalization of environmental externalities into prices;
- *regulations* for land use to reconcile environment and development and to avoid degradation of coastal zones through unsuitable leisure activities;
- *extending the length of visits and spreading visits throughout the year* to reduce pollution from transport, create more permanent jobs and increase local incomes.

To make such an alternative development path come true, tourism-specific economic and regulatory tools will have to be developed, especially the definition of maximum *accommodation* capacities at destinations and the development of quality charters and labels. An alternative development also assumes the conditionality of public aid and better contribution from tourists to maintain the Mediterranean environment (applying the user-pays principle). Formulas with the *visitor pay back principle*, which would draw lessons from the recent failed experiences in the Balearic Islands, could be developed and applied successfully. For example, duties imposed on boat transport towards islands and at airports could be conceived and implemented. These formulas have several advantages compared with taxes paid only by professional accommodation providers: they encourage longer stays, so reducing pollution from transport, allow significant resources for developing destinations in a sustainable way, and do not discriminate between categories of tourists and accommodation providers.

Successful decoupling also requires a greater contribution from tourism towards enhancing Mediterranean diversity and in-depth development of countries. In this scenario, strategies no longer aim at purely quantitative goals (indicators of the number of beds or tourist flows) but at sustainable development goals (indicators of quality and added value created for local populations and countries).

Reducing internal imbalances and enhancing the Mediterranean heritage

In the alternative scenario, domestic disparities are reduced, social cohesion is maintained and the strong points and specificities of the region in a globalizing world are enhanced.

Reducing domestic disparities and relieving coastal zones

Reducing domestic disparities relies on rural and urban development to alleviate poverty, desertification and social segregation, and relieving coastal zones to reduce the currently proliferating pressures and rebalance tourist flows.

The alternative scenario first assumes that *fragile rural areas* in the developing and transition countries are upgraded.

The issues are social, economic and environmental. At the social level, the goal defined for this alternative development path is to succeed between now and 2025 in reducing most of the disparities currently observed between cities and rural areas. It is assumed that rural illiteracy will be wiped out, and that rates of access to basic services (such as water, roads, electricity, education and health) will be generally improved, approaching the averages in urban areas. At the economic level, a rural revival is assumed to start by improving agricultural productivity (mainly in rain-fed agriculture, too long ignored in agricultural developments despite its major importance) and diversifying activities (for example tourism, handicrafts, industry and services). This will contribute to a growth of domestic markets and national economies, which would, otherwise, remain strongly impeded by keeping a large part of the population marginalized. Maintaining a large rural population also gives the SEMC the possibility to revive rural areas rather than abandon them, so avoiding the developments currently observed in northern countries where revival was started too late to prevent devitalization and abandonment of many mountainous areas. And finally, at the environmental level, it is assumed that combating desertification will be possible through rural development.

Such a scenario implies shifting from what is still a top-down administered development to an accountable development in which adapted rural policies are put in place. This requires basic public services to be upgraded (roads, drinking water and sanitation, electricity, education and health). This could be achieved by structuring rural centres (villages and small towns), stimulating and supporting the involvement of local and professional actors in integrated local development processes, and diversifying the economy. Chapter 5 discussed the importance and difficulties of such policies. Successful examples stress that one of the main conditions is encouraging local development and training. NGOs can play an important part in this.

The objective of reducing domestic disparities also concerns *cities*. The alternative scenario assumes specific preventive efforts (to avoid an explosion of informal residential neighbourhoods where possible, the re-absorption of which is always difficult and expensive) and upgrading actions (improving the living environment in existing poor neighbourhoods). This means ending, as far as possible, the current trends of segregation and exclusion, and making cities 'all-inclusive'. Chapter 4 highlighted the need for more specific support to small and medium-sized cities, which in many countries still lack the capacities for coping with the challenges facing them.

To enable this balanced and in-depth development, the alternative scenario assumes that *coastal zones* will be relieved in favour of inland areas. Chapter 6 assumes a *transfer to inland areas of a third of the additional tourist flows* projected for the coastal regions up to 2025. This transfer would make it possible to reduce the current pressures on coastal zones and coastal agricultural plains and assist in dynamic local development in rural and urban areas. It will require considerable diversification in tourism and policies for territorial management (for example on infrastructures and the development of inland areas).

Enhancing the strong points of the Mediterranean region

Developing the exceptional wealth and diversity of the Mediterranean heritage can be an essential contribution to the economic renewal of the region.

With its bio-geographical, cultural and historic characteristics, the Mediterranean has great potential for another kind of development. The sea, its common element, has always provided many benefits, especially through the trade it facilitates and the exceptional attractiveness of its coastlines. The convergence of geography and history have endowed the region with an exceptionally *rich heritage*: species diversity, unique plant life communities, agricultural products, diet, the beauty and variety of its landscapes, the architectural quality of its historic cities, its incomparable archaeological remains, the diversity of its villages and the wealth of its rural traditions. This cultural heritage has become richer through the contributions of its varied civilizations. Its climate, its landscapes and its lifestyles still make it one of the most pleasant places on earth to live.

Many examples illustrate the need for a heritage approach. Entire regions on the northern shore have managed to emerge from serious economic crises through new strategies based on enhancing heritage and local specificities and adapting to new markets. The wine-cultivating regions are a good example, where renewal has enabled companies and regions to engage in quality policies. Many other examples exist of urban and mountain renewal based on heritage strategies. Some regions have used a high quality of living as a basis for developing technology centres. The northern countries have not nearly exploited all their possibilities; those in the south and east would gain much it they too were to engage in the development of their rich potential.

The challenges are important for cities and for agriculture and rural areas. Under the alternative scenario, which could be called the 'Tuscan development' (after the controlled development of the Tuscany region of Italy), the garden vocation of the Mediterranean would be confirmed in its cities, coastal zones, countrysides and mountains in response to new market demands. The scenario assumes the development of new synergies as sources of added value between tourism, agriculture, industry and handicrafts, which would make it possible to encourage a more qualitative rather than a quantitative development.

Summary and Call for Action

This scenario assumes a *reorientation of agricultural, tourism and food policies* based on local certifications of quality, diversification of agriculture and agro-industries towards unique local products, promoting the Mediterranean diet and cultural, rural, agri-, and eco-tourism as well as tourism around conferences. The benefits would be considerable with a changing Mediterranean tourism image through forms of tourism that are complementary to those of coastal resorts, increased incomes from agriculture and tourism for local people, the emergence of new generations of local entrepreneurs who contribute to safeguarding and restoring the cultural and natural heritage, improved satisfaction of domestic and international tourist demand, recognition of the economic value of high-quality landscapes and Mediterranean products, a shift towards healthy diets, and, finally, decreased pressures on the environment. All these aspects would put the Mediterranean in an advantageous position on a world scale.

The alternative scenario in particular assumes a widespread adoption of heritage strategies in the various specific territories. Concerted projects of *urban renewal* involving local people, associations and companies and relying on the heritage identity of Mediterranean cities would increase, like the many experiences already underway (from Tunis to Aleppo, Naples and Genoa to Barcelona and Valencia). In rural areas a reorientation would highlight the unique village lands (*terroir* in French) and natural sites by exploiting all the diversity and typical local heritage. These strategies could result in significant development of territorial dynamics in special project areas. B*iosphere Reserves* and N*atural Regional Parks* could be promoted in natural areas by providing proper technical and financial tools and resources (investment budgets, training programmes and subsidies for projects by local actors).

Striving for more sustainable development

Moving towards the alternative scenario that has just been described will require important changes. It will benefit from *technological advances*, which will certainly increase in the next 20 years, but it will not be possible without basic progress in *governance*, both in individual countries and at the level of regional cooperation.

The 1989 Blue Plan scenarios called for (1) the implementation of national policies that would integrate the environment with development and (2) policies of north–south and south–south regional cooperation. Although all countries have subsequently set up environmental strategies, policies and plans and, increasingly, sustainable development policies, and although an ambitious Euro-Mediterranean Partnership was inaugurated in 1995 between the EU and 12 partners in the southern and eastern Mediterranean, this progress has not yet made it possible to reverse, even marginally, the strong current trends.

There is, therefore, a need to examine the *constraints* that explain this lack of success and the conditions that need to be created to overcome them. The analyses in the previous chapters highlight a number of weaknesses and paths to progress.

The main weakness is most certainly the *insufficient mobilization and accountability of the various professional and local actors* for sustainable development. This elicits questions about the 'governance' of sustainable development in the countries. The issue of governance calls, in particular, for new forms of dialogue and *participation*, expressed by more fluid relationships between governments and public institutions, and the private sector and civil society. This opening up should be manifest both in the early stages, through transparency in deliberation and decision making, and further down the line in the effective implementation of policies. Even in the countries where the decision-making process has really been opened up, much remains to be done to improve participation and public debate, for example by increasing opportunities for consultation, diversifying forms of consultation, strengthening pluralistic expertise and through the framing of procedures. It should also be stressed that participation is only one side of the coin, the other being *responsibility*, which calls for defining common rules and commitments that each side should adhere to.

The alternative scenario also assumes significant progress in *regional cooperation*, which is currently insufficient and has not yet really integrated the principles of sustainable development.

Obstacles to be overcome

A slack economy that lacks innovation

The region suffers from an *entrepreneurial deficit* and a *lack of innovation*. This weakness owes much to the slack and 'mining' nature of the Mediterranean economy: seeking maximum income sources (land/real estate, commercial monopolies) rather than added value from directly productive activities; exporting manpower and executives (brain drain); and excessive polarization of the economy by exploiting natural resources (agriculture/water, tourism/coastal areas, energy/hydrocarbons).

The *financial sector* in many developing countries is dominated by a banking system that has a near-monopoly of savings and has remained fairly traditional despite slight progress made since the 1990s. This is expressed in high real interest rates, poor access to financing for small and medium size enterprises, little encouragement for

3 A Sustainable Future for the Mediterranean

innovation and difficulties in financing sustainable development (for example energy savings that often require an immediate overcharge for a deferred benefit and innovative relay devices). Expenditure on *research and development* is particularly feeble, and *training institutions* are not sufficiently committed to creating enterprises and local development.

Declining and undervalued links between society and the environment

The specificities of the Mediterranean environment and culture have, over the centuries, favoured the development of extensive well-adapted know-how. The past few decades, however, have seen a clear reduction in this know-how. The decline has been especially marked in regions that are undergoing various rapid changes, including rural-to-urban migration and urbanization, accelerated development of standardized tourism and a residential economy, and conflicts and population displacements. There are many indicators of this decline, including buildings that are poorly adapted to the climate and risks; abandonment of town planning practices that had created the beauty of Mediterranean cities; a regression of collective rules for sustainable management of land and water, soil, grazing-land and forest resources; and a retreat from the Mediterranean diet just when its qualities are being recognized and promoted elsewhere in the world.

The move of economic and technological centres towards the northern hemisphere has contributed much to this. For several centuries, the models for consumption, development and administrative organization have been designed in the north. For the past few decades the Mediterranean has copied these exogenous models, which are poorly adapted to the Mediterranean's specificities. At the same time, insufficient benefit has been derived from its particularities to carry out research and development that is adapted to those characteristics, or to assume a significant place in worldwide innovation.

Poorly adapted systems of public administration

Despite their economic, environmental and social advantages in the mid-term, priority actions identified in the alternative scenario are often difficult to finance in the real world of either the NMC or the SEMC. One of the reasons is that the benefits are not always monetary, they are deferred to the future or do not benefit the investor.

The room for manoeuvre for *governments* in budgetary terms is limited, even in the more developed countries of the northern shore. Many obstacles can paralyse actions by governments: efforts to try to satisfy demands from diverse categories; a natural tendency to add levels, structures and services without ever removing any; reducing expenses in all sectors in proportion to the budget without a real budgetary choice; the permanence in some parts of the Mediterranean of conflict situations; and the difficulty some governments have with relinquishing tasks that can be addressed by the private sector. All these obstacles limit the abilities of governments to play their fundamental role of being 'strategists' for sustainable development.

Budgetary problems are particularly troublesome in the SEMC, where countries will face enormous requirements in the coming decades for their social and economic development and environmental protection with only limited public financial resources. In particular, the very heavy weight of sovereignty expenses and debt servicing are worth recalling. Tensions and conflicts in the region, moreover, explain the large proportion of national budgets devoted to security and defence (between 3 per cent and 7 per cent of GDP compared with a world average of 2.3 per cent). The weight of public sector salaries is also a major element of expenditure. On the other hand, capital expenditure (from 5–8 per cent of the GDP) and research and development expenditure (0.6 per cent in the SEMC compared with a world average of 2.3 per cent) are limited.

Another important obstacle is the *lack of integrated policies* and policy inconsistency. Economic policies have remained highly sectoral (tourism, transport, agriculture, etc.) and in general continue to favour traditional supply approaches (major irrigation, roads, seaside resorts, hydrocarbons), with little integration of environmental issues. Environmental policies basically remain top-down, command-and-control and regulatory; implementation is often hard and beleaguered by a lack of resources (e.g. number of inspectors) or by effective and dissuasive sanctions. Initiated by new administrations with weak authority and inter-ministerial power, policies have little encouraged local mobilization and have not succeeded in promoting the upfront integration of environmental concerns into consumer and production patterns or into economic and sectoral policies. The incentive, taxation and regulatory mechanisms for sustainable development are insufficient, which encourages overexploitation of natural resources, especially water, soils and the coastal zone.

Local authorities in most developing countries in the region do not have all the necessary capacities to act effectively for sustainable development. In the southern and eastern Mediterranean, the share of expenditure by local authorities compared with total state expenditure is one of the lowest in the world. In rural areas and coastal zones, many participatory and undeniably successful projects have only been able to operate through outside support and ad hoc facilities placed outside the usual administration systems (for example the GEFRIF project in Morocco). In many countries protected areas do not have the resources to act as a laboratory for development.

The more developed *decentralized systems* that exist in several EU-Med countries do not necessarily address long-term issues. All too often, local authorities have developed strategies driven by short-term financial and electoral gains (land tenure opportunism and vote-catching), which have only accelerated a wild urbanization of coastal zones, reduced productive activities and degraded the living environment and identity of specific territories. Decentralization policies have also led to unnecessarily multiplied levels of administrative decision-making procedures (particularly in France).

International and regional cooperation that is not up to the challenges

International and regional cooperation is not yet up to the challenges to be met. It has had the merit, however, of signalling, both at the international and regional level, the need for a change towards sustainable development. Nevertheless, the vision remains insufficiently shared and is seldom translated into strategies or action programmes. The gap between speech and reality remains considerable.

The major global actors are encountering difficulties developing a convergent approach for tackling the vital problems of the planet, whether these concern the loss of biodiversity, mitigating the greenhouse effect, reducing poverty or resolving conflicts in the Mediterranean region.

Donors and the major actors in regional and international cooperation, although increasingly making reference to international or regional commitments relating to sustainable development (for example those formulated in Rio, Kyoto, Monterrey, Johannesburg, the EU sustainable development strategy), are still having difficulties in changing their strategies and methods. For example, economic reforms undertaken at the advice of international financial institutions and the EU have mostly put the accent on reducing the role of states, trade liberalization (without lifting the barriers to the free circulation of individuals), removing subsidies, and privatization, rather than improving the performance and competitiveness of local and professional actors or facilitating better relationships with governments to engage in processes of sustainable development together.

An analysis of the official development assistance to the Mediterranean countries shows that the share devoted to the environment, urban and rural development, fisheries, agriculture and forestry, rail transport, human capital and democracy remains relatively limited: on average 19 per cent of the total over a period of 30 years.[8]

In this general context, the EU is a unique example of regional integration. It nevertheless still has difficulties in integrating sustainability into its actions.

Insufficient awareness and information

With increasing problems, awareness has improved significantly, so that sustainable development has become a popular subject in the media and in economic and political speeches. Socio-economic, environmental as well as ethical considerations are contributing to this and reminding us that our societies rely on value systems that are powerful determinants of future developments.

However, the concept of sustainable development has remained vague, largely due to a *poor understanding of the issues and the potential benefits* of sustainable development policies, especially those of decoupling. One of the weaknesses is derived from the reluctance of decision-makers to evoke and analyse certain failures of consumption patterns in the developed countries. As for the capability to develop and structure information that truly helps in decision making, mobilizing various actors and getting them to become accountable, it remains insufficient.

A first weakness is related to the *available data*. Of the 130 indicators for sustainable development adopted by the Mediterranean Commission on Sustainable Development and selected on the basis of criteria of relevance and availability, only 60 are regularly documented by countries and used by international institutions.[9] This reveals a paradoxical situation: there is a considerable amount of data available but of little use, while few but essential data are lacking. The many gaps concern in particular:

- There is very little information that is comparable over time and between countries on the state of the environment (for example water and air quality, biodiversity) and land use. In most developing and transition countries, laboratories, monitoring stations and appropriate methodologies are largely lacking. The environment is also a new field for statistical departments. They are, however, structuring and acquiring experience in this field, especially due to the strong impulse given since 1999 by the Euro-Mediterranean Medstat Environment Programme for the 12 partner countries of the EU.
- There is poor understanding of negative externalities (the costs of the impacts of degradation) and positive externalities (amenities) of economic activities on the environment. Heritage accounting, which takes non-commercial capital and degradation into account, is either non-existent or insufficient. The costs and effects of environmental degradation on health remain poorly measured. And, similarly, the real effects of tourism on local destinations are inadequately quantified.

- Specific territorial information is inadequate, even in the most developed countries. There is not enough information to follow up strategic questions about the environmental and socio-economic evolution of cities, coastal zones and various kinds of rural areas such as mountains. Follow-up on 'Mediterranean' issues should also call for detailed data on the scale of coastal regions. This is rarely the case because of the absence of an official and operational definition of these areas or because of different definitions in countries, which does not facilitate international comparisons.

A second weakness relates to a lack of *policy assessment* in the light of sustainable development (ex ante and ex post assessments of public policies, impact assessments of plans and programmes,[10] monitoring methods and periodic re-examinations). Adapted analytical tools, such as strategic environmental assessments of sectoral plans and programmes and cost–benefit analyses of different options, carried out beforehand, would make it possible to direct choices towards win–win strategies, but such tools are seldom applied. In any major sectoral programme (transport, energy, industry), planned measures result in positive and negative impacts in other fields that need to be understood if they are to be corrected. The present inconsistency results from a lack of integrating analytical tools, a reluctance to effectively open up deliberations (if not decisions) despite a growing recognition of the need for participation, and a lack of real political will to change the ways things are done.

Strengthened by the first two weaknesses, the third relates to the lack of capacity to structure and disseminate summarized information for sustainable development. Of course all countries, and many companies, are producing reports and strategies on the environment and increasingly on sustainable development. Some southern and eastern countries have managed to implement their first environmental and sustainable development observatories or equivalent agencies, which have produced worthwhile reports. But monitoring and assessment facilities generally remain weak. It often appears difficult to shift from a static or purely statistical approach in environmental observation towards a dynamic and prospective approach relying on a new generation of indicators and producing information more likely to be of interest to, reach and convince political and economic decision-makers.

Towards governance for sustainable development in the countries

This report identifies three major levers of progress in governance for sustainable development:

1. decompartmentalization of policies and integration of the environment in economic and sectoral policies;
2. promotion of specific territorial approaches of sustainable development;
3. an enabling framework with new tools and methodologies.

Decompartmentalizing sectoral policies and integrating the environment into economic policies

The alternative scenario assumes a strong integration of environmental concerns and sustainability into the various economic policies.

This integration has started in EU fisheries and agricultural policies. Indeed, after an initial phase of agri-environmental measures, the entire common agricultural policy is being reformed with the introduction of eco-conditionality principles in assistance and the launch of the second pillar relating to rural development.

Integration is also making progress in developing and transition countries. Examples are the reorientation of agricultural policy in Tunisia based on a water-saving strategy in the irrigation sector, and the success of Egypt that has become the eighth leading country in the world in using natural gas in vehicles (52,000 vehicles).[11]

Shifting developments in the region to those described in the alternative (decoupling) scenario requires a strong acceleration of integration efforts and an evolution towards *innovative and integrated supply and demand policies*. Challenges are especially important in the sectors of water (and thus agriculture), energy (thus housing) and transport. Particular attention should be paid to tourism policies given their importance in the region and the possible benefits of their reorientation.

The alternative scenario also assumes a reorientation of *environmental policies* in the following three directions:

1. First, environmental policies would gain much from a better integration of cultural, social and economic aspects, especially when their success depends on economic and social progress and on the know-how of local people, as is the case in rural areas. Such integration will also be important in policies on waste management and public transport, where developing countries would gain by avoiding European-like developments, which are often costly and do not create much employment.

Summary and Call for Action

2. Second, they would gain by acquiring a more strategic dimension. The struggle against climate change and the sustainable management of water resources require long-term visions and strategies that could be included in each sectoral policy.
3. Third, they will have to ensure risk prevention and protection of environmental common goods through powerful measures (protecting coastal zones, agricultural land and valuable ecosystems, construction bans in flood-prone areas). It is essential to ensure that the scarce resources which are protected by environmental policies are not sacrificed on the altar of sustainable development.

As for *land-use planning and regional development policies*, they should gain in importance because of the specificities of the Mediterranean area. This calls for both a Basin-wide planning vision and a reconsideration of planning in the light of sustainable development, by reorienting transport services to succeed in decoupling, revitalizing inland areas, protecting and sustainably managing coastal zones and agricultural land, and developing new participatory methods for territorial planning and development.

One path of progress towards integration is the *mutual strengthening of policies in pairs* (environment and health, transport and regional planning, energy and global warming, agriculture and sustainable water management, tourism and sustainable rural development), with specific strategies and programmes. This path can increase the effectiveness of policies much more than juxtaposing single-discipline analyses and sectoral policies, as remains too often the case in certain 'integrated' plans and some early strategies for sustainable development that have had no results.

Strengthening the territorial approach and sustainable local development

Questions of sustainable development first emerged in a context very far from local concerns: global risks and north–south relationships. Today several reasons argue for a special place for *local territories* in sustainable development strategies. The *territory* can be defined as a space of affiliation: a neighbourhood, city or town, an urban area, a unique village land or countryside, an island, a catchment area, a bay, a mountain valley. This specific territory can become a project area for local development or for management of a commonly-held asset. Hence, its limits can coincide (or not) with administrative or political boundaries.

The specific territory level is inescapable when engaging in *heritage development strategies*. Each territory has its own wealth of culture, nature, landscapes, products, know-how and skills, traditions and enterprises. A local project can be defined on the scale of a coastal area, a mountain range, a small region or an urban area. On such a scale, a project will be able to enhance and dynamically conserve this heritage, isolate added values thanks to synergies between activities, and organize the mobilization of actors for its implementation.

For many environmental and sustainable development issues, solutions can only be found at the appropriate territorial scale, requiring the cooperation of several administrative units and the participation of the people and enterprises concerned. Such is the case for cleaning up a polluted bay, sustainably managing a body of water or a catchment area, organizing public transport, managing waste in an urban area, preventing natural risks, combating desertification, and implementing integrated rural development or sustainable management of coastal fisheries.

Even the conservation of regional or international public commons requires local approaches. For example, reducing greenhouse gas emissions or cleaning up the Mediterranean Sea needs the mobilization of cities that can monitor emissions and set measurable progress goals. Protecting biodiversity will not be achieved without more local processes that can reconcile the environment and development.

The territorial scale corresponds with *very substantial realities*. It is the scale at which problems of sustainable development are better perceived and where fair solutions can be found originating from public debate. It is also the only level that can guarantee a minimum of integration, the issue at the heart of sustainable development. The need to find solutions to substantive problems is a positive incentive to de-compartmentalize institutional rationales and build bonds between the social, economic and ecological dimensions of sustainable development in a more democratic way. At the scale of cities or rural areas, sustainable development processes provide a real opportunity to bring socio-economic and ecological cultures that traditionally ignore each other closer together. Such local action probably has a greater chance of being effective than at the national or global level since responsibilities are easier to establish, the interdependence of actors easier to factor in, and the projects easier to formulate and implement.

Developing *participatory approaches for sustainable development* that mobilize actors can be easier on smaller scales, dealing with a limited area at a time. One of the main characteristics of local strategies for sustainable development, in particular Local Agendas 21 and charters for natural parks, is a reliance on different procedures of dialogue with interest groups, associations, enterprises or even the entire population. They also provide an occasion for strengthening democratic legitimacy.

The alternative scenario indeed assumes a *reorientation of Mediterranean development for the benefit of the local and territorial processes* of sustainable development, so as to move away from the current situation where much of the

A Sustainable Future for the Mediterranean

experience and committed action remains at a brainstorming or 'declaration' level. Structured initiatives over time will increase, such as Local Agenda 21 processes, strategic plans at the scale of urban areas or regions, urban travel plans, programmes for integrated coastal area management, regional natural parks, biosphere reserves, projects on integrated rural development and to combat desertification, etc.

This reorientation will require action at several territorial levels while implementing the subsidiarity principle, and should not automatically result in ever more administrative levels. On the contrary, flexibility needs to be maintained and actors brought together in *projects*, depending on the various terrirorial scales required. It can lead to the joint formulation of plans (master plans, management plans) making it possible to balance and blend protection and development, share activities in the area and set rules that the various actors in the area will respect over time in order to ensure the conservation of their common good.

Promoting such territory-specific approaches will raise the question of the evolving role and patterns of *state* intervention related to *decentralization* or *deconcentration*. It will be up to each country to find paths for progress adapted to its own context and specificities. No sustainable progress will be achieved without strengthening the capacities, responsibilities and resources of local authorities (fiscal decentralization) and without a greater focus in public policies on specific territories (in particular focusing on rural public policies for sustainable development and management of rural areas). Decentralizing official development assistance (by encouraging decentralized cooperation) would help the developing countries to follow this path.

Another problem to be solved relates to the contrast between short-term political mandates and long-term environmental issues. Decentralization does not guarantee commitment to sustainable development. One solution would be to organize *guidance by upper government levels* (the regional and national echelons) to encourage virtuous processes and penalize processes that do not comply with the principles of sustainable development. A good example is to be found in French regional natural parks: for each park a ten-year development plan has to be developed, which becomes the 'charter' of the park. This guarantees the commitment of the various actors and is supported by granting a quality label (the 'regional park' label) and financial resources that are formalized by contract.

This orientation also assumes a participatory approach: contracts and charters bring social actors together for the specific projects and set objectives measured through *indicators*, while progress is assessed at periodic meetings.

An enabling framework

High-level involvement at various scales

Engaging in a sustainable development process generally implies *breaking* with 'business as usual' trends. But such a break will not be possible if there is no involvement at the highest decision-making levels. This is as true for enterprises as it is for local authorities or national governments.

It is important for governments (guarantors of the public interest and thus also of inter-generational equity) to develop a framework that favours commitment to sustainable development by everyone. Such an enabling framework should in particular lead to the confirmation, either constitutionally or legally, of basic rights and principles – the right to a healthy environment for current and future generations – and the derived principles, including the precautionary principle, the principle of integrating the environment into sectoral politics, and the principles of the polluter pays and the user pays that take negative and positive externalities into account. To make the implementation of territory-specific and local processes possible, this framework also implies de-concentrated or decentralized approaches. It requires a confirmation of the subsidiarity principle, while guiding local development through rules and incentives that ensure long-term consideration.

Long-term strategies and goals; repositioning administrations in charge of the environment

Apart from advocating principles, an enabling framework should lead to organized action. Setting mid- and long-term goals of sustainable development is needed as well as implementing action programmes to achieve these goals. The Johannesburg World Summit, in particular, called on countries to formulate and implement sustainable development strategies. The Mediterranean sustainable development strategy currently under development will only have an impact if it is extended by national strategies that can best be defined along sectoral lines (strategies for sustainable transport, sustainable tourism, sustainable rural development, etc.) and integrated into economic and social development plans. The latter plans would gain when becoming 'sustainable development plans', so emphasizing the will to evolve in this direction.

To make such progress, it is necessary that governments (and local authorities) renew their prospective, strategic orientation and assessment tools so that they can better determine their mid- and long-term priorities and monitor progress on a regular basis through measurable indicators. The elaboration by each Mediterranean country of a national outlook report on the environment and development would contribute to the adaptation of global and regional considerations on countries' specific conditions.

The challenge of sustainable development could lead to a repositioning of administrations in charge of the environment, land planning and development. These administrations would gain by focusing on their functions of monitoring and assessment, orientation and regulation rather than on management, which should be more the concern of territory-specific authorities and other, more sectoral ministries. It is vital to give these administrations a more strategic role to orient and facilitate the implementation of environmental and sustainable development policies by all actors concerned, in particular the other sectoral ministerial departments and local authorities and enterprises. It is also of major importance to strengthen assessment procedures by independent agencies outside conventional administrative hierarchies. This would ensure cross-cutting approaches and consideration of all the interests and dimensions of projects and policies.

Developing new tools and partnerships

An enabling framework for sustainable development will lead to the setting up of new packages of tools and partnerships. There is no 'one model' to follow, and progress will probably come more from the 'periphery' rather than the 'centre'. This emphasizes the need to experiment and for increased active partnerships (companies, professional associations, syndicates, educational institutions, the media, local authorities, rural communities).

Economic tools are indispensable for correcting market failures, driving economic agents to adopt environment-friendly attitudes, and developing less polluting and less natural resource-depleting practices and technologies. Success often requires simultaneously fiscal incentives (tax breaks, accelerated writing off, and certain exemptions), economic incentives (subsidies, low-interest loans, buying polluting rights or rights for consuming resources, price guarantees), and a gradual internalization in prices of environmental costs generated by economic activity (the externalities). This internalization occurs mainly through increased tariffs for natural resources (in particular water, which is not being paid for at its real value) and by developing new taxation in conformity with the principles of the polluter pays and the user pays. Taxes could be based on damage (air and water pollution), on polluting products and potentially polluting equipment, and the contribution by tourists to maintenance of the Mediterranean environment. The tax issue is especially important in the south and east, where new systems are developing. Dismantling protection measures, other than reducing customs income, is going to change relative prices and the diversity of products (one could, for example, expect lower car and packaging prices). It will be in the interest of countries to anticipate such changes to avoid serious environmental impacts.

Internalization may, however, run up against considerable reluctance. Equity and social acceptance aspects need to be taken into account (several countries, for example, have developed very gradual pricing systems for drinking water). Specific mechanisms would be needed (the implementation of interim financing, subsidies for water-saving irrigation equipment, tax breaks for energy efficiency, free public transport on certain tourist destinations and cities, the restoration of old buildings for tourism), all to achieve sustainability objectives (resource savings, reduced congestion, development of integrated tourism that enhances cultural heritage, development of organic agriculture, etc.).

Incentives (such as micro-credit or subsidies for pilot projects) are also needed in integrated rural development and to enable the emergence of new generations of enterprises. Developing or transition countries would gain, for example, by promoting local tourism enterprises that can respond better to changes in international and domestic demand (this appears not to be the case in several countries that are focused only on international tourism) and by creating positive synergies with other sectors (construction, agriculture, handicrafts), all for the benefit of local development.

Apart from economic tools, there are other very effective tools for sustainable development. Standards, for example, are powerful vectors for implementing innovative technologies and protecting the environment, as can be seen in current developments in the EU. Voluntary tools such as charters, labels, geographic labelling systems, and mandatory labelling (for example on energy efficiency or risks of flooding) can also be far-reaching when reorienting production and consumption models. The same is true for training and awareness-raising campaigns. Eco-conditionality of assistance is another path to develop, especially in agriculture and tourism, where any subsidies should be on the condition that the rules and objectives of sustainable development are respected.

Land tenure is another important issue for sustainable development and justifies specific tools. The development of European agriculture since the 1950s owes much not only to economic tools (especially adjustment and market regulation funds), which have provided a certain financial security, but also to improved land property structures, which have provided long-term management guarantees to farmers (especially through the right to tenant farming), a condition for all sustainable investment. Concession systems in the public maritime domain have made it possible to mobilize local actors and render them accountable (by giving them rights and duties) for developments in shellfish raising.

In developing countries threatened by desertification, clarifying the rights of access to grazing grounds and forest resources and the rules for collective management of pastureland and forests are fundamental conditions for winning the fight against poverty and desertification. Protecting coastal zones requires adapted tools such as coastal agencies (enabling property acquisition or land auditing). Combating urban sprawl calls for the design of new tools, not just regulatory and economic but also for property issues, particularly to maintain green, agricultural and forest belts between agglomerations.

New action methods to involve civil society

Sustainable development will be based on an active civil society that participates in the decision-making process and is involved in the formulation as well as the implementation of sustainable development strategies. The groups of actors identified by Agenda 21 should be called on to play their part in this regard, most particularly women, educators, the media, NGOs, and, of course, business and rural communities.

The implementation of demand-driven management policies and integrated territory-specific development calls for a shift from hard to soft approaches. It is a new art of governing and administering that often still needs to be developed. In particular, administrations could learn how to move away from overly interventionist and technical approaches of the command-and-control type to become partners of local and professional actors and innovative in implementing new tools.

This shift will not occur by itself. It requires an evolution in the role and methods used by consultancy firms. These often advise very extreme technical solutions, which are the very opposite of what is required when trying to find local actors or new partners. Also, in research a profound evolution is needed (towards multidisciplinary and action-oriented research), as well as in training programmes at universities and engineering and administrative schools (towards skills rather than knowledge, development of personal training and social sciences, teaching through projects close to the 'ground', trainee programmes), and in local development organizations. Moves to put mediation facilities in place should be promoted, facilities capable of ensuring interfaces between disciplines and mobilizing and enhancing the different expertises required, including those of relevant local or professional actors.

Mobilizing technological progress

Technological progress, well controlled and targeted to exploit the strong points of the Mediterranean region, can play a major role in shifting towards more sustainable development. This is especially true when combating pollution and promoting the rational use of natural resources. In many countries important efforts, including normative measures, would be necessary in order to introduce and disseminate advanced technologies in the field of vehicles, construction, thermal plants, cleaner technologies in industry, waste and wastewater treatment and prevention of noise from infrastructures. It is also important to develop technologies able to exploit the strong points of the region (solar and wind power, the quality and diversity of unique products) and that are adapted to the specific constraints of the eco-region (relief, climate, urbanization, unique village lands).

Financing sustainable development

The alternative scenario assumes strengthening certain public expenditures: research and development, training and the capitalization of knowledge, tackling environmental liabilities (waste and wastewater treatment) and other infrastructures (public transport, rural services), subsidies, user awareness-raising campaigns, and financing positive externalities in agriculture and forestry for sustainable management of the Mediterranean territory.

Financing these public expenditures in the developing and transition countries requires a considerable strengthening of international solidarity in conformity with the commitments made at the conferences of Rio, Monterrey and Johannesburg, and in the framework of the Millennium Development Goals. However, financing also requires national administrative and tax reforms and the mobilization of private savings. In the developing and transition countries where the general context is one of deregulation and economic liberalization, which implies a loss of tax income related to the dismantling of tariffs, fundamental reforms are required. Partly already committed, these reforms go in complementary directions: a reform of tax revenues and the rationalization of public expenditure.

In tax matters, much effort will have to be made, especially in direct taxation where progress in reform is less than in indirect taxation. One necessary step forward will be the broadening of the tax base. In some countries, taxation at source still mainly concerns civil servants, employees of large companies and the financial sector. It needs to be expanded to other categories of economic agents (other salaried employees, the private sector and wealthy individuals). The IMF[12] stresses that broadening the tax base (and modernizing the collection system) could enable the Mashrek countries to see their direct tax revenues come close to those recorded in Maghreb countries (2–2.5 per cent of GDP). All would gain from getting closer to the levels attained in Turkey and the eastern European countries (4–6 per cent). Another important objective for many countries would be the search for mechanisms that would enable a shift of the current mass of small, private, informal businesses to the formal economy. This implies very refined actions in order to offer them sufficient benefits in exchange. Subcontracting for large firms and a wider application of micro credits could contribute to this.

A suppression of many exemptions would also make it possible to increase government revenues. As FEMISE has shown, exemptions aimed at attracting foreign capital have not clearly proved their worth.

Increasing the potential of local taxation will also be needed. This could be accompanied by improved management of local public services. The financial resources of local authorities remain relatively feeble in several countries and often rely on special ad hoc taxes (label rights, quarry permits) and local taxation (building taxes, slaughtering taxes), which are insufficient for funding expenditure needs. Some paths have already been successfully explored; they should now be encouraged.

Improved rationalization of public expenditure would leave significant room for manoeuvre. This report has illustrated that preliminary cost–benefit analyses on public investments need to be made more systematic. A few and too little known examples have shown the relevance of such analyses in the water and energy sectors. In some cases, the cost of a cubic metre of water saved thanks to good water management can be between three and ten times less expensive for the community than the cost of the water mobilized by supply policies. There is no doubt that some public expenses for agriculture or tourism, for example, would benefit from being reoriented.

Last, but not least, is the importance of a political and military relaxation in the region. The FEMISE (2003) Report estimates that up to US$9.5 thousand million could be saved in seven southern and eastern Mediterranean countries[13] (this is the equivalent of the total amount of direct foreign investment) if they reduced just the defence part of their total expenditures from 10.8 per cent to 5 per cent (the French average).

Towards Mediterranean governance for sustainable development

The alternative scenario assumes that regional co-operation will be strengthened at several levels and that a better synergy will be created between the various existing cooperation frameworks.

More operational eco-regional cooperation

The Mediterranean level is the relevant scale to understanding the environmental issues specific to the eco-region and drawing up a Mediterranean vision of sustainable development that is common in its design and differentiated in its objectives and implementation.

Because it was faced with common problems and issues, the Mediterranean eco-region has managed for several decades to put in place various frameworks for cooperation on sustainable development issues. ICSEMS, GFCM and ICAMAS are very valuable bodies in the fields of science, fisheries and agronomic training respectively.

The Mediterranean Action Plan, in addition to its legal tools, has always had the commitment of all Mediterranean-rim countries and the EU, and permanent action facilities (MEDPOL and the Regional Activity Centres, including the Blue Plan). These action facilities ensure continuing regional cooperation for protecting marine and coastal environments, managing coastal areas, and monitoring and prospective studies for environment and development. The MAP secretariat comes under the political authority of the member countries and is financed by contributions from member countries, which ensures its autonomy.

MAP has integrated sustainable development as a guiding principle in its thinking and activities. Since the publication of *Futures for the Mediterranean Basin. The Blue Plan* in 1989, and the development of *Agenda 21 for the Mediterranean* in 1994, considerable effort has been made in the context of MAP to tackle the issue of development from a sustainability point of view. It has relied especially on the work carried out on a Mediterranean-wide basis by MAP various components and by institutions such as METAP, ICAMAS, FEMISE, MEDCOAST, MEDFORUM, MIO-ECSDE, MEDCITIES and OME. Each institution works according to its own mandate on building a common regional vision of sustainable development. The Mediterranean vision of sustainable development has been built on this invaluable input around which an 'epistemic community' of specialist networks, governmental and local authority representatives and NGOs has developed.[14]

Thus, the Mediterranean is not just an eco-region in the bio-geographical and cultural sense of the word but also an area of multidimensional cooperation around a common vision of sustainable development.

In 1996 the governments of the region and the European Community put in place a Mediterranean Commission on Sustainable Development (MCSD) with a very broad and ambitious mandate in terms of a sustainable development strategy for the region. Like many comparable institutions, especially the United Nations Commission for Sustainable Development, MCSD has produced some excellent, innovative work, in particular relating to sustainable tourism, water-demand management and the relationships between trade and the environment. Nevertheless it has been criticized on the following points:

- socio-economic organizations, and to a lesser degree local authority organizations and representatives, have not participated greatly;
- many Contracting Parties have been represented by their Ministries of the Environment, which, added to

- the active presence of NGOs, has put a strong environmental stamp on the Commission, thus creating a feeling of imbalance to the detriment of the socio-economic aspects;
- feedback on the work of the Commission has been insufficient;
- follow-up by the actors and bodies concerned has remained weak.

MAP undertook to carry out an evaluation of MCSD and gave it a new impetus. The European Commission has shown its confidence in MCSD, and the Euro-Mediterranean Partnership[15] asked that the Mediterranean Strategy for Sustainable Development, which was announced at the Johannesburg World Summit, be developed within the framework of MAP, and MCSD in particular. The process for drawing up this strategy began in 2003.

It is hoped that the future Mediterranean Strategy for Sustainable Development will be subject to a validation and implementation procedure that would give it the political backing required for this process (as seen in the Baltic 21 Agenda, approved by a conference of Prime Ministers). It is also hoped that MCSD will become a framework for effectively engaging all administrations and IGOs involved in the region, not only the environmental actors. Its adoption at the Euro-Mediterranean level, for example by the Conference of the Ministers of Foreign Affairs, also deserves consideration.

In this context, a special place should be reserved for the sub-regional level. RAMOGE, the Ionian and Adriatic initiatives, and the Arab Maghreb Union are all sub-regional cooperation frameworks, which should take up sustainable development issues. Similarly it would be advisable to better structure Mediterranean cooperation for sustainable development around specific territories such as islands, mountains and coastal zones.

Lastly, the search for better synergy between the various regional and sub-regional cooperation frameworks is seen as a necessary and inescapable path to making progress towards sustainable development. In the water sector for example, operational synergy is needed between the Euro-Mediterranean Partnership and the specialized and active networks in the region (EU Water Initiative, GWP Med, MCSD).

More sustainability in EU enlargement towards the Mediterranean

It is important that the gradual enlargement of the EU to the eastern Mediterranean countries of the northern shore occurs within a perspective of sustainable development for the entire Mediterranean region. This requires new ideas and appropriate cooperation frameworks.

Concerning the EU itself, lessons could be learned from the past about the impacts of enlargement in terms of sustainable development, which have been both positive and negative for the countries involved.

For new member countries, the pace of development generated by enlargement is an important issue, as are the effects of their integration into a large competitive market. Under the influence of their accession to the single market and the policy of convergence, Spain and Greece have and still are experiencing growth rates higher than the European average. This leads to rapid economic convergence but also to an acceleration of hard-to-control phenomena, such as increases in household waste production, the car fleet and freight transport by road. This rapid pace does not give these countries enough time to adjust their infrastructures and capacities to cope with growing demand. Consequently the OECD has observed a considerable delay in waste management in the European Mediterranean countries.

In addition, the EU does not yet appear to have sufficiently included Mediterranean specificities in its vision and actions. Inadequate support has sometimes been given in the sectors of agriculture, forests, tourism and infrastructures. A shared vision of the multiple roles of Mediterranean agriculture would be useful, as well as on a development model for islands and mountains in Mediterranean Europe, and on the possibilities for encouraging a better reconciliation of tourism with sustainable development.

In the alternative scenario lessons are drawn from past but recent experiences that may be transferable to the new and future members, namely to better integrate the objective of sustainability in enlargement policies: preserving the positive aspects (especially the allocation of EU funds for curbing pollution, urban public transport, and sustainable agriculture and rural development) and reducing negative aspects as far as possible, particularly those related to the enlarged market. This implies better internalization of environmental costs generated by the expanded market, and putting in place more proactive European and national policies on sustainable development matters in the sectors concerned (energy, transport, tourism, urban development, waste, agriculture and rural development). To do so, Mediterranean specificities need to be taken into account, for which the structural, cohesion and agricultural funds could be used.

A new Euro-Mediterranean Partnership, a laboratory for sustainable development

The Mediterranean region has been the cradle of European civilizations, and the future of Europe is again being played out for better or (perhaps) worse in the Mediterranean.

This is already the case for its environment with the growing risks of extreme climate events, degradation of the common sea, the effects in time of desertification and the shift northwards of water stress problems. It is

Summary and Call for Action

even more the case for its future economic weight in the world and its security and stability. Successfully implementing a common commitment to sustainable co-development between the two shores would be far reaching.

Relaunching the Euro-Mediterranean Partnership through sustainable development

Launched in Barcelona in 1995, the Euro-Mediterranean Partnership was a political response to this immense challenge. Its instigators had hoped for rapid peace in the Near East in the post-Oslo days, but in vain. They thought that putting a Euro-Mediterranean free-trade zone in place would be enough to drive economic revival and prosperity in the region. While recognizing the merits of free trade, many renowned experts now voice the opinion that the EU should recognize its limitations, not to mention risks, and that there is an urgent need to considerably strengthen its intervention in the region so as to contribute to its development.

The experience of European integration shows how much a regional partnership can be a factor in sustainable progress if, in addition to the institution of a large market, it is also based on a discovery and recognition of each other, on a well-understood effort of solidarity towards the less developed (the economic and environmental progress of which benefits all partner countries), and on a shared political commitment to certain values, principles, concrete goals and common rules. However, the MEDA budget remains limited, the problem of free circulation of individuals persists, and current Euro-Mediterranean association agreements only contain real commitments on dates for dismantling tariffs. Other aspects dealt with, including the integration of the environment with development, generally remain very vague.

The question is whether, after nearly 50 years of unison and enlargement, the EU will make its relations with the rest of the world its new major project. The first challenge will lie in its relations with its direct neighbours in the eastern and southern Mediterranean with whom it shares an evident common destiny.

The new neighbourhood policy, which should lead to neighbourhood agreements starting in 2005, along with a new financial instrument in 2007 (the neighbourhood tool) could be the occasion for the EU to make this move. This is on the condition that the scale and nature of the challenges to be met are properly gauged, while relying on a common and differentiated vision, giving a new dimension to the commitments of partners on both shores, and providing sufficient resources to this policy.

In doing this, the EU should not ignore the scale of the disparities between the two shores or the seriousness and specificity of unsustainable development problems, in the Mediterranean region in general and in the southern and eastern countries in particular. In this way, the EU would respond to the appeals being sent out worldwide.

When looking at the action plan adopted in Johannesburg (2002) it can be noted that its philosophy is particularly relevant in the Mediterranean, since it is aimed at developing poor countries, equity, resolving the hardest social issues, controlling energy, protecting and managing water resources in a sustainable way, combating desertification, etc. Its joint application would make the Mediterranean a living laboratory for sustainable development on the confines of the developed countries and the developing ones of northern Africa and western Asia.

The Partnership would gain much if it gradually turned in this direction while relying on various available documents and ongoing studies.[16] Sustainable development could effectively become an excellent vector of association and co-development; determined commitment by partners would make it possible to orient the region towards a win–win scenario (north and south, environment and development). An alternative development path would thus see sustainable development erected as a guiding principle in the neighbourhood policy, in line with the position already assumed on the Euro-Mediterranean level.[17]

Strengthening south–south cooperation

The definitive resolution of many problems and the stability of the region will probably occur through sub-regional political and economic cooperation in the south and east, as called for by the Blue Plan in 1989.

The Partnership would benefit from contributing to this development. Association agreements already call for strengthened south–south trade. Bilateral and especially multilateral south–south agreements (for example the Agadir initiative) are called on to expand and develop. Beyond trade agreements themselves, an alternative development path would see the implementation of south–south initiatives, especially on a sub-regional level (Maghreb, Mashrek). These initiatives may cover multiple fields, including a coordinated development of transport infrastructures while integrating the objective of sustainability, the circulation of individuals, water management, research and development, etc.

The Partnership regional programmes could lead to identifying and initiating sub-regional projects, as has occurred in some cooperation programmes on statistics. A significant portion of funding should be allocated to such programmes.

The Partnership would also benefit much if the southern and eastern countries play a more active role in formulating proposals. These countries are of course largely centred on the Mediterranean Basin, but often they are also part of other regional structures. Special habits all too often inscribed in the tradition of centralized assistance grants from state to state. It is advisable to evolve towards assistance that results in a new kind of partnership that mobilizes local and professional actors (for example to set up natural regional parks and Agendas 21). Pilot projects could be carried out to demonstrate the validity of these new foundations, followed by more general application as part of Mediterranean-wide programmes that ensure a close relationship with the realities of the field.

Options could be explored among the EU, other donors and states to innovate in this direction: broadening the 'project' rationale (major infrastructures) to a 'programme' rationale for mobilizing and supporting projects of a large number of actors; extend formulas already proved valid in the EU to the entire Mediterranean, as done with the ERDF, INTERREG and LEADER programmes; set up financial twinning between local authorities on both shores to enable access to credit for those on the southern shore with guarantees from those on the northern side.

It would also be in the interest of the Partnership to use the typical Mediterranean character as a lever for partnership and regional development. Identifying common values between the European and Mediterranean worlds is to rediscover and recreate the Mediterranean identity together, a condition for better regional cohesion, successful cooperation and new, better positions in the face of globalization.

For their part, the SEMC could review their national development strategies and institutional set up with regard to the principles established and adopted in Barcelona, the neighbourhood policy, the common vision of sustainable development, and the search for a gradual convergence of their standards with certain EU standards. They would also gain from playing a more active role in the Partnership, especially by developing new south–south cooperation initiatives, by getting their EU partners to understand their specificities and their national issues, and by giving a more strategic and innovative dimension to the programmes financed by the Partnership. They could also commit to facilitating a decentralized cooperation between their local authorities and those on the northern shore.

The EU and the SEMC would also gain from sending strong signals to the market about their determination to make sustainable development a guiding principle in their Partnership and as a consequence the evolution of Mediterranean development. A signal on economic policies would, for example, be the adoption of a Euro-Mediterranean agreement on investments containing sustainability rules (adhering to social and environmental rules, transferring the best available technology). Signals could also be sent on various sectoral policies. In the tourism sector, for example, it is in the interest of all countries to show a common determination to commit the region to sustainable tourism and correct market failures. A regional cooperation mechanism would contribute to this. In the transport sector clear favouring of rail and maritime transport would help the companies involved to take risks and invest in the 'seaside highways' and rail freight.

To advance in this direction, it is vital that public opinion in general, and economic and political decision-makers of the Partnership in particular, have a greater awareness of sustainable development issues in the region, the obstacles to its implementation, the progress being made and the diversity and wealth of examples of good practice available. The Euro-Mediterranean Ministers of Economy could, for example, make it a topic of common concern with a view to a certain reorientation of policies. The new bodies established within the Euro-Mediterranean Partnership, the Cooperation and Dialogue Committee for the Facility for Euro-Mediterranean Investment and Partnership (FEMIP) and the Economic Dialogue for MEDA, offer opportunities to mobilize Ministries of Economy for win–win sustainable development solutions.

Finally, the need for a stronger, networked regional environmental information system and greater dissemination of information should be emphasized. Stronger regional input through better synergy between MAP, the Euro-Mediterranean Partnership and the European Environment Agency could contribute to this. It would bolster national and territorial environment and sustainable development observatories or equivalent agencies in the developing and transition countries, and develop 'benchmark' approaches for monitoring sustainable development. Comparisons would be all the more useful since many environmental and developmental problems in the Mediterranean countries have many traits in common. Monitoring through sustainable development indicators, analysing obstacles, highlighting good practices, ex ante and ex post assessments, and exchanging experiences are all important tools for shared progress.

As for the various *donors*, they should adopt a better evaluation, from a sustainable development point of view, of the efficiency and effectiveness of the assistance they provide.

Concluding remarks

Comparing the 1989 Blue Plan report with this 2005 one may give the impression that little has changed: the responses that have been implemented have been insufficient, and developments are rather close to what was feared in the relatively bleak trend scenario. The main global threats (climate change, mass losses of biodiversity, desertification, cultural standardization, exclusion and impoverishment, violence) have only

Summary and Call for Action

become stronger, and are at least as present in the Mediterranean as in other parts of the world.

And yet, is the future so dark? The analyses of this report suggest that an alternative development path is more credible than it was 15 years ago. There are even many elements for reasonable optimism.

Demographic transition in the southern and eastern Mediterranean is rapid. The successful integration of three large Mediterranean countries – Greece, Spain and Portugal – into the EU demonstrates that success is possible if resources are provided. These countries have experienced profound transformations of their economies and societies, although, especially in the field of environment, all remains far from perfect.

Through the concept of sustainable development, increasing numbers of leaders, experts and businessmen and -women have become aware that preserving the environment, social justice and economic effectiveness are the three interdependent pillars of the same process of development.

Many environmental and developmental innovations have emerged in fields such as technology, specific territory management, and forms of social organization and lifestyles. They demonstrate that solutions adapted to specific problems and cultures can be found, but they are often isolated and not well-known.

There are thus many elements for hope. What, then, would be the conditions? Three basic keys for success along an alternative development path are here highlighted.

Linking the environment to social justice. This was too neglected in the initial generation of environmental policies, which may partially explain their mixed success. It is generally the poorest who suffer most from degraded environmental conditions. Poverty and unclear rights and rules on access to resources, often lead to practices that are very harmful to the environment, especially in rural areas.

Linking environment to economic efficiency. The economy is not the enemy of the environment. Indeed, saving resources is an essential dimension of their proper management. Policies should aim at precise objectives and be accompanied by well-defined indicators of goals, resources and results to enable assessment of the efficiency and effectiveness of environmental and developmental policies. The basic thrust of these policies should be the development of better production and consumption patterns.

Devising new social relationships and international cooperation to manage the environment and sustainable development. The words participation, partnership and governance are increasingly used, although they are not always clearly defined. The fundamental idea that underlies them is that sustainable development can only succeed if all social actors take part in a 'positive sum game'. This is true for countries and at some international levels where new north–south solidarity around the concepts of common and differentiated responsibility and sustainable development needs to be devised or put in place urgently. Europe could play a pioneering role in this direction by evolving its relations with its Mediterranean neighbours within the framework of a renewed Euro-Mediterranean Partnership.

More than ever, the question of how to reconcile individual interests and the common wealth of nations is being put forward in the Mediterranean. Coercive systems have been ineffective everywhere. The laissez-faire of absolute free trade is increasing social inequalities and impacts on the environment. Between these two extremes, methods and institutions are needed to make individual interests work in favour of the common good. This is the main key for achieving sustainable development. Relying on the immense heritage of ancient cultures, philosophical legacies and modern science, the Mediterranean countries are well placed to meet this challenge.

Notes

1. The Sustainability Impact Assessment of the EMFTA was launched late in 2004.
2. From 2000 to 2020 for all the SEMC, including Jordan (source: FEMISE, 2003).
3. Differences between the baseline scenario and the IFRI 'Europe-Russia-Mediterranean' scenario (see Part 1, Box 7 p36).
4. In particular the authoritative reports by FEMISE, CEPII, the World Bank, OECD, UNDP, IFRI.
5. IFRI, 2002.
6. Or a stabilization in the Mediterranean region at 7 per cent of the global CO_2 emissions (compared with 9 per cent projected under the baseline scenario).
7. The baseline scenario already includes a decoupling assumption similar to the one integrated in the European Union Goteborg strategy. This is the reason why in the alternative scenario traffic is not halved compared with the baseline scenario.
8. Estimate from the OECD's DAC database.
9. The maximum number of indicators calculated for a country is 103 (France).
10. The European directive 2001/42 of June 2001 requires that plans and programmes likely to have a notable influence on the environment should be submitted to an environmental impact assessment.
11. International Association for Natural Gas Vehicles, April 2004 (www.iangv.org).
12. Nashashibi, 2002.
13. Egypt, Israel, Jordan, Lebanon, Morocco, Turkey and Tunisia.
14. Haas, 1990.
15. Meeting of the Euro-Mediterranean Ministers of the Environment, Athens, 2002.
16. In particular the current sustainability analysis of the effects of the implementation of the Euro-Mediterranean free-trade zone, the future Mediterranean strategy for sustainable development, and this report.
17. Euro-Mediterranean Conference of the Ministers of Foreign Affairs, Valencia, 2002 and the 'Athens Statement' of the Conference of the Euro-Mediterranean Ministers of the Environment, 2002.
18. The Athens Euro-Mediterranean Conference of Environment Ministers in 2002 stressed the need to establish integration strategies in each sector and each regional programme.

19 As decided by the Valencia Euro-Mediterranean Conference of Ministers of Foreign Affairs, 2002.

References

FEMISE (2003) *Report on the Euro-Mediterranean Partnership* 2003. Marseille: FEMISE (www.femise.org)

Haas, P. (1990) *Saving the Mediterranean. The Politics of International Environmental Cooperation.* Columbia: Columbia University Press

IFRI (2002) *Le commerce mondial au XXIe siècle.* French Institute of International Relations, Paris

Nashashibi, K. (2002) *Fiscal Revenues in South Mediterranean Arab Countries: Vulnerabilities and Growth Potential.* Washington, DC: IMF. IMF Working Paper WP/02/67, April

Statistical annex

Geographical framework
A1 Countries and Mediterranean coastal regions specificities

Population (Permanent and Seasonal)
A2 Population of Mediterranean countries, 1960–2025
A3 Population age distribution, 1970–2025
A4 Average household size
A5 Urban population in Mediterranean countries, 1970–2025
A6 Number of tourists in the Mediterranean countries, 1990–2025

Economy
A7 GDP growth rates in the Mediterranean countries, 1970–2025
A8 Unemployment rate, 1980–2003
A9 Net capital flows to the Mediterranean
A10 Net official capital flows to the Mediterranean
A11 Net capital flows per country compared with the Mediterranean average

Rural areas and agriculture
A12 Agricultural and rural populations in the Mediterranean, 1960–2025
A13 Land use in Mediterranean countries
A14 Agriculture and intensification
A15 Trade in food and agricultural products, and agricultural added value

Fisheries and aquaculture
A16 Mediterranean fisheries
A17 Mediterranean aquaculture

Energy and transport
A18 Energy intensity, per country, 1971–2025
A19 Primary energy supply sources per country, baseline and alternative scenarios, 2000–2025
A20 Primary commercial energy demand, baseline and alternative scenarios, 1971–2025
A21 Electricity production by source, per country, baseline and alternative scenarios, 2000–2025
A22 Electricity demand, baseline and alternative scenarios, 1971–2025
A23 Electricity production potential from renewable energies in the SEMC in 2020
A24 The fluctuation of low-tension electricity prices (residential sector) in a number of SEMC
A25 Motorization rate

Coastal regions
A26 Population of Mediterranean coastal regions, 1970–2025
A27 Density of the Mediterranean coastal regions compared with the density of the country, 1970–2025
A28 Urban population in Mediterranean coastal regions, 1970–2025
A29 Number of tourists in the Mediterranean coastal regions, 1990–2025
A30 Human pressures on coastal regions and on the coastline
A31 Human settlements and infrastructures along the coast
A32 Type and treatment of industrial sewage

Water
A33 Water resources per country, 2000–2025
A34 Water demand by country, baseline and alternative scenarios, 2000-2025

Emissions in the atmosphere
A35 CO_2 emissions related to energy activities per country, baseline and alternative scenarios

Biodiversity
A36 Biosphere reserves in the Mediterranean in 2003
A37 Protected Areas in Mediterranean countries

A Sustainable Future for the Mediterranean

This statistical annex holds tables showing the basic data and the major quantitative assumptions or results of the scenarios worked out in this report.

Most of these data are extracted from the databases or statistical reports of international agencies such as FAO, World Bank, OECD, etc. The use of these international sources, themselves based on national statistical sources, ensures data comparability and consistency of the analyses carried out by the *Plan Bleu* experts.

For certain topics such as inland waters, urban agglomerations, *Plan Bleu* uses some databases compiled by well-known experts.

Nevertheless, some data from national sources, considered as essential or notably different from the international data used by *Plan Bleu*, have been added as a note to a table or in the text of the related chapter.

Conventions

'—':	Non available
AAGR:	Annual Average Growth Rate
Mt:	million tonnes
Mtoe:	million tonnes of oil equivalent
koe:	kilo of oil equivalent
kWh:	kilowatt hour (1000 Wh)
TWh:	terawatt hour (1 thousand million kilowatt hours = 10^{12} Wh)
GWP:	Global Warming Potential

Groups of countries

NMC (Northern Mediterranean countries): Spain, France, Italy, Greece, Monaco, Malta, Cyprus, Slovenia, Croatia, Bosnia-Herzegovina, Serbia-Montenegro, Albania

- 4 EU-Med (before EU enlargement in May 2004): Spain, France, Italy, Greece
- EAC (eastern Adriatic countries): Slovenia, Croatia, Bosnia-Herzegovina, Serbia-Montenegro, Albania

SEMC (Southern and Eastern Mediterranean countries):

- East: Turkey, Syria, Lebanon, Israel, Palestinian Territories
- South: Egypt, Libya, Tunisia, Algeria, Morocco.

Statistical annex

Geographical framework

Table A1 Countries and Mediterranean coastal regions specificities

Country	Area km²	Mediterranean coastal regions	Area km²	Coastline (km) % of the country area	Total	Of which Islands
Spain	504,782	12 provinces / 50	95,504	19	2580	910[1]
France	543,965	9 départements / 96	46,248	9	1703	802[2]
Italy	301,333	57 provinces / 103	165,846	55	7375	3766[3]
Greece	131,626	41 nomes / 51	100,975	77	15,021	7700
Monaco	2	1 (the whole country)	2	100	4	
Malta	316	2 regions (the whole country)	316	100	180	180[4]
Cyprus	9251	6 districts (the whole country)	9251	100	782	782
Slovenia	20,273	1 Statistical region / 12	1044	5	47	
Croatia	56,542	7 counties / 21	26,150	46	5835	4058
Bosnia-Herzegovina	51,129	1 canton[5]	4401	9	23	
Serbia-Montenegro	102,173	1 group of communes[6]	6508	6	294	
Albania	28,748	11 districts / 36	7833	27	418	
Turkey	814,578	10 provinces / 80	122,612	15	5191	809
Syria	185,180	2 mohafazats / 14	4189	2	183	
Lebanon	10,452	4 governorates / 5	4892	47	225	
Israel	22,145	5 districts / 6	19,865	90	179	
Palestinian Territories	6020	1 group of governorates[7]	363	6	55	
Egypt	1,001,450	11 governorates[8] / 27	114,767	11	955	
Libya	1,759,540	16 baladyats[9] / 25	274,911	16	1770	
Tunisia	163,610	14 governorates / 24	45,712	28	1298	301
Algeria	2,381,741	15 wilayates / 48	47,027	2	1200	
Morocco	710,850[10]	4 provinces + 3 préfectures / 67[11]	41,950	3	512	
Total Med	**8,805,706**	**234 administrative units**	**1,118,690**	**13**	**45,830**	**19,308**

Source: National Statistical yearbooks and *Plan Bleu* estimates, 2003

Notes:
1. Balearics
2. Corsica
3. of which Sicily 1126km and Sardinia 1387km
4. Including Gozo 43km
5. Bosnia: 10 cantons; Herzegovina: 5 regions
6. Montenegro: coastal and central municipalities
7. Gaza includes 5 out of 14 governorates of the Palestinian Authority
8. of which 1 partial (desert excluded)
9. of which 1 non-coastal and 2 partials (desert excluded)
10. with Western Sahara
11. Morocco: 42 provinces and 25 préfectures

Population (permanent and seasonal)

Table A2 Population of Mediterranean countries, 1960–2025

Country	1970	1985	2000	2025	AAGR 1970–2000	AAGR 2000–2025
	(1000 inhabitants)				(%)	
Spain	34,027	38,156	39,815	40,769	0.5	0.1
France	50,569	55,216	59,412	64,177	0.5	0.3
Italy	53,758	56,951	57,456	53,925	0.2	−0.3
Greece	8716	9905	10,558	10,393	0.6	−0.1
Monaco	24	28	34	41	1.2	0.8
Malta	319	344	389	430	0.7	0.4
Cyprus	615	645	785	900	0.8	0.5
Slovenia	1670	1913	1965	2029	0.5	0.1
Croatia	4406	4662	4473	4193	0.1	−0.3
Bosnia-Herzegovina	3564	4129	3972	4324	0.4	0.3
Serbia-Montenegro	8691	9948	10,856	12,217	0.7	0.5
Albania	2184	2978	3114	3820	1.2	0.8
Turkey	35,666	50,267	65,627	87,303	2.1	1.1
Syria	6277	10,298	15,936	24,003	3.2	1.7
Lebanon	2177	2790	3206	4147	1.3	1.0
Israel	2935	4097	5851	7861	2.3	1.2
Palestinian Territories	1134	1510	3150	6072	3.5	2.7
Egypt	32,364	46,140	66,007	94,895	2.4	1.5
Libya	1986	3719	6038	8832	3.8	1.5
Tunisia	5127	7182	9615	12,892	2.1	1.2
Algeria	13,623	21,492	30,332	42,329	2.7	1.3
Morocco	15,081	21,579	28,505	38,174	2.1	1.2
Total NMC	168,542	184,877	192,829	197,218	0.4	0.1
Total east	48,189	68,962	93,770	129,386	2.2	1.3
Total south	68,181	100,112	140,497	197,122	2.4	1.4
Total SEMC	116,370	169,074	234,267	326,508	2.4	1.3
Total Med	284,912	353,951	427,096	523,726	1.4	0.8

Source: Attané and Courbage, 2001

Statistical annex

Table A3 Population age distribution, 1970–2025

Shore	Age group (years)	Size (Millions inhabitants)			Share of total population (%)			AAGR (%)	
		1970	2000	2025	1970	2000	2025	1970–2000	2000–2025
North	<15	43	32	30	26	17	15	−0.96	−0.25
	15–65	106	128	123	64	67	62	0.63	−0.19
	>65	18	31	43	16	16	22	1.80	1.33
	Total	167	192	196	100	100	100	0.44	0.08
South and east	<15	50	75	73	43	32	22	1.33	−0.16
	15–65	61	146	221	52	62	67	2.94	1.66
	>65	5	13	34	4	6	10	3.31	3.80
	Total	117	235	328	100	100	100	2.35	1.33
Total Med		**284**	**427**	**523**				**1.35**	**0.81**

Source: Plan Bleu, Attané and Courbage, 2001

Table A4 Average household size

Country	1985	2000	2025
Spain	3.7	3.1	3.2
France	2.7	2.5	2.2
Italy	3.0	2.5	2.3
Greece	3.2	2.7	2.4
Malta	3.3	2.9	2.5
Cyprus	3.7	3.9	3.8
Slovenia	3.2	2.7	2.6
Croatia	3.1	2.8	2.5
Serbia-Montenegro	3.5	3.2	3.0
Albania	5.6	4.8	4.8
Turkey	5.3	4.2	3.3
Syria	6.5	6.2	4.7
Israel	3.6	3.5	3.1
Egypt	5.0	4.9	4.0
Libya	6.9	7.7	7.5
Tunisia	5.4	4.8	4.2
Algeria	7.0	6.1	4.9
Morocco	5.9	5.3	4.4
Total NMC	**3.2**	**2.8**	**2.5**
Total SEMC	**5.6**	**5.0**	**4.1**
Total Med	**4.0**	**3.7**	**3.3**

Source: UN-Habitat, 2001

Table A5	Urban population in Mediterranean countries, 1970–2025					
Country	1970	1985	2000	2025	AAGR 1970–2000	AAGR 2000–2025
	(1000 inhabitants)				(%)	
Spain	23,774	29,125	31,851	32,941	1.0	0.1
France	31,395	34,709	36,982	39,820	0.5	0.3
Italy	37,252	40,788	41,135	39,095	0.3	−0.2
Greece	4504	5782	6289	6717	1.1	0.3
Monaco	24	28	34	41	1.2	0.8
Malta	235	252	292	333	0.7	0.5
Cyprus	238	374	501	699	2.5	1.3
Slovenia	435	602	706	887	1.6	0.9
Croatia	1613	2147	2370	2659	1.3	0.5
Bosnia-Herzegovina	645	1192	1877	2702	3.6	1.5
Serbia-Montenegro	3136	4501	5815	7839	2.1	1.2
Albania	518	816	1112	1877	2.6	2.1
Turkey	12,901	25,664	43,517	66,424	4.1	1.7
Syria	2629	5433	9342	17,273	4.3	2.5
Lebanon	1193	1713	2462	3320	2.4	1.2
Israel	2316	3368	4986	6861	2.6	1.3
Palestinian Territories	632	816	1683	3570	3.3	3.1
Egypt	18,016	28,999	42,485	78,777	2.9	2.5
Libya	862	2447	4449	6798	5.6	1.7
Tunisia	1814	3495	5608	8828	3.8	1.8
Algeria	4772	9132	15,321	25,852	4.0	2.1
Morocco	4948	9078	15,037	25,053	3.8	2.1
Total NMC	103,769	120,317	128,963	135,611	0.7	0.2
Total east	19,670	36,995	61,990	97,448	3.9	1.8
Total south	30,412	53,151	82,900	145,308	3.4	2.3
Total SEMC	50,082	90,146	144,890	242,756	3.6	2.1
Total Med	153,850	210,463	273,853	378,367	1.9	1.3

Source: Plan Bleu, Attané and Courbage, 2001

Note: Urban population is understood as the population living in towns with more than 10,000 inhabitants.

Statistical annex

Table A6 Number of tourists in the Mediterranean countries, 1990–2025, (in thousands)

Country	International tourism 1990	2000	2025	National tourism 1990	2000	2025	TOTAL 1990	2000	2025
Spain	37,441	47,898	80,759	17,917	20,704	28,538	55,358	68,602	109,297
France	52,497	75,595	116,333	33,405	42,777	48,133	85,902	118,372	164,466
Italy	26,679	41,181	57,345	29,366	31,601	38,287	56,045	72,782	95,631
Greece	8873	12,500	18,708	4072	4751	6755	12,945	17,251	25,463
Monaco	245	300	594	—	—	—	245	300	594
Malta	872	1216	2022	125	156	249	997	1372	2271
Cyprus	1561	2686	4405	143	314	540	1704	3000	4945
Slovenia	—	1090	3993	799	786	1400	799	1876	5393
Croatia	—	5831	11,786	1423	1118	2348	1423	6949	14,135
Bosnia-Herzegovina	—	110	594	871	199	1384	871	309	1978
Serbia-Montenegro	—	239	2689	2599	1086	4887	2599	1325	7576
Yugoslavia SFR	7880	—	—	—	—	—	7880	—	—
Albania	30	32	215	98	125	879	128	157	1093
Turkey	4799	9586	33,991	11,295	16,407	41,032	16,094	25,993	75,023
Syria	562	1416	4351	1212	1594	6961	1774	3010	11,311
Lebanon	—	742	5442	—	962	2074	—	1704	7515
Israel	1063	2417	4423	2327	3043	5110	3390	5460	9533
Palestinian Territories	—	330	—	—	63	1336	—	393	1336
Egypt	2411	5116	23,983	6800	6601	21,826	9211	11,717	45,809
Libya	96	174	1738	318	604	3533	414	778	5271
Tunisia	3204	5057	10,603	1600	2115	5028	4804	7172	15,630
Algeria	1137	866	1482	4973	6066	11,006	6110	6932	12,488
Morocco	4024	4113	10,962	3843	4561	9544	7867	8674	20,505
Total NMC	**136,078**	**188,678**	**299,441**	**90,818**	**103,615**	**133,400**	**226,896**	**292,293**	**432,841**
Total SEMC	**17,296**	**29,817**	**96,975**	**32,368**	**42,015**	**107,448**	**49,664**	**71,832**	**204,423**
Total Med	**53,374**	**218,495**	**396,416**	**123,185**	**145,630**	**240,848**	**276,559**	**364,125**	**637,263**

Source: *Plan Bleu* and WTO, 2001

Notes:
The number of international tourist arrivals is assessed by surveys in borders (WTO).
The number of national tourists is estimated by the *Plan Bleu* from assumptions on the rate of holiday departure.

Economy

Table A7 GDP growth rates in the Mediterranean countries, 1970–2025

	AAGR (%) 1970–1985	AAGR (%) 1985–2000	Baseline scenario assumptions AAGR (%) 2000–2025
Spain	2.9	3.3	
France	2.8	2.3	
Italy	3.0	2.0	
Greece	3.1	2.0	
4 EU-Med	**2.9**	**2.3**	**2.4**
Monaco	—	1.6[2]	
Malta	7.4	5.3	
Cyprus	8.7[1]	5.1	
Islands	**—**	**5.1**	**4.0**
Slovenia	—	1.8[2]	
Croatia	—	–1.4[2]	
Bosnia-Herzegovina	—	25.6	
Serbia-Montenegro	—	—	
Albania	—	0.9	
EAC	**—**	**—**	**4.6**
Turkey	**4.3**	**4.2**	**4.7**
Israel	4.8	4.9	
Palestinian territories	—	1.6[4]	
Israel & Palestinian Territories			**3.6**
Syria	7.5	3.9	
Lebanon		3.2[3]	
Egypt	6.6	4.3	
Mashrek	**6.8**	**4.2**	**4.7**
Libya	—	–1.2[2]	
Tunisia	6.3	4.1	
Algeria	5.6	1.4	
Morocco	4.6	3.0	
Maghreb	**5.3**	**2.4**	**4.0**
Total NMC	**2.7**	**2.6**	**2.5**
Total east	**4.7**	**4.8**	**4.2**
Total south	**4.4**	**3.1**	**4.2**
Total SEMC	**4.6**	**4.1**	**4.2**
Total Med	**2.9**	**2.5**	**2.7**
	1970–1985	1985–2000	
Sub-Saharan Africa	2.7	2.3	
Latin America and the Caribbean	4.0	2.8	
South East Asia and Pacific	6.8	7.6	
High income OECD countries	3.1	2.8	
US	3.2	3.2	
World	**3.4**	**3.0**	

Source: World Bank for 1970–2000; *Plan Bleu* 2001 from background papers by E. Fontela Mahjoub M. for 2025

Notes:
GDP in US$1995
1   period 1975–1985
2   period 1990–2000
3   period 1998–2000
4   period 1994–2000

Table A8 Unemployment rate, 1980–2003 (in % of labour force)

	1980	1985	1990	1995	2000	2001	2002	2003
Spain	11.1	21.0	16.0	22.7	13.9	10.5	11.4	11.3
France	6.1	10.2	9.2	11.6	10.0	8.8	8.9	9.7
Italy	7.6	10.3	11.4	11.4	10.5	9.5	9.0	8.7
Greece	2.4	7.8	7.0	9.1	11.1	10.2	9.6	—
Malta	—	8.1	3.9	3.7	6.5	6.5	6.8	7.9
Cyprus	2.0	3.3	1.8	2.6	4.9	4	3.3	4.1
Slovenia	—	—	4.7	7.4	7.2	5.9	5.9	6.6
Croatia	—	6.5	8.2	14.5	16.1	15.8	14.8	14.3
Bosnia-Herzegovina	—	—	—	—	—	—	—	—
Serbia-Montenegro	—	—	—	—	21.2	22.3	24.7	—
Albania	5.6	6.7	9.5	12.9	16.8	16.4	15.8	15.0
Turkey	—	11.2	8.0	7.5	6.6	8.5	10.6	—
Syria	3.9	—	—	—	—	11.2	11.7	—
Lebanon	—	—	—	8.6[1]	—	—	—	—
Israel	4.8	6.7	9.6	6.9	8.8	9.3	10.3	10.7
Palestinian Territories	—	—	—	18.2	14.2	25.5	31.3	25.6
Egypt	5.2	—	8.6	11.3	9.0	9.2	—	—
Libya	—	—	—	—	—	—	—	—
Tunisia	—	—	15.3[2]	15.6[3]	15.6	15.0	14.9	14.3
Algeria	—	—	19.8	27.9	29.8	27.3	—	—
Morocco	—	—	15.8	22.9	21.5	19.5	18.3	—

Source: World Bank, World Development Indicators 2004; ILO (http://laborsta.ilo.org/)

Notes
1   1997
2   1989
3   1994

Table A9 — Net capital flows to the Mediterranean – Annual average (in constant 1995 US$)

	1971–1980		1981–1990		1991–2000	
	Million $	$ per cap.	Million $	$ per cap.	Million $	$ per cap.
Total Net Flows (TNF)	12,820	89	12,719	70	16,805	70
– Official Flows	7941	55	9137	50	9638	40
– Private Flows	4878	34	3582	20	7167	30
GDP	193,627		322,718		517,095	
TNF/GDP	6.6%		3.9%		3.2%	

Source: OECD/DAC, 2004

Notes:
Excluding the 4 EU Mediterranean countries
For each year, population (and GDP) includes population (and GDP) of the Mediterranean countries having benefited from capital contributions.

Table A10 — Net official capital flows to the Mediterranean – Share of the multilateral funds (%)

	1971–1980		1981–1990		1991–2000	
Share of the multilateral funds in the net official flows	15.7		17.2		16.5	
– of which European Commission	1.6	*10*	3.2	*19*	13.2	*80*
– of which other multilateral	6.2	*40*	4.5	*26*	6.4	*39*
– of which IBRD	7.8	*50*	9.5	*55*	–3	*–19*

Source: OECD/DAC, 2004

Notes:
Excluding the 4 EU Mediterranean countries
Numbers in italics indicate share of each donor in the total of the multilateral aid.
'Other multilateral' includes IBRD (International Bank for Reconstruction and Development), African Development Bank., Arab Agencies. EBRD, GEF.

Statistical annex

Table A11 Net capital flows per country compared with the Mediterranean average

		Net official flows per inhabitant from EU[1] (%)			Total Foreign Direct Investment from DAC[2] countries per inhabitant
		1971–1980	1981–1990	1991–2000	1991–2000
Islands	Cyprus	117	154	141	1361
	Malta	464	302	323	1014
EAC	Slovenia	—	—	135	389
	Croatia	—	—	39	390
	Bosnia-Herzegovina	—	—	572	–6
	Serbia-Montenegro	—	—	100	7
	Albania	—	14	307	15
SEMC	Turkey	68	30	12	123
	Syria	21	32	26	29
	Israel	201	99	15	959
	Palestinian Territories	—	—	397	2
	Lebanon	58	100	127	96
	Egypt	45	70	66	40
	Libya	57	33	4	43
	Tunisia	250	150	115	91
	Algeria	98	58	134	78
	Morocco	81	119	86	26
Mediterranean[3] average (base = 100)		**100**	**100**	**100**	**100**
in US$/inhabitant		**8**	**15**	**21**	**113**

Source: OECD/DAC, 2004

Notes:
1 The average is calculated by dividing the total net flows coming from the EU and received by the Mediterranean countries by their population.
2 DAC members: Australia, Austria, Belgium, Canada, Denmark, Finland, France, Germany, Greece, Ireland, Italy, Japan, Luxembourg, Netherlands, New Zealand, Norway, Portugal, Spain, Sweden, Switzerland, United Kingdom, United States, Commission of the European Communities.
3 Excluding the four EU Mediterranean countries.

This table should be read as follows: during the decade 1971–1980, the net official flow per inhabitant received by Morocco accounted for 81% of the Mediterranean average level of the net official flow received from EU.

Rural areas and agriculture

Table A12 Agricultural and rural populations in the Mediterranean, 1960–2025

Countries	Total population[1] (thousands)	Rural population[1] (thousands)				Rural population[1] (% of total population[1])	Rural population[2] (thousands)
	2000	1960	1990	2000	2025	2000	2000
Spain	39,815	11,941	8700	7964	7828	20	8936
France	59,412	19,372	21,187	22,430	24,357	41	14,588
Italy	57,456	17,693	16,199	16,321	14,830	26	19,018
Greece	10,558	4784	4134	4269	3676	35	4234
Monaco	34	0	0	0	0	0	0
Malta	389	83	93	97	97	25	35
Cyprus	785	379	266	284	201	26	236
Slovenia	1965	1243	1351	1259	1142	58	1010
Croatia	4473	3053	2438	2103	1534	34	1967
Bosnia-Herzegovina	3972	2752	2979	2095	1622	41	2269
Serbia-Montenegro	10,856	5887	5548	5041	4378	40	5109
Albania	3114	1213	2292	2002	1943	62	1808
Turkey	65,627	20,375	24,537	22,110	20,879	32	22,824
Syria	15,936	2932	5444	6594	6730	42	7865
Lebanon	3206	855	1067	744	827	26	359
Israel	5851	573	765	865	1000	17	505
Palestinian Territories	3150	599	769	1467	2502	79	
Egypt	66,007	12,905	18,196	23,522	16,118	24	38,914
Libya	6038	991	1399	1589	2034	34	654
Tunisia[3]	9615	2994	3760	4007	4064	42	3261
Algeria	30,332	7614	13,563	15,011	16,477	54	12,980
Morocco	28,505	8463	13,161	13,468	13,121	46	13,307
NMC	192,829	68,398	65,188	63,866	61,607	32	59,210
SEMC	234,267	58,304	82,662	89,377	83,752	35	100,669
MED	427,096	126,702	147,850	153,243	145,359	34	159,879
Med. France[4]	7082	1265	1610	1775	1998	28	
NMC revised[5]	140,499	50,292	45,611	43,211	39,248	28	
Mediterranean[6]	374,766	108,595	128,273	132,588	123,000	32	

Source: Plan Bleu from Géopolis, World Population Prospects, Attané and Courbage, 2001; FAO (http://faostat.fao.org), AGRESTE (www.agreste.agriculture.gouv.fr), ILO (http://laborsta.ilo.org/)

Notes:
1   Data from Géopolis
2   Data from FAO
3   For Tunisia, data from the 2004 population census: total population: 9.9 million, rural population: 3.48 million, rural population/total population: 35 per cent
4   Med. France = Mediterranean regions of France
5   NMC revised = NMC, excluding the non-Mediterranean regions of France
6   Mediterranean = NMC revised + SEMC

Table A12   Agricultural and rural population in the Mediterranean; 1960–2025 (continuation)

Countries	Agricultural population[2] (thousands)					(% of rural pop.[1])	Agricultural labour force[2] (thousands)					(% of total active population[2])			
	1960	1990	2000	2010	2025	2000	1960	1990	2000	2010	2025	1960	1990	2000	2010
Spain	12,224	4635	2780	1765	1087	35	4727	1892	1293	796	490	41.1	11.9	7.4	4.5
France	10,095	3116	1985	1244	806	9	4243	1356	857	555	359	22.1	5.5	3.3	2.0
Italy	14,965	4878	2911	1828	1075	18	6157	2101	1352	795	467	30.8	8.6	5.3	3.2
Greece	3750	1901	1381	1002	728	32	1735	963	775	563	409	52.2	23.0	16.8	11.9
Monaco	4	1	0	0	0	2	1	0	0	0					
Malta	29	9	6	4	4	6	9	3	2	2	2	9.5	2.6	1.4	1.3
Cyprus	240	92	65	45	31	23	98	44	33	23	16	41.9	13.6	8.6	5.4
Slovenia	1049	119	34	13	5	3	464	55	20	7	2	63.8	5.7	2.0	0.7
Croatia	2759	730	370	203	104	18	1220	338	186	95	49	64.2	16.1	8.5	4.4
Bosnia-Herzegovina	1913	475	194	97	46	9	846	220	96	46	22	64.2	11.3	5.2	2.3
Serbia-Montenegro	5195	3025	2015	1340	888	40	2297	1400	1008	668	443	64.4	29.7	19.9	12.9
Albania	1175	1796	1496	1383	805	75	521	857	751	729	425	71.2	54.6	48.2	41.8
Turkey	18,227	20,950	20,365	18,650	12,244	92	11,043	13,012	14,426	14,570	9565	78.7	53.6	46.2	39.0
Syria	2832	4104	4535	4930	3744	69	899	1147	1434	1815	1378	60.7	33.1	27.8	23.7
Lebanon	697	198	123	74	49	17	197	62	47	31	20	38.3	7.3	3.7	1.9
Israel	304	186	163	129	109	19	110	73	70	59	50	14.4	4.1	2.7	1.8
Palestinian Territories	—	—	—	—	—	—	54	85	—	—	—	24.2	20.2	—	—
Egypt	19,586	24,664	24,805	23,418	22,237	105	6651	7899	8591	9064	8607	65.9	40.3	33.3	26.6
Libya	801	471	303	209	148	19	261	140	107	76	54	59.5	10.9	6.0	3.2
Tunisia	2592	2295	2319	2224	1674	58	838	805	942	1013	763	61.6	28.1	24.6	20.9
Algeria	7594	6396	7307	7463	6593	49	2310	1827	2545	3212	2838	71.1	26.1	24.3	21.3
Morocco	8676	11,110	10,877	10,383	9619	81	3124	4073	4251	4376	4054	73.1	44.7	36.1	29.0
NMC	**53,399**	**20,778**	**13,237**	**8924**	**5579**	**21**	**22,319**	**9230**	**6373**	**4279**	**2684**	**35.7**	**11.4**	**7.4**	**4.9**
SEMC	**61,309**	**70,374**	**70,797**	**67,480**	**56,417**	**81**	**25,487**	**29,123**	**32,413**	**34,216**	**27,329**	**70.2**	**41.3**	**34.5**	**28.2**
MED	**114,708**	**91,152**	**84,034**	**76,404**	**61,995**	**55**	**47,806**	**38,353**	**38,786**	**30,495**	**013**	**48.0**	**25.3**	**21.6**	**18.5**
Med. France[3]	2289	678	450	282	183	25	743	226	150	97	63				
NMC revised[4]	593	18,340	11,702	7962	4956	27	18,819	8100	5666	3821	2388				
Mediterranean[5]	**106,902**	**88,714**	**82,499**	**75,442**	**61,372**	**63**	**44,306**	**37,138**	**38,079**	**38,037**	**29,716**				

Source: Plan Bleu from Géopolis, World Population Prospects, Attané and Courbage, 2001; FAO (http://faostat.fao.org), AGRESTE (www.agreste.agriculture.gouv.fr), ILO (http://laborsta.ilo.org/)

Notes:
1. Data from Géopolis
2. Data from FAO
3. Med. France = Mediterranean regions of France
4. NMC revised = NMC, excluding the non-Mediterranean regions of France
5. Mediterranean = NMC revised + SEMC

Table A13 Land use in Mediterranean countries (1000ha)

Countries	Total area	Arable land				Permanent crops				Permanent meadows and pastures			
	2000	1961	1990	2000	AAGR 1990–2000 (%)	1961	1990	2000	AAGR 1990–2000 (%)	1961	1990	2000	AAGR 1990–2000 (%)
Spain	50,599	16,246	15,335	13,256	−1.4	4484	4837	4965	0.3	12,500	10,300	11,462	1.1
France	55,150	19,606	17,999	18,440	0.2	1799	1191	1142	−0.4	13,134	11,380	10,124	−1.2
Italy	30,134	12,862	9012	8292	−0.8	2746	2960	2841	−0.4	5075	4868	4353	−1.1
Greece	13,196	2794	2899	2741	−0.6	906	1068	1113	0.4	5210	5255	4675	−1.2
Malta	32	17	12	8	−4.0	1	1	1	0.0				
Cyprus	925	100	106	101	−0.5	100	51	42	−1.9	5	5	4	−2.2
Slovenia	2025	—	200	173	−1.8	—	36	31	−1.9	—	328	314	−0.5
Croatia	5654	—	1212	1458	2.3	—	113	128	1.6	—	1079	1570	4.8
Bosnia-Herzegovina	5120	—	800	660	−2.4	—	240	150	−5.7	—	1200	1030	−1.9
Serbia-Montenegro	10,217	—	3720	3406	0.0	—	356	330	−0.9	—	2112	1851	−1.6
Yugoslavia SFR		7691				691				6570			
Albania	2875	432	579	578	0.0	47	125	121	−0.3	753	417	445	0.7
Turkey	77,482	23,013	24,647	24,138	−0.2	2154	3030	2534	−1.8	11,350	12,000	12,378	0.3
Syria	18,518	6146	4885	4542	−0.7	235	741	810	0.9	8560	7869	8359	0.6
Lebanon	1040	172	183	190	0.4	90	122	142	1.5	7	12	16	2.9
Israel	2106	318	348	338	−0.3	79	88	86	−0.2	114	148	142	−0.4
Palestinian Territories	38	8	9	11	2.0	12	10	8	−2.2				
Egypt	100,145	2499	2284	2821	2.1	69	364	466	2.5				
Libya	175,954	1700	1805	1815	0.1	270	350	335	−0.4	9200	13,300	13,300	0.0
Tunisia	16,361	3100	2909	2864	−0.2	1150	1942	2105	0.8	4398	3793	4561	1.9
Algeria	238,174	6472	7081	7662	0.8	594	554	530	−0.4	38,405	31,041	31,829	0.3
Morocco	44,655	6590	8707	8767	0.1	380	736	967	2.8	16,400	20,900	21,000	0.0
NMC[1]	175,927	59,748	51,874	49,113	−0.5	10,774	10,978	10,864	−0.1	43,247	36,944	35,828	0
SEMC	674,473	50,018	52,858	53,148	0.1	5033	7887	7972	0.1	87,286	88,630	90,755	0.2
MED	850,400	109,766	104,732	102,261	−0.2	15,807	18,868	18,771	−0.1	130,533	125,74	834,126	0.1
Med. France[2]	6828		551	517	−0.6		209	494	9.0		1540	1398	−1.0
NMC revised[3]	127,605		34,426	31,190	−1.0		9996	10,216	0.2		27,104	27,102	0.0
Mediterranean[4]	802,078		87,284	84,338	−0.3		17,883	18,188	0.2		115,734	117,857	0.2

Source: FAO, Eurostat, Montgolfier, 2002; French Ministry of Agriculture

Notes:
1 The NMC sum does not include the Yugoslavia SFR data
2 Med. France = Mediterranean regions of France
3 NMC revised = NMC, excluding the non-Mediterranean regions of France
4 Mediterranean = NMC revised + SEMC

Table A13 Land use in Mediterranean countries (continuation) (1000ha)

Countries	Total area	Total forest area	Natural Wood-lands	Forest Plantations	Total forests	AAGR Total forests (%)	Forest Plantations	Total forests	Forest area yearly damaged by fire (average)		Number of fire outbreaks (average)	
	2000	1990		2000		1990–2000	2000	2000	1981–1988	1989–1998	1981–1988	1989–1988
Spain	50,599	13,510	12,466	1,904,981	14,370	0.6	12,611	26	219	192	9373	18,217
France	55,150	14,725	14,380	961	15,341	0.4	1833	17,174	37	29	4555	6135
Italy	30,134	9708	9870	133	10,003	0.3	985	10,988	157	96	12,759	11,233
Greece	13,196	3299	3479	120	3599	0.9	3154	6753	56	48	1355	1727
Malta	32	—	—	—	—	—	—	—	—	—	—	—
Cyprus	925	119	172	0	172	3.8	214	386	4	0.4	72	33
Slovenia	2025	1085	1106	1	1107	0.2	67	1174	—	0.8	—	93
Croatia	5654	1763	1736	47	1783	0.1	330	2113	—	12	—	282
Bosnia-Herzegovina	5120	2273	2216	57	2273	0.0	433	2706	—	0.9	—	139
Serbia-Montenegro	10,217	2901	2848	39	2887	0.0	586	3473	—	—	—	—
Albania	2875	1069	886	102	991	–0.8		991	0.3	0.5	112	366
Turkey	77,482	10,005	8371	1854	10,225	0.2	10,695	20,920	11	12	1402	1934
Syria	18,518	461	232	229	461	0.0	35	496	—	—	—	—
Lebanon	1040	37	34	2	36	–0.3	35	71	—	—	—	—
Israel	2106	82	41	91	132	4.9	46	178	4	6	907	969
Palestinian Territories	38	—	—	—	—	—	—	—	—	—	—	—
Egypt	100,145	52	0	72	72	3.3	—	72	—	—	—	—
Libya	175,954	311	190	168	358	1.4	446	804	—	—	—	—
Tunisia	16,361	499	308	202	510	0.2	328	838	2	1.5	443	128
Algeria	238,174	1879	1427	718	2145	1.3	1662	3807	43	14	—	—
Morocco	44,655	3037	2491	534	3025	0.0	1265	4290	4	3.5	217	277
NMC	**175,927**	**50,452**	**49,159**	**3364**	**52,526**	**0.4**	**20,213**	**72,739**	**473**	**380**	**28,226**	**38,225**
SEMC	**674,473**	**16,363**	**13,094**	**3870**	**16,964**	**0.4**	**14,512**	**31,476**	**63**	**37**	**2969**	**3308**
MED	**850,400**	**66,815**	**62,253**	**7234**	**69,490**	**0.4**	**34,725**	**104,215**	**536**	**417**	**31,195**	**41,533**
Med. France[1]	6828	2211			2458	1.1	325	3783	31	20	2822	2668
NMC revised[2]	127,605	37,938			39,643	0.4	19,705	59,348	467	371	26,493	34,758
Mediterranean[3]	**802,078**	**54,301**			**56,607**	**0.4**	**34,217**	**90,824**	**530**	**408**	**29,462**	**38,066**

Source: FAO, Eurostat, Montgolfier, 2002; French Ministry of Agriculture

Notes:
1. Med. France = Mediterranean regions of France.
2. NMC revised = NMC, excluding the non-Mediterranean regions of France.
3. Mediterranean = NMC revised + SEMC.

Table A14 Agriculture and intensification

Countries	Arable land and permanent crops	Fertilizer consumption in kg/ha of arable land and permanent crops		Pesticide consumption in kg/ha of arable land and permanent crops		Organic agriculture (certified and conversion)	
	2000 1000 ha	2000 1000 ha	AAGR 1990–2000 (%)	2000 1000 ha	AAGR 1990–2000 (%)	2003 1000 ha	(% of arable land + permanent crops)
Spain	18,221	118.0	1.9	1.989	0.1	485	2.7
France	19,582	211.7	−3.3	4.978	−0.2	420	2.1
Italy	11,133	155.6	−0.4	13.707	−1.7	1230	11.0
Greece	3854	118.3	−3.9	2.759	4.3	31	0.8
Malta	9	83.3	3.4	36.100	−4.6	—	—
Cyprus	143	100.8	−3.6	—	—	0.052	0.0
Slovenia	204	370.4	4.1	13.862	12.1	5	2.6
Croatia	1586	140.6	0.2	—	—	0.120	0.0
Bosnia-Herzegovina	810	51.6	17.5	—	—	—	—
Serbia-Montenegro	3736	51.1	7.8	0.899	0.6	15	0.4
Albania	699	26.8	−15.5	—	—	—	—
Turkey	26,672	78.3	1.4	2.357	1.3	57	0.2
Syria	5352	68.3	2.4	1.156	16.7	0.074	0.0
Lebanon	332	157.4	6.6	5.571	—	0.250	0.1
Israel	424	221.7	−0.7	5.959	—	7	1.7
Palestinian Territories	19	—	—	—	—	—	—
Egypt	3287	382.8	0.6	2.256	−37.3	15	0.5
Libya	2150	25.6	−3.4	—	—	—	—
Tunisia	4969	22.3	2.6	0.364	—	18	0.4
Algeria	8192	11.3	−3.8	0.331	−13.9	—	—
Morocco	9734	37.7	0.2	—	—	12	0.1
NMC	**59,977**	**149**	**−1.7**			**2187**	**3.6**
SEMC	**61,131**	**73**	**1.3**			**110**	**0.2**
MED	**121,108**	**111**	**−0.9**			**2296**	**1.9**
Med. France[1]	1011					32	3.2
NMC revised[2]	41,406					1799	4.3
Mediterranean[3]	**102,537**					**1908**	**1.9**

Source: FAO, Eurostat, Montgolfier, 2002; Ministère de l'Agriculture et de la Pêche (France)

Notes:
1   Med. France = Mediterranean regions of France.
2   NMC revised = NMC, excluding the non-Mediterranean regions of France.
3   Mediterranean = NMC revised + SEMC.

Table A14 Agriculture et intensification (continuation)

Countries	Irrigated areas				Cereals				Olive groves				Vineyards			
	1961	1990	2000	AAGR 1990–2000	1961	1990	2002	AAGR 1990–2002	1961	1990	2003	AAGR 1990–2003	1961	1990	2003	AAGR 1990–2003
	(1000ha)			(%)	(1000ha)			(%)	(1000ha)			(%)	(1000ha)			(%)
Spain	1950	3402	3655	0.7	6930	7551	6717	−1.0		2064	2400	1.2	1742	1402	1166	−1.4
France	360	1970	2634	2.9	9140	9060	9307	0.2	45	15	17	1.0	1418	908	852	−0.5
Italy	2400	2711	2700	0.0	6387	4413	4305	−0.2	1229	1134	1141	0.0	1691	1024	868	−1.3
Greece	430	1195	1451	2.0	1773	1470	1296	−1.0	0	691	765	0.8	247	146	129	−0.9
Malta	1	1	2	7.2	4	2	3	1.7	0	0	0	−8.8	1	1	1	−0.7
Cyprus	30	36	40	1.1	142	58	56	−0.2	20	6	8	1.7	35	25	17	−3.0
Slovenia	—	2	3	5.2	—	119	98	−1.9	—	0.2	1	13.9	—	20	15	−2.6
Croatia	—	2	3	5.2	—	593	715	1.9	—	15	15	0.0	—	56	57	0.2
Bosnia-Herzegovina	—	2	3	5.2	—	304	399	2.8	—	—	—	—	—	5	4	−3.3
Serbia-Montenegro	—	78	20	−15.6	—	2399	2094	−1.3	—	2.7	1.5	−5.2	—	89	69	−2.3
Yugoslavia SFR	121				5455								272			
Albania	156	423	340	−2.2	348	321	175	−4.9	19	45	29	−3.3	10	14	7	−5.4
Turkey	1310	3800	4500	1.7	12,865	13,640	13,981	0.2	392	537	597	0.8	775	580	565	−0.2
Syria	558	693	1211	5.7	2116	4138	2974	−2.7	80	391	499	1.9	69	109	50	−5.8
Lebanon	41	86	104	1.9	90	41	54	2.3	27	43	58	2.3	24	29	14	−5.6
Israel	136	206	194	−0.6	155	114	87	−2.2	10	13	19	3.0	12	5	6	2.3
Palestinian Territories	8	11	12	0.9	—	2	2	0.0	0	0	0	—	—	0	8	29.5
Egypt	2568	2648	3219	2.2	1724	2283	2723	1.5	3	9	50	14.0	9	38	64	4.1
Libya	121	470	470	0.0	489	404	342	−1.4	0	60	90	3.2	3	7	8	0.7
Tunisia	100	300	380	2.4	1125	1443	430	−9.6	540	1392	1500	0.6	46	29	27	−0.5
Algeria	229	384	560	3.8	2709	2366	1850	−2.0	0	170	178	0.3	349	88	65	−2.3
Morocco	875	1258	1305	0.4	3773	5603	4955	−1.0	155	365	500	2.5	76	50	50	0.0
NMC[1]	**5448**	**9822**	**10,851**	**1.0**	**30,178**	**26,291**	**25,164**	**−0.4**	**1313**	**3974**	**4377**	**0.7**	**5416**	**3691**	**3185**	**−1.1**
SEMC	**5946**	**9856**	**12,027**	**2.0**	**25,046**	**30,034**	**27,398**	**−0.8**	**1207**	**2980**	**3490**	**1.2**	**1364**	**935**	**857**	**−0.7**
MED	**11,394**	**19,678**	**22,878**	**1.5**	**55,224**	**56,325**	**52,562**	**−0.6**	**2519**	**6954**	**7868**	**1.0**	**6780**	**4626**	**4042**	**−1.0**
Med. France[2]		334	304	−0.9		259	226	−1.1		16	17	0.5		557	410	−2.3
NMC revised[3]		8186	8521	0.4		17,490	16,083	−0.7		3974	4377	0.7		3340	2743	−1.5
Mediterranean[4]		**18,042**	**20,548**	**1.3**		**47,524**	**43,481**	**−0.7**		**6955**	**7867**	**1.0**		**4275**	**3600**	**−1.3**

Source: FAO, Eurostat, Montgolfier, 2002; Ministère de l'Agriculture et de la Pêche (France)

Notes:
1 The NMC sum includes the Yugoslavia SFR data for 1961.
2 Med. France = Mediterranean regions of France.
3 NMC revised = NMC, excluding the non-Mediterranean regions of France.
4 Mediterranean = NMC revised + SEMC.

Table A15 Trade in food and agricultural products, and agricultural added value

Countries	Agricultural export[1] (1000 US$ 1995)					Agricultural imports[1] (1000 US$ 1995)				
	1962	1970	1980	1991	2001	1962	1970	1980	1991	2001
Spain	386,785	728,375	3,504,169	8,731,636	14,985,267	436,564	904,975	4,110,284	9,091,596	11,573,627
France	1,299,096	3,168,208	17,250,124	34,163,489	33,078,925	2,308,853	3,272,555	13,991,269	23,642,467	23,903,173
Italy	757,265	1,248,316	5,783,387	12,059,651	16,248,078	1,701,090	3,462,174	13,944,250	24,302,973	21,571,702
Greece	208,258	326,365	1,288,725	2,820,511	2,502,988	117,817	254,399	1,158,265	3,141,371	3,367,604
Malta	4345	6147	30,236	40,336	61,152	30,340	43,101	161,475	209,157	262,090
Cyprus	26,552	60,746	180,025	326,965	352,268	22,245	37,539	169,973	271,750	613,382
Slovenia	—	—	—	448,682	362,217	—	—	—	611,121	745,022
Croatia	—	—	—	537,820	428,526	—	—	—	575,535	792,653
Bosnia-Herzegovina	—	—	—	40,946	41,631	—	—	—	138,903	638,934
Serbia-Montenegro	—	—	—	494,851	343,276	—	—	—	400,438	468,133
Yugoslavia SFR	*214,495*	*331,065*	*1,036,553*			*253,619*	*380,214*	*1,546,412*		
Albania	5861	19,604	110,708	46,025	23,426	12,787	10,592	37,923	180,218	262,733
Turkey	328,505	490,494	1,949,121	3,429,215	3,730,336	113,212	98,183	267,423	1,866,501	2,907,760
Syria	138,857	143,925	273,871	667,816	792,086	51,865	106,434	613,933	731,265	762,318
Lebanon	33,480	70,723	205,593	129,567	139,819	115,412	158,821	587,450	889,318	1,157,159
Israel	92,618	202,145	870,641	1,189,001	991,609	120,253	224,154	936,048	1,250,347	1,859,489
Palestinian Territories		11,367	81,198	49,820	100,052	—	7853	37,315	42,838	581,600
Egypt	338,303	526,911	674,545	406,407	637,652	242,023	250,392	2,551,398	2,717,815	3,418,643
Libya	3942	1196	29	46,631	28,529	26,637	125,500	1,223,768	1,256,263	1,283,268
Tunisia	73,459	62,255	190,761	365,592	424,476	58,043	85,255	518,450	569,281	874,275
Algeria	267,628	169,863	119,202	60,063	33,899	217,796	182,524	2,097,893	2,429,119	2,737,170
Morocco	157,583	231,893	515,821	632,511	736,173	129,374	157,354	931,889	931,954	1,692,595
NMC	**2,902,657**	**5,888,825**	**29,183,928**	**59,710,912**	**68,427,754**	**4883,315**	**8,365,548**	**35,119,852**	**62,565,528**	**64,199,053**
SEMC	**1,434,374**	**1,910,772**	**4,880,782**	**6,976,624**	**7,614,631**	**1,074,617**	**1,396,470**	**9,765,567**	**12,684,700**	**17,274,277**
MED	**4,337,031**	**7,799,598**	**34,064,710**	**66,687,536**	**76,042,385**	**5,957,932**	**9,762,018**	**44,885,419**	**75,250,228**	**81,473,330**
NMC – FR[2]	1,603,561	2,720,617	11,933,804	25,547,423	35,348,830	2,574,462	5,092,993	21,128,583	38,923,061	40,295,880
MED – FR[3]	2,868,749	3,821,545	9,761,565	13,953,249	15,229,261	2,149,234	2,792,940	19,531,135	25,369,401	34,548,555

Source: FAOSTAT, 2003 (http://faostat.fao.org/)

Notes:
1 The data in this table correspond to three-year averages of the imports and of the exports (calculated by *Plan Bleu*).
2 NMC – FR = NMC excluding France.
3 MED – FR = MED excluding France.

Statistics annex

Table A15 Trade in food and agricultural products, and agricultural added value (continuation)

Countries	Agricultural trade balance[1] (1000 US$ 1995)					Agricultural added value per active agricultural worker (US$ 1995)			
	1962	1970	1980	1991	2001	1971	1981	1991	2001
Spain	−49,779	−176,600	−606,114	−359,960	3,411,640	4720	7431	14,666	22,580
France	−1,009,757	−104,347	3,258,854	10,521,022	9,175,752	12,149	19,961	32,482	60,468
Italy	−943,824	−2,213,859	−8,160,863	−12,243,322	−5,323,624	7766	11,416	16,128	27,654
Greece	90,441	71,966	130,461	−320,859	−864,616	6034	8977	11,643	13,850
Malta	−25,995	−36,954	−131,240	−168,820	−200,938	—	—	—	—
Cyprus	4306	23,207	10,052	55,216	−261,114	—	—	—	—
Slovenia	—	—	—	−162,439	−382,805	—	—	16,608	39,172
Croatia	—	—	—	−37 715	−364,127	—	—	6303	10,098
Bosnia-Herzegovina	—	—	—	−97 957	−597,303	—	—	—	—
Serbia-Montenegro	—	—	—	94,413	−124,857	—	—	—	—
Yugoslavia SFR	*−39,124*	*−49,149*	*−509,859*						
Albania	−6926	9012	72,785	−134,193	−239,307	—	1183	989	1900
Turkey	215,293	392,311	1,681,698	1,562,715	822,576	1717	1840	1832	1787
Syria	86,991	37,491	−340,061	−63,449	29,768	933	2468	2103	2669
Lebanon	−81,932	−88,098	−381,857	−759,751	−1,017,340	—	—	—	30,832
Israel	−27,635	−22,009	−65,407	−61,346	−867,880	—	—	—	—
Palestinian Territories	—	3513	43,883	6982	−481,548	—	—	—	—
Egypt	96,279	276,519	−1,876,853	−2 311,408	−2,780,992	618	736	1059	1332
Libya	−22,695	−124,304	−1,223,739	−1,209,632	−1,254,739	—	—	—	—
Tunisia	15,416	−23,000	−327 689	−203,689	−449,799	1377	1863	3055	3072
Algeria	49,832	−12,661	−1,978,691	−2,369,056	−2,703,271	663	1,408,914	1988	2013
Morocco	28,209	74,539	−416,068	−299,443	−956,422	1415	914	2162	1693
NMC	**−1,980,658**	**−2,476,723**	**−5,935,924**	**−2,854,616**	**4,228,701**				
SEMC	**359,757**	**514,302**	**−4,884,785**	**−5,708,076**	**−9,659,647**				
MED	**−1,620,900**	**−1 962,420**	**−10,820,709**	**−8,562,692**	**−5,430,945**				
NMC − FR[2]	−970,901	−2,372,376	−9,194,778	−13,375,638	−4,947,050				
MED − FR[3]	719,515	1,028,605	−9,769,570	−11,416,152	−19,319,293				

Source: FAOSTAT, 2003 (http://faostat.fao.org/)

Note:
1 The data of the trade balance in this table correspond to a 3-year average (calculated by *Plan Bleu*).
2 NMC − FR = NMC excluding France.
3 MED − FR = MED excluding France.

Fisheries and aquaculture

Table A16 Mediterranean fisheries

Fishing area	Captures in the Mediterranean (excluding Black Sea, Azov and Marmara seas) in tonnes						
	1970	1975	1980	1985	1990	1995	2000
The Adriatic sea	144,516	195,378	216896	220,359	149,686	144,563	155,271
The Aegean sea	44,043	62,866	91,392	109,236	133,430	172,663	114,551
The Balearic islands	128,668	190,507	218,758	232,823	249,853	277,625	261,373
Golfe du Lion	40,971	35,514	34,983	36,381	40,790	30,636	41,635
The Ionian	83,884	147,059	156,256	245,860	242,003	273,841	199,617
The Levant	22,265	21,241	34,141	37,789	70,668	79,153	80,915
Sardinia	149,436	93,647	75,917	102,997	59,386	84,740	77,902
Unknown (GFCM area)	—	—	—	—	—	24	—
Tunas (GFCM area)	34,093	26,924	44,998	66,568	69,233	75,459	71,577
Total	647,876	773,136	873,342	1,052,012	1,015,049	1,138,702	1,002,841
of which the Atlantic red tunny	4694	11,266	10,040	19,296	17,238	37,560	23,092
Lobsters, spiny-rock lobsters	3553	4363	4045	7473	8758	7539	4057
Marine fishes not identified	65,821	119,766	116,098	113,148	102,124	110,210	67,690
River eels	5494	3183	3156	3042	2638	849	464
Squids, cuttlefishes, octopuses	44,587	47,024	51,659	65,722	72,129	62,526	53,805
Clams, cockles, arkshells	17,813	55,320	30,321	26,954	24,141	38,493	37,403
Herrings, sardines, anchovies	244,829	274,138	345,178	364,026	339,871	380,416	372,518
Miscellaneous pelagic fishes	56,989	50,021	55,177	87,224	68,863	95,066	71,913
Miscellaneous demersal fishes	9203	8026	10,412	17,735	16,952	17,731	17,280
Cods, hakes, haddocks	26,749	32,895	41,936	60,458	46,037	73,466	42,943
Flounders, halibuts, soles	6221	6948	8634	11,031	12,295	12,624	7724
Miscellaneous coastal fishes	91,381	105,804	115,801	140,874	158,652	155,934	142,674
Mussels	14,290	5565	9560	11,897	20,409	34,249	44,831
Shrimps, prawns	10,613	14,677	14,896	29,701	33,718	26,181	32,872

Source: FAO GFCM capture production 1970–2001 (www.fao.org/fi/statist/FISOFT/FISHPLUS.asp)

Note: GFCM = General Fisheries Commission for the Mediterranean.

Table A17 Mediterranean aquaculture

Aquaculture: production (aquatic plants excluded) in tonnes in the Mediterranean – Black Sea (marine + brackish waters)							
	1970	1975	1980	1985	1990	1995	2000
Mollusc – Total	16,304	39,254	55,926	82,613	135,798	185,885	190,618
France	4004	8732	8520	10,454	25,782	29,000	17,181
Italy	12,000	30,000	46,948	70,001	100,910	144,000	147,000
Greece	—	—	18	200	3686	10,889	24,356
Fish – Total	1693	2483	3782	4979	13,083	51,737	177,070
of which total Bass+ sea bream	10	110	375	902	7788	42,864	145,344
France	—	—	—	85	330	3640	4200
Italy	10	110	370	700	1900	6800	14,100
Greece	—	—	5	67	3550	18,926	65,240
Malta	—	—	—	—	—	900	1746
Cyprus	—	—	—	1	52	322	1683
Slovenia	—	—	—	—	—	49	73
Croatia	—	—	—	—	—	337	2100
Serbia-Montenegro	—	—	—	—	—	386	2173
Albania	—	—	—	—	—	—	—
Turkey	—	—	—	—	1133	7620	33,337
Israel	—	—	—	—	84	930	2701
Egypt	—	—	—	—	—	1817	18,893
Tunisia	—	—	—	20	368	390	611
Algeria	—	—	—	4	4	10	20
Morocco	—	—	—	—	157	1123	640
of which total grey mullet	1270	1820	2500	3205	3006	7246	26,793
Italy	1270	1820	2500	3200	3000	3000	3,000
Greece	—	—	—	—	—	443	526
Egypt	—	—	—	—	—	3800	23,265
Tunisia	—	—	—	5	6	3	2
Crustaceans – Total	—	—	—	17	78	130	119
Total	**18,297**	**46,995**	**59,728**	**87,936**	**148,983**	**238,052**	**367,820**

Aquaculture: value (aquatic plants excluded) in 1000 US$ in the Mediterranean – Black Sea (marine + brackish waters)							
	1970	1975	1980	1985	1990	1995	2000
North				87,252	291,760	526,338	68,275
South and east				274	18,865	120,750	372,353
Mediterranean countries				**87,527**	**310,625**	**647,088**	**1,058,627**

Source: FAO, Fishstat 2003 (www.fao.org/fi/statist/FISOFT/FISHPLUS.asp)

Energy and transport

Table A18 Energy intensity, per country, trends and scenarios to 2025 (in toe/US$ constant 95)

Countries	1971	1980	1992	2000	Baseline scenario 2025	AAGR (%) 1980–2000	AAGR (%) 1992–2000	AAGR (%) 2000–2025
Spain	0.143	0.168	0.165	0.172	0.107	0.13	0.51	−1.88
France	0.177	0.165	0.148	0.140	0.102	−0.80	−0.66	−1.27
Italy	0.196	0.170	0.149	0.140	0.106	−0.96	−0.74	−1.12
Greece	0.129	0.155	0.195	0.194	0.174	1.14	−0.03	−0.45
Malta	0.426	0.237	0.273	0.204	0.272	−0.74	−3.53	1.16
Cyprus	—	0.241	0.226	0.222	0.171	−0.42	−0.25	−1.05
Slovenia	—	—	—	0.263	0.110	—	—	−3.44
Croatia	—	—	—	0.329	0.196	—	—	−2.04
Bosnia-Herzegovina	—	—	—	0.689	0.238	—	—	−4.17
Serbia-Montenegro	—	—	—	1.020	0.795	—	—	−0.99
Albania	—	—	—	0.513	0.393	—	—	−1.06
Turkey	0.213	0.297	0.307	0.345	0.390	0.75	1.49	0.49
Syria	1.013	0.818	1.115	1.346	0.791	2.52	2.39	−2.10
Lebanon	—	—	0.308	0.394	0.287	—	3.12	−1.26
Israel	0.221	0.192	0.184	0.190	0.168	−0.05	0.36	−0.49
Palestinian Territories	—	—	—	0.277	0.292	—	—	0.21
Egypt	0.440	0.508	0.631	0.576	0.363	0.63	−1.15	−1.83
Libya	—	—	0.302	0.477	0.319	—	5.87	−1.60
Tunisia	0.240	0.294	0.294	0.282	0.210	−0.22	−0.53	−1.16
Algeria	0.232	0.391	0.640	0.594	0.569	2.12	−0.93	−0.17
Morocco	0.177	0.212	0.235	0.251	0.205	0.85	0.80	−0.80
NMC	0.176	0.167	0.153	0.154	0.113	−0.39	0.08	−1.25
SEMC	0.255	0.323	0.371	0.393	0.354	0.99	0.72	−0.42
TOTAL	0.182	0.180	0.178	0.184	0.156	0.13	0.42	−0.66
					Alternative scenario			
NMC	0.176	0.167	0.153	0.154	0.101	−0.39	0.08	−1.66
SEMC	0.255	0.323	0.371	0.393	0.273	0.99	0.72	−1.44
TOTAL	0.182	0.180	0.178	0.184	0.133	0.13	0.42	−1.31

Source: World Bank, *World Development Indicators*, 2002; OME (2002)

Statistical annex

Table A19 Primary energy supply sources per country, baseline and alternative scenarios, 2000-2025

Baseline scenario	Primary commercial energy total demand[1] (Mtoe)		Share of renewable energy in % of the total energy balance sheets (TPES)			
	2000	2025	A 2000	B 2000	A 2025	B 2025
Spain	121	157	6.0	2.5	7.0	4.2
France	246	317	6.8	2.4	7.6	4.1
Italy	169	214	5.3	4.0	6.0	5.0
Greece	27	50	5.3	1.6	7.0	5.1
Malta	1	3	0.0	0.0	0.9	0.9
Cyprus	2	5	2.0	1.5	1.5	1.3
Slovenia	6	8	12.0	5.0	10.2	4.6
Croatia	7	15	11.3	6.5	7.7	5.3
Bosnia-Herzegovina	4	4	14.2	10.0	15.8	11.8
Serbia-Montenegro	13	28	9.4	7.6	7.8	7.0
Albania	2	4	29.3	25.7	19.8	18.3
Turkey	71	251	12.5	4.1	6.4	3.9
Syria	18	39	4.6	4.4	2.6	2.4
Lebanon	5	12	3.4	0.9	2.3	1.2
Israel	20	37	3.4	2.9	2.3	2.1
Palestinian Territories	1	3	25.6	9.8	12.1	5.3
Egypt	45	83	5.4	2.7	4.4	2.8
Libya	16	29	0.9	0.0	0.6	0.1
Tunisia	7	16	15.8	0.1	9.3	2.2
Algeria	29	69	0.3	0.0	0.5	0.4
Morocco	10	20	5.0	0.7	6.4	4.3
NMC	**599**	**806**	**6.4**	**3.1**	**7.1**	**4.6**
SEMC	**222**	**560**	**7.0**	**2.6**	**4.6**	**2.8**
TOTAL	**821**	**1365**	**6.6**	**3.0**	**6.1**	**3.9**
Alternative scenario						
NMC	**599**	**724**	**6.4**	**3.1**	**19.1**	**16.4**
SEMC	**222**	**433**	**7.0**	**2.6**	**10.2**	**7.9**
TOTAL	**821**	**1157**	**6.6**	**3.0**	**15.8**	**13.2**

Source: OME from the International Energy Agency from 1971 to 2000, energy balances of the OECD and non-OECD countries, 2001.

Notes:
1   total demand of energy, excluding CWR

A: Share of total renewables in TPES (%);(Hydro+REn+CWR)/TPES.
B: Share of renewables excluding combustible renewables and waste in TPES (%); (Hydro+REn)/TPES.
CWR: Combustible Renewables and Waste: Solid biomass, gas/liquid from biomass, Municipal and Industrial waste.
TPES: Total Primary Energy Supply (methodology IEA); TPES = Total+CWR.

For availability and comparability reasons, the tables only concern primary commercial energy balance sheets; they exclude non-commercial primary energy, in particular the biomass; they therefore have a tendency to underestimate the shares of renewable energy in the energy balance sheet.

For example, France would have had only 2.4 per cent of renewable energies (REns+HYDRO) in the primary commercial energy balance sheet in 2000 whereas with the biomass, this share should be nearly 6.8 per cent of the total primary energy balance sheet (columns A and B). Likewise, by including the biomass in the renewable energies, Turkey would have already had over 12 per cent of its total primary energy demand in the form of renewable energies in 2000, whereas they were in fact under 5 per cent of its primary commercial energy balance sheet.

To convert electricity production and trade into primary energy, the coefficient of 0.086 toe/MWh was used, in accordance with the method recommended by the IEA.

Trend scenarios built by the countries on the basis of different assumptions may lead to other projections. For example, according to Syria's projections the share of renewable energy in TPES in 2025 would be of 3 per cent (figure from the Ministry of Local Administration and Environment).

Table A19 Primary energy supply sources per country, baseline and alternative scenarios, 2000–2025 (continuation)

Baseline scenario

	Supply sources (%)											
	Coal		Oil		N. Gas		Nucl.		Hydro		REn	
	2000	2025	2000	2025	2000	2025	2000	2025	2000	2025	2000	2025
Spain	17	11	54	50	13	27	13	8	2	2	0.5	2.3
France	6	4	35	38	14	19	44	35	2	2	0.1	1.9
Italy	7	8	52	42	34	43	0	0	2	2	1.8	2.7
Greece	34	29	58	54	6	11	0	0	1	2	0.5	3.6
Malta	0	0	100	98	0	1	0	0	0	0	0.0	0.9
Cyprus	1	5	99	90	0	4	0	0	0	0	1.5	1.3
Slovenia	21	21	40	40	14	22	20	14	5	4	0.0	0.5
Croatia	6	3	53	52	30	36	0	0	7	5	0.0	0.3
Bosnia-Herzegovina	64	18	22	39	5	30	0	0	10	12	0.0	0.4
Serbia-Montenegro	55	46	23	24	12	23	0	0	8	7	0.0	0.5
Albania	1	0	67	25	1	56	0	0	27	18	0.0	1.1
Turkey	33	38	44	29	18	30	0	0	4	3	0.7	0.5
Syria	0	0	71	56	25	41	0	0	4	2	0.0	0.2
Lebanon	3	1	94	53	0	45	0	0	1	1	0.1	0.6
Israel	33	29	65	35	0	34	0	0	0	0	3.0	2.1
Palestinian Territories	0	0	71	39	0	55	0	0	0	0	11.6	5.7
Egypt	2	1	59	50	36	46	0	0	3	2	0.0	1.3
Libya	0	0	74	41	26	59	0	0	0	0	0.0	0.1
Tunisia	1	1	58	41	41	56	0	0	0	0	0.0	2.2
Algeria	2	1	29	30	69	68	0	0	0	0	0.0	0.4
Morocco	26	21	70	51	0	23	0	0	1	2	0.1	2.3
NMC	**12**	**10**	**45**	**43**	**19**	**27**	**21**	**16**	**3**	**3**	**1**	**2**
SEMC	**15**	**20**	**54**	**37**	**27**	**40**	**0**	**0**	**2**	**2**	**1**	**1**
TOTAL	**13**	**14**	**48**	**40**	**21**	**33**	**15**	**9**	**2**	**2**	**1**	**2**

Alternative scenario

	Coal		Oil		N. Gas		Nucl.		Hydro		REn	
	2000	2025	2000	2025	2000	2025	2000	2025	2000	2025	2000	2025
NMC	**12**	**8**	**45**	**35**	**19**	**24**	**21**	**16**	**3**	**3**	**1**	**14**
SEMC	**15**	**18**	**54**	**32**	**27**	**41**	**0**	**0**	**2**	**3**	**1**	**5**
TOTAL	**13**	**12**	**48**	**34**	**21**	**30**	**15**	**10**	**2**	**3**	**1**	**11**

Source: OME from the International Energy Agency from 1971 to 2000, Energy balances of the OECD and non-OECD countries.

Notes:
Hydro: energy from hydro power plants. It includes large as well as small hydro.
REn: Geothermal energy, solar energy and wind energy.

Statistical annex

Table A20 Primary commercial energy demand, baseline and alternative scenarios, 1971–2025

Baseline scenario

	Total consumption of primary energy (Mtoe)					AAGR (%)		Primary energy consumption per capita (koe/capita)			AAGR (%)	
	1971	1980	1992	2000	2025	1971 –2000	2000 –2025	1971	2000	2025	1971 –2000	2000 –2025
Spain	43	69	93	121	157	3.6	1.1	1267	3031	3857	3.1	0.8
France	155	190	223	246	317	1.6	1.0	3058	4146	4941	1.1	0.6
Italy	115	140	157	169	214	1.3	1.0	2147	2940	3974	1.1	1.0
Greece	9	16	22	27	50	3.8	2.5	1050	2549	4809	3.1	2.2
Malta	0	0	1	1	3	3.8	5.2	860	2095	6734	3.1	4.1
Cyprus	1	1	2	2	5	4.6	2.9	1036	3010	5379	3.7	2.0
Slovenia	—	—	—	6	8	—	1.0	—	3102	3835	—	0.7
Croatia	—	—	—	7	15	—	2.9	—	1656	3605	—	2.7
Bosnia-Herzegovina	—	—	—	4	4	—	0.1	—	1053	1002	—	−0.2
Serbia-Montenegro	—	—	—	13	28	—	3.0	—	1239	2289	—	2.1
Yugoslavia SFR	*21*	*33*	*30*					*1119*				
Albania	—	—	—	2	4		3.7	0	506	1029	—	2.5
Turkey	13	26	48	71	251	6.0	5.2	364	1076	2879	3.8	3.5
Syria	3	5	11	18	39	6.8	3.1	431	1147	1623	3.4	1.2
Lebanon	2	3	3	5	12	3.4	3.7	866	1538	2935	2.0	2.3
Israel	6	9	13	20	37	4.1	2.4	2165	3451	4647	1.6	1.0
Palestinian Territories	—	0	0	1	3	—	3.9	—	356	477	—	1.0
Egypt	7	15	34	45	83	6.5	2.5	222	684	879	4.0	0.9
Libya	2	7	11	16	29	8.4	2.3	787	3081	3288	4.8	0.2
Tunisia	1	3	5	7	16	5.6	3.6	266	692	1255	3.3	2.1
Algeria	4	12	27	29	69	7.4	3.5	266	956	1627	4.5	1.9
Morocco	2	5	8	10	20	4.9	3.0	162	346	537	2.7	1.5
NMC	**344**	**450**	**528**	**599**	**806**	**1.9**	**1.2**	**2039**	**3105**	**4084**	**1.5**	**0.9**
SEMC	**40**	**84**	**158**	**222**	**560**	**6.1**	**3.8**	**345**	**951**	**1715**	**3.6**	**2.1**
TOTAL	**384**	**534**	**686**	**821**	**1365**	**2.7**	**2.1**	**1347**	**1925**	**2607**	**1.2**	**1.1**

Alternative scenario

NMC	**344**	**450**	**528**	**599**	**724**	**1.9**	**0.8**	**2039**	**3105**	**3673**	**1.5**	**0.6**
SEMC	**40**	**84**	**158**	**222**	**433**	**6.1**	**2.7**	**345**	**951**	**1326**	**3.6**	**1.2**
TOTAL	**384**	**534**	**686**	**821**	**1157**	**2.7**	**1.4**	**1347**	**1925**	**2210**	**1.2**	**0.5**

Source: OME from the International Energy Agency from 1971 to 2000, Energy balances of OECD and non-OECD countries

Table A21 Electricity production by source, per country, baseline and alternative scenarios, 2000–2025

Baseline scenario

	Electricity production (TWh)		Electricity sources (%)											
			Coal		Oil		N. Gas		Nucl.		Hydro.		REn	
	2000	2025	2000	2025	2000	2025	2000	2025	2000	2025	2000	2025	2000	2025
Spain	222	363	36	28	10	7	9	28	28	14	13	10	3	12
France	535	669	3	0	2	1	4	8	78	67	13	13	1	10
Italy	270	381	11	4	32	9	38	61	0	0	16	15	3	11
Greece	53	102	64	43	17	11	11	18	0	0	7	9	1	20
Malta	2	5	0	34	100	60	0	0	0	0	0	0	0	6
Cyprus	3	9	0	0	100	97	0	0	0	0	0	0	0	3
Slovenia	14	18	34	33	1	3	2	11	35	27	28	23	1	3
Croatia	11	22	14	5	16	27	15	25	0	0	55	42	0	2
Bosnia-Herzegovina	10	12	51	44	0	4	0	0	0	0	49	50	0	2
Serbia-Montenegro	32	75	56	56	3	3	3	11	0	0	38	29	0	2
Albania	5	11	0	0	1	5	0	15	0	0	99	75	0	5
Turkey	125	538	31	33	8	7	36	39	0	0	25	19	0	2
Syria	23	65	0	0	22	22	36	62	0	0	41	15	0	2
Lebanon	8	18	0	0	94	8	0	82	0	0	6	5	0	4
Israel	43	72	69	56	31	0	0	42	0	0	0	0	0	3
Palestinian Territories	0	9	—	0	—	2	—	93	—	0	—	0	—	4
Egypt	76	197	0	0	16	7	65	79	0	0	19	8	0	7
Libya	15	44	0	0	78	2	22	97	0	0	0	0	0	1
Tunisia	11	40	0	0	12	10	87	79	0	0	1	1	0	11
Algeria	25	72	0	0	3	6	97	90	0	0	0	0	0	4
Morocco	14	49	58	41	36	0	0	37	0	0	5	10	1	11
NMC	**1157**	**1664**	**16**	**13**	**12**	**6**	**13**	**26**	**42**	**30**	**15**	**14**	**2**	**11**
SEMC	**340**	**1104**	**22**	**22**	**20**	**7**	**41**	**56**	**0**	**0**	**16**	**12**	**0**	**4**
TOTAL	**1497**	**2768**	**18**	**16**	**14**	**6**	**20**	**38**	**32**	**18**	**15**	**13**	**1**	**8**

Alternative scenario

	2000	2025	2000	2025	2000	2025	2000	2025	2000	2025	2000	2025	2000	2025
NMC	1157	1524	16	7	12	1	13	17	42	31	15	16	2	29
SEMC	340	736	22	15	20	2	41	51	0	0	16	18	0	14
TOTAL	1497	2260	18	9	14	1	20	28	32	21	15	16	1	24

Source: OME (2002)

Notes:
Hydro: energy from hydro power plants. It includes large as well as small hydro.
REn: geothermal energy, solar energy and wind energy, excluding bio-masse.

Trend scenarios built by the countries on the basis of different assumptions may lead to other projections. For example, Syria projects a total electricity production of 76.6 TWh in 2025 (figure from the Ministry of Local Administration and Environment).

Statistical annex

Table A22 Electricity demand, baseline and alternative scenarios, 1971–2025

Baseline scenario

	Total electricity consumption							Electricity consumption per capita				
	TWh					AAGR (%)		kWh/capita			AAGR (%)	
	1971	1980	1992	2000	2025	1971–2000	2000–2025	1971	2000	2025	1971–2000	2000–2025
Spain	62	110	156	222	363	4.5	2.0	1810	5568	8892	4.0	1.6
France	156	258	458	535	669	4.3	0.9	3082	9006	10,424	3.8	0.5
Italy	121	186	223	270	381	2.8	1.4	2256	4698	7060	2.6	1.4
Greece	12	23	37	53	102	5.4	2.6	1327	5060	9834	4.7	2.3
Malta	0	0	1	2	5	6.5	3.9	958	4928	11,628	5.8	3.0
Cyprus	1	1	2	3	9	5.8	3.8	1081	4293	9556	4.9	2.8
Slovenia	—	—	—	14	18	—	1.0	0	6933	8674	—	0.8
Croatia	—	—	—	11	22	—	2.8	0	2399	5128	—	2.7
Bosnia-Herzegovina	—	—	—	10	12	—	0.6	0	2626	2775	—	0.2
Serbia-Montenegro	—	—	—	32	75	—	3.5	0	2938	6114	—	2.6
Yugoslavia SFR	*30*	*60*	*83*					*1610*				
Albania	—	—	—	5	11	—	3.1	0	1587	2775	—	1.9
Turkey	9	23	67	125	538	9.7	6.0	242	1904	6162	7.4	4.1
Syria	1	4	12	23	65	10.2	4.3	214	1420	2708	6.7	2.3
Lebanon	1	3	4	8	18	6.2	3.4	631	2445	4389	4.8	2.0
Israel	8	13	24	43	72	6.1	2.1	2603	7339	9159	3.6	0.8
Palestinian Territories	—	—	—	0	9	—	—	0	0	1482	—	—
Egypt	8	19	46	76	197	8.1	3.9	247	1151	2076	5.4	2.1
Libya	1	5	11	15	44	11.4	4.2	340	2911	4925	7.7	1.8
Tunisia	1	3	6	11	40	8.8	5.5	181	1101	3103	6.4	3.6
Algeria	2	7	18	25	72	8.8	4.3	164	838	1708	5.8	2.5
Morocco	2	5	10	14	49	6.5	5.0	152	497	1270	4.2	3.3
NMC	**381**	**638**	**961**	**1157**	**1664**	**3.9**	**1.5**	**2259**	**6000**	**8439**	**3.4**	**1.2**
SEMC	**33**	**81**	**199**	**340**	**1104**	**8.4**	**4.8**	**284**	**1455**	**3380**	**5.8**	**2.9**
TOTAL	**414**	**719**	**1160**	**1497**	**2768**	**4.5**	**2.5**	**1452**	**3511**	**5285**	**3.1**	**1.4**

Alternative scenario

NMC	**381**	**638**	**961**	**1157**	**1524**	**3.9**	**1.1**	**2259**	**6000**	**7712**	**3.4**	**0.9**
SEMC	**33**	**81**	**199**	**340**	**736**	**8.4**	**3.1**	**284**	**1455**	**2696**	**5.8**	**2.1**
TOTAL	**414**	**719**	**1160**	**1497**	**2260**	**4.5**	**1.7**	**1452**	**3511**	**4802**	**3.1**	**1.1**

Source: OME, *Plan Bleu*, 2002

Table A23 Electricity production potential from renewable energies in the SEMC in 2020

	Potential (MW)	Electricity produced (TWh/yr)
Wind energy	10,000	20
Photovoltaic	2500	5
Solar thermal	6000	15
Biomass	8000	48
Geothermal	2900	17
Total	**29,400**	**105**

Source: OME (2002)

Table A24 The fluctuation of low-tension electricity prices (residential sector) in a number of SEMC

Country	Date Available price	1st block kWh per month	Price in 1/100 Euro	2nd Block kWh per month	Price in 1/100 Euro	3rd block kWh per month	Price in 1/100 Euro	4th block kWh per month	Price in 1/100 Euro	5th block kWh per month	Price in 1/100 Euro
Malta	11/93	<33	5.56	<67	5.44	>67	6.92				
Turkey	2000		3.64								
Syria	1996	<50	0.60	<100	0.84	<200	1.20	<300	1.81	>300	3.61
Lebanon	1993	<100	2.84	<300	4.46	<400	6.49	<500	9.73	>500	16.22
Palestinian Territories	2001	Const.	11.06								
Egypt	1/5/94	<50	1.42	<200	2.36	<350	3.13	<650	4.26	<1000	5.97
Tunisia	1996	<50	5.25	>50	7.14						
Algeria	1997	<41.7	2.18	>41.7	4.73						
Morocco	5-2001	<100	8.61	<200	9.26	<500	10.08	>500	13.77		

Source: Cornut, Ademe, 2001

Note: Prices have been converted into euros by using the exchange rates of May 2001 (infoeuro site: http://europa.eu.int/comm/budget/inforeuro):

Country	1 EURO =
Malta	0.40440 MTL
Turkey	1 050 000 TRL
Syria	41.5058 SYP
Lebanon	1356.6 LBP
Palestinian Territories	3.75041 ILS
Egypt	3.5184 EGP
Tunisia	1.27 9 TND
Algeria	66.8524 DZD
Morocco	9.7754 MAD

Statistical annex

Table A25 Motorization rate (Number of vehicles per 1000 inhabitants)

	1985	1990	1995	2000	2003
Spain	337	424	485	573	611
France	506	541	584	646	652
Italy	477	555	597	642	679
Greece	222	289	340	366	513
Monaco	645	694	764	830	—
Malta	316	407	513	705	750
Cyprus	291	414	527	631	615
Slovenia	—	372	429	528	500
Croatia	—	196	175	246	324
Bosnia-Herzegovina	—	142	19	37	299
Serbia-Montenegro	—	166	210	277	170
Albania	8	17	36	52	84
Turkey	43	59	92	133	134
Syria	43	39	53	60	76
Lebanon	269	513	544	587	519
Israel	293	333	420	463	482
Palestinian Territories	—	84	103	156	—
Egypt	41	44	49	58	64
Libya	385	388	464	414	367
Tunisia	70	75	95	127	109
Algeria	75	99	94	107	140
Morocco	57	63	78	77	84
NMC	**386**	**463**	**503**	**562**	**592**
SEMC	**65**	**81**	**100**	**120**	**124**
MED	**260**	**307**	**330**	**362**	**373**

Source: Plan Bleu estimates 2004, from IRF, World Road Statistics (www.irfnet.org/wrs.asp) and SMMT (www.smmt.co.uk), CCFA (www.ccfa.fr), ACEA (www.acea.be)

Coastal regions

Table A26 Population of Mediterranean coastal regions, trends and baseline scenario to 2025					
Country	1970	2000	2025	AAGR 1970–2000	AAGR 2000–2025
	(1000 inhabitants)			(%)	
Spain	12,245	15,560	16,464	0.8	0.2
France	4770	6265	7151	0.9	0.5
Italy	30,339	32,837	30,291	0.3	−0.3
Greece	7703	9482	9412	0.7	0.0
Monaco	24	34	41	1.2	0.8
Malta	319	389	430	0.7	0.4
Cyprus	615	785	900	0.8	0.5
Slovenia	90	101	106	0.4	0.2
Croatia	1434	1529	1480	0.2	−0.1
Bosnia-Herzegovina	203	226	246	0.4	0.3
Serbia-Montenegro	215	266	289	0.7	0.3
Albania	844	1193	1468	1.2	0.8
Turkey	6606	13,691	19,514	2.5	1.4
Syria	692	1533	2040	2.7	1.1
Lebanon	1884	2906	3766	1.5	1.0
Israel	2623	5045	6585	2.2	1.1
Palestinian Territories	442	1119	1996	3.1	2.3
Egypt	11,262	22,929	32,633	2.4	1.4
Libya	1694	5179	7784	3.8	1.6
Tunisia	3545	6762	9154	2.2	1.2
Algeria	5794	12,271	17,669	2.5	1.5
Morocco	1678	3233	4586	2.2	1.4
Total NMC	58,801	68,667	68,278	0.5	0.0
Total east	12,247	24,294	33,901	2.3	1.3
Total south	23,973	50,374	71,826	2.5	1.4
Total SEMC	36,220	74,668	105,727	2.4	1.4
Total Med	95,021	143,335	174,005	1.4	0.8

Source: Plan Bleu, Attané and Courbage, 2001

Statistical annex

Table A27 Density of the Mediterranean coastal regions compared with the density of the country, trends and baseline scenario to 2025 (inhabitants/km^2)

	1970		2000		2025	
	Country	Med coastal regions	Country	Med coastal regions	Country	Med coastal regions
Spain	67	128	79	163	81	172
France	93	103	109	135	118	155
Italy	178	183	191	198	179	183
Greece	66	76	80	94	79	93
Monaco	12,000	12,000	17,000	17,000	20,500	20,500
Malta	1009	1009	1231	1231	1361	1361
Cyprus	66	66	85	85	97	97
Slovenia	82	86	97	97	100	102
Croatia	78	55	79	58	74	57
Bosnia-Herzegovina	70	46	78	51	85	56
Serbia-Montenegro	85	33	106	41	120	44
Albania	76	108	108	152	133	187
Turkey	44	54	81	112	107	159
Syria	34	165	86	366	130	487
Lebanon	208	385	307	594	397	770
Israel	133	132	264	254	355	331
Palestinian Territories	188	1218	523	3083	1009	5499
Egypt	32	98	66	200	95	284
Libya	1	6	3	19	5	28
Tunisia	31	78	59	148	79	200
Algeria	6	123	13	261	18	376
Morocco	22	83	40	159	54	226
Total NMC	97	127	110	148	113	147
Total SEMC	17	55	33	114	46	162
Total Med	33	85	49	128	59	156

Source: Plan Bleu, Attané and Courbage, 2001

Table A28 Urban population in Mediterranean coastal regions, trends and baseline scenario to 2025

Country	1970	2000	2025	AAGR 1970–2000	AAGR 2000–2025
	(1000 inhabitants)			(%)	
Spain	9743	13,162	14,399	1.0	0.4
France	3447	4490	5153	0.9	0.6
Italy	22,259	25,399	24,543	0.4	−0.1
Greece	4266	6022	6434	1.2	0.3
Monaco	24	34	41	1.2	0.8
Malta	235	292	333	0.7	0.5
Cyprus	238	501	699	2.5	1.3
Slovenia	17	36	42	2.6	0.6
Croatia	519	794	929	1.4	0.6
Bosnia-Herzegovina	55	121	174	2.7	1.5
Serbia-Montenegro	109	276	365	3.2	1.1
Albania	209	480	809	2.8	2.1
Turkey	2631	9,181	14,399	4.3	1.8
Syria	186	634	1167	4.2	2.5
Lebanon	1086	2355	3176	2.6	1.2
Israel	2036	4215	5600	2.5	1.1
Palestinian Territories	416	1049	1890	3.1	2.4
Egypt	6702	14,748	26,954	2.7	2.4
Libya	796	4172	6451	5.7	1.8
Tunisia	1636	4582	7075	3.5	1.8
Algeria	2749	6744	10,138	3.0	1.6
Morocco	450	1582	2620	4.3	2.0
Total NMC	**41,118**	**51,607**	**53,920**	**0.8**	**0.2**
Total east	**6355**	**17,434**	**26,232**	**3.4**	**1.6**
Total south	**12,332**	**31,828**	**53,238**	**3.2**	**2.1**
Total SEMC	**18,687**	**49,262**	**79,470**	**3.3**	**1.9**
Total Med	**59,805**	**100,869**	**133,390**	**1.8**	**1.1**

Source: Plan Bleu, Attané and Courbage, 2001

Note: Urban population understood as the population living in urban areas with more than 10,000 inhabitants.

Statistical annex

Table A29 Number of tourists in the Mediterranean coastal regions, trends and baseline scenario to 2025 (in thousands)

Country	International tourism			National tourism			TOTAL		
	1990	2000	2025	1990	2000	2025	1990	2000	2025
Spain	26,209	33,529	56,531	7167	8282	11,415	33,375	41,810	67,946
France	10,499	15,119	23,267	6013	7700	8664	16,512	22,819	31,931
Italy	17,341	26,768	40,141	20,556	22,121	26,801	37,898	48,888	66,942
Greece	8429	11,875	17,772	3665	4276	6080	12,094	16,151	23,852
Monaco	245	300	594	—	—	—	245	300	594
Malta	872	1216	2022	125	156	249	997	1372	2271
Cyprus	1561	2686	4405	143	314	540	1704	3000	4945
Slovenia	—	273	1397	160	197	350	160	469	1747
Croatia	—	5423	10,608	825	805	1667	825	6228	12,275
Bosnia-Herzegovina	—	11	89	87	20	138	87	31	227
Serbia-Montenegro	—	24	672	780	163	1466	780	187	2138
Yugoslavia SFR	5122	—	—	—	—	—	5122	—	—
Albania	15	16	107	49	62	439	64	78	547
Turkey	2879	6231	23,794	4518	6563	16,413	7398	12,794	40207
Syria	225	142	870	363	478	2784	588	620	3654
Lebanon	—	482	3809	—	769	1659	—	1252	5468
Israel	532	1692	3096	1862	2434	4088	2393	4126	7184
Palestinian Territories	—	33	—	—	13	401	—	46	401
Egypt	482	512	2398	2380	2310	7639	2862	2822	10,037
Libya	91	165	1651	270	513	3003	362	679	4654
Tunisia	3044	4804	10,072	1440	1904	4525	4484	6708	14,598
Algeria	227	260	593	2238	3033	5503	2465	3293	6096
Morocco	604	617	1644	1153	1368	2863	1756	1985	4507
NMC	**70,294**	**97,238**	**157,605**	**39,569**	**44,094**	**57,810**	**109,863**	**141,333**	**215,415**
SEMC	**8084**	**14,938**	**47,928**	**14,224**	**19,386**	**48,877**	**22,308**	**34,323**	**96,806**
MED	**78,378**	**112,176**	**205,533**	**53,794**	**63,480**	**106,688**	**132,171**	**175,656**	**312,221**

Source: Plan Bleu and WTO, 2001

Notes:
The number of international tourist arrivals is assessed by border surveys (WTO).
The number of national tourists is estimated by the *Plan Bleu* from assumptions on the rate of holiday departure.

Table A30 Human pressures on coastal regions and on the coastline, trends and baseline scenario to 2025

Country	International and national tourists in coastal regions during the peak month (1000)			Population and tourists density in coastal regions during the peak month (inhabitants/km²)		Population of urban areas located along the coastline (1000)		
	1990	2000	2025	2000	2025	1990	2000	2025
Spain	6675	8362	13,589	250	315	9357	9747	10,663
France	3302	4564	6386	234	293	3288	3492	4007
Italy	7580	9778	13,388	257	263	17,992	18,009	17,402
Greece	2419	3230	4770	126	140	5078	5345	5710
Monaco	49	60	119	47,000	79,853	30	34	41
Malta	199	274	454	2099	2798	227	253	288
Cyprus	341	600	989	150	204	246	297	414
Slovenia	32	94	349	187	436	35	36	42
Croatia	165	1246	2455	106	150	727	741	867
Bosnia-Herzegovina	17	6	45	53	66	—	—	—
Serbia-Montenegro	156	37	428	47	110	52	60	80
Albania	13	16	109	154	201	212	245	414
Turkey	1480	2559	8041	133	225	3801	5221	8188
Syria	118	124	731	396	661	428	616	1134
Lebanon	—	250	1094	645	993	1721	2151	2901
Israel	479	825	1437	296	404	2462	3264	4337
Palestinian Territories	—	9	80	3108	5719	524	1049	1890
Egypt	572	564	2007	205	302	4994	5927	10,833
Libya	72	136	931	19	32	2395	3514	5433
Tunisia	897	1342	2920	177	264	2982	3859	5958
Algeria	493	659	1219	275	402	4017	4979	7485
Morocco	351	397	901	117	173	872	1204	1994
NMC	**21,973**	**28,267**	**43,083**	**209**	**240**	**37,244**	**38,259**	**39,928**
SEMC	**4462**	**6865**	**19,361**	**122**	**188**	**24,197**	**31,785**	**50,153**
MED	**26,434**	**35,131**	**62,444**	**158**	**209**	**61,441**	**70,043**	**90,082**

Source: WTO; *Plan Bleu*, 2003; Attané and Courbage, 2001; Géopolis

Statistical annex

Table A30 Human pressures on coastal regions and on the coastline, trends and baseline scenario to 2025 (continuation)

Country	Number of inhabitants of urban areas located along the coastline per kilometre of coast (inhabitant/km of coast)			Number of tourists per kilometre of coast during the peak month (persons/km of coast)[1]		
	1990	2000	2025	1990	2000	2025
Spain	3627	3778	4133	2587	3241	5267
France	1931	2050	2353	1939	2680	3750
Italy	2440	2442	2360	1028	1326	1815
Greece	338	356	380	161	215	318
Monaco	7475	8500	10,250	12,250	15,000	29,676
Malta	1263	1404	1601	1107	1524	2523
Cyprus	314	380	530	436	767	1265
Slovenia	745	773	901	—	2013	7500
Croatia	125	127	149	—	213	421
Bosnia-Herzegovina	—	—	—	—	268	1978
Serbia-Montenegro	178	205	271	—	127	1455
Albania	507	587	990	31	37	262
Turkey	732	1006	1577	285	493	1549
Syria	2338	3367	6197	643	677	3994
Lebanon	7651	9559	12,892	—	1113	4861
Israel	13,756	18,237	24,229	2674	4610	8027
Palestinian Territories	9530	19,073	34,364	—	166	1457
Egypt	5229	6206	11,343	599	591	2102
Libya	1353	1985	3070	41	77	526
Tunisia	2298	2973	4590	691	1034	2249
Algeria	3347	4150	6238	411	549	1016
Morocco	1704	2352	3895	686	775	1761
NMC	**1087**	**1094**	**1160**	**641**	**825**	**1257**
SEMC	**2092**	**2407**	**3400**	**386**	**593**	**1674**
MED	**1341**	**1528**	**1966**	**577**	**767**	**1363**

Source: WTO; *Plan Bleu*, 2003; Attané and Courbage, 2001; Géopolis

Note:

1 Calculated as the number of tourists in the coastal regions divided by the length of the coast.

Table A31 Human settlements and infrastructures along the coast

Country	Urban areas of more than 10,000 inhabitants located along the coastline						Airports in coastal urban areas and small islands	Commercial ports	Desalination plants	Yachting		Part of the coast under artificial land cover[1] (%)
	1950	1960	1970	1980	1990	1995				Port	Berths	
Spain	43	45	51	63	68	71	9	30	37	157	65,637	75
France	14	14	16	17	21	21	9	18		147	80,195	63
Italy	154	162	173	184	196	196	23	76	61	334	87,995	71
Greece	27	28	28	33	36	37	32	48	13	15	6300	26
Monaco	1	1	1	1	1	1	1	1		2	860	100
Malta	2	2	2	2	3	3	2	2	15			98
Cyprus	3	3	3	4	4	4	2	8	9	2	680	34
Slovenia	0	1	1	1	2	2		3		5	1436	100
Croatia	6	6	6	9	10	10	4	13		60	12,863	58
Bosnia-Herzegovina												41
Serbia-Montenegro	0	0	0	2	4	4		2				52
Albania	3	3	3	4	5	6		4				8
Turkey	12	17	25	30	46	49	7	19		15	10,000	73
Syria	3	3	4	4	4	4	1	3				69
Lebanon	4	4	5	5	11	11	1	6	5			88
Israel	5	7	8	8	9	10	1	4	1	5	2730	100
Palestinian Territories	4	4	4	4	4	4	1					100
Egypt	14	17	31	44	74	76	4	8	5			45
Libya	3	6	10	14	17	21	2	14	59			32
Tunisia	9	14	19	23	25	25	5	10	1	6	2200	42
Algeria	7	12	14	16	23	23	4	15	32			35
Morocco	4	4	4	4	6	6	4	2		2		20
NMC	253	265	284	320	350	355	82	205	135	722	255,966	
SEMC	65	88	124	152	219	229	30	81	103	28	14,930	
MED	318	353	408	472	569	584	112	286	238	750	270,896	42

Source: Géopolis, UNEP/MAP/MEDPOL, 2001; Lloyds' lists – Ports of the World 2003, various national sources

Note:

1 Estimates of coast under artificial land cover derived from night-time light radiation surveys (NOAA).

Statistical annex

Table A32 Type and treatment of industrial sewage (in m³ per year)

	Process wastewater		Cooling water		Domestic water (industry)		Total		
	Not processed	Processed	Not processed	Processed	Not processed	Processed	Not processed	Processed	Total
Total industrial discharge	71,995,300	77,802,400	529,302,954	4,929,900	38,041,159	3,385,145	639,339,413	86,117,445	725,456,858
		52%		1%		8%		12%	
of which in the Med. Sea	13,728,600	6,280,500	512,815,380	1,999,500	22,074,000	49,500	548,617,980	8,329,500	556,947,480
		31%		<1%		<1%		>1%	
of which in the river	13,582,800	6,458,900	6,996,300	2,930,400	1,286,932	554,375	21,866,032	9,948,675	31,814,707
		32%		30%		30%		31%	

Source: UNEP/MAP/MEDPOL, 2002

Water

Table A33 Water resources per country, 2000–2025

Countries and territories in the Mediterranean basin	Average yearly flows (surface and ground water) in km³/year			Average natural renewable resources (surface and ground water) in km³/year		
	Internal inflow (effective precipitations) (1)	External inflow (from neighbouring countries) (2)	Of which non-Med countries	Yearly average (1)+(2)	Regular[1] resources (from surface and underground	Annual inflow in decennial dry year[2]
Spain	28	0.35	0.1	28.35	10	10
France	64	8.5	8.5	72.5	35	53
Italy	182.5	8.8	2	191.3	30	
Greece	58	16.25	10.2	74.25	10	
Malta	0.05			0.05		0.03
Cyprus	0.78			0.78	0.3	
Slovenia	4.21			4.21		
Croatia	18	13.65		31.65		
Bosnia-Herzegovina	14			14		
Serbia-Montenegro	16			16		
Albania	26.9	14.8	2	41.7	5	13 to 30
Turkey	66	3.45	2.8	69.45	20	
Syria	5	0.96		5.96	2.5	
Lebanon	4.8			4.8	2.5	1.4
Israel	0.63	0.38		1.01	1	1
Palestinian Territories	0.616	0.01		0.626		
Egypt	0.8	55.5	55.5	56.3	23	65
Libya	0.7			0.7	0.6	
Tunisia	3.7	0.32		4.02	1.0	0.97
Algeria	11.97	0.03		12	2.3	5
Morocco	5			5	1.4	1.5
Total Mediterranean (Without double counting)	**511.656**	**123**	**81.1**	**592.756**	**144.3**	**136.9**

Source: Plan Bleu, Margat, 2004

Notes:
1 Annual inflow, in an average year, which means ensured for 11 months out of 12 (in practice, low-water discharge of rivers being equivalent to ground flow).
2 Which means ensured for nine years out of ten.

Statistical annex

Table A34 Water demand by country, baseline and alternative scenarios, 2000–2025

Baseline scenario (km³)	Total demand		Sectors							
			Domestic water		Irrigation		Industries		Energy	
	2000	2025	2000	2025	2000	2025	2000	2025	2000	2025
Spain	18.2	21.9	2.07	2.90	11.86	13.00	0.85	1.00	3.39	5.00
France	16.7	12.6	1.71	1.50	1.78	2.00	1.09	1.10	12.10	8.00
Italy	42.0	37.0	8.00	7.00	20.00	21.00	8.00	4.00	6.00	5.00
Greece	8.7	8.3	0.87	1.00	7.60	6.90	0.11	0.20	0.12	0.20
Malta	0.0	0.0	0.04	0.04	0.01	0.01	0.00	0.00	0.00	0.00
Cyprus	0.3	0.3	0.10	0.10	0.24	0.19	0.00	0.00	0.00	0.00
Slovenia	0.0	0.0	0.03	0.04	0.00	0.00	0.00	0.00	0.00	0.00
Croatia	0.2	0.5	0.18	0.45	0.00	0.00	0.00	0.02	0.00	0.00
Bosnia-Herzegovina	0.1	0.1	0.03	0.04	0.06	0.06	0.00	0.00	0.00	0.00
Serbia-Montenegro	0.8	0.8	0.50	0.70	0.20	0.00	0.10	0.10	0.00	0.00
Albania	1.4	2.8	0.40	0.80	1.00	1.70	0.00	0.30	0.00	0.00
Turkey	11.7	18.6	3.37	4.50	7.60	13.00	0.70	1.10	0.00	0.00
Syria	3.9	4.2	0.35	0.50	3.19	3.50	0.31	0.15	0.00	0.00
Lebanon	1.3	1.8	0.35	0.52	0.90	1.10	0.07	0.14	0.00	0.00
Israel	1.8	2.2	0.50	1.00	1.20	1.00	0.10	0.20	0.00	0.00
Palestinian Territories	0.3	0.8	0.10	0.47	0.16	0.30	0.00	0.04	0.00	0.00
Egypt	72.8	85.0	4.54	6.00	60.73	65.00	7.53	14.00	0.00	0.00
Libya	2.2	3.7	0.51	1.00	1.63	2.50	0.10	0.20	0.00	0.00
Tunisia	2.3	2.2	0.34	0.47	1.88	1.60	0.06	0.17	0.00	0.00
Algeria	2.9	4.3	1.25	2.00	1.05	1.40	0.40	0.92	0.20	0.00
Morocco	1.9	2.7	0.20	0.40	1.70	2.30	0.00	0.00	0.00	0.00
NMC	88	84.3	13.93	14.57	42.75	44.86	10.15	6.72	21.61	18.20
SEMC	101	125.5	11.50	16.86	80.04	91.70	9.27	16.92	0.20	0.00
MED	189	209.8	25.44	31.43	122.79	136.56	19.42	23.64	21.81	18.20
Alternative scenario										
NMC	88	65	13.9	11.3	42.7	32.0	10.2	3.6	21.6	18.2
SEMC	101	92	11.5	12.1	80.0	70.7	9.3	8.7	0.2	0.0
MED	189	157	25.4	23.4	122.8	102.8	19.4	12.3	21.8	18.2

Source: Plan Bleu, Margat, 2004

Emissions in the atmosphere

Table A35 CO_2 emissions related to energy activities per country, baseline and alternative scenarios

| | CO_2 emissions (Mt CO_2) | | | | | | Alt. S | Emission of CO_2 (kg per capita) | | Emission of CO_2 (kg GWP per 1000 $ constants 95) | |
| | | | | | Baseline scenario | | | Base. S | Alt. S | Base. S | Alt. S |
	1971	1990	1998	2000	2010	2025	2025	2025	2025	2025	2025
Spain	121	212	254	298	364	382	287	9367	7049	260	195
France	441	364	372	384	446	531	401	8272	6251	170	129
Italy	304	397	420	429	474	531	404	9841	7494	262	199
Greece	25	69	81	83	128	145	107	13,986	10,278	506	372
Monaco	—	—	—	—	—	—	—	—	—	—	—
Malta	1	2	2	2	5	8	5	18,653	11205	755	453
Cyprus	2	4	6	7	10	14	10	15,089	11211	479	356
Slovenia		13	15	14	16	19	14	9280	6881	265	197
Croatia	—	—	19	18	26	37	28	8757	6710	477	365
Bosnia-Herzegovina	—	—	5	14	7	11	8	2514	1,896	596	449
Serbia-Montenegro		60	55	41	58	84	61	6854	4989	2382	1734
Yugoslavia SFR	65	137									
Albania	4	7	2	3	6	8	6	2058	1609	786	615
Turkey	44	138	185	208	454	743	520	8512	5952	1152	805
Syria	8	32	46	47	72	98	75	4099	3138	1999	1530
Lebanon	5	6	15	13	24	31	22	7477	5252	732	514
Israel	17	35	56	63	76	106	75	13,431	9557	486	346
Palestinian Territories	—	0	0	2	—	7	5	1127	892	689	546
Egypt	21	83	106	115	159	209	156	2203	1642	908	677
Libya	4	27	41	44	54	73	56	8231	6372	799	618
Tunisia	4	13	17	17	30	40	33	3087	2562	517	429
Algeria	10	57	65	72	116	169	125	3998	2954	1398	1033
Morocco	6	19	26	30	48	57	44	1500	1148	573	439
NMC	**963**	**1191**	**1231**	**1292**	**1540**	**1768**	**1332**	**8967**	**6753**	**247**	**186**
SEMC	**118**	**410**	**557**	**611**	**1033**	**1533**	**1111**	**4695**	**3404**	**968**	**702**
MED	**1081**	**1601**	**1788**	**1900**	**2573**	**3302**	**2443**	**6304**	**4665**	**378**	**280**
World	14,753	22,984	22,982		30,083	36,680	36,680				
Mediterranean	7.3%	7.0%	7.8%		8.6%	9.0%	6.7%				

Source: IEA, 2001 for 1971–1999; OME, 2002 estimates for 2000–2025

Biodiversity

Table A36 Biosphere reserves in the Mediterranean in 2003

Morocco
1 Arganeraie
2 South Moroccos Oasis

Algeria
3 Chrea
4 Djurdjura
5 Tassili N'Ajjer
6 El Kala

Tunisia
7 Djebel Chambi
8 Djebel Bou-Hedma
9 Ichkeul
10 Zembra and Zembretta Islands

Egypt
11 Omayed
12 Wadi Allaqi

Israel
13 Carmel Mount

Greece
14 Gorge of Samaria
15 Mount Olympus

Serbia and Montenegro
16 Golija-Studenica
17 Tara River Basin

Croatia
18 Velebit Mountain

Slovenia
19 Julian Alps

Italy
20 Miramare
21 Collemeluccio-Montedimezzo
22 Cilento and Vallo di Diano
23 Somma-Vesuvio and Miglio d'Oro
24 Circeo
25 Tuscan Islands
26 Valle del Ticino

France
27 Fango Valle
28 Luberon
29 Mount Ventoux
30 Camargue
31 Cevennes

Spain
32 Menorca
33 Montseny
34 Ordesa Viñamala
35 Bardenas Reales
36 Valles del Jubera, Leza, Cidacos y Alhama
37 Urdaibai
38 Picos de Europa
39 Redes
40 Somiedo
41 Terras do Miño
42 Muniellos
43 Valle de Laciana
44 Cuenca Alta del Rio Manzanares
45 Mancha Humeda
46 Las Sierras de Cazorta y Segura
47 Cabo de Gata-Nijar
48 Sierra Nevada
49 Sierra de las Nieves y su Entorno
50 Grazzalema
51 Doñana
52 Marismas del Odiel
53 Las Desehas de Sierra Morena
54 Monfragüe

Source: UNESCO MAB, 2004 (www.unesco.org/mab/)

Table A37 Protected areas in Mediterranean countries

	Protected areas 1970–2004 (1000 hectares)						Protected area as percentage of the national (marine and terrestrial) territory 2004	
	According to the IUCN categories (Ia–VI)					All categories (IUCN & national)		
Country	1970	1980	1990	2000	2004	2004	IUCN	Total
Spain	904	1567	3657	4240	4240	4807	6.8	7.7
France	1815	4288	5532	7226	7226	7319	11.6	11.7
Italy	271	480	1442	1878	1878	5724	4.1	12.5
Greece	37	167	232	491	491	688	2.0	2.8
Monaco	0	0.05	0.05	0.05	0.05	0.05	25.5	25.5
Malta	0.01	0.01	1.32	4.90	5.86	5.85	1.4	1.4
Cyprus	67	67	69	78	78	92	3.4	4.0
Slovenia	87	89	128	150	150	150	7.3	7.3
Croatia	50	91	450	572	572	572	6.5	6.5
Bosnia-Herzegovina	27	27	27	27	27	27	0.5	0.5
Serbia-Montenegro	96	188	323	338	338	387	3.3	3.8
Albania	58	58	60	103	103	103	2.9	2.9
Turkey	291	474	1039	1256	1256	3353	1.5	3.9
Syria	0	0	0	0	0	357	0.0	1.9
Lebanon	0	0	4	4	4	8	0.3	0.5
Israel	33	53	263	295	295	408	11.7	16.2
Egypt	48	48	253	9744	11,812	12,767	11.2	12.1
Libya	0	157	157	173	173	221	0.1	0.1
Tunisia	0	41	46	46	46	258	0.2	1.3
Algeria	13	32	11,949	11,957	11,957	11,970	5.0	5.0
Morocco	330	340	340	373	373	567	0.8	1.2
NMC	**3411**	**7022**	**11,920**	**15,108**	**15,109**	**19,876**	**6.6**	**8.7**
SEMC	**715**	**1146**	**14,050**	**23,848**	**25,916**	**29,909**	**3.7**	**4.3**
MED	**4126**	**8168**	**25,970**	**38,956**	**41,025**	**49,785**	**4.4**	**5.3**
Med. France[1]					1519	1533	22.2	22.5
NMC revised[2]					9402	14,090	5.5	8.2
Mediterranean[3]					**35,318**	**43,999**	**4.0**	**5.0**

Source: UNEP-WCMC/WDBPA v2.03 (World Database on Protected Areas (sea.unep-wcmc.org/wdbpa)), Plan Bleu, 2005

Notes:
1   Med. France = Mediterranean regions of France (Languedoc-Roussillon, PACA and Corse).
2   NMC revised = NMC excluding non Mediterranean part of France.
3   Mediterranean = NMC revised + SEMC.

This table includes the Marine and Coastal Protected Areas.

Statistical annex

Table A37 Protected areas in Mediterranean countries (continuation)

Country	Number of protected areas in 2004			Protected areas distribution by IUCN categories (in percentage of total country area) in 2004							
	IUCN	Total	Of which Marine	Strict Nature Reserve Ia	Wilderness Area Ib	National Park II	Natural Monument III	Habitat/Species Management Area IV	Protected Landscape/sea scape V	Managed Resources Protected area VI	Non Classified NC
Spain	327	602	72	0.0	0.1	4.6	0.0	35.9	47.6	0.0	11.8
France	1262	1327	103	0.0	0.0	3.6	0.0	4.8	90.4	0.0	1.3
Italy	324	752	79	0.0	3.9	7.3	0.0	5.8	15.8	0.0	67.2
Greece	123	147	25	0.0	0.1	44.9	2.3	18.8	5.2	0.0	28.7
Monaco	2	2	2	0.0	0.0	0.0	0.0	100.0	0.0	0.0	0.0
Malta	93	93	5	0.1	63.3	0.5	5.5	3.5	4.7	22.3	0.0
Cyprus	10	19	5	0.0	0.0	9.9	0.0	74.1	1.1	0.0	14.9
Slovenia	45	46	1	0.0	0.0	55.9	0.5	0.0	43.6	0.0	0.0
Croatia	200	200	19	5.8	0.0	16.5	0.1	1.9	75.7	0.0	0.0
Bosnia-Herzegovina	21	31	1	0.0	0.1	63.7	0.0	11.0	24.3	0.0	1.0
Serbia-Montenegro	103	178	7	0.2	0.6	36.7	0.5	22.7	26.7	0.0	12.6
Albania	52	52	11	14.1	0.0	20.2	3.4	42.1	2.4	17.7	0.0
Turkey	85	474	61	0.8	0.0	11.8	0.0	14.4	4.1	6.3	62.5
Syria	0	28	2	0.0	0.0	0.0	0.0	0.0	0.0	0.0	100.0
Lebanon	2	24	8	0.0	0.0	44.8	0.0	6.4	0.0	0.0	48.8
Israel	185	288	41	0.0	0.0	0.8	0.5	68.9	2.4	0.0	27.5
Egypt	34	49	28	2.7	0.0	26.1	0.0	26.5	30.0	5.4	9.3
Libya	8	12	6	0.0	0.0	23.1	0.0	55.2	0.0	0.0	21.7
Tunisia	7	42	5	0.0	0.0	17.6	0.0	0.0	0.0	0.0	82.3
Algeria	18	25	9	0.3	0.0	98.5	0.0	0.4	0.7	0.0	0.1
Morocco	13	34	13	1.1	0.5	0.0	42.0	22.2	34.2	0.0	0.0
NMC	**2562**	**3447**	**330**	**0.3**	**1.2**	**7.9**	**0.1**	**13.8**	**52.6**	**0.1**	**24.0**
SEMC	**352**	**976**	**173**	**1.5**	**0.0**	**51.1**	**0.8**	**15.4**	**14.9**	**3.1**	**13.3**
MED	**2914**	**4423**	**503**	**1.0**	**0.5**	**34.3**	**0.5**	**14.8**	**29.6**	**1.9**	**17.5**
Med. France[1]	225	232	50	0.0	0.0	7.6	0.1	5.4	57.6	0.0	29.4
NMC revised[2]	1525	2352	277	0.3	1.6	10.0	0.2	17.1	34.5	0.1	36.1
Mediterranean[3]	**1877**	**3328**	**450**	**1.1**	**0.5**	**38.0**	**0.6**	**15.9**	**21.2**	**2.1**	**20.6**

Source: UNEP-WCMC/WDBPA v2.03 (World Database on Protected Areas, http://sea.unep-wcmc.org/wdbpa), Plan Bleu, 2005

Notes:
1   Med. France = Mediterranean regions of France (Languedoc-Roussillon, PACA and Corse).
2   NMC revised = NMC excluding non Mediterranean part of France.
3   Mediterranean = NMC revised + SEMC.

This table includes the Marine and Coastal Protected Areas.

References

Attané I., Courbage, Y. (2001) *La Démographie en Méditerranée. Situation et Projections*. Les Fascicules du Plan Bleu, no 11. Paris: Economica

Cornut, B.; ADEME. (2001) *Les tarifs des énergies dans la région méditerranéenne et leurs impacts sur le développement des énergies renouvelables*. Forum pour le développement des énergies renouvelables dans la région méditerranéenne, Marrakech, May

Fontela, E. (2001) *Éléments pour le cadrage économique du rapport sur l'environnement et le développement en Méditerranée*. Report for the Blue Plan

IEA. *World energy outlook*. (1999, 2000, 2001, 2002) Paris: IEA

Mahjoub, A. (2001) *La Méditerranée: une rétrospective économique, sociale, environnementale et une brève prospective*. Report for the Blue Plan

Margat, J.; Plan Bleu. (2004) *L'eau des Méditerranéens: situation et perspectives*. MAP Technical Report Series no 158. Athens: MAP (www.unepmap.gr)

Montgolfier Jean de et al. (2002) *Les espaces boisés méditerranéens. Situation et perspectives*. Les Fascicules du Plan Bleu, no 12. Paris: Economica

OME. (2002) *Energie en Méditerranée*. Report for the Blue Plan, May

UNEP-MAP-MEDPOL. (2001) *Seawater desalinisation in Mediterranean countries: assessment of environmental impacts and proposed guidelines for the management of brine*. Athens: MAP

UNEP/MAP/MEDPOL. (2002) *Transboundary Diagnostic Analysis for the Mediterranean Sea (TDA) Draft*. Athens: MAP

UN-Habitat. Global Urban Observatory. *Statistical Annexes to the Global Report on Human Settlements 2001* (www.unhabitat.org)

World Tourism Organization. (2001) *Tourism 2020 Vision. Vol. 7: Global forecasts and profiles of market segments*. Madrid: WTO

List of Figures, Tables and Boxes

Part 1 The Mediterranean and its Development Dynamics

Figures

0	Structure of the report	xii
1	A multidimensional Mediterranean region	4
2	Population of countries and coastal regions, 1970–2025	6
3	The bio-climatic features of the Mediterranean	7
4	The Mediterranean, a fragmented area	8
5	Constraints to Mediterranean soil fertility	9
6	Seisms, probable maximal intensity	10
7	Volcanic activity and earthquakes in Mediterranean cities	11
8	Zones with high levels of endemic plant biodiversity in the Mediterranean bio-climatic area	11
9	GDP growth rates, 1970–2002	13
10	Income gaps: SEMC income per capita compared with four EU-Med average income	14
11	The social importance of agriculture in 2001	15
12	GDP structure and services structure	16
13	Share of the receipts from international tourists as a percentage of goods and services exports	17
14	Human Development Index (HDI) and ecological footprint	18
15	The coastal overdevelopment	24
16	Evolution of the population in the coastal zone cities of the Latin Arc, 1910/1911 – 1999/2000/2001	25
17	The Mediterranean area by night	26
18	Fertility rate (per woman), N1, 1950–2025	31
19	Convergence of demographic parameters	32
20	Number and size of households, 1985–2025, all Mediterranean countries together	32
21	Population age distribution, N1, 1970–2025	33
22	Net entries and withdrawals of the 20 to 64-year-old bracket	33
23	Urban and rural populations in the Mediterranean countries, N1, 1970–2025	34
24	Foreign and Mediterranean emigrant peoples	37
25	Number and origin of international tourists in the Mediterranean countries, 1999	38
26	Tourist visits (domestic and international) N1, N3 trends and projections to 2025	39
27	SEMC and EAC trade in goods with the EU	40
28	Net amounts of the EU's main financial transfers in 2002	47
29	Debt and debt service	58
30	Current account balance	59

Tables

1	List of Mediterranean countries and their abbreviations, N1	5
2	Average annual damage cost of environmental degradation in some Mediterranean countries	27
3	The Mediterranean share in the world's economy, 1980–2002	35
4	FDI in the Mediterranean countries	40
5	Assumptions for the baseline scenario (2025)	63

Boxes

1	Share of the 'Mediterranean coastal regions' population in the countries, 2000 (N3/N1)	4
2	Ecological footprint of Mediterranean populations (N1)	18
3	The concept of decoupling between economic growth and pressure on the environment	19
4	Sustainable development indicators. The example of HDI	20
5	Resilience of Mediterranean societies towards impoverishment	20
6	The Mediterranean Region and the Millennium Development Goals	22
7	The Euro-Mediterranean pole in two world economy scenarios to 2050	35
8	Illegal immigration to the EU	36
9	Trade in goods between Mediterranean countries: 1998–2000	41
10	Trade in services between the SEMC, the EAC and the European Union	42
11	Origin and destination of capital flows to the Mediterranean	44
12	Spain and Greece in the EU: A success story clouded by environmental impacts	45
13	The Euro-Mediterranean Partnership	46
14	The Euro-Mediterranean Partnership and sustainable development	47
15	The Barcelona system (Barcelona Convention, MAP, MCSD)	49
16	Agencies for environmental protection	52
17	Framework laws for protecting the environment	52

18	Disputes over applying Community environmental law	54
19	Company commitment to sustainable development	56
20	Key sectors of increasingly exposed SEMC economies	61
21	Democratic governance for human development in the Arab countries	62

Chapter 1 Water

Figures

1.1	Mediterranean catchment area, NV	71
1.2	Average rainfall distribution in the Mediterranean basin	72
1.3	Inter-annual variation in river discharge	74
1.4	Total demand per country, baseline scenario, N1, 1980–2025	75
1.5	The structure of the water demand per sector and per country groups, NV, 2000–2025	76
1.6	Exploitation indices per basin, NV, 2000–2025	77
1.7	Main water transfers, NV	79
1.8	Urban and industrial wastewater, NV, 2000–2025	84
1.9	Main rivers subjected to chronic pollution, NV	85
1.10	People not having access to an improved source of drinking water or improved sanitation, N1	87
1.11	Shared Mediterranean basins	88
1.12	Total demand, baseline and alternative scenarios, NV, 2000–2025	94
1.13	Official Development Assistance in the water sector in the Mediterranean, cumulative 1973–2001	105

Tables

1.1	Unsustainable production indices, NV, 2000	81
1.2	Percentage of population connected to a sewage collection network and benefiting from treatment, N1	90
1.3	Estimate of recoverable losses by Mediterranean Basin regions, NV, 2000	93
1.4	Use of economic instruments for WDM in a few Mediterranean countries	99
1.5	Cost recovery and agricultural water prices in some Mediterranean countries	100

Boxes

1.1	What is the outlook for renewable natural resources?	73
1.2	What is the 'exploitable part' of renewable natural resources?	78
1.3	Supply policies, still dominated by major project programming	79
1.4	Overexploitation of Mediterranean coastal aquifers	80
1.5	Water regulated by dams – an unsustainable resource?	80
1.6	Examples of underground water pollution in the Mediterranean Basin	85
1.7	Examples of lower levels observed in water tables	86
1.8	The Framework Directive on water, a major stimulus for integrated management	89
1.9	Examples of savings made by clean production techniques	91
1.10	Some options for better resource management in arid climates	92
1.11	Infrastructure savings through WDM, Rabat-Casablanca	95
1.12	Tunisian national strategy for managing water demand	96
1.13	The benefits of processes that involve users	97
1.14	Tariffs compatible with social goals	102
1.15	Large variability in consumer reactions to water prices	103

Chapter 2 Energy

Figures

2.1	Primary energy demand, baseline scenario to 2025	112
2.2	Final consumption per sector, 1971–2000	113
2.3	Primary energy demands by source, baseline scenario, 1971–2025	114
2.4	Electricity production by source, baseline scenario, 1971–2025	114
2.5	Primary energy production by NMC and SEMC source, baseline scenario	117
2.6	Growing energy-dependence index of a number of Mediterranean countries	117
2.7	CO_2 emissions related to energy activities, baseline scenario, 1971–2025	120
2.8	CO_2 emissions from the energy sector in the four EU-Med countries, 1970–2025	120
2.9	The doubling of intra-Mediterranean gas trade, 2000–2025	121
2.10	The development of gas infrastructures in the Mediterranean	122
2.11	Electricity interconnections in the Mediterranean	123
2.12	Map of average solar radiation, April, 1981–1990	129
2.13	Learning curves, the increase in installed capacity strongly reduces costs	134
2.14	Pump price of super-grade petrol	135
2.15	ODA total amount intended for the Mediterranean countries in the energy sector, 1973–2001	139
2.16	Map of the Euro-Mediterranean energy trade (gas, oil, electricity), 2000	142

2.17 Feasible energy savings with the alternative scenario, 2025   144
2.18 Total CO_2 emissions according to the two scenarios   144

Tables

2.1 Risks and effects on the environment related to energy consumption and production   119
2.2 Electricity production potential from renewable energies in the SEMC in 2020   128
2.3 The distribution of the ODA intended for the Mediterranean countries in the energy sector, Cumulative, 1973–2001   139

Boxes

2.1 The risks to the worldwide supply of hydrocarbons   117
2.2 The main electricity interconnections in the Mediterranean, 2002   122
2.3 Energy savings by clean production technologies   126
2.4 Bio-climatic building to save energy   127
2.5 The situation and development perspectives for wind energy in the SEMC   128
2.6 Examples of energy efficiency and renewable energy strategies in the NMC   131
2.7 Examples of energy efficiency and renewable energy policies in the SEMC   132
2.8 Energy prices and costs   136
2.9 The main recommendations of the Johannesburg Plan of Action in the energy field, 2002   137
2.10 MEDENER, a Mediterranean network for energy efficiency agencies   138
2.11 Some energy efficiency and renewable energy projects in the Mediterranean region   139

Chapter 3 Transport

Figures

3.1 Passenger traffic (air, rail and road), 1970–2000   150
3.2 The car fleet in the Mediterranean, 2002   150
3.3 The modal split of domestic passenger traffic in the Mediterranean countries, 1999   151
3.4 Air passenger traffic between the SEMC and the EU, main regular connections, 2001   152
3.5 Port passenger traffic, 1985–2000   153
3.6 Freight traffic (air, road and rail), 1970–2000   153
3.7 Maritime freight traffic, 1985–2000   154
3.8 The modal split of domestic freight traffic (excluding maritime) in the Mediterranean countries, 1999   155
3.9 Modal split of international freight traffic (excluding sea) in the Mediterranean countries, in 1999   157
3.10 Intra-southern Europe trade in 1998   157
3.11 Trade opening and road freight traffic in some Mediterranean countries, 1970–2000   161
3.12 Passenger traffic (rail+air+road), baseline scenario, 2025   162
3.13 Passenger port traffic in 2025, baseline scenario   162
3.14 Freight traffic (air+rail+road), baseline scenario, 2025   163
3.15 Maritime freight traffic, baseline scenario, 2025   163
3.16 The population exposed to road noise over 55 decibels, 1970–2000   164
3.17 People exposed to road noise above 55 decibels, baseline scenario, 2025   165
3.18 VOC emissions, 1970–2000   166
3.19 NOx emissions, 1970–2000   166
3.20 CO_2 emissions, 1971–2000   167
3.21 VOC emissions, baseline scenario, 2025   167
3.22 NOx emissions, baseline scenario, 2025   168
3.23 CO_2 emissions, baseline scenario, 2025   168
3.24 Areas occupied by land transport infrastructures, 1984–2000   169
3.25 Areas occupied by land transport infrastructures, baseline scenario, 2025   170
3.26 Number of road deaths per 1000 vehicles, 1984–2000   171
3.27 Number of road deaths, baseline scenario, 2025   171
3.28 Share of transport in household budgets, 1985–2000   172
3.29 Share of transport in household budgets, baseline scenario, 2025   172
3.30 Costs of congestion, 1995–2000   173
3.31 Congestion costs, baseline scenario, 2025   173
3.32 The Mediterranean-flag merchant fleet, breakdown by freight type, 2001   180
3.33 The Mediterranean-flag merchant fleet, breakdown by country, 2001   180
3.34 Accidental and operational polluting emissions into the sea related to maritime traffic, 1985–2000   182
3.35 Accidental and operational polluting emissions into the sea related to maritime traffic, baseline scenario 2025   182
3.36 Corridors of International Euro-Mediterranean networks in the eastern Mediterranean   188
3.37 Corridor of International Euro-Mediterranean Maghreb Networks   189
3.38 Priority infrastructure network for the Balkans   190
3.39 Official Development Assistance in the transport sector, 1973–2001   190
3.40 Passenger traffic, alternative scenario, 2025   191
3.41 Freight traffic, alternative scenario, 2025   191
3.42 People exposed to road noise above 55 decibels, alternative scenario, 2025   191
3.43 CO_2 emissions, alternative scenario, 2025   192

3.44 VOC emissions, alternative scenario, 2025   192
3.45 NOx emissions, alternative scenario, 2025   192
3.46 The number of road accident deaths, alternative scenario, 2025   192
3.47 Congestion costs, alternative scenario, 2025   192
3.48 The share of transport in household budgets, alternative scenario, 2025   193
3.49 Polluting emissions into the sea related to maritime traffic, alternative scenario, 2025   193

Tables

3.1 Plans for investment in rail infrastructures in the north-western Mediterranean   174

Boxes

3.1 The hub system   152
3.2 High speed vessels   154
3.3 Railways in the SEMC   156
3.4 Traffic across the Alps and the Pyrenees   157
3.5 The logistical difficulties of a German investor in Tunisia, Leoni Tunisie   159
3.6 Maritime traffic in the Mediterranean   160
3.7 The development of high-speed rail lines (LGV) in Italy, Spain and France   175
3.8 The prospects for re-opening the Lebanese rail system   176
3.9 Examples of successful combined transport   177
3.10 The Franco-Italian short-sea shipping project   177
3.11 Examples of difficulties in assessing externalities   184
3.12 Liberalization of road transport, examples of impact in the city of Batna, Algeria   187

Chapter 4 Urban Areas

Figures

4.1 Urban population in the Mediterranean countries, trends and projections   199
4.2 Urban population in the countries and their Mediterranean coastal regions, projections 2000–2025   200
4.3 Urbanization rate in the Mediterranean countries, 1950, 2000 and projections to 2030   200
4.4 Cities with 10,000 or more inhabitants, 1950 and 1995   201
4.5 Population in cities according to town sizes   203
4.6 Metropolises with more than one million inhabitants   203
4.7 Land use changes in Marseilles, Padua-Mestre, Nicosia and Istanbul   208
4.8 Land use in Great Sfax, Tunisia. Trends and scenarios   209
4.9 Passenger travel modes in the Middle East, North Africa and in western Europe   215
4.10 Production of household waste   228
4.11 Local level expenditure as a percentage of total government expenditure, 1997–2000   232
4.12 World distribution of public–private partnerships (per cent)   237

Tables

4.1 Growth rates of cities, per town class   204
4.2 Estimates of the extent of unregulated housing   207
4.3 Balance of recent natural disasters in the southern and eastern Mediterranean   213
4.4 Motorized travel share in six very large cities of the SEMC   216
4.5 Changes in transport modal share in Cairo   216
4.6 Share of informal systems in total public transport   217
4.7 Examples of solidarity measures for ensuring access to drinking water   227
4.8 Two scenarios for waste production through 2025   230
4.9 Authorities at sub-national levels in south Mediterranean countries   233
4.10 Responsibilities of local authorities in the Mediterranean   234
4.11 Share of expenditure by local authorities   235
4.12 Proportion of transfers to local authorities from the Government budget   235
4.13 Local authority resources   235
4.14 Capital expenditure by local authorities   236

Boxes

4.1 Land speculation and pressures on the agricultural plains in Lebanon   209
4.2 New towns in Egypt   210
4.3 Development scenarios in the French Riviera conurbation   210
4.4 Gated cities in the Mediterranean   211
4.5 The industrial accident in Toulouse (France)   213
4.6 Underground and tramways in six Mediterranean metropolises   217
4.7 Urban Travel Plans in French cities   218
4.8 Integrated urban and metropolitan transport in Naples   219
4.9 Urban regeneration and renewal in Barcelona, Naples, Tunis and Aleppo   221
4.10 Air and health, developments in epidemiological approaches   224
4.11 Air pollution and health in Casablanca   225
4.12 Access to water in different quarters in Rabat-Salé, Morocco   226
4.13 Competition for access to water in Algiers   226
4.14 Ecological management of household waste in Barcelona   229
4.15 Household waste in Cairo and Alexandria: The zabaleens and foreign companies   231
4.16 Public regulation of concessions in Morocco   238

4.17 Local Agendas 21 239
4.18 Integrated urban development policies in Italy 241
4.19 The URBAN programme: Giving a dynamic boost to European urban areas in crisis 243
4.20 The Medcities network 243

Chapter 5 Rural Areas

Figures

5.1 The Mediterranean's main agricultural and natural systems 251
5.2 Total agricultural population in the Mediterranean countries, trends and projections to 2025 254
5.3 The population of the Languedoc-Roussillon, France, in 1901 and in 1999 255
5.4 Urban, rural and 'remote' rural populations in three EU Mediterranean regions 256
5.5 The rural population in Mediterranean countries: trends and projections to 2025 257
5.6 Drought-related fluctuations in agricultural and total GDP in Morocco, 1992–1999 260
5.7 The relative percentage of forest, cultivated land and pasturelands 262
5.8 Arable land per capita 263
5.9 Wood cover in the Mediterranean region 263
5.10 Percentage of irrigated lands in the Mediterranean in 1995 266
5.11 Fertilizer consumption 268
5.12 Growth in protected area coverage in the Mediterranean 272
5.13 Agriculture value-added per worker 274
5.14 Agricultural trade balance in the Mediterranean countries: 1970–2001 274
5.15 The evolution of food intake in the northern Mediterranean countries 275
5.16 Mediterranean biosphere reserves 288

Tables

5.1 Farm size distribution in a small number of EU Mediterranean countries 254
5.2 Agrarian structures in selected southern and eastern Mediterranean countries 258
5.3 Land and water used by agriculture, 2000–2025 267
5.4 Arid lands affected by desertification 269
5.5 Annual losses in agricultural production on degraded lands 272
5.6 The evolution of production indices per capita in a selection of SEMC 279
5.7 Technological trend and agricultural production growth in SEMC 279

Boxes

5.1 Four possible scenarios for the rural areas in the developed Mediterranean countries 257
5.2 Rural poverty in the Mediterranean developing countries, the magnitude of the disparities with urban dwellers 259
5.3 The impact of NAFTA on Mexican agriculture and the environment 260
5.4 The impact of free trade on rain-fed agriculture in Morocco 261
5.5 Changes in forest land and irrigated areas in Tunisia over a century 264
5.6 Shrub encroachment 264
5.7 Some data on the aggravation of soil salinization 267
5.8 Tunisia's 1000 hillside-lakes programme 267
5.9 The development of olive-growing and the environment in the Mediterranean 268
5.10 The magnitude of soil degradation in various countries 270
5.11 Loss of high quality agricultural land and rural character in the Mediterranean region 270
5.12 Organic agriculture in Europe and the Mediterranean. An example of success in Egypt: The Sekem Group 277
5.13 Oasis: Laboratories for land reform and agricultural renewal 278
5.14 The entry of Spain and Greece into the common agricultural market, EU aid and sustainability 280
5.15 A necessary assessment of the possible impacts of GMOs in the Mediterranean 281
5.16 The agro-pastoral development programme and local initiatives in Tunisia 284
5.17 Massive migration from mountain areas, Albania as an example 285
5.18 Mountain revival in Cyprus, the Iberian dehesas and France's Cévennes 286
5.19 The GEFRIF project in Morocco 287
5.20 The value of the multiple roles of Mediterranean agriculture and forests 290
5.21 Developing forestry in Calabria 294
5.22 Forest products other than wood in North Africa 294
5.23 The European LEADER programme 296
5.24 Rural development in Morocco, Turkey and Lebanon 297
5.25 The weakness of official development assistance in targeting the Mediterranean rural sector 299
5.26 Networks on the Mediterranean rural areas 300

Chapter 6 Coastal Areas

Figures

6.1 Sedimentary coasts and their ecosystems 306
6.2 The 100 Mediterranean historical coastal sites 308
6.3 Pressures on the coastal zone 309
6.4 Coastal city population, 1950–2025 310

A Sustainable Future for the Mediterranean

6.5	Number of coastal cities with population more than 10,000, 1950–1995	310
6.6	Cumulative density of resident and tourist populations in the peak month in coastal regions, 2000–2025	311
6.7	Road and airport infrastructure along coasts	312
6.8	Mediterranean commercial ports	313
6.9	Marinas in Italy	313
6.10	Mediterranean coastal energy infrastructure	314
6.11	Coastal construction from 1975 to 1990 in Malaga province (Spain)	315
6.12	The increase in artificial land covering a few Mediterranean coastal regions (%)	316
6.13	One hundred and one priority pollution hot spots	318
6.14	Urban areas with more than 150,000 population equivalent in compliance with the EU Waste Water Directive	322
6.15	Wastewater treatment by coastal cities with more than 10,000 inhabitants, 2002	323
6.16	Sites with groundwater overexploitation and saline intrusion	326
6.17	Main ways non-indigenous marine species are introduced and spread of Caulerpa	327
6.18	Protected coastal areas in the Mediterranean, 1950–1995	329
6.19	Specially Protected Areas of Mediterranean Interest (SPAMI) and wetlands classified through the RAMSAR Convention	329
6.20	Fishing: Catches	331
6.21	Bass and sea bream, aquaculture production in the Mediterranean and the Black Sea, 1990–2000	333
6.22	Tourism system in a specific territory: actors and operational levels	346

Tables

6.1	Urbanization of the coastal strip in 1995 and 2025 in a number of selected countries	317
6.2	The flow of nutrients carried by river water into the Mediterranean Sea	319
6.3	The Blue Flag as applied in Mediterranean-rim countries	345

Boxes

6.1	The coastal zone: Official definitions	305
6.2	Islands: Assets and drawbacks of insularity	306
6.3	Traditional techniques for exploitation of the Mediterranean coastal waters	308
6.4	A convergence of pressures in the coastal zone of Algiers	309
6.5	Increase in risks and costs of flooding due to coastal urbanization in France's Languedoc-Roussillon region	317
6.6	Strong tourism-related urbanization on the Croatian coast and measures developed to reconcile environment and development	317
6.7	Eutrophication and its impacts on the northern Adriatic coastline	319
6.8	Factoring in environmental benefits in calculating the treatment tax in Morocco	324
6.9	Examples of savings made by cleaner production techniques in the industrial sector	324
6.10	Factors in the degradation of Posidonia meadows	328
6.11	The Strategic Action Programme on Biodiversity (SAP-BIO)	330
6.12	EU mobilization for integrated management of coastal areas	335
6.13	Coastal agencies and action plans in selected countries	335
6.14	Regulation to limit construction and roads in coastal zones	337
6.15	Examples of civil society initiatives for the Mediterranean	337
6.16	MAP Coastal Area Management Programmes: Location and some results	339
6.17	Countering the intolerable increase in urbanization along the coast of Algeria	340
6.18	Socio-economic imperatives of tourism, some figures	341
6.19	Planned investments in the tourism sector in the Mediterranean	341
6.20	Tourism puts considerable pressure on island environments	343
6.21	The contribution of tourism to the water sector in Tunisia	344
6.22	Examples of a tourist eco-tax in Spain and France	346
6.23	The Calvia Local Agenda 21 (Balearics, Spain)	347
6.24	Towards a reorientation of Greek policies for tourism development	347
6.25	Islands as new sustainable development laboratories?	350

List of Acronyms and Abbreviations

AAGR	Annual Average Growth Rate	CARDS	Assistance Programme for the Western Balkans
AC 10	accession countries to the European Union in 2004 (Cyprus, Malta, Slovenia, Czech Republic, Estonia, Hungary, Latvia, Lithuania, Poland, Slovakia)	CCE	Commission of the European Communities
ACCOBAMS	Agreement on the Conservation of Cetaceans of the Black Sea, Mediterranean Sea and Contiguous Atlantic Area	CDER	Centre de Développement des Énergies Renouvelables (Morocco)
		CDM	Clean Development Mechanism
		CEC	Commission for Environmental Cooperation of North America
ADEM	Association Nationale de Développement de l'Économie de Marché (Algeria)	CEDARE	Centre for Environment and Development for the Arab Region and Europe
ADEME	Agency for the Environment and Energy Management (France)	CEEC	Central and Eastern European Countries
AEP	drinking water supply/alimentation en eau potable	CEMAGREF	National Institute for Agricultural and Environmental Engineering (France)
AEWA	Agreement on the conservation of African-Eurasian migratory waterbirds	CEMT/ECMT	European Conference of Ministers of Transport
AFIT	Agence française d'ingénierie touristique	CENEAP	National Centre of Studies and Analyses for Planning (Algeria)
AFSSE	Agence française de sécurité sanitaire environnementale	CEPII	Centre d'études prospectives et d'informations internationales (France)
ANDRA	National Agency for Radioactive Waste Management (France)	CERTU	Research Centre for Networks, Transport, Urban Planning and Public Infrastructure (France)
ANER	National Agency for Renewable Energies (Tunisia)		
		CFCE	French Foreign Trade Centre
ANPA	National Agency for the Protection of the Environment (Italy)	CGPM/GFCM	General Fisheries Commission for the Mediterranean
ANPE	National Agency for the Protection of the Environment (Tunisia)	CH_4	methane
		CIESM	International Commission for the Scientific Exploration of the Mediterranean Sea
APAL	Agency for coastal protection and development (Tunisia)		
APAT	Agency for the Protection of the Environment and Technical Services (Italy)	CIHEAM	International Centre for Advanced Mediterranean Agronomic Studies
		CIS	Commonwealth of Independent States
APHEIS	Air Pollution and Health European Information System	CITES	Convention on International Trade in Endangered Species of Wild Fauna and Flora
ARPA	Regional Agency for the Protection of the Environment (Italy)		
		CITET	International Centre for Environmental Technologies of Tunisia
ASEAN	Association of Southeast Asian Nations		
BAT	best available technology	CMS	Convention on the conservation of migratory species of wild animals
BOD	biochemical oxygen demand		
BOD	biological oxygen demand	CO	carbon monoxide
BOD_5	biological oxygen demand at five days	CO_2	carbon dioxide
CAMP	Coastal Area Management Programme	CODATU	cooperation for urban mobility in the developing world
CAP	Common Agricultural Policy		

CORRIMED	International Euro-Mediterranean Network Corridors	FDI	foreign direct investment
COV/VOC	volatile organic compound	FEEE	Foundation for Environmental Education in Europe
CP RAC	Regional Activity Centre for Cleaner Production (MAP)	FEMIP	Facility for the Euro-Mediterranean Investment and Partnership
CRAB	Coastal Resources Advisory Board (Malta)	FEMISE	Euro-Mediterranean Network of Economic Institutes
CSD	Commission on Sustainable Development (UN)	FFEM	French Global Environment Facility
CWR	combustible renewables and waste	FIDEME	French Investment Fund for the Environment and Energy Resource
DAC	Development Assistance Committee (OECD)	FMCU	World Federation of United Cities
DATAR	Delegation for Economic Development and Regional Planning (France)	FOGIME	French Guarantee Fund for Energy Resource Investment
DDT	dichloro-diphenyl-trichloroethane	G	giga (or thousand million)
DRS	erosion control and land reclamation	G7	Group of Seven: Canada, France, Germany, Italy, Japan, UK, US
EAC	eastern Adriatic countries	G8	Group of Eight: Canada, France, Germany, Italy, Japan, Russian Federation, UK, US
EBRD	European Bank for Reconstruction and Development		
EC	European Commission	GDP	gross domestic product
ECMT	European Conference of Ministers of Transport	GEF	Global Environment Facility
		GFCM	General Fisheries Commission for the Mediterranean
EDF	Électricité de France		
EE	energy efficiency	GHG	greenhouse gas
EEA	European Environment Agency	GMO	genetically modified organism
EEAA	Egyptian Environmental Affairs Agency	GNI	gross national income
EEC	European Economic Community	GNP	gross national product
EIB	European Investment Bank	GRP	Piano regolatore generale/master plan (Italy)
EMAS	eco-management and audit scheme		
EMFTA	Euro-Mediterranean Free Trade Area	GSR	guaranteed solar results
EMP	Euro-Mediterranean Partnership	GTZ	Deutsche Gesellschaft für Technische Zusammenarbeit
EMWIS	Euro-Mediterranean Information System on the know-how in the Water sector		
		GWP	Global Water Partnership
		ha	hectare
ENEA	National Agency for New Technologies, Energy and the Environment (Italy)	HACCP	Hazard Analysis and Critical Control Point
EPR	European pressurised reactor	HCB	hexachlorobenzene
ERDF	European Regional Development Fund	HCH	hexachlorocyclohexane
ERS RAC	Environment Remote Sensing Regional Activity Centre (MAP)	HDI	Human Development Index
		HSB	high speed boat
ESCWA	Economic and Social Commission for Western Asia (UN)	IAMF	International Association for Mediterranean Forests
ESHA	European Small Hydraulic Association	IAURIF	Institute for Urban Planning and Development of the Paris Ile-de-France Region
ESRI	Environmental Systems Research Institute		
ETC	European Topic Centre	IBRD	International Bank for Reconstruction and Development
ETC IW	European Topic Centre for Inland Waters		
		ICAEN	Catalan Institute of Energy (Spain)
ETP	evapotranspiration	ICALPE	International Centre for Alpine Environments
EU	European Union		
EuroMeSCo	Euro-Mediterranean network of foreign policy institutes	ICAM	Integrated Coastal Areas Management
		ICAMAS	International Centre for Advanced Mediterranean Agronomic Studies
EUWI	European Union Water Initiative		
FAO	Food and Agriculture Organization of the United Nations	ICCAT	International Commission for the Conservation of Atlantic Tunas

List of Acronyms and Abbreviations

ICLEI	International Council for Local Environmental Initiatives	JRC	European Joint Research Centre
ICRAM	Central Institute for scientific and technological research as applied to the sea (Italy)	k	kilo
		kg	kilogram
		km	kilometre
IDAL	Investment Development Authority of Lebanon	koe	kg of oil equivalent
		kV	kilo volt
IEA	International Energy Agency	kWh	kilowatt hour
IFAD	International Fund for Agricultural Development	LAG	Local action groups
		LBS	Land-based Sources
IFEN	French Institute for the Environment	LEADER	Community Initiative for Rural Development (EU)
IFOAM	International Federation of Organic Agriculture Movements	LGV	high speed rail line
		LNG	liquefied natural gas
IFRI	French Institute of International Relations	LOTI	French law on the organization of public transport
IHEI	International Hotel Environment Initiative	LPG	liquefied petroleum gas
		M	mega (or million)
ILO	International Labour Organization	MAB	Man and Biosphere
IMF	International Monetary Fund	MAP	Mediterranean Action Plan
IMO	International Maritime Organization	MARPOL	International Convention for the Prevention of Pollution from Ships
INAT	Institut National d'Agronomie de Tunisie	MATE	Ministère de l'Aménagement du Territoire et de l'Environnement (Algeria)
INE	National Statistics Institute (Spain)		
INRA	National Institute for Agronomic Research (France)	MCF	Mediterranean Carbon Fund
		MCSD	Mediterranean Commission on Sustainable Development
INRETS	National Institute for Transport and Safety Research (France)	MED POL	Programme for the Assessment and Control of Pollution in the Mediterranean Region (MAP)
INSEE	National Institute of Statistics and Economic Studies (France)		
INSULEUR	Network of the Insular Chambers of Commerce and Industry of the European Union	MEDA	main EU financial instrument for the implementation of the Euro-Mediterranean Partnership
INTERREG	Community initiative for interregional cooperation (EU)	MEDENER	Mediterranean Association of the National Agencies for Energy Conservation
IOPC	International Oil Pollution Compensation Funds	MEDREC	Mediterranean Renewable Energy Centre (Tunisia)
IPCC	Intergovernmental Panel on Climate Change	MEDREP	Mediterranean Renewable Energy programme
IPP	independent power producer	MEDWET	Mediterranean Wetlands Initiative
IPPC	integrated pollution prevention and control	MEFTA	Middle East Free Trade Agreement
		MENA	Middle East and North Africa
IPTS	European Institute for Prospective Technological Studies	MEPI	Middle East Partnership Initiative
		MERCOSUR	Southern Common Market (South America)
IRF	International Road Federation		
ISO	International Organization for Standardization	METAP	Mediterranean Environment Technical Assistance Programme
ISPA	Instrument for Structural Policies for Pre-Accession (EU)	MFN	most favoured nation
		MIO-ECSDE	Mediterranean Information Office for Environment, Culture and Sustainable Development
ISTAT	National Institute of Statistics (Italy)		
ISTED	Institut des Sciences et des Techniques de l'Équipement et de l'Environnement pour le Développement (France)		
		MIT	Massachusetts Institute of Technology
IUCN	World Conservation Union	MSW	municipal solid waste
IULA	International Union of Local Authorities	Mt	million tonnes

Mtoe	million tonnes of oil equivalent	PHARE	Programme of Pre-accession Aid for Central and Eastern Europe
MWRI	Ministry of Water Resources and Irrigation (Egypt)	PICTE	Integrated plan for quality tourism in Spain
N	nitrogen	PM	particulate matter
NACA	Network of Aquaculture Centres in Asia-Pacific	PNAE-DD	National Action Plan for the Environment and Sustainable Development (Algeria)
NAFTA	North American Free Trade Agreement	PNUD/UNDP	United Nations Development Programme
NCIT	new communication and information technologies	PNUE/UNEP	United Nations Environment Programme
NEAP	National Environmental Action Plan	POGAR	Programme on Governance in the Arab Region
NECC	National Energy Conservation Centre (Turkey)	POP	persistent organic pollutants
NEPAD	New Partnership for Africa's Development	ppp	purchasing power parity
NGO	non-governmental organization	PPP	public–private partnership
NMC	Northern Mediterranean Countries	PRU	urban renewal programme (Italy)
NOAA	National Oceanic and Atmospheric Administration (US)	PSIRU	Public Services International Research Unit (UK)
NO_2	nitrogen dioxide	PTS	persistent toxic substances
NO_X	nitrogen oxide	PUT/UTP	urban traffic plan (Italy)
NUTS	Nomenclature of Territorial Units for Statistics	PV	solar photovoltaic energy
OACI	International Civil Aviation Organization	RAMOGE	Agreement between the Governments of France, Monaco and Italy to constitute a pilot coastal zone for preventing and combating pollution of the marine environment
ODA	Official Development Assistance		
OECD	Organisation for Economic Cooperation and Development		
OME	Mediterranean Energy Observatory		
ONAS	National Sanitation Agency (Tunisia)	RAMSAR	Convention on Wetlands of International Importance Especially as Waterfowl Habitat
ONEM	Observatoire National de l'environnement du Maroc		
ONEP	National Agency Drinking Water (Morocco)	R&D	Research and Development
		RED	Régie autonome de distribution d'eau et d'électricité (Morocco)
ONF	National Forestry Office (France)		
OPEC	Organization of the Petroleum Exporting Countries	REMPEC	Regional Marine Pollution Emergency Response Centre for the Mediterranean Sea
OPRC	International Convention on Oil Pollution Preparedness, Response and Cooperation		
		REn	geothermal, solar and wind energy
		REs	renewable energy sources
OTEDD	Observatoire Tunisien de l'Environnement et du Développement Durable	RIVM	National Institute for Public Health and the Environment (The Netherlands)
O_3	tropospheric ozone	SAP	Strategic Action Programme
P	phosphorus	SAP-BIO	Strategic Action Programme for Conserving Marine and Coastal Biological Diversity in the Mediterranean
PAC/CAMP	Coastal Area Management Programme		
PAGER	grouped potable water supply plan for the rural population		
PAH	polycyclic aromatic hydrocarbons	SAPARD	Special Accession Programme for Agriculture and Rural Development (EU)
PAP RAC	Priority Actions Programme Regional Activity Centre (MAP)		
		SBA	Sustainable Business Associates
PCB	polychlorinated biphenyl	SD	Sustainable Development
PDU	urban transport plan (France)	SDI	Indicators for Sustainable Development
PECO	Central and Eastern European Countries		
		SEMC	Southern and Eastern Mediterranean Countries
pH	potential hydrogen (level of acidity or alkalinity)		

List of Acronyms and Abbreviations

SMAP	Short and Medium Term Action Programme for the Environment (EU-MEDA)
SO_2	sulphur dioxide
SO_X	sulphur oxides
SPA	Specially Protected Areas
SPA RAC	Regional Activity Centre for Specially Protected Areas (MAP)
SPAMI	Specially Protected Areas of Mediterranean Importance
SRU	French law on solidarity and urban renewal
SSS	short sea shipping
STEP	water treatment plant
TDA	Transboundary Diagnostic Analysis
TDW	deadweight ton
TEA	tourist expansion area
TEU	twenty-foot equivalent unit
TFD	tourism for development
TGV	high speed train
TIRS	Transportation Inter Regional Survey
TNF	total net flow
toe	tonne of oil equivalent
TPES	Total Primary Energy Supply
TSEZ	tourist sites and expansion zones (Algeria)
TWh	terawatt hour
UAA	usable agricultural area
UIC	International Union of Railways
UK	United Kingdom
UMA	Arab Maghreb Union
UN	United Nations
UNCCD	United Nations Convention to Combat Desertification
UNCED	United Nations Conference on the Environment and Development
UNCHS	United Nations Centre for Human Settlements
UNCLOS	United Nations Convention on the Law of the Sea
UNCSD	United Nations Commission on Sustainable Development
UNCTAD	United Nations Conference on Trade and Development
UND	Road hauliers association (Turkey)
UNDP	United Nations Development Programme
UNECE	United Nations Economic Commission for Europe
UNEP	United Nations Environment Programme
UNESCO	United Nations Educational, Scientific and Cultural Organization
UNHSP	United Nations Human Settlements Programme
UNICEF	United Nations International Children's Emergency Fund
UNIDO	United Nations Industrial Development Organization
UNSD	United Nations Statistics Division
US	United States of America
USRR	Former Union of Soviet Socialist Republics
VAT	value added tax
VOC	volatile organic compound
WCMC	World Conservation Monitoring Centre
WDI	World Development Indicators
WDM	water demand management
WHO	World Health Organization
WRI	World Resources Institute
WTO	World Tourism Organization
WTO	World Trade Organization
WTTC	World Travel and Tourism Council
WWF	World Wide Fund for Nature

ISO country codes

Albania	AL
Algeria	DZ
Bosnia-Herzegovina	BA
Croatia	HR
Cyprus	CY
Egypt	EG
France	FR
Greece	GR
Israel	IL
Italy	IT
Lebanon	LB
Libya	LY
Malta	MT
Monaco	MC
Morocco	MA
Palestinian Territories	PS
Serbia and Montenegro	CS
Slovenia	SI
Spain	ES
Syria	SY
Tunisia	TN
Turkey	TR

Index

access
  to drinking water 23, 27, 86–8, 225–7, 259
  to education vii, 21
  to electricity 112, 118, 259
  to health services 259
  to Internet 24, 59
  to sanitation 23, 27, 259
Africa 3, 6, 7, 9, 23, 50, 158, 161
ageing population 31–3, 253
Agenda-21 51, 53, 239, 244, 289, 296, 347, 381
agricultural employment, decrease in by 2025 257
agricultural water prices 100
agriculture vii, xi, 7, 8, 10, 15, 17, 29, 35, 43, 46, 47, 50, 51, 52, 56, 58, 60, 61, 63, 75, 76, 82, 83, 85, 86, 88, 91, 93, 96, 98, 99, 103, 104, 105, 202, 209, 225, 249–53, 257, 258, 260, 261, 265, 267, 268, 273, 274, 276–8, 280–2, 284, 286, 289–93, 295, 299, 309, 310, 319, 330, 336, 338, 346, 347, 348, 350, 351, 360, 362, 363, 364, 367, 368, 370, 371, 372, 373, 374, 375, 376, 377, 379, 380, 381, 382, 384, 385
  and agri-food industry 293–4
  awareness of many roles of 290–1
  better water management and soil conservation 370–1
  coupling/decoupling 291–2
  free trade 279–81
  organic 277
  outlook 273–82
  risks of unsustainable baseline scenario 281–2
  south and east
    droughts and environmental degradation 278–81
    production growth 278–81
    trade deficits and north/south productivity 273–4
    transition towards 'rational' agriculture 276–8
agro-pastoral development programme 284
air pollution see pollution
air passenger traffic 151–2, 312
Albania
  call for action 359, 361, 362
  coastal areas 311, 316, 322, 339, 341, 345
  energy 113
  Mediterranean development dynamics xiii, 3, 5, 14, 15, 19, 21, 23, 24, 36, 39, 43, 45, 51, 52, 53, 54
  rural areas 253, 255, 259, 265, 273, 276, 285, 288
  transport 154, 158, 180
  urban areas 207, 228, 232, 242, 244
  water 71, 73, 78, 89, 99, 100
Algeria
  call for action 359, 362, 368, 385
  coastal areas 305, 307, 311, 312, 328, 331, 335, 336, 337, 340, 341, 342
  energy 111, 113, 114, 115, 116, 117, 122, 123, 130, 132, 135, 136, 137
  Mediterranean development dynamics 4, 5, 6, 9, 14, 16, 17, 19, 20, 21, 22, 23, 24, 27, 31, 38, 39, 48, 51, 52, 53, 54, 55, 57, 58, 59
  rural areas 258, 259, 262, 263, 264, 265, 266, 267, 270, 271, 273, 274, 278, 279, 280, 282, 284, 285, 288, 289, 294, 299
  transport 156, 158, 162, 167, 169, 172, 176, 179, 180, 184, 187
  urban areas 202, 207, 210, 212, 226, 228, 233, 234, 235, 236, 237, 242
  water 71, 76, 77, 78, 79, 80, 81, 82, 84, 86, 90, 96, 105
Alpine traffic 157
alternative scenarios vii, xii, 36, 72, 94, 120, 124, 179, 185, 209, 229–31, 253, 268, 287, 289, 295, 325, 330, 347, 360, 365–74, 376, 377, 380, 382
  coastal areas 348–52
  energy 143–5
  principles and objectives 349–51, 365
  rural areas 291–3
  transport 190–3
  urban areas 238–42, 244–5
  water 93–5
  see also baseline scenario
animal production 275, 278
aquaculture 307, 308, 319, 325, 326, 330–4, 336, 350, 352, 366
  sustainable management of 330–4
Arab Maghreb Union 46, 176, 189
archaeological sites 12, 308, 338, 372
Asia vii, 3, 6, 22, 23, 35, 36, 47, 120, 159, 160, 184, 280, 359, 362, 383
automobile transition in south and east 214–16
awareness, insufficient 375–6

baseline scenarios vii, xii, 29, 30, 31, 35, 37, 39, 40, 43, 50, 57, 63–4, 361–5, 369, 370
  coastal areas 305, 310, 311, 316, 317, 321, 348-49
  energy 111–23
  general assumptions 63–4
  risks in 361–4
    internal disparities 362–3
    neglected environment 363–4
    north–south gap 361–2
  rural areas 257, 267, 279
  transport 161, 162, 163, 165, 168, 170, 171, 174, 182
  urban areas 200, 205, 209, 215, 230
  water 75–8, 82, 83, 84, 89, 94
  world economy 35
  see also alternative scenario
bio-climatic building 127, 371
bio-climatic region xiii, 4, 5, 6–7, 11, 199, 266
biodiversity xi, 3, 9–11, 30, 45, 48, 49, 53, 170, 249, 252, 261, 269, 273, 274, 276, 278, 280, 281, 282, 285, 286, 288, 290, 291, 292, 298, 299, 300, 305, 306, 307, 309, 319, 325, 334, 346, 349, 352, 360, 363, 364, 365, 369, 375, 377, 386
  agricultural 10
  animal biodiversity 10
  coastal zones 325–30
  conserving 366–7
  ecosystems 30
  habitats 6, 9, 10, 24
    degradation of 326–8
  landscapes 271–2
  marine biodiversity 10, 30, 50
  plant biodiversity 9, 11
  rural landscapes 250–1
  terrestrial biodiversity 9, 10, 30, 366–7
biomass 111, 115, 128, 131, 139, 265, 270, 290, 292, 319, 364
Blue Flag ecological label 344–5
Bosnia-Herzegovina
  call for action 359, 361
  coastal areas 336, 337
  energy 113
  Mediterranean development dynamics 4, 5, 15, 19, 23, 24, 39, 45, 51
  rural areas 255, 259, 263, 265, 273, 285, 288
  transport 163, 166, 173
  urban areas 228, 242
  water 71, 73
budget deficit 57, 226

call for action 359–87
  facing sustainable development challenges 364–73
  risks in baseline scenario 361–4
  striving for more sustainable development 373–86
Canada 38, 47, 260, 268, 282
capital flows see trade
car ownership 214
catchment areas viii, xiii, 4, 24, 71, 72, 73, 78, 82, 83, 84, 89, 92, 93, 95, 266, 287, 291, 298, 306, 319, 325, 336, 350, 351, 363, 367, 377
CDM (Clean Development Mechanism) 140–2
Central America 19
central Europe 39, 187, 294
child mortality 23
China 35, 36, 64, 142, 282
cities see urban areas
civil society 62, 104, 278, 296, 328, 335, 337, 340, 351, 360, 373, 380
climate change xiii, 10, 29–30, 73, 74, 118, 120, 121, 124, 127, 138, 212, 298, 328, 360, 364, 369, 384, 386
  baseline scenario 63
climate hazards and droughts, response to 283
CO_2 emissions 19, 23, 27, 29, 30, 119–20, 132, 144, 165, 167, 168, 169, 191, 214, 225, 362, 364, 369, 370
  alternative scenario 144, 192
  baseline scenario 120, 168
coastal over-development 24, 24, 309–18
coastal regions/areas xiii, 3, 4, 5, 6, 7, 8, 12, 18, 24–6, 27, 30, 31, 34, 49, 51, 52, 53, 61, 63, 80, 97, 118, 149, 199, 200, 203, 303–52
  alternative scenario 348–52
  baseline scenario 305, 310, 311, 316, 317, 321, 348–9
  conditions for better governance 351–2

444

Index

convergence of pressures 309
coveted place for development 307–8
cultural heritage 308
definitions 305–6
degradation of habitats and species 326–8
description 305–18
direct pressures on 309
ecosystems 306–7
energy infrastructure 314
erosion and salt water intrusion 325
increasingly built-up 309–18
industrial concentration 312–15
mobilizing for integrated coastal area management 335–40
pollution 318–25
    chemical contamination 320
    combating 321–5
    land-based 318–21
    instruments for reducing 314–15, 345–8
    organic and microbiological 320–1
    solid and hazardous waste 321
population growth 310–11
protecting ecosystems and biodiversity 325–30, 366
regional development polices 340
relieving 371–2
risks of high-impact biological invasions 326
sustainable coastal management scenario 348–52
    objectives of alternative scenario 349–51
    risks of baseline scenario 348–9
tourism and sustainable development 341–8
under artificial cover by 2025 315–18
urbanization 309–18
coastal shipping 174, 177–8, 187, 188, 189, 194
Common Agricultural Policy (CAP) 257, 268, 276, 280, 298, 376
common goods viii, 364, 367, 377, 384, 385
conflicts, persistence of 12–13
congestion 27, 61, 152, 155, 164, 172, 173, 174, 184, 185, 190, 191, 192, 194, 290, 317, 347, 364, 369, 370, 379
consumption patterns 30, 38, 50, 57, 63, 276, 310, 387
conventions 52, 53, 104, 118, 149, 194, 240, 270, 321, 354–6, 367
    ACCOBAMS Agreement 328
    Barcelona Convention vii, viii, xiii, 48, 49, 104, 182, 244, 288, 305, 321, 328, 330, 335, 344, 352, 360, 366
    Convention on Biological Diversity 330
    Kyoto Protocol 30, 44, 104, 118, 120, 132, 140, 142, 144, 218, 367, 384
    Land-Based Sources (LBS) Protocol 321, 322
    MARPOL 73/78 Convention 181, 183
    OPRC 90 Convention 182
    Protocol Concerning Specially Protected Areas (SPA) 328, 330
    RAMOGE Agreement 328
    RAMSAR Convention on wetlands 328, 329
    UN Convention to Combat Desertification and Drought (UNCDD) 270
    UN Convention on the Law of the Sea (UNCLOS) 328
    UN Framework Convention on Climate Change 30, 44, 118, 137, 138, 139, 298, 369, 384
cooperation viiii, xi, 3, 13, 34, 36, 43, 51, 54, 63, 89, 96, 99, 137–40, 142, 143, 145, 182, 183, 186–90, 193, 194, 199, 214, 218, 223, 236–40, 244, 245, 288, 289, 295, 298–300, 325, 328, 330, 332, 334, 335, 336, 338, 340, 342, 344, 346, 348, 352, 359, 360, 361, 367, 373, 375, 377, 378, 381–7
    environmental cooperation 49–50, 321–2, 381–2
    international and regional, insufficient 375
    north–south cooperation 46–9, 64, 383
    rural development 298–300
    south–south cooperation 45–6, 64, 383–4
    strong political commitment for 385–6
    urban dimension 242–3
    water management 104–6
corruption 62
Corsica 12, 252, 263, 271, 285, 306, 310, 335
Crete 6, 86, 188, 271, 275, 306, 347, 350
Croatia
    call for action 359 361, 362
    coastal areas 312, 313, 317, 318, 323, 333, 336, 337, 339, 341, 345
    energy 113, 118, 126
    Mediterranean development dynamics 4, 5, 13, 14, 15, 19, 20, 24, 39, 43, 45, 51, 53, 64
    rural areas 251, 253, 255, 265, 273, 277, 288, 289
    transport 154, 158, 163, 166, 173, 180
    urban areas 228, 232, 244
    water 71, 73, 89, 92, 99
current account balance 60
Cyprus
    call for action 359, 361
    coastal areas 306, 307, 312, 314, 316, 331, 332, 333, 334, 338, 342, 343, 345
    energy 135
    Mediterranean development dynamics xiii, 4, 5, 6, 8, 9, 12, 14, 17, 43, 45, 51
    rural areas 253, 254, 255, 265, 270, 271, 286, 288
    transport 151, 156, 166, 173, 180, 181, 184
    urban areas 202, 228, 244
    water 71, 75, 76, 77, 79, 80, 81, 82, 90, 95, 99, 100, 103

Dardanelles 6
debt 259, 385
    debt service 24, 58, 59, 374
    external debt 58
decentralization
    administrative without financial means 232–6
    financial 233–6
    regional development planning and local governance 296–7
    and urban governance 231–6
decompartmentalizing sectoral policies 376–7
decoupling economic growth from environmental pressures 19, 190–1, 367–8
demand management 93–8, 266, 324, 344, 368
democratic governance 62, 340
demographic convergence, north–south 39–4
desertification 24, 45, 48, 73, 118, 249, 250, 252, 259, 263, 269–70, 272, 273, 282, 291, 298, 300, 340, 362, 363, 367, 370, 371, 372, 377, 378, 380, 382, 383, 386
domestic disparities, reducing 371–2
drought 6–7, 9, 21, 30, 59, 63, 73, 74, 89, 96, 103, 249, 250, 251, 259, 260, 267, 270, 278, 279, 281, 282, 283, 294, 298, 363
    and climate hazards, response to 283

earthquakes 9, 10, 11, 24, 212, 213, 214, 239, 242, 349, 363
East and Pacific Asia 13, 21
Eastern Adriatic countries (EAC)
    call for action 362, 382
    coastal areas 317, 319, 326, 331, 332, 333
    energy 113, 115
    Mediterranean development dynamics 5, 7, 8, 9, 17, 18, 21, 22, 24, 40, 43, 45, 47, 54, 59
    rural areas 249, 252, 253, 255, 259, 265, 273, 274, 275, 276, 292, 299
    transport 149, 151, 155, 158, 159, 163, 170, 174, 175, 176, 177, 183, 188, 191, 192
    urban areas 207, 228, 239, 242, 243
    water 75, 76, 77, 89, 105
eastern Europe 39, 54, 142, 158, 186, 188, 232, 277, 298, 385
eco-efficiency 174, 163, 239, 369
    of development patterns 18
    of transport modes 178
ecological deficit 17
ecological footprint 17, 347
    and HDI 18
economic development vii, xii, 12, 29, 48, 49, 56, 116, 124, 125, 142, 149, 155, 176, 189, 200, 212, 233, 234, 238, 282, 286, 289, 294, 312, 333, 340, 344, 347, 349, 362, 364, 374
    based on exploitation of declining natural resources 15–19
    decoupling from pressures on environment 367–8
economic diversification 254, 261, 264, 287, 292
economic handicaps 57–9
economic instruments 55, 99–104, 135–7, 184–5, 345–6, 379
economic performance
    lack of innovation 373–4
    poor 13–15
economic polarization 35
economic policies, integrating environmental concerns 376–7
economic specialization 193, 218, 244
eco-regional cooperation 381–2
ecosystems 24, 30, 71, 72, 77, 78, 82–6, 87, 89, 94, 95, 119, 120, 121, 214, 222, 251, 253, 266, 271, 286, 287, 288, 290, 292, 305, 306, 307, 310, 317, 318, 325, 327, 331, 332, 334, 348, 349, 351, 359, 364, 366, 367, 377
    coastal zones 325–30
    degradation 82–6
    marine ecosystems 30
    terrestrial ecosystems 30
education vii, viii, 12, 18, 19, 21, 22, 23, 31, 34, 46, 59, 62, 105, 118, 139, 179, 225, 233, 240, 243, 259, 292, 294, 306, 344, 350, 360, 372, 379
    primary education and literacy 23
EEA 54, 91, 164, 169
Egypt
    call for action 359, 362, 364, 366, 369, 376
    coastal areas 305, 307, 308, 310, 311, 312, 313, 316, 317, 322, 323, 324, 325, 327, 328, 333, 334, 336, 337, 339, 340, 342, 347
    energy 111, 113, 115, 116, 117, 118, 121, 123, 128, 129, 130, 132, 135, 136, 140, 142
    Mediterranean development dynamics 3, 4, 5, 7, 14, 15, 16, 18, 19, 20, 21, 22, 23, 24, 27, 33, 36, 38, 39, 40, 43, 46, 48, 51, 52, 53, 54, 57, 58, 59
    rural areas 250, 258, 259, 261, 262, 264, 265, 266, 267, 268, 270, 273, 274, 277, 278, 279, 281, 289, 294, 299
    transport 151, 153, 156, 158, 160, 162, 163, 164, 170, 171, 172, 173, 174, 180, 181, 184, 186, 189
    urban areas 200, 202, 207, 209, 210, 212, 214, 227, 228, 229, 231, 232, 233, 234, 235, 236, 242, 244
    water 71, 73, 74, 76, 77, 78, 79, 81, 82, 83, 84, 86, 88, 89, 91, 97, 99, 100, 101, 102, 105

445

electricity 36, 112–15, 118–33, 135–9, 141–4, 178, 237, 238, 259, 284, 292, 295, 297, 320, 324, 362, 371, 372,
   infrastructures 121–3
   production 114
   growth in demand 112
   trade 141
EMAS 55, 58
employment rate 33
enabling framework 378–81
endemic species 9, 250, 271, 307
energy efficiency (EE) and renewable energy vii, 115, 123–42, 143, 145, 379
   agencies to promote 139
   conditions for developing 130–7
   examples 131–3
   funding 133–5, 137–8
   immediate costs and deferred benefits 133–4
   need for public support 134–5
   potential in Mediterranean 125–8
   price signals 135–7
   projects 139–40
   regional cooperation 138–43
      EU's role 142–3
   strategic priority 368–9
energy sector 17, 109–45
   alternative scenario to 2025 143–5
      energy savings 144
   baseline scenario 111–23
   committing to rational use 124
   consequences for environment 118–21
   energy demand management 123
   EU's role in cooperation 142–3
   expensive access 118
   growth in demand 111–15
   infrastructures, impact and risks 121
   increasing dependence 116–17
   increasing risks 116–23
   public initiatives and national strategies 131–2
   regional cooperation 137–42
   towards sustainability 123–45
   and transport and residential sectors 112, 119, 168–9
   unsustainable energy development 111–23
enforcement and compliance *see* law enforcement
enhancing strong points of Mediterranean region 372–3
environment-related infractions 55
environmental agencies 51–2
environmental assessment 53–4
environmental degradation, costs 24–7, 272–3
environmental governance 50–7
   baseline scenario 63
environmental policies 51, 52
   integrating into economic policies 376–7
environmental pressures 19, 54–7
   decoupling from economic growth 19, 190–1, 367–8
environmental protection 53, 101, 154, 187, 220
   framework laws 53
epidemiological studies 87–8, 167, 224–5
EU-Med subgroup
   call for action 375
   coastal areas 331
   energy 118, 120, 129, 131, 136, 138, 144
   Mediterranean development dynamics 5, 13, 14–15, 18, 19, 21, 22, 40, 43, 45, 47, 51, 56, 57, 58, 64
   rural areas 253, 266, 267, 272, 273, 275, 277, 285
   transport 149, 155, 157, 161, 166, 173, 179, 186
   urban areas 232, 236, 239, 240
   water 84, 86, 91, 101

Euro-Mediterranean Partnership (EMP) xi, 46–8, 49, 382–6
   Euro-Mediterranean free-trade area (EMFTA) 46, 49
   Facility for Euro-Mediterranean Investment and Partnership (FEMIP) 46, 47, 186, 385
   MEDA funding 46, 47, 104, 105, 138, 139, 140, 142, 183, 242, 298, 299, 300, 361, 366, 383, 385, 386
   relaunching through sustainable development 383
   SMAP (Short and Medium Term Environment Action Programme) 48
Euro-Mediterranean pole in two world economy scenarios 36
European Commission 40, 43, 44, 49, 50, 55, 89, 105, 140, 180, 194, 244, 272, 332, 334, 335, 348, 352, 359, 382
European Court of Justice 55
European Investment Bank (EIB) 46, 47, 48, 50, 106, 139, 186, 242, 244, 299
European Leader programme 296
European Neighbourhood Policy 48
European Union (EU) xi, 4, 15, 23, 36–51, 53, 63, 64, 89, 105, 119, 127, 129, 135–7, 140, 142–3, 149, 151, 152, 155, 157–9, 162–4, 170, 172, 173, 175, 178, 180–3, 185–90, 206, 240, 242–5, 253, 260, 261, 267–9, 273, 274, 277, 278, 280–2, 286, 287, 292, 295, 298–300, 320, 321, 322, 332, 334, 335, 338, 340, 344, 348, 360, 361, 366, 369, 370, 373, 375, 376, 379, 381, 382, 383, 384, 385, 386, 387
   EU enlargement 43–5, 59, 273, 278
      more sustainability 382
   EU farm subsidies 45
   EU structural funds 58, 64, 205, 243
   role in Mediterranean energy cooperation 142–3
eutrophication 319–20

FAO 21, 76, 253, 261, 264, 267, 291, 300, 332, 367
farm size distribution 254
farmland, conservation of 289
fauna 6, 10, 256, 257, 272, 307, 319,
FEMIP *see* Euro-Mediterranean Partnership
FEMISE 59, 381
fertility rate 5, 38, 63, 259, 359
   reduced 30–1
financing 40, 57, 58, 143, 184–6, 189, 217, 218, 220, 222, 237, 241–4, 270, 287, 324, 342, 347, 365, 373, 374, 379, 380, 384, 385
   strong political commitment for 385–6
   water management 98–9
fisheries
   regulating development of fish farming 333–4
   risks and regulatory measures 334–5
   sustainable management of 330–4
   unsustainable development 331–2
floods xi, 6–7, 9, 27, 30, 63, 92, 170, 212, 213, 242, 249, 252, 272, 282, 294, 317, 349, 363
flora 6, 9, 250, 251, 271, 307
food 6, 12, 15, 41, 84, 88, 91, 96, 161, 250, 253, 255, 256, 261, 268, 286, 290, 292, 293, 295, 299, 300, 307, 314, 319, 320, 334, 350, 368, 373
   and agriculture outlook 273–82
      paradox baseline scenario 276
   Mediterranean food regime 274–5
   security 21
foreign direct investments (FDI) 40, 43, 57, 60, 62, 64, 126
forest fires 8, 24, 30, 249, 251, 252, 263–5, 290, 292, 363, 367, 371

forestry 249, 251, 265, 283, 286, 287, 291, 292, 293, 294, 295, 297, 300, 367, 375, 380
   and pastoralism 294–5
France
   call for action 359, 364, 366, 375
   coastal areas 305, 307, 308, 310, 312, 313, 314, 315, 317, 320, 321, 325, 327, 328, 330, 332, 333, 334, 335, 336, 337, 338, 341, 342, 344, 345, 346, 348, 350
   energy 112, 113, 115, 116, 122, 124, 126, 128, 129, 130, 131, 133, 134, 135, 136, 139, 143
   Mediterranean development dynamics xiii, 3, 4, 5, 12, 14, 16, 19, 20, 21, 22, 23, 24, 36, 37, 39, 40, 42, 43, 46, 51, 52, 53, 54, 55, 57, 58, 59
   rural areas 251, 252, 253, 254, 255, 256, 257, 263, 265, 267, 268, 270, 272, 273, 274, 275, 277, 282, 285, 286, 287, 296
   transport 149, 151, 152, 154, 155, 157, 158, 162, 163, 164, 167, 169, 173, 174, 175, 176, 178, 179, 183, 184, 187, 190
   urban areas 202, 206, 207, 212, 213, 218, 223, 227, 228, 229, 230, 232, 234, 235, 236, 237, 238, 239, 240, 242, 243, 244
   water 71, 73, 75, 77, 78, 79, 83, 84, 85, 87, 88, 90, 97, 99, 100, 101, 103, 105
free trade *see* trade
freight traffic 154–61
   combined rail/road/sea/river 176–8
   domestic, dominated by road 155
   international 155–61
   projections to 2025 163

gas
   infrastructures 121, 122
   trade 141
GDP 13–21, 27, 35, 36, 40, 45, 54, 57, 59–61, 64, 98, 111, 112, 116, 118, 125, 132, 136, 152, 155, 161–4, 173, 174, 191–3, 229–31, 233, 235, 260, 273, 280, 291, 306, 324, 328, 341, 359, 362, 364, 365, 374, 380
   growth 13
   structure 16
gender equality 23, 360
geothermal energy 130
Germany 36, 40, 42, 46, 105, 115, 129, 139, 159, 175, 234, 242, 277, 342
GFCM (General Fisheries Commission for the Mediterranean) 50, 332, 334, 352, 381
Gibraltar 6, 37, 154, 158, 160, 161, 180, 181, 183
Global Compact 58
Global Environment Facility (GEF) 49, 140
global warming 7, 29–30, 63, 73, 111, 123, 127, 132, 149, 265, 305, 328, 349, 363, 367
globalization xi, 29, 34–50, 56, 57, 63, 125, 193, 205, 220, 237, 249, 250, 285, 290, 300, 349, 359, 360, 370, 386
   baseline scenario 63
governance viii, xii, 48, 50, 57, 59, 62, 63, 90, 96, 104, 105, 131, 193, 231, 232, 251, 287, 296, 297, 330, 337, 340, 344, 348, 349, 350, 351, 365, 373, 376, 381, 387
   for sustainable development 376–81
Greece
   call for action 359, 361, 368, 382, 387
   coastal areas 305, 307, 313, 316, 317, 320, 321, 323, 325, 327, 331, 332, 333, 334, 335, 336, 337, 339, 341, 342, 343, 345, 347
   energy 113, 115, 116, 120, 123, 125, 127, 128, 130, 131, 134, 135
   Mediterranean development dynamics xiii, 3, 4, 5, 7, 9, 13, 14, 15, 16, 18, 19, 21, 23, 24, 36, 37, 39, 43, 45, 51, 53, 55, 60
   rural areas 253, 254, 255, 263, 265, 267,

Index

269, 271, 274, 277, 280, 282, 285, 288, 296, 298
transport 149, 151, 154, 155, 157, 158, 160, 163, 170, 173, 176, 177, 180, 183, 184, 190
urban areas 199, 202, 212, 223, 227, 228, 234, 235, 236, 243, 244
water 71, 77, 78, 79, 80, 86, 90, 91, 99, 100, 101, 102, 105
Greek islands 8
groups of Mediterranean countries 5
Gulf countries 6, 35, 36
Gulf wars 12, 36

harbours viii, 8, 119, 170, 222, 309, 310, 313, 316, 327
hazardous waste *see* waste
health vii, viii, xi, 12, 19, 20–1, 23, 27, 32, 34, 49, 52, 62, 81, 82, 84, 86–90, 116, 118, 120, 122, 136, 164, 167, 176, 179, 210, 217, 220, 223–5, 229, 234, 235, 242–4, 259, 268, 273, 274–8, 292, 293, 297, 306, 307, 318, 319, 320, 321, 324, 328, 334, 349, 351, 360, 364, 371, 372, 373, 375, 377, 378
facilities 22
model 275–6
high-productivity agricultural model 276–8
high-speed vessels 154
historical sites 12
UNESCO World Heritage List 12
households 22
households size 31–2
household waste *see* waste
housing 22, 125, 126, 130, 133, 142, 192, 194, 202, 205–7, 209–12, 220, 222, 226, 227, 234, 239–41, 244, 256, 310, 335, 362, 370, 371, 376
development and urban sprawl 207–10
Human Development Index (HDI) 18, 19, 76
and ecological footprint 18
human health 82, 116, 120, 136, 224, 229, 242, 320, 324, 349
hydrocarbons
gas 113–15, 117
oil 117
hydrological regime 73
hydropower plants 97, 128

illiteracy 21, 360, 372
immigration, illegal 37
impact assessment 52, 53, 361, 376
impoverishment 20, 61, 193, 261, 273, 281, 386
income inequality 19
income per capita 14
industry 16
cleaner production 91–2, 126, 324
concentration on coastal areas 312-5
extraction industry 16
industrial accidents 213–4
residential development and new technologies 295
infant mortality 21
informal sector 15, 55
information, insufficient 375–6
inspections (environmental) 53–5
International Centre for Advanced Mediterranean Agronomic Studies (ICAMAS) 50, 381
International Commission for the Scientific Exploration of the Mediterranean Sea (ICSEMS, CIESM) 50, 381
International Energy Charter 137
International Monetary Fund (IMF) 34
international terrorism 39
Iraq 12, 117, 156, 266
irrigation
rapid growth of 75–7
and salinization 265–8
islands 8, 113, 116, 124, 154, 172, 173, 177, 194, 222, 250, 251, 258, 267, 271, 305, 306, 307, 308, 310, 312, 317, 318, 325, 327, 333, 336, 337, 341, 342, 343, 346, 347, 348, 349, 350, 352, 371, 382
ISO 14001 58, 345
Israel
call for action 359, 368
coastal areas 306, 311, 312, 313, 316, 325, 331, 333, 335, 336, 337, 339
energy 113, 115, 116, 125, 130, 135
Mediterranean development dynamics 3, 4, 5, 13, 14, 19, 20, 21, 22, 23, 24, 31, 38, 39, 40, 42, 43, 46, 48, 51, 54, 57, 59, 64
rural areas 258, 259, 262, 264, 265, 267, 268, 270, 274, 277, 279, 280, 281
transport 155, 158, 164, 166, 169, 170, 172, 173, 176, 182, 189
urban areas 205, 227, 228, 232, 234, 235, 236, 243, 244
water 71, 75, 76, 77, 78, 80, 81, 82, 84, 85, 88, 89, 90, 91, 95, 99, 100, 101, 102
Israeli–Palestinian conflict 12
Italy
call for action 359, 366, 370, 372, 385
coastal areas 308, 310, 312, 313, 314, 315, 317, 319, 320, 321, 322, 325, 328, 331, 332, 333, 334, 336, 337, 341, 342, 343, 345, 346, 348
energy 114, 115, 116, 123, 127, 128, 130, 131, 134, 135, 136, 139, 142
Mediterranean development dynamics 3, 4, 5, 9, 14, 15, 19, 20, 21, 24, 30, 36, 37, 39, 40, 42, 51, 52, 53, 54, 55, 57, 58
rural areas 253, 254, 255, 256, 257, 262, 263, 265, 267, 269, 271, 272, 273, 274, 275, 285, 294, 296
transport 149, 151, 154, 155, 157, 158, 159, 160, 161, 162, 163, 164, 169, 170, 172, 174, 175, 176, 177, 178, 181, 183, 184, 186, 190
urban areas 202, 212, 218, 220, 223, 227, 228, 229, 232, 234, 235, 236, 237, 239, 240, 241, 243, 244
water 71, 73, 75, 77, 78, 82, 83, 85, 86, 90, 91, 99, 100, 101, 105

Japan 40, 46
Jordan 46, 48

land
cultivated: importance of permanent crops 262–3
degradation 27, 264–5, 269–72
costs of 272–3
dynamics 261–73
inappropriate technical response 282–3
use planning 209–10
landscape, Mediterranean viii, xi, xiii, 3, 10, 12, 24, 61, 78, 82, 119, 121, 122, 123, 124, 249, 250, 251, 252, 257, 262, 263, 265, 268, 269, 271, 272, 273, 276, 278, 285, 286, 288, 291, 292, 293, 295, 298, 307, 308, 317, 318, 336, 338, 342, 343, 347, 348, 349, 350, 351, 364, 366, 367, 369, 371, 372, 373, 377
biodiversity 271–2
conserving 366-7
pastureland as major component 263–4
Latin America 13, 23, 24, 58
law enforcement 54–6, 367
League of Arab States 46, 384
Lebanon
call for action 362
coastal areas 311, 312, 316, 317, 322, 323, 324, 335, 336, 337, 342
energy 116, 123, 125, 127, 136, 140
Mediterranean development dynamics 3, 4, 5, 7, 12, 13, 14, 15, 18, 19, 20, 21, 23, 24, 27, 36, 38, 39, 43, 51, 53, 57, 59
rural areas 251, 258, 259, 262, 264, 265, 266, 267, 270, 271, 274, 277, 278, 279, 285, 288, 297
transport 149, 154, 156, 158, 166, 170, 172, 178, 180, 181, 184, 189
urban areas 207, 209, 211, 212, 214, 228, 229, 232, 233, 234, 236, 244
water 71, 77, 82, 99, 100, 101, 102
legislation 43, 48, 51, 53, 55, 62, 89, 90, 127, 182, 183, 187, 222, 228, 229, 238, 284, 322, 338
Libya
call for action 362, 363
coastal areas 310, 312, 336, 337, 341, 342
energy 111, 116, 117, 121, 123, 125, 132, 135
Mediterranean development dynamics 1, 3, 4, 5, 6, 7, 16, 17, 19, 21, 23, 31, 34, 38, 39, 51, 57, 58
rural areas 250, 258, 259, 262, 263, 264, 265, 266, 267, 274, 278, 279, 288
transport 156, 160, 176, 181, 184
urban areas 244
water 71, 74, 77, 78, 79, 80, 81, 82, 86
life expectancy 18, 21, 31, 39, 81, 272, 275
lifestyles, Mediterranean 12
Local Agendas-21 239
local authorities
control and supervision 236
duties 232–3
expenditure 235–6
inter-municipal cooperation 236–7
local development, sustaining 377–8
long-term strategies and goals 378–9
loss of agricultural land 24, 206, 207–9, 249, 265, 270–1, 289, 309, 315, 316, 336, 340, 363, 367

Maghreb countries
call for action 359, 361, 380, 382, 383, 384
coastal areas 327
energy 122, 128
Mediterranean development dynamics 5, 15, 17, 22, 23, 24, 36, 37, 39, 42, 43, 57, 62, 64
rural areas 150, 263, 264, 265, 267, 268, 279, 283, 284, 294, 297
transport 151, 158, 164, 168, 176, 189, 190
urban areas 202, 205, 226, 228, 232, 239
water 105
Malta
call for action 359, 361, 363
coastal areas 306, 316, 324, 331, 333, 334, 336, 337, 339, 341, 342, 343, 344, 345, 350
Mediterranean development dynamics xiii, 3, 4, 5, 8, 14, 18, 19, 24, 40, 43, 45, 51, 64
rural areas 253, 254, 262, 265, 270, 271
transport 151, 154, 156, 159, 164, 166, 170, 173, 180, 181, 184
urban areas 202, 227, 244
water 71, 75, 76, 77, 80, 81, 82, 85, 90, 95, 99, 100, 102
MAP (Mediterranean Action Plan) 49, 51, 300, 318, 321, 322, 324, 326, 328, 330, 334, 335, 337, 338, 339, 344, 350, 352, 366, 367, 381, 382, 385, 386
maritime traffic 152–3
improving safety 182–3
projections to 2025 163
risks, and reducing pollution 179–83
market economy 35, 62, 63, 273, 278, 340, 361, 365

Mashrek
  call for action 359, 361, 380, 383
  Mediterranean development dynamics 17, 22, 36, 42, 43, 57, 64
  rural areas 279
  transport 151, 164, 191
  urban areas 202, 232, 236, 239
maternal health 23
MCSD (Mediterranean Commission for Sustainable Development) 49, 97, 335, 344, 345, 381–2
MEDA see Euro-Mediterranean partnership
MEDCITIES 244, 381
MEDCOAST 337, 381
MEDENER (Mediterranean network for energy efficiency agencies) 138
MEDFORUM 381
MEDPOL 318, 324, 381
Mediterranean area, definition xi, 3–5
Mediterranean diet, benefits and decline 274–6
Mediterranean merchant fleet 180–1
Mediterranean Sea 6, 30, 49, 50, 71, 120, 157, 159, 183, 194, 290, 306, 307, 319, 328, 333, 377, 384
  main maritime routes 160
  maritime oil traffic 160
  protecting 384
  transit trade through 159–61
Medwet programme 50, 328, 329, 337
megapolization 309–18
MEPI (Middle East Partnership Initiative) 46
METAP (Mediterranean Environment Technical Assistance Programme) 50, 381
meteorological events, extreme 30
metropolitan and regional development 205, 236–7
Mexico 47, 260
micro-hydroelectricity 128
Middle East conflicts 39
migration between SEMC and EU 35–8
Millennium Development Goals (MDG) vii, 22–4
military expenditure 13
MIO-ECSDE 381
mobility, increased 206
Monaco
  coastal areas 311, 313, 314, 316, 328, 336, 337
  energy 118
  Mediterranean development dynamics 3, 4, 5, 24, 38
  transport 151, 166, 170, 173, 183, 190
  urban areas 228, 244
  water 71
Morocco
  call for action 359, 362, 367, 374, 385
  coastal areas 311, 312, 313, 314, 322, 323, 324, 325, 331, 333, 334, 335, 337, 341, 342, 344, 347
  energy 111, 112, 113, 115, 116, 118, 122, 123, 125, 126, 128, 129, 130, 134, 136, 140, 142, 144
  Mediterranean development dynamics 3, 4, 5, 7, 14, 18, 19, 20, 21, 22, 23, 24, 27, 38, 39, 43, 46, 48, 51, 53, 54, 57, 58, 59, 60, 61
  rural areas 251, 258, 259, 260, 261, 262, 263, 264, 265, 266, 267, 269, 270, 271, 272, 273, 274, 277, 279, 280, 281, 283, 284, 285, 287, 288, 290, 291, 294, 295, 297, 299
  transport 153, 156, 158, 161, 163, 169, 176, 178, 179, 181, 183, 186
  urban areas 199, 200, 202, 205, 207, 212, 213, 216, 225, 226, 227, 228, 229, 232, 233, 234, 235, 236, 237, 238, 242, 244
  water 71, 76, 77, 78, 79, 81, 82, 84, 86, 90, 91, 94, 95, 96, 100, 101, 102, 103, 105
motorization see car ownership
mountains 7, 8, 12, 250, 251, 252, 253, 254, 268, 282, 285–9, 291, 292, 295, 296, 362, 367, 372, 376, 382
  degradation or conservation and restoration 285–9
  importance and fragility of Mediterranean mountain system 285–6
  reconciling development and conservation 286–9
municipal waste see waste
municipalities 232–3

National Commissions for Sustainable Development 56
natural gas 114–15
natural hazards
  natural disasters 212–14
  reducing 367
  vulnerability to natural hazards 213–14
natural resources, exploitation of 17
Near East 12, 46, 47
NGOs 49, 50, 54, 131, 183, 289, 296, 297, 300, 328, 330, 337, 339, 344, 348, 366, 372, 380, 381, 382
Nile river/valley 7, 8, 30, 71, 73, 78, 79, 80, 81, 83, 89, 97, 210, 216, 250, 307, 308, 319, 325, 328
noise pollution 164–5, 191, 348
North Africa 23, 30, 120, 122, 123, 189, 214, 215, 238, 243, 251, 262, 264, 270, 271, 294, 383
North America 47, 48, 120, 180, 362
North American Free Trade Agreement (NAFTA) 36, 47, 48, 60, 161, 260
north–south cooperation see cooperation
north–south demographic convergence 39–4
north–south gap 14, 361–2
Northern Mediterranean countries (NMC)
  call for action 364, 368, 369, 374
  energy 111, 112, 113, 114, 115, 116, 117, 118, 120, 122, 125, 126, 128, 129, 131, 143
  Mediterranean development dynamics 3, 5, 17, 21, 22, 31, 32, 33, 38
  rural areas 253, 262, 263, 265, 269, 271, 272
  transport 151, 52, 153, 154, 155, 162, 163, 164, 165, 170, 171, 172, 178
  urban areas 199, 202, 203, 204, 229, 230, 231, 236
  water 71, 73, 76, 77, 83, 84, 89, 91, 94
NOx emissions 166
  baseline scenario 168

OECD 12, 21, 22, 38, 44, 54, 105, 120, 242, 253, 260, 275, 348, 382
Official Development Assistance (ODA) 23, 46, 105–6, 137, 190, 242
oil
  exporting countries 116
  trade 141
OME 112, 128, 381
organic agriculture 277

Palestinian Territories
  coastal areas 311
  energy 116, 130, 135
  Mediterranean development dynamics 4, 14, 19, 23, 40, 48, 57, 64
  rural areas 259, 262
  urban areas 205, 236, 242
  water 71, 76, 77, 82, 91, 105
parking control 219
participatory management of natural resources in arid areas 283–5
partnerships, developing new 379–80
passenger mobility, projections 215
passenger traffic 149–54
  domestic 151
  international 151–4
  projections to 2025 162
plains viii, 7, 8, 24, 74, 83, 85, 206, 209, 249, 250, 252, 255, 257, 264, 267, 271, 280, 282, 285, 287, 289, 292, 306, 308, 317, 325, 327, 336, 340, 349, 363, 364, 372
polluter-pays 51, 53, 184
pollution 10
  air 27, 218–19, 223–5
  agricultural 268–9
  atmospheric 165–8
  coastal zones 318–25, 344–5
  marine pollution 83–4
    accidental by ships 181
    land-based sources 318–21
    operational from ships 181–2
  rivers 85
  underground water 85
  water resources 83–4, 90
population
  in coastal regions 4, 310–11
  growth 5–6
    baseline scenario 31, 63
  rural, projections to 2025 33-4, 257
  urban, projections to 2025 33–4, 199–202
Portugal 45, 60, 229, 252, 265, 280, 286, 294, 298, 387
poverty 8, 12, 15, 19, 21, 22, 24, 43, 48, 51, 118, 137, 202, 209, 222, 244, 249, 253, 258, 259, 260, 261, 264, 270, 276, 282, 284, 286, 287, 291, 294, 296, 297, 298, 340, 360, 362, 371, 375, 380, 387
  MDG 22
private sector 21, 46, 49, 104, 131, 134, 140, 186, 216, 221, 227, 231, 241, 294, 342, 373, 380
protected areas 271, 272, 278, 330, 335, 336, 351, 352, 367, 375
  natural parks 288, 296, 297, 346, 377, 378
  specially protected areas (SPA) 49, 328, 329, 330
public administration, poorly adapted systems 374–5
public expenditure 21, 57, 62, 131, 176, 232, 233, 235, 343, 380, 381
  military expenditure 13
  on education 21
  on human health 21
public–private partnerships 237–8
public transport 216–23
  formal and informal 216–17
Pyrenean traffic 157

rail transport 149–51, 153–8, 161–4, 167, 169, 170, 173–9, 183–91, 193, 210, 216–19, 221, 225, 312, 350, 363, 369, 370, 375, 386
rainfall 7, 8, 29, 30, 72, 170, 212, 249, 283, 285
RAMOGE 382
regional cooperation 298–300
regional development 50, 56, 63, 199, 202, 205, 210, 244, 289, 296, 308, 340, 350, 351, 377, 386
regional disparities 280
  and decline of health model 275–6
regional integration 29, 34–50, 64, 194, 375
  baseline scenario 63
renewable energy see energy efficiency and renewable energy
renewable natural resources (RNS) viii, 51, 73, 75, 77, 78, 82, 92, 363
residential sector and energy 112–13
resource management 90–2, 93, 104, 284, 287,

Index

294, 296, 300
Rhone, river 8, 9, 30, 73, 79, 83, 176, 177, 307, 319, 320
Rio Declaration 51
rivers 4, 8, 71, 73, 74, 77, 78, 79, 82, 84, 85, 106, 119, 161, 176, 2150, 266, 269, 272, 306, 307, 319, 320, 323, 325, 363
    discharge 74
    pollution 85
road deaths 170, 171, 370
road safety 171, 174, 178–9, 183, 192
rural areas 247–300
    alternatives for 289–300
        objectives of alternative scenario 291–3
    awareness of many roles of 290–1
    baseline scenario 257, 267, 279
    developing fragile 292
    enhancing quality 292–3
    in north
        collapse of agricultural manpower 253–4
        rural migration to rural revival 254–5
    in south and east
        large agricultural populations and internal disparities 258
        rural poverty 259–61
        stabilized rural population to 2025 261
    integrated development 293
    originality, fragility 249–53
    rural landscapes 250
        and biodiversity 250–1
    suburban, conservation of 289
    sustainable development, towards 290–3
        territory-specific approach 295–7
        towards policies for 297–8
    towards new social structure 255–6
    unsustainable developments 253–61
rural populations 34, 249, 254, 256, 259, 296, 362
rural–urban migration 61, 118, 199, 205, 290, 310, 362, 374

salinization viii, 9, 251, 252, 265–70, 279, 281, 327, 363, 370
salt water intrusion 325
Sardinia 6, 80, 123, 177, 253, 271, 306, 331, 341
sea and trade 6
sea level rise 29, 30, 328, 363
seisms *see* earthquakes
Serbia and Montenegro
    call for action 359, 361, 362
    coastal areas 341, 344
    energy 113
    Mediterranean development dynamics 4, 5, 15, 23, 24, 39, 45
    rural areas 255, 265, 285
    transport 151, 158, 166, 173, 180
    water 71, 73
services sector 16, 17
sewage collection and treatment 90
Sicily 6, 55, 80, 177, 306, 327
Silva Mediterranea 50, 295, 300, 367
Slovenia
    call for action 359, 361
    coastal areas 308, 311, 313, 316, 323, 331, 336, 337, 345
    energy 113, 115, 118
    Mediterranean development dynamics 4, 5, 14, 19, 20, 21, 24, 39, 40, 43, 45, 51, 53, 54
    rural areas 253, 255, 265, 273, 277, 285
    transport 158, 166, 173, 180
    urban areas 223, 228, 232, 244
    water 71, 90, 99, 101
SMAP *see* Euro-Mediterranean Partnership
social change 221,

indicators 22
social disparities 19–24
society and environment, declining links 374
soil(s)
    conserving 366–7, 370–1
    erosion 9, 118, 268, 269–70, 287, 367, 370
        coastal 325
    fertility 9
solar photovoltaic (PV) energy 129
solar radiation 111, 129
solar thermal energy 129
South America 19, 40
southeast Asia 22, 23, 24, 40, 59, 185
Southern and Eastern Mediterranean countries (SEMC)
    call for action 361, 362, 363, 364, 365, 368, 369, 370, 372, 374, 384, 385, 386
    coastal areas 320, 323, 324, 334
    energy 111, 112, 113, 114, 115, 116, 117, 118, 120, 122, 123, 124, 125, 126, 127, 128, 129, 132, 133, 134, 135, 137, 138, 139, 140, 142, 143
    free trade 59–61
    key sectors of increasingly exposed economies 61
    Mediterranean development dynamics 3, 5, 13, 14, 15, 16, 17, 19, 21, 22, 23, 30–1, 32, 34, 35–8, 39–40, 42, 43, 45–6, 47, 49, 51, 54, 56, 57, 58, 59, 60, 61, 62, 63, 64
    multiple development challenges 57–63
    reforms 62–3
    rural areas 258, 259, 262, 263, 264, 269, 271, 272, 274, 276, 279, 280, 281, 298, 299
    transport 151, 152, 154, 155, 156, 157, 158, 162, 163, 164, 165, 167, 168, 169, 170, 171, 172, 178, 179, 183, 185, 186, 187, 188, 190, 192
    urban areas 199, 202, 203, 204, 205, 207, 212, 214, 216, 225, 227, 228, 229, 230, 231, 232, 233, 236, 237, 238, 239, 240, 242
    water 71, 73, 74, 75, 76, 77, 78, 83, 84, 89, 90, 91, 94, 95, 98, 101, 106
Spain
    call for action 359, 361, 363, 366, 368, 370, 382, 385, 387
    coastal areas 305, 307, 308, 311, 312, 313, 314, 315, 316, 317, 320, 321, 324, 325, 327, 332, 333, 334, 335, 336, 337, 341, 342, 343, 345, 346, 347
    energy 113, 115, 116, 120, 123, 125, 126, 127, 128, 130, 131, 132, 134, 143
    Mediterranean development dynamics 3, 4, 5, 7, 12, 13, 14, 18, 19, 20, 22, 24, 30, 37, 39, 43, 45, 47, 51, 53, 55, 58, 60
    rural areas 253, 254, 255, 256, 262, 263, 265, 267, 269, 271, 273, 274, 277, 280, 282, 283, 286, 292, 294, 296, 300
    transport 149, 151, 153, 154, 155, 157, 158, 162, 170, 174, 175, 176, 181, 183, 184, 190
    urban areas 199, 202, 212, 223, 227, 228, 229, 232, 234, 235, 236, 239, 240, 242, 243, 244
    water 71, 73, 77, 78, 79, 80, 81, 82, 83, 84, 86, 88, 90, 91, 95, 97, 99, 100, 101, 102, 103
specially protected areas *see* protected areas
sub-Saharan Africa 22, 23
Suez Canal 6, 154, 160, 161, 183, 187, 266, 326
sustainability vii, viii, 12, 43, 48, 142, 199, 239, 252, 253, 257, 260, 277, 278, 279, 280, 291, 293, 296, 308, 330, 331, 333, 339, 347, 350, 351, 352, 370, 375, 376, 379, 381, 382, 383, 386
    impact assessment 52, 53, 332, 336, 337,

376
    indicators 19
    sustainable development vii, viii, xi, 3, 12, 13, 18–19, 22, 24, 29, 43, 45, 47–51, 54, 56–9, 63, 90, 104, 111, 115, 118, 135, 136, 145, 179, 186, 187, 189, 190, 210, 211, 213, 222, 238, 239, 244, 272, 281, 289–91, 293, 295, 296, 299, 300, 330, 331, 334, 338, 341–4, 346–51, 360, 361, 364, 365, 368, 371–86
        financing 380–1
        towards Mediterranean governance for 381–6
    unsustainable development 12, 19–24, 80–1, 111–23, 331–2
Syria
    call for action 362
    coastal areas 311, 312, 322, 327, 331, 339
    energy 111, 113, 116, 117, 118, 123, 125, 132, 135, 136, 140
    Mediterranean development dynamics xiii, 4, 5, 12, 13, 14, 15, 16, 18, 19, 21, 22, 23, 24, 27, 34, 38, 39, 43, 51, 54, 57, 58, 59
    rural areas 251, 258, 259, 261, 263, 264, 265, 266, 267, 268, 270, 271, 273, 274, 278, 279, 282, 284, 285, 288, 300
    transport 155, 156, 170, 173, 176, 180, 181, 184, 189
    urban areas 200, 205, 207, 222, 228, 229, 233, 236, 237, 244
    water 71, 75, 76, 77, 78, 79, 81, 82, 84, 86, 90, 105

taxation 57, 100, 135, 136, 137, 143, 172, 184, 185, 190, 206, 220, 235, 292, 374, 379, 380, 381
technological and scientific progress 35, 86, 111, 113, 118, 130, 134, 137, 145, 279, 365
    mobilizing 380
territorial approaches, strengthening 377–8
territorial challenges in Mediterranean 282–9
territorial disparities 24–6
textiles/clothing sector 61
tourism vii, viii, xiii, 10, 15, 17, 24, 29, 38, 39, 42, 45, 49, 61, 63, 74–6, 78, 96, 149, 151, 154, 172, 181, 191, 202, 211, 221, 222, 239, 255, 256, 276, 278, 279, 285–93, 295, 296, 305, 307–9, 311, 312, 315–19, 321, 325, 327, 328, 330, 332, 336–8, 340–52, 359, 361–5, 368, 371–8, 381, 382, 385, 386
    baseline scenario 38–9, 310–1
    coastal zones and sustainable development 341–8
        forms of governance 344–8
    domestic tourists 39
    evaluating benefits 343–4
    international tourists xiii, 38–9
    reconciling with environment 371
    system 346
trade
    baseline scenario 63
    capital flows 40–3, 44
    EU and Mediterranean countries 158
    free trade 46, 59–63, 261
        agricultural 279–81
    in goods 39–41
    intensification 39–43
    liberalization 186–90
    in services 40, 42
traffic management 179
    see also transport
transparency 62, 96, 97, 103, 238, 373
transport 147–94
    air 151–2, 190
    baseline scenario 161, 162, 163, 165, 168, 170, 171, 174, 182
    alternative scenario 190–4

boom benefiting road sector 149–64
breaking current vicious cycle 349–70
coastal shipping 177–8
congestions costs 173–4
and energy consumption 112–13, 168–9
funding 186
impacts of land and air transport 164–79
  environmental 164–73
    on land 169–70
    social 170–3
  improving management in urban areas 217–20
international cooperation 186–90
lack of competitiveness of land systems 158–9
liberalization 186–90
maritime transport 152–4, 159–61, 179–83
rail 155–6, 174–6
rationalizing taxes and subsidies 184–6
road 149–51
  safety 1789
share of, in household budgets 172–3
sustainability 183–94
  urban transport systems 219–20
traffic projections to 2025 161–4
urban 214–23
Tunisia
  call for action 362, 368, 370, 376
  coastal areas 305, 307, 308, 311, 312, 314, 316, 321, 322, 324, 325, 328, 331, 333, 335, 336, 337, 339, 341, 342, 344, 348
  energy 111, 116, 122, 123, 125, 128, 129, 130, 132, 140, 142
  Mediterranean development dynamics xiiii, 4, 5, 14, 15, 16, 18, 19, 20, 21, 22, 23, 24, 27, 38, 39, 43, 46, 48, 51, 52, 54, 57, 59
  rural areas 258, 259, 262, 264, 265, 266, 267, 268, 270, 273, 276, 277, 278, 279, 280, 283, 284, 289, 294, 295, 299
  transport 155, 156, 158, 159, 166, 176, 177, 178, 184, 186
  urban areas 200, 202, 207, 209, 216, 222, 224, 227, 228, 232, 233, 234, 235, 236, 237, 239, 242, 244
  water 71, 74, 77, 78, 79, 80, 81, 82, 83, 84, 86, 90, 91, 92, 95, 96, 98, 99, 100, 101, 103, 105
Turkey
  call for action 359, 361, 362, 364, 380, 384
  coastal areas 305, 307, 310, 311, 312, 313, 314, 319, 322, 323, 327, 332, 333, 334, 336, 337, 338, 339, 341, 342, 343, 345, 347
  energy 111, 112, 113, 115, 116, 120, 123, 124, 125, 126, 128, 129, 130, 132, 134, 136, 138
  Mediterranean development dynamics 3, 4, 5, 9, 12, 13, 14, 15, 18, 19, 20, 21, 22, 23, 24, 31, 33, 34, 36, 38, 39, 40, 42, 43, 45, 51, 53, 54, 57, 58, 59, 64
  rural areas 258, 259, 261, 262, 263, 264, 265, 266, 267, 268, 270, 271, 273, 274, 277, 278, 279, 280, 281, 285, 288, 289, 295, 297
  transport 149, 151, 153, 155, 156, 158, 159, 161, 164, 166, 169, 170, 171, 173, 174, 176, 177, 178, 180, 181, 183, 184,
  urban areas 199, 200, 202, 205, 207, 209, 212, 213, 226, 227, 228, 232, 233, 234, 235, 236, 239, 242, 243, 244
  water 71, 73, 75, 76, 77, 79, 80, 82, 84, 85, 86, 87, 90, 91, 97, 99, 100, 191, 102, 105,

unemployment VII, 13, 14–15, 19, 22, 43, 57, 60, 61, 202, 243, 256, 276, 306, 360,
  youth unemployment 22–3
UK (United Kingdom) 20, 42, 43, 105, 176, 342
UNCED (United Nations Conference on the Environment and Development) 51
underground transport 217
UNECE 54
UNEP (United Nations Environment Programme) 49
unique development model, avoiding paradigm 385
United States (US) 23, 34, 35, 36, 38, 40, 41, 43, 44, 46, 47, 59, 105, 139, 158, 159, 178, 242, 260, 268, 281, 293, 385
urban areas 197–245
  air pollution and health 223–5
  alternative scenario 238–42, 244–5
  baseline scenario 200, 205, 209, 215, 230
  cities as candidates for decoupling 360
  cities sizes 202–5
  decentralization and urban governance 231–6
    possibilities for action by cities 231–40
  Euro-Med cooperation 242–5
  household waste 228–31
  increased expansion 206–23
  informal housing 207
  loss of agricultural lands 207–9, 270–1
  metropolitan development 205
  public–private partnerships 237–8
  town planning 209–11, 218, 219, 220–3, 238–9
  social and ecological inequalities 210–12
  sustainable urban development 238–40
    insufficient city-to-city cooperation 243–5
    need for national support 240–2
    possibilities for action by cities 231–42
  unequal access to drinking water 225–7
  urban environment 223–31
  urban regeneration and renewal, examples 220–3
  urban sprawl 206–10
  urban transport 214–23
    pollution
    sustainable transport systems 219–20
  vulnerability to natural and technological hazards 212–14
urban population, projections 33–4, 199, 200
  coastal regions and coastline 310
URBAN programme 243
urbanization 10, 34, 199–202

VOC emissions 166
  baseline scenario 168
volcanic activity 9, 11

warming see global warming
waste
  hazardous waste 321
  household waste 227–31
    baseline and alternative scenarios 229–31
    collection 22
    production 228–31

recycling 228–9
reduction at source 229
uncontrolled dumpsites 228
waste management 27
waste water treatment 90–1, 323
water resources
  access difficult 86–8
  alternative scenario 93–5
  baseline scenario 75–8, 82, 83, 84, 89, 94
  better water management in agriculture 370–1
  changes to regime 82–3
  combating pollution 90
  conflicts over 88–9
  conserving 366-7
  degradation in quality 84–6
  demand management (WDM) 93–8
    barriers to 95–8
    infrastructure savings 95
    lack of commitment to 95
    problem of funding 98–9
  desalination 81–2
  drinking water, reduced availability 86–8
  enhancing supply by managing demand 368
  exploitable 78
  Framework Directive 89, 95, 101, 105
  growing demands in south and east 74–7
    agricultural 75–6
    drinking 76
    energy and industrial sectors 76–7
    structure 76
    total demand per country 75
  lower levels in water tables 86
  main transfers 79
  pollutant emissions 83–4
  pressures on 72–89
    increasing withdrawals and waterworks 77–9
  rare and fragile 72–4
  recycling 81–3
  sustainable resource management 89–106
  rising cost 86
  supply policies 79
  tariffs 102–3
  underground water pollution 85
  unequal access and resource wastage 225–7
  unsustainable exploitation 80–1
wetlands 9, 24, 50, 73, 86, 252, 289, 307, 312, 327, 328, 329, 333, 349, 350, 363, 366
wind power 128, 129, 130, 140, 380
women 15, 22, 31, 171, 239, 259, 275, 284, 285, 287, 294, 348, 380, 387
  empowerment 23
woodlands 128, 212, 249, 262, 264–5, 270, 283
  degraded in south 264–5
  increasing wooded areas in north and risk of forest fires 263
wood cover 263
working-age population 31–3
World Bank 21, 34, 50, 51, 54, 97, 239, 243, 244, 273, 289, 336, 339, 341
World Conservation Union 50, 300
World Trade Organization (WTO) 34, 46, 61, 261, 273

Yugoslavia, former 12, 158, 177, 189

SF-36 PHYSICAL & MENTAL HEALTH SUMMARY SCALES:
A Manual for Users of Version 1, Second Edition

John E. Ware, Jr., Ph.D.
Mark Kosinski, M.A.

QualityMetric, Inc. · Lincoln, Rhode Island
March, 2001

Copyright © 1994

Copyright © 1994 – 2001, John E. Ware, Jr., Ph.D.

All rights reserved. No part of this manual covered by the copyrights hereon may be reproduced or transmitted in any form or by any means—electronic, mechanical, including photocopy, recording, or any information storage or retrieval system—without permission of the copyright holder. Permission to reproduce the SF-36™ Health Survey and scoring algorithms is hereby granted to the holder of this manual for his/her own personal use.

ISBN: 1-891810-09-X

The SF-36™ Health Survey is reproduced with permission of the Medical Outcomes Trust (MOT), a non-profit organization that ensures the availability of the SF-36™ Health Survey, while preserving standardization of its content, scoring, and labeling. Permission to use the SF-36™ Health Survey is routinely granted by the MOT, upon receipt of a request, to individuals and organizations for their own non-commercial use on a royalty-free basis.

Requests for permission to reproduce parts of this manual should be sent to John E. Ware, Jr., Ph.D., QualityMetric Incorporated, 640 George Washington Highway, Lincoln, RI 02865, info@qmetric.com.

Permission to reproduce sample SF-36 items and to make reference to specific normative data from this manual for purposes of documenting or interpreting a specific published study is hereby granted to the holder of this manual with the understanding that proper citation will be made to this manual as the source.

SF-36 is a registered trademark of the Medical Outcomes Trust.

Suggested citation:

Ware JE and Kosinski M. SF-36 PHYSICAL & MENTAL HEALTH SUMMARY SCALES: A MANUAL FOR USERS OF VERSION 1,
Second Edition, Lincoln, RI: QualityMetric, 2001

Table of Contents

Copyright © 1994 i

Table of Contents ii

Acknowledgements iv

Preface vi
    SECOND EDITION VI
    FIRST EDITION VII

Chapter 1: How To Use This Manual 1

Chapter 2: Introduction 3

Chapter 3: Construction Of Summary Measures 5
    SF-36 MEASUREMENT TOOL 5
    METHODOLOGICAL ISSUES 12
    SF-12 HEALTH SURVEY 16

Chapter 4: Norm-Based Scoring And Interpretation 18
    ADVANTAGES OF NORM-BASED SCORING 18

Chapter 5: Norm-Based Scoring Of SF-36 Scales 22
    IMPORTANCE OF STANDARDIZATION 22
    NORM-BASED SCORING OF SCALE SCORES, STANDARD FORM (4-WEEK RECALL) 23
    NORM-BASED SCORING OF SCALE SCORES, ACUTE FORM (1-WEEK RECALL) 25

Chapter 6: Norm-Based Scoring For Physical & Mental Summary Measures 28
    STANDARDIZATION OF SCALES (Z-SCORES), ACUTE FORM 32
    MISSING DATA ESTIMATION 33
    FEATURES OF PCS AND MCS SCORES 33

Chapter 7: Reliability And Statistical Power 36
    ESTIMATION OF RELIABILITY 37
    SUMMARY OF RELIABILITY ESTIMATES 37
    STATISTICAL POWER 39

Chapter 8: Empirical Validation 45
    INITIAL VALIDATION STUDIES 46
    CORRELATIONS WITH SPECIFIC SYMPTOMS 53
    CORRELATIONS WITH OTHER MOS SCALES 54

Chapter 9: Interpretation: Content And Criterion-Based 57
    CONTENT-BASED INTERPRETATION 57
    CRITERION-BASED INTERPRETATION 64
    LIFE SATISFACTION AND MCS 69
    BURDEN OF CHRONIC CONDITIONS 70
    EFFECTS OF AGING 74
    SUMMARY OF EFFECT SIZES 75

EVALUATION OF CHANGES IN PCS AND MCS SCORES	83
FINAL COMMENT	86

Chapter 10: Norms: 1998 General U.S. Population 88

BACKGROUND	88
HOW THE SF-36 WAS NORMED	90

Chapter 11: Applications: Outcomes Research 120

HEART VALVE REPLACEMENT	121
DUODENAL ULCER TREATMENT	127
METHOD FOR RELATING PCS AND MCS SCALE SCORES TO "CRITERION" SCORES	128

Chapter 12: Applications: Clinical Practice 131

CHARTS FOR LONGITUDINAL DATA	132
PROFILES FOR INDIVIDUAL PATIENTS	134
PATIENT PROFILES: MORE EXAMPLES	136
MCS AND CLINICAL DEPRESSION	138
PCS SCORES AND PHYSICAL DISEASE	139

Chapter 13: Individual License Information 141

TYPES OF LICENSES OFFERED	142
SINGLE-USER, NON-COMMERCIAL LICENSE	142
DISCLAIMER OF WARRANTY	143

References 144

PCS/MCS References 154

Appendix A: Definitions Of Criterion Variables 158

Appendix B: Table Of Results 162

Appendix C: Scoring Algorithms: 1990 Norm-Based Scoring of PCS and MCS 171

STEPS IN SCORING	172
FEATURES OF PCS AND MCS SCORES	173
SCORING CHECKS	174

Appendix D: 1990 Norms for PCS and MCS 176

GENERAL POPULATION DATABASE	178
MOS DATABASE	179
EFFECTS OF DATA COLLECTION METHOD	180
U.S. GENERAL POPULATION NORMS	181
NORMS FOR CHRONIC CONDITIONS, GENERAL U.S. POPULATION	182
NORMS FOR CHRONIC CONDITIONS: GENERAL U.S. POPULATION	183

Acknowledgements

Development and validation of the SF-36 Health Survey was supported by a grant (#89-6515) from the Henry J. Kaiser Family Foundation, Menlo Park, California, to The Health Institute at New England Medical Center Hospitals (J.E. Ware, Jr., Ph.D., Principal Investigator).

Development of the physical (PCS) and mental (MCS) component summary health measures for the SF-36 was supported by unrestricted research grants for the International Quality of Life Assessment (IQOLA) Project from Glaxo Research Institute, Research Triangle Park, North Carolina, and Schering-Plough Corporation, Kenilworth, New Jersey, and by The Health Institute at New England Medical Center Hospitals (NEMCH) from its own research funds.

The national norming of the SF-36 Health Survey in the US population in 1998 was supported, in part, by the Musculoskeletal Education and Research Institute of the American Academy of Orthopedic Surgeons and by QualityMetric, Inc., Lincoln, RI, from its own research funds. The 1990 national norming of the SF-36 in the US population, also documented here, was sponsored by the Functional Outcomes Program of the Henry J. Kaiser Family Foundation, The Health Institute, NEMCH, Boston, MA.

Several publications documenting the SF-36 were very useful in preparing the second edition of this manual. Most useful was the first edition entitled, *SF-36 Physical and Mental Health Summary Scales: A User's Manual* (Ware, Kosinski and Keller, 1994). We gratefully acknowledge the contributions of Susan D. Keller, PhD, our co-author, for her many contributions to that manual. Other useful documents include: *How To Score the MOS 36-Item Short-Form Health Survey (SF-36)* (Ware, 1988; Medical Outcomes Trust, 1991); *Scoring Exercise for the SF-36 Health Survey* (Medical Outcomes Trust, 1994); and *SF-36 Health Survey Manual and Interpretation Guide* (Ware, Snow, Kosinski, et al., 1993).

We also gratefully acknowledge our IQOLA Project collaborators in 10 countries for the very informative studies of the PCS and MCS, as summarized here, including: the Director of the IQOLA Project (Barbara Gandek, MS), and other collaborators from Denmark (Jakob Bjorner, Md, PhD), France (Alan Leplege, MD, PhD), Germany (Monika Bullinger, PhD), Italy (Giovanni Apolone, MD), The Netherlands (Neil K Aaronson, PhD), Norway (Stein Kaasa, Md,PhD), Spain (Jordi Alonso, MD, PhD), Sweden (Marianne Sullivan, PhD), and the United Kingdom (John Brazier, PhD). In addition to sharing their results, which are also published elsewhere (Gandek and Ware, 1998), our IQOLA Project collaborators provided valuable comments regarding the scoring and interpretation of the PCS and MCS summary health measures from the SF-36.

We gratefully acknowledge: William H. Rogers, PhD for his invaluable assistance with our analyses of mortality data from the Medical Outcomes Study (MOS) and for providing statistical consultation throughout the development of the SF-36 and other tools in the "SF" family of measures; Colleen A. McHorney, PhD, for her contribution to initial studies of the factor structure of the SF-36 in the U.S.; Martha Bayliss, MSc for her efforts during the first analyses of the PCS and MCS in the MOS; and both Marie Perrone and Rosamaria Amoros-Cherry for their assistance in preparing the second edition of this manual.

Finally, we acknowledge and thank our many colleagues in the US and around the world who have applied and evaluated the SF-36 PCS and MCS summary measures in population studies, clinical trials, and in clinical practice and who have provided very constructive and useful feedback regarding the strengths and weaknesses of these summary measures. We also thank those among them who granted permission to reanalyze their data and present their results for purposes of the examples discussed here.

Preface

Second Edition

This manual was written as a companion to the *SF-36 Health Survey Manual and Interpretation Guide*, which documents the development, psychometric evaluation and norming of the eight scales in the SF-36 Health Survey, Version 1 (v1). This manual provides comparable information for the physical (PCS) and mental (MCS) component summary measures from the standard (4-week recall) and acute (1-week recall) forms of the SF-36 (v1).

This manual also documents more up-to-date norms for SF-36 scales and summary measures based on 1998 surveys of the general US population and introduces norm-based scoring (NBS) for the eight SF-36 scales and two summary measures scored from the SF-36 (v1). Results from the original psychometric evaluations of the PCS and MCS summary measures are reproduced here. Results from more recent psychometric evaluations (e.g., reliability estimates from 1998 studies) are also documented and those evaluations have been expanded (e.g., replications of studies of the factor structure of the SF-36 in 10 countries) in this manual. This manual also documents, for the first time, the scoring of PCS and MCS summary measures based on the acute (1-week recall) version of the SF-36 (v1). More than 40 studies about the PCS and MCS summary measures, published after the first edition of this manual have been added to the reference list.

Because health status and outcomes monitoring efforts underway are using means and variances from the original 1990 SF-36 norming study, we have reproduced here (from the first edition of this manual) the entire chapter documenting 1990 scoring algorithms (see Appendix C) and the entire chapter documenting 1990 norms (see Appendix D). These appendices also serve the purpose of documenting the methods and results cited in published studies based on the original 1990 scoring and norm-based interpretation guidelines for the SF-36 (v1) Health Survey.

Although this manual was written to "stand alone," it does not contain complete background information about the development of the SF-36 or explanations of the psychometric methods used in evaluating assumptions underlying the construction and scoring of the eight SF-36 scales that are used to estimate the PCS and MCS summary scores. Those topics are more completely covered in the original *SF-36 Health Survey Manual and Interpretation Guide* (Ware, Snow, Kosinski and Gandek, 1993). The most up-to-date summary of the history of the development and usage of the SF-36 is published in Ware JE "SF-36 Health Survey Update," *Spine*, 2000, 25 (4), 3130-3139.

Health status and outcomes scores based on the two SF-36 summary measures are compared with scores based on the eight-sclae profile in "Comparison of Methods for the Scoring and Statistical Analysis of the SF-36 Health Profile and Summary Measures: Summary of Results from the Medical Outcomes Study," *Medical Care*, 1995, 33 (Suppl 4), AS264-A5279 by JE Ware, M Kosinski, M Bayliss et al.

Because of the complexity of the algorithms required to estimate the SF-36 PCS and MCS summary measures, which are linear composites of standardized scale scores weighted using factor score coefficients, this manual includes a test data set and documented results for use in checking the accuracy of computations. We strongly recommend the use of this test data (also available at sfinfo@qmetric.com) to verify the accuracy of scores before they are interpreted or compared with results published in this manual.

First Edition

This manual was written as a companion to the *SF-36 Health Survey Manual and Interpretation Guide*, which documents the development of the SF-36 Health Survey and summarizes information about the eight SF-36 scales, including assumptions underlying their construction and scoring; reliability, precision, and data quality; and validation and interpretation guidelines based on content-, criterion-, and norm-based strategies.

This second manual provides the same information for the SF-36 physical and mental health summary measures, which were developed subsequently, and extends the normative data and other interpretation guidelines substantially. Although this manual was written to "stand alone," it does not contain complete background information about the development of the SF-36 or explanations of psychometric methods used in evaluating scaling assumptions, reliability, or validity. Those topics are more completely explained in the original *SF-36 Health Survey Manual*.

Results for the two summary measures are compared with previous results for the eight-scale profile in "Comparison of Methods for the Scoring and Statistical Analysis of the SF-36 Health Profile and Summary Measures: Summary of Results from the Medical Outcomes Study" by Ware, Kosinski, Bayliss, and others, which is scheduled for publication in *Medical Care* early in 1995.

Because of the complexity of the algorithms required to estimate the SF-36 Physical and Mental Component Summary measures, which are linear composites aggregated using norm-based scoring methods, this manual includes a computer diskette with the code for scoring algorithms and also a test data set for checking the accuracy of computations.

As with the original SF-36 manual, we invite suggestions for improvements in our documentation of the SF-36. Many were received in response to the first manual, resulting in numerous additions to the content of this manual, including a computer disk, norms for change scores in the MOS, and expanded empirical tests of validity and interpretation guidelines. The result is a lengthier manual than we had originally intended. We have tried to organize it in a way that makes desired information easy to find and use.

We encourage users of the manual to fill out and return the information form for our mailing list (Appendix C) to ensure receipt of new information in a timely manner (e.g., an errata sheet). Among the updates scheduled are U.S. population norms for change scores, additional estimates of test-retest reliability and confidence intervals, and much improved algorithms for handling missing item and scale data.

Chapter 1: How To Use This Manual

This chapter offers suggestions on how to use this manual and how to find information quickly.

Introduction

Chapter 2 provides an introductory explanation of how the discovery of the two-dimensional factor structure underlying the eight SF-36 scales led to the construction of physical and mental health summary measures.

Construction Of Summary Measures

Chapter 3 presents the SF-36 measurement model and reviews the psychometric methods used in the development of the SF-36 Physical (PCS) and Mental (MCS) Component Summary scales. The generalizability of results across populations is demonstrated and important methodological issues are discussed, including how the PCS and MCS led to construction of a new 12-item survey (SF-12)

Norm-Based Scoring

Chapter 4 provides an explanation of norm-based scoring methods and illustrates their advantages in interpreting data.

Scoring of Scales

Chapter 5 explains how to score SF-36 scales using norm-based scoring for both standard and acute forms.

Scoring of Summaries

Chapter 6 explains how to score the physical and mental summary scales using norm-based scoring for both standard and acute forms.

Reliability And Statistical Power

Chapter 7 presents numerous reliability estimates for the SF-36 summary scales, explains the internal-consistency and test-retest methods used in estimating reliability, provides estimates of sample sizes required to achieve statistical power for various study designs, and of confidence intervals for interpreting individual patient scores.

Empirical Validity

Chapter 8 summarizes results from the first two rounds of empirical tests of the validity of the PCS and MCS summary scales and compares results with those using the eight-scale SF-36 profile in the same tests. Correlations with specific symptoms and with numerous other MOS health scales are presented.

Content And Criterion-Based Interpretation

Chapter 9 explains the meaning of very high and low PCS and MCS scores and presents 19 tables of content-based and criterion-based interpretation guidelines and explains their use. Effect sizes from numerous cross-sectional and longitudinal analyses are ordered from largest to smallest.

Norms

Chapter 10 presents 42 tables of norms for a representative sample of the 1998 general U.S. population for both standard and acute forms by age, gender, and chronic conditions.

Applications: Outcomes Research

Chapter 11 illustrates the use of norms and content- and criterion-based interpretation guidelines in interpreting the summary measures in outcomes research and compares results with those using SF-36 profiles published by others.

Applications: Clinical Practice	Chapter 12 presents examples of how SF-36 summary measures can be used in monitoring individual patients over time, in screening patients, and illustrates different display formats.
Scoring Algorithms	Appendix D documents the algorithms for aggregating the eight SF-36 scales to score the PCS and MCS summary scales. Using 1990 norm-based scoring methods.
Norm-Based Interpretation	Appendix E presents 50 tables of norms for a representative sample of the 1990 general U.S. population (for seven age groups, men and women), as well as for MOS patients, including norms for one-year change scores. This chapter also describes sampling methods and sample characteristics.
References And Bibliography	Complete citations for more than 125 referenced publications, including more than 87 publications about the SF-36, are included.

Chapter 2: Introduction

The cycle of measurement development involves a model of health status components and the search for the most valid operational definitions of those components. The cycle must grapple with both uncertainty about the underlying conceptual framework of health and the validity of assessment methods. Out of necessity, the process is circular and tests of hypotheses about the structure of health are vulnerable to the limits of the measures used in hypothesis testing. Advances in understanding of the structure of health can lead to breakthroughs in measurement strategies.

Physical And Mental Health

Discovery of the physical and mental components of health has substantial implications for the construct validation of health measures and a new strategy for creating more useful summary measures of the information they contain. The observation that the eight SF-36 scales define distinct physical and mental health clusters in factor analytic studies of both patients participating in the Medical Outcomes Study (MOS) (McHorney, Ware, Raczek, et al., 1993; Ware, Gandek, & the IQOLA Project Group, 1994) and in the general U.S. population (Ware, Snow, Kosinski, et al., 1993) constitutes considerable support for the construct validity of the SF-36. The generalizability of these findings is underscored by results showing that the factor content of each SF-36 scale -- the extent to which each scale measures a physical and/or mental health component -- is also very similar across populations (Ware, Snow, Kosinski, et al., 1993; Ware, Kosinski, Bayliss, et al., 1995). These results have proven very useful in establishing interpretation guidelines for each of the SF-36 scales, as documented in the original *SF-36 Health Survey Manual and Interpretation Guide* (Ware, Snow, Kosinski, et al., 1993).

The discovery that from 80 to 85 percent of the reliable variance in the eight SF-36 scales is accounted for by physical and mental components of health opens the door for a breakthrough in scoring and interpretation. Specifically, this result suggests that psychometrically-based summary measures have the potential to reduce the number of statistical comparisons required in analyzing SF-36 data from eight to two without substantial loss of information. Preliminary tests of their potential have yielded promising results (Ware, Kosinski, Bayliss, et al., 1995).

Psychometric-Based Summary Measures

In support of aggregating highly related health measures using psychometric methods, initial results from the MOS suggested that scales with the same factor content are much more likely to lead to the same conclusions about health outcomes than scales with different factor content (McHorney, Ware, & Raczek, 1993). Thus, measures having the same factor content are good candidates for aggregation, because less information is likely to be lost. Factor analytic methods are likely to prove useful in deriving the weights used in aggregation. Although factor analytic methods have been used in construct validation (Ware, 1976; Goldberg & Hillier, 1979; Ware, Brook, Davies-Avery, et al., 1980; Veit & Ware, 1983; Derogatis, 1986; Wiklund, Lindvall, Swedberg, et al., 1987; Mason, Anderson, & Meenan, 1988; Hall, Epstein, & McNeil, 1989; Hays & Stewart 1990; Schag, Heinrich, Aadland, et al., 1990), the scoring of higher-order factors to achieve summary measures of health status has been pursued rarely (Davies & Ware, 1981; Veit & Ware, 1983). As a result, little is known about the tradeoffs between the simplicity of fewer statistical comparisons with aggregate summary measures versus the greater sensitivity of a scale to effects concentrated in a particular dimension of health (Ware, 1984). In theory, to the extent that results are the same across conceptually related scales, a summary measure that aggregates them should be *more* useful than any one of them in detecting a difference in health status at a point in time or a change in health over time. To the extent that differences are concentrated in a particular subscale, a summary measure is *less* likely to capture that difference or change (Ware, Kosinski, Bayliss, et al., 1995).

Standardized Scoring

Following several years of evaluating the factor structure of the SF-36 among those with various chronic conditions and among the "well" in general populations in the U.S. and in other countries, it is clear that physical and mental factors account for the great majority of the variance in SF-36 scales across populations. It follows that the standardized scoring of measures of physical and mental factors will add a useful option for the scoring of SF-36 data with advantages in interpreting and presenting results.

This user's manual summarizes the empirical work leading to the development of the SF-36 Physical Component Summary (PCS) and Mental Component Summary (MCS) scales and provides the documentation necessary to score and interpret those measures.

Chapter 3: Construction Of Summary Measures

This chapter documents the factor analytic methods used to evaluate the construct validity of the SF-36 in relation to a two-factor -- physical and mental -- model of health across populations. The construction of the PCS and MCS summary measures of physical and mental health based on those methods and some of the more important methodological issues are also explained. The PCS and MCS summary measures are referred to as "component" measures (Ware, Kosinski, Bayliss, et al., 1995) because they were derived and scored using a factor analytic method called principal *components* analysis (Harman, 1976).

Factor analyses of correlations among the eight SF-36 scales have consistently identified two factors, as documented below. On the strength of the pattern of their correlations with the eight scales, they have been interpreted as "physical" and "mental" dimensions of health status. These physical and mental components accounted for 81.5% of the reliable variance in SF-36 scales in the general U.S. population (Ware, Kosinski, Bayliss, et al., 1995), and 82.4% in the MOS (McHorney, Ware, Raczek, 1993). Similar physical and mental components have been observed for other comprehensive surveys, including the Health Insurance Experiment Medical History Questionnaire (Ware, Brook, Davies-Avery, et al., 1980), the MOS Functioning and Well-Being Profile (Hays & Stewart, 1990), and the Sickness Impact Profile (Bergner, Bobbitt, Carter, et al., 1981). In further support of the generalizability of this two-dimensional model of health, we would expect factor analytic studies of other comprehensive surveys to yield recognizably similar physical and mental components.

The psychometric approach to summarizing health measures illustrated here is in contrast to a utility index, in which measures are aggregated without regard to their inter-relationships. A utility index achieves a single summary score at the expense of sensitivity and specificity to physical versus mental components of health status. A strength of the PCS and MCS summary scales described here is their value in distinguishing a physical from a mental health outcome.

SF-36 Measurement Model

Figure 3.1 illustrates the measurement model underlying the construction of SF-36 multi-item scales and summary measures. This model has three levels: (1) items, (2) scales that aggregate items, and (3) summary measures that aggregate scales. All but one of the 36 items (self-reported health transition) are used to score the eight SF-36 scales.

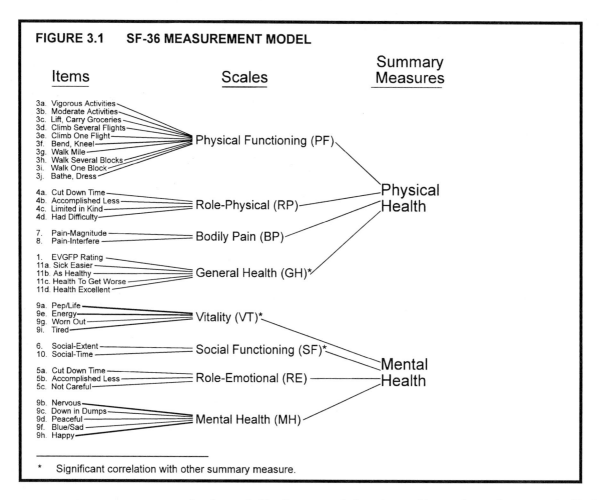

* Significant correlation with other summary measure.

As shown in the figure, each item is used in scoring only one scale. Tests of assumptions underlying the algorithms used in scoring the eight scales have been strongly supported in the U.S. (Ware, Snow, Kosinski, et al., 1993; McHorney, Ware, Lu, et al., 1994) and in other countries (Ware, Keller, Gandek, et al., 1998). The eight scales also form two distinct higher-ordered clusters. Three scales (Physical Functioning, Role-Physical, and Bodily Pain) correlate most highly with the physical component and contribute most to scoring of the PCS measure of that component. The mental component correlates most highly with the Mental Health, Role-Emotional, and Social Functioning scales, which contribute most to the scoring of the MCS measure of that component. Three of the scales have noteworthy correlations with both components: the Vitality scale correlates substantially with both; General Health correlates with both but higher with the physical component; and Social Functioning correlates much higher with the mental component. Reasons for these patterns of factor loadings are discussed elsewhere (Ware, Snow, Kosinski, et al., 1993).

Criteria And Hypotheses

Criteria commonly used to evaluate factor analyses using the principal components method were routinely applied (Harman, 1976) using SAS software (SAS Institute, 1989). First, we examined whether eigenvalues for the first two components were both greater than unity, as required for rotation. Other criteria, previously used, including the scree test, five percent rule, and common factor test, were also satisfied (Ware, Davies-Avery, & Brook, 1980). Second, we evaluated the pattern of correlations between the two rotated components and eight SF-36 scales to determine the basis for their interpretations as physical and mental components.

We hypothesized that the PF scale would have the highest and a very strong loading ($r > .80$) on the "physical" component followed by the RP and BP scales. The MH scale was hypothesized to have the strongest loading ($r > .80$) on the "mental" component, followed by the RE and SF scales. Because they are general measures, the GH and VT scales were hypothesized to correlate with both components. Lastly, we evaluated the adequacy of the two components in explaining the variation in each SF-36 scale. We hypothesized that a large proportion of the reliable variance in each scale (three-fourths) would be explained by the two components.

To evaluate the generalizability of the two-dimensional SF-36 model of health across different patient groups, we conducted principal component analyses in 23 subgroups of MOS patients differing in demographic characteristics and medical conditions. These are the same subgroups analyzed in tests of scaling assumptions for the eight SF-36 scales, as described elsewhere (McHorney, Ware, Lu, et al., 1994). These tests are extended here to include 12 subgroups differing in demographic characteristics from the general U.S. population, as well as general population samples from ten countries participating in the International Quality of Life Assessment (IQOLA) Project. In all analyses, we used the same method to extract two principal components from the correlations among the SF-36 scales and rotated them to orthogonal simple structure to facilitate interpretation.

Results

Tables 3.1 through 3.3 present results for the principal component analyses of SF-36 scales for the total 1998 general U.S. population standard (4-week recall) and acute (1-week recall) forms; the total patient sample from the MOS (McHorney, Ware, & Raczek, 1993); the total 1990 general U.S. population; and general populations in Denmark, France, Germany, Italy, Netherlands, Norway, Spain, Sweden, and the United Kingdom (Ware, Kosinski, Gandek, et al., 1998). Results for the 12 subgroups from the general U.S. population are summarized in Table 3.4. Results for the 23 patient subgroups from the MOS are summarized in Table 3.5. Results for age and gender subgroups in 9 countries are summarized in Table 3.6.

TABLE 3.1 CORRELATIONS BETWEEN SF-36 SCALES AND ROTATED PRINCIPAL COMPONENTS IN THE 1998 GENERAL U.S. POPULATION, STANDARD (N=1,982) AND ACUTE RECALL (N=1,646) FORMS

Scales	Standard Form (4-Week Recall)			Acute Form (1-week Recall)		
	PCS	MCS	h^2/r_{tt}	PCS	MCS	h^2/r_{tt}
Physical Function	.87	.13	.82	.87	.16	.83
Role Physical	.83	.28	.88	.82	.30	.88
Bodily Pain	.77	.31	.74	.79	.29	.84
General Health	.61	.50	.76	.64	.47	.75
Vitality	.43	.70	.76	.43	.69	.75
Social Function	.55	.65	.84	.55	.66	.86
Role Emotion	.25	.70	.70	.30	.73	.75
Mental Health	.10	.91	.98	.11	.90	.99
Reliable Variance[1]	81.3%			82.9%		

h^2/r_{tt} Variance in each SF-36 scale explained by the two principal components (h^2) divided by the reliability of each SF-36 scale (r_{tt}).

[1] Percent of the total reliable variance in SF-36 scales explained by the two principal components.

As hypothesized, eigenvalues for the first two components were both greater than unity in all analyses, strongly supporting the two-dimensional model of health.

TABLE 3.2 CORRELATIONS BETWEEN SF-36 SCALES AND ROTATED PRINCIPAL COMPONENTS IN THE MEDICAL OUTCOMES STUDY (MOS) AND THE GENERAL U.S. POPULATION

SF-36 SCALES	MOS (N = 3,445)			U.S. POPULATION (N = 2,474)		
	PCS	MCS	h^2/r_{tt}	PCS	MCS	h^2/r_{tt}
Physical Functioning (PF)	.88	.04	.84	.85	.12	.78
Role Physical (RP)	.78	.30	.83	.81	.27	.82
Bodily Pain (BP)	.77	.24	.79	.76	.28	.72
General Health (GH)	.68	.32	.72	.69	.37	.78
Vitality (VT)	.59	.57	.77	.47	.64	.75
Social Functioning (SF)	.44	.71	.82	.42	.67	.92
Role Emotional (RE)	.19	.81	.83	.17	.78	.78
Mental Health (MH)	.12	.90	.91	.17	.87	.92
Reliable Variance[1]			82.4%			81.5%

h^2/r_{tt} Variance in each SF-36 scale explained by the two principal components (h^2) divided by the reliability of each SF-36 scale (r_{tt}).

[1] Percent of the total reliable variance in SF-36 scales explained by the two principal components.

The pattern of correlations between SF-36 scales and the two rotated components in Tables 3.1 and 3.2 strongly support the physical and mental health interpretation of the two components. In each analysis, the PF scale loaded strongest on the "physical" component and weakest on the "mental" component. Also, the RP and BP scales both had stronger loadings on the "physical" component than the "mental" component. The MH scale had the highest loading and the RE and SF scales had stronger loadings on the "mental" component than the "physical" component. The GH and VT scales had noteworthy loadings (r > .30) on both physical and mental components, as expected for general measures.

TABLE 3.3 CORRELATIONS BETWEEN SF-36 SCALES AND ROTATED PRINCIPAL COMPONENTS IN 9 COUNTRIES

Scales	Denmark		France		Germany		Italy		Netherlands		Norway		Spain		Sweden		United Kingdom	
	PC	MC	PC	MC	PC	MC	PC	MC	PC	MC	PC	MC	PC	MC	PC	MC	PC	MC
PF	0.86	0.10	0.85	0.12	0.85	0.16	0.86	0.13	0.86	0.16	0.83	0.15	0.81	0.19	0.84	0.15	0.84	0.20
RP	0.78	0.30	0.78	0.29	0.79	0.28	0.64	0.42	0.69	0.44	0.82	0.26	0.76	0.24	0.78	0.30	0.80	0.27
BP	0.69	0.32	0.78	0.32	0.82	0.24	0.73	0.37	0.80	0.20	0.80	0.25	0.74	0.26	0.76	0.28	0.80	0.19
GH	0.70	0.41	0.60	0.49	0.72	0.38	0.75	0.34	0.75	0.31	0.74	0.39	0.73	0.34	0.66	0.50	0.73	0.39
VT	0.39	0.75	0.41	0.73	0.52	0.64	0.44	0.70	0.52	0.66	0.40	0.71	0.51	0.61	0.43	0.74	0.54	0.64
SF	0.33	0.72	0.37	0.78	0.33	0.75	0.27	0.81	0.49	0.68	0.33	0.78	0.50	0.62	0.32	0.78	0.55	0.62
RE	0.27	0.70	0.28	0.71	0.20	0.71	0.23	0.76	0.13	0.84	0.21	0.78	0.18	0.78	0.25	0.71	0.20	0.78
MH	0.11	0.89	0.09	0.91	0.16	0.87	0.25	0.81	0.25	0.86	0.14	0.90	0.22	0.84	0.15	0.90	0.18	0.87
Reliable Variance[1]	81.3		84.7		82.9		84.0		82.1		83.8		76.3		82.6		82.0	

Abbreviations: PC = Physical Component; MC = Mental Component; PF = Physical Functioning; RP = Role Physical; BP = Bodily Pain; GH = General Health; VT = Vitality; SF = Social Functioning; RE = Role Emotional; MH = Mental Health.

Source: Ware JE, Kosinski M, Gandek BG, et al.; 1998.

Table 3.4 summarizes the range of factor loadings observed in 12 analyses of subgroups of the general U.S. population differing in age, gender, race, and education. The two components explained from 77% to 87% (median = 81%) of the reliable variance in the eight scales, and the range of correlations observed for each scale and each component is strikingly similar to that observed in Table 3.1 for the general U.S. population.

Table 3.5 summarizes results for analyses of 23 subgroups of MOS patients differing in sociodemographic characteristics, diagnosis, and other clinical variables defined elsewhere (McHorney, Ware, Lu, et al., 1994). From 75% to 83% (median = 80%) of the reliable variance in the eight scales was explained. Again, the amount of reliable variance explained and the pattern of correlations between scales and the two components replicates other results to date.

Table 3.6 presents therange of correlations between SF-36 scales and rotated principal components for five subgroups differing in age (three groups) and gender (two groups) within each of the nine countries. The pattern of correlations observed are consistent with the patterns observed for the total populations in each country (see Table 3.3). Only in Italy are departures observed, particularly among the 65 and older group.

The results presented in Tables 3.1-3.6 also demonstrate the adequacy of the two components in explaining the reliable variance in each SF-36 scale. In all but four of the 40 factor analyses, the total reliable variance in SF-36 scales explained by the two components exceeded 81%. As hypothesized, the two components explained at least three-fourths of the reliable variance in each SF-36 scale, with few exceptions.

TABLE 3.4 RANGE OF CORRELATIONS BETWEEN SF-36 SCALES AND ROTATED PRINCIPAL COMPONENTS, 12 SUBGROUPS OF THE GENERAL U.S. POPULATION[1] (N = 2,474)

SF-36 SCALES	PCS	MCS
Physical Functioning (PF)	.77 - .88	.05 - .30
Role Physical (RP)	.67 - .82	.16 - .43
Bodily Pain (BP)	.70 - .84	.17 - .46
General Health (GH)	.53 - .76	.29 - .69
Vitality (VT)	.31 - .73	.44 - .82
Social Functioning (SF)	.34 - .62	.46 - .73
Role Emotional (RE)	.06 - .48	.57 - .83
Mental Health (MH)	.11 - .27	.84 - .90

[1] Results above come from analyses of 12 subgroups in the general U.S. population: 3 age, 2 gender, 3 race, and 4 education groups.

TABLE 3.5 RANGE OF CORRELATIONS BETWEEN SF-36 SCALES AND ROTATED PRINCIPAL COMPONENTS, 23 MOS SUBGROUPS[1] (N = 3,445)

SF-36 SCALES	PCS	MCS
Physical Functioning (PF)	.83 - .90	.01 - .16
Role Physical (RP)	.69 - .81	.20 - .41
Bodily Pain (BP)	.65 - .83	.12 - .42
General Health (GH)	.56 - .76	.17 - .51
Vitality (VT)	.43 - .79	.27 - .66
Social Functioning (SF)	.30 - .72	.37 - .82
Role Emotional (RE)	.11 - .38	.66 - .88
Mental Health (MH)	.01 - .34	.75 - .92

[1] Results above come from principal components analyses of 23 subgroups in the MOS: 3 age, 2 gender, 3 race, 4 education, 2 poverty status, 6 diagnoses, and 3 clinical groups.

TABLE 3.6 RANGE OF CORRELATIONS BETWEEN SF-36 SCALES AND ROTATED PRINCIPAL COMPONENTS ACROSS AGE AND GENDER GROUPS IN 9 COUNTRIES

	Denmark		France		Germany		Italy		Netherlands	
Scale	Physical	Mental	Physical	Mental	Physical	Mental	Physical	Mental	Physical	Mental
PF	.79-.86	.04-.16	.82-.86	.06-.19	.81-.86	.10-.24	.81-.86	.04-.20	.81-.86	.09-.20
RP	.72-.80	.25-.39	.68-.78	.25-.41	.73-.81	.26-.31	.60-.83	.29-.42	.60-.75	.30-.47
BP	.68-.79	.23-.37	.73-.81	.27-.34	.76-.86	.17-.32	.43-.77	.31-.63	.76-.85	.18-.26
GH	.58-.78	.33-.56	.56-.65	.44-.48	.62-.75	.33-.46	.53-.77	.33-.57	.58-.77	.22-.49
VT	.29-.56	.64-.79	.32-.51	.69-.76	.34-.69	.46-.76	.30-.48	.66-.82	.37-.70	.46-.74
SF	.22-.39	.69-.75	.33-.47	.71-.80	.27-.38	.70-.78	.26-.29	.76-.81	.38-.58	.56-.75
RE	.13-.44	.55-.79	.15-.41	.63-.76	.14-.24	.68-.73	.19-.61	.40-.79	.09-.24	.73-.91
MH	.06-.16	.86-.91	.07-.17	.90-.92	.12-.25	.82-.89	.18-.27	.80-.86	.11-.19	.70-.89

	Norway		Spain		Sweden		United Kingdom	
Scale	Physical	Mental	Physical	Mental	Physical	Mental	Physical	Mental
PF	.77-.85	.12-.22	.69-.81	.07-.23	.81-.85	.08-.29	.76-.86	.15-.26
RP	.76-.83	.21-.35	.62-.80	.14-.40	.75-.78	.26-.33	.76-.81	.21-.33
BP	.79-.83	.19-.27	.67-.77	.19-.35	.74-.80	.26-.30	.78-.81	.14-.27
GH	.66-.79	.36-.41	.49-.83	.19-.42	.61-.69	.47-.54	.57-.78	.35-46
VT	.37-.52	.66-.72	.19-.53	.59-.76	.42-.62	.64-.74	.37-.59	.59-.70
SF	.29-.46	.65-.80	.43-.53	.55-.66	.28-.35	.77-.79	.43-.56	.61-.74
RE	.10-.28	.75-.84	.13-.19	.66-.90	.11-.45	.64-.77	.10-.27	.76-.81
MH	.12-.20	.86-.91	.03-.36	.70-.86	.13-.19	.88-.91	.11-.29	.82-.89

Methodological Issues

Factor analysis has proven to be very useful in testing hypotheses about the structure of health and in evaluating the construct validity of the SF-36 and other health surveys (Ware, 1976; Goldberg & Hillier, 1979; Ware, Brook, & Davies-Avery, 1980; Veit & Ware, 1983; Derogatis, 1986; Hall, Epstein, & McNeil, 1989; Wiklund, Lindvall, Swedberg, et al., 1987; Mason, Anderson, & Meenan, 1988; Schag, Heinrich, Aadland, et al., 1990; Hays & Stewart, 1990; McHorney, Ware, & Raczek, 1993). We have given considerable attention to the implications of different methods of factor extraction and rotation. In many cases, conclusions do not vary across methods. When they do, the choice among methods depends on the purpose(s) of the factor analysis (Nunnally & Bernstein, 1994).

The choice of method for the SF-36 studies follows from the first author's earlier work and considerations of work published by others (Ware, Miller, & Snyder, 1973; Snyder & Ware, 1974; Ware & Snyder, 1975; Ware, Davies-Avery, & Brook, 1980). Consistent with guidelines suggested by Harris & Harris (1971), conclusions about the factor structure of the SF-36 were not of great consequence in choosing a method. That structure is robust across methods and populations. In fact, a good test of a structural model is its robustness across factor analytic methods (Harris & Harris, 1971). For example, comparisons across methods for the same matrices were often employed during the development of the Health Perceptions Questionnaire (Ware, Miller, & Snyder, 1973; Ware, 1976; Ware & Karmos, 1976), from which items were selected for the SF-36 GH Scale. Those studies also demonstrated the advantages of homogeneous, short, multi-item scales over single-item measures as the unit of analysis in factor analytic studies. These advantages are also well-documented in empirical studies of personality variables (Comrey, 1973).

The two-dimensional factor structure of the eight SF-36 scales has also been shown to satisfy criteria for "simple structure" (Nunnally & Bernstein, 1994) across patient and general population samples in the U.S. and across other countries (as summarized above). To facilitate re-analyses by others, the matrices of correlations for the SF-36 in general and patient populations are reproduced at the end of this section (see Tables 3.5 and 3.6).

Principal Components

As noted above, the interpretation of the first two factors as dimensions of physical and mental health has been straightforward and robust across methods. Thus, the choice of factor analytic method was not governed by considerations in interpreting the factors. The choice of the principal components method was based on other considerations, including the ease of estimation of factor scores for the two summary measures, estimation of the factor content of the eight SF-36 scales in relation to physical and mental health status, the explanatory power of the factors, and their validity in discriminating between physical and mental dimensions of health status. These considerations are discussed briefly below.

The advantages of components analysis over principal factor analysis are noteworthy for the purposes of achieving: (1) a simple additive model of factor content facilitating the interpretation of each scale; (2) summary measures that explain as much of the variance in the eight scales as possible; (3) summary scales that are easy to estimate statistically; and (4) summary scores that are interpretable as physical and mental dimensions of health. The goal in constructing the PCS and MCS scores was to explain as much of the variance in the eight SF-36 variables as possible with only two summary measures (Ware, Kosinski, Bayliss, et al., 1995). Components analysis attempts to explain as much of the variance as possible, in contrast to principal factors, which attempt to reproduce the original correlation matrix (Harman, 1976). Second, as explained in Chapter 4, the computation of scores for each principal component is a straightforward estimation using scores for the observed variables (i.e., the eight SF-36 scale scores) in contrast to approximations involved in estimating scores for principal factors. These differences and the advantages of components analysis are discussed in numerous texts on factor analysis and psychometric methods (e.g., Harman, 1976; Nunnally & Bernstein, 1994).

There are good theoretical arguments for both orthogonal and oblique factor rotations (Harman, 1976; Nunnally & Bernstein, 1994). As argued in numerous texts, orthogonal components are ideal for our purposes. Our initial objective in factor analyzing the correlations among SF-36 scales was to test the construct validity of the SF-36 and to establish guidelines for interpreting each scale, on the basis of its physical and mental health factor content (McHorney, Ware, & Raczek, 1993; Ware, Snow, Kosinski, et al., 1993; Ware, Kosinski, Bayliss, et al., 1995). For this purpose, orthogonal components, which are not correlated, have clear advantages. For example, "factor loadings," which are product moment correlations between scales and factors, can be squared and summed across factors to estimate the amount of variance in each scale accounted for by each factor and the amount of variance in each scale that is explained by all factors (i.e., the communality). As a result, factor content and implications for the interpretation of each scale are more straightforward (Ware, Snow, Kosinski, et al., 1993).

Orthogonal Components

To provide a visual image of the contribution of the eight SF-36 scales to the physical and mental components, their factor loadings are plotted in Figure 3.2 for the general U.S. population. This plot reveals a progression from the upper left corner, with the MH scale correlating most highly with the mental component (MCS) and least with the physical component (PCS), to the lower right corner, where the highest correlation with the PCS is observed for the PF scale, which also correlates the least with the MCS. In between is a progression of loadings from the lower right (PF) to the upper left (MH). This ordering of scales from PF in the lower right to MH in the upper left corresponds to their ordering in the SF-36 profile.

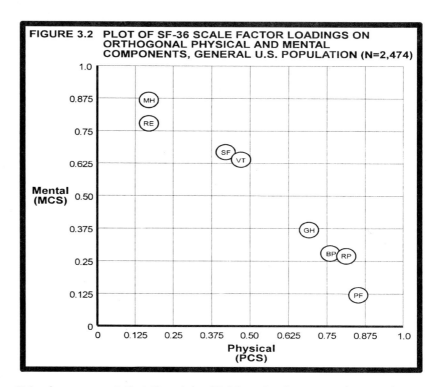

FIGURE 3.2 PLOT OF SF-36 SCALE FACTOR LOADINGS ON ORTHOGONAL PHYSICAL AND MENTAL COMPONENTS, GENERAL U.S. POPULATION (N=2,474)

It is also apparent that the eight SF-36 scales form two distinct clusters with four scales (MH, RE, SF, VT) correlating highest with the MCS and lowest with the PCS, and a second cluster (PF, RP, BP, GH) correlating highest with the PCS and lowest with the MCS. As shown in Chapter 4, the factor score coefficients used to score the PCS and MCS correspond to these two clusters. Specifically, the highest positive coefficients (0.42 to 0.25) are used to weight the four best physical scales in scoring the PCS, and the highest positive coefficients (0.49 to 0.24) are used to weight the four best mental scales in scoring the MCS. Because the factor score coefficients take into account the correlations *among* the eight scales, they differ from the factor loadings; some are negative. Negative factor score coefficients were also observed in oblique principal factor solutions and are not unique to the principal components method (data not reported). Oblique solutions, which can allow substantial correlations between health dimensions (factors), can facilitate the identification of factors but they complicate understanding of the factor content of scales because loadings are not additive in an oblique solution. Correlations among factors and factor loadings must both be taken into account in interpreting an oblique solution, complicating the interpretation of each scale.

If the PF and MH scales had proven to be substantially correlated, or if the PCS and MCS were substantially correlated on cross-validation, there would be good reason to favor an oblique solution. However, it is clear that physical and mental health are only weakly positively correlated. Correlations between the PF and MH scales, the best physical and mental health measures among the eight SF-36 scales, are low, with medians ranging from only 0.22 to 0.30 in 39 patient and general population studies in the U.S., Germany, Sweden, and the U.K. Cross validation of the orthogonal two-dimensional model (using U.S. factor score coefficients) in these samples has demonstrated very low empirical correlations between the PCS and MCS scores (medians of -0.01 to 0.07 across 39 estimates in studies to date). These correlations would be much larger upon cross-validation if the orthogonal solution and scoring were a gross distortion.

Additional convincing empirical evidence favoring the scoring of orthogonal principal components in summarizing SF-36 information about physical and mental health is their superiority in discriminating between physical and mental health outcomes in empirical tests. Comparisons of alternative scoring strategies revealed that much of the advantage in interpretation gained with the SF-36 PCS and MCS is lost when the physical and mental components are scored with an oblique solution (data not shown).

TABLE 3.7 PRODUCT MOMENT CORRELATIONS AND RELIABILITY COEFFICIENTS (IN PARENTHESES), SF-36 SCALES (4-WEEK RECALL) FOR THE 1998 GENERAL U.S. POPULATION

	PF	RP	BP	GH	VT	SF	RE	MH
PF	(.94)							
RP	.66	(.88)						
BP	.57	.65	(.85)					
GH	.57	.54	.57	(.84)				
VT	.44	.52	.54	.63	(.88)			
SF	.54	.64	.59	.56	.60	(.86)		
RE	.35	.46	.38	.39	.42	.58	(.82)	
MH	.26	.34	.38	.51	.63	.59	.51	(.85)

N=1,982

TABLE 3.8 PRODUCT MOMENT CORRELATIONS AND RELIABILITY COEFFICIENTS (IN PARENTHESES), SF-36 SCALES (1-WEEK RECALL) FOR THE 1998 GENERAL U.S. POPULATION

	PF	RP	BP	GH	VT	SF	RE	MH
PF	(.94)							
RP	.69	(.88)						
BP	.62	.66	(.86)					
GH	.58	.57	.60	(.84)				
VT	.46	.54	.53	.61	(.88)			
SF	.55	.66	.59	.58	.60	(.86)		
RE	.42	.52	.41	.43	.48	.64	(.83)	
MH	.29	.35	.39	.50	.61	.59	.55	(.83)

N=1,646

TABLE 3.9 PRODUCT MOMENT CORRELATIONS AND RELIABILITY COEFFICIENTS (IN PARENTHESES), SF-36 SCALES FOR THE 1990 GENERAL U.S. POPULATION

	PF	RP	BP	GH	VT	SF	RE	MH
PF	(.93)							
RP	.65	(.89)						
BP	.52	.61	(.90)					
GH	.55	.55	.56	(.81)				
VT	.44	.50	.52	.58	(.86)			
SF	.45	.52	.49	.47	.51	(.68)		
RE	.30	.42	.32	.35	.44	.53	(.82)	
MH	.28	.35	.39	.46	.63	.56	.54	(.84)

N=2,474

TABLE 3.10 PRODUCT MOMENT CORRELATIONS AND RELIABILITY COEFFICIENTS (IN PARENTHESES), SF-36 SCALES FOR THE MOS (N = 3,445)

	PF	RP	BP	GH	VT	SF	RE	MH
PF	(.93)							
RP	.60	(.84)						
BP	.56	.62	(.82)					
GH	.53	.48	.50	(.78)				
VT	.49	.57	.50	.57	(.87)			
SF	.45	.52	.53	.47	.57	(.85)		
RE	.29	.48	.34	.37	.51	.55	(.83)	
MH	.20	.34	.38	.41	.58	.66	.60	(.90)

SF-12 Health Survey

The construction of the PCS and MCS and evidence to date supporting their usefulness in cross-sectional and longitudinal tests provides the foundation for the construction of a health survey that is much shorter than the SF-36. The number of items in a survey is, at least in part, a function of the number of health dimensions for which separate scores are to be estimated with precision. If two summary scores are useful for most purposes, as suggested by results reported here and elsewhere (Ware, Kosinski, Bayliss, et al., 1995), it is likely that the two summary scores can be estimated with fewer items well enough for some purposes.

Following discovery of the PCS and MCS summary scales, we began evaluating how well these scales could be reproduced using much shorter questionnaires. Those studies identified ten items from six of the eight SF-36 scales that reproduced at least 90% of the variance in both the PCS and MCS, as defined using the SF-36 scales. Addition of two more items created a 12-item short-form yielding satisfactory estimates of the PCS and MCS, as well as scores for the two additional concepts necessary to represent the profile of eight SF-36 concepts. We refer to this new short form as the SF-12. Four of the eight concepts (physical functioning, role-physical, role-emotional, and mental health) are estimated using two items each. The remaining scales (bodily pain, general health, vitality, and social functioning) are measured using one item each. The eight-concept profile based on SF-12 items (calibrated to reproduce the original eight SF-36 scales) appears to be very similar, on average, to the original SF-36 profile, although each score is estimated with less precision. This disadvantage of single-item scales and very short multi-item scales has been demonstrated in previous studies (McHorney, Ware, Rogers, et al., 1992).

Results from preliminary tests of empirical validity, like those described in Chapter 6 for the SF-36 PCS and MCS summary scales, suggest that 12-item versions of the PCS and MCS will correlate with the SF-36 versions in the 0.93 to 0.97 range, upon cross-validation. However, the SF-12-based PCS and MCS define fewer levels and should be expected to yield less reliable assignments of individuals to those levels. For large group studies these differences in measurement reliability are not as important, because confidence intervals for group averages are more determined by sample size. Given that the SF-12 can be self-administered in about two to three minutes, the trade-off may be worthwhile for many purposes.

We have prepared documentation of items and scoring algorithms for the SF-12 (Ware, Kosinski, & Keller, 1995), and results of the preliminary evaluation of empirical validity for the SF-12 have been published (Ware, Kosinski & Keller, 1995). The SF-12 should be most useful for purposes of large group and population monitoring of health status and changes in health over time. Because SF-12 items are a subset of the original SF-36, complete comparability can be maintained across data sets using SF-12 scoring algorithms and data for either short-form survey.

Chapter 4: Norm-Based Scoring And Interpretation

The interpretation of results has been made much easier with the norm-based scoring of SF-36 scales and summary measures. Specifically, norm-based scoring has proven to be very useful when interpreting differences across scales in the SF-36 profile and the physical and mental summary measures. As originally documented in the first edition of this manual (Ware et al, 1994), linear transformations were performed to transform scores to a mean of 50 and standard deviations of 10, in the general US population for both the SF-36 physical and mental health summary measures. This same transformation can be applied to all eight SF-36 scales to simplify the interpretation of scores on each scale.

We use 1998 norms to introduce norm-based scoring (NBS) for the eight-dimension profiles because these norms are much more up-to-date and reflect substantial differences in most "physical health" scales, in comparison with 1990 norms. These differences from 1990 to 1998 should be taken into account when drawing conclusions about population health, disease burden and treatment benefits. 1998 norms and associated scoring algorithms for SF-36, SF-12 and SF-8 also make it possible to estimate scores for the eight dimensions of health, which they all have in common, and for their two summary measures using either SF-36, SF-12 or SF-8. Finally, the 1998 norms and NBS scoring algorithms provide the long awaited linkage necessary to compare results across studies relying on the eight-scale profile or two summary measures based on any of these three forms.

Advantages Of Norm-Based Scoring

The advantage of norm-based scoring can be illustrated by comparing the SF-36 profile scored using the original 0 – 100 algorithms with the profile based on the norm-based scoring algorithms for the same sample. For purposes of this comparison, we scored SF-36 profiles both ways for a sample of asthmatic patients who participated in a clinical trial (Okamoto et al, 1996). The original SF-36 0-100 scoring produced the profile shown in Figure 4.1. The shape of this profile – the peaks and valleys due to higher and lower scores across scales – reflect both the impact of asthma on SF-36 health concepts, as well as arbitrary differences in the ceilings and floors of the SF-36 scales. Three scales, namely General Health (GH), Vitality (VT), and Mental Health (MH), measure relatively wide score ranges and set the ceiling relatively high by measuring very favorable levels of those health concepts (Ware et al, 1993). Other scales, such as Physical Functioning (PF), and Role-Physical (RP), assess a narrower range based on a lower ceiling. For these scales the most favorable levels (scored 100 using the original SF-36 algorithms) represent the absence of limitations and do not extend the range into well being. Thus, the average score for each scale differs substantially across the profile for reasons that have nothing to do with asthma, using the original SF-36 0 – 100 scoring (see "norm" in Figure 4.1). Ignoring these norms, a reasonable inference from the profile in Figure 4.1 is that asthma has a greater impact on Physical Functioning than on Vitality; this inference is incorrect. (See the two shaded scales).

General population norms provide a basis for meaningful comparisons across scales (see Figure 4.1). For example, the PF scale averages between 80 and 90 while the VT average score is below 60 (on the 100-point score range) in the general population. In relation to these norms, the impact of asthma is actually, much larger on the Physical Functioning Scale than on the Vitality Scale, although both are statistically significant. Using the original 0 – 100 scoring, these differences in norms must be kept in mind when interpreting a profile. Differences in standard deviations, which are also substantial across some scales, must also be considered for purposes of comparing results across scales.

0 – 100 SCORING

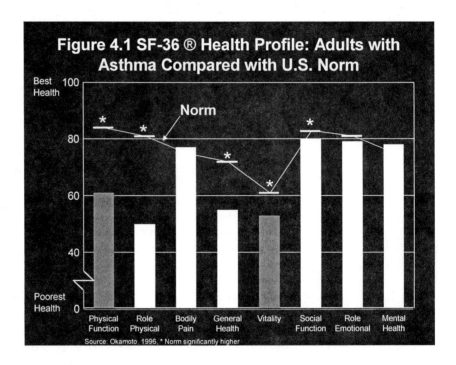

Figure 4.1 SF-36 ® Health Profile: Adults with Asthma Compared with U.S. Norm
Source: Okamoto, 1996. * Norm significantly higher

In norm-based scoring, each scale is scored to have the same average (50) and the same standard deviation (10 points). Without referring to tables of norms, it is clear with this method that anytime a scale score is below 50, health status is below average, and each point is one-tenth of a standard deviation. As shown in Figure 4.2, with norm-based scoring, differences in scale scores much more clearly reflect the impact of the disease, in this example the impact of asthma. Clinicians can more quickly and appropriately interpret the effect of asthma on a SF-36 health profile.

Norm-Based Scoring

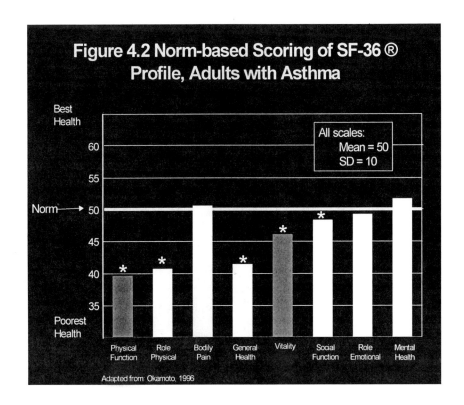

Other advantages of norm-based scoring are shown in Figure 4.3. Results for physical and mental summary measurers, which have always been scored in relation to norms, can be compared directly with results for the eight scales when all are standardized in relation to populations norms.

Because the Physical (PCS) and Mental (MCS) component summary measures take into account the correlations among the eight SF-36 scales, it is clear from the example in Figure 4.3 that asthma has a very broad impact on the physical component of health (from the profile in Figure 4.2 with five significant differences) and from the physical summary measure (see Figure 4.3).

The application of norm-based scoring to a clinical trial of treatment effects is also illustrated in Figure 4.3. Patients treated using an inhaler showed statistically significant improvements (white bars with *) relative to baseline (shaded bars) after 16 weeks of treatment on three of the eight SF-36 scales, those most closely associated with physical functioning. As expected, the physical summary measure also improved significantly.

Norm-Based Scoring

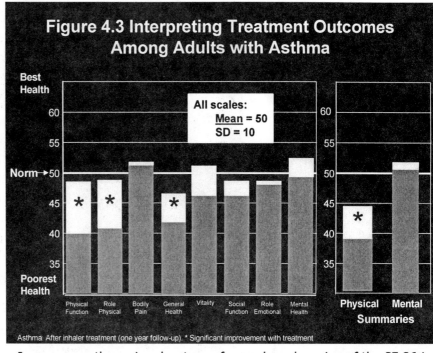

Figure 4.3 Interpreting Treatment Outcomes Among Adults with Asthma

Asthma: After inhaler treatment (one year follow-up). * Significant improvement with treatment

Conclusion

In summary, the main advantage of norm-based scoring of the SF-36 is easier interpretation. When interpreting the scores you no longer have to remember the norms for 8 scales. In norm-based scoring, the general population norm is built into the scoring algorithm. All scores above or below 50 can be interpreted as above or below the general population norm. And because the standard deviations for each scale are equalized at 10, it's easier to see exactly how far above (or below) the mean the score is in standard deviation units. Furthermore, since all 8 scales are scored to have the same standard deviation, comparisons in scores across scales and summary measures can be made directly.

Norm-based scoring has another very important advantage. It provides a basis for direct comparisons between scores from the original version of the SF-36 (v1) & Version 2 (v2). To facilitate such comparisons, we have made available a service for Version 1 scoring of scales and summary measures using the 1998 norms. In addition to providing much more up-to-date norms for use in interpreting Version 1 scales and summary measures, this scoring service assures the comparability of results for both versions of the SF-36 Health Survey. For more information, visit www.SF-36.com.

Again, we take this opportunity to underscore the importance of carefully documenting the SF-36 form (v1), norms and scoring algorithms used in reports of "Study Methods" accompanying results based on the 1998 SF-36 (v1) U.S. population norms. Further, because tables and figures are sometimes distributed separately, we also recommend inclusion of explicit references to "1998 SF-36 (v1) U.S. population norms" and to "norm-based scoring (NBS)" in tables and figures presenting results based on the new 1998 U.S. population norms.

Chapter 5: Norm-Based Scoring Of SF-36 Scales

This chapter provides instructions for the norm-based scoring of the eight multi-item scales of the SF-36 Health Survey. Algorithms are provided for both standard (4-week) and acute (1-week) recall survey forms. The instructions provided in this chapter assume that each of the eight SF-36 scales have already been scored on the 0-100 scale according to standard formulas for item aggregation and transformation of scores to a 0-100 scale published in the SF-36 Manual and Interpretation (Ware, Snow, Kosinski, Gandek, 1993).

Following the 0-100 score transformation is the norm-based scoring of each scale to have a mean of 50 and a standard deviation of 10 in the 1998 general U.S. population. Norm-based scoring methods of scales from standard (4-week recall) and acute (1 week recall) forms are presented separately. Finally, formal checks for errors in scoring are explained. A test dataset is available upon request (info@qmetric.com) to confirm successful reproduction of the scoring methods outlined in this chapter.

Importance Of Standardization

As with all standardized tests, standardization of content and scoring is what makes interpretation of SF-36 scales possible. The content of the SF-36 form and the scoring algorithms were selected and standardized following careful study of many options. The algorithms described in this chapter were chosen to be as simple as possible, satisfy the assumptions of the methods used to construct SF-36 scales, and to achieve the nearly perfect linear association with the original SF-36 scores, necessary to preserve their original interpretations.

Changes in the content of the survey or in the scoring method may compromise the reliability and validity of scores. Changes are also likely to bias scores sufficiently to invalidate normative comparisons and to prevent comparisons of results across studies.

There are at least two good reasons to adhere to the standards of content and scoring described in this manual. First, they are most likely to produce scores with the same reliability and validity as those previously reported for SF-36 scales. Second, comparisons of results across studies are made possible to the benefit of all who use these content and scoring standards. Third, differences in scores will have the same interpretations.

Prior to using the SF-36 scoring rules, it is essential to verify that the questionnaires being scored, including the questions asked (item stems), response choices and numbers assigned to response choices at the time of data entry, have been reproduced exactly. The scoring rules documented in the SF-36 Manual and Interpretation Guide are appropriate for the standard SF-36 survey questions, response choices, and numbers assigned to response choices.

Norm-Based Scoring Of Scale Scores, Standard Form (4-Week Recall)

Once each SF-36 scale has been scored on a 0-100 scale, the next step involves the norm-based scoring of each scale using the formulas shown below. Table 5.1 provides the information necessary to apply these formulas to each SF-36 scale from the standard form (4 week recall). The means and standard deviations (Table 5.1) used in norm-based scoring come from the 1998 general U.S. population. A linear z-score transformation is used so that all eight SF-36 scales have a mean of 50 and a standard deviation of 10 in the 1998 general U.S. population.

The advantage of the standardization and norm-based scoring of the 8 SF-36 scales is that results for one scale can be meaningfully compared with the other scales and their scores have a direct interpretation in relation to the distribution of scores in the 1998 general U.S. population. Specifically, all scores above or below 50 are above or below the average, respectively, in the 1998 general U.S. population. Because the standard deviation is 10 for all 8 scales, each one point difference or change in scores also has a direct interpretation. A one point difference or change is one-tenth of a standard deviation unit or an effect size of 0.10. Lastly, norm-based scoring provides the basis for comparing scale scores across Version 1.0 and Version 2.0 standard forms.

The first step in norm-based scoring consists of standardizing each SF-36 scale using a z-score transformation. A z-score for each scale is computed by subtracting the 1998 general U.S. population mean (Table 5.1) for each SF-36 scale and dividing the difference by the corresponding scale standard deviation (Table 5.1) from the 1998 general U.S. population. Formulas are listed below.

STEP 1. FORMULAS FOR Z-SCORE STANDARDIZATION OF SF-36 SCALES (STANDARD FORM)

$$PF\_Z = (PF - 82.96845) / 23.83795$$
$$RP\_Z = (RP - 77.93107) / 35.34865$$
$$BP\_Z = (BP - 70.22865) / 23.35310$$
$$GH\_Z = (GH - 70.10060) / 21.35900$$
$$VT\_Z = (VT - 56.99917) / 21.12677$$
$$SF\_Z = (SF - 83.56494) / 23.02758$$
$$RE\_Z = (RE - 83.10276) / 31.64149$$
$$MH\_Z = (MH - 75.21913) / 17.60698$$

Means and standard deviations are from Table 5.1.

TABLE 5.1 1998 GENERAL U.S. POPULATION MEANS AND STANDARD DEVIATIONS USED TO DERIVE SF-36 Z-SCORES (STANDARD FORM)

SF-36 SCALE	MEAN*	STANDARD DEVIATION
PHYSICAL FUNCTIONING	82.96845	23.83795
ROLE PHYSICAL	77.93107	35.34865
BODILY PAIN	70.22865	23.35310
GENERAL HEALTH	70.10060	21.35900
VITALITY	56.99917	21.12677
SOCIAL FUNCTIONING	83.56494	23.02758
ROLE EMOTIONAL	83.10276	31.64149
MENTAL HEALTH	75.21913	17.60698

*Note: The means and standard deviations for each SF-36 scale are based on the 0-100 scoring.

The second step involves transforming each SF-36 z-score to the norm-based (50, 10) scoring. This is accomplished by multiplying each z-score from Step 1 by 10 and adding the resulting product to 50. Formulas are listed below.

STEP 2. NORM-BASED TRANSFORMATION OF SF-36 Z-SCORES (STANDARD FORM):

Norm-Based PF: PF = 50 + (PF_Z * 10)
Norm-Based RP: RP = 50 + (RP_Z * 10)
Norm-Based BP: BP = 50 + (BP_Z * 10)
Norm-Based GH: GH = 50 + (GH_Z * 10)
Norm-Based VT: VT = 50 + (VT_Z * 10)
Norm-Based SF: SF = 50 + (SF_Z * 10)
Norm-Based RE: RE = 50 + (RE_Z * 10)
Norm-Based MH: MH = 50 + (MH_Z * 10)

Scoring Checks

Because of errors in reproducing a form, entering data, programming or processing, which can lead to inaccurate scale scores, we strongly recommend formal scoring checks prior to using the scales. Any discrepancies observed during the following checks should be investigated for scoring errors:

1. Calculate SF-36 scale scores by hand for several respondents and compare the results to those produced by your scale scoring computer software.

2. After items have been coded into their final item values, inspect the frequency distributions for the items to verify that only the final item values shown in Tables 6.1 through 6.8 in the SF-36 Manual and Interpretation Guide are observed. Discrepancies should be limited to respondents with values estimated for missing data as described in the manual.

3. After items have been recoded and scale scores have been computed, inspect the correlation between each scale and its component items to verify that all correlations are positive in direction and substantial in magnitude (0.30 or higher).

4. Check correlation's between the General Health scale and the other seven scales to verify that all are positive; with rare exceptions they should also be substantial in magnitude (0.30 or higher).

5. For those familiar with principal factor or components analysis, inspect correlations between the eight scales and the first unrotated factor or component extracted from the correlation's among those scales. Regardless of extraction method, these correlations should be positive and substantial in magnitude (0.30 or higher).

Norm-Based Scoring Of Scale Scores, Acute Form (1-Week Recall)

After scoring the acute version SF-36 scales on a 0-100 scale, the next step involves the norm-based scoring (50/10) using the formulas shown below. Table 5.2 provides the information necessary to apply these formulas to each SF-36 scale from the Acute form (1 week recall). The means and standard deviations (Table 5.2) used in norm-based scoring come from the 1998 general U.S. population. A linear z-score transformation is used so that all eight SF-36 scales have a mean of 50 and a standard deviation of 10 in the 1998 general U.S. population.

The advantages of the norm-based scoring of the standard recall form above also apply to the acute version. The methods for transforming 0-100 scores on the acute scales to norm-based scores are also similar to those used for the standard form. The first step in norm-based scoring consists of standardizing each SF-36 scale using a z-score transformation. A z-score for each scale is computed by subtracting the 1998 general U.S. population mean (Table 5.2) for each acute form SF-36 scale and dividing the difference by the corresponding scale standard deviation (Table 5.2) from the 1998 general U.S. population. Formulas are listed below.

STEP 1. FORMULAS FOR Z-SCORE STANDARDIZATION OF SF-36 SCALES (ACUTE FORM)

$$PF\_Z = (PF - 82.34934) / 24.55363$$
$$RP\_Z = (RP - 78.37950) / 35.25259$$
$$BP\_Z = (BP - 73.37809) / 24.09701$$
$$GH\_Z = (GH - 70.69938) / 21.60431$$
$$VT\_Z = (VT - 57.64847) / 21.81907$$
$$SF\_Z = (SF - 85.61860) / 22.83605$$
$$RE\_Z = (RE - 82.44273) / 32.52898$$
$$MH\_Z = (MH - 76.23511) / 17.25352$$

Means and standard deviations are from Table 5.2.

TABLE 5.2 1998 GENERAL U.S. POPULATION MEANS AND STANDARD DEVIATIONS USED TO DERIVE SF-36 Z-SCORES (ACUTE FORM)

SF-36 SCALE	MEAN*	SD
PHYSCIAL FUNCTIONING	82.34934	24.55363
ROLE PHYSICAL	78.37950	35.25259
BODILY PAIN	73.37809	24.09701
GENERAL HEALTH	70.69938	21.60431
VITALITY	57.64847	21.81907
SOCIAL FUNCTIONING	85.61860	22.83605
ROLE EMOTIONAL	82.44273	32.52898
MENTAL HEALTH	76.23511	17.25352

*Note: The means and standard deviations for each SF-36 scale are based on the 0-100 scoring.

Step 2. Norm-Based Transformation Of SF-36 Z-Scores (Acute Form)

The second step involves transforming each SF-36 z-score from Step 1 to the norm-based (50, 10) scoring. This is accomplished by multiplying each z-score by 10 and adding the resulting product to 50. Formulas are listed below.

STEP 2. NORM-BASED TRANSFORMATION OF SF-36 Z-SCORES (ACUTE FORM):

Norm-Based PF: $PF = 50 + (PF\_Z * 10)$
Norm-Based RP: $RP = 50 + (RP\_Z * 10)$
Norm-Based BP: $BP = 50 + (BP\_Z * 10)$
Norm-Based GH: $GH = 50 + (GH\_Z * 10)$
Norm-Based VT: $VT = 50 + (VT\_Z * 10)$
Norm-Based SF: $SF = 50 + (SF\_Z * 10)$
Norm-Based RE: $RE = 50 + (RE\_Z * 10)$
Norm-Based MH: $MH = 50 + (MH\_Z * 10)$

Scoring Checks

Because of errors in reproducing a form, entering data, programming or processing, which can lead to inaccurate scale scores, we strongly recommend formal scoring checks prior to using the scales. Any discrepancies observed during the following checks should be investigated for scoring errors.

1. Calculate SF-36v2 scale scores by hand for several respondents and compare the results to those produced by your scale scoring computer software.

2. After items have been coded into their final item values, inspect the frequency distributions for the items to verify that only the final item values shown in Tables 6.1 through 6.8 in the SF-36 Manual and Interpretation Guide are observed. Discrepancies should be limited to respondents with values estimated for missing data as described in the manual.

3. After items have been recoded and scale scores have been computed, inspect the correlation between each scale and its component items to verify that all correlations are positive in direction and substantial in magnitude (0.30 or higher).

4. Check correlation's between the General Health scale and the other seven scales to verify that all are positive; with rare exceptions they should also be substantial in magnitude (0.30 or higher).

5. For those familiar with principal factor or components analysis, inspect correlations between the eight scales and the first unrotated factor or component extracted from the correlations among those scales. Regardless of extraction method, these correlations should be positive and substantial in magnitude (0.30 or higher).

Chapter 6: Norm-Based Scoring For Physical & Mental Summary Measures

This chapter provides scoring instructions for the SF-36 physical (PCS) and mental (MCS) component summary measures. Scoring of the SF-36 PCS and MCS summary measures involves three steps. First, the eight SF-36 scales are standardized using means and standard deviations from the 1998 general U.S. population. Second, they are aggregated using weights (factor score coefficients) from the 1990 general U.S. population. These are the same weights as those used to score PCS and MCS from the SF-36 Version in the first edition of this manual (Ware 1994). Finally, aggregate PCS and MCS scores are standardized using a linear T-score transformation to have a mean of 50 and a standard deviation of 10, in the 1998 general U.S. population. Norm-based scoring of the standard (4-week recall) and acute (1 week recall) forms is presented separately.

General U.S. population statistics used in the standardization and in the aggregation of SF-36 scale scores are presented in Table 6.1 for the standard form and in Table 6.2 for the acute form. Note that for both standard and acute forms the same factor score coefficients are used to score PCS and MCS. To make sure that the original coefficients used in this scoring "recipe" have the same affect in 1998, as in 1990, 1998 variances are utilized. Detailed information including formulas for scale aggregation and transformation of scores are presented below. Formal checks using a test dataset available upon request (info@qmetric.com) can be performed to confirm the successful reproduction of PCS and MCS scales, as discussed later in the chapter. We strongly recommend these tests.

Importance Of Standardization

As with the 1998 NBS scoring of SF-36 scales, which are aggregated to score the summary measures, standardization of the scoring of the PCS and MCS scales is vital to their interpretation. Any changes in scoring of the SF-36 scales or the algorithms for the summary measures may compromise their reliability and validity. Changes in scoring may also invalidate normative comparisons, based on the 1998 norms documented here.

The methods documented here achieve 1998 scores that are a nearly perfect transformation of the 1990 scores. For scales and summary measures, NBS based on 1998 norms only shifts the score distribution to better reflect the health of the US population in 1998. Otherwise, 1998 scores have the same interpretations as 1990 scores.

Norm-Based Scoring Of PCS And MCS, Standard Form (4-Week Recall)

Standard form PCS and MCS scales are scored using norm-based methods. The means and standard deviations used in scoring come from the 1998 general U.S. population and the factor score coefficients come from the 1990 general U.S. population (Ware et al. 1994). A linear T-score transformation method is used so that both the PCS and MCS have a mean of 50 and a standard deviation of 10 in the 1998 general U.S. population.

The advantage of the standardization and norm-based scoring of the PCS and MCS is that results for one can be meaningfully compared with the other and their scores have a direct interpretation in relation to the distribution of scores in the general U.S. population. Specifically, all scores above and below 50 are above and below the average, respectively, in the 1998 general U.S. population. Because the standard deviation is 10 for both PCS and MCS measures, each one point difference in scores also has a direct interpretation. A one point difference is one-tenth of a standard deviation.

TABLE 6.1 1998 GENERAL U.S. POPULATION MEANS, STANDARD DEVIATIONS AND 1990 FACTOR SCORE COEFFICIENTS USED TO DERIVE PCS AND MCS SCALE SCORES, STANDARD FORM

SF-36 SCALE	MEAN*	STANDARD DEVIATION	Factor Score Coefficients	
			PCS	MCS
PF	82.96845	23.83795	0.42402	-0.22999
RP	77.93107	35.34865	0.35119	-0.12329
BP	70.22865	23.35310	0.31754	-0.09731
GH	70.10060	21.35900	0.24954	-0.01571
VT	56.99917	21.12677	0.02877	0.23534
SF	83.56494	23.02758	-0.00753	0.26876
RE	83.10276	31.64149	-0.19206	0.43407
MH	75.21913	17.60698	-0.22069	0.48581

*Note: The means and standard deviations for each SF-36 scale are based on the 0-100 scoring.

It should be noted that mean scores for most of the scales vary substantially form those from the 1990 norms. For those scales that have been changed in any way from 1990 to 1998, several explanations are being explored (e.g., sampling bias in the 1990 study, downward temporal trends in physical health). Regardless of the explanations, these differences underscore the importance of using up-to-date norms and also the importance of using the same normative data when equating Versions 1 and 2 of the SF-36. 1998 norms for both make these comparisons possible.

Steps In Scoring

Following the scoring of the eight scales according to the standard SF-36 scoring algorithms (0-100 scale) explained in the SF-36 Manual and Interpretation Guide (Ware et al., 1993, PCS and MCS are scored in three steps as explained below:

Standardization Of Scales (Z-Scores), Standard Form

The first step in computing PCS and MCS consists of standardizing each of the 8 SF-36 scales using a z-score transformation. This is the same as Step 1 used in the norm-based scoring of the 8 SF-36 scales explained in Chapter 5. A z-score for each scale is computed by subtracting the mean 0-100 1998 general US population score (see Table 6.1) for each SF-36 scale and dividing the difference by the corresponding scale standard deviation. Note that the SF-36 scales scored on the 0-100 scale are used in step 1. Norm-based SF-36 scale scores are not used in this step. Formulas are listed below.

STEP 1. Formulas For Z-Score Standardization Of SF-36 Scales, (Standard Form)

$$PF\_Z = (PF - 82.96845) / 23.83795$$
$$RP\_Z = (RP - 77.93107) / 35.34865$$
$$BP\_Z = (BP - 70.22865) / 23.35310$$
$$GH\_Z = (GH - 70.10060) / 21.35900$$
$$VT\_Z = (VT - 56.99917) / 21.12677$$
$$SF\_Z = (SF - 83.56494) / 23.02758$$
$$RE\_Z = (RE - 83.10276) / 31.64149$$
$$MH\_Z = (MH - 75.21913) / 17.60698$$

Means and standard deviations are from Table 6.1.

Aggregation Of Scale Scores (Standard Form)

After a z-score has been computed for each SF-36 scale, the second step involves computation of aggregate scores for the physical and mental components using the physical and mental factor score coefficients from the 1990 general U.S. population as given in Table 6.1.

Computation of an aggregate physical component score consists of multiplying each SF-36 scale z-score by its respective physical factor score coefficient and summing the eight products, as shown below. Similarly, an aggregate mental component score is obtained by multiplying each SF-36 scale z-score by its respective mental factor score coefficient and summing the eight products.

STEP 2. Formulas For Aggregating Scales In Estimating Aggregate Physical And Mental Component Scores (Standard Form)

$$AGG\_PHYS = (PF\_Z*.42402) + (RP\_Z*.35119) + (BP\_Z*.31754) + (GH\_Z*.24954) + (VT\_Z * .02877) + (SF\_Z*-.00753) + (RE\_Z*-.19206) + (MH\_Z*-.22069)$$

$$AGG\_MENT = (PF\_Z*-.22999) + (RP\_Z*-.12329) + (BP\_Z*-.09731) + (GH\_Z*-.01571) + (VT\_Z*.23534) + (SF\_Z*.26876) + (RE\_Z*.43407) + (MH\_Z*.48581)$$

Transformation Of Summary Scores

The third step involves transforming each component score to the norm-based (50, 10) scoring. This is accomplished by multiplying each aggregate component scale score by 10 and adding the resulting product to 50. Formulas are shown in step 3.

STEP 3. FORMULAS FOR T-SCORE TRANSFORMATION OF COMPONENT SCORES (STANDARD FORM)

Transformed Physical (PCS) = 50 + (AGG_PHYS * 10)
Transformed Mental (MCS) = 50 + (AGG_MENT * 10)

Norm-Based Scoring Of PCS And MCS, Acute Form

Acute form PCS and MCS scales are scored using norm-based methods. The means and standard deviations used in scoring come from the 1998 general U.S. population and the factor score coefficients come from the 1990 general U.S. population documented in the first edition of this manual (Ware et al. 1994). A linear T-score transformation method is used so that both the PCS and MCS have a mean of 50 and a standard deviation of 10 in the 1998 general U.S. population.

The advantage of the standardization and norm-based scoring of the PCS and MCS is that results for one can be meaningfully compared with the other and their scores have a direct interpretation in relation to the distribution of scores in the general U.S. population. Specifically, all scores above and below 50 are above and below the average, respectively, in the 1998 general U.S. population. Because the standard deviation is 10 for both PCS and MCS measures, each one point difference in scores also has a direct interpretation. A one point difference is one-tenth of a standard deviation.

TABLE 6.2 1998 GENERAL U.S. POPULATION MEANS, STANDARD DEVIATIONS AND 1990 FACTOR SCORE COEFFICIENTS USED TO DERIVE PCS AND MCS SCALE SCORES, ACUTE FORM

SF-36 SCALE	MEAN*	STANDARD DEVIATION	FACTOR SCORE COEFFICIENTS	
			PCS	MCS
PF	82.34934	24.55363	0.42402	-0.22999
RP	78.37950	35.25259	0.35119	-0.12329
BP	73.37809	24.09701	0.31754	-0.09731
GH	70.69938	21.60431	0.24954	-0.01571
VT	57.64847	21.81907	0.02877	0.23534
SF	85.61860	22.83605	-0.00753	0.26876
RE	82.44273	32.52898	-0.19206	0.43407
MH	76.23511	17.25352	-0.22069	0.48581

*Note: The means and standard deviations for each SF-36 are based on the 0-100 scoring.

Steps In Scoring

Following the scoring of the eight scales according to the standard SF-36 scoring algorithms (0-100 scale) explained earlier in SF-36 Manual and Interpretation Guide (Ware et al., 1993), PCS and MCS are scored in three steps as explained below:

Standardization Of Scales (Z-Scores), Acute Form

The first step in computing PCS and MCS consists of standardizing each of the 8 SF-36 scales using a z-score transformation. This is the same as Step 1 used in the norm-based scoring of the 8 SF-36 scales explained in Chapter 5. A z-score for each scale is computed by subtracting the mean 0-100 1998 general US population score (see Table 6.2) for each SF-36 scale and dividing the difference by the corresponding scale standard deviation. Note that the SF-36 scales scored on the 0-100 scale are used in step 1. Norm-based SF-36 scale scores are not used in this step. Formulas are listed below.

STEP 1.   FORMULAS FOR Z-SCORE STANDARDIZATION OF SF-36 SCALES (ACUTE VERSION)

$PF\_Z = (PF - 82.34934) / 24.55363$
$RP\_Z = (RP - 78.37950) / 35.25259$
$BP\_Z = (BP - 73.37809) / 24.09701$
$GH\_Z = (GH - 70.69938) / 21.60431$
$VT\_Z = (VT - 57.64847) / 21.81907$
$SF\_Z = (SF - 85.61860) / 22.83605$
$RE\_Z = (RE - 82.44273) / 32.52898$
$MH\_Z = (MH - 76.23511) / 17.25352$

Means and standard deviations are from Table 6.2

Aggregation Of Scale Scores (Acute Form)

After a z-score has been computed for each SF-36 scale, the second step involves computation of aggregate scores for the physical and mental components using the physical and mental factor score coefficients from the 1990 general U.S. population as given in Table 6.2.

Computation of an aggregate physical component score consists of multiplying each SF-36 scale z-score by its respective physical factor score coefficient and summing the eight products, as shown in step 2. Similarly, an aggregate mental component score is obtained by multiplying each SF-36 scale z-score by its respective mental factor score coefficient and summing the eight products.

STEP 2.   FORMULAS FOR AGGREGATING STANDARDIZED SCALES IN ESTIMATING AGGREGATE PHYSICAL AND MENTAL COMPONENT SCORES (ACUTE FORM)

$AGG\_PHYS = (PF\_Z * .42402) + (RP\_Z * .35119) + (BP\_Z * .31754) + (GH\_Z * .24954) + (VT\_Z * .02877) + (SF\_Z * -.00753) + (RE\_Z * -.19206) + (MH\_Z * -.22069)$

$AGG\_MENT = (PF\_Z * -.22999) + (RP\_Z * -.12329) + (BP\_Z * -.09731) + (GH\_Z * -.01571) + (VT\_Z * .23534) + (SF\_Z * .26876) + (RE\_Z * .43407) + (MH\_Z * .48581)$

Transformation Of Summary Scores

The third step involves transforming each component score to the norm-based (50, 10) scoring. This is accomplished by multiplying each aggregate component scale score by 10 and adding the resulting product to 50. Formulas are shown in step 3.

STEP 3. FORMULAS FOR T-SCORE TRANSFORMATION OF COMPONENT SCORES

Transformed Physical (PCS) = 50 + (AGG_PHYS * 10)
Transformed Mental (MCS) = 50 + (AGG_MENT * 10)

Missing Data Estimation

Results from ongoing evaluations of options to score PCS and MCS when a respondent is missing any one of the eight SF-36 scales has shown considerable promise. Evaluation of missing data rates across general and clinical populations has shown that 50% of those who have missing PCS and MCS scores are missing PCS and MCS because of missing data on one SF-36 scale. QualityMetric's Missing Data Estimator (MDE) will enable you to score PCS and MCS with data for 7 of the 8 SF-36 scales. PCS and MCS scores from MDE have proven to be reliable and valid (Ware et al., forthcoming). Further efforts are underway to evaluate options for scoring PCS and MCS when more than 1 SF-36 scale score is missing. More information will be available at these websites www.sf-36.com and www.qmetric.com.

Until QualityMetric's MDE is made available, it is recommended that component scale scores be set to missing if the respondent is missing any one of the eight SF-36 scales. To minimize the number of component scores missing, we recommend estimating each of the eight scale scores if half or more of the items are complete, as documented earlier.

Features Of PCS And MCS Scores

The PCS and MCS were constructed and scored to achieve a number of advantages, in addition to reducing the SF-36 from an eight-scale profile to two summary measures without substantial loss of information. Features of the PCS and MCS scores, including their reliability, confidence intervals (CI), skewness (percent ceiling and floor), and number of levels observed in the general U.S. population, are summarized in Table 6.3. These results confirm some of the theoretical advantages of the two summary measures as compared to the eight SF-36 scales, including a very large increase in the number of levels defined, smaller confidence intervals relative to each of the eight scales, as well as the elimination of both floor and ceiling effects.

TABLE 6.3 COMPARISON OF FEATURES OF SF-36 SCALES AND SUMMARY MEASURES, 1998 GENERAL U.S. POPULATION

	SF-36 SCALES[b]	SUMMARY MEASURES	
		PCS	MCS
Reliability	.84 - .95	.95	.93
95% CI (±)	12 – 17	4.5	5.2
% Floor	0.0 – 2.1	0	0
% Ceiling	2.1 – 60.2	0	0
# of Levels[a]	8 - 20	486	494

[a] Scores rounded to first decimal place
[b] Statistics are presented as the range of results found across the eight SF-36 scales (standard and acute forms) in the 1998 general U.S. population.

Scoring Checks

Because errors can lead to inaccurate scale scores, we strongly recommend formal scoring checks of SF-36 scales prior to computing the SF-36 component summary scales. These formal scoring checks are explained in full detail in the SF-36 Manual and Interpretation Guide (Ware et al., 1993).

The following scoring checks are also strongly recommended for the SF-36 component summary scales. Any discrepancies should be investigated for scoring errors:

Check correlation's between the eight SF-36 scales and the PCS and MCS scales. The PF, RP, and BP scales should correlate highest with the PCS and lowest with the MCS. The MH, RE, and SF scales should correlate highest with the MCS and lowest with the PCS. The GH and VT scales should correlate moderately with both physical and mental component scales.

Check the correlation between the physical and mental component summary scales. The correlation should be very low, generally lower than 0.30.

Scoring Exercise

The algorithms for scoring the SF-36 scales and summary measures for use with SAS software (along with a test data set) are available in electronic format to purchasers of this manual – by sending an email request to info@qmetric.com. QualityMetric also offers a Scoring & Data Quality Analysis service (information available at www.qmetric.com).

Table 6.4 presents descriptive statistics for the eight SF-36 scales (0-100 and norm-based) and the physical and mental component summary measures from the test dataset for the standard form. Table 6.5 presents descriptive statistics for the eight SF-36 scales (0-100 and norm-based) and the physical and mental component summary measures from the test dataset for the acute form.

After scoring the test dataset, you should observe the same means, standard deviations, and minimum and maximum observed values as those presented in Table 6.4 for the standard form and in Table 6.5 for the acute form.

TABLE 6.4 TEST DATASET DESCRIPTIVE STATISTICS*: SF-36 SCALES AND SUMMARY MEASURES (N = 100), STANDARD FORM

	NO. OF CASES	0-100 SCORES				NORM-BASED SCORES			
		MEAN	STANDARD DEVIATION	MIN. VALUE	MAX. VALUE	MEAN	STANDARD DEVIATION	MIN. VALUE	MAX. VALUE
Physical Functioning	100	82.55	23.74	5	100	49.82	9.96	17.29	57.14
Role Physical	100	78.25	35.82	0	100	50.09	10.13	27.95	56.24
Bodily Pain	100	70.34	23.11	0	100	50.05	9.90	19.93	62.75
General Health	100	65.32	22.72	0	100	47.76	10.63	17.18	63.99
Vitality	100	58.22	22.36	0	94	50.58	10.59	23.02	70.35
Social Functioning	100	83.00	22.79	0	100	49.75	9.90	13.71	57.14
Role Emotional	100	80.33	35.17	0	100	49.12	11.11	23.74	55.34
Mental Health	100	73.74	22.15	0	100	49.16	12.58	7.28	64.07
Physical Component Summary (PCS)	100	N/A	-	-	-	49.76	9.78	20.92	63.12
Mental Component Summary (MCS)	100	N/A	-	-	-	49.34	11.70	8.67	63.46

*All descriptive statistics are rounded to the second decimal point.

TABLE 6.5 TEST DATASET DESCRIPTIVE STATISTICS*: SF-36 SCALES AND SUMMARY MEASURES (N = 100), ACUTE FORM

	NO. OF CASES	0-100 SCORES				NORM-BASED SCORES			
		MEAN	STANDARD DEVIATION	MIN. VALUE	MAX. VALUE	MEAN	STANDARD DEVIATION	MIN. VALUE	MAX. VALUE
Physical Functioning	100	77.75	27.44	0	100	48.13	11.17	16.46	57.19
Role Physical	100	71.25	39.79	0	100	47.98	11.29	27.77	56.13
Bodily Pain	100	69.85	24.42	0	100	48.54	10.13	19.55	61.05
General Health	100	68.75	22.42	5	100	49.10	10.38	19.59	63.56
Vitality	100	59.98	22.51	0	100	51.07	10.32	23.58	69.41
Social Functioning	100	84.88	23.52	0	100	49.67	10.30	12.51	56.30
Role Emotional	100	83.67	31.96	0	100	50.38	9.82	24.66	55.40
Mental Health	100	78.35	15.85	8	100	51.23	9.19	10.45	63.77
Physical Component Summary (PCS)	100	47.49	10.87	19.00	63.13	47.49	10.87	19.00	63.13
Mental Component Summary (MCS)	100	51.76	8.61	13.43	62.06	51.76	8.61	13.44	62.07

*All descriptive statistics are rounded to the second decimal point.

SF-36 algorithms have been made available to computer software vendors and other organizations providing scoring and analysis services for the SF-36. Look for the symbol below.

This symbol is your assurance that computer software products and data processing services produce results that are comparable with this Manual and with other normative data and interpretation guidelines for the SF-36 Health Survey.

Chapter 7: Reliability And Statistical Power

Reliability estimates for the PCS and MCS scales were calculated using data from the Medical Outcomes Study (MOS), the 1990 and 1998 general U.S. populations, and data from general population surveys in Denmark, France, Italy, Netherlands, Norway, Spain, Sweden, and the U.K. As documented below, reliability estimates based on both internal consistency and test-retest methods have been high, ranging from 0.89 to 0.94 for the PCS and from 0.74 to 0.91 for the MCS studies of patients in the U.S. and in general population studies across countries. While in previous studies of the eight SF-36 scales (McHorney, Ware, Lu, et al., 1994), reliability estimates were slightly lower for more disadvantaged study participants, PCS and MCS reliability estimates were similar for disadvantaged and more advantaged groups.

Background

Indices of reliability give an indication of the extent to which the scores produced by a particular measurement procedure are consistent and reproducible. A measurement procedure is reliable to the extent that items within the same scale give the same results or to the extent that an individual scores the same across repeated administrations of the scale (Nunnally & Bernstein, 1994). A reliability coefficient is an estimate of how much of the variation in a score is real or truth, as opposed to chance or random errors. For example, a reliability coefficient of 0.80 indicates that 80% of the total measured variance is true score. Suggested levels of reliability are 0.70 or greater for scales used in group-level analyses and 0.90 or greater for scales used in decisions at the individual level (Nunnally & Bernstein, 1994).

There are several methods that can be used to estimate reliability. Reliability can be estimated by correlating scores from one administration with scores from another (test-retest reliability), or by correlating scores and testing the equivalence of individual answers across alternative forms of an instrument (alternative forms reliability), or by examining the equivalence of responses within the same test from a single administration (internal consistency reliability) (Nunnally & Bernstein, 1994). Most studies to date have used the internal consistency method and Cronbach's coefficient alpha to estimate the reliability of SF-36 scales, although all three methods listed above have been used in one or more studies (Ware, Snow, Kosinski, et al., 1993; McHorney & Ware, 1995).

Estimation Of Reliability

Both test-retest and internal consistency methods have been used in estimating the reliability of the PCS and MCS scales. Because the PCS and MCS scales are a linear combination of eight scales measuring distinct health constructs, it is necessary to take into account the reliability of each scale as well as the covariances among them in estimating reliability using the internal consistency method (Nunnally & Bernstein, 1994). Using the covariance matrix of SF-36 scales from each sample, and the physical and mental factor score coefficients from the general U.S. population, reliability is estimated in the following steps: (1) each off-diagonal covariance is multiplied by the product of its respective factor score coefficient, summed, and multiplied by two (two sides of the matrix); (2) the observed score variance is calculated by multiplying each diagonal of the covariance matrix by the squared factor score coefficient; (3) total score variance is calculated by summing the products of steps one and two; (4) true score variance is calculated by multiplying each diagonal entry computed in step #2 by the respective SF-36 scale reliability; and (5) each component summary reliability is computed by subtracting the true score variance (step #4) from the observed variance (step #2) divided by the total score variance (step #3), the result of which is subtracted from one.

Summary Of Reliability Estimates

General Population

Reliability coefficients for the PCS and MCS scales have been estimated using the internal consistency method for respondents to general population surveys in ten countries (Denmark, France, Germany, Italy, Netherlands, Norway, Spain, Sweden, The U.K., and the U.S.). Table 7.1 summarizes these results. The sampling methods for these surveys and the characteristics of respondents are documented elsewhere (Ware, Keller, Gandek, et al., 1995). Return rates ranged from 62.5% to 83.1%. Table 7.1 also presents test-retest reliability estimates for PCS and MCS in the general U.S. and U.K. populations.

TABLE 7.1 RELIABILITY ESTIMATES FOR PCS AND MCS SCALES IN GENERAL POPULATION FROM TEN COUNTRIES

COUNTRIES	(N)	PCS	MCS
Denmark	4,084	.91	.85
France	3,656	.90	.88
Germany	2,914	.94	.87
Italy	1,483	.90	.86
Netherlands	1,771	.92	.89
Norway	2,323	.92	.89
Spain	9,151	.94	.90
Sweden	8,930	.92	.88
U.K.	2,056	.93	.88
U.S. (1990)	2,474	.92	.89
U.S. (1998) 4-week recall	1,982	.94	.89
U.S. (1998) 1-week recall	1,646	.94	.89
U.K. Test-Retest	180	.89	.80
U.S. Test-Retest	540	.87	.74

Internal consistency reliability coefficients ranged from 0.90 to 0.94 for the PCS scale and 0.85 to 0.90 for the MCS scale. These results suggest that the underlying two dimensions of health measured by the SF-36 Health Survey can be reliably scored for the purpose of summarizing SF-36 scale score data in general populations studied to date.

Test-retest reliability for an interval of two weeks between administrations has also been estimated using data from the U.K. (Brazier, Harper, Jones, et al., 1992). The correlations between scores from administrations two weeks apart were 0.89 for the PCS and 0.80 for the MCS (N = 180). In the U.S., these correlations were 0.87 and 0.74, respectively (N=540).

These data demonstrate that the component summary measures have reliabilities that generally equal or exceed those of the eight scales.

TABLE 7.2 RELIABILITY ESTIMATES FOR PCS AND MCS SCALES: MOS PATIENT SUBGROUPS (N = 3,445)

		N	PCS	MCS
TOTAL SAMPLE			.92	.90
AGE	<65	2456	.91	.91
	65 – 74	700	.92	.86
	75+	287	.91	.86
GENDER	Female	2126	.92	.90
	Male	1319	.91	.90
RACE	White	2625	.92	.90
	Black	481	.90	.88
	Other	221	.90	.89
EDUCATION	<8	209	.94	.89
	9 – 11	313	.93	.90
	12	1005	.91	.89
	>12	1791	.91	.91
POVERTY STATUS	Poverty	253	.93	.91
	Non-Poverty	2864	.93	.90
DIAGNOSIS	Hypertension	2089	.92	.86
	Diabetes	624	.92	.86
	CHF	216	.91	.88
	MI	107	.89	.84
	Clinical Depression	503	.92	.88
	Symptomatic Depression	785	.91	.87
DISEASE SEVERITY	Uncomplicated Medical	1136	.90	.85
	Complicated Medical	289	.91	.86
	Psychiatric and Medical	300	.89	.86

NOTE: Estimates based on internal consistency method (see text).

Patient Subgroups

The reliability of the PCS and MCS scales was also estimated using the internal consistency method for 23 subgroups of patients participating in the MOS and for the total MOS sample. These patients (N = 3,445) differed in sociodemographic characteristics, diagnosis, and disease severity, as defined elsewhere (McHorney, Ware, Lu et al., 1994). Table 7.2 summarizes these results. Estimates of reliability for the SF-36 component summary scale scores varied very little across the groups, with a range of coefficients from 0.89 to 0.94 for the PCS and from 0.84 to 0.91 for the MCS. Estimates of reliability tended to be higher for the PCS relative to the MCS. Minimum standards of reliability for purposes of group comparisons (r > .70) were satisfied in all 23 patient subgroups for both physical and mental component scales. Minimum reliability standards required for comparisons of individual patients (r > .90) were met in 21 out of 23 patient groups for the physical component scale and for one-third of the patient groups for the mental component scale.

General Population Subgroups

The internal consistency reliability of the PCS and MCS scales was also estimated for 12 subgroups of respondents from the general U.S. population sample. As shown in Table 7.3, reliability coefficients varied very little across the 12 subgroups, with a range of 0.90 to 0.94 for the physical component scale and a range of 0.85 to 0.89 for the mental component scale.

TABLE 7.3 RELIABILITY ESTIMATES FOR PCS AND MCS SCALES: GENERAL U.S. POPULATION SUBGROUPS (N = 2,474)

		N	PCS	MCS
TOTAL SAMPLE			.93	.88
AGE	<65	1757	.92	.87
	65 – 74	442	.93	.85
	75+	264	.91	.89
GENDER	Female	1414	.93	.88
	Male	1055	.93	.87
RACE	White	2077	.93	.88
	Black	223	.94	.85
	Other	174	.90	.88
EDUCATION	<8	215	.93	.88
	9 – 11	277	.93	.86
	12	820	.93	.89
	>12	1162	.92	.87

NOTE: Estimates based on internal consistency method (see text).

Minimum standards of reliability for group comparisons were satisfied for both component scales across all subgroups. Minimum reliability standards recommended for individual respondents were met for the physical component summary scale for all subgroups; and were approached, though not achieved, for the mental component summary.

Statistical Power

Statistical power, the probability that a difference will be found when there is one, is largely determined by features of the sample design, such as the effect size under study and sample size. Statistical power is also determined by measurement reliability, because noisy measures have greater error variance relative to systematic variance, and thus less statistical power. Results of previous studies (McHorney, Ware, Rogers, et al., 1992) have shown that measures that reliably define more levels of health are more precise in detecting differences between groups of patients differing in health. Since the PCS and MCS measures define *many* more levels of health than any of the SF-36 scales and have been shown to be as reliable or more reliable, one can expect them to achieve greater statistical power in detecting differences in physical and mental health, with some exceptions discussed in Chapter 8.

Tables 7.4 through 7.8 present estimates of sample sizes necessary to detect average group differences in the PCS and MCS scores equal to 1, 2, 5, 10, and 20 points. We relied upon formulas published by Cohen (1988) and variance estimates from general U.S. population studies in estimating these sample sizes. We estimated sample sizes for five different study designs beginning with the most powerful design, an experimental comparison between two randomly formed groups with comparisons between repeated assessments over time, to the least powerful, a simple comparison between two group means. A non-directional hypothesis (two-tailed test) with a false rejection rate of 5%, and with a statistical power of 80%, was assumed for all estimates.

In comparing these sample size tables to those presented for the eight scales in the *SF-36 Health Survey Manual and Interpretation Guide* (Ware, Snow, Kosinski, et al., 1993), one should keep in mind that the number of points used to define each difference are not comparable (in standard deviation units) between the components and the eight scales. For example, a 10-point difference represents one standard deviation unit for both the PCS and MCS, but only half to one-third that amount for each of the eight scales. Thus, 10 points, as measured by the PCS or MCS is comparable to approximately 20 points or 30 points as measured by each of the eight scales.

Experimental Studies

Tables 7.4 and 7.5 present sample size estimates for two experimental study designs: two randomly formed groups with repeated assessments and two randomly formed groups with post-intervention assessments only. The repeated measures experimental design and all other repeated measures designs discussed in the following section assume a correlation of 0.70 between administrations. This is a conservative assumption given that correlations for the PCS and MCS scales are 0.89 and 0.80, respectively, in a test-retest study (Brazier, Harper, Jones, et al., 1992). The MOS has observed correlations for the PCS and MCS scales of 0.73 and 0.69, respectively, between repeated administrations six months apart.

Table 7.4 presents sample size estimates for a two-group randomized groups experiment with repeated SF-36 measures. As this table illustrates, it takes many more subjects to detect a one-point difference than to detect a very large difference of 10 points. In comparison to the best SF-36 scales measuring physical (PF) and mental (MH) health, the component summary scales reduce considerably the number of subjects required to detect small and large differences. For example, a sample size of 1,364 subjects in each randomly formed group is required to detect a two-point difference ($p < 0.05$, 80% power) in the SF-36 PF scale, in comparison with a sample size of 201 subjects in each randomly formed group for the PCS scale. However, this comparison is misleading. Two points on the PF scale is much smaller than two points on the PCS, in standard deviation (SD) terms. Taking into account this difference, it would be more appropriate to compare the power of the PCS in detecting a one-point difference (i.e., 0.1 SD unit) relative to the power of the PF scale to detect a two-point difference, which is much closer to a difference of 0.1 SD units. More than 1,300 people per group would be required to detect a two-point difference in the PF scale and the two other best SF-36 physical health scales (RP, BP) (see Table 7.4, Ware, Snow, Kosinski, et al., 1993). A one-point difference (0.1 SD unit) in the PCS is detectable with 801 people per group.

TABLE 7.4 SAMPLE SIZES NEEDED TO DETECT DIFFERENCES BETWEEN POST-INTERVENTION SCORES OF TWO EXPERIMENTAL GROUPS WITH PRE-INTERVENTION SCORES AS COVARIATES

	NUMBER OF POINTS DIFFERENCE				
	1	2	5	10	20
Physical Component Summary	1571	393	64	17	5
Mental Component Summary	1571	393	64	17	5

NOTE: Estimates assume alpha = 0.05, two-tailed test, power = 80% (Cohen, 1988), and an intertemporal correlation of .70.

Table 7.5 presents sample size estimates for comparisons between two experimental groups with post-intervention PCS and MCS measures only. Comparisons between the sample sizes in Table 7.5 and Table 7.4 reveal gains in statistical power from a repeated measures experimental design relative to one with post-intervention measures only. About twice as many subjects are required to detect the same difference in scores with a post-intervention design than with a repeated measures design. For example, 1,571 subjects are required to detect the smallest difference (one point) in the PCS and MCS scores in Table 7.5 compared with only 801 subjects in Table 7.4.

TABLE 7.5 SAMPLE SIZES NEEDED TO DETECT DIFFERENCES BETWEEN TWO EXPERIMENTAL GROUPS, POST-INTERVENTION SCORES ONLY

	NUMBER OF POINTS DIFFERENCE				
	1	2	5	10	20
Physical Component Summary	1571	393	64	17	5
Mental Component Summary	1571	393	64	17	5

NOTE: Estimates assume alpha = 0.05, two-tailed test, power = 80% (Cohen, 1988), and an intertemporal correlation of .70.

Non-Experimental Studies

Tables 7.6 to 7.8 present sample size estimates for three non-experimental comparisons involving SF-36 component summary scales: (1) comparisons between two self-selected groups with administrations before and after intervention(s) (Table 7.6); (2) repeated measures over time for a single group (Table 7.7); and (3) a comparison between a group mean score and a fixed score, such as the general population norm (Table 7.8).

The sample size estimates in Table 7.6 for a non-experimental, two-group study with repeated measures assumes that difference scores will be analyzed to maximize the internal validity of the study design. Comparisons between the sample sizes presented in Table 7.6 and Table 7.4 illustrate the power gained from an experimental versus a non-experimental two-group comparison. The power gained is approximately 29% for the smallest difference (one point) in the PCS and MCS scores.

Table 7.7 presents the sample sizes required to detect differences in the PCS and MCS scores over time within one group. As Table 7.7 illustrates, the sample size required to detect a change in the PCS and MCS scale scores over time within one group is smaller than the sample sizes required in the other study designs. However, the results are more difficult to interpret than the results for study designs that compare scores between two groups receiving different interventions (Cook & Campbell, 1979).

Estimates of sample sizes required to compare average PCS and MCS scores to a fixed norm, such as the general population, are presented in Table 7.8. As illustrated, a difference of five points on the PCS and MCS scales between a sample mean and a norm can be detected with only 32 subjects in the sample, compared with 197 subjects for a difference of two points. These estimates of sample sizes are less than those required for most of the SF-36 scales for similar comparisons (in terms of SD units). For example, to detect a 10-point difference in a PF scale score (i.e., one-half SD unit) between a sample and a norm requires 44 subjects.

Power Advantages Of PCS And MCS

Comparable effect size comparisons between the sample sizes needed to detect differences with the eight SF-36 scales, as compared to the PCS and MCS scales show a consistent advantage of the summaries. This is due to the summaries' generally higher or equal reliabilities in combination with the increased precision from defining many more scale levels. The power advantage of the components relative to the eight scales will be largest for SF-36 scales with the highest standard deviations (e.g., RP, RE) and less for others (e.g., MH, GH, VT). However, these analyses have not addressed another important consideration, namely, validity (see Chapter 8). It is also important to keep in mind that comparisons between SF-36 scales and the PCS and MCS should be made in comparable units, taking into account differences in their standard deviations.

TABLE 7.6 SAMPLE SIZES NEEDED TO DETECT DIFFERENCES BETWEEN TWO SELF-SELECTED GROUPS, REPEATED MEASURES DESIGN

	NUMBER OF POINTS DIFFERENCE				
	1	2	5	10	20
Physical Component Summary	1122	281	46	12	4
Mental Component Summary	1122	281	46	12	4

TABLE 7.7 SAMPLE SIZES NEEDED TO DETECT DIFFERENCES OVER TIME WITHIN ONE GROUP

	NUMBER OF POINTS DIFFERENCE				
	1	2	5	10	20
Physical Component Summary	561	140	23	6	2
Mental Component Summary	561	140	23	6	2

TABLE 7.8 SAMPLE SIZES NEEDED TO DETECT DIFFERENCES BETWEEN A GROUP MEAN AND A FIXED NORM

	NUMBER OF POINTS DIFFERENCE				
	1	2	5	10	20
Physical Component Summary	786	197	32	9	3
Mental Component Summary	786	197	32	9	3

NOTE: Estimates assume alpha = 0.05, two-tailed test, power = 80% (Cohen, 1988), and an intertemporal correlation of .70.

Confidence Intervals For Individual Scores

Confidence intervals provide valuable information about the amount of fluctuation that can be expected in a single score due to measurement error. A confidence interval around an individual score is a function of the standard deviation (SD) of the score distribution and the standard error of measurement (SEM). The size of the SEM is a function of the reliability of that score (Nunnally, 1978).

Two attributes of the SF-36 PCS and MCS scales (relatively small standard deviations and high reliability) lead to reductions in confidence intervals around individual scores of about one-half to one-fifth relative to those for the eight SF-36 scales. With smaller confidence intervals, fluctuations in an individual patient score due to chance are less likely, facilitating their use in monitoring individual patients in clinical practice.

Table 7.9 compares estimates of confidence intervals for the eight SF-36 scales and the PCS and MCS scales for an individual respondent's score. These estimates are based on the reliability and standard deviation (SD) of each scale in the general U.S. population. If either the reliability or SD varies substantially in a sample, confidence intervals should be re-estimated using published formulas (e.g., Nunnally & Bernstein, 1994). It is also important to keep in mind that estimates of confidence intervals are not always symmetrical around the *observed* score.

The confidence intervals in Table 7.9 can be used to take into account fluctuations due to measurement error when interpreting scores for one patient or other respondents. Intervals for three levels of confidence are presented in Table 7.9: 68% (1 SEM), 90% (1.64 SEM), and 95% (2 SEM). Examples of how confidence intervals can be used to interpret individual scores are presented below. Chapter 13 presents norms for changes in scores for patients participating in the MOS. Chapter 12 discusses their use in clinical practice.

According to Table 7.9, individual patient scores on the PCS and MCS would be expected to fall within 2.8 and 3.2 points, respectively, about 68% of the time. To be much more certain about an individual's PCS and MCS score, use the 95% confidence interval, which is 5.7 points for the PCS and 6.3 points for the MCS. Compared to the SF-36 scales, the PCS and MCS scales make it possible to monitor the scores of individual patients with a much higher degree of confidence. As discussed in the preceding section, however, the gains are not as large (in SD units) as they appear to be.

TABLE 7.9 CONFIDENCE INTERVALS FOR INDIVIDUAL PATIENT SCORES

SCALE	LABEL	CONFIDENCE INTERVALS (CI)		
		68%[a]	90%[b]	95%[c]
Physical Functioning	PF	6.2	10.2	12.3
Role Physical	RP	11.3	18.7	22.6
Bodily Pain	BP	7.5	12.4	15.0
General Health	GH	8.8	14.7	17.6
Vitality	VT	7.8	13.0	15.6
Social Functioning	SF	12.8	21.3	25.7
Role Emotional	RE	14.0	23.2	28.0
Mental Health	MH	7.2	12.0	14.0
Physical Component Summary	PCS	2.8	4.6	5.7
Mental Component Summary	MCS	3.2	5.2	6.3

[a] 68% CI equals 1 SEM
[b] 90% CI equals 1.64 SEM
[c] 95% CI equals 2 SEM

NOTE: These estimates are based on reliability estimates and standard deviations for SF-36 scales and the PCS and MCS in the general U.S. population

Individual scores on the PCS and MCS can be compared to general U.S. population norms or to norms for diagnostic groups by using the confidence levels presented in Table 7.9. Suppose that a clinician wanted to know whether a PCS score of 44 for a 40 year old male patient was below that of the general U.S. population. Because the mean and SD are 50 and 10, respectively, for the PCS in the general U.S. population, it is obvious that the patient is well below the norm. Using the norms presented in Table 14.5 for males age 40, it is apparent that a score of 44 is well below the norm of 52.1 for a 40 year old male in the general U.S. population (52.1 - 44 = 8.1). Because 8.1 is greater than the 95% confidence interval of 5.7 (from Table 5.9), the clinician can be confident that the patient's score of 44 is below the norm for men of a similar age in the general U.S. population, more than would be expected due to measurement error. These examples and other considerations are discussed in Chapter 12.

Chapter 8: Empirical Validation

Studies of validity are about the meaning of scores and whether or not they have their intended interpretations. The methods we have used in studies of the SF-36 have followed guidelines recommended for use in validating psychological and educational measures by the American Psychological Association, the American Educational Research Association, and the National Council on Measurement in Education (American Psychological Association, 1985). The same methods were used to study validity and to establish interpretation guidelines for the eight SF-36 scales, as discussed in the original *SF-36 User's Manual* (Ware, Snow, Kosinski, et al., 1993). That manual explains the methods, beginning with comparisons of the content of the SF-36 with other widely used measures. The content analysis, which has been subsequently updated and extended (Ware, 1995) revealed that the SF-36 includes eight of the health concepts most frequently represented in widely used health status measures. The SF-36 differs from most other measures in that it attempts to represent a wider range of levels for most of those concepts.

This chapter is divided into three sections, beginning with a summary of results from the first published tests of the validity of the PCS and MCS scales relative to the eight-scale SF-36 profile. The PCS and MCS scales performed in the 80-100 percent range relative to the best SF-36 scale in empirical tests of validity. These results, which are summarized elsewhere, (Ware, Kosinski, Bayliss, et al., 1995), constitute strong support of the usefulness of the PCS and MCS in summarizing and interpreting results. Detailed results, including nine tables not previously published, are presented here.

The second section extends the empirical evaluation of the validity of the PCS and MCS scales by replicating the original "four-group" tests that demonstrated the validity of the eight SF-36 scales in discriminating among MOS patient groups known to differ in severity of physical and/or mental condition, as defined clinically (McHorney, Ware, & Raczek, 1993). When the same tests were applied to the PCS and MCS, the summary scales never missed a difference captured by an SF-36 scale and performed in the 70-100 percent range in tests of empirical validity, relative to the best SF-36 scale. These tests call attention to the tradeoffs involved with the simplicity of two summary measures relative to the richness of the eight-scale SF-36 profile.

A third section summarizes results of correlational analyses of the validity of the PCS and MCS in relation to 33 health status scales and summary scales developed during the course of the MOS (Stewart & Ware, 1992) and measures of the frequency of 19 specific symptoms in four categories (Ware, Snow, Kosinski, et al., 1993). These results provide information useful in deciding which measures are most likely to add information beyond what can be known from the two summary measures, and demonstrate the sensitivity of the PCS and/or MCS to most specific symptoms.

Initial Validation Studies

The first round of validation studies of the two summary measures focused on their empirical validity relative to that of the profile of eight scales constructed from the SF-36. Although there has been some debate regarding the tradeoffs involved with these two approaches (Ware, 1984; Bergner & Rothman, 1987; Patrick & Erickson, 1993), their implications have not been explored empirically, with two exceptions based on single criteria (Katz, Larson, Phillips, et al., 1992; Beaton, Bombardier, & Hogg-Johnson, 1994). To evaluate these tradeoffs, we used the logic of "known groups" validity (Kerlinger, 1964) and defined "criterion" groups of patients differing in ways that impact on physical or mental health status. To test the generalizability of results, 16 tests were performed involving both cross-sectional and longitudinal study designs. Results are presented elsewhere and are summarized here; nine tables of specific findings not included in the original publication are appended (see Appendix B).

Comparisons between the summary measures and the SF-36 eight-scale profile were designed to approximate as closely as possible their intended use and circumstances that might affect conclusions. Several considerations guided selection of criteria used in defining groups reported here: (a) a strong theoretical foundation for hypotheses, including both direction of differences and whether physical or mental dimensions of health should be most affected; and (b) replication across both cross-sectional and longitudinal designs. Conclusions about different methods should be based on multiple tests. Criteria known to involve physical more than mental health differences, as well as the reverse pattern, were selected. Finally, data from both cross-sectional and longitudinal study designs were analyzed. Although measures that do best in discriminating the effects of differences at a point in time should also be most responsive to the impact of those changes over time, this principal has been questioned (Guyatt, Walter, & Norman, 1987).

Seven categories of comparisons were performed, involving groups of patients differing in: (1) presence of chronic medical conditions (four conditions); (2) severity of hypertension (two levels), diabetes (four levels), and severity of congestive heart failure (two levels); (3) the presence of any one of 16 comorbid conditions (and a count of 10 others); (4) frequency of acute symptoms in four clusters (ear, nose, and throat (ENT), central nervous system (CNS), musculoskeletal, and gastrointestinal and genitourinary (GI/GU)); (5) cross-sectional and longitudinal comparisons of age effects among the most well group available, patients with uncomplicated hypertension; (6) longitudinal comparisons of patients classified at one-year follow-up according to self-reported changes in physical, mental, and general health status (five categories each); and (7) cross-sectional comparison of patients with and without clinical depression and longitudinal comparison of patients recovering from depression. All tests were based on clinical data used in previous MOS reports (Wells, Hays, Burnam, et al., 1989; Wells, Stewart, Hays, et al., 1989; Kravitz, Greenfield, Rogers, et al., 1992; McHorney, Ware, & Raczek, 1993). Specific clinical definitions are documented in Appendix A.

Hypotheses

A strong theoretical foundation for generating hypotheses makes it easier to draw conclusions about measurement validity (Kerlinger, 1964). The result that would be expected for a valid measure must be known in advance for each test. We hypothesized that: (a) patients with more severe conditions, such as congestive heart failure, would score worse, particularly in physical health, than those with uncomplicated hypertension; (b) patients at more advanced levels of disease severity or with comorbid conditions would score lower; (c) scores would be lower for patients reporting a greater frequency of acute symptoms; (d) self-reported changes in physical, mental, and general health at one-year follow-up would be most related to changes estimated from repeated measures of the same concepts; (e) physical health (but not mental health) would decline with age; and (f) mental health would be better for patients without clinical depression and improve with clinical recovery from major depression.

Analysis Plan

Analyses of groups in the first three categories used ordinary least squares multiple regression techniques (SAS Institute, 1989) with statistical adjustment for differences in age, gender, race, poverty, study site, health care setting, and season of the year, to maintain comparability with previous MOS analyses (Wells, Burnam, Rogers, et al., 1992; Rogers, Wells, Meredith, et al., 1993). Longitudinal analyses and other cross-sectional analyses used the same statistical methods but without adjustments for baseline patient characteristics. All analyses of SF-36 profiles used multivariate analysis of variance (MANOVA) (Stevens, 1992), which provides an overall test of whether average scores (cross-sectional) or average changes in scores (longitudinal) for any of the eight SF-36 scales differed across any of the groups being compared. For those tests that yielded a significant MANOVA F-ratio, regression models were estimated to test the relative validity (RV) of each scale. Thus, according to this convention, only those scales that met two statistical criteria were considered valid: (a) significant overall MANOVA F for the set of criterion variables (defining patient groups) in relation to the profile of eight scales; and (b) significant univariate F for the same set of criterion variables and the scale in question.

To estimate the RV for each scale in relation to the best of the eight SF-36 scales, ratios of F-statistics were compared as in previous MOS studies (McHorney, Ware, & Raczek, 1993). The F-statistic for each measure is a ratio of the variance in scores due to differences among independent groups or between repeated assessments for the same group, relative to the within group (error) variance. The F-statistic for a given measure in a given test would be larger when the measure yields a larger average separation in mean scores being compared and/or a smaller error variance. The RV estimate for each measure in each test indicates in proportional terms the empirical validity of the scale in question relative to the most valid scale in that test (Liang, Larsen, Cullen, et al., 1985; Winer, Brown, & Michels, 1991; McHorney, Ware, & Raczek, 1993).

When one measure performs exceptionally well, estimates (based on RV) of the usefulness of other measures sometimes appear relatively low to the point of being misleading (Ware, Snow, Kosinski, et al., 1993). Therefore, standardized estimates of effect size (ES) (Kazis, Anderson, &

Summary Of Results: First Round

Meenan, 1989) were also computed by dividing the average difference for each measure by the standard deviation (SD) for that measure, using SD estimates from the general U.S. population, as published elsewhere (Ware, Snow, Kosinski, et al., 1993).

Table 8.1 summarizes RV coefficients for the two summary measures and the eight-scale SF-36 profile across all 16 tests, as published elsewhere (Ware, Kosinski, Bayliss, et al., 1995). The first 12 columns include criterion variables defining differences in physical health, and the last three columns are tests for differences in mental health. (An exception is the ninth column (GI/GU symptom cluster) shown previously to impact most on mental health (Stewart, Greenfield, Hays, et al., 1989).) For each test, the "best" of the eight SF-36 scales (with the highest F-ratio) is labeled RV = 1.00 and is boldfaced. Horizontal lines in Table 8.1 distinguish SF-36 scales hypothesized to be most valid in measuring physical health (PF, RP, and BP) versus mental health (MH, RE, and SF). The two more general scales (GH and VT) are in the middle grouping. F-ratios for the eight-scale MANOVA

Tests are presented for each criterion, followed by results for the two summary measures (PCS and MCS). Blanks in Table 8.1 indicate a nonsignificant MANOVA F-statistic for the eight scales in that test or a nonsignificant univariate F-statistic for a particular scale in that test. RV coefficients were not estimated for measures with nonsignificant F-statistics.

The two summary measures did well in these tests relative to the profile of eight SF-36 scales. In comparisons that involved physical health "criteria" and yielded significant differences for any of the three physical scales (PF, RP, and BP), statistical conclusions based on the PCS agreed nine of nine times (RV coefficients for the PCS ranged from 0.20 to 0.89, median = 0.79).

Differences in either or both of the two "general" scales (GH and VT) were significant in 13 of the 16 tests, including five with RV = 1.00. In nearly all instances, the PCS and/or MCS also performed well. The PCS captured 12 of 13 of the significant differences captured by a general scale (range of RV coefficients 0.01 to 0.89, median = .72); the MCS detected 11 of 13 (RV = 0.03-1.47, median = 0.35).

The three best mental health scales (MH, RE, and SF) yielded one or more significant results in 13 of the 16 tests (RV coefficients ranged from 0.03 - 1.00, median = 0.40). For 12 of 13, the MCS also produced statistically significant results (RV = .03 - 1.47, median = .43). Some coefficients were low because both physical and mental tests were included. For the four mental health tests (three right-hand columns in Table 8.1 and GI/GU symptoms), RV was 0.98 - 1.00 for the MH scale. For the MCS in these four tests, RV's of 1.02, 1.03, 1.47, and 0.93, respectively, were observed.

Table 8.1 also documents instances in which statistical conclusions varied across analyses of profiles and summary measures. Effects of GI/GU symptoms detected by the PF and BP scales were missed by the PCS; they were detected by the MCS. PCS scores differed significantly across levels of severity of diabetes, but were missed by the eight-scale profile. Differences in the SF-36 profile were significant across levels of hypertension severity, but the univariate F-statistics were not significant for any of the eight scales.

Results from tests involving the severity of acute symptoms suggest that the validity of the two summary measures varies substantially across symptom clusters. For example, the impact of ENT symptoms was best detected by the PCS. The impact of the CNS symptoms was detected by both summary measures, with the MCS performing better than the PCS. Musculoskeletal symptoms were reflected in scores for both summary measures, with the PCS clearly more affected than the MCS.

TABLE 8.1 SUMMARY OF RELATIVE VALIDITY COEFFICIENTS FOR SF-36 PROFILES AND SUMMARY MEASURES, SIXTEEN COMPARISONS USED TO TEST VALIDITY

MEASURES	SEVERITY OF DISEASE					SYMPTOM CLUSTERS					AGE DIFFERENCES		SELF-REPORTED CHANGE			CLINICAL DEPRESSION	
	CHRONIC CONDITIONS	HYPER-TENSION	DIABETES	CHF	COMORBID CONDITIONS	EAR, NOSE & THROAT	CENTRAL NERVOUS SYSTEM	MUSCULO SKELETAL	GI/GU[a]		CROSS-SECTIONAL	LONGI-TUDINAL	PHYSICAL	GENERAL	MENTAL	CROSS-SECTIONAL	LONGI-TUDINAL
PF	.87***			1.00***	.57***	1.00***	.75***	.28***	.43*		1.00***	.62***	1.00***	.79***	.19***	.00	
RP	.43***			.82***	.44***	.32*	.36***	.27***			.42***		.39***	.40***	.11***	.05***	.23***
BP	.12*			.25*	1.00***		.23***	1.00***	.39*		.14**		.20***	.21***	.09***	.06***	
GH	1.00***			.82***	.87***	.42**	.38***	.11***			.08*	1.00***	.74***	1.00***	.39***	.07***	.16***
VT	.46***			.64***	.53***		1.00***	.13***				.41***	.30***	.49***	.34***	.24***	1.00***
SF	.21***			.27*	.25***		.44***	.08***	.82***				.61***	.65***	.53***	.58***	.52***
RE	.12*			.34**	.13*		.38***	.03***	.36*				.11***	.11***	.46***	.41***	.84***
MH	.12*			.25*	.30***	.91***	.84***		1.00***		.20***		.26***	.46***	1.00***	1.00***	.98***
MANOVA F 8 SCALES	6.61***	2.17*	NS[b]	3.42***	5.14***	3.57***	9.96***	3.07***	28.14***		7.65***	2.60**	9.35***	12.91***	10.97***	165.85***	4.89***
PCS	.71***		1.00*	.83***	.89***	.86***	.61***	.55***			.82***	.20**	.79***	.74***	.05**	.01*	
MCS				.24*	.13*	.51***	.86***	.03***	.93***		.35***		.18***	.29***	1.02***	1.03***	1.47***

Summary Of Results: Second Round

Tables 8.2 - 8.4 present unpublished analyses of tests of the validity of the PCS and MCS in discriminating among four mutually exclusive groups of MOS patients known to differ in severity of medical (physical) and psychiatric conditions. The top panels of Tables 8.2 - 8.4 repeat results from analyses of the eight-scale SF-36 profile in the same tests as reported elsewhere (McHorney, Ware, & Raczek, 1993), and the bottom panel extends the comparisons to include the PCS and MCS scales. Tables 8.3 and 8.4 add columns to standardize effect size (ES) estimates in addition to mean differences, F-ratios, and RV estimates. Data in the top panel differ in some instances from the previous publication due to slight improvements in the definitions of clinical status.

TABLE 8.2 COMPARISON OF MEANS FOR SF-36 SCALES AND SUMMARY MEASURES, MOS PATIENTS DIFFERING IN MEDICAL AND PSYCHIATRIC CONDITIONS (N = 1,150)

SCALE	COMPARISON GROUPS			
	MINOR MEDICAL (N = 697)	SERIOUS MEDICAL (N = 162)	PSYCHIATRIC ONLY (N = 242)	PSYCHIATRIC & SERIOUS MEDICAL (N = 49)
Physical Functioning	80.14 (0.8)	57.07 (2.1)	79.90 (1.5)	45.28 (4.1)
Role Physical	69.55 (1.4)	40.28 (3.1)	56.20 (2.5)	22.95 (4.5)
Bodily Pain	76.01 (0.8)	65.10 (2.0)	66.02 (1.5)	48.71 (3.5)
General Health	67.26 (0.7)	47.62 (1.7)	57.65 (1.4)	40.61 (2.5)
Vitality	62.19 (0.7)	47.45 (1.7)	44.28 (1.4)	37.41 (3.1)
Social Functioning	91.62 (0.6)	78.63 (2.0)	67.04 (1.7)	60.71 (3.7)
Role Emotional	84.12 (1.2)	72.63 (3.1)	45.04 (2.6)	44.22 (5.7)
Mental Health	82.06 (0.5)	77.18 (1.3)	54.01 (1.3)	54.67 (3.1)
Physical Component Summary	46.37 (0.4)	36.27 (0.9)	47.95 (0.7)	33.54 (1.4)
Mental Component Summary	54.29 (0.3)	52.23 (0.8)	37.62 (0.8)	41.69 (1.6)

NOTE: Table entries are means (standard errors), comparison groups are defined as in previous MOS validity studies (McHorney, Ware, & Raczek, 1993)

The results in Tables 8.3 and 8.4 largely replicate results and lead to conclusions similar to those from the first round of validity studies of the PCS and MCS. Comparisons between groups known (by clinical diagnosis) to differ from the minor medical group in terms of serious medical (physical) morbidity (the first set of columns in Table 8.3 and the second set of columns in Table 6.4) involve the most pure *physical* differences. In these tests, the PCS yielded very large differences (RV = 0.94 and 0.82) relative to the best SF-36 scale (GH and PF, respectively). In both of these tests, the MCS yielded significant, but small, group differences as well as small RV and ES estimates, as hypothesized (RV = 0.06, ES = 0.21 and RV = 0.05, ES = 0.41).

In the two validity tests involving relatively pure mental health comparisons (second set of columns in Table 8.3 and first set of columns in Table 8.4) large differences in group means, RV and ES estimates were observed for the MCS (RV = 1.06, ES = 1.67 and RV = 0.69 and ES = 1.05, respectively), relative to the best SF-36 scale (MH in both tests). In both tests for differences in mental health criteria, PCS yielded very small differences and RV and ES estimates in the first set (0.01 and 0.16, respectively) and insignificant differences in the second.

When physical and mental differences are confounded (third sets of columns in both Table 8.3 and 8.4), both the PCS and MCS revealed significant differences with large ES estimates (1.17 to 1.46 SD units), and with RV estimates ranging from 0.48 to 1.07, relative to the best SF-36 scale (SF and MH, respectively).

TABLE 8.3 SUMMARY OF CLINICAL VALIDITY TESTS INVOLVING MINOR MEDICAL PATIENTS

	SERIOUS MEDICAL vs. MINOR MEDICAL				PSYCHIATRIC vs. MINOR MEDICAL				BOTH SERIOUS MEDICAL & PSYCHIATRIC vs. MINOR MEDICAL			
	MEAN DIFFERENCE	ES[1]	F	RV[2]	MEAN DIFFERENCE	ES	F	RV	MEAN DIFFERENCE	ES	F	RV
PF	-23.06	0.99	130.19[a]	0.91	-0.23	0.01	0.02	0.00	-34.86	1.50	109.41[a]	0.64
RP	-29.27	0.86	83.36[a]	0.58	-13.35	0.39	23.72	0.04	-46.59	1.37	77.97[a]	0.46
BP	-10.91	1.29	10.90[a]	0.08	-9.99	0.42	35.40	0.07	-27.30	1.15	70.06[a]	0.41
GH	-19.65	0.97	144.00[a]	1.00	-9.60	0.47	44.62	0.08	-26.65	1.31	100.60[a]	0.59
VT	-14.75	0.70	69.39[a]	0.48	-17.91	0.85	139.24	0.26	-24.78	1.18	69.89[a]	0.41
SF	-13.00	0.59	71.91[a]	0.50	-24.58	1.08	315.04	0.59	-30.91	1.36	169.52[a]	1.00
RE	-11.49	0.35	16.48[a]	0.11	-39.08	1.18	243.67	0.46	-39.90	1.21	74.65[a]	0.44
MH	-4.88	0.27	14.36[a]	0.10	-28.05	1.55	530.38	1.00	-27.39	1.52	153.02[a]	0.90
PCS	-10.10	1.01	135.26[a]	0.94	1.58	0.16	4.45[c]	0.01	-12.83	1.28	80.82[a]	0.48
MCS	-2.06	0.21	8.24[b]	0.06	-16.67	1.67	561.69[c]	1.06	-12.60	1.26	110.46[a]	0.65

[a] $p < .001$ [b] $p < .01$ [c] $p < .05$
[1] Effect Size (ES) = means difference/SD, where SD comes from the general U.S. population
[2] RV = relative validity (see text)

TABLE 8.4 SUMMARY OF CLINICAL VALIDITY TESTS INVOLVING CHRONICALLY ILL PATIENTS

	PSYCHIATRIC AND SERIOUS MEDICAL vs. SERIOUS MEDICAL ONLY				PSYCHIATRIC AND SERIOUS MEDICAL vs. PSYCHIATRIC ONLY				PSYCHIATRIC vs. SERIOUS MEDICAL			
	MEAN DIFFERENCE	ES[1]	F	RV[2]	MEAN DIFFERENCE	ES	F	RV	MEAN DIFFERENCE	ES	F	RV
PF	-11.80	0.51	6.76[b]	0.11	-34.62	1.49	85.75[a]	1.00	22.10	0.95	82.62[a]	0.59
RP	-17.32	0.51	7.78[b]	0.13	-33.23	0.98	31.92[a]	0.37	15.92	0.47	16.00[a]	0.11
BP	-16.39	0.69	15.76[a]	0.26	-17.31	0.73	20.70[a]	0.24	0.92	0.04	0.14	0.00
GH	-7.01	0.34	4.28[c]	0.07	-17.05	0.84	25.00[a]	0.29	10.05	0.49	19.80[a]	0.14
VT	-10.04	0.48	8.12[b]	0.13	-6.87	0.33	4.20[c]	0.05	-3.17	0.15	2.10	0.01
SF	-17.91	0.79	18.23[a]	0.30	-6.33	0.28	2.40	0.03	-11.58	0.51	19.45[a]	0.14
RE	-28.41	0.86	19.36[a]	0.32	-0.82	0.02	0.02	0.00	-27.59	0.83	45.43[a]	0.32
MH	-22.51	1.25	60.37[a]	1.00	0.65	0.04	0.04	0.00	-23.17	1.28	140.66[a]	1.00
PCS	-2.73	0.27	2.31	0.04	-14.41	1.44	70.06[a]	0.82	11.68	1.17	105.06[a]	0.75
MCS	-10.54	1.05	41.47[a]	0.69	4.08	0.41	4.24[c]	0.05	-14.61	1.46	150.55[a]	1.07

[a] $p < .001$ [b] $p < .01$ [c] $p < .05$
[1] Effect Size (ES) = means difference/SD, where SD comes from the general U.S. population
[2] RV = relative validity (see text)

Advantages And Disadvantages Of PCS And MCS

The two summary measures reduced the number of statistical analyses to 32 from 128 (8 scales times 16 tests), illustrating their advantage in reducing the number of statistical comparisons. The summary measures proved to be very useful in most of these tests. In both cross-sectional and longitudinal tests, the PCS rarely missed a difference it was expected to capture, and it was the only measure to detect the impact of differences in the severity of diabetes.

The PCS consistently performed below the best physical health scale in the physical health tests, although usually with an RV · 0.80, relative to the best of the eight scales. The simplicity of a single measure of physical health appears to go hand-in-hand with an empirical validity that is about 80% of that achieved by the best SF-36 scale. The MCS consistently performed as well as or better than the best scale in mental health tests. Thus, these analyses revealed little or no tradeoff involved in relying on the MCS in testing hypotheses about mental health.

The PCS and MCS were also expected to have an advantage in interpretation as physical versus mental health measures, respectively. Because most of the eight scales are substantially intercorrelated and most have complicated physical and mental factor content, most have more than one interpretation. Interpretation has been shown to be least complicated for the PF and MH scales and most complicated for the VT scale (McHorney, Ware, & Raczek, 1993; Ware, Snow, Kosinski, et al., 1993; see also Chapter 3). A difference or change in VT scores could reflect changes in physical or mental health or both. The problem this presents in interpretation was illustrated in the first round of validity studies (see Appendix B, Table B.1). Those studies revealed a large difference in average VT scores favoring patients with uncomplicated hypertension over those with CHF ($\Delta = 14.1$, $t = 5.97$, $p < 0.001$), a comparison of groups known to differ in physical health. The average change in VT scores for patients who recovered from clinical depression was also large ($\Delta = 14.4$, $t = 8.6$, $p < 0.001$). Are these differences in physical or mental health? Because the pattern of differences was the same for VT in both tests, it is not clear whether they have the same interpretation. However, the PCS and MCS produced different results suggesting that changes in physical and mental health were involved, respectively. The impact of CHF was significant for the PCS ($\Delta = 8.9$, $t = 6.96$, $p < 0.001$), but not for the MCS ($\Delta = 1.6$, $t = 1.55$, $p > 0.05$), suggesting a *physical* health difference. The MCS improved ($\Delta = 9.8$, $t = 9.42$, $p < 0.001$) with recovery from depression and the PCS did not ($\Delta = 1.4$, $t = <1$, $p > 0.05$), suggesting a *mental* health difference. These results illustrate an advantage of the PCS and MCS in interpreting health outcomes.

Correlations With Specific Symptoms

Table 8.5 presents correlations between the PCS and MCS and self-reported frequency of symptoms in the MOS. Symptoms were reported for the prior four week period. Symptoms are grouped into four categories: (1) those correlating highly (0.30 or better) with the PCS only; (2) those that correlate 0.30 or better with the MCS scale; (3) those correlating 0.30 or better with both the PCS and MCS scales; and (4) symptoms correlating less than 0.30 with both the PCS and MCS. All table entries are product-moment correlations.

A number of overall patterns of results are apparent in Table 8.5. The symptoms most strongly associated with the PCS scores include shortness of breath, stiffness and pain in muscles, backaches or lower back pain, and chest pain. Correlations were highest with the MCS for headaches more than usual, waking up early/unable to sleep, and being dizzy when standing up. Other symptoms correlating highest with the MCS (lightheaded while on feet and feeling drowsy or sedated) also correlated equally highly with the PCS. It is interesting that the same four groupings of symptoms was formed when these criteria were applied to the PF and MH scales in analyses of these symptoms reported in the original *SF-36 User's Manual* (Ware, Snow, Kosinski, et al., 1993; Table 6.16).

Other correlations in the last category are noteworthy, including significant associations between the PCS and sudden weakness, urinating more than usual, acid indigestion after meals, and coughing that produced sputum. Sudden weakness and acid indigestion also correlated highest with the MCS.

The results presented in Table 8.5 are useful in speculating about the symptoms most likely to be underlying differences in the PCS and MCS scores. These results suggest that the PCS and/or MCS are sensitive to differences in self-reports of the frequency of a wide range of different symptoms, including all 19 symptoms shown in Table 8.5.

TABLE 8.5 CORRELATIONS BETWEEN THE PCS AND MCS AND SPECIFIC SYMPTOMS IN FOUR CATEGORIES

SPECIFIC SYMPTOMS	MEAN[a]	SD	COMPONENT SUMMARIES	
			PCS[b]	MCS[b]
Shortness of breath (climbing stairs)	1.89	1.2	-.55	-.19
Stiffness, pain in muscles	2.89	1.4	-.53	-.12
Backaches or lower back pain	2.41	1.4	-.42	-.16
Chest pains brought on by activities	1.48	0.9	-.40	-.18
Pins and needles in your feet	1.75	1.2	-.38	-.14
Dry mouth	2.16	1.3	-.37	-.26
Heart pounding or palpitations	1.58	1.0	-.32	-.27
Blurred vision	1.53	1.0	-.30	-.24
Headaches more than usual	1.69	1.1	-.23	-.38
Waking up early, unable to sleep	2.34	1.3	-.29	-.35
Dizzy when standing up	1.61	0.9	-.27	-.32
Lightheaded while on feet	1.63	0.9	-.38	-.32
Drowsy or sedated	1.87	1.1	-.33	-.38
Urinating more than usual	1.78	1.2	-.28	-.15
Sudden weakness relieved by eating	1.39	0.8	-.28	-.26
Acid indigestion after meals	2.20	1.2	-.25	-.25
Coughing that produced sputum	1.82	1.2	-.22	-.15
Trouble passing urine	1.22	0.7	-.15	-.14
Fainting or passing out	1.03	0.2	-.09	-.09

[a] 1 = never, 2 = once or twice, 3 = a few times, 4 = fairly often, 5 = very often
[b] Note: short form versions of PF, BP, and SF scales were used to construct the PCS and MCS

Correlations With Other MOS Scales

Correlations between the PCS and MCS scales and other measures of known validity can be useful in evaluating their validity. Table 8.6 summarizes previously unpublished associations for 33 measures studied in the MOS. These measures are grouped into 10 different categories. The labels used to identify them, as well as their construction and scoring, are documented elsewhere (Stewart & Ware, 1992, specifically in Chapter 20 [Tables 20-2 and 20-3, pp. 350-360]). It is important to keep in mind when interpreting this table that in many instances one or more SF-36 items are also included in the MOS measure being correlated with the PCS or MCS. The resulting correlations, which are inflated, are labeled in the table.

The correlations in Table 8.6 indicated how well the PCS and MCS scales reproduce longer-form measures (e.g., MHI 32) and how well they reflect measures not directly represented in the SF-36 (e.g., sexual functioning). The scoring of the MOS scales is indicated in parentheses after each variable name. As would be expected, correlations between unfavorably scored measures and the PCS and MCS, which are scored positively, are negative.

The entries in Table 8.6 can be very useful in judging the value of adding measures of other concepts to supplement the SF-36. For example, based on substantial correlations (0.53 to 0.63), it is clear that the PCS well reflects overall satisfaction with physical ability and mobility. Both the PCS and MCS correlate substantially with the summary of eight physical and psycho-physiologic symptoms measured in the MOS (-0.55 and -0.41, respectively).

Among the measures not included in the SF-36, the MOS cognitive functioning scale had a very high correlation with the MCS ($r = 0.70$). Also, variations in sleep, as measured by the MOS sleep problems index, correlate substantially with the MCS ($r = -0.57$). In contrast, correlations between the SF-36 scales and sexual functioning (problems) tended to be low. Given that the same pattern was observed for the eight SF-36 scales (Ware, Snow, Kosinski, et al., 1993), these results suggest that variations in sexual functioning are not well represented in the SF-36 scales. Thus, sexual functioning is a candidate for inclusion in a generic health battery designed to supplement the SF-36. Accordingly, two large NIH-sponsored clinical trials using the SF-36 to monitor outcomes of treatment for women at high risk of breast cancer and men with prostate disease supplement the SF-36 with the MOS Sexual Problems Scale (see Stewart & Ware, 1992).

TABLE 8.6 CORRELATIONS BETWEEN THE PCS AND MCS AND MOS FUNCTIONING AND WELL-BEING MEASURES

MEASURE		k^1	PCS[b]	MCS[b]
Physical Functioning	Satisfaction w/ Physical Ability (+)[2]	1	0.63	0.34
	Mobility (+)	2	0.53	0.23
Role Functioning	Role Limitations due to Physical Health (-)	7	-0.77	-0.34
	Role Limitations due to Emotional Problems (-)	3	-0.15	-0.81
	Unable to do Work due to Health (-)	1	-0.23	-0.06
	Unable to do Housework due to Health (-)	1	-0.23	-0.22
Social, Family, Sexual Functioning	Social Activity Limitations due to Health (+)	4	0.41	0.67
	Sexual Problems (-)	5	-0.13	-0.28
	Satisfaction with Family Life (+)	3	0.02	0.48
	Overall Happiness w/ Family Life	1	0.02	0.52
	Marital Functioning (+)	6	0.03	0.43
Psychological Distress/Well-Being	Anxiety (-)	6	-0.11	-0.78
	Depression/Behavioral-Emotional Control (-)	13	-0.07	-0.88
	Positive Affect (+)	7	0.12	0.83
	Feelings of Belonging (+)	3	0.02	0.38
	Psychological Well-Being (+)	10	0.09	0.81
	Mental Health Index (+)	32	0.09	0.90
	Mental Health Index II (+)	17	0.08	0.90
Cognitive Functioning	Cognitive Functioning (+)	6	0.18	0.70
Health Perceptions	Current Health (+)	7	0.65	0.45
	Prior Health (+)	3	0.36	0.14
	Health Outlook (+)	6	0.41	0.22
	Health Concern (-)	4	-0.22	-0.12
	Resistance to Illness (+)	4	0.28	0.32
	General Health Rating Index (+)	19	0.60	0.42
	Health Distress (-)	6	-0.45	-0.57
Sleep	Sleep Problems Index (-)	9	-0.34	-0.57
Pain	Effects of Pain (-)	6	-0.67	-0.38
	Pain Severity (-)	5	-0.61	-0.24
	Days Pain Interfered (-)	1	-0.56	-0.26
	Overall Pain Index (-)	12	-0.70	-0.34
Physical/Psycho-Physiologic Symptoms	Physical/Psycho-Physiologic Symptoms (-)	8	-0.55	-0.41
Quality of Life	Life Satisfaction (+)	1	0.11	0.68

[1] K = number of items

[2] (+) scale scores from low to high reflect positive status, (-) scale scores from low to high reflect negative status

[a] Correlations between scales and summary measures are not significant (p< .05)

[b] Correlation is inflated because measure includes one or more SF-36 items

Chapter 9: Interpretation: Content And Criterion-Based

Background

Traditional analyses of validity include empirical tests of "whether" and "how" valid a measure is, and results are most often expressed in terms of correlation coefficients. In contrast to such correlational analyses, analyses presented here were designed to yield interpretation guidelines for differences in PCS and MCS scores of specific amounts. Two kinds of results are presented. The first includes plots of responses to specific SF-36 items to establish content-based interpretation guidelines for differences throughout the range of PCS and MCS scores. The second includes results based on analyses of external "criteria," such as comparisons between groups differing in chronic conditions, and analyses using PCS and MCS scores to predict job loss due to health problems, utilization of health care services, and five-year mortality rates. Analyses of chronic conditions yield interpretation guidelines based on the hypothesized impact of specific conditions on PCS and MCS scores. Plots of results from predictive tests of validity yield interpretation guidelines based on the consequences of PCS and MCS scores and particularly the social relevance of differences in scores.

Content-Based Interpretation

Content-based interpretation guidelines are based on analyses of the content of SF-36 items as a way of understanding the meaning and interpretation of differences in PCS and MCS scores in between the extremes. This is accomplished by plotting specific item responses across levels of the PCS and MCS as they were plotted for the eight SF-36 scales (see Ware, Snow, Kosinski, et al., 1993 for a further discussion). For example, it is useful to know that about 90% of those within or below a PCS score of 40-44 in the general U.S. population report health-related limitations in their performance of vigorous physical activities.

Meaning Of High And Low Scores

Table 9.1 presents content-based descriptions of the health states associated with very high and very low scores on the PCS and MCS scales. These descriptions, which are based on the known contributions of the eight SF-36 scales to the definition of those health states (from Chapters 3 and 4), can be used in summarizing what the PCS and MCS measure.

Very high or very low scores for the PCS and MCS reflect a combination of physical and mental *function* and *well-being*, the extent of social and role *disability*, and *personal evaluation* of health status. The lowest state of health reflects substantial functional limitation, severe social and role disability, distress, *and* very unfavorable evaluations of health status and outlook. Very high scores are earned only in the absence of limitations, in the absence of disability in social or usual role activities, and with high levels of well-being and very favorable personal health evaluations.

TABLE 9.1 DESCRIPTION OF VERY HIGH AND VERY LOW PCS AND MCS HEALTH STATUS LEVELS

SCALE	VERY LOW	VERY HIGH
PCS	Substantial limitations in self care, physical, social, and role activities; sever bodily pain; frequent tiredness; health rated "poor"	No physical limitations, disabilities, or decrements in well-being; high energy level; health rated "excellent"
MCS	Frequent psychological distress, substantial social and role disability due to emotional problems; health in general rated "poor"	Frequent positive affect; absence of psychological distress and limitations in usual social/role activities due to emotional problems; health rated "excellent"

Although operational definitions are similar for some of the physical and mental health items, they are distinct both conceptually and empirically. The PCS reflects *physical morbidity* and *etiology*, whereas the MCS reflects *psychological or mental morbidity* and *etiology*. It is important to note that a very high PCS score requires more than freedom from physical limitations and social and role disability; it requires an evaluation of current health as "excellent." Likewise, the most favorable personal evaluation of health as "excellent" is not enough for a very high score; PCS scores are lower with limitations or disabilities in the physical spectrum, reflecting the consequences of such limitations and disabilities in physical health. The same logic is reflected in the scoring of the MCS. Both the PCS and MCS place high weights on both the personal and the social implications of different health states. For these reasons, the PCS and MCS are unique in their comprehensiveness as summary health measures.

How To Use These Tables

To facilitate the interpretation of tables of results presented in Chapter 9, the same format is used for all tables. This format is explained in detail for Table 9.2, which presents 10 content-based interpretation guidelines for scores at eight levels of the PCS scale. The range of scores defining each of the eight levels, the mean PCS score for each level, and the sample size are presented in the left-most columns. The eight levels represent five-point intervals throughout the range of PCS scores observed to date in the general U.S. population. Scores at the highest and lowest levels have been collapsed to maintain sample sizes above 100.

Content-based interpretation guidelines were prepared in several steps. First, items with good face validity, and representing scales most highly correlated with the PCS, such as the 10 items in Table 9.2, were selected from the SF-36. Second, responses to each item were dichotomized in a way that is meaningful and that reveals differences across levels of the scale in the score ranges of interest. Third, the percentage of responses to each dichotomous item at each PCS level being interpreted was estimated and plotted.

All table entries are percentages. The first pair of columns headed "%" presents the percentage of the general U.S. population (N = 2,474) at each level of PCS scale scores who reported any limitations in "vigorous activities." The second column in the first pair of columns, headed "Δ/Δ," presents "difference ratios," defined as the percentage point change in each "criterion" item associated with a one-point change in PCS scores, from one particular level to an adjacent level. For example, from level one to level two (an average change of about five on the PCS scale), limitations in vigorous activities increased by 8 percentage points, or about 1.6 percentage points for each PCS point (8.2/5.1 = 1.6). Comparison of difference ratios across levels of PCS illustrate that the vigorous activity item is most useful for interpreting differences in the PCS scores at the top levels, whereas "walking one block" is most useful at the lower levels.

When one or more scores being compared across levels are close to the extremes of the range defining a particular level, or where difference ratios vary inconsistently, it is necessary to interpolate, using the Δ/Δ column entries across adjacent rows. For example, a 3-point change in PCS in the 45-54 range is associated with about 13.8% (3 x 4.6) change in the probability of a limitation in vigorous activities.

As hypothesized (Ware, Snow, Kosinski, et al., 1993), these scale scores have monotonic relationships with the variables used to establish validity and interpretation guidelines. Accordingly, the Δ/Δ columns nearly always reflect a smooth monotonic trend showing increasing and decreasing unit changes without inconsistent reversals. Most of the exceptions are apparent in Table 9.2. We have chosen not to collapse adjacent levels to "smooth out" these trends because for nearly all other interpretation guidelines substantial information would be lost. Complete documentation makes it possible to choose the most appropriate way to interpolate in handling such situations. Numerous examples are explained in Chapter 9.

SF-36™ Summary Measures Manual

Interpretation: Content and Criterion-Based, Pg. 60

TABLE 9.2 PERCENTAGE OF ADULTS ENDORSING SELECTED SF-36 ITEMS AT EIGHT LEVELS OF PCS SCORES: GENERAL U.S. POPULATION

| | PCS SCORES | | | PHYSICAL LIMITATIONS | | | | | | ROLE DISABILITY | | | | PAIN | | VITALITY | | | | GENERAL HEALTH | | | |
| | | | | 1 Vigorous Activities | | 2 Walking 1 Block | | 3 Climbing Stairs | | 4 Difficulty at Work | | 5 Cut Down Worktime | | 6 Very Severe | | 7 Have Lot of Energy | | 8 Feeling Tired | | 9 Excellent Health | | 10 Fair/Poor Health | |
LEVELS	RANGE	MEAN	(N)	%	Δ/Δ	%	Δ/Δ	%	Δ/Δ	%	Δ/Δ	%	Δ/Δ	%	Δ/Δ	%	Δ/Δ	%	Δ/Δ	%	Δ/Δ	%	Δ/Δ
1	60 - 73	61.9	157	20.0	1.6	0.6	-0.1	0.6	0.0	0.0	0.2	0.0	0.1	0.0	0.0	40.1	-4.1	10.2	0.9	47.4	2.6	1.9	0.0
2	55 - 59	56.8	691	28.2	7.5	0.3	0.1	0.6	0.6	1.2	0.7	0.4	0.8	0.0	0.2	61.2	2.7	5.5	0.3	34.2	4.6	2.0	0.7
3	50 - 54	52.3	565	61.9	4.6	0.9	1.5	3.2	1.9	4.4	2.4	4.1	1.1	1.0	0.2	48.9	3.0	6.7	0.4	13.4	1.8	5.3	1.5
4	45 - 49	47.3	304	85.1	1.2	8.2	1.8	12.9	2.9	16.6	6.0	9.5	3.0	2.3	0.3	33.9	1.7	8.6	1.1	4.3	0.0	12.8	1.9
5	40 - 44	42.2	194	91.0	-0.2	17.3	4.1	27.7	3.0	47.2	4.0	24.9	4.9	4.1	1.6	25.3	1.3	14.5	2.3	4.2	-0.1	22.3	3.0
6	35 - 39	37.1	161	89.9	1.1	38.4	1.2	43.2	4.7	67.5	4.2	49.7	3.1	12.4	1.8	18.7	1.8	26.1	2.3	3.7	0.7	37.5	4.6
7	30 - 34	32.1	131	95.3	0.3	44.5	4.2	66.9	2.8	88.5	0.7	65.1	1.9	21.4	3.1	9.9	0.5	37.4	0.9	0.0	-0.1	60.5	1.8
8	8 - 29	23.3	190	97.9		81.6		92.0		94.7		81.9		48.4		5.3		45.0		0.5		76.7	

NOTE: All table entries are percentages
KEY TO COLUMN DEFINITIONS

1 % any limitations in vigorous activities (PF01)
2 % any limitations in walking one block (PF09)
3 % any limitations in climbing one flight of stairs (PF05)
4 % reporting difficulty performing at work due to physical health (RP4)
5 % reporting cutting down amount of time spent on work due to physical health (RP1)
6 % reporting very severe or severe bodily pain (BP1)
7 % reporting having a lot of energy all or most of the time (VT2)
8 % reporting feeling tiered all or most of the time (VT4)
9 % reporting excellent health (GH1)
10 % reporting fair or poor health (GH1)

Vigorous Physical Activities And PCS

As shown in the first column of percentages, from 90-98% of those at the lowest four PCS levels (levels 5-8) reported limitations in vigorous physical activities. Across the four higher PCS levels (levels 1-4) these percentages decline from 85%, 62%, 28%, to 20%. Thus, a five-point increase in the PCS from the midpoint of Level 4 (45-49), which is just below the mean, to the midpoint of the next level (50-54) represents a substantial decline (from 85.1% to 61.9%) in the percentage who are limited in vigorous activities.

Walking One Block/ Climbing Stairs And PCS

The second and third pairs of columns of Table 9.2 present the percentage of the general U.S. population who are limited in walking one block and climbing one flight of stairs at each of eight levels of the PCS. The pattern of results across levels is similar for both items. At the bottom of the PCS range (levels 6-8) 38% - 82% reported limitations in walking one block and 43% - 92% reported limitations in climbing one flight of stairs. Very few respondents (3.2% or less) reported either limitation at the top of the range (levels 1-3 of the PCS scores). Thus, these two items appear to be most useful for interpreting and explaining differences in scale scores at the middle and lower levels of the PCS score distribution.

ROle Disability And PCS

Pairs of columns numbered (4) and (5) present the percentage of the general U.S. population who reported difficulty performing at work and the need to cut down the amount of time at work because of physical health problems. For both items, these limitations were absent at the top levels of the PCS and were very prevalent at the bottom levels of the PCS.

Bodily Pain And PCS

Column (6) presents the percentage of the general U.S. population experiencing "severe" or "very severe" pain at each level of the PCS. At the bottom two levels of the PCS, 21% and 48% of the general U.S. population experienced severe or very severe bodily pain, whereas pain was more rare at the top levels of the PCS.

The percentages of the general U.S. population reporting "a lot of energy" and "feeling tired" all or most of the time at each level of the PCS are presented in the pairs of columns labeled (7) and (8). These results are most useful in interpreting the top and bottom of the PCS scale range, respectively.

Health Evaluations And PCS

Results in columns (9) and (10) show the percentage of the general U.S. population that evaluated their health as "excellent" versus "fair or poor" at each level of the PCS. "Excellent" evaluations are very rare except at the very top of the scale range. Evaluations of "fair" or "poor" increase progressively from the top to the bottom of the PCS scale range, reaching very high percentages at the bottom two levels of the PCS scale.

Table 9.3 presents data for eight SF-36 items used in content-based interpretations of nine levels of MCS scores. This table follows the same format as described above for the PCS (see How to Use These Tables, p. 67).

Downhearted/Blue And MCS

Column (1) of Table 9.3 presents the percentage of the general U.S. population feeling "downhearted or blue" all or most of the time at nine levels of the MCS scores. As illustrated, the downhearted or blue item clearly defines the bottom of the MCS scale: it is only at the lowest level of the MCS that a large percentage (44%) of respondents endorsed this item.

Feeling Happy And MCS

As shown in column (2) of Table 9.3, there is a more consistent increase in the percentage reporting being happy from the lower to the higher MCS scale levels (from 96.4% to 5.7%, lowest to highest).

Role Disability And MCS

Columns (3), (4), and (5) show the percentages of the general U.S. population reporting limitations of various kinds in everyday role activities *due to emotional problems*. For each of the items, there is a progressive increase in the percentage reporting limitations from the higher to the lower levels of the MCS scores. As the rate of increase diminishes at the higher levels, the Δ/Δ columns become smaller again, as would be expected.

Social Disability And MCS

Column (6) (Social) presents the percentage of the general U.S. population who report interference in usual social activities all or most of the time. At the bottom five levels of the MCS scale, 14.0% to 38.5% reported such limitations.

Feeling Tired/Having Energy And MCS

The last two pairs of columns in Table 9.3, (7) and (8), present percentages of the general U.S. population "feeling tired" and "having a lot of energy" all or most of the time at each level of the MCS. As with the role disability items, a systematic relationship between MCS scores and reports of tiredness and energy are apparent, with large percentages of individuals reporting tiredness at the lower MCS levels and large percentages of individuals reporting energy at the higher MCS levels.

TABLE 9.3 PERCENTAGE OF ADULTS ENDORSING SELECTED SF-36 ITEMS AT NINE LEVELS OF MCS SCORES: GENERAL U.S. POPULATION

| | | | | DISTRESS/WELL-BEING | | | | | ROLE DISABILITY | | | | | | SOCIAL | | VITALITY | | | |
| | MCS SCORES | | | 1 Downhearted Blue | | 2 Happy Person | | 3 Cut Down Worktime | | 4 Accomplished Less at Work | | 5 Worked Less Carefullly | | 6 Activity Limitation | | 7 Tired | | 8 Lot of Energy | |
LEVELS	RANGE	MEAN	(N)	%	Δ/Δ	%	Δ/Δ	%	Δ/Δ	%	Δ/Δ	%	Δ/Δ	%	Δ/Δ	%	Δ/Δ	%	Δ/Δ
1	65 - 74	67	28	0.0	0.0	96.4	-0.2	0.0	0.1	0.0	0.3	0.0	0.1	0.0	0.3	7.1	-0.6	64.3	-0.8
2	60 - 64	61.5	269	0.0	0.0	97.4	1.7	0.7	-0.1	1.5	0.1	0.4	0.1	1.5	0.2	3.7	-0.1	68.8	1.4
3	55 - 59	57.1	729	0.1	0.2	89.8	5.4	0.4	0.7	1.9	2.0	1.0	0.9	2.5	1.2	3.3	0.9	62.4	5.4
4	50 - 54	52.3	512	1.0	0.1	63.7	5.0	3.7	1.7	11.5	3.8	5.3	3.0	8.2	1.1	7.8	1.3	36.3	3.0
5	45 - 49	47.2	287	1.4	0.2	38.3	1.1	12.2	3.8	31.0	4.9	20.6	3.0	14.0	0.1	14.3	2.2	21.0	1.1
6	40 - 44	41.9	190	2.7	1.2	32.6	0.9	32.6	5.7	57.1	5.1	36.5	5.6	14.3	0.8	25.8	0.8	15.3	0.3
7	35 - 39	37.2	153	8.6	0.4	28.1	4.3	59.5	3.3	81.0	2.2	62.7	3.3	18.3	0.3	29.4	2.9	13.8	1.5
8	30 - 34	32.1	102	10.8	4.0	5.9	0.0	76.5	1.1	92.2	0.2	79.4	0.8	19.8	2.2	44.1	1.2	5.9	0.0
9	9 - 29	23.8	123	43.9		5.7		85.4		94.3		86.2		38.5		53.7		5.7	

NOTE: All table entries are percentages

KEY TO COLUMN DEFINITIONS

1 % report being downhearted or blue all or most of the time (MH4)
2 % reporting being happy all or most of the time (MH5)
3 % cut down amount of time spent at work due to emotional problems (RE1)
4 % accomplished less than would like due to emotional problems (RE2)
5 % didn't do work as carefully due to emotional problems (RE3)
6 % physical or emotional problems interfere with social activities (SF2)
7 % reporting feeling tiered all or most of the time (VT4)
8 % reporting having a lot of energy all or most of the time (VT2)

Criterion-Based Interpretation

Criterion-based tests of validity are based on analyses of relationships between the measures in question and other variables, referred to as "criteria," measured either concurrently or after some period of time. Criteria relied upon here were chosen to be conceptually related to the PCS and MCS and, in the absence of a "gold standard," to provide the most useful interpretation guidelines. Specifically, these criteria include variables that: (1) were important both clinically and socially (e.g., clinical diagnosis and job loss, respectively); (2) represented plausible outcomes of the variations in physical and mental health; and (3) were measured independently of the PCS and MCS scales.

Current Work Status And PCS

An important social consequence of poor physical health status is that it may prevent one from working at a paying job. For each of four levels of PCS scores, Table 9.4 presents the percentage of MOS participants who were working at a paying job when the MOS began (N = 2,069). Work status was determined concurrently with the PCS scale. The percentage of MOS patients who were eligible to work and could not work ranged from a high of 57.6% for PCS scores below 35 (level 4) to a low of 5.2% for PCS scores greater than or equal to 55 (level 1). For each one-point difference in PCS scale scores below 45, a two-point increase was observed in the percentage unable to work. Much less difference in the percentage reporting inability to work was observed between the two top scale levels (levels 1-2).

TABLE 9.4 PERCENTAGE OF MOS PATIENTS[1] UNABLE TO WORK AT FOUR LEVELS OF THE PCS

	PCS SCORES			CANNOT WORK	
LEVELS	RANGE	MEAN	(N)	%	Δ/Δ
1	55-72	57.5	326	5.2	
					0.1
2	45-54	49.9	715	6.3	
					2.1
3	35-44	39.8	502	27.1	
					2.2
4	8-34	26.2	526	57.6	

[1] MOS baseline sample eligible to work with complete SF-36 data (N = 2,069).

Subsequent Job Loss And PCS

Table 9.5 presents estimates of the percentages of MOS patients (working initially) who had lost their jobs one or two years later because of their health, at four levels of baseline PCS scores. Only patients who reported that they were working at a paying job at "baseline" and who were available at one- and two-year follow-ups were included in the analyses. There is a perfect ordering (from the top to the bottom PCS score levels) of the percentage reporting job loss at follow-up. The change was substantial, approximately a 10-fold increase in job loss, from the top to the bottom PCS levels. As shown in the Δ/Δ columns, approximately a one percentage point change in the probability of job loss is apparent (at one and two years) for each one-point change in PCS scores.

TABLE 9.5 PERCENTAGE OF WORKING MOS PATIENTS[1] WHO REPORTED JOB LOSS AT THE ONE AND TWO-YEAR FOLLOW-UP, FOUR LEVELS OF THE PCS

	PCS SCORES			FOLLOW-UP (%)			
				ONE YEAR		TWO YEAR	
LEVELS	RANGE	MEAN	(N)	%	Δ/Δ	%	Δ/Δ
1	55-72	57.6	130	3.1		3.8	
					0.9		0.9
2	45-54	49.9	321	10.0		11.0	
					0.8		1.1
3	35-44	40.4	184	17.9		21.7	
					1.1		1.1
4	8-34	27.2	115	32.2		35.8	

[1] MOS sample working at baseline and who reported not being able to work due to health at one year (N = 750) and two year (N = 735) follow-up assessment

Hospital Stays And PCS

Table 9.6 presents the percentage of MOS patients reporting one or more overnight stays in a hospital during the six-month period following completion of the SF-36. Percentages were estimated for four levels of PCS scores. The percent hospitalized overnight increases with lower PCS scale scores. From the top PCS level to the bottom, the percentage of patients hospitalized overnight nearly tripled. In the 8-44 point range on the PCS, each one-point difference was associated with approximately a 0.4 percentage-point difference in hospitalization rates.

Doctor Visits And PCS

Table 9.7 presents results for two criteria: the percentage of the general U.S. population reporting a doctor visit in the past month and the percentage with one or more chronic conditions across eight levels of the PCS scale (rates of chronic conditions are discussed below). Information about doctor visits was collected by self-report concurrently with the SF-36. The percentage of the general U.S. population reporting one or more doctor visits in the past month increases gradually from a low of 12% at the highest PCS level to a high of 53% at the lowest PCS level, more than a four-fold increase.

TABLE 9.6 PERCENTAGE OF MOS PATIENTS[1] WHO WERE HOSPITALIZED OVERNIGHT WITHIN SIX MONTHS, FOUR LEVELS OF THE PCS

	PCS SCORES			HOSPITALIZED OVERNIGHT	
LEVELS	RANGE	MEAN	(N)	%	Δ/Δ
1	55-72	57.8	271	4.4	
					0.1
2	45-54	50.1	584	5.3	
					0.3
3	35-44	40.1	364	8.0	
					0.4
4	8-34	26.7	302	12.9	

[1] MOS longitudinal sample with data at baseline and six-month follow-up assessments, (N = 1,521).

TABLE 9.7 PERCENTAGE OF THE GENERAL U.S. POPULATION REPORTING A RECENT VISIT TO THE DOCTOR AND ONE OR MORE PHYSICAL CONDITIONS, EIGHT LEVELS OF THE PCS

	PCS SCORES			RECENT DR. VISIT[1]		1+ PHYSICAL CONDITIONS[2]	
LEVELS	RANGE	MEAN	(N)	%	Δ/Δ	%	Δ/Δ
1	60-73	61.9	157	12.0		26.1	
					0.6		1.4
2	55-59	56.8	691	15.0		33.1	
					1.1		4.9
3	50-54	52.3	565	20.0		55.2	
					1.2		4.0
4	45-49	47.3	304	26.0		75.3	
					1.4		1.3
5	40-44	42.2	194	33.0		81.9	
					1.6		1.1
6	35-39	37.1	161	41.0		78.6	
					1.4		1.4
7	30-34	32.0	131	48.0		94.7	
					0.6		0.2
8	8-29	23.3	190	53.0		96.8	

[1] Percentage reporting visit to a medical doctor within the past month (N=637)
[2] Percentage reporting one or more of the following conditions: hypertension, congestive heart failure, myocardial infarction, diabetes, angina, chronic lung disease, arthritis, back/sciatica, or weakness/limitations in arms or legs (N=1,419)

Probability Of A Chronic Condition And PCS

Table 9.7 also presents the percentage of the general U.S. population reporting one or more chronic conditions across the eight levels of PCS scores. The percentage reporting one or more chronic conditions ranged from 26.1% at the top of the scale to 96.8% at the bottom of the scale, a 3.7-fold increase.

Five-Year Survival And PCS

An important test of the validity of the PCS scale scores is their usefulness in predicting five-year mortality. Ongoing MOS studies of the PCS scale scores and five-year mortality rates suggest that differences in scores have substantial implications for survival. As illustrated in Figure 7.1, findings-to-date indicate that the percentage who died within five years increased from only 1.8% at the top of the PCS scale to 21.5% at the bottom scale level, nearly a 12-fold increase (see Table 7.8). Predictions based on logistic regression yielded very large differences in odds-ratios of dying across PCS score levels, with and without statistical adjustment for differences in age. As illustrated in Table 7.8, patients scoring at the bottom scale level were nearly seven times more likely to die within the following five years than patients scoring at the top scale level, with adjustment for differences in age. Patients at scale levels two and three were twice as likely to die as patients scoring at the top level.

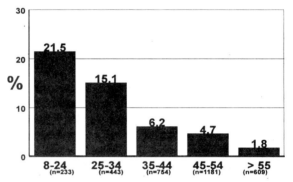

TABLE 9.8 FIVE-YEAR MORTALITY RATES FOR MOS PATIENTS[1] AT FIVE LEVELS OF THE PCS

	PCS Scores			Died		Odds-Ratio of Dying	
Levels	Range	Mean	(N)	%	Δ/Δ	Unadjusted	Age Adjusted[2]
1	55-72	57.8	609	1.8		1	1
					0.4		
2	45-54	50.0	1181	4.7		2.7**	2.0*
					0.1		
3	35-44	39.9	754	6.2		3.6**	2.2*
					0.9		
4	25-34	29.9	443	15.1		9.7**	4.8**
					0.7		
5	8-24	20.4	233	21.5		14.8**	6.8**

[1] MOS longitudinal sample with complete SF-36 data at baseline assessment (N = 3,220).
[2] Results from a logistic regression model predicting death and controlling for age at baseline assessment.
** $p < .001$ * $p < .05$

Depression Screener and MCS

Results useful in interpreting MCS scores are presented in Table 7.9, including rates of positive screening for likelihood of depression, substantial life stress, and life satisfaction observed across the MCS levels measured concurrently in the general U.S. population. Definitions for these "criteria" are documented in Appendix A. Table 9.9 first presents the percentage of the general U.S. population who screened positive for likelihood of depression at each of nine levels of the MCS scale. These percentages increased consistently from a low of 10.7% at the highest MCS scale level to a high of 93.5% at the lowest scale level, an 8.7-fold increase. These results underscore the clinical relevance of differences in MCS scores in the general U.S. population and offer guidelines for interpreting specific score differences in clinical terms throughout the scale range. The implications of these results for using the MCS as a screening tool are addressed in Chapter 12 on clinical applications.

TABLE 9.9 RATES OF POSSIBLE DEPRESSION, STRESS, AND LIFE SATISFACTION IN THE GENERAL U.S. POPULATION, NINE LEVELS OF THE MCS

MCS Scores				Screen for Depression[1]		Stress[2]		Life Satisfaction[3]	
Levels	Range	Mean	(N)	%	Δ/Δ	%	Δ/Δ	%	Δ/Δ
1	65-74	67	28	10.7		0.0		78.6	
					0.3		1.2		0.9
2	60-64	61.5	269	12.3		6.7		73.4	
					1.4		1.1		1.6
3	55-59	57.1	729	18.4		11.5		66.3	
					2.2		1.4		3.9
4	50-54	52.3	512	29.1		18.4		47.5	
					3.2		2.4		3.1
5	45-49	47.2	287	45.6		30.5		31.9	
					2.5		1.3		-0.1
6	40-44	41.9	190	58.9		37.6		32.4	
					3.2		1.0		3.0
7	35-39	37.2	153	73.9		42.5		18.4	
					3.0		2.4		2.2
8	30-34	32.1	102	89.2		54.9		7.0	
					0.5		2.9		0.2
9	9-29	23.8	123	93.5		78.9		5.7	

[1] Respondents who answered YES to one or more of the following: two or more weeks in the last year feeling sad, blue, or depressed; two or more years in your life you felt depressed most days; felt depressed much of the time in the last year.
[2] Quite a bit or a great deal of stress experienced in your daily living in the past four weeks.
[3] How happy, satisfied, pleased with your personal life (% extremely happy or very happy).

Life Stress And MCS

The second of the three "criteria" presented in Table 9.9 is the percentage of the general U.S. population experiencing a great deal of stress in daily living across nine levels of the MCS scale. None reported such stress at the highest MCS level and the percentages ranged from a low of 6.7% at the second MCS level to a high of 78.9% at the lowest level. A perfect ordering of these percentages is apparent across the nine MCS levels.

Life Satisfaction And MCS

The third "criteria" analyzed in Table 9.9 is a widely-used measure of "quality of life," specifically the percentage of the general U.S. population who are happy, satisfied, or pleased with their life all or most of the time. These percentages are reported for nine levels of the MCS scale. A 13.8-fold increase in the percent satisfied or pleased with their life is apparent at the top level of the MCS (78.6%) relative to the bottom level (only 5.7%). The percentages decline nearly consistently with an aberration observed only between levels 5 and 6. Not surprisingly, the MCS correlates substantially with evaluations of quality of life in both the general U.S. population ($r = 0.57$, $p < 0.001$) as shown here, and among MOS patients ($r = 0.68$, $p < 0.001$) as documented further in Chapter 8 (see Table 8.6).

Diagnosis Of Clinical Depression And MCS

Table 9.10 presents results for two clinical "criteria" — a clinical diagnosis of clinical depression and the percentage receiving mental health specialty care — at nine levels of the MCS scale. A perfect ordering of the percentage of patients with a diagnosis of clinical depression is apparent across levels of the MCS scale, from a low of 0.0% at the top level to a high of 59.4% at lowest level. As indicated in the Δ/Δ column, these percentages tend to increase at an increasing rate (from less than 1% to more than 2% per MCS point) from the top to bottom. These results are very similar to results reported above for the general U.S. population, although a much lower prevalence was observed in the MOS, as would be expected for more stringent diagnostic criteria (e.g., a three-item self-report screener in the general population study versus a diagnostic interview schedule in the MOS).

Mental Health Treatment And MCS

Table 9.10 also presents estimates of the percentage of MOS patients receiving mental health treatment from a formally trained mental health specialist during the six-month period after administration of the SF-36. These percentages are reported for eight levels of the MCS. At the top two MCS levels (collapsed) the percentage receiving mental health specialty treatment from a mental health specialist was only 2.9% compared with 69.3% at the bottom MCS scale levels. The percentages increased consistently across scale levels and the largest differences in the percentage treated were observed at the lower three levels of the MCS (2.0 - 3.2 percentage-point difference in treatment rate per MCS point).

TABLE 9.10 PERCENTAGE OF MOS PATIENTS DIAGNOSED WITH DEPRESSIVE DISORDER AND MENTAL HEALTH SPECIALTY CARE, NINE LEVELS OF THE MCS

MCS SCORES				CLINICAL DEPRESSION[1]		RECEIVING MENTAL HEALTH SPECIALITY CARE[2]	
Levels	Range	Mean	(N)	%	Δ/Δ	%	Δ/Δ
1	65-72	66.2	78	0.0		--[a]	
					0.4		--
2	60-64	61.4	421	1.9		2.9	
					0.3		0.7
3	55-59	57.1	861	3.0		5.8	
					0.9		1.5
4	50-54	52.2	519	7.3		13.0	
					0.8		0.6
5	45-49	47.1	348	11.2		16.2	
					1.0		1.3
6	40-44	42.1	259	16.2		22.7	
					2.5		2.5
7	35-39	37.2	234	28.6		35.1	
					2.2		3.2
8	30-34	32.1	209	39.7		51.4	
					2.2		2.0
9	3-29	23.2	291	59.4		69.3	

[1] MOS baseline sample with complete SF-36 data (N = 3,220)
[2] MOS longitudinal sample with SF-36 data at baseline and six-month follow-up (N = 1,739)
[a] Level 9 (65-72) was collapsed with level 8 due to insufficient sample size.

Burden Of Chronic Conditions

Estimates of the health burden associated with clinically-defined groups known to differ in physical and mental health have proven very useful in interpreting health status scores and changes in scores over time (Brook, Ware, Rogers, et al., 1984; Ware, Brook, Rogers, et al., 1986). Because many adults with a clinical condition have more than one (Stewart, Greenfield, Hays, et al., 1989), it is necessary to take "co-morbidity" into account along with other confounding factors (e.g., age differences) in making estimates of the health burden associated with a specific condition.

Statistical Method

To estimate the impact of each of the five MOS tracer conditions alone on PCS and MCS scores and the added effects of 16 comorbid conditions, we used multivariate statistical methods. These methods insure that the formula used to estimate the impact of each condition adjusts as much as possible for differences in demographic and socioeconomic characteristics, and for other conditions.

Overall F-tests confirmed that the set of variables defining the five tracer conditions and 16 comorbid conditions in the MOS differed significantly from zero (see Table footnotes). An independent regression model was estimated for the PCS and MCS scales. The same statistical approach was followed in estimating the impact of self-reported conditions in the general U.S. population.

Sixteen of the 27 comorbid conditions were each common enough to estimate their unique effects on the PCS and MCS. The other 12 were "controlled for" in the model, but results are not reported. Results to date support an additive model; the unique impact of each did not vary much across other conditions as observed in previous MOS analyses (Stewart, Greenfield, Hays, et al., 1989).

Disease Impact: MOS

Table 9.11 presents estimates of the burden of chronic conditions for use in interpreting differences in PCS and MCS scores. These are estimates of the unique effects of the five "tracer" conditions studied in the MOS along with 16 comorbid conditions, adjusting for sociodemographic variables (as described above).

As shown in Table 9.11, the largest effects on PCS scores in the MOS were observed for rheumatoid arthritis (-7.61, ± 1.9), hip impairment (-6.09, ± 1.4), congestive heart failure (-5.43, ± 1.2), osteoarthritis (-5.19, ± 1.2), and ulcers (-5.01, ± 2.1). All of these effects exceeded five points, which is one-half of a standard deviation, and all involve physical conditions. Other significant effects are documented in Table 9.11, including differences for all five "tracer" conditions and for 10 of 16 comorbid conditions, for which differences were estimated.

Only two conditions were associated with significant effects on the MCS scores. The largest difference was observed for clinical depression (-12.72, ± 1.1), a difference of more than a standard deviation. The only other significant difference in the MCS scores was observed for asthma (-6.20, ± 2.0).

Disease Impact: General Population

In the general U.S. population, the largest effects observed on the PCS, which were in the 6 - 8 point range, were observed for limitations in the use of an arm or leg (-7.15, ± 0.9) and for congestive heart failure (-6.72, ± 1.3) (see Table 9.12). Other substantial differences were observed for (in decreasing order of impact): back pain/sciatica (-3.75, ± 0.5), angina (-3.67, ± 1.2) diabetes/low blood sugar (-3.44, ± 0.8), chronic lung disease (-3.12, ± 0.9), myocardial infarction in the past year (-2.75, ± 1.2), and arthritis or rheumatism (-2.77, ± 0.5). Differences for hypertension and allergies were smaller, but statistically significant.

TABLE 9.11 ESTIMATES OF THE UNIQUE EFFECTS OF CHRONIC CONDITIONS ON THE PCS AND MCS, MOS PATIENTS (N = 1,790)

Comparison Group Means	N	PCS 53.27[a]	(1.2)	MCS 48.95[a]	(1.2)
Tracer Conditions					
Clinical Depression	262	-2.34[c]	(1.0)	-12.72[a]	(1.1)
Congestive Heart Failure (CHF)	218	-5.43[a]	(1.2)	-.98	(0.9)
Diabetes Type II	442	-3.48[a]	(0.7)	.58	(0.6)
Hypertension	1293	-1.85[c]	(0.8)	.61	(0.7)
Recent Myocardial Infarction (MI)	107	-3.24[b]	(1.0)	-.87	(1.0)
Comorbid Conditions					
Anemia	76	-3.05[c]	(1.5)	.62	(1.2)
Angina (No MI)	288	-4.02[a]	(0.8)	-.39	(0.7)
Asthma	50	-.86	(1.7)	-6.20[b]	(2.0)
Back Pain/Sciatica	502	-4.38[a]	(0.7)	-.83	(0.7)
COPD	117	-3.14[c]	(1.3)	-1.15	(1.0)
Dermatitis	356	-.85	(0.8)	-.71	(0.8)
Hip Impairment	75	-6.09[a]	(1.4)	-2.28	(1.5)
Irritable Bowel Disease	81	-2.92	(1.6)	-1.25	(1.2)
Kidney Disease	24	-1.89	(2.7)	-1.23	(3.0)
Musculoskeletal Complaints	341	-2.10[b]	(0.7)	-.16	(0.7)
Osteoarthritis	164	-5.19[a]	(1.2)	1.26	(1.0)
Other Lung Disease	26	2.04	(1.9)	-.16	(2.0)
Past MI	165	-3.08[b]	(1.0)	-.92	(1.2)
Rheumatoid Arthritis	39	-7.61[a]	(1.9)	3.01	(1.6)
Ulcer	53	-5.01[c]	(2.1)	-1.05	(1.6)
Urinary Tract Infection	127	-.89	(0.9)	-.75	(1.1)
Adjusted R2		0.3365		0.2858	
F (model)		19.53[a]		15.62[a]	

[a] $p < 0.001$
[b] $p < 0.01$
[c] $p < 0.05$

Comparison group is a hypothetical MOS female patient age 18-44 years, with 12 years education and no chronic conditions.
Numbers in parentheses are standard errors.
Note: These models are adjusted for main effects of age, sex, race, and education.

The largest effect on the MCS in the general U.S. population was observed for those who screened positive for likelihood of depression using three screening items from the MOS screener (Burnam, Wells, Leake, et al., 1988). Those with depression scored nearly one standard deviation (-9.30, ± 0.5) lower on the MCS. Six other conditions impacted the MCS a statistically significant amount, including: chronic lung disease (-3.03, ± 0.9), visual impairments (-2.92, ± 0.7), and dermatitis (-2.00, ± 0.8); significant differences in the 1 - 2 point range were back pain/sciatica, hearing impairments, and dermatitis. A noteworthy difference in MCS scores (-2.36, ± 1.2) for MI was not statistically significant at conventional levels due to a small sample size.

TABLE 9.12 ESTIMATES OF THE UNIQUE EFFECTS OF CHRONIC CONDITIONS ON THE PCS AND MCS, GENERAL U.S. POPULATION (N = 2,393)

	N	PCS		MCS	
Comparison Group Means		54.45[a]	(0.5)	52.46[a]	(0.5)
Chronic Conditions					
Allergy	842	-.82[c]	(0.4)	.04	(0.4)
Angina	112	-3.67[b]	(1.2)	.18	(1.0)
Arthritis	862	-2.77[a]	(0.5)	-.92	(0.5)
Back Pain/Sciatica	531	-3.75[a]	(0.5)	-1.58[b]	(0.5)
Cancer	108	-.83	(0.9)	-.31	(1.1)
Chronic Lung Disease	194	-3.12[a]	(0.9)	-3.03[b]	(0.9)
Congestive Heart Failure (CHF)	93	-6.72[a]	(1.3)	-1.36	(1.2)
Depression Screen	700	-.42	(0.4)	-9.30[a]	(0.5)
Dermatitis	224	-.41	(0.6)	-2.00[c]	(0.8)
Diabetes	156	-3.44[a]	(0.8)	.30	(0.9)
Hearing Impairment	405	-.94	(0.6)	-1.16[c]	(0.6)
Hypertension	701	-1.53[a]	(0.4)	-.10	(0.5)
Limitation in use of Arm(s)/Leg(s)	274	-7.15[a]	(0.9)	-.16	(0.8)
Myocardial Infarction (MI)	69	-2.75[c]	(1.2)	-2.36	(1.2)
Vision Impairment	280	-1.11	(0.7)	-2.92[a]	(0.7)
Adjusted R^2		0.4679		0.3039	
F (model)		73.54[a]		37.00[a]	

[a] $p < 0.001$
[b] $p < 0.01$
[c] $p < 0.05$

Comparison group is white females, age 18-44 years, with 12 years education. Results adjust for age, sex, race, and education.
Numbers in parentheses are standard errors.
Note: These models are adjusted for main effects of age, sex, race, and education.

Comparison Of Estimates

Although different methods were used to determine "tracer" conditions (physician report) and comorbid conditions (physician and patient reports) in the MOS, in comparison with the general U.S. population (self-report), estimates for most conditions included in both investigations showed considerable agreement. Of those significant in both studies, hypertension showed the smallest effect on the PCS in both the MOS and U.S. studies (-1.85 and -1.53, respectively). Congestive heart failure had one of the largest effects on PCS among conditions common to both studies (-5.43 and -6.72, respectively). Diabetes estimates fell in the middle range (-3.48 and -3.44, respectively) in both studies.

Among the largest discrepancies in results for the PCS were estimates for rheumatoid arthritis and osteoarthritis in the MOS in comparison with "arthritis" in the general population survey (-7.61 and -5.19 versus -2.77, respectively). Both studies showed the largest effects on the MCS for depression, although the impact of depression on the MCS was about one-third larger in the MOS, which used more extensive clinical criteria. Visual impairment and hearing impairment, which were not studied in the MOS, showed significant negative effects on the MCS in the general U.S. population survey.

Effects Of Aging

Given the substantial differences in PCS scores observed across age groups for both men and women in cross-sectional analyses (see Chapter 8), it is reasonable to expect change scores for the PCS scale to favor younger adults. If confirmed in longitudinal analyses, age-related differences in the impact of a year of aging might offer another basis for interpreting changes in PCS scores.

Changes in PCS scores were estimated for three age groups among MOS patients with only uncomplicated hypertension. This group of patients was selected for purposes of estimating age effects because they were the most "well" of those followed longitudinally in the MOS; thus age differences in health transitions were least likely to be confounded with medical comorbidity.

As shown in Table 9.13, the average change in PCS scores over a two-year follow-up differed significantly across the three age groups ($F = 4.5$, $p < 0.01$). This significant result was accounted for by the significant average decline in PCS scores ($\Delta = -2.0$, $p < 0.001$) among those 65 years of age and older (mean age = 71.4). The decline of two PCS points during the two-year follow-up period amounts to one PCS point per year. Thus, for example, it is reasonable to explain that each one-point decline in PCS scores observed in a clinical trial of alternative treatments is equal to the age-related decline in physical health among those 65 and older.

TABLE 9.13 ESTIMATES OF THE EFFECTS OF AGING TWO YEARS ON PCS SCORES, MOS PATIENTS WITH UNCOMPLICATED HYPERTENSION (N = 581)

Age Groups	Mean Age	(N)	Difference Score (Δ)		Δ/year
18-44	38.4	93	1.1	(0.8)	0.5
45-64	56.7	225	-0.8	(0.6)	-0.4
65+	71.4	263	-2.0*	(0.6)	-1.0*
F for Effect of Age			4.5[a]		

Significance of mean change in PCS scores for each age group: * $p < .001$
Significance test for the difference in mean change scores across age groups: [a] $p < .01$
Note: Uncomplicated hypertension defined as documented for the "minor medical" group in McHorney, Ware, & Raczek, 1993. Numbers in parentheses are standard errors.

Summary Of Effect Sizes

Among the results discussed in Chapter 9 are those from numerous cross-sectional and longitudinal group-level comparisons of average PCS and MCS scores. Because these comparisons involve groups known to differ in meaningful ways, results from these comparisons should be useful in: (a) interpreting score differences in future studies; (b) explaining results to others; and (c) planning future studies. To facilitate their use for these and other purposes, differences observed from group comparisons are summarized below.

PCS Scores

Tables 9.14 and 9.15 summarize results from cross-sectional and longitudinal comparisons, respectively, of average PCS scores for groups presented in this and other chapters. Each table orders average differences from the largest to the smallest "effect size." Because the PCS (and MCS) have a standard deviation (SD) of 10 in the general U.S. population, table entries can be easily interpreted in SD units by dividing each difference by 10 (moving the decimal one place to the left). Thus, the first entry in Table 9.14, which is 14.41, represents a difference of 1.44 SD units between groups with and without serious physical morbidity. Table 9.14 presents results from 62 cross-sectional comparisons of PCS scores. Differences that were significant statistically under the conditions of the study are indicated using three conventional levels for two-tailed tests. However, because sample sizes and other unique features of study designs and analytic methods influence statistical conclusions, they should be interpreted cautiously. For each comparison, the definition of the groups or the intervention that occurred between repeated measurements is defined very briefly, under the column headed "comparison," and the table notes where in this manual results are displayed and/or a reference if results have been published.

TABLE 9.14 RANK ORDER OF AVERAGE GROUP DIFFERENCES IN PCS SCORES OBSERVED IN CROSS-SECTIONAL STUDIES

Difference	Comparison	Sample	Source
14.41[a]	Patients with and w/out Serious Physical Morbidity	MOS Depressed Patients	Table 8.4
12.83[a]	Serious Physical and Mental (vs) Minor Physical	MOS Patients	Table 8.3
11.68[a]	Serious Mental (vs) Serious Physical	MOS Patients	Table 8.3
10.10[a]	Serious Physical (vs) Minor Physical	MOS Patients	Table 8.3
8.29[a]	CHF Severity 2 (vs) Severity 1	MOS CHF Patients	Ware, et al., 1995, Appendix Table B2
7.65[a]	CHF (vs) Hypertension	MOS Patients	Ware, et al., 1995, Appendix Table B1
7.61[a]	Rheumatoid Arthritis, Unique Effect	MOS Patients	Table 9.13
7.60[a]	Age 18-44 (vs) Age 65+, Uncomplicated Hypertension	MOS Hypertension Patients	Ware, et al., 1995, Table 7.15
7.15[a]	Limitations in Use of Arm/Leg, Unique Effect	General U.S. Population	Table 9.12
6.72[a]	CHF, Unique Effect	General U.S. Population	Table 9.12
6.09[a]	Hip Impairment, Unique Effect	MOS Patients	Table 9.11
5.57[a]	CHF (vs) Diabetes	MOS Patients	Ware, et al., 1995, Appendix Table B1
5.43[a]	CHF, Unique Effect	MOS Patients	Table 9.11
5.22[b]	CHF (vs) MI	MOS Patients	Ware, et al., 1995, Appendix Table B1
5.20[c]	Diabetes Severity 4 (vs) Severity 2	MOS Diabetic Patients	Ware, et al., 1995, Appendix Table B2
5.19[a]	Osteoarthritis, Unique Effect	MOS Patients	Table 9.11
5.01[c]	Ulcers, Unique Effect	MOS Patients	Table 9.11
4.80[c]	Diabetes Severity 4 (vs) Severity 1	MOS Patients	Ware, et al., 1995, Appendix Table B2
4.38[a]	Back Pain/Sciatica, Unique Effect	MOS Patients	Table 9.11
4.02[a]	Angina, Unique Effect	MOS Patients	Table 9.11
4.00[c]	Age 45-64 (vs) Age 65+, Uncomplicated Hypertension	MOS Hypertension Patients	Ware, et al., 1995, Table 7.15
3.75[a]	Back Pain/Sciatica, Unique Effect	General U.S. Population	Table 9.12
3.67[b]	Angina, Unique Effect	General U.S. Population	Table 9.12
3.60[c]	Age 18-44 (vs) Age 45-64, Uncomplicated Hypertension	MOS Hypertension Patients	Ware, et al., 1995, Table 7.15
3.48[a]	Diabetes, Unique Effect	MOS Patients	Table 9.11
3.44[a]	Diabetes, Unique Effect	General U.S. Population	Table 9.12
3.24[b]	MI, Unique Effect	MOS Patients	Table 9.11
3.14[c]	COPD, Unique Effect	MOS Patients	Table 9.11
3.12[a]	Chronic Lung Disease, Unique Effect	General U.S. Population	Table 9.12
3.08[b]	Past MI, Unique Effect	MOS Patients	Table 9.11

TABLE 9.14 RANK ORDER OF AVERAGE GROUP DIFFERENCES IN PCS SCORES OBSERVED IN CROSS-SECTIONAL STUDIES (CONTINUED)

Difference	Comparison	Sample	Source
3.05[c]	Anemia, Unique Effect	MOS Patients	Table 9.11
2.97	Diabetes Severity 4 (vs) Severity 3	MOS Diabetic Patients	Ware, et al., 1995, Appendix Table B2
2.92	Irritable Bowel Disease, Unique Effect	MOS Patients	Table 9.11
2.77[a]	Arthritis, Unique Effect	General U.S. Population	Table 9.12
2.75[c]	MI, Unique Effect	General U.S. Population	Table 9.12
2.73	Patients with and w/out Serious Mental	MOS Serious Physical Patients	Table 8.4
2.43[c]	Hypertension (vs) MI	MOS Patients	Ware, et al., 1995, Appendix Table B1
2.34[c]	Clinical Depression, Unique Effect	MOS Patients	Table 9.11
2.23	Diabetes Severity 3 (vs) Severity 2	MOS Patients	Ware, et al., 1995, Appendix Table B2
2.10[b]	Musculoskeletal Complaints, Unique Effect	MOS Patients	Table 9.11
2.08[c]	Diabetes (vs) Hypertension	MOS Patients	Ware, et al., 1995, Appendix Table B1
2.04	Other Lung Disease, Unique Effect	MOS Patients	Table 9.11
1.89	Kidney Disease, Unique Effect	MOS Patients	Table 9.11
1.85[c]	Hypertension, Unique Effect	MOS Patients	Table 9.11
1.83	Diabetes Severity 3 (vs) Severity 1	MOS Patients	Ware, et al., 1995, Appendix Table B2
1.80	Depression (vs) Minor Physical	MOS Patients	Ware, et al., 1995, Appendix Table B8
1.58[c]	Depression (vs) Minor Physical	MOS Patients	Table 8.3
1.53[a]	Hypertension, Unique Effect	General U.S. Population	Table 9.12
1.11	Vision Impairment, Unique Effect	General U.S. Population	Table 9.12
1.03	Hypertension Severity 2 (vs) Severity 1	MOS Patients	Ware, et al., 1995, Appendix Table B2
.94	Hearing Impairment, Unique Effect	General U.S. Population	Table 9.12
.89	UTI, Unique Effect	MOS Patients	Table 9.11
.86	Asthma, Unique Effect	MOS Patients	Table 9.11
.85	Dermatitis, Unique Effect	MOS Patients	Table 9.11
.83	Cancer, Unique Effect	General U.S. Population	Table 9.12
.82[c]	Allergies, Unique Effect	General U.S. Population	Table 9.12
.50	Ulcer Treatment — Maintenance (vs) Intermittent Therapy	Duodenal Ulcer Patients	Rampal, et al., 1994 (Figure 9.4)
.42	Depression, Unique Effect	General U.S. Population	Table 9.12
.41	Dermatitis, Unique Effect	General U.S. Population	Table 9.12
.40	Diabetes Severity 2 (vs) Severity 1	MOS Patients	Ware, et al., 1995, Appendix Table B2
.35	Diabetes (vs) MI	MOS Patients	Ware, et al., 1995, Appendix Table B1

[a] $p < .001$ [b] $p < .01$ [c] $p < .05$

PCS Cross-Sectional Differences

The largest differences in PCS scores observed to date were observed in comparisons between groups with serious physical conditions relative to patients without serious physical conditions.

The largest cross-sectional difference in PCS scores (-14.41, $p < 0.001$) was observed between depressed patients with and without serious physical comorbidity, as defined above (e.g., CHF). Cross-sectional differences greater than 10 points on the PCS summary scale (i.e., more than one SD unit) were also observed for three other comparisons involving groups of patients known to differ in the seriousness of their *physical* morbidity as clinically defined (see Table 9.14).

PCS Longitudinal Differences

Table 9.15 presents 27 average changes in PCS scores observed in longitudinal comparisons. Another finding that may be useful in interpreting changes in PCS scores is the decline ($\Delta = -2.0$) associated with two years of aging among patients with uncomplicated hypertension who were 65 and older (mean age = 71.4). Each year of aging for this group was associated with a one-point decline in PCS scores, on average.

Many ongoing studies will contribute to the understanding of the PCS and MCS scores that should be considered clinical and socially relevant. Guidelines for judging the clinical and social relevance of differences in such differences must await the result of those studies.

Although these are very large differences in terms of SD units (Cohen, 1988), even larger differences between groups would be expected, for example, in analyses of groups at even more severe levels of either physical or mental (emotional) morbidity. The differences reported above and any difference of 0.8 SD units or larger would be considered large according to conventional statistical standards (Cohen, 1988). Results from many ongoing studies will further contribute to the understanding of the PCS and MCS scores and the sizes of differences that should be considered clinically and socially relevant. Chapter 9 documents a wide range of differences that should be useful in interpreting the PCS and MCS scores and in explaining those differences to others. More definitive guidelines for establishing the importance of differences must await the results of future studies.

The three largest average changes in PCS scores were estimated from studies comparing scores before and after treatment for physical conditions. The treatment included hip replacement ($\Delta = 9.55$), therapy for low back pain ($\Delta = 7.66$), and heart valve replacement ($\Delta = 7.64$). These three studies are discussed further in Chapter 11 (Applications: Outcomes Research) and in the articles referenced. Another study of treatment effects yielded significant average changes in PCS scores, including maintenance drug therapy for ulcers ($\Delta = 4.60$) and intermittent drug therapy for ulcers ($\Delta = 3.20$) (see Rampal, Martin, Marquis, et al., 1994)

TABLE 9.15   RANK ORDER OF AVERAGE CHANGE IN PCS SCORES OBSERVED IN LONGITUDINAL STUDIES

Average Change	Comparison	Sample	Source
9.55[a]	Hip Replacement, Before and After	Hip Replacement Patients	Katz, et al., 1992 (Figure 9.2)
7.66[a]	Low Back Pain Therapy, Before and After	Spine Center Patients	Lansky, et al., 1992 (Figure 9.3)
7.64[a]	Heart Valve Replacement, Before and After	AVR and MVR Patients	Phillips & Lansky, 1992 (Figure 9.1)
-6.60[b]	Average Change in 1 Year for Patients Reporting "Lot More Limited"	MOS Patients	Appendix Table B7
4.60[c]	Ulcer, Before and After Maintenance Treatment, 1 Year	Duodenal Ulcer Patients	Rampal, et al., 1994 (Figure 9.4)
-3.90[c]	Average Change in 1 Year for Patients Reporting "Somewhat More Limited"	MOS Patients	Appendix Table B7
3.80[c]	Average Change in 1 Year for Patients Reporting "Less Limited"	MOS Patients	Appendix Table B7
-3.30[b]	COPD, 1 Year Change	MOS Patients	Table 14.45
3.20[c]	Ulcer, Before and After Intermittent Treatment, 1 Year	Duodenal Ulcer Patients	Rampal, et al., 1994 (Figure 9.4)
-2.09[c]	Osteoarthritis, 1 Year Change	MOS Patients	Table 14.48
-2.00[c]	Aging 2 Years, Age Group 65+ — Uncomplicated Hypertension	MOS Patients	Table 9.15
-1.96	Musculoskeletal Complaints, 1 Year Change	MOS Patients	Table 14.47
-1.69	CHF, 1 Year Change	MOS Patients	Table 14.39
1.40	Average Change in 1 Year for Patients Reporting "Less Limited"	MOS Patients	Appendix Table B7
-1.24	Angina, 1 Year Change	MOS Patients	Table 14.49
-1.23	Recovery from Depression	MOS Patients	Ware, et al., 1995 Appendix Table B9
1.10	Aging 2 Years, Age Group 18-44 — Uncomplicated Hypertension	MOS Patients	Table 9.15
-.80	Aging 2 Years, Age Group 45-64 — Uncomplicated Hypertension	MOS Patients	Table 9.15
-.70	Varicosities, 1 Year Change	MOS Patients	Table 14.50
-.70	BPH, 1 Year Change	MOS Patients	Table 14.44
-.64	Clinical Depression, 1 Year Change	MOS Patients	Table 14.38
.44	MI, 1 Year Change	MOS Patients	Table 14.41
-.40	Hypertension, 1 Year Change	MOS Patients	Table 14.40
-.33	Back Pain/Sciatica, 1 Year Change	MOS Patients	Table 14.43
.22	Diabetes, 1 Year Change	MOS Patients	Table 14.42
-.20	Dermatitis, 1 Year Change	MOS Patients	Table 14.46
-.10	Average Change in 1 Year for patients Reporting "Stayed the Same"	MOS Patients	Appendix Table B7

[a] $p < .001$ [b] $p < .01$ [c] $p < .05$

TABLE 9.16 RANK ORDER OF AVERAGE GROUP DIFFERENCES IN MCS SCORES OBSERVED IN CROSS-SECTIONAL ANALYSES

Difference	Comparison	Sample	Source
16.67[a]	Serious Mental (vs) Minor Physical	MOS Patients	Table 8.3
14.61[a]	Serious Mental (vs) Serious Physical	MOS Patients	Table 8.3
12.72[a]	Depression, Unique Effect	MOS Patients	Table 9.11
12.60[a]	Serious Physical + Mental (vs) Mental + Minor Physical	MOS Patients	Table 8.4
10.54[a]	Mental + Serious Physical (vs) Serious Physical	MOS Patients	Table 8.4
9.30[a]	Depression, Unique Effect	General U.S. Population	Table 9.12
6.20[b]	Asthma, Unique Effect	MOS Patients	Table 9.11
4.67[c]	CHF Severity 2 (vs) Severity 1	MOS Patients	Ware, et al., 1995 Appendix B2
4.08[c]	Mental (vs) Mental + Serious Physical	MOS Patients	Table 8.4
3.03[b]	Chronic Lung Disease, Unique Effect	General U.S. Population	Table 9.12
3.01	Rheumatoid Arthritis, Unique Effect	MOS Patients	Table 9.11
2.92[a]	Vision Impairment, Unique Effect	General U.S. Population	Table 9.12
2.57	Diabetes Severity 2 (vs) Severity 3	MOS Patients	Ware, et al., 1995 Appendix B2
2.36	MI, Unique Effect	General U.S. Population	Table 9.12
2.28	Hip Impairment, Unique Effect	MOS Patients	Table 9.11
2.10	Diabetes Severity 4 (vs) Severity 3	MOS Patients	Ware, et al., 1995 Appendix B2
2.06[b]	Serious Physical (vs) Minor Physical	MOS Patients	Table 8.3
2.00[c]	Dermatitis, Unique Effect	General U.S. Population	Table 9.12
1.97	CHF (vs) Hypertension	MOS Patients	Ware, et al., 1995 Appendix B1
1.71	MI (vs) Hypertension	MOS Patients	Ware, et al., 1995 Appendix B1
1.61	CHF (vs) Diabetes	MOS Patients	Ware, et al., 1995 Appendix B1
1.58[b]	Back Pain/Sciatica, Unique Effect	General U.S. Population	Table 9.12
1.36	CHF, Unique Effect	General U.S. Population	Table 9.12
1.35	MI (vs) Diabetes	MOS Patients	Ware, et al., 1995 Appendix B1
1.33	Diabetes Severity 2 (vs) Severity 1	MOS Patients	Ware, et al., 1995 Appendix B2
1.26	Osteoarthritis, Unique Effect	MOS Patients	Table 9.11
1.25	Irritable Bowel Disease, Unique Effect	MOS Patients	Table 9.11
1.24	Diabetes Severity 3 (vs) Severity 1	MOS Patients	Ware, et al., 1995 Appendix B2

TABLE 9.16 RANK ORDER OF AVERAGE GROUP DIFFERENCES IN MCS SCORES OBSERVED IN CROSS-SECTIONAL ANALYSES (CONTINUED)

Difference	Comparison	Sample	Source
1.23	Kidney Disease, Unique Effect	MOS Patients	Table 9.11
1.16[c]	Hearing Impairment, Unique Effect	General U.S. Population	Table 9.12
1.16	Hypertension Severity 1 (vs) Severity 2	MOS Patients	Ware, et al., 1995 Appendix B2
1.15	COPD, Unique Effect	MOS Patients	Table 9.11
1.05	Ulcers, Unique Effect	MOS Patients	Table 9.11
.98	CHF, Unique Effect	MOS Patients	Table 9.11
.92	Arthritis, Unique Effect	General U.S. Population	Table 9.12
.92	Past MI, Unique Effect	MOS Patients	Table 9.12
.90	Ulcer Treatment — Intermittent (vs) Maintenance	Duodenal Ulcer Patients	Rampal, et al., 1994 (Figure 9.4)
.87	MI, Unique Effect	MOS Patients	Table 9.11
.86	Diabetes Severity 4 (vs) Severity 1	MOS Patients	Ware, et al., 1995 Appendix B2
.83	Back Pain/Sciatica, Unique Effect	MOS Patients	Table 9.11
.75	UTI, Unique Effect	MOS Patients	Table 9.11
.71	Dermatitis, Unique Effect	MOS Patients	Table 9.11
.62	Anemia, Unique Effect	MOS Patients	Table 9.11
.61	Hypertension, Unique Effect	MOS Patients	Table 9.11
.58	Diabetes, Unique Effect	MOS Patients	Table 9.11
.47	Diabetes Severity 4 (vs) Severity 2	MOS Patients	Ware, et al., 1995 Appendix B2
.39	Angina, Unique Effect	MOS Patients	Table 9.11
.36	Diabetes (vs) Hypertension	MOS Patients	Ware, et al., 1995 Appendix B1
.31	Cancer, Unique Effect	General U.S. Population	Table 9.12
.30	Diabetes, Unique Effect	General U.S. Population	Table 9.12
.26	CHF (vs) MI	MOS Patients	Ware, et al., 1995 Appendix B1
.18	Angina, Unique Effect	General U.S. Population	Table 9.12
.16	Other Lung Disease, Unique Effect	MOS Patients	Table 9.11
.16	Limitations in Use of Arm/Leg, Unique Effect	General U.S. Population	Table 9.12
.16	Musculoskeletal Complaints, Unique Effect	MOS Patients	Table 9.11
.10	Hypertension, Unique Effect	General U.S. Population	Table 9.12
.04	Allergies, Unique Effect	General U.S. Population	Table 9.12

[a] $p < .001$ [b] $p < .01$ [c] $p < .05$

TABLE 9.17 RANK ORDER OF AVERAGE CHANGES IN MCS SCORES OBSERVED IN LONGITUDINAL STUDIES

Average Change	Comparison	Sample	Source
10.93[a]	Recovery From Depression	MOS Depression Patients	Ware, et al., 1995 Appendix B9
-7.30[b]	Average Change in 1 Year for Patients Reporting "A Lot More Limited"	MOS Patients	Appendix B7
7.20[b]	Average Change in 1 Year for Patients Reporting "A Lot Less Limited"	MOS Patients	Appendix B7
6.40[c]	Ulcer — Before and After Maintenance Treatment, 1 Year	Duodenal Ulcer Patients	Rampal, et al., 1994 (Figure 9.4)
5.40[c]	Ulcer — Before and After Intermittent Treatment, 1 Year	Duodenal Ulcer Patients	Rampal, et al., 1994 (Figure 9.4)
4.78[c]	COPD, 1 Year Change	MOS Patients	Appendix E45
-4.00[c]	Average Change in 1 Year for Patients Reporting "Somewhat More Limited"	MOS Patients	Appendix B7
3.86[a]	Clinical Depression, 1 Year Change	MOS Patients	Appendix E38
3.73[b]	Hip Replacement, Before and After	Hip Replacement Patients	Katz, et al., 1992 (Figure 9.2)
3.18[c]	Heart Valve Replacement, Before and After	AVR and MVR Patients	Phillips & Lansky, 1992 (Figure 9.1)
2.50[c]	Average Change in 1 Year for Patients Reporting "Somewhat Less Limited"	MOS Patients	Appendix B7
1.57	MI, 1 Year Change	MOS Patients	Appendix E41
1.31	Low Back Pain Therapy, Before and After	Spine Center Patients	Lansky, et al., 1992 (Figure 9.3)
-1.24	Musculoskeletal Complaints, 1 Year Change	MOS Patients	Appendix E47
-1.00	Varicosities, 1 Year Change	MOS Patients	Appendix E50
.91	CHF, 1 Year Change	MOS Patients	Appendix E39
.56	Back Pain/Sciatica, 1 Year Change	MOS Patients	Appendix E43
-.21	Osteoarthritis, 1 Year Change	MOS Patients	Appendix E48
.20	Hypertension, 1 Year Change	MOS Patients	Appendix E40
.18	Diabetes, 1 Year Change	MOS Patients	Appendix E42
-.17	Dermatitis, 1 Year Change	MOS Patients	Appendix E46
-.13	BPH, 1 Year Change	MOS Patients	Appendix E44
.10	Average Change in 1 Year for Patients Reporting "Stayed the Same"	MOS Patients	Appendix B7
.10	Angina, 1 Year Change	MOS Patients	Appendix E49

[a] $p < .001$ [b] $p < .01$ [c] $p < .05$

MCS Scores

The largest differences in MCS scores from cross-sectional studies involved comparisons between groups differing in seriousness of mental disorder (see Table 9.16). Other large differences include the effects of asthma (-6.20), congestive heart failure (-4.67), and chronic lung disease (-3.03). Interestingly, the estimate of the unique effect of vision impairment on MCS scores (-2.92) is large and significant in the general U.S. population. Table 9.16 also includes significant and unique negative effects of dermatitis and back pain/sciatica on MCS scores.

Changes in MCS scores summarized in Table 9.17 were statistically significant for all differences greater than two points. The largest average change ($\Delta = 10.93$) was observed for the MOS patients who (according to clinical criteria) recovered from clinical depression. Patients who reported being "more distressed" at one-year follow-up also had large average declines in their MCS scores, as would be expected ($\Delta = 7.30$).

Significant changes in MCS scores were also reported in response to drug therapy for ulcers ($\Delta = 6.40$ or 5.40, depending on regimen) and following hip replacement ($\Delta = 3.73$) and heart valve replacement ($\Delta = 3.18$).

Evaluation of Changes In PCS And MCS Scores

The scoring and interpretation of changes in PCS and MCS scores assumes much about how those changes are evaluated by those who experience them. Analyses of personal evaluations of actual changes in PCS and MCS scores over a one-year period tend to support these assumptions (Ware, Kosinski, Bayliss, et al., 1995; see also Chapter 8 and Appendix A). To address this issue further and to expand the SF-36 interpretation guidelines to include the values of patients, self-evaluated changes over a one-year interval were used as "criteria" for interpreting measured changes in PCS and MCS scores among MOS patients. MOS patients were asked two questions about physical functioning and mental health during their first year follow-up survey: "Compared to one year ago, are you more or less limited now in your everyday physical activities because of your health?" with five response choices ranging from "a lot more limited now" to "a lot less limited now." A similar question was asked for changes in mental health.

Table 9.18 presents the percentages evaluating their physical status as "more" to "less" limited (five categories of evaluation) at each of nine levels of measured change in PCS scores over the one-year period. Trends in results are also illustrated in Figures 9.2-9.4, which charts the percentages giving favorable evaluations of change (Figure 9.2), about the same (Figure 9.3), and unfavorable evaluations (Figure 9.4) at different levels of change in PCS scale scores. Overall, the figures and Table 7.18 support the hypothesized interpretation of changes in PCS scores. Specifically, measured declines (change levels 6-9) are most likely to be evaluated unfavorably (more limited or somewhat more limited) and measured improvements (change levels 1-4) are most likely to be evaluated favorably. Those scoring the same at both assessments (level 5) are most likely to evaluate their status as the same (73.7%). The percentage evaluating their health as the same declines with the magnitude of improvement or worsening in PCS scores over time, to slightly more than one-third at the highest and lowest PCS levels.

TABLE 9.18 PERCENTAGE OF MOS PATIENTS REPORTING MORE OR LESS PHYSICAL LIMITATIONS AFTER ONE YEAR, NINE CATEGORIES OF PCS CHANGE SCORES (N = 1,539)

	Measured PCS Changes			Self-Evaluated Physical Transition[1]				
Level of Change	Range	Mean Change	(N)	A lot more limited	Somewhat more limited	About the same	Somewhat less limited	A lot less limited
1	16 to 34	20.5	55	1.8	7.3	38.2	21.8	30.9
2	11 to 15	12.6	80	2.5	11.2	50.0	15.0	21.3
3	6 to 10	7.6	166	4.2	10.8	57.2	13.2	14.6
4	1 to 5	2.8	381	4.5	7.3	63.8	10.8	13.6
5	0	0.0	95	5.2	4.2	73.7	9.5	7.4
6	-5 to -1	-2.6	405	5.4	13.5	64.9	8.1	8.1
7	-10 to -6	-7.7	193	6.8	21.2	53.9	11.4	6.7
8	-15 to -11	-12.8	83	12.0	27.8	43.4	12.0	4.8
9	-43 to -16	-21.2	72	29.2	26.4	34.7	6.9	2.8

1 Self-evaluated transition: Compared to a year ago, are you more or less limited now in your everyday physical activities because of your health?

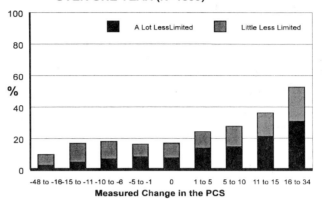

FIGURE 9.2 PLOT OF SELF-EVALUATED PHYSICAL HEALTH TRANSITIONS AND CHANGES IN PCS SCORES OVER ONE YEAR (N=1539)

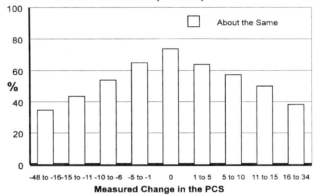

FIGURE 9.3 PLOT OF SELF-EVALUATED PHYSICAL HEALTH TRANSITIONS AND CHANGES IN PCS SCORES OVER ONE YEAR (N=1539)

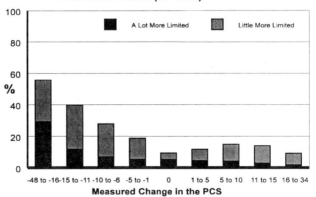

FIGURE 9.4 PLOT OF SELF-EVALUATED PHYSICAL HEALTH TRANSITIONS AND CHANGES IN PCS SCORES OVER ONE YEAR (N=1539)

TABLE 9.19 PERCENTAGE OF MOS PATIENTS REPORTING MORE OR LESS EMOTIONAL PROBLEMS AFTER ONE YEAR, NINE CATEGORIES OF MCS CHANGE SCORES (N = 1,539)

Measured MCS Changes				Self-Evaluated Mental Transition[1]				
Level of Change	Range	Mean Change	(N)	A lot more often	Somewhat more often	About the same	Somewhat less often	A lot less often
1	16 to 45	22.5	111	0.0	2.7	19.8	22.5	55.0
2	11 to 15	12.7	89	2.2	6.7	36.0	20.3	34.8
3	6 to 10	7.6	158	3.2	6.3	41.8	24.0	24.7
4	1 to 5	2.6	358	2.2	7.3	58.1	16.8	15.6
5	0	0.0	109	0.0	9.2	65.2	12.8	12.8
6	-5 to -1	-2.7	384	2.1	9.9	59.9	15.1	13.0
7	-10 to -6	-7.7	131	7.7	16.0	51.1	17.6	7.6
8	-15 to -11	-13.0	61	11.5	22.9	42.6	11.5	11.5
9	-37 to -16	-21.0	58	19.0	34.5	29.3	13.8	3.4

[1] Self-evaluated transition: Compared to one year ago, how often do you feel bothered by emotional problems, such as feeling anxious, depressed, or irritable now?

As shown in Table 9.19 (and Figure 9.3), patients who worsened in MCS scores were much more likely to evaluate their current mental health unfavorably in comparison with one year ago. Likewise, those whose MCS scores improved were much more likely to evaluate their current mental health favorably in comparison to one year ago.

Final Comment

Generally, smaller average changes were less likely to be significant, as would be expected. All average *changes* greater than two points on the PCS were statistically significant and none below two points were significant in the 27 longitudinal studies summarized here in Table 9.15. With only two exceptions out of 62 studies, the same pattern held true for cross-sectional studies involving the PCS, as summarized in Table 9.15. However, statistical significance should not be equated with clinical and social relevance, which are addressed earlier in this chapter.

FIGURE 9.5 PLOT OF SELF-EVALUATED MENTAL HEALTH TRANSITIONS AND CHANGES IN THE MCS SCORES OVER ONE YEAR (N=1539)

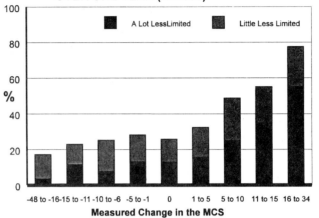

FIGURE 9.6 PLOT OF SELF-EVALUATED MENTAL HEALTH TRANSITIONS AND CHANGES IN THE MCS SCORES OVER ONE YEAR (N=1539)

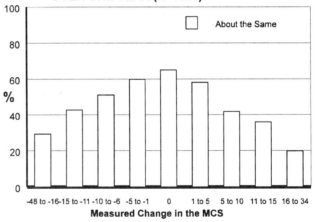

FIGURE 9.7 PLOT OF SELF-EVALUATED MENTAL HEALTH TRANSITIONS AND CHANGES IN THE MCS SCORES OVER ONE YEAR (N=1539)

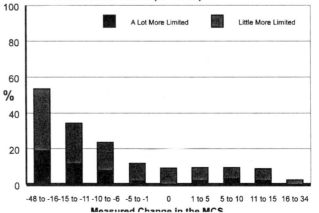

Chapter 10: Norms: 1998 General U.S. Population

This chapter presents norms for SF-36 scales and summary measures from both standard and acute forms in the 1998 general U.S. population. The chapter is divided into three sections. The first section explains norm-based interpretations and provides background information. The remaining two major sections present: (1) 1998 general U.S. population norms for the standard form (4-week recall scales and summary measures; and (2) 1998 general U.S. population norms for the acute form (1-week recall) scales and summary measures.

Background

Normative data makes it possible to interpret SF-36 scale and summary measure scores for an individual respondent or the average for a group of respondents by comparing them with the distribution of scores for other individuals. Scores can be understood as departures from expected or typical scores; these expected or typical scores are called norms. Norms can be computed at the individual or group level. At the individual level, norms are scores that are typical of the individual under stable conditions. At the group level, norms are average values for a group and can be calculated based on a sample from the general population. Norm-based interpretation answers the questions of whether or not an observed score is typical: Is the score one would expect to find for this individual or group of individuals?

Because estimates of scores for the scales and summary measures of the SF-36 are "norm-based", they have the advantage of a direct interpretation in this regard. For example, it is useful to know that scores above and below 50 are above and below the mean, respectively, in the general U.S. population for all SF-36 scales and summary measures. However, because health status varies with age, gender and with the presence of diseases, specific norms for subgroups of the general U.S. population are presented in this chapter to facilitate the interpretation of SF-36 scale scores and summary measure scores for these subgroups.

Norm-based comparisons require valid norms for a well-defined and representative sample of the population of interest. This chapter presents such data for the SF-36 standard and acute forms. Some view the publication of such norms as more than just a convenience. Accepted guidelines for the standardization, scoring, and the documentation of widely used psychological tests emphasize the publication of norms prior to their widespread use (APA, 1985).

This chapter presents SF-36 scale and summary measure norms for representative samples of the 1998 non-institutionalized general U.S. population. Norms are presented separately for standard (N=1,982) and acute (N=1,646) forms. Because health status scores for some concepts differ significantly across age groups and for men and women, norms are presented for the total population and separately for seven age groupings and for males and females. In addition, norms are presented for individuals with various chronic conditions (self-reported) using data from the 1998 general U.S. population.

Figure 10.1   Guide to Tables Presenting Normative Data (Standard Form)

Norms for SF-36 Scales and Summary Measures

SF-36 Norms for the General U.S. Population

Table 10.1	Total Sample, Frequency Distribution.....	94
Table 10.2	Total Sample, Descriptive Statistics........	95
Table 10.3	Males..	95
Table 10.4	Females..	95
Table 10.5	Males & Females, by Age Groups............	96
Table 10.6	Males, by Age Group..................................	98
Table 10.7	Females, by Age Group.............................	100
Table 10.8	No Chronic Conditions.............................	102
Table 10.9	Anemia..	102
Table 10.10	Back Pain/Sciatica.....................................	102
Table 10.11	Cancer (except skin)..................................	103
Table 10.12	Depression...	103
Table 10.13	Diabetes..	103
Table 10.14	Heart Disease...	104
Table 10.15	Hypertension (high blood pressure).......	104
Table 10.16	Kidney Disease...	104
Table 10.17	Liver Disease..	105
Table 10.18	Lung Disease..	105
Table 10.19	Osteoarthritis..	105
Table 10.20	Rheumatoid Arthritis................................	106
Table 10.21	Ulcers..	106

Figure 10.2 Guide to Tables Presenting Normative Data (Acute Form)

Norms for SF-36 Scales and Summary Measures

SF-36 Norms for the General U.S. Population

Table 10.22	Total Sample, Frequency Distribution.....	107
Table 10.23	Total Sample, Descriptive Statistics........	108
Table 10.24	Males...	108
Table 10.25	Females...	108
Table 10.26	Males & Females, by Age Groups............	109
Table 10.27	Males, by Age Group................................	111
Table 10.28	Females, by Age Group............................	113
Table 10.29	No Chronic Conditions.............................	114
Table 10.30	Anemia...	115
Table 10.31	Back Pain/Sciatica....................................	115
Table 10.32	Cancer (except skin)...............................	115
Table 10.33	Depression..	116
Table 10.34	Diabetes..	116
Table 10.35	Heart Disease...	116
Table 10.36	Hypertension (high blood pressure).......	117
Table 10.37	Kidney Disease...	117
Table 10.38	Liver Disease..	117
Table 10.39	Lung Disease..	118
Table 10.40	Osteoarthritis...	118
Table 10.41	Rheumatoid Arthritis...............................	118
Table 10.42	Ulcers..	119

How The SF-36 Was Normed

U.S. general population norms were estimated from responses to the 1998 National Survey of Functional Health Status, which included the SF-36 Health Survey. Sampling and data collection methods are described below.

Sampling

Households were drawn from the sampling frames, maintained by National Family Opinion (NFO) Research. The NFO panel includes 550,000 households representative of the non-institutionalized adult U.S. population and is matched to the U.S. Census data on geographical region, market size, age, income (SES) and household size. Panel households are balanced demographically to the four Census regions and the nine Census divisions and are also in the correct proportion by state within each of the nine divisions. Households are identified by frequently used geographical classifications to provide complete sampling and identification flexibility including 1) Metropolitan Statistical Areas (MSA), county or groups of counties with at least one city of 50,000 of more inhabitants, with the exception of New England MSA's which consist of towns or cities rather than counties. NFO Panel members are located in more than 300 MSA's; 2) Consolidated Metropolitan Statistical Areas (CMSA), a group of two or more adjacent primary metropolitan statistical areas with combined populations of more than one million. NFO households are located in all 20 CMSA's; 3) Counties, NFO households are located in 99.4% of the more than 3,000 counties located in contiguous states; 4) Federal Information Processing Standards (FIP Place Codes) used to designate all cities and towns; and 5) Census tracts, geographical areas of approximately 4,000 people each used for statistical analysis of Census data. The NFO panel is managed to maintain current demographic information about household members.

The NFO used a two-stage area probability sample design. In the first stage, quota sampling was used based on age, sex, and income. The primary sampling units (PSUs) used were Standard Metropolitan Statistical Areas or non-metropolitan counties. These PSUs were stratified by region, market size, age, income and household size before selection. The units of selection at the second stage were households stratified according to age, sex, and race.

To better represent the demographic characteristic's of the US population in 1998, weights based on the 1998 census were applied to equate the ages and gender of respondents, in comparison with that census.

Data Collection

Data were collected by the National Research Corporation (NRC) using a single wave of questionnaires mailed to randomly selected members of the NFO panel. Pre-notification announcements, replacement surveys, reminders and follow-up postcards, were not utilized. Data were collected for 12 weeks between October and December 1998. At the end of the data collection period, the unlocatable rate was 8% for respondents consistent with previous NFO surveys using this strategy, and the overall response rate for the survey was 67.8%.

A total of 1,982 completed the Standard form and 1,646 completed the Acute form. The mean age of respondents was 50.7 years (18-96) and 23% were elderly (65 and older). More than half of the respondents were female (59.6%) and the majority were white (84.2%), with at least 12 years of education (79.8%). Sampling weights were applied to adjust the sample to match the age, gender, and age by gender distribution of the 1998 census.

U.S. NORMS FOR SF-36 SCALES AND SUMMARY MEASURES: STANDARD (4-WEEK RECALL) AND ACUTE FORMS (1-WEEK RECALL)

Normative data for SF-36 scales and summary measures for standard and acute forms are presented separately in this chapter. Tables 10.1 through 10.21 present descriptive statistics for each SF-36 scale and summary measure for the standard form. Tables 10.22 through 10.42 present descriptive statistics for each SF-36 scale and summary measure for the acute form.

Tables presenting normative data include descriptive statistics for each SF-36 scale and summary measure in the 1998 general U.S. population. These include the mean, median (50th percentile), 25th and 75th percentiles, standard deviation, and observed range of scores.

NORMS FOR THE SF-36 SCALES AND SUMMARY MEASURES

Tables 10.2 (standard form) and 10.23 (acute form) present descriptive statistics for each SF-36 scale and summary measure in the total general U.S. population (males and females combined and separately). The mean and standard deviations for the total sample are 50 and 10, respectively, for all scales and summary measures because of the scoring method used in their standardization (see chapters 5 and 6). The medians (50 percentile score) for each scale are generally higher than the mean scores.

This reflects some skewness of the score distributions in the 1998 general US population, with more respondents scoring above the mean.

To determine whether the score for an individual or the average (mean) for a particular group is above or below the average for the U.S. general population, look at the first row of Table 10.1 (standard form) or Table 10.22 (acute form). For example, the mean Physical Functioning (PF) scale score in the general U.S. population is 50.00 (see the upper left-hand corner of Table 10.1). Scores above and below 50.00 are above and below average, respectively.

SCALE AND SUMMARY MEASURE NORMS FOR MALES AND FEMALES

Tables 10.3- 10.4 (standard form) and Tables 10.24-10.25 (1-week recall) present SF-36 scale and summary measure norms separately for males and females in the 1998 U.S. general population. These tables are useful in determining whether a score for males or females is above or below the average score for males and females in the 1998 general U.S. population.

SCALE AND SUMMARY MEASURE NORMS FOR AGE GROUPS

Tables 10.5 (standard form) and 10.26 (acute form) present SF-36 scale and summary measure norms for six different age groups among males and females combined. These age groupings were selected (1) to be large enough to satisfy minimum standards for precision; (2) to correspond with standard practices for defining age-specific groups; and (3) to correspond with age groupings used by others in reporting norms for the SF-36 (Brazier et al., 1992; Jenkinson et al., 1993, Ware et al., 1993, Ware et al., 1994) and that are forthcoming from other countries. Comparing results across age groups clearly shows that health status, in particular physical health, is related to age. Generally, the scores for all physical health scales and summary measures decline with age. For example, whereas the mean PF score for the total sample is 50.00, the mean for the 18 to 34 group is higher (54.07) and the mean for the 75 and older group is lower (41.77) (Table 10.5).

SCALE AND SUMMARY MEASURE NORMS FOR MALES AND FEMALES BY AGE GROUP

Tables 10.6-10.7 (standard form) and Tables 10.26-10.27 (1-week recall) present SF-36 scale and summary measure norms by age group separately for males and females in the 1998 U.S. general population. These tables are useful in determining whether a score for males or females is above or below the average score for males and females in a particular age group in the 1998 general U.S. population.

SCALE AND SUMMARY MEASURE NORMS FOR CHRONIC CONDITIONS

The 1998 general U.S. population survey included a checklist of 13 chronic conditions. This checklist asked respondents if their doctor had ever told them they had any of the following conditions or if they now had any of the following conditions: heart disease, hypertension, lung disease, diabetes, ulcers, kidney disease, liver disease, anemia, cancer, depression, osteoarthritis, rheumatoid arthritis, or low back pain or sciatica. Tables

10.8-10.21 (standard form) and Tables 10.29-10.42 (1-week recall) present SF-36 scale and summary measure norms for individuals with these various chronic conditions (self reported) and for a "well" group of respondents who did not report any of the chronic conditions on the chronic condition checklist. The norms in these tables are not adjusted for differences in sociodemographic characteristics and comorbid conditions. It is important to keep this in mind when making comparisons. For example, in interpreting general population norms for those who self-reported heart disease (Table 10.14 and 10.35), note that the majority also suffered from other conditions on the checklist and that the average age was generally higher than that of the total general population.

STANDARD FORM

TABLE 10.1 Frequency Distribution for PCS and MCS Scale Scores (4-Week Recall): 1998 General U.S. Population (N=1,982)

	PCS Scale				MCS Scale		
Score	f	%	Cum. %	Score	f	%	Cum. %
69	1	0.05	100.00	73	1	0.05	100.00
68	1	0.05	99.95	71	1	0.05	99.95
67	2	0.10	99.9	68	3	0.15	99.90
66	5	0.25	99.8	67	2	0.10	99.75
65	7	0.35	99.55	66	7	0.35	99.65
64	6	0.30	99.19	65	11	0.55	99.29
63	15	0.76	98.89	64	9	0.45	98.74
62	13	0.66	98.13	63	23	1.16	98.28
61	36	1.82	97.48	62	37	1.87	97.12
60	87	4.39	95.66	61	47	2.37	95.26
59	105	5.30	91.27	60	105	5.30	92.89
58	104	5.25	85.97	59	119	6.00	87.59
57	144	7.27	80.73	58	151	7.62	81.58
56	123	6.21	73.46	57	123	6.21	73.97
55	116	5.85	67.26	56	111	5.60	67.76
54	97	4.89	61.40	55	115	5.80	62.16
53	103	5.20	56.51	54	115	5.80	56.36
52	80	4.04	51.31	53	106	5.35	50.55
51	77	3.88	47.28	52	68	3.43	45.21
50	79	3.99	43.39	51	71	3.58	41.78
49	62	3.13	39.40	50	75	3.78	38.19
48	37	1.87	36.28	49	63	3.18	34.41
47	53	2.67	34.41	48	45	2.27	31.23
46	48	2.42	31.74	47	42	2.12	28.96
45	38	1.92	29.31	46	43	2.17	26.84
44	39	1.97	27.40	45	36	1.82	24.67
43	33	1.66	25.43	44	44	2.22	22.86
42	30	1.51	23.76	43	36	1.82	20.64
41	36	1.82	22.25	42	35	1.77	18.82
40	22	1.11	20.43	41	21	1.06	17.05
39	23	1.16	19.32	40	26	1.31	15.99
38	26	1.31	18.16	39	25	1.26	14.68
37	33	1.66	16.85	38	17	0.86	13.42
36	20	1.01	15.19	37	21	1.06	12.56
35	29	1.46	14.18	36	27	1.36	11.50
34	19	0.96	12.71	35	23	1.16	10.14
33	33	1.66	11.76	34	17	0.86	8.98
32	25	1.26	10.09	33	16	0.81	8.12
31	19	0.96	8.83	32	19	0.96	7.32
30	19	0.96	7.87	31	12	0.61	6.36
29	20	1.01	6.91	30	13	0.66	5.75
28	15	0.76	5.90	29	14	0.71	5.10
27	9	0.45	5.15	28	13	0.66	4.39
26	11	0.55	4.69	27	10	0.50	3.73
25	10	0.50	4.14	26	12	0.61	3.23
24	9	0.45	3.63	25	7	0.35	2.62
23	6	0.30	3.18	24	7	0.35	2.27
22	13	0.66	2.88	23	6	0.30	1.92
21	17	0.86	2.22	22	5	0.25	1.61
20	10	0.50	1.36	21	6	0.30	1.36
19	5	0.25	0.86	20	4	0.20	1.06
18	1	0.05	0.61	19	2	0.10	0.86
17	1	0.05	0.55	18	2	0.10	0.76
16	2	0.10	0.50	17	2	0.10	0.66
15	5	0.25	0.40	16	2	0.10	0.55
14	2	0.10	0.15	15	3	0.15	0.45
13	1	0.05	0.05	9-13	6	0.30	0.30
Total	1982			Total	1982		

Norms For The General U.S. Population: (4-Week Recall)

TABLE 10.2   NORMS FOR THE GENERAL U.S. POPULATION, TOTAL SAMPLE

Total Sample (N=1,982)		PF	RP	BP	GH	VT	SF	RE	MH	PCS	MCS
	Mean	50.00	50.00	50.00	50.00	50.00	50.00	50.00	50.00	50.00	50.00
	25th Percentile	46.65	49.18	42.19	43.87	44.32	46.28	44.81	44.19	45.46	45.07
	50th Percentile (median)	55.05	56.24	51.61	50.89	51.42	57.14	55.34	52.71	53.17	52.89
	75th Percentile	57.14	56.24	55.90	57.91	58.52	57.14	55.34	57.26	57.05	57.19
	Standard Deviation	10.00	10.00	10.00	10.00	10.00	10.00	10.00	10.00	10.00	10.00
	Range	15-58	27-57	19-63	17-64	20-71	13-58	23-56	7-65	12-69	8-73

TABLE 10.3   NORMS FOR THE GENERAL U.S. POPULATION, MALE SAMPLE

Males (N=784)		PF	RP	BP	GH	VT	SF	RE	MH	PCS	MCS
	Mean	51.50	50.85	50.63	50.19	50.69	50.79	50.37	50.96	50.92	50.49
	25th Percentile	50.85	49.17	46.04	46.21	44.32	46.28	55.34	45.90	47.53	46.06
	50th Percentile (median)	55.04	56.24	51.61	52.29	51.42	57.14	55.34	52.71	53.54	53.32
	75th Percentile	57.14	56.24	55.90	57.91	58.52	57.14	55.34	57.26	57.16	57.14
	Standard Deviation	8.91	9.58	9.65	9.76	9.54	9.48	9.86	9.41	9.28	9.25
	Range	12-58	27-57	19-63	17-64	23-71	13-58	23-56	7-65	15-69	8-68

TABLE 10.4   NORMS FOR THE GENERAL U.S. POPULATION, FEMALE SAMPLE

Females (N=1,198)		PF	RP	BP	GH	VT	SF	RE	MH	PCS	MCS
	Mean	48.61	49.21	49.41	49.82	49.35	49.27	49.66	49.11	49.15	49.54
	25th Percentile	44.56	42.09	41.77	43.87	41.95	40.85	44.80	43.63	43.46	44.01
	50th Percentile (median)	52.94	56.24	50.76	50.89	49.05	57.13	55.34	50.44	52.56	52.29
	75th Percentile	57.14	56.24	55.89	57.91	56.15	57.13	55.34	57.25	56.92	57.22
	Standard Deviation	10.73	10.32	10.28	10.22	10.37	10.40	10.12	10.43	10.60	10.40
	Range	15-58	27-57	19-63	17-64	23-71	13-58	23-56	7-65	12-69	12-73

Norms For The General U.S. Population: (4-Week Recall)

TABLE 10.5 NORMS FOR FIVE AGE GROUPS, MALES AND FEMALES COMBINED: THE GENERAL U.S. POPULATION

Ages 18-34 (N=366)		PF	RP	BP	GH	VT	SF	RE	MH	PCS	MCS
	Mean	54.07	52.45	52.05	51.36	49.14	50.45	50.02	49.05	53.76	48.00
	25th Percentile	52.95	56.24	46.47	47.61	41.95	46.28	44.80	43.62	51.14	43.08
	50th Percentile (median)	57.14	56.24	51.61	53.23	49.05	57.13	55.34	50.44	55.76	50.84
	75th Percentile	57.14	56.24	57.60	57.91	56.15	57.13	55.34	54.98	58.06	55.37
	Standard Deviation	6.09	7.85	8.94	8.95	9.75	9.09	9.54	9.66	7.04	9.57
	Range	15-58	27-57	19-63	17-64	23-66	19-58	23-56	14-65	20-67	12-68

Ages 35-44 (N=435)		PF	RP	BP	GH	VT	SF	RE	MH	PCS	MCS
	Mean	52.06	51.50	50.73	50.85	50.29	50.76	50.94	49.33	51.81	49.61
	25th Percentile	50.85	49.17	46.04	46.20	44.32	46.28	55.34	43.63	49.21	45.54
	50th Percentile (median)	55.04	56.24	51.61	53.23	51.42	57.13	55.34	52.71	54.29	52.67
	75th Percentile	57.14	56.24	55.90	57.91	58.52	57.13	55.34	57.25	57.59	56.71
	Standard Deviation	7.95	9.01	9.46	9.75	9.74	9.10	9.10	10.47	8.54	9.88
	Range	17-58	27-57	19-63	17-64	23-70	13-58	23-56	7-65	14-69	8-66

Ages 45-54 (N=417)		PF	RP	BP	GH	VT	SF	RE	MH	PCS	MCS
	Mean	49.40	50.10	49.19	49.31	50.40	50.07	50.60	49.44	49.37	50.32
	25th Percentile	46.65	49.17	41.76	43.86	44.32	46.28	55.34	43.62	44.40	45.76
	50th Percentile (median)	52.94	56.24	50.75	50.88	51.42	57.13	55.34	52.71	52.23	53.91
	75th Percentile	57.14	56.24	55.89	57.91	58.52	57.13	55.34	57.25	56.68	57.24
	Standard Deviation	10.01	9.94	10.19	10.70	10.49	10.09	9.53	10.71	10.37	10.27
	Range	15-58	27-57	19-63	17-64	23-71	13-58	23-56	14-65	12-69	9-73

Norms For The General U.S. Population: (4-Week Recall)
TABLE 10.5 (CONTINUED)

Ages 55-64 (N=310)		PF	RP	BP	GH	VT	SF	RE	MH	PCS	MCS
	Mean	47.80	48.98	48.69	49.66	51.30	50.68	50.06	52.08	47.77	52.29
	25th Percentile	42.46	42.09	41.76	43.86	44.32	46.28	55.34	48.17	41.76	48.50
	50th Percentile (median)	52.94	56.24	50.75	50.88	53.78	57.13	55.34	54.98	51.07	55.43
	75th Percentile	55.04	56.24	55.89	56.97	58.52	57.13	55.34	59.53	55.50	58.35
	Standard Deviation	10.30	10.65	10.80	10.17	10.25	10.13	10.63	9.35	10.34	9.39
	Range	15-58	27-57	19-63	19-64	23-71	13-58	23-56	9-65	14-64	17-67

Ages 65+ (N=454)		PF	RP	BP	GH	VT	SF	RE	MH	PCS	MCS
	Mean	41.77	44.04	46.89	47.26	49.95	47.63	48.08	51.83	42.72	52.38
	25th Percentile	31.97	27.95	37.48	39.18	44.32	40.85	44.80	45.89	33.40	47.62
	50th Percentile (median)	44.55	49.17	46.47	48.54	51.42	51.70	55.34	54.98	44.61	55.24
	75th Percentile	52.94	56.24	55.89	55.57	56.15	57.13	55.34	59.53	52.40	59.62
	Standard Deviation	12.40	11.78	10.82	10.72	10.04	12.02	11.64	9.25	11.55	9.55
	Range	15-58	27-57	19-63	17-64	23-71	13-58	23-56	14-65	14-65	15-69

Norms For The General U.S. Population: (4-Week Recall)
TABLE 10.6 NORMS FOR MALES BY AGE GROUPS: THE GENERAL U.S. POPULATION

Ages 18-34 (N=132)

	PF	RP	BP	GH	VT	SF	RE	MH	PCS	MCS
Mean	55.36	53.17	52.65	52.01	50.32	51.66	50.63	50.91	54.40	49.32
25th Percentile	55.04	56.24	46.47	48.54	44.32	46.28	55.34	47.60	52.39	44.72
50th Percentile (median)	57.14	56.24	51.61	53.23	51.42	57.13	55.34	52.71	56.13	51.54
75th Percentile	57.14	56.24	59.32	57.91	58.12	57.13	55.34	57.25	57.97	55.46
Standard Deviation	4.38	7.22	8.33	7.97	8.84	7.74	9.32	8.21	6.13	8.16
Range	17-58	27-57	19-63	26-64	23-66	24-58	23-56	18-65	25-66	13-61

Ages 35-44 (N=203)

	PF	RP	BP	GH	VT	SF	RE	MH	PCS	MCS
Mean	52.78	52.02	51.07	50.91	51.06	51.65	51.81	50.68	51.97	50.80
25th Percentile	52.48	56.24	46.47	46.20	44.32	46.28	55.34	45.89	49.92	48.06
50th Percentile (median)	55.04	56.24	51.61	53.23	51.42	57.13	55.34	54.98	53.75	53.31
75th Percentile	57.14	56.24	55.89	57.91	58.52	57.13	55.34	57.25	57.43	56.88
Standard Deviation	6.76	8.83	9.28	9.83	9.26	8.82	8.36	9.87	7.99	8.92
Range	19-58	27-57	19-63	17-64	23-71	13-58	23-56	7-65	23-66	8-64

Ages 45-54 (N=167)

	PF	RP	BP	GH	VT	SF	RE	MH	PCS	MCS
Mean	51.00	50.99	50.45	49.62	51.29	50.44	50.48	49.37	50.90	49.93
25th Percentile	48.75	49.17	46.04	43.86	44.32	46.28	55.34	41.35	46.95	45.37
50th Percentile (median)	55.04	56.24	51.61	50.88	53.78	57.13	55.34	54.98	53.89	54.31
75th Percentile	57.17	56.24	55.89	57.91	60.88	57.13	55.34	59.53	57.10	57.58
Standard Deviation	8.74	9.37	9.95	10.45	10.81	10.38	9.65	11.42	9.67	11.05
Range	17-58	27-57	19-63	19-64	23-71	13-58	23-56	14-65	15-69	9-68

Norms For The General U.S. Population: (4-Week Recall)
TABLE 10.6 (CONTINUED)

Ages 55-64 (N=106)		PF	RP	BP	GH	VT	SF	RE	MH	PCS	MCS
	Mean	49.16	49.03	49.13	49.02	51.57	51.40	50.17	53.32	48.05	52.48
	25th Percentile	44.55	42.09	42.19	43.86	46.68	46.28	55.34	50.44	42.36	50.13
	50th Percentile (median)	52.94	56.24	51.18	50.88	53.78	57.13	55.34	57.25	51.48	55.99
	75th Percentile	55.04	56.24	55.89	55.57	58.52	57.13	55.34	59.53	54.71	58.40
	Standard Deviation	9.04	10.86	10.19	10.22	9.77	9.44	11.00	8.31	9.73	8.98
	Range	21-58	27-57	25-63	19-64	23-71	19-58	23-56	20-65	24-64	20-66

Ages 65+ (N=176)		PF	RP	BP	GH	VT	SF	RE	MH	PCS	MCS
	Mean	43.11	44.96	46.71	46.49	49.59	47.37	47.54	51.59	43.51	51.48
	25th Percentile	34.07	35.02	37.48	39.18	44.32	40.85	44.80	45.89	33.88	46.19
	50th Percentile (median)	46.65	49.17	46.47	48.54	51.42	51.70	55.34	54.98	46.32	54.41
	75th Percentile	52.94	56.24	55.89	55.57	56.15	57.13	55.34	59.53	52.47	58.48
	Standard Deviation	12.65	11.78	10.89	10.96	9.73	11.98	11.88	9.11	11.28	9.61
	Range	15-58	27-57	19-63	19-64	23-70	13-58	23-56	23-65	18-65	15-67

SF-36™ Summary Measures Manual — Norms: 1998 General U.S. Population, Pg. 100

Norms For The General U.S. Population: (4-Week Recall)
TABLE 10.7 NORMS FOR FEMALES BY AGE GROUPS: THE GENERAL U.S. POPULATION

Ages 18-34 (N=234)

	PF	RP	BP	GH	VT	SF	RE	MH	PCS	MCS
Mean	52.76	51.72	51.44	50.69	47.96	49.22	49.39	47.15	53.10	46.66
25th Percentile	50.85	49.17	46.47	46.20	41.95	40.85	44.80	41.35	49.35	39.76
50th Percentile (median)	55.04	56.24	54.18	53.23	49.05	54.42	55.34	49.02	55.05	49.96
75th Percentile	57.14	56.24	57.60	57.91	56.15	57.13	55.34	54.98	58.26	55.29
Standard Deviation	7.19	8.38	9.50	9.82	10.48	10.15	9.73	10.61	7.81	10.65
Range	15-58	27-57	29-63	17-64	23-66	19-58	23-56	14-65	20-67	12-68

Ages 35-44 (N=232)

	PF	RP	BP	GH	VT	SF	RE	MH	PCS	MCS
Mean	51.36	50.97	50.38	50.79	49.52	49.88	50.09	47.99	51.65	48.44
25th Percentile	50.85	49.17	46.04	46.20	44.32	46.28	44.80	41.35	47.44	42.23
50th Percentile (median)	55.04	56.24	51.61	53.23	49.05	51.70	55.34	49.30	54.42	51.43
75th Percentile	57.14	56.24	55.89	57.91	56.15	57.13	55.34	57.25	58.09	56.57
Standard Deviation	8.93	9.17	9.654	9.68	10.14	9.31	9.72	10.90	9.07	10.63
Range	17-58	27-57	19-63	17-64	23-71	13-58	23-56	7-65	14-69	13-66

Ages 45-54 (N=250)

	PF	RP	BP	GH	VT	SF	RE	MH	PCS	MCS
Mean	47.87	49.25	47.98	49.01	49.55	49.73	50.72	49.50	47.90	50.69
25th Percentile	43.16	42.09	41.76	42.93	41.95	46.28	55.34	43.62	41.66	46.23
50th Percentile (median)	52.94	56.24	46.47	50.88	51.42	57.13	55.34	52.71	50.63	53.30
75th Percentile	57.14	56.24	55.89	57.91	56.15	57.13	55.34	57.25	56.36	57.06
Standard Deviation	10.89	10.40	10.29	10.95	10.13	9.81	9.44	10.00	10.82	9.47
Range	15-58	27-57	19-63	17-64	23-71	19-58	23-56	16-65	12-65	18-73

Norms For The General U.S. Population: (4-Week Recall)
TABLE 10.7 (CONTINUED)

Ages 55-64 (N=204)		PF	RP	BP	GH	VT	SF	RE	MH	PCS	MCS
	Mean	46.56	48.93	48.29	50.25	51.05	50.03	49.96	50.95	47.51	51.79
	25th Percentile	41.93	42.09	41.76	43.86	44.32	46.28	50.07	45.89	41.47	47.99
	50th Percentile (median)	50.85	56.24	50.75	53.23	51.42	57.13	55.34	52.71	50.86	55.16
	75th Percentile	55.04	56.24	55.89	57.91	60.88	57.13	55.34	59.53	56.06	58.04
	Standard Deviation	11.21	10.50	11.34	10.12	10.69	10.71	10.32	10.10	10.90	9.75
	Range	15-58	27-57	19-63	19-64	25-71	13-58	23-56	9-65	14-62	17-66

Ages 65+ (N=278)		PF	RP	BP	GH	VT	SF	RE	MH	PCS	MCS
	Mean	40.82	43.39	47.02	47.81	50.20	47.82	48.46	52.00	42.17	53.02
	25th Percentile	29.87	27.95	37.48	39.18	41.95	40.85	44.80	48.17	33.27	48.68
	50th Percentile (median)	44.55	42.09	46.47	48.54	51.42	51.70	55.34	52.71	42.68	55.78
	75th Percentile	50.85	56.24	55.89	55.57	58.52	57.13	55.34	59.53	51.87	59.85
	Standard Deviation	12.15	11.77	10.79	10.54	10.27	12.08	11.47	9.35	11.72	9.48
	Range	15-58	27-57	19-63	17-64	23-71	13-58	23-56	14-65	14-63	20-69

Norms For The General U.S. Population: (4-Week Recall)

TABLE 10.8 NORMS FOR "HEALTHY" GROUP WITH NO CHRONIC CONDITIONS: GENERAL U.S. POPULATION

Healthy (N=571)		PF	RP	BP	GH	VT	SF	RE	MH	PCS	MCS
	Mean	54.67	54.68	56.27	55.37	54.36	54.06	52.85	53.04	55.83	52.48
	25th Percentile	55.04	56.24	51.61	50.88	49.05	51.70	55.34	50.44	53.75	49.90
	50th Percentile (median)	57.14	56.24	55.89	56.97	56.15	57.13	55.34	54.98	56.86	54.26
	75th Percentile	57.14	56.24	62.74	60.25	60.88	57.13	55.34	59.53	59.21	57.57
	Standard Deviation	5.61	4.98	6.89	7.05	8.05	6.10	7.11	7.56	5.34	7.25
	Range	15-58	27-57	24-63	28-64	25-71	13-58	23-56	18-65	27-67	22-68

TABLE 10.9 NORMS FOR ANEMIA: GENERAL U.S. POPULATION

Anemia (N=82)		PF	RP	BP	GH	VT	SF	RE	MH	PCS	MCS
	Mean	47.09	45.47	44.99	45.75	44.22	45.11	44.74	46.75	46.12	45.24
	25th Percentile	43.16	35.02	37.48	36.84	37.22	35.42	34.27	39.08	37.54	38.77
	50th Percentile (median)	52.94	49.17	42.19	48.54	46.68	46.28	44.80	48.17	48.69	47.11
	75th Percentile	57.14	56.24	51.61	53.23	51.42	57.13	55.34	54.98	56.26	52.94
	Standard Deviation	11.94	11.30	10.35	10.78	9.10	12.25	11.92	10.34	11.70	10.37
	Range	15-58	27-57	19-63	19-64	23-66	13-58	23-56	25-65	14-66	18-65

TABLE 10.10 NORMS FOR BACK PAIN, SCIATICA: GENERAL U.S. POPULATION

Back Pain (N=766)		PF	RP	BP	GH	VT	SF	RE	MH	PCS	MCS
	Mean	46.60	46.44	44.55	46.46	46.52	46.87	47.60	47.63	45.60	47.95
	25th Percentile	40.36	35.02	37.48	39.18	39.58	40.85	44.80	41.35	37.99	41.82
	50th Percentile (median)	50.85	49.17	46.47	48.54	46.68	51.70	55.34	50.44	48.56	50.86
	75th Percentile	55.04	56.24	51.61	53.23	53.78	57.13	55.34	57.25	54.43	56.16
	Standard Deviation	11.31	11.43	9.28	10.55	10.21	11.16	11.25	10.88	10.84	10.97
	Range	15-58	27-57	19-63	17-64	23-71	13-58	23-56	7-65	14-65	8-73

Norms For The General U.S. Population: (4-Week Recall)

TABLE 10.11   NORMS FOR CANCER (EXCEPT SKIN): GENERAL U.S. POPULATION

Cancer (N=70)	PF	RP	BP	GH	VT	SF	RE	MH	PCS	MCS
Mean	41.01	43.25	43.32	43.40	46.86	43.23	44.17	49.95	41.14	48.54
25th Percentile	29.87	27.95	33.63	36.84	39.58	35.42	34.27	43.62	32.14	41.36
50th Percentile (median)	42.46	49.17	41.76	43.86	46.68	46.28	55.34	52.71	41.49	51.32
75th Percentile	52.94	56.24	51.61	50.88	53.78	57.13	55.34	57.25	49.41	59.27
Standard Deviation	13.34	12.09	10.64	10.94	10.96	12.60	13.15	10.17	11.52	11.24
Range	15-58	27-57	19-63	19-64	23-68	13-58	23-56	20-65	17-61	20-67

TABLE 10.12   NORMS FOR DEPRESSION: GENERAL U.S. POPULATION

Depression (N=256)	PF	RP	BP	GH	VT	SF	RE	MH	PCS	MCS
Mean	43.54	43.04	42.93	40.86	40.51	39.01	39.27	36.51	45.13	36.78
25th Percentile	34.07	27.95	33.63	31.22	32.48	29.99	23.73	27.72	34.46	28.30
50th Percentile (median)	48.75	42.09	41.76	41.52	41.95	40.85	34.27	36.81	47.26	36.53
75th Percentile	55.04	56.24	51.61	50.88	46.68	46.28	55.34	45.89	56.07	45.47
Standard Deviation	12.72	11.91	10.74	11.41	9.99	11.95	12.37	11.28	12.54	11.60
Range	15-58	27-57	19-63	17-63	23-68	13-58	23-56	7-62	12-69	8-66

TABLE 10.13   NORMS FOR DIABETES: GENERAL U.S. POPULATION

Diabetes (N=169)	PF	RP	BP	GH	VT	SF	RE	MH	PCS	MCS
Mean	42.26	44.60	44.89	41.19	45.90	44.93	46.33	48.10	42.05	48.24
25th Percentile	29.87	27.95	37.48	33.56	37.22	35.42	34.27	41.35	31.76	43.22
50th Percentile (median)	46.65	49.17	46.47	41.52	46.68	46.28	55.34	50.44	45.02	50.86
75th Percentile	55.04	56.24	55.89	50.88	56.15	57.13	55.34	57.25	51.94	57.90
Standard Deviation	13.30	12.00	10.61	10.91	11.68	12.30	12.83	11.99	11.49	12.00
Range	15-58	27-57	19-63	17-62	23-66	13-58	23-56	7-65	14-60	8-67

Norms For The General U.S. Population: (4-Week Recall)

TABLE 10.14 NORMS FOR HEART DISEASE: GENERAL U.S. POPULATION

Heart Dis. (N=184)	PF	RP	BP	GH	VT	SF	RE	MH	PCS	MCS
Mean	39.39	42.23	43.18	40.47	45.63	43.95	45.88	48.10	39.36	48.84
25th Percentile	29.87	27.95	37.48	31.81	39.58	35.42	34.27	41.35	30.22	41.62
50th Percentile (median)	42.46	42.09	41.76	41.52	46.68	46.28	55.34	50.44	39.57	51.49
75th Percentile	50.85	56.24	51.61	48.54	53.78	57.13	55.34	57.25	49.02	58.23
Standard Deviation	12.80	12.04	10.42	10.98	10.28	12.29	12.80	11.62	11.31	11.44
Range	15-58	27-57	19-63	19-64	23-66	13-58	23-56	14-65	14-62	17-67

TABLE 10.15 NORMS FOR HYPERTENSION: GENERAL U.S. POPULATION

Hypertension (N=503)	PF	RP	BP	GH	VT	SF	RE	MH	PCS	MCS
Mean	43.80	46.07	46.50	45.68	48.34	47.75	48.84	49.60	44.08	50.63
25th Percentile	36.16	35.02	37.48	38.24	41.95	40.85	44.80	43.62	35.91	44.79
50th Percentile (median)	46.65	49.17	46.47	47.61	49.05	51.70	55.34	52.71	46.47	53.42
75th Percentile	55.04	56.24	55.89	53.23	56.15	57.13	55.34	57.25	53.54	58.05
Standard Deviation	11.93	11.75	10.49	10.43	10.28	11.32	11.12	10.41	11.56	10.00
Range	15-58	27-57	19-63	17-64	23-71	13-58	23-56	9-65	12-65	14-73

TABLE 10.16 NORMS FOR KIDNEY DISEASE: GENERAL U.S. POPULATION

Kidney Dis. (N=45)	PF	RP	BP	GH	VT	SF	RE	MH	PCS	MCS
Mean	37.35	40.19	40.38	36.56	42.22	40.52	45.11	44.48	36.79	46.08
25th Percentile	24.51	27.95	33.20	24.98	37.22	35.42	34.27	39.08	26.40	38.49
50th Percentile (median)	36.16	35.02	37.91	36.84	41.95	40.85	55.34	41.35	37.00	45.00
75th Percentile	52.94	56.24	46.04	43.86	51.42	51.70	55.34	54.98	48.58	55.89
Standard Deviation	14.61	12.27	8.28	10.30	8.69	12.20	12.66	10.39	11.43	11.29
Range	15-58	27-57	25-56	19-54	23-52	19-58	23-56	24-60	21-58	25-67

Norms For The General U.S. Population: (4-Week Recall)

TABLE 10.17 NORMS FOR LIVER DISEASE: GENERAL U.S. POPULATION

Liver Dis. (N=35)	PF	RP	BP	GH	VT	SF	RE	MH	PCS	MCS
Mean	45.65	42.01	41.63	38.36	45.27	40.49	40.93	44.59	42.65	42.75
25th Percentile	38.26	27.95	33.20	28.88	39.58	24.56	23.73	36.81	33.40	37.03
50th Percentile (median)	50.85	42.09	41.76	39.18	49.05	46.28	44.80	50.44	41.39	43.66
75th Percentile	52.94	56.24	55.89	48.54	53.78	57.13	55.34	57.25	55.42	56.26
Standard Deviation	12.16	13.55	13.28	12.29	11.49	16.30	15.32	15.40	13.96	15.20
Range	19-58	27-57	19-63	17-63	23-61	13-58	23-56	7-65	19-60	8-62

TABLE 10.18 NORMS FOR LUNG DISEASE: GENERAL U.S. POPULATION

Lung Dis. (N=100)	PF	RP	BP	GH	VT	SF	RE	MH	PCS	MCS
Mean	37.82	39.29	42.70	38.60	41.84	40.58	42.95	44.74	38.26	44.94
25th Percentile	27.77	27.95	33.63	28.88	37.22	29.99	23.73	34.54	30.53	33.92
50th Percentile (median)	38.26	35.02	41.76	38.24	41.95	40.85	44.80	48.17	37.00	47.36
75th Percentile	46.65	49.17	50.78	50.88	46.68	51.70	55.34	54.98	47.35	55.18
Standard Deviation	12.35	11.68	10.71	12.34	9.40	12.47	13.90	12.20	11.09	12.54
Range	15-58	27-57	19-63	17-64	23-64	13-58	23-56	14-65	18-62	15-67

TABLE 10.19 NORMS FOR OSTEOARTHRITIS: GENERAL U.S. POPULATION

OA (N=286)	PF	RP	BP	GH	VT	SF	RE	MH	PCS	MCS
Mean	38.97	41.20	40.77	42.99	45.31	43.69	45.57	47.56	38.85	48.72
25th Percentile	27.77	27.95	33.20	35.90	39.58	35.42	34.27	41.35	29.19	41.24
50th Percentile (median)	40.36	42.09	41.76	43.86	46.68	46.28	55.34	48.17	39.01	51.27
75th Percentile	50.85	56.24	46.47	50.88	51.42	57.13	55.34	57.25	48.64	57.77
Standard Deviation	12.75	11.96	9.86	10.70	10.07	12.54	12.72	10.64	11.81	10.98
Range	15-58	27-57	19-63	17-64	23-71	13-58	23-56	7-65	14-62	13-73

Norms For The General U.S. Population: (4-Week Recall)

TABLE 10.20 NORMS FOR RHEUMATOID ARTHRITIS: GENERAL U.S. POPULATION

RA (N=133)		PF	RP	BP	GH	VT	SF	RE	MH	PCS	MCS
	Mean	42.52	43.17	42.19	43.24	47.11	44.75	45.40	47.46	41.66	48.11
	25th Percentile	34.07	27.95	33.20	33.56	39.58	35.42	34.27	41.35	33.05	41.36
	50th Percentile (median)	46.65	49.17	41.76	46.20	46.68	46.28	55.34	48.17	42.68	50.14
	75th Percentile	52.94	56.24	50.75	53.23	56.15	57.13	55.34	57.25	51.04	56.56
	Standard Deviation	12.65	12.37	10.25	11.97	10.67	12.28	13.15	11.25	11.37	11.22
	Range	15-58	27-57	19-63	17-64	23-71	13-58	23-56	7-65	14-58	8-65

TABLE 10.21 NORMS FOR ULCERS: GENERAL U.S. POPULATION

Ulcers (N=145)		PF	RP	BP	GH	VT	SF	RE	MH	PCS	MCS
	Mean	43.40	43.33	42.77	41.62	43.93	43.84	44.81	44.21	42.62	45.03
	25th Percentile	34.07	27.95	33.63	33.56	34.85	35.42	34.27	34.54	33.88	36.29
	50th Percentile (median)	46.65	49.17	41.76	41.52	44.32	46.28	55.34	48.17	44.16	47.92
	75th Percentile	55.04	56.24	51.61	50.88	53.78	57.13	55.34	54.98	52.84	55.30
	Standard Deviation	12.87	12.05	10.53	11.44	11.22	12.28	12.79	12.97	11.68	12.71
	Range	15-58	27-57	19-63	17-64	23-64	13-58	23-56	7-65	14-65	8-66

ACUTE FORM

TABLE 10.22 Frequency Distribution for PCS and MCS Scale Scores (1-Week Recall): 1998 General U.S. Population (N=1,646)

	PCS Scale				MCS Scale		
Score	F	%	Cum. %	Score	F	%	Cum. %
70	1	0.06	100.00	73	1	0.06	100.00
69	2	0.12	99.94	71	1	0.06	99.94
68	1	0.06	99.82	70	1	0.06	99.88
67	7	0.43	99.76	67	4	0.24	99.82
66	2	0.12	99.33	66	6	0.36	99.57
65	4	0.24	99.21	65	6	0.36	99.21
64	1	0.06	98.97	64	13	0.79	98.85
63	8	0.49	98.91	63	16	0.97	98.06
62	8	0.49	98.42	62	32	1.94	97.08
61	32	1.94	97.93	61	28	1.70	95.14
60	53	3.22	95.99	60	58	3.52	93.44
59	78	4.74	92.77	59	97	5.89	89.91
58	114	6.93	88.03	58	113	6.87	84.02
57	109	6.62	81.11	57	114	6.93	77.16
56	104	6.32	74.48	56	106	6.44	70.23
55	102	6.20	68.17	55	101	6.14	63.79
54	103	6.26	61.97	54	104	6.32	57.65
53	75	4.56	55.71	53	94	5.71	51.34
52	73	4.43	51.15	52	66	4.01	45.63
51	60	3.65	46.72	51	61	3.71	41.62
50	69	4.19	43.07	50	63	3.83	37.91
49	48	2.92	38.88	49	48	2.92	34.08
48	34	2.07	35.97	48	49	2.98	31.17
47	34	2.07	33.90	47	42	2.55	28.19
46	30	1.82	31.83	46	22	1.34	25.64
45	30	1.82	30.01	45	23	1.40	24.30
44	29	1.76	28.19	44	34	2.07	22.90
43	25	1.52	26.43	43	22	1.34	20.84
42	29	1.76	24.91	42	21	1.28	19.5
41	23	1.40	23.15	41	32	1.94	18.23
40	24	1.46	21.75	40	23	1.40	16.28
39	30	1.82	20.29	39	16	0.97	14.88
38	22	1.34	18.47	38	24	1.46	13.91
37	12	0.73	17.13	37	26	1.58	12.45
36	26	1.58	16.4	36	14	0.85	10.87
35	24	1.46	14.82	35	19	1.15	10.02
34	22	1.34	13.37	34	12	0.73	8.87
33	25	1.52	12.03	33	17	1.03	8.14
32	17	1.03	10.51	32	15	0.91	7.11
31	16	0.97	9.48	31	10	0.61	6.20
30	11	0.67	8.51	30	13	0.79	5.59
29	14	0.85	7.84	29	6	0.36	4.80
28	20	1.22	6.99	28	5	0.30	4.43
27	21	1.28	5.77	27	8	0.49	4.13
26	10	0.61	4.50	26	6	0.36	3.65
25	7	0.43	3.89	25	10	0.61	3.28
24	12	0.73	3.46	24	5	0.30	2.67
23	13	0.79	2.73	23	7	0.43	2.37
22	10	0.61	1.94	22	3	0.18	1.94
21	3	0.18	1.34	21	5	0.30	1.76
20	3	0.18	1.15	20	5	0.30	1.46
19	9	0.55	0.97	19	6	0.36	1.15
18	2	0.12	0.43	18	3	0.18	0.79
17	1	0.06	0.30	16	2	0.12	0.61
16	1	0.06	0.24	14	3	0.18	0.49
15	1	0.06	0.18	13	1	0.06	0.30
12	1	0.06	0.12	12	1	0.06	0.24
11	1	0.06	0.06	3-11	3	0.18	0.18
Total	1646	100		Total	1646	100	

Norms For The General U.S. Population: Acute Form

TABLE 10.23 NORMS FOR THE GENERAL U.S. POPULATION, TOTAL SAMPLE

Total Sample (N=1646)		PF	RP	BP	GH	VT	SF	RE	MH	PCS	MCS
	Mean	50	50	50	50	50	50	50	50	50	50
	25th Percentile	47.01	49.04	44.86	43.66	44.20	45.35	45.15	45.23	45.40	45.80
	50th Percentile (median)	55.15	56.13	50.26	52.92	51.08	56.30	55.40	52.18	53.48	53.06
	75th Percentile	57.19	56.13	61.05	57.55	57.95	56.30	55.40	56.82	57.21	57.00
	Standard Deviation	10	10	10	10	10	10	10	10	10	10
	Range	16-58	27-57	19-62	17-64	23-70	12-57	24-56	5-64	11-71	3-74

TABLE 10.24 NORMS FOR THE GENERAL U.S. POPULATION, MALE SAMPLE

Males (N=659)		PF	RP	BP	GH	VT	SF	RE	MH	PCS	MCS
	Mean	51.71	51.45	51.42	51.17	51.43	51.28	51.41	51.26	51.46	51.17
	25th Percentile	51.08	49.04	45.28	45.97	46.49	50.82	55.40	47.55	48.62	48.04
	50th Percentile (median)	55.15	56.13	54.41	52.92	53.37	56.30	55.40	54.50	54.24	53.66
	75th Percentile	57.19	56.13	61.05	57.55	60.24	56.30	55.40	59.14	57.67	57.25
	Standard Deviation	8.65	9.07	9.21	9.54	9.65	9.13	8.77	9.03	8.86	8.59
	Range	16-58	27-57	19-62	17-64	23-70	12-57	24-56	5-64	16-67	11-72

TABLE 10.25 NORMS FOR THE GENERAL U.S. POPULATION, FEMALE SAMPLE

Females (N=987)		PF	RP	BP	GH	VT	SF	RE	MH	PCS	MCS
	Mean	48.42	48.66	48.69	48.92	48.67	48.82	48.70	48.84	48.65	48.91
	25th Percentile	42.93	41.95	40.71	43.66	41.91	45.35	45.15	42.91	41.91	43.50
	50th Percentile (median)	53.12	56.13	50.26	50.60	48.79	56.30	55.40	52.18	52.16	52.47
	75th Percentile	57.19	56.13	61.05	56.62	57.95	56.30	55.40	56.82	56.72	56.69
	Standard Deviation	10.87	10.62	10.52	10.30	10.14	10.61	10.86	10.69	10.86	10.88
	Range	16-58	27-57	19-62	17-64	23-70	12-57	24-56	5-64	11-71	3-74

Norms For The General U.S. Population: Acute Form

TABLE 10.26 NORMS FOR FIVE AGE GROUPS, MALES AND FEMALES COMBINED: THE GENERAL U.S. POPULATION

Ages 18-34 (N=293)

	PF	RP	BP	GH	VT	SF	RE	MH	PCS	MCS
Mean	53.47	51.69	51.70	51.83	49.71	50.57	49.81	48.98	53.31	48.30
25th Percentile	53.12	49.04	45.28	48.29	41.91	45.35	45.15	42.91	50.23	43.87
50th Percentile (median)	57.19	56.13	54.41	54.30	51.08	56.30	55.40	52.18	55.97	51.73
75th Percentile	57.19	56.13	61.05	58.93	57.95	56.30	55.40	56.82	58.39	55.60
Standard Deviation	7.21	8.85	9.49	9.47	9.74	9.40	10.21	9.81	8.10	10.08
Range	18-58	27-57	19-62	17-64	23-70	12-57	24-56	5-64	24-71	3-63

Ages 35-44 (N=371)

	PF	RP	BP	GH	VT	SF	RE	MH	PCS	MCS
Mean	52.89	51.85	51.11	50.66	50.19	50.86	51.74	50.07	52.04	50.05
25th Percentile	51.08	49.04	45.28	45.97	44.20	45.35	55.40	45.22	49.63	47.31
50th Percentile (median)	55.15	56.13	54.41	52.92	51.08	56.30	55.40	52.18	54.46	53.41
75th Percentile	57.19	56.13	61.05	57.55	57.95	56.30	55.40	56.82	57.25	56.53
Standard Deviation	7.28	8.41	9.19	9.32	9.68	8.98	8.49	10.52	7.99	9.80
Range	16-58	27-57	19-62	17-64	23-68	17-57	24-56	5-64	16-68	11-67

Ages 45-54 (N=332)

	PF	RP	BP	GH	VT	SF	RE	MH	PCS	MCS
Mean	50.41	50.75	49.63	48.91	49.97	49.82	50.56	49.97	49.95	50.03
25th Percentile	49.04	49.04	41.13	43.66	44.20	45.35	45.15	45.23	46.20	45.74
50th Percentile (median)	55.15	56.13	50.26	50.60	51.08	56.30	55.39	52.18	52.92	53.29
75th Percentile	57.19	56.13	61.05	55.23	57.95	56.30	55.39	57.98	56.59	57.13
Standard Deviation	9.53	9.49	9.78	10.29	10.47	10.53	9.38	10.23	9.69	9.83
Range	16-58	27-57	19-62	17-64	23-70	12-57	24-56	8-64	12-70	6-65

Norms For The General U.S. Population: Acute Form
TABLE 10.26 (CONTINUED)

Ages 55-64 (N=248)		PF	RP	BP	GH	VT	SF	RE	MH	PCS	MCS
	Mean	47.04	49.58	48.21	49.43	50.67	49.95	49.69	50.30	47.90	51.07
	25th Percentile	42.93	41.95	40.71	41.34	44.20	45.35	45.15	45.23	42.88	47.42
	50th Percentile (median)	51.08	56.13	49.43	51.99	51.08	56.30	55.40	52.18	51.55	54.67
	75th Percentile	55.15	56.13	54.41	57.55	60.24	56.30	55.40	59.14	55.29	57.72
	Standard Deviation	10.98	10.15	10.51	10.13	10.34	10.39	9.99	10.36	10.71	9.90
	Range	16-58	27-57	19-62	19-64	23-70	12-57	24-56	15-64	15-67	15-67

Ages 65+ (N=402)		PF	RP	BP	GH	VT	SF	RE	MH	PCS	MCS
	Mean	41.32	43.97	46.94	47.19	49.90	48.04	47.74	51.66	42.61	52.36
	25th Percentile	32.30	27.77	36.56	39.03	41.91	39.88	40.03	47.55	33.19	47.69
	50th Percentile (median)	42.93	49.04	49.43	48.29	48.79	56.30	55.40	54.50	43.67	54.48
	75th Percentile	51.08	56.13	54.41	55.23	57.95	56.30	55.40	59.14	53.35	58.86
	Standard Deviation	11.51	11.85	10.91	10.67	10.21	11.25	11.50	8.96	11.42	9.24
	Range	16-58	27-57	19-62	17-64	23-70	12-57	24-56	17-64	11-62	17-74

Norms For The General U.S. Population: Acute Form

TABLE 10.27 NORMS FOR MALES BY AGE GROUPS: THE GENERAL U.S. POPULATION

Ages 18-34 (N=106)		PF	RP	BP	GH	VT	SF	RE	MH	PCS	MCS
	Mean	54.63	52.59	52.91	54.19	52.02	52.06	51.34	50.92	54.42	50.32
	25th Percentile	55.15	56.13	45.28	50.60	46.49	50.82	55.40	47.55	53.24	47.37
	50th Percentile (median)	57.19	56.13	54.41	57.08	53.37	56.30	55.40	54.50	56.45	53.40
	75th Percentile	57.19	56.13	61.05	59.86	60.24	56.30	55.40	56.82	58.65	56.18
	Standard Deviation	5.91	8.38	8.49	7.93	8.89	8.62	9.24	8.03	6.82	8.09
	Range	18-58	27-57	28-62	26-64	28-70	23-57	24-56	28-62	29-67	23-60

Ages 35-44 (N=178)		PF	RP	BP	GH	VT	SF	RE	MH	PCS	MCS
	Mean	53.14	52.75	51.65	50.79	50.71	51.44	52.72	50.79	52.33	50.89
	25th Percentile	53.12	56.13	45.28	45.97	44.20	50.82	55.40	45.23	50.10	47.69
	50th Percentile (median)	57.19	56.13	54.41	52.92	52.22	56.30	55.40	52.18	54.73	53.18
	75th Percentile	57.19	56.13	61.05	57.55	57.95	56.30	55.40	59.14	57.59	57.25
	Standard Deviation	7.26	7.62	9.10	9.13	9.45	8.59	7.19	10.03	7.69	8.97
	Range	16-58	27-57	19-62	24-64	23-68	17-57	24-56	5-64	16-63	11-67

Ages 45-54 (N=130)		PF	RP	BP	GH	VT	SF	RE	MH	PCS	MCS
	Mean	53.17	52.91	51.79	49.70	51.55	51.83	52.13	51.57	52.14	51.29
	25th Percentile	52.10	56.13	49.43	43.66	46.49	50.82	55.40	47.55	49.79	48.11
	50th Percentile (median)	55.15	56.13	54.41	52.92	53.37	56.30	55.40	54.50	53.76	53.67
	75th Percentile	57.19	56.13	61.05	56.62	60.24	56.30	55.40	59.14	56.94	57.16
	Standard Deviation	5.92	7.23	8.34	9.22	9.92	8.34	7.39	8.92	6.97	8.19
	Range	26-58	27-57	19-62	21-64	23-70	12-57	24-56	15-64	18-64	14-65

Norms For The General U.S. Population: Acute Form
TABLE 10.27 (CONTINUED)

Ages 55-64 (N=82)		PF	RP	BP	GH	VT	SF	RE	MH	PCS	MCS
	Mean	48.12	50.94	49.46	50.30	51.44	50.69	50.27	51.79	49.02	51.88
	25th Percentile	44.97	49.04	44.86	43.66	44.20	50.82	45.15	47.55	45.61	49.27
	50th Percentile (median)	51.08	56.13	54.41	52.92	53.37	56.30	55.40	54.50	51.85	55.20
	75th Percentile	55.15	56.13	54.41	57.55	60.24	56.30	55.40	59.14	56.16	58.32
	Standard Deviation	10.11	9.46	10.51	9.75	10.29	10.18	9.56	9.97	10.13	9.71
	Range	18-58	27-57	19-62	26-64	23-70	17-57	24-56	22-64	19-67	19-67

Ages 65+ (N=163)		PF	RP	BP	GH	VT	SF	RE	MH	PCS	MCS
	Mean	43.82	45.51	48.70	47.32	51.08	49.05	49.51	51.94	44.42	52.87
	25th Percentile	34.78	34.85	40.71	40.41	44.20	45.35	45.15	47.54	35.47	48.77
	50th Percentile (median)	47.00	49.04	50.25	49.67	53.36	56.29	55.39	54.50	46.39	54.40
	75th Percentile	53.11	56.13	54.40	56.61	60.24	56.29	55.39	59.13	54.51	58.75
	Standard Deviation	11.48	11.71	10.17	11.63	10.85	10.80	10.40	9.04	11.22	8.53
	Range	16-58	27-57	24-62	17-64	23-70	12-57	24-56	17-64	19-62	17-72

Norms For The General U.S. Population: Acute Form
TABLE 10.28 NORMS FOR FEMALES BY AGE GROUPS: THE GENERAL U.S. POPULATION

Ages 18-34 (N=187)		PF	RP	BP	GH	VT	SF	RE	MH	PCS	MCS
	Mean	52.30	50.78	50.48	49.44	47.36	49.06	48.27	47.01	52.18	46.26
	25th Percentile	51.07	49.04	41.12	43.65	39.61	45.35	45.15	40.59	49.08	40.82
	50th Percentile (median)	55.15	56.13	54.40	50.60	48.78	56.29	55.39	49.86	55.18	50.11
	75th Percentile	57.18	56.13	61.04	57.54	55.66	56.29	55.39	56.81	58.08	54.47
	Standard Deviation	8.17	9.23	10.29	10.29	10.03	9.93	10.93	11.08	9.10	11.42
	Range	20-58	27-57	19-62	17-64	23-65	12-57	24-56	5-64	24-71	3-63

Ages 35-44 (N=193)		PF	RP	BP	GH	VT	SF	RE	MH	PCS	MCS
	Mean	52.63	50.96	50.58	50.54	49.67	50.28	50.77	49.35	51.75	49.23
	25th Percentile	51.07	49.04	45.27	45.97	44.20	45.35	55.39	45.22	48.43	47.04
	50th Percentile (median)	55.15	56.13	50.25	52.91	51.07	56.29	55.39	52.18	54.38	53.46
	75th Percentile	57.18	56.13	61.04	57.54	57.95	56.29	55.39	56.81	57.17	56.20
	Standard Deviation	7.31	9.06	9.26	9.53	9.90	9.34	9.52	10.96	8.29	10.51
	Range	24-58	27-57	28-62	17-64	23-65	17-57	24-56	10-64	23-68	13-61

Ages 45-54 (N=202)		PF	RP	BP	GH	VT	SF	RE	MH	PCS	MCS
	Mean	47.77	48.69	47.57	48.15	48.45	47.89	49.05	48.43	47.86	48.83
	25th Percentile	42.93	41.94	40.71	41.34	41.91	39.87	45.15	40.59	40.28	40.84
	50th Percentile (median)	53.11	56.13	49.42	50.60	48.78	56.29	55.39	52.18	51.51	52.79
	75th Percentile	57.18	56.13	54.40	55.23	57.95	56.29	55.39	56.81	56.19	56.92
	Standard Deviation	11.41	10.86	10.60	11.19	10.79	11.97	10.76	11.14	11.33	11.06
	Range	16-58	27-57	19-62	17-64	23-70	12-57	24-56	8-64	12-70	6-64

Norms For The General U.S. Population: Acute Form
TABLE 10.28 (CONTINUED)

Ages 55-64 (N=166)		PF	RP	BP	GH	VT	SF	RE	MH	PCS	MCS
	Mean	46.06	48.34	47.07	48.64	49.95	49.27	49.16	48.93	46.88	50.32
	25th Percentile	38.86	41.94	40.71	41.34	44.20	45.35	45.15	42.90	41.01	44.78
	50th Percentile (median)	51.07	56.13	49.42	50.60	51.07	56.29	55.39	52.18	50.39	53.67
	75th Percentile	55.15	56.13	54.40	57.54	57.95	56.29	55.39	56.81	55.14	57.19
	Standard Deviation	11.67	10.63	10.42	10.45	10.37	10.57	10.38	10.57	11.15	10.06
	Range	16-58	27-57	19-62	19-64	23-68	12-57	24-56	15-64	15-66	15-66

Ages 65+ (N=239)		PF	RP	BP	GH	VT	SF	RE	MH	PCS	MCS
	Mean	39.55	42.88	45.69	47.10	49.06	47.31	46.49	51.45	41.32	51.98
	25th Percentile	30.71	27.76	36.56	38.10	41.91	39.87	34.90	47.54	31.43	46.45
	50th Percentile (median)	39.37	41.94	45.27	48.28	48.78	56.29	55.39	52.18	41.50	54.47
	75th Percentile	49.04	56.13	54.40	55.23	57.95	56.29	55.39	59.13	51.84	58.94
	Standard Deviation	11.22	11.85	11.24	9.95	9.66	11.51	12.06	8.92	11.40	9.71
	Range	16-58	27-57	19-62	19-64	23-70	12-57	24-56	22-64	11-62	23-74

TABLE 10.29 NORMS FOR "HEALTHY" GROUP WITH NO CHRONIC CONDITIONS: GENERAL U.S. POPULATION

Healthy (N=480)		PF	RP	BP	GH	VT	SF	RE	MH	PCS	MCS
	Mean	54.63	54.43	55.60	55.21	53.98	53.78	53.19	53.16	55.37	52.64
	25th Percentile	55.15	56.13	54.40	50.60	48.78	56.29	55.39	49.86	53.75	49.79
	50th Percentile (median)	57.18	56.13	54.40	56.61	55.66	56.29	55.39	54.50	56.70	54.51
	75th Percentile	57.18	56.13	61.04	59.85	60.24	56.29	55.39	59.13	58.69	57.35
	Standard Deviation	5.38	5.42	6.71	6.59	8.43	6.03	6.40	7.44	5.49	7.00
	Range	18-58	27-57	28-62	28-64	28-70	17-57	24-56	19-64	26-67	16-67

Norms For The General U.S. Population: Acute Form

TABLE 10.30 NORMS FOR ANEMIA: GENERAL U.S. POPULATION

Anemia (N=63)		PF	RP	BP	GH	VT	SF	RE	MH	PCS	MCS
	Mean	45.89	44.30	44.45	43.09	45.19	42.19	41.61	44.43	45.53	42.72
	25th Percentile	36.82	27.76	36.56	34.40	37.32	34.40	24.65	35.95	38.33	32.06
	50th Percentile (median)	51.07	49.04	45.27	45.97	44.20	39.87	45.15	47.54	48.61	47.91
	75th Percentile	57.18	56.13	50.25	52.91	55.66	56.29	55.39	56.81	54.90	55.26
	Standard Deviation	12.08	11.75	10.49	11.83	11.56	12.94	13.90	13.51	11.25	13.92
	Range	16-58	27-57	19-62	17-64	23-65	12-57	24-56	15-64	21-66	14-63

TABLE 10.31 NORMS FOR BACK PAIN, SCIATICA: GENERAL U.S. POPULATION

Back Pain (N=615)		PF	RP	BP	GH	VT	SF	RE	MH	PCS	MCS
	Mean	46.83	46.51	44.46	46.39	46.53	47.03	47.49	46.98	45.84	47.58
	25th Percentile	38.86	34.58	36.56	39.03	39.61	39.87	34.90	40.59	38.13	40.40
	50th Percentile (median)	51.07	49.04	45.27	48.28	46.49	50.82	55.39	49.86	49.41	51.03
	75th Percentile	55.15	56.13	50.25	55.23	55.66	56.29	55.39	54.50	54.55	55.86
	Standard Deviation	11.50	11.31	9.92	10.83	10.31	11.60	11.53	11.14	11.09	11.25
	Range	16-58	27-57	19-62	17-64	23-70	12-57	24-56	5-64	12-70	6-74

TABLE 10.32 NORMS FOR CANCER (EXCEPT SKIN): GENERAL U.S. POPULATION

Cancer (N=49)		PF	RP	BP	GH	VT	SF	RE	MH	PCS	MCS
	Mean	43.86	46.89	46.16	42.77	48.01	47.69	48.62	48.66	43.80	49.94
	25th Percentile	34.78	34.85	36.56	36.75	39.61	39.87	45.15	40.59	38.08	41.32
	50th Percentile (median)	49.04	56.13	49.42	43.65	51.07	56.29	55.39	52.18	44.76	52.25
	75th Percentile	53.11	56.13	54.40	50.60	57.95	56.29	55.39	59.13	52.13	57.46
	Standard Deviation	11.73	11.44	10.96	11.06	11.33	12.16	11.27	10.54	10.76	10.78
	Range	16-58	27-57	19-62	17-63	23-65	12-57	24-56	17-64	18-59	17-74

Norms For The General U.S. Population: Acute Form

TABLE 10.33 NORMS FOR DEPRESSION: GENERAL U.S. POPULATION

Depression (N=206)		PF	RP	BP	GH	VT	SF	RE	MH	PCS	MCS
	Mean	43.46	42.68	43.28	40.46	39.95	39.04	38.23	35.87	45.31	35.93
	25th Percentile	32.75	27.76	36.56	32.08	30.45	28.92	24.65	28.99	34.59	27.25
	50th Percentile (median)	47.00	41.94	40.71	39.03	39.61	39.87	34.90	35.95	46.83	35.83
	75th Percentile	55.15	56.13	54.40	50.60	46.49	50.82	55.39	45.22	55.19	43.95
	Standard Deviation	12.52	12.03	11.67	11.69	10.29	12.75	12.55	11.99	12.53	11.99
	Range	16-58	27-57	19-62	17-64	23-65	12-57	24-56	5-60	18-71	3-65

TABLE 10.34 NORMS FOR DIABETES: GENERAL U.S. POPULATION

Diabetes (N=145)		PF	RP	BP	GH	VT	SF	RE	MH	PCS	MCS
	Mean	43.97	45.36	45.87	40.94	45.20	46.01	47.18	47.88	43.14	48.05
	25th Percentile	34.78	34.85	36.97	33.47	37.32	39.87	45.15	40.59	34.59	41.49
	50th Percentile (median)	45.87	49.04	45.27	40.41	46.49	50.82	55.39	49.86	43.83	50.71
	75th Percentile	53.11	56.13	54.40	50.60	53.36	56.29	55.39	56.81	53.23	56.46
	Standard Deviation	11.05	11.48	10.62	10.88	10.33	12.08	11.83	11.34	10.86	10.98
	Range	16-58	27-57	19-62	19-63	23-70	12-57	24-56	5-64	15-67	11-65

TABLE 10.35 NORMS FOR HEART DISEASE: GENERAL U.S. POPULATION

Heart Dis. (N=146)		PF	RP	BP	GH	VT	SF	RE	MH	PCS	MCS
	Mean	39.26	40.93	44.12	41.15	44.97	44.28	45.86	48.29	39.26	48.95
	25th Percentile	30.71	27.76	36.56	33.47	37.32	34.40	34.90	42.90	31.50	41.61
	50th Percentile (median)	38.86	41.94	41.12	41.34	44.20	45.35	55.39	52.18	38.72	51.26
	75th Percentile	49.04	56.13	52.72	49.67	51.07	56.29	55.39	56.81	48.18	57.76
	Standard Deviation	11.32	11.66	10.59	10.07	10.19	12.43	12.37	10.71	10.62	10.95
	Range	16-58	27-57	19-62	17-64	23-70	12-57	24-56	15-64	15-59	14-67

Norms For The General U.S. Population: Acute Form

TABLE 10.36 NORMS FOR HYPERTENSION: GENERAL U.S. POPULATION

Hypertension (N=414)	PF	RP	BP	GH	VT	SF	RE	MH	PCS	MCS
Mean	44.63	46.09	46.76	45.15	47.36	47.63	48.35	48.94	44.60	49.62
25th Percentile	34.78	34.85	36.55	38.10	39.61	39.87	45.15	42.90	36.46	43.62
50th Percentile (median)	48.13	49.04	45.27	45.97	48.02	56.29	55.39	49.86	47.80	52.70
75th Percentile	55.15	56.13	54.40	52.91	55.66	56.29	55.39	56.81	53.35	57.16
Standard Deviation	11.38	11.42	10.59	10.44	10.38	11.25	11.02	10.43	10.95	10.20
Range	16-58	27-57	19-62	17-64	23-70	12-57	24-56	10-64	12-67	14-67

TABLE 10.37 NORMS FOR KIDNEY DISEASE: GENERAL U.S. POPULATION

Kidney Dis. (N=42)	PF	RP	BP	GH	VT	SF	RE	MH	PCS	MCS
Mean	44.64	42.15	44.05	40.86	48.38	42.71	45.40	47.82	42.16	47.53
25th Percentile	36.82	27.76	36.56	33.47	41.91	34.40	34.90	38.27	33.92	36.65
50th Percentile (median)	44.97	41.94	40.71	38.10	44.20	45.35	55.39	47.54	44.10	48.38
75th Percentile	55.15	56.13	50.25	50.60	57.95	56.29	55.39	59.13	54.40	57.93
Standard Deviation	11.28	12.11	10.72	11.91	10.42	13.48	12.43	13.19	12.12	12.92
Range	18-58	27-57	24-62	17-63	30-68	17-57	24-56	19-64	21-59	24-74

TABLE 10.38 NORMS FOR LIVER DISEASE: GENERAL U.S. POPULATION

Liver Dis. (N=39)	PF	RP	BP	GH	VT	SF	RE	MH	PCS	MCS
Mean	39.43	42.00	42.20	41.32	47.27	46.19	46.83	49.53	38.73	51.03
25th Percentile	31.73	27.76	32.82	36.71	44.20	34.40	34.90	45.22	27.83	47.76
50th Percentile (median)	36.82	41.94	36.56	43.65	48.78	50.82	55.39	49.86	38.88	54.30
75th Percentile	47.00	56.13	61.04	48.28	55.66	56.29	55.39	59.13	51.83	59.35
Standard Deviation	11.29	12.90	12.80	9.87	10.10	12.34	11.97	11.77	12.78	12.44
Range	24-58	27-57	28-62	17-53	30-65	23-57	24-56	26-64	21-58	24-74

Norms For The General U.S. Population: Acute Form

TABLE 10.39 NORMS FOR LUNG DISEASE: GENERAL U.S. POPULATION

Lung Dis. (N=64)	PF	RP	BP	GH	VT	SF	RE	MH	PCS	MCS
Mean	38.82	40.48	43.31	38.71	43.11	42.20	46.13	46.31	38.40	47.38
25th Percentile	28.67	27.76	36.56	31.16	37.32	34.40	34.90	38.27	28.51	41.06
50th Percentile (median)	38.86	34.85	44.86	36.71	44.20	45.35	55.39	47.54	38.63	52.06
75th Percentile	51.07	56.13	50.25	48.28	51.07	56.29	55.39	56.81	46.17	56.01
Standard Deviation	11.75	11.71	10.03	9.77	9.76	12.62	12.24	11.91	10.93	11.94
Range	16-58	27-57	19-62	17-63	23-61	12-57	24-56	19-64	19-61	19-66

TABLE 10.40 NORMS FOR OSTEOARTHRITIS: GENERAL U.S. POPULATION

OA (N=246)	PF	RP	BP	GH	VT	SF	RE	MH	PCS	MCS
Mean	39.14	41.36	39.67	40.99	44.09	42.78	43.50	45.46	38.97	46.35
25th Percentile	28.67	27.76	32.41	31.16	35.80	34.40	34.90	35.95	30.98	35.47
50th Percentile (median)	38.86	41.94	36.97	41.34	44.20	45.35	45.15	47.54	39.15	48.74
75th Percentile	49.04	56.13	45.27	50.60	51.07	56.29	55.39	54.50	48.18	56.34
Standard Deviation	11.82	11.87	9.91	11.51	10.56	12.75	12.85	11.68	11.24	12.27
Range	16-58	27-57	19-62	17-64	23-70	12-57	24-56	10-64	11-65	13-74

TABLE 10.41 NORMS FOR RHEUMATOID ARTHRITIS: GENERAL U.S. POPULATION

RA (N=147)	PF	RP	BP	GH	VT	SF	RE	MH	PCS	MCS
Mean	39.85	41.64	40.29	41.68	43.98	43.13	45.51	46.52	39.11	47.54
25th Percentile	28.67	27.76	32.82	33.47	37.32	34.40	34.90	40.59	28.80	38.95
50th Percentile (median)	40.89	41.94	40.29	43.65	44.20	45.35	55.39	57.54	40.28	50.23
75th Percentile	51.07	56.13	45.27	50.60	51.07	56.29	55.39	54.50	49.15	55.74
Standard Deviation	12.85	12.13	10.62	11.53	10.01	12.94	12.52	10.29	12.23	10.83
Range	16-58	27-57	19-62	17-64	23-65	12-57	24-56	15-64	11-61	14-74

Norms For The General U.S. Population: Acute Form
TABLE 10.42   NORMS FOR ULCERS: GENERAL U.S. POPULATION

Ulcers (N=125)	PF	RP	BP	GH	VT	SF	RE	MH	PCS	MCS
Mean	44.85	44.04	42.59	43.53	44.44	43.88	44.98	43.23	44.10	44.32
25th Percentile	36.82	27.76	36.56	35.79	37.32	34.40	34.90	35.95	37.24	37.22
50th Percentile (median)	49.04	49.04	41.12	43.65	44.20	50.82	55.39	45.22	45.28	47.31
75th Percentile	55.15	56.13	50.25	52.91	53.36	56.29	55.39	52.18	53.81	53.13
Standard Deviation	12.40	11.50	10.52	11.61	10.44	13.04	12.09	11.88	11.35	11.55
Range	16-58	27-57	19-62	17-64	23-65	12-57	24-56	5-64	11-62	10-67

Chapter 11: Applications: Outcomes Research

Background

Alternative treatments and health care in general should be judged in terms of "how closely the result approaches the fundamental objectives of prolonging life, relieving distress, restoring function, and preventing disability" (Lembcke, 1952). Accordingly, outcomes research broadens the definition of outcome beyond traditional clinical endpoints to represent the implications of disease and treatment in terms of what people are able to do and how they feel. As reviewed elsewhere (Ware, 1995), substantial advances have been made in the validity and practicality of patient-based methods for assessing these outcomes. The reason for using these methods is that much is known about the dollar costs of health care and little is known about the health benefits. It is becoming increasingly accepted that patients are the best source of information about health benefits.

Results summarized in Chapter 6 illustrate advantages of both the SF-36 profile of eight scales and the PCS and MCS summary scales. Other surveys that offer both kinds of scaling options include the Sickness Impact Profile (Bergner, Bobbitt, Carter, et al., 1981) and the Duke Health Profile (Parkerson, Broadhead, & Tse, 1990). Measures that yield a single score without the option of analyzing individual dimensions include the Index of Well-Being (Kaplan & Anderson, 1988) and the EuroQOL (EuroQol Group, 1990).

This chapter summarizes results from re-analyses of four published studies that compared SF-36 health profiles to estimate health outcomes following: a) heart valve replacement surgery (Phillips & Lansky, 1992); b) hip replacement surgery (Katz, Larson, Phillips, et al., 1992); c) treatment for low back pain (Lansky, Butler, & Waller, 1992); and d) drug therapy for duodenal ulcers (Rampal, Martin, Marquis, et al., 1994). The first two of these studies were also discussed in the original *SF-36 User's Manual* (Ware, Snow, Kosinski, et al., 1993). Numerous other studies -- 85 at last count -- are cited in the references.

Published results from these four longitudinal studies were used to estimate the PCS and MCS scores and to compare findings and conclusions with those for the eight-scale SF-36 profile. These comparisons were made with two objectives: a) to address the issue of what is gained and what is lost with reliance on the SF-36 summary measures; and b) to illustrate the use of the interpretation guidelines presented in Chapters 7 and 8 to interpret results published by others. Are interpretation guidelines useful in understanding the implications and consequences of differences in outcomes for patients measured using the PCS and MCS?

For each study, mean scores for the eight SF-36 scales were aggregated using the formulas for the PCS and MCS scales in Chapter 4. The resulting estimates represent the mean scores that would have been observed for the PCS and MCS, if those summaries had been computed at the time of the original analyses. In the place of variance estimates, which cannot be estimated for the PCS and MCS from published descriptive statistics for SF-36 scales, estimates of standard deviations from the general U.S. population were relied upon. (From the published variance estimates for the SF-36 scales across the four studies, it appears that the general population standard deviations used here provide good approximations for purposes of significance testing for differences in PCS and MCS scores.)

In the examples presented below, we use interpolation to estimate the "criterion" value associated with different PCS and MCS scores, based on data provided in the tables in Chapter 9 and 14. A detailed explanation of how this was done is presented at the end of the chapter.

Heart Valve Replacement

Phillips & Lansky (1992) reported SF-36 profiles for 62 patients before and after heart valve replacement. Among the conclusions from their study were: a) substantial decrements in all eight SF-36 health status scales before surgery, particularly for five of the scales, (PF, RP, VT, SF, and RE); and b) significant improvements in all eight SF-36 scales after surgery.

Figure 11.1 compares SF-36 profiles before surgery with those at six-month follow-up. (The "normative" profile originally published is not reproduced.) The estimates for the PCS and MCS, which we have added to Figure 11.1, are consistent with results for the profile of eight scores. The average score of 35.1 for the PCS before surgery is well below the mean of 50 for the general U.S. population, and is below the mean observed for the PCS in norms for nearly all chronic conditions studied in the MOS and general population (Appendix E). These results indicate that the summaries would have led to very similar conclusions about the physical and mental health burden of these patients just prior to heart surgery.

The improvements in the eight SF-36 scales observed at the six-month follow-up after surgery were also captured by the PCS and MCS. Because the two summary scales take into account the correlations among the eight scales, they help to clarify that the burden of heart valve disease is concentrated in the physical dimension of health and that, although improvements in both the PCS and MCS were significant, improvement following heart valve replacement was concentrated in the physical dimension ($\Delta = 7.6$ for the PCS and 3.2 for the MCS).

The average improvement of 7.6 points in the PCS following surgery (from 35.1 to 42.7) can be interpreted using the norms in Chapter 14, as well as the content- and criterion-based guidelines presented in Chapter 9. Both the initial score of 35.1 and the amount of change observed must be taken into account because the interpretation guidelines depend on both, as explained in Chapter 9. Examples of results useful in interpreting an improvement of 7.6 from a pre-treatment PCS score of 35.1 are listed below (source tables from Chapters 9 and 14 are given in parentheses):

FIGURE 11.1 MEAN SF-36 SCORES BEFORE AND AFTER HEART VALVE REPLACEMENT SURGERY (N=62)

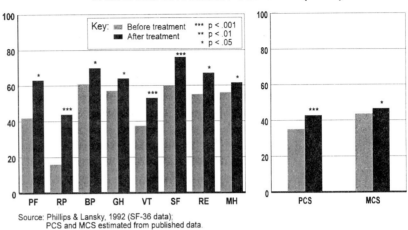

Source: Phillips & Lansky, 1992 (SF-36 data); PCS and MCS estimated from published data.

Norm-based interpretations using tables in Chapter 8 reveal that:

A. The average PCS score of 35.1 before surgery is 1.5 SD units below the mean (Table 8.1) and specifically just below the 11th percentile.

B. The change in the PCS scale from 35.1 to 42.7 represents an improvement from the 11th to the 20th percentile (Table 8.2).

C. Changes in the MCS, an average of 43.3 to 46.5 (Δ = +3.2), represents an improvement from approximately the 22nd to the 28th percentile in the general U.S. population (Table 8.2).

Content-based interpretations include specific estimates of percentage-point reductions in the likelihood of limitations in:

D. Walking one block (reduced from 37.3% to 19.1%) (Table 9.2).

E. Climbing one flight of stairs (reduced from 56.1% to 28.7%) (Table 9.2).

Criterion-based interpretation guidelines include:

F. Lowered probability of overnight hospitalization (reduced from 9.0% to 6.5%) (Table 7.6).

One-third decrease in the predicted probability of death within five years (from approximately 12.4% to 8.5%) (Table 7.8 with interpolation).

Although the average improvement in the MCS scores from 43.3 to 46.5 (Δ = +3.2) might seem numerically small, the clinical and social implications appear noteworthy from results presented in Chapter 7, including:

H. A 21% reduction (from 15.0% to 11.8%) in the likelihood of clinical depression (Table 9.10).

I. A 20% reduction (from 21.1% to 16.9%) in the likelihood of mental health treatment (Table 9.10).

J. A 12% reduction in the probability of reporting substantial life stress (from 35.8% to 31.4%) (Table 9.9).

Numerous other norm-based and criterion-based interpretations of these results are possible using the tables presented in Chapters 8 and 7, respectively.

Hip Replacement

The top and bottom panels in figure 11.2 compare unstandardized (0-100 scoring) and standardized (SD units) estimates of changes in the eight SF-36 scales with average changes in PCS and MCS scores for 54 adults following total hip arthroplasty (from Katz, Larson, Phillips, et al., 1992). Note that the PCS and MCS change scores maintain the same relationship across panels because both are scored in SD units. Because the standardization of change scores following treatments is a linear transformation, standardization does not change conclusions from statistical tests for significance. However, standardization does change some conclusions about which sf-36 scales changed most. The top panel (unstandardized) suggests that RP improved substantially more than GH; however, as shown in the second panel (standardized scores), they changed nearly the same amount in SD units and GH is more significant, statistically. Likewise, the top panel suggests that re improved more than MH; however, as shown in the bottom panel, the reverse is true when change scores are expressed in SD units.

Although all eight SF-36 scales improved a statistically significant amount following total hip arthroplasty, comparison of the estimates of changes in the PCS and MCS scales indicates that improvement was much greater in physical health. The PCS improved more than twice as much as the MCS.

Examples of norm-based interpretation using the tables presented in Chapter 8 include:

A. An average improvement of 9.5 in the PCS (from 39.0 to 48.5) represents an improvement in physical health large enough to move from about the 15th to the 33rd percentile of the distribution of scores in the general U.S. population (Table 8.2).

B. An average improvement in MCS scores of 3.8 (from 39.7 to 43.5) represents an improvement in mental health large enough to move from about the 17th percentile to the 22nd percentile (Table 8.2).

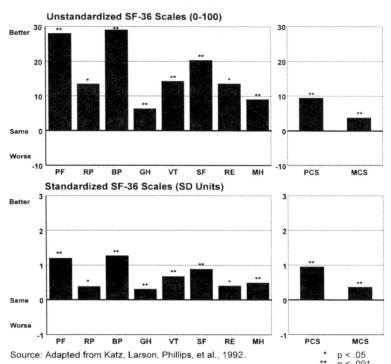

FIGURE 11.2 COMPARISON OF UNSTANDARDIZED AND STANDARDIZED CHANGES IN SF-36 SCALE SCORES AND SUMMARY MEASURES FOLLOWING HIP REPLACEMENT (N=54)

Source: Adapted from Katz, Larson, Phillips, et al., 1992.
* $p < .05$
** $p < .001$

The tables presented in Chapter 9 can be used to interpret the average improvement in the PCS and MCS based on item content. Content-based interpretations indicate that the observed improvement in the PCS score reflects reductions in the percentage of people likely to report on the SF-36:

C. That they had to cut down on work time (87% fewer people had to cut down on work time three months after hip replacement; percent cutting down work time dropped from 42.3 before the operation to 5.25 three months after the operation) (Table 9.2).

D. That they had severe pain (90% reduction in those reporting severe pain from 10.5% before the operation to 1% three months after the operation) (Table 9.2).

Criterion-based interpretations express improvement in the PCS scores in terms of reductions in the percentage of people likely to:

E. Be unable to work (75% reduction in probability of reporting inability to work from 28.8% to 7.1%) (Table 9.5).

F. Have a doctor visit in the past month (38% reduction in the probability of visiting the doctor in the past month from 38.3% to 23.6%) (Table 9.7).

Content-based interpretations associated with the 3.8 point improvement on the MCS scores observed for hip replacement include:

G. A 44% reduction in percentage that worked less carefully due to emotional problems (from 48.7% to 27.4%) (Table 11.3).

H. A 28% reduction in the percentage that accomplish less at work due to emotional problems (from 68.3% to 48.9%) (Table 11.3).

Criterion-based interpretation guidelines include:

I. A 46% increase in the percentage being extremely or very happy with their personal life (from 25.9% to 37.3%) (Table 9.9).

J. A 32% reduction in the likelihood of mental health treatment (from 28.9% to 19.4%) (Table 11.10).

Treatment For Low Back Pain

Lansky, Butler, & Waller (1992) reported SF-36 profiles for 113 spine center patients before and after outpatient treatment of low back pain. Conclusions reached from their study included: (a) substantial deficits were observed during initial assessment of two of the eight SF-36 scales, specifically patients' ability to function in their physical roles and pain; (b) patients reported significant improvement in both physical roles and pain at the 90-day follow-up assessment; and (c) SF-36 scales measuring general health perceptions and emotional functioning did not change significantly during the course of treatment.

Figure 11.3 compares mean scores for the SF-36 profile and the PCS and MCS summary scales before treatment for low back pain and at 90-day follow-up after treatment. Significant improvements were reported for five of the eight SF-36 scales (PF, RP, BP, VT, and SF) and for the PCS summary scale. There were no significant declines. In standard deviation units, improvements were largest for the PF, RP, BP, and SF scales, which is consistent with an improvement in the physical component of health. This interpretation is supported by the large and significant improvement in the PCS scale, in the absence of a significant improvement in the MCS scale. Interestingly, despite this substantial improvement in physical functioning, on average, these patients did not evaluate their health status more favorably after surgery, as indicated by the general health scale scores before and after treatment.

Examples of norm-based interpretations of the PCS and MCS using the tables presented in Appendix E include:

A. An average improvement of 7.6 points on the PCS (from 32.3 to 39.9) represents an improvement in physical health large enough to move from about the 8th to the 17th percentile of the distribution of scores in the general U.S. population (Appendix E2).

B. A before treatment score of 32.3 on the PCS represents 1.8 standard deviation units below the mean of the general U.S. population. A post treatment score of 39.9 on the PCS represents 1.0 standard deviation units below the mean of the general U.S. population (Appendix E2).

C. Before and after treatment scores on the MCS, 47.2 and 48.5 represent 0.28 and 0.15 standard deviation units, respectively, below the mean of the general U.S. population (Appendix E2).

Examples of content-based interpretations using the tables presented in Chapter 7 include:

D. A 36% reduction in the percentage having difficulty performing at work due to physical limitations (from 87.6% to 55.7%) (Table 9.2).

E. A 64% reduction in the likelihood of having severe or very severe bodily pain (from 21.0% to 7.4%) (Table 9.2).

FIGURE **11.3** MEAN SF-36 SCORES BEFORE AND AFTER TREATMENT FOR LOW BACK PAIN (N=113)

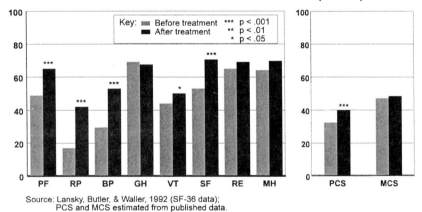

Source: Lansky, Butler, & Waller, 1992 (SF-36 data); PCS and MCS estimated from published data.

Criterion-based interpretation guidelines include:

F. A 25.7% reduction in job loss over one year (from 26.6% to 18.3%) (Table 9.5).

G. A 22.2% reduction in the likelihood of having seen the doctor in the last month (from 47.6% to 37.0%) (Table 9.7).

The MCS scores after treatment did not differ significantly from those before treatment and, therefore, were not interpreted further.

Duodenal Ulcer Treatment

Rampal, Martin, Marquis, et al. (1994) conducted a study that compared SF-36 profiles over a one-year follow-up period for 581 duodenal ulcer patients, randomized into two drug treatment regimens: maintenance treatment versus intermittent treatment. Among the conclusions from their study were: a) all SF-36 scores improved over time in both treatment regimens; and b) the improvement in SF-36 scores over one year was greater for patients in the maintenance treatment regimen than for patients in the intermittent treatment regimen.

Figure 11.4 compares average *changes* in SF-36 scores and the two summary measures over one year for both treatment regimens. Significant improvements in all eight scales were observed with both treatment regimens, except for the GH scale for those treated intermittently. The estimates for the PCS and MCS, which we have added to Figure 9.4, are consistent with results for the profile of eight scores. Like the SF-36 profile, significant improvements on PCS and MCS scores are observed for both treatment regimens, and the improvement is greater in the maintenance treatment regimen for both summary scales.

Examples of norm-based interpretations for the PCS and MCS using the tables presented in Appendix E include:

A. An average improvement on the PCS from 47.9 to 52.6 for the maintenance treatment regimen represents an improvement in physical health large enough to move from about the 31st to the 46th percentile of the distribution of scores in the general U.S. population (Appendix E2).

B. An average improvement on the MCS from 45.1 to 51.5 for the maintenance treatment regimen represents an improvement large enough to move from nearly one-half a standard deviation (SD) below the general U.S. population mean to 0.15 SD units above that mean.

Examples of content-based interpretations using the tables presented in Chapter 9 include:

C. For those in the maintenance treatment regimen, the improvement in PCS meant a 74% reduction in difficulty performing at work due to physical limitations (from 15.2% to 3.9%) (Table 11.2); and the improvement in MCS meant a 62% reduction in accomplishing less at work due to emotional problems (from 39.0% to 14.7%) (Table 11.3).

D. For those in the intermittent treatment regimen, the improvement in PCS meant a 38% reduction in reports of fair or poor health (from 12.5% to 7.7%) (Table 11.2); and the improvement in MCS meant a 53% reduction in feeling tired all or most of the time (from 21.0% to 9.7%) (Table 11.3).

Examples of criterion-based interpretations using the tables presented in Chapter 9 include:

E. For those in the maintenance treatment regimen, the improvement in PCS meant a 34% reduction in the likelihood of job loss over one year (from 11.5% to 7.5%) (Table 11.5); and the improvement in PCS meant a 38% reduction in the likelihood of clinical depression (from 12.7% to 7.8%) (Table 11.10).

F. For those in the intermittent treatment regimen, the improvement in PCS meant a 15% reduction on the likelihood of having seen the doctor in the last month (from 25.7% to 21.8%) (Table 11.7); and the improvement in MCS meant a 35% reduction in the likelihood of mental health treatment (from 20.0% to 13.0%) (Table 11.10).

Method For Relating PCS And MCS Scale Scores To "Criterion" Scores

Scale scores are interpreted by relating a difference or change in scale score to other specific results as explained in Chapters 7 and 8. Since only ranges and means within ranges for PCS and MCS scale scores and "criteria" are printed in the tables in Chapter 9 and Appendix E, one must calculate ratios of differences and interpolate to estimate the "criterion" value that is associated with a particular score. The numbers presented in the difference ratio column (Δ/Δ) on the tables in Chapter 9 are for difference ratios *between* levels. These can be used to relate differences in scale scores to differences in criterion values for differences that occur *between* the levels on the table. To relate differences in scale scores to differences in criterion values *within* a level, it is necessary to calculate the difference ratio *within* a level. This ratio is then used to associate the scale scores with the "criterion" value. This method is illustrated in an example below in which the death rate associated with scores on the PCS is estimated based on interpolation using data from a portion of Table 9.8.

Death Rate in Five Years Associated with Heart Valve Replacement. Patients had an average PCS score of 35.1 before heart valve replacement and 42.7 after heart valve replacement. How is the change in the PCS score related to predicted differences in the death rate? Portions of Table 9.8 reproduced below in Table 11.1 show the percent of people who died within five years at each of five score ranges on the PCS and the mean PCS score within each of those ranges. The mean scores on PCS before (35.1) and after (42.7) heart valve replacement lie within one level (level 3, Table 9.1). If either of these scores were equivalent to the mean for that level (39.9), the percent that would die in five years (6.2) could be read directly from the table. However, since the percentages associated with these scores are not presented in the table, they must be estimated by interpolation. This requires two steps. First, the appropriate difference ratio (change in criterion per unit change in PCS or MCS) must be calculated. Next, this ratio is used to estimate the criterion value for a particular score. To estimate the predicted change in the percent dying within five years corresponding to the change in PCS scores before and after heart surgery, do the following:

1. *Identify the numbers to use to calculate the change in the criterion (e.g., death rate) per unit change in PCS (i.e., the difference ratio).* This is done by looking at the column of means for each level. Choose those levels whose means are lesser and greater than the lower and higher scores, respectively. On Table 9.1, the scale scores for before (35.1) and after surgery (42.7) fall within the mean PCS scores for levels 4 (29.9) and 2 (50.0).

2. *Calculate the percent change on the criterion (e.g., death rate) per unit change in PCS.* At level 2 (Table 11.1), the mean PCS is 50 and the percent that would die within five years is 4.7. At level 4 (Table 11.1), the mean PCS is 29.9 and the percent that would die within five years is 15.1. To calculate the percent change in death rate per unit change in the PCS, take the difference in the percent who died associated with the mean PCS scores at levels 4 and 2 (15.1%-4.7%=10.4%) and divide it by the difference in the means at levels 2 and 4 (50.0-29.9=20.1). The percent death rate change per unit change in PCS is 0.52 (10.4%/20.1).

TABLE 11.1   PORTION OF TABLE 9.8 USED FOR INTERPRETATION: FIVE-YEAR MORTALITY RATES FOR MOS PATIENTS AT FIVE LEVELS OF THE PCS SCALE

	PCS Scores			Died	
Levels	Range	Mean	(N)	%	Δ/Δ
1	55-72	57.8	609	1.8	
					0.4
2	45-54	50	1181	4.7	
					0.1
3	35-44	39.9	754	6.2	
					0.9
4	25-34	29.9	443	15.1	
					0.7
5	8-24	20.4	233	21.5	

3. *Calculate the percent on the criterion (e.g., the percent who die) at one score value (e.g., the before surgery score).* The percent who die at the before surgery score of 35.1 should be less than those who die at the mean for level 4 (29.9), because the trend in the data is for the death rate to go down as scores go up (higher scores indicate better health). To estimate how many fewer people should die, subtract the mean PCS score at level 2 (29.9) from the before PCS score (35.1) and multiply this result by the percent change in death rate per unit change in PCS [(35.1-29.9)*0.52=2.7%]. Subtract this result (2.7%) from the percentage who die at a score of 29.9 (15.1%) to get the percent who would die at a score of 35.1 (12.4%).

4. *Calculate the percent on the "criterion" (e.g., the percent who die) at the other score value (e.g., the after surgery score).* To get the percent who would die at the average after surgery score of 42.7, multiply the difference between the before (35.1) and after surgery (42.7) PCS scores of 7.6 (42.7-35.1=7.6) by the percent change in death rate per unit change in PCS (0.52). This result (3.95%) should be subtracted from the percent dying at the before surgery score (12.4%-3.95%=8.5%). Thus, the percent who would die at a score of 42.7 is estimated to be 8.5 percent.

It should be understood that the interpretations offered in this chapter are approximations. They are based on data from single studies. Also, simpler rather than more complex calculations are used to promote better understanding. For example, the difference ratios presented in this manual are based on simple averaging and so assume a linear relationship between score levels. When values associated with each score level differ greatly, a more accurate approach would be to put greater weight on the values that are closer to the score of interest. We have calculated simple averages for difference ratios for each of several levels and thereby capture some of the variation in change in criterion associated with change in scores at different levels of scale scores. For example, on Table 9.1, the difference ratio between levels 4 and 3 (0.9) is much larger than that between levels 3 and 2 (0.1). Table 9.2 shows a wide variation in difference ratios for each of the 10 criteria listed.

Final Comment

Outcomes research seeks to inform decision-makers about health benefits in terms of changes in what people are able to do and how they feel. *Generic* measures of these outcomes, such as the PCS and MCS summary scales, provide a common yardstick for purposes of estimating health burden across diseases and for comparing benefits across treatments. The interpretation of group differences detected by these measures can be greatly facilitated using normative data as well as content- and criterion-based interpretation guidelines.

The estimates of changes in PCS and MCS scores discussed above were prepared to illustrate the use of norm-, content-, and criterion-based interpretation guidelines from Chapters 9 and 14. Results for the PCS and MCS were also compared with published results for the eight-scale SF-36 profile to provide preliminary tests of whether the summaries do a good job of reflecting profile changes. The studies re-analyzed here are among the first longitudinal SF-36 studies published. The first three studies reported very large changes following treatment (see Chapter 9, Summary of Effect Sizes). Whether the satisfactory performance of the PCS and MCS in summarizing SF-36 profiles and in detecting hypothesized changes in these studies will generalize to others remains to be determined. In the meantime, documentation of changes in the SF-36 profile along with changes in the PCS and MCS summary scales in the same graph, as illustrated above, facilitates comparisons of results.

Chapter 12: Applications: Clinical Practice

Background

Standardized health surveys have the potential to become the new "laboratory tests" of medical practice (Deyo & Carter, 1992). Without them, it appears that patient functioning and well-being are unlikely to be discussed during a typical medical visit. Two-thirds to three-fourths of U.S. adults reported that physicians rarely or never ask about the extent of their limitations in performing everyday activities, even in the presence of chronic conditions (Schor, Lerner, & Malspeis, 1995). As a result, practicing physicians are unaware of relatively concrete impairment manifested by observable limitations in physical, social, and role functioning (Rubenstein, McCoy, Cope, et al., 1991). Differences in severity of psychological distress also are often not apparent to treating physicians (Wells, Hays, Burnam, et al., 1989). Severely psychologically-distressed patients suffering from psychiatric disorders often go unrecognized and untreated even when mental health treatment is generously covered by health insurance (Ware, Manning, Duan, et al., 1984). It has been suggested that more widespread use of standardized health measures may improve clinical practice (American College of Physicians, 1988; Berwick, Murphy, Goldman, et al., 1991).

Standardization of functional status and well-being assessments in everyday medical practice may be useful in: (1) ensuring that all important dimensions of functional status and well-being are considered consistently; (2) detecting, explaining, and tracking changes in functional capacity over time; (3) making it possible to better consider the patient's total functioning when choosing among therapies; (4) guiding the efficient use of community resources and social services; and (5) predicting more accurately the course of chronic disease. Such data also would make it possible for physicians to better inform patients about the clinical and functional tradeoffs involved in alternative treatments (Fowler, Wennberg, Timothy, et al., 1988). Clearly, a great potential exists for standardized measures of functional status and well-being administered routinely as part of the clinical database

Confidence Intervals

As with all health status scales, the interpretation of individual patient scores for the PCS and MCS must take into account the amount of "noise" in the scores that they yield. The "noise level" can be quantified and displayed visually as a confidence interval (CI) around a patient's score. The size of the CI is a function of both the reliability of a score and the standard deviation of the score distribution in the population of interest (see Chapter 7). Because score reliability determines the size of confidence intervals most when *individual* scores are interpreted, psychometricians recommend a much higher standard of score reliability for measures to be interpreted at the individual patient level, as opposed to *average* scores for large groups of patients (Guilford, 1954). For example, lowering the reliability of the PCS from an actual 0.93 to a hypothetical 0.50 would increase the 95% confidence interval for an individual patient PCS score by more than 250% (from ± 5.7 to ± 14.4).

Estimates for the 95% confidence interval for the PCS and MCS are 5.7 and 6.3, respectively, in the general U.S. population. Other estimates for larger intervals are presented in Chapter 7. To the extent that either score reliability or score variability differ in the patient population of interest, relative to those for the general U.S. population used here, confidence intervals should be re-estimated using coefficients of reliability and standard deviations from the population of interest. As documented in Chapter 7 and Appendix E (for reliability and standard deviations, respectively), such differences are sometimes an issue for the patients used as examples here.

Monitoring Individual Patients

Soon after the SF-36 became available, began New England Medical Center (NEMC) testing patient-based systems for monitoring and improving health outcomes in various outpatient clinics. One project included quarterly administrations of the SF-36 to expand the definition of the "adequacy" or "quality" of dialysis beyond traditional laboratory test values among patients with end-stage renal disease (ESRD) (Kurtin, Davies, Meyer, et al., 1992; Meyer, Espindle, DeGiacomo, et al., 1994). Reported results for individual hemodialysis patients illustrate both the feasibility and usefulness of periodic health assessments in managing patients during the progression from advanced renal failure to end-stage renal disease (Meyer, Espindle, DeGiacomo, et al., 1994).

Charts For Longitudinal Data

Figure 12.1 plots the results of 12 quarterly administrations of the SF-36 for one patient over a three-year period that included his seventh and eighth years on dialysis (from Meyer, Espindle, DeGiacomo, et al., 1994). The patient is a middle-aged married male, who is an employed parent. He completed SF-36 forms at the time of regularly scheduled outpatient visits for hemodialysis. Results for each of the eight SF-36 scales for the first two years have been published. Analyses are extended here to include results for the PCS and MCS scales. The solid horizontal line in each panel defines a stable "norm" for a general U.S. population male of the same age. These norms vary considerably across scales, as documented in Appendix E. Visit one scores (1/91) were at or above the norm for six of eight scales (all except VT and PF) and were within the 95% CI for the PCS and MCS. It should be noted that the 95% CIs, which are indicated by the vertical lines above and below the score for each observation, are much narrower for the PCS and MCS (top panel) than for the eight scales (lower panel). As explained and further documented in Chapter 7 and Appendix D, this is an advantage of the two summary measures in monitoring individual patients.

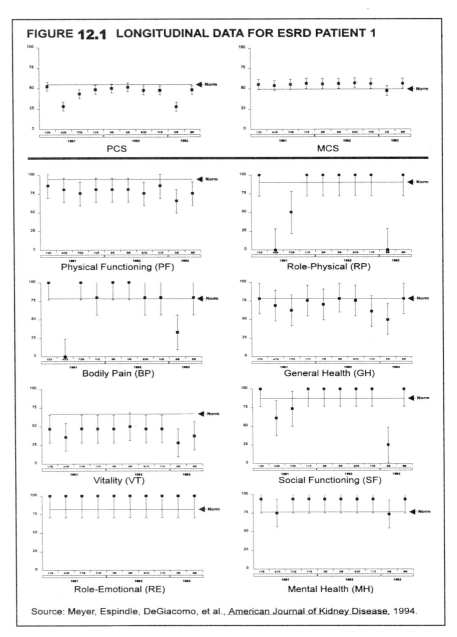

FIGURE 12.1 LONGITUDINAL DATA FOR ESRD PATIENT 1

Source: Meyer, Espindle, DeGiacomo, et al., *American Journal of Kidney Disease*, 1994.

In conjunction with an adverse medical event, this patient's PCS score declined dramatically by visit two (4/91). Underlying the decline in the PCS were drops in the BP, RP, and SF scales. Recovery to levels high enough that the 95% CI included the general population norm were observed for the BP scale by the third observation period (7/91); followed by recovery to the norm for the RP and SF scales by the fourth period (11/91). Improvements in these scales are reflected in the PCS, which came within two standard errors of normal by the fifth and sixth visits. The PCS did not completely reach the norm because of the VT scale score, which remained well below the norm across all visits. This pattern of low VT scores is often observed for patients with chronic renal failure. Longitudinal monitoring revealed that this patient's functional health and well-being remained stable throughout most of the observation period.

Figure 12.1 illustrates a format for displaying longitudinal results for an individual patient that NEMC clinicians have found useful. The figure also illustrates several important lessons. First, the PCS and MCS reflect changes that are observed in specific SF-36 scales. Second, general population norms can be useful in understanding individual patient scores, which were below the norm for physical health, but above the norm for mental health. Third, this example calls attention to the much smaller size of confidence intervals around individual patient scores for the PCS and MCS relative to the eight SF-36 scales. CIs were particularly large for the RP and SF scales relative to the PCS and MCS. These differences are explained in Chapters 4 and 5. Finally, this example illustrates the value of establishing a personal norm for each individual patient. Changes over time for an individual patient may be best judged in relation to what is "normal" for that patient.

The display of longitudinal observations for an individual patient is simplified by the PCS and MCS scales. Because longitudinal results for only one dimension of health can be effectively displayed in the same chart, eight charts were required for a display of longitudinal data for the SF-36 scales (Meyer, Espindle, DeGiacomo, et al., 1994). If the PCS and MCS are used to summarize that information, as illustrated here, only two displays are required. Results for the eight scales provide useful back-up for use in better understanding what is underlying a change in a PCS or MCS summary score.

Profiles For Individual Patients

Along with short forms like the SF-36 and systems that rapidly process survey forms with a high degree of reproducibility, acceptance by clinicians also requires that the display of a profile be user-friendly. Figure 12.2 illustrates another format for displaying SF-36 profiles and the PCS and MCS scales for individual patients that is currently being evaluated.

Figure 12.2 shows the SF-36 profile and the PCS and MCS summary scores for the patient with ESRD, discussed above. "Initial" (I) and "follow-up" (F) visits correspond to the first and second visits discussed above. The bar graphs for the SF-36 profile and for the PCS and MCS, along with the scores printed, reveal that this patient had worsened substantially between SF-36 administrations. Significant declines, which are larger than the 95% confidence interval, are indicated by downward arrows () in Figure 10.2. Significant declines were observed for four of the scales (RP, BP, SF, and MH) and for the PCS. The decline of 24 points (from 52 to 28) in the PCS scores represents a decline from just below the median (50th percentile) to well below the 25th percentile for a 38-year-old male (see norms in Appendix E5).

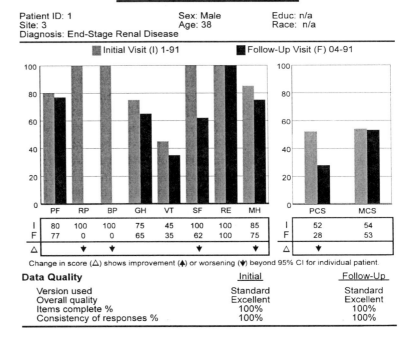

FIGURE **12.2** INITIAL AND FOLLOW-UP PROFILES FOR ESRD PATIENT 1

Data Quality

The printout shown in Figure 12.2 also summarizes data quality. Specifically, it indicates that the Standard SF-36 form and scoring were used for both initial and follow-up administrations. Thus, it is appropriate to compare them. Further, the overall data quality was "excellent." Specifically, 100% of the items were complete, and the Response Consistency Index (RCI), which is an indicator of the patient's consistency in responding across 15 pairs of SF-36 items, yielded 100% consistency scores for both administrations. In the general population (N = 2,474) and in the MOS (N = 3,434), 90.3% and 94.5% had no inconsistent responses, respectively; 6.1% and 3.4% had one inconsistent response. The RCI and normative data for interpreting the RCI are explained in the original *SF-36 User's Manual* (Ware, Snow, Kosinski, et al., 1993).

Processing Systems

Processing system for the SF-36 have the features discussed above, including the RCI for monitoring data quality, along with other options. With advances in data processing systems, a doctor and patient can monitor the patient's functional health and well-being on a regular basis, inexpensively, without delay, and with highly reproducible results. Such systems remove the practical barriers to monitoring patient health outcomes routinely in everyday clinical practice.

Patient Profiles: More Examples

Figures 12.3 and 12.4 present additional examples of the longitudinal monitoring of PCS and MCS scores for an individual patient. The patients involved were selected from among those followed during the first year of the MOS. The first patient is a 37-year old female who, according to clinical criteria (documented in Appendix A), recovered from a depressive disorder that was diagnosed when the MOS began. As was the case for the great majority of these patients (McHorney, Ware, Lu, et al., 1994), data quality was more than satisfactory for both assessments nearly one year apart. As indicated by the upward arrows (⇧), improvements in her scores were greater than would be expected by chance for six of the eight SF-36 scales. Comparison of scores for the PCS and MCS, which take correlations among the eight scales into account, clearly indicates that this patient's improvement was concentrated in the mental component of health status. Her initial (I) MCS score of 19 was below that of nearly 99% of women her age in the general U.S. population at baseline and improved to a score of 44, which is above the 25th percentile (see norms in Appendix E4). This improvement in the MCS scores of 25 points is about seven times greater than the average improvement observed for depressed patients in the MOS (see norms in Appendix E38). In addition to the decrease in suffering associated with a MCS score change from only 19, which indicates substantial psychological distress, to a score of 44, interpretation guidelines in Chapter 9 suggest that this patient's probability of receiving mental health specialty care (an important cost consideration) was substantially reduced as a result of this improvement (Chapter 9, Table 9.10).

FIGURE **12.3** INITIAL AND FOLLOW-UP PROFILES FOR A PATIENT WITH CLINICAL DEPRESSION

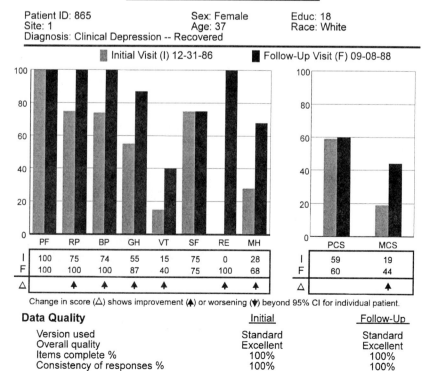

Figure 12.4 presents the last example of the longitudinal monitoring of PCS and MCS scores and the SF-36 profile for an individual patient. The patient is a 50-year old male who was diagnosed as having congestive heart failure (CHF) upon enrollment into the Medical Outcomes Study (MOS) (see criteria in Appendix B). The follow-up assessment was performed approximately eight months later. Comparison of his initial scores with the norms for patients with congestive heart failure indicate that this patient's PCS score was within two or three points of average for those patients with CHF in the MOS and scored well above average (at the 75th percentile) on the MCS (see norms in Appendix E25).

As was the case for the average MOS patient with CHF (see norms for change scores Appendix E39), the patient in Figure 12.4 declined nine points on the PCS scale (from 37 to 28) during the eight-month follow-up period. (It should be noted that norms for the PCS change scores reported in Appendix E cover a one-year period.) This amount of decline is more than four times the average for patients with CHF and is well below the 25th percentile of change scores for such patients. Underlying this change in PCS scores were significant declines (larger than the 95% confidence level) for the PF, RP, and VT scales. Scores for three other scales known to be most indicative of mental health status (MH, RE, and SF) did not decline more than would be expected by chance during the follow-up period. Consistent with these results, the MCS summary measure showed no significant change.

FIGURE **12.4** INITIAL AND FOLLOW-UP PROFILES FOR A PATIENT WITH CHF

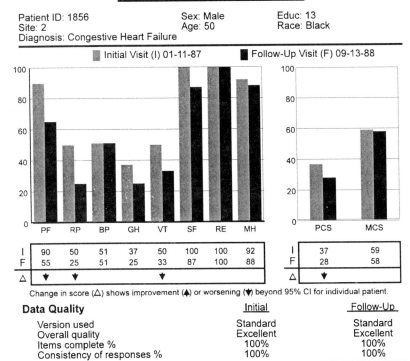

Patient Screening

Another promising clinical application of the PCS and MCS summary scales is that of patient screening. Because more than half of the patients with depressive disorders go undetected in primary care practices (Berwick, Murphy, Goldman, et al., 1991; Wells, Hays, Burnam, et al., 1989), for example, there is much to be gained from using short-form surveys proven to be valid for purposes of patient screening. The SF-36 is an especially good candidate because it is very short — requiring only about five to ten minutes to complete — and because it is widely used for monitoring health status outcomes. Thus, the survey used in monitoring outcomes may serve the dual purpose of patient screening. In support of this application, the SF-36 MH scale, which is also often referred to as the MHI-5, has performed well in tests of sensitivity and specificity relative to other screening tools for depression and other mental disorders (Berwick, Murphy, Goldman, et al., 1991).

MCS And Clinical Depression

To evaluate the MCS scale as a screening tool for clinical depression, we used the same statistical methods as used elsewhere (Berwick, Murphy, Goldman, et al., 1991). SF-36 surveys were self-administered independently but in conjunction with a two-stage screening process that relied upon a short-form of the CESD and then the NIMH Diagnostic Interview Schedule (DIS) criteria for clinical depression, as described elsewhere (Burnam, Wells, Leake, et al., 1988; Wells, Hays, Burnam, et al., 1989). Using these methods, 503 of the 3,445 MOS patients (14.6%) were diagnosed with major depression and/or dysthymia.

Using receiver operating curve (ROC) analysis, we evaluated the sensitivity and specificity, as well as the ROC curve, which combines both, across the full range of MCS scores. The MH scale was also evaluated using the well-documented cut-off of 52 or below for detecting depression (Berwick, Murphy, Goldman, et al., 1991; Ware, Snow, Kosinski, et al., 1993). (Incidentally, the optimal MH cut-off of 52 was confirmed in these analyses.)

As summarized in Table 12.1, the best all-around cut-off for the MCS is at a score of 42 or below, which achieves an area under the ROC curve of 0.77 and a sensitivity and specificity of 73.7% and 80.6%, respectively. These results compare favorably with those for the MH scale in the same test. The area under the ROC curve for the MH scale was 0.76 with a sensitivity of 66.8% and a specificity of 86.2% (data not reported). For purposes of screening applications requiring larger or smaller rates of false positives or false negatives, other MCS scale cut-offs can be used as documented in Table 12.1.

The evaluation of the eight SF-36 scales and the MCS in screening patients for depression and other mental conditions is ongoing at NEMC and other sites. Both the SF-36 and SF-12 versions of the MCS (see Chapter 3) are being evaluated. Work in progress indicates that the three additional screening items shown in previous MOS studies to improve the performance of the CESD short-form (Burnam, Wells, Leake, et al., 1988) do not improve the performance of the SF-36. Thus, a 39-item questionnaire that adds three additional screening questions (about a two-week episode during the past year, depression over the past one- and two-year periods) does not appear to be worth the additional respondent burden for purposes of patient screening as described above.

TABLE 12.1 SUMMARY OF THE PERFORMANCE OF THE MCS (AT VARIOUS CUTPOINTS) IN SCREENING FOR CLINICAL DEPRESSION[1], MOS PATIENTS (N = 3,445)

	MCS SCALE CUTPOINTS					
	·23	·32	·37	·42	·47	·52
Sensitivity (%)	17.9	47.3	62.2	73.7	81.1	9.5
Specificity (%)	98.5	93.4	87.6	80.6	70.5	56.4
False Positive Rate (%)	1.5	6.6	12.4	19.4	29.5	43.6
False Negative Rate (%)	82.1	52.7	37.8	26.3	18.9	10.5
Area Under ROC Curve	0.58	0.70	0.75	0.77	0.76	0.73

[1] N = 503 (14.6%) patients from the MOS baseline sample were diagnosed with clinical depression; NIMH (DIS) criteria met for major depression and/or dysthymia (Wells, Hays, Burnam, et al., 1989).

PCS Scores And Physical Disease

Previous studies have shown that the probability of a diagnosable physical condition increases substantially with declines in general health status. For example, at the top levels of the General Health Rating Index (Davies & Ware, 1981; Ware, 1992), from which the SF-36 GH scale was constructed, only about five percent were diagnosed with a physical condition. That number increased to about 95% at the lower scale levels.

To evaluate the PCS scale as a screening tool for physical disease, ROC analyses (as described above) were performed across the full range of the PCS scores using data from a 1990 general population survey used to norm the SF-36 (see Chapter 8). Chronic physical conditions, as reported by patients, were based on definitions in Appendix B. A physical condition was counted if one or more of the following was reported: arthritis, angina, back pain/sciatica, chronic lung disease, CHF, diabetes, hypertension, limitation in use of arm(s)/leg(s), or myocardial infarction. In the general U.S. population (N = 2,474), 57.3% reported one or more of these conditions.

As summarized in Table 12.2, the best all around cut-off for the PCS is at a score of 50 or below, which achieves an area under the ROC curve of 0.72, a sensitivity of 60%, and a specificity of 84.8%. (Note that 50 is the average PCS score in the general U.S. population.) As shown in Table 10.2, the PCS cut-offs of 55 and 45 substantially increase sensitivity (to 83% and 98%) and decrease specificity (55% and 10%), respectively.

These results, which are consistent with estimates of the unique effects of each of these physical conditions on PCS scores (see Chapter 9, Table 9.14), suggest that there is a very high probability of physical disease underlying low PCS scores, and that the PCS may prove useful as a first-stage screener for such conditions.

TABLE 12.2 SUMMARY OF THE PERFORMANCE OF THE PCS (AT VARIOUS CUTPOINTS) IN SCREENING FOR A PHYSICAL CONDITION[1], GENERAL U.S. POPULATION (N = 2,474)

	PCS SCALE CUTPOINTS						
	< 30	< 35	< 40	< 45	< 50	< 55	< 60
Sensitivity (%)	12.7	21.0	31.3	42.4	60.0	83.4	97.7
Specificity (%)	99.4	98.6	96.3	92.7	84.8	55.3	9.5
False Positive (%)	0.6	1.5	3.7	40.07.3	15.3	44.7	90.5
False Negative (%)	87.3	79.0	68.7	57.6	39.9	16.6	2.3
ROC	0.56	0.60	0.64	0.68	0.72	0.69	0.54

[1] Chronic conditions — presence of one or more of the following: arthritis, angina, back pain/sciatica, chronic lung disease, CHF, diabetes, hypertension, limitation in use of arm(s)/leg(s), or myocardial infarction.

Final Comment

There are many issues involved in using health status assessments in everyday clinical practice (Nelson, Wasson, Kirk, et al., 1987; Deyo & Carter, 1992; Ware, 1992). Practicality is essential. Measurement precision — a narrow confidence interval around each patient score — is a prerequisite for interpretation. Despite the "noise level" inherent in short-form scales, repeated assessments of the SF-36 yield interpretable estimates of changes in patient health status that would otherwise have been unknown to clinicians (Meyer, Espindle, DeGiacomo, et al., 1994).

The PCS and MCS scores increased measurement precision beyond that achieved by any of the eight SF-36 scales in psychometric evaluations documented in Chapter 7. As illustrated above, the PCS and MCS appear to yield reproducible and useful summaries of results in patient-level analyses. These examples also illustrate the gains in precision expected for both the PCS and MCS.

Another rate-limiting factor in clinical applications is the extent of understanding of what the scores mean and the availability of user-friendly documentation of that information. The PCS and MCS clearly have straightforward interpretations as measures of physical and mental health status, respectively. However, despite the fact that the PCS and MCS aggregate the most highly-related SF-36 scales, which are most likely to yield the same results, some information of value to clinicians may be lost in the "averaging" process. The display formats for presenting the PCS and MCS summary scales, in tandem with the richness of the SF-36 profile, should help to make the best of both apparent to clinicians. It is also hoped that interpretation guidelines documented in this manual will facilitate the use of the summary measures in everyday clinical practice.

Chapter 13: Individual License Information

Our goals in licensing the SF-8 Health Survey are the same as our goals for the SF-36 and SF-12 health surveys. Our policies are intended to assure the royalty-free availability of the SF-8 Health Survey, and the documentation necessary to use and interpret it properly, to individuals and organizations for their own use. Further, we are committed to meeting the needs of those seeking licenses to reproduce the SF-8, translations of the SF-8, scoring algorithms and norms for commercial applications. Finally, we are continuing our ongoing collaboration and financial support of our academic and other not-for-profit partners in developing, validating and norming the SF-8. These partners include the Medical Outcomes Trust (MOT), which recently merged with the Health Assessment Lab (HAL) in Boston, and the International Quality of Life Assessment (IQOLA) Project's international network of investigators. We ask our licensees and other industry sponsors to support these partners and projects from the proceeds of their license fees and by way of other industry grants for research and development efforts.

QualityMetric Inc. began its financial support of the MOT, HAL and other organizations when it was incorporated in 1997. At that time, QualityMetric entered into an agreement with the MOT, the other co-copyright holder of the SF-36 and SF-12 health surveys, to be the exclusive source of commercial licenses for these surveys and to provide ongoing financial support to the MOT.

The types of licenses offered to single- and multi-users for commercial and non-commercial applications are described below. According to all licenses, use shall not modify, abridge, condense, translate, adapt, recast, or transform the SF-8 or the Basic Scoring Algorithms in any manner or form, including but not limited to any minor or significant change in wording or organization of the SF-8. We welcome proposals for other arrangements, which should be directed to www.qmetric.com/marketplace/index.cgi.

Types Of Licenses Offered

Single-User, Non-Commercial License

This category of license is granted to individual owners of copies of the SF Survey scoring manuals (SF-36, SF-36v2, SF-12, SF-12v2, SF-8, HIT-6, AOMS, HQLQ). Owners of the manuals are authorized to type the items from the surveys, the scoring algorithms and normative data into a single computer for their own personal use. Licensee is not authorized to transfer or distribute machine-readable copies to any other person, to use the scoring algorithms or normative data on more than one computer, or to distribute executable programs containing the scoring algorithms or normative data. This license covers non-commercial use only. It does not entitle the licensee to market, re-sell, sub license, or otherwise distribute or use the SF Surveys, scoring algorithms, or normative data as part of a product or service offering (regardless of whether or not a fee is charged). This single-user license does not allow reproduction or transmission of the survey instrument(s), scoring algorithm(s), and/or normative data on a computer network, Intranet, Internet server, Interactive Voice Recognition (IVR) system or Personal Digital Assistant (PDA). This single-user, non-commercial license is the only royalty-free SF license.

Multi-User, Non-Commercial License

This category of license is granted to owners of copies of the SF Surveys scoring manuals (SF-36, SF-36v2, SF-12, SF-12v2, SF-8, HIT-6, AOMS, HQLQ) to use the survey instrument(s), scoring algorithm(s), and/or normative data on more than one computer, on their corporate computer network, or restricted-access Intranet, Internet server, Interactive Voice Recognition (IVR) system or Personal Digital Assistant (PDA), for non-commercial purposes. This license does not entitle the licensee to market, re-sell, sub license, or otherwise distribute or use the SF Surveys forms, scoring algorithms or normative data outside their organization, as part of a product or service offering (whether or not a fee is charged). This non-commercial license requires payment of a usage fee based on the number of SF Surveys anticipated to be used over a one-year license period. For more information and/or to obtain a multi-user, non-commercial license agreement, please go to:

www.qmetric.com/marketplace/index.cgi.

Commercial License

Holders of this category of license are authorized to market, re-sell, sublicense, or otherwise distribute and use the SF Surveys (SF-36, SF-36v2, SF-12, SF-12v2, SF-8, HIT-6, AOMS, HQLQ), scoring algorithms, and/or normative data as part of a product or service. This commercial license requires an annual fee based on the size of the organization (as measured by previous year's revenue) as well as a royalty fee based on the number of SF Surveys anticipated to be used during a single calendar year license period. For more information and/or to obtain a commercial license agreement, please go to:

www.qmetric.com/marketplace/index.cgi.

Note: Licenses are subject to change at any time.

Disclaimer Of Warranty

We make no warranties, express or implied, that the algorithms contained in this volume are free of error, or are consistent with any particular standard of merchantability, or that they will meet your requirements for any particular application. They should not be relied on for solving a problem whose incorrect solution could result in injury to a person or loss of property. If you do use the programs in such a manner, it is at your own risk. The authors and publisher disclaim all liability for direct or consequential damages resulting from your use of the algorithms.

References

Aaronson NK, Acquadro C, Alonso J, Apolone G, Bucquet D, Bullinger M, Bungay K, Fukuhara S, Gandek B, Keller S, Razavi D, Sanson-Fisher R, Sullivan M, Wood-Dauphinee S, Wagner AK, & Ware JE. International quality of life assessment (IQOLA) project. *Quality of Life Research* 1992; 1:349-351.

American College of Physicians. Comprehensive functional assessment for elderly patients. *Annals of Internal Medicine* 1988; 109:70-72.

American Psychological Association. *Standards for Educational and Psychological Testing*. Washington, D.C.: American Psychological Association, 1985.

Beaton DE, Bombardier C, & Hogg-Johnson S. Choose your tool: A comparison of the psychometric properties of five generic health status instruments in workers with soft tissue injuries. *Quality of Life Research* 1994; 3:50-56.

Bergner L, Bergner M, Hallstrom AP, Eisenberg M, & Cobb L. Health status of survivors of out-of-hospital cardiac arrest six months later. *American Journal of Public Health* 1984; 74(5):508-510.

Bergner M, Bobbitt RA, Carter WB, & Gilson BS. The Sickness Impact Profile: Development and final revision of a health status measure. *Medical Care* 1981; 19:787-805.

Bergner M & Rothman ML. Health status measures: An overview and guide for selection. *Annual Review of Public Health* 1987; 8:191-210.

Berwick DM, Murphy JM, Goldman PA, Ware JE, Barsky AJ, & Weinstein MC. Performance of a five-item mental health screening test. *Medical Care* 1991; 29:169-176.

Bonfils S. Les questionnaires de Qualité de View: Irremplaçables instruments d'approximation. *Impertinente Psychosomatique*. Montrouge: John Libbey Eurotext, 1993:293-300.

Bousquet J, Knani J, Dhivert H, Richard A, Chichoye A, Ware JE, & Michel FB. Quality-of-life in asthma: I. Internal consistency and validity of the SF-36 questionnaire. *American Journal of Respiratory and Critical Care Medicine* 1994; 149:371-375.

Brazier J. The SF-36 health survey questionnaire - a tool for economists. *Health Economics* 1993; 2.

Brazier J, Jones N, & Kind P. Testing the validity of the EuroQOL and comparing it with the SF-36 health survey questionnaire. *Quality of Life Research* 1993; 2:169-180.

Brazier JE, Harper R, Jones NMB, O'Cathain A, Thomas KJ, Usherwood T, & Westlake L. Validating the SF-36 health survey questionnaire: New outcome measure for primary care. *British Medical Journal* 1992; 305:160-164.

Brook RH, Ware JE, Rogers WH, *et al. The Effect of Coinsurance on the Health of Adults: Results from the RAND Health Insurance Experiment*. Santa Monica, CA: The RAND Corporation, 1984; R-3055-HHS.

Bullinger M. German translation and psychometric testing of the SF-36 Health Survey: Preliminary results from the IQOLA Project. *Social Science and Medicine* 1995.

Burnam MA, Wells KB, Leake B, & Landsverk J. Development of a brief screening instrument for detecting depressive disorders. *Medical Care* 1988; 26(8):775-789.

Cleary PD, Greenfield S, & McNeil BJ. Assessing quality of life after surgery. *Controlled Clinical Trials* 1991; 12:189S-203S.

Cohen J. *Statistical Power Analysis for the Behavioral Sciences -- Second Edition*. Hillsdale, NJ: Lawrence Erlbaum Associates, 1988.

Comrey AL. *A First Course in Factor Analysis*. New York: Academic Press, 1973.

Cook TD & Campbell DT. *Quasi-Experimentation: Design and Analysis Issues for Field Settings*. Chicago, IL: Rand McNally College Publishing Co., 1979.

Davies AR & Ware JE. *Measuring Health Perceptions in the Health Insurance Experiment*. Santa Monica, CA: Rand Corporation, 1981; R-2711-HHS. (Health Insurance Experiment Series).

Derogatis LR. The Psychosocial Adjustment to Illness Scale (PAIS). *Journal of Psychosomatic Research* 1986; 30(1):77-91.

Deyo RA & Carter WB. Strategies for improving and expanding the application of health status measure in clinical settings: A researcher-developer viewpoint. *Medical Care* May 1992; 30(5 Suppl.):MS176-MS186.

Dixon P, Heaton J, Long A, & Warburton A. Reviewing and applying the SF-36. *Outcomes Briefing* August 1994; 4:3-25.

EuroQOL Group. EuroQOL - a new facility for the measurement of health-related quality of life. *Health Policy* 1990; 16:199-208.

Fowler FJ, Wennberg JE, Timothy RP, Barry MJ, Mulley AG, & Henley D. Symptom status and quality of life following prostatectomy. *Journal of the American Medical Association* 1988; 259:3018-3022.

Fryback DG, Dasbach EJ, Klein R, Klein BE, Dorn N, Peterson K, & Martin PA. The Beaver Dam health outcomes study: Initial catalog of health state quality factors. *Medical Decision Making* 1993; 13:89-102.

Garratt AM, MacDonald LM, Ruta DA, Russell IT, Buckingham JK, & Krukowski ZH. Towards measurement of outcome for patients with varicose veins. *Quality in Health Care* 1993; 2:5-10.

Garratt AM, Ruta DA, Abdalla MI, & Russell IT. The SF-36 Health Survey Questionnaire: II. Responsiveness to changes in health status for patients with four common conditions. *Quality in Health Care* 1995.

Garratt AM, Ruta DA, Abdalla MI, Buckingham JK, & Russell IT. The SF-36 Health survey questionnaire: An outcome measure suitable for routine use within the NHS? *British Medical Journal* 1993; 306:1440-1444.

Goldberg DP & Hillier V. A scaled version of the General Health Questionnaire. *Psychological Medicine* 1979; 9:139-145.

Greenfield S, Nelson EC, Zubkoff M, Manning WG, Rogers W, Kravitz RL, Keller A, Tarlov AR, & Ware JE. Variations in resource utilization among medical specialties and systems of care: Results from the Medical Outcomes Study. *Journal of the American Medical Association* 1992; 267:1624-1630.

Guilford JP. *Psychometric Methods*. New York: McGraw-Hill, 1954.

Guyatt G, Walter S, & Norman G. Measuring change over time: Assessing the usefulness of evaluative instruments. *Journal of Chronic Diseases* 1987; 40(2):171-178.

Haley SM, McHorney CA, & Ware JE. Evaluation of the MOS SF-36 Physical Functioning scale (PF-10): I. Unidimensionality and reproducibility of the Rasch item scale. *Journal of Clinical Epidemiology* 1994; 47:671-684.

Hall JA, Epstein AM, & McNeil BJ. Multidimensionality of health status in an elderly population: Construct validity of a measurement battery. *Medical Care* 1989; 27(3 Suppl..):S168-S177.

Hall JA, Epstein AM, DeCiantis ML, & McNeil BJ. Physicians' liking for their patients: More evidence for the role of affect in medical care. *Health Psychology* 1993; 12(2):140-146.

Harman HH. *Modern Factor Analysis*, 3rd edition, revised. Chicago: University of Chicago Press, 1976.

Harris ML & Harris CW. A factor analytic interpretation strategy. *Educational and Psychological Measurement* 1971; 31:589-606.

Hays RD, Sherbourne CD, & Mazel RM. The RAND 36-Item Health Survey 1.0. *Health Economics* 1993; 2:217-227.

Hays RD & Stewart AL. The structure of self-reported health in chronic disease patients. *Psychological Assessment: A Journal of Consulting and Clinical Psychology* 1990; 2:22-30.

Hill S & Harries U. Assessing the outcome of health care for the older person in community settings: Should we use the SF-36. *Outcomes Briefing* 1994; 4:26-27.

Idler EL & Angel RJ. Age, chronic pain, and subjective assessments of health. *Advances in Medical Sociology* 1990; 1:131-152.

Jacobson AM, de Groot M, & Samson JA. The evaluation of two measures of quality of life in patients with Type I and Type II Diabetes Mellitus. *Diabetes Care* April 1994; 17(4):267-274.

Jenkinson C, Coulter A, & Wright L. Short form 36 (SF-36) Health Survey Questionnaire: Normative data for adults of working age. *British Medical Journal* 1993; 306:1437-1440.

Jenkinson C & Wright L. The SF-36 health survey questionnaire. *Auditorium* 1993; 2:7-12.

Jenkinson C, Wright L, & Coulter A. Criterion validity and reliability of the SF-36 in a population sample. *Quality of Life Research* 1994; 3:7-12.

Jenkinson C, Wright L, & Coulter A. *Quality of Life Measurement in Health Care: A Review of Measures and Population Norms for the UK SF-36*. Oxford, England: Health Services Research Unit, Department of Public Health and Primary Care, University of Oxford, June 1993.

Jette DU & Downing J. Health status of individuals entering a cardiac rehabilitation program as measured by the Medical Outcomes Study 36-Item Short-Form Survey (SF-36). *Physical Therapy*, 1994; 74(6):521-527.

Kantz ME, Harris WJ, Levitsky K, Ware JE, & Davies AR. Methods for assessing condition-specific and generic functional status outcomes after total knee replacement. *Medical Care* 1992; 30(Suppl.):MS240-MS252.

Kaplan RM & Anderson JP. The Quality of Well-Being Scale: Rational for a single quality of life index. in: Walker SR & Rosser RM. *Quality of Life: Assessment and Application*. Lancaster: MTP Press Limited, 1988:51-77.

Kappelle LJ, Adams HP, Heffner ML, Torner JC, Gomez F, & Biller J. Prognosis of young adults with ischemic stroke: A long-term follow-up study assessing recurent vascular events and functional outcome in the Iowa Registry of Stroke in Young Adults. *Stroke* 1994; 25(7):1360-1365.

Katz JN, Harris TM, Larson MG, Krushell RJ, Brown CH, Fossel AH, Liang MH. Predictors of functional outcomes after arthroscopic partial meniscectomy. *Journal of Rheumatology* 1992; 19(12):1938-1942.

Katz JN, Larson MG, Phillips CB, Fossel AH, & Liang MH. Comparative measurement sensitivity of short and longer health status instruments. *Medical Care* October 1992; 30(10):917-925.

Kazis LE, Anderson JJ, & Meenan RF. Effect sizes for interpreting changes in health status. *Medical Care* March 1989; 27(3 Suppl.):S178-S189.

Kerlinger FN. *Foundations of Behavioral Research*. New York: Holt, Rinehart and Winston, 1964.

Kravitz RL, Greenfield S, Rogers WH, Manning WG, Zubkoff M, Nelson E, Tarlov AR, & Ware JE. Differences in the mix of patients among medical specialties and systems of care: Results from the Medical Outcomes Study. *Journal of the American Medical Association* 1992; 267:1617-1623.

Krousel-Wood MA & Re RN. Health status assessment in a hypertension section of an internal medicine clinic. *American Journal of the Medical Sciences* 1994; 308(4):211-217.

Krousel-Wood MA, McCune TW, Abdoh A, & Re RN. Predicting work status for patients in an occupational medicine setting who report back pain. *Archives of Family Medicine* 1994; 3(4): 349-355.

Kurtin PS, Davies AR, Meyer KB, DeGiacomo JM, & Kantz ME. Patient-based health status measures in outpatient dialysis: Early experiences in developing an outcomes assessment program. *Medical Care* 1992; 30(Suppl.):MS136-MS149.

Kutner NG, Schechtman KB, Ory MG, & Baker DI. Older adult's perceptions of their health and functioning in relation to sleep disturbance, falling, and urinary incontinence. *Journal of the American Geriatrics Society* 1994; 42(7):757-62.

Lancaster TR, Singer DE, Sheehan MA, Oertel LB, Maraventano SW, Hughes RA, & Kistler JP. The impact of long-term warfarin therapy on quality of life: Evidence from a randomized trial. *Archives of Internal Medicine* 1991; 151:1944-1949.

Lansky D, Butler JBV, & Waller FT. Using health status measures in the hospital setting: From acute care to 'outcomes management'. *Medical Care* 199 2; 30(Suppl.):MS57-MS73.

Lembcke PA. Measuring the quality of medical care through vital statistics based on hospital service areas: 1. Comparative study of appendectomy rates. *American Journal of Public Health* 1952; 42:276-286.

Lerner DJ, Levine S, Malspeis S, & D'Agostino RB. Job strain and health-related quality of life in a national sample. *American Journal of Public Health* 1994; 84(10):1580-1585.

Levin NW, Lazarus JM, & Nissenson AR. National cooperative rHu erythropoietin study in patients with chronic renal failure -- an interim report. *American Journal of Kidney Disease* 1993; 22:3-12.

Liang J, Wu SC, Krause NM, Chiang TL, & Wu HY. The structure of the mental health inventory among Chinese in Taiwan. *Medical Care* 1992; 30(5):659-676.

Liang MH, Larson MG, Cullen KE, & Schwartz JA. Comparative measurement efficiency and sensitivity of five health status instruments for arthritis research. *Arthritis and Rheumatism* 1985; 28:542-547.

Lyons RA, Perry HM, & Littlepage BN. Evidence for the validity of the Short-Form 36 Questionnaire (SF-36) in an elderly population. *Age and Ageing* 1994; 23(3):182-184.

Mangione CM, Marcantonio ER, Goldman L, Cook EF, Donaldson MC, Sugarbaker DJ, Poss R, & Lee TH. Influence of age on measurement of health status in patients undergoing elective surgery. *Journal of the American Geriatrics Society* 1993; 41(4):377-383.

Martin C, Marquis P, & Bonfils S. A 'quality of life questionnaire' adapted to duodenal ulcer therapeutic trials. *Scandinavian Journal of Gastroenterology* 1994; 206(29 Suppl.):40-43.

Martin C, Marquis P, Ware JE, & Bonfils S. A Quality of life study of nizatidine in 581 DU patients: Maintenance treatment versus intermittent treatment. *Gasatroenterology* 1993; 104:A141.

Mason JH, Anderson JJ, & Meenan RF. A model of health status for rheumatoid arthritis: A factor analysis of the Arthritis Impact Measurement Scales. *Arthritis and Rheumatism* 1988; 31(6):714-720.

McCallum J. *The New "SF-36" Health Status Measure: Australian Validity Tests*. National Centre for Epidemiology and Population Health, The Australian National University, July 1994. (Record Linkage Pilot Study).

McHorney CA, Kosinski M, & Ware JE. Comparisons of the costs and quality of norms for the SF-36 Health Survey collected by mail versus telephone interview: Results from a national survey. *Medical Care* 1994; 32:551-567.

McHorney CA & Ware JE. Construction of an alternate form general mental health scale for the MOS SF-36 Health Survey. *Medical Care* 1995; 33:15-28.

McHorney CA, Ware JE, & Raczek AE. The MOS 36-Item Short-Form Health Survey (SF-36): II. Psychometric and clinical tests of validity in measuring physical and mental health constructs. *Medical Care* 1993; 31:247-263.

McHorney CA, Ware JE, Lu JFR, & Sherbourne CD. The MOS 36-Item Short-Form Health Survey (SF-36): III. Tests of data quality, scaling assumptions and reliability across diverse patient groups. *Medical Care* 1994; 32:40-66.

McHorney CA, Ware JE, Rogers WH, Raczek A, & Lu JFR. The validity and relative precision of MOS Short- and Long-Form Health Status Scales and Dartmouth COOP Charts: Results from the Medical Outcomes Study. *Medical Care* 1992; 30(Suppl.):MS253-MS265.

Medical Outcomes Trust. *How To Score the SF-36 Short-Form Health Survey*. Boston, MA: Medical Outcomes Trust, 1991.

Medical Outcomes Trust. *Scoring Exercise for the SF-36 Health Survey, 2nd edition*. Boston, MA: Medical Outcomes Trust, 1994.

Medical Outcomes Trust. *SF-36 Health Survey Scoring Manual for English Language Adaptations: Australia/New Zealand, Canada, United Kingdom*. Boston, MA: Medical Outcomes Trust, 1994.

Meyer KB, Espindle DM, DeGiacomo JM, Jenuleson CS, Kurtin PS, & Davies AR. Monitoring dialysis patients' health status. *American Journal of Kidney Diseases* 1994; 24(2):267-279.

Nelson EC, Wasson J, Kirk J, Keller A, Clark D, Dietrich A, Stewart A, & Zubkoff M. Assessment of function in routine clinical practice: Description of the COOP chart method and preliminary findings. *Journal of Chronic Disease* 1987; 40(Suppl. 1):55S-63S.

Nerenz DR, Repasky DP, Whitehouse FW, & Kahkonen DM. Ongoing assessment of health status in patients with diabetes mellitus. *Medical Care* 1992; 30(Suppl.):MS112-MS124.

Nunnally JC. *Psychometric Theory* 2nd edition. New York: McGraw-Hill, 1978.

Nunnally JC & Bernstein IH. *Psychometric Theory 3rd edition*. New York: McGraw-Hill, 1994.

Okun MA & Stock WA. Correlates and components of subjective well-being among the elderly. *Journal of Applied Gerontology* 1987; 6:95-112.

Orley J & Kuyken W (eds.). *Quality of Life Assessment: International Perspectives*. Berlin: Springer-Verlag, 1994.

Osterhaus JT, Townsend RJ, Gandek B, & Ware JE. Measuring the functional status and well-being of patients with migraine headaches. *Headache* June 1994; 34(6):337-343.

Park Nicollet Medical Foundation & The Johns Hopkins University. *A Comparison of Alternative Approaches to Risk Management*. Washington, D.C.: Physician Payment Review Commission, 1994. (Selected External Research Series).

Parkerson GR, Broadhead WE, Tse CKJ. The Duke health profile: A 17-item measure of health and disfunction. *Medical Care* 1990; 28:1056-1072.

Patrick DL & Erickson P. *Health Status and Health Policy: Allocating Resources to Health Care*. New York: Oxford University Press, 1993.

Phillips RC & Lansky DJ. Outcomes management in heart valve replacement surgery: Early experience. *Journal of Heart Valve Disease* 1992; 1:42-50.

Rampal P, Martin C, Marquis P, Ware JE, & Bonfils S. A quality of life study in five hundred and eighty-one duodenal ulcer patients: Maintenance versus intermittent treatment with nizatidine. *Scandinavian Journal of Gastroenterology* 1994; 206(29 Suppl.):44-51.

Rector TS, Ormaza SM, & Kubo SM. Health status of heart transplant recipients versus patients awaiting heart transplantation: A preliminary evaluation of the SF-36 Questionnaire. *Journal of Heart and Lung Transplantation* 1993; 12(6 Pt. 1): 983-986.

Rogers WH, Wells KB, Meredith LS, Sturm R, & Burnam A. Outcomes for adult outpatients with depression under prepaid or fee-for-service financing. *Archives of General Psychiatry* 1993; 50:517-525.

Rubenstein LV, McCoy JM, Cope DW, Barrett PA, Hirsch SH, Messer KS, & Young RT. Improving patient functional status: A randomized trial of computer-generate resource and management suggestions. Abstract presented at the American Federation of Clinical Research 1991.

Ruta DA, Abdalla MI, Garratt AM, Coutts A, & Russell IT. The SF-36 Health Survey Questionnaire: I. Reliability in two patient-based studies. *Quality in Health Care* 1995.

Ruta DA, Garratt AM, Wardlaw D, & Russell IT. Developing a valid and reliable measure of health outcome for patients with low back pain. *Spine* 1994; 19(17):1887-1896.

SAS Institute Inc. *SAS/STAT User's Guide, Version 6, Fourth Edition, Volume 1*. Cary, NC: SAS Institute Inc., 1989.

Schag CAC, Heinrich RL, Aadland RL, & Ganz PA. Assessing problems of cancer patients: Psychometric properties of the Cancer Inventory of Problem Situations. *Health Psychology* 1990; 9:83-102.

Schor EL, Lerner DJ, & Malspeis S. Physician's assessment of functional health status and well-being: The patient's perspective. *Archives of Internal Medicine* 1995.

Siu AL, Hays RD, Ouslander JG, Osterwell D, Valdez RB, Krynski M, & Gross A. Measuring functioning and health in the very old. *Journal of Gerontology* 1993; 48(1):M10-M14.

Snyder MK & Ware JE. *A Study of Twenty-two Hypothesized Dimensions of Patient Attitudes Regarding Medical Care*. Springfield, Virginia: National Technical Information Service, 1974; MHC 74-10, PB-239-518/AS 65.

Stevens J. *Applied Multivariate Statistics for the Social Sciences* 2nd edition. Hillsdale, NJ: Lawrence Erlbaum Associates, 1992.

Stewart AL, Greenfield S, Hays RD, Wells KB, Rogers WH, Berry SD, McGlynn EA, & Ware JE. Functional status and well-being of patients with chronic conditions: Results from the Medical Outcomes Study. *Journal of the American Medical Association* August 1989; 262(7):907-913.

Stewart AL, Hays RD, & Ware JE. Health perceptions, energy/fatigue, and health distress measures. In: Stewart AL & Ware JE. *Measuring Functioning and Well-Being: The Medical Outcomes Study Approach*. Durham, NC: Duke University Press, 1992:143-172.

Stewart AL & Ware JE. *Measuring Functioning and Well-Being: The Medical Outcomes Study Approach*. Durham, NC: Duke University Press, 1992.

Stoller EP. Self-assessments of health by the elderly: The impact of informal assistance. *Journal of Health and Social Behavior* September 1984; 25:260-270.

Street RL, Gold WR, & McDowell T. Using health status surveys in medical consultations. *Medical Care* 1994; 32(7):732-744.

Sullivan M. LIVSKVALITESMÄTNING. Nytt generellt och nytt tumörspecifikt formulär för utvärdering och planering. *Läkartidningen* 1994; 91:1340-1342.

Sullivan M, Karlsson J, & Ware JE. *SF-36 Hälsoenkät: Svensk Manual Och Tolkningsguide (Swedish Manual and Interpretation Guide)*. Gothenburg: Sahlgrenska University Hospital, 1994.

Sullivan M, Karlsson J, & Ware JE. The Swedish SF-36 health survey: I. Evaluation of data quality, scaling assumptions, reliability, and construct validity across general populations in Sweden. *Social Science and Medicine* 1995.

Tarlov AR, Ware JE, Greenfield S, Nelson EC, Perrin E, & Zubkoff M. The Medical Outcomes Study: An application of methods for monitoring the results of medical care. *Journal of the American Medical Association* 1989; 262:925-930.

Thalji L, Haggerty CC, Rubin R, Berckmans TR, & Pardee BL. *1990 National Survey of Functional Health Status: Final Report*. Chicago, IL: NORC, 1991.

Usherwood T & Jones N. Self-perceived health status of hostel residents: use of the SF-36D Health Survey Questionnaire. *Journal of Public Health Medicine* 1994; 25(4):311-314.

van Tulder MW, Aaronson NK, & Bruning PF. The quality of life of long-term survivors of Hodgkin's disease. *Annals of Oncology* 1994; 5:153-158.

Veit CT & Ware JE. The structure of psychological distress and well-being in general populations. *Journal of Consulting and Clinical Psychology* 1983; 51:730-742.

Vickrey BG, Hays RD, Graber J, Rausch R, Engel J, & Brook RH. A health-related quality of life instrument for patients evaluated for epilepsy surgery. *Medical Care* 1992; 30:299-319.

Viramontes JL & O'Brien B. Relationship between symptoms and health-related quality of life in chronic lung disease. *Journal of General Internal Medicine* 1994; 9(1):46-48.

Wagner AK, Keller SD, Kosinski M, Baker GA, Jacoby A, Hsu M, Chadwick DW, & Ware JE. Advances in methods for assessing the impact of epilepsy and antiepileptic drug therapy on patients' health-related quality of life. *Quality of Life Research* 1995.

Ware JE. Captopril, enalapril, and quality of life (letter to the editor). *New England Journal of Medicine* 1993; 329(7):506-507.

Ware JE. Comments on the use of health status assessment in clinical settings. *Medical Care* 1992; 30:MS205-MS209.

Ware JE. Evaluating measures of general health concepts for use in clinical trials. in: Furberg CD & Schuttinga JA. *Quality of Life Assessment: Practice, Problems, and Promise*. U.S. Department of Health and Human Services, Public Health Service, National Institutes of Health, 1990: 51-63.

Ware JE. *How to Score the Revised MOS Short Form Health Scale (SF-36)*. Boston, MA: The Health Institute, New England Medical Center Hospitals, 1988.

Ware JE. Measuring patients' views: The optimum outcome measure. *British Medical Journal* 1993; 306:1429-1430.

Ware JE. Methodological considerations in the selection of health status assessment procedures. in: Wenger NK, Mattson ME, Furberg CD, & Elinson J (eds.). *Assessment of Quality of Life in Clinical Trials of Cardiovascular Therapies*. New York: Le Jacq Publishing Co., 1984:87-111.

Ware JE. The MOS 36-Item Short-Form Health Survey (SF-36). in: Sederer LI & Dickey B. *Outcomes Assessment in Clinical Practice*. Baltimore, MD: Williams and Wilkins, 1995.

Ware JE. Scales for measuring general health perceptions. *Health Services Research* Winter 1976; 11(4):396-415.

Ware JE. The status of health assessment 1994. *Annual Review of Public Health* 1995; 16:327-354.

Ware JE, Brook RH, Rogers WH, Keeler EB, Davies AR, Sherbourne CD, Goldberg GA, Camp P, & Newhouse JP. Comparison of health outcomes at a health maintenance organization with those of fee-for-service care. *Lancet* 1986; 1017-1022.

Ware JE, Brook RH, Davies-Avery A, Williams KN, Stewart AL, Rogers WH, Donald CA, & Johnston SA. *Model of Health and Methodology*. Santa Monica, CA: Rand Corporation, May 1980; R-1987/1-HEW. 49. (Conceptualization and Measurement of Health for Adults in the Health Insurance Study; vol I).

Ware JE, Davies-Avery A, & Brook RH. *Analysis of Relationships Among Health Status Measures*. Santa Monica: Rand Corporation, 1980; R-1987/6-HEW. 61. (Conceptualization and Measurement of Health for Adults in the Health Insurance Study; vol VI).

Ware JE, Gandek B, & the IQOLA Project Group. The SF-36 Health Survey: Development and use in mental health research and the IQOLA Project. *International Journal of Mental Health* 1994; 23(2):49-73.

Ware JE & Karmos AH. Scales for measuring general health perceptions. *Health Services Research* 1976; 11:396-415.

Ware JE, Keller SD, Gandek B, Brazier JE, Sullivan M, & the IQOLA Project team. Evaluating translations of health status questionnaires: Methods from the IQOLA Project. *International Journal of Technology Assessment in Health Care* 1995.

Ware JE, Kosinski M, Bayliss MS, McHorney CA, Rogers WH, & Raczek A. Comparison of methods for the scoring and statistical analysis of SF-36 health profile and summary measures: Summary of results from the Medical Outcomes Study. *Medical Care* 1995.

Ware JE, Manning WG, Duan N, Wells KB, & Newhouse JP. Health status and the use of outpatient mental health services. *American Psychologist* 1984; 39:1090-1100.

Ware JE, Miller WG, & Snyder MK. *Comparison of Factor Analytic Methods in the Development of Health Related Indexes from Questionnaire Data*. Rockville, MD: Health Services Research Methods Branch, 1973; PB-239 517. 58.

Ware JE & Sherbourne CD. The MOS 36-item short-form health survey (SF-36): I. Conceptual framework and item selection. *Medical Care* 1992; 30:473-483.

Ware JE, Snow KK, Kosinski M, & Gandek B. *SF-36 Health Survey Manual and Interpretation Guide*. Boston, MA: The Health Institute, New England Medical Center, 1993.

Ware JE & Snyder MK. Dimensions of patient attitudes regarding doctors and medical care services. *Medical Care* August 1975; 13(8):669-682.

Weinberger M, Kirkman MS, Samsa GP, et al. The relationship between glycemic control and health-related quality of life in patients with non-insulin-dependent diabetes mellitus. *Medical Care* 1994; 32(12):1173-1181.

Weinberger M, Samsa GP, Hanlon JT, Schmader KE, Doyle ME, Cowper PA, Utlech KM, Cohen HJ, & Feussner JR. An evaluation of a brief health status measure in elderly veterans. *Journal of the American Geriatrics Society* 1991; 39:691-694.

Wells KB, Burnam MA, Rogers W, et al. The course of depression in adult outpatients: Results from the Medical Outcomes Study. *Archives of General Psychiatry* 1992; 49:788-794.

Wells KB, Hays RD, Burnam MA, Rogers WH, Greenfield S, & Ware JE. Detection of depressive disorder for patients receiving prepaid or fee-for-service care: Results from the Medical Outcomes Study. *Journal of the American Medical Association* 1989; 262:3298-3302.

Wells KB, Stewart A, Hays RD, Burnam MA, Rogers W, Daniels M, Berry S, Greenfield S, & Ware JE. The functioning and well-being of depressed patients: Results from the Medical Outcomes Study. *Journal of the American Medical Association* 1989; 262(7):914-919.

Wiklund I, Lindvall K, Swedberg K, & Zupkis RV. Self-assessment of quality of life in severe heart failure: An instrument for clinical use. *Scandinavian Journal of Psychology* 1987; 28:220-225.

Winer BJ, Brown DR, & Michels KM. *Statistical Principles in Experimental Design* 3rd Edition. New York: McGraw-Hill, Inc., 1991.

Wood W, Rhodes N, & Whelan M. Sex differences in positive well-being: A consideration of emotional style and marital status. *Psychological Bulletin* 1989; 106:249.

PCS/MCS References

The following references are representative of studies that present results using the PCS and MCS:

Alemao Evo A; Larson Rodney A., and Cady Paul S. Examining the Role of Health Status and Age in Patient Satisfaction with Prescription Services. *American Association of Pharmacy-Projects & Abstracts* 1999 Jul 3.

Alter H J; Braun R, and Zazzali J L. Health status disparities among public and private emergency department patients. *Academy of Emergency Medicine.* 1999 Jul; 6, (7): 736-43.

Benzo R; Flume P A; Turner D, and Tempest M. Effect of pulmonary rehabilitation on quality of life in patients with COPD: the use of SF-36 summary scores as outcomes measures [In Process Citation]. *Journal of Cardiopulminary Rehabilitation.* 2000 Jul-2000 Aug 31; 20, (4): 231-4.

Bobes J; Gutierrez M; Gibert J; Gonzalez M P; Herraiz L, and Fernandez A. Quality Of Life In Schizophrenia: Long-Term Follow-Up in 362 Chronic Spanish Schizophrenic Outpatients Undergoing Risperidone Maintenance Treatment. *European Psychiatry.* 1998; 13, ( ): 158-63.

Byles J E; Mishra G, and Schofield M. Factors associated with hysterectomy among women in Australia. *Health Place.* 2000 Dec 1; 6, (4): 301-308.

Cass A R; Volk R J, and Nease D E. Health-related quality of life in primary care patients with recognized and unrecognized mood and anxiety disorders. *International Journal of Psychiatry Medicine.* 1999; 29, (3): 293-309.

DeOreo P B. Hemodialysis patient-assessed functional health status predicts continued survival, hospitalization, and dialysis-attendance compliance. *American Journal of Kidney Medicine.* 1997 Aug; 30, (2): 204-12.

Friedman A W; Alarcon G S; McGwin Jr G; Straaton K V; Roseman J; Goel N, and Reveille J D. Systemic lupus erythematosus in three ethnic groups. IV. Factors associated with self-reported functional outcome in a large cohort study. LUMINA Study Group. Lupus in Minority Populations, Nature versus Nurture. *Arthritis Care Research.* 1999 Aug; 12, (4): 256-66.

Funk G F; Karnell L H; Dawson C J; Means M E; Colwill M L; Gliklich R E; Alford E L, and Stewart M G. Baseline and post-treatment assessment of the general health status of head and neck cancer patients compared with United States population norms. *Head Neck.* 1997 Dec; 19, (8): 675-83.

Gandek B; Ware J E; Aaronson N K; Apolone G; Bjorner J B; Brazier J E; Bullinger M; Kaasa S; Leplege A; Prieto L, and Sullivan M. Cross-validation of item selection and scoring for the SF-12 Health Survey in nine countries: results from the IQOLA Project. International Quality of Life Assessment. *Journal of Clinical Epidemiology.* 1998 Nov; 51, (11): 1171-8.

Gartsman G M; Khan M, and Hammerman S M. Arthroscopic repair of full-thickness tears of the rotator cuff. *Journal of Bone Joint Surgery (American).* 1998 Jun; 80, (6): 832-40.

Jenkinson C. The SF-36 physical and mental health summary measures: an example of how to interpret scores. *Journal of Health Service Research Policy.* 1998 Apr; 3, (2): 92-6.

Jenkinson C; Gray A; Doll H; Lawrence K; Keoghane S, and Layte R. Evaluation of index and profile measures of health status in a randomized controlled trial. Comparison of the Medical Outcomes Study 36-Item Short Form Health Survey, EuroQol, and disease specific measures. *Medical Care.* 1997 Nov; 35, (11): 1109-18.

Jenkinson C and Layte R. Development and testing of the UK SF-12 (short form health survey). *Journal of Health Service Research Policy.* 1997 Jan; 2, (1): 14-8.

Jenkinson C; Layte R; Jenkinson D; Lawrence K; Petersen S; Paice C, and Stradling J. A shorter form health survey: can the SF-12 replicate results from the SF-36 in longitudinal studies? *Journal of Public Health Medicine.* 1997 Jun; 19, (2): 179-86.

Jenkinson C; Stewart-Brown S; Petersen S, and Paice C. Assessment of the SF-36 version 2 in the United Kingdom. *Journal of Epidemiology Community Health.* 1999 Jan; 53, (1): 46-50.

Assessment of the SF-36 version 2 in the United Kingdom [see comments]. 1999 Jan; 53, (1): 46-50. Notes: Comment in: *Journal of Epidemiology Community Health* 1999 Oct,53(10):651-2

Jenkinson C; Stradling J, and Petersen S. How should we evaluate health status? A comparison of three methods in patients presenting with obstructive sleep apnoea. *Quality of Life Research.* 1998 Feb; 7, (2): 95-100.

Johnson J A and Coons S J. Comparison of the EQ-5D and SF-12 in an adult US sample. *Quality of Life Research.* 1998 Feb; 7, (2): 155-66.

Johnson J A and Pickard A S. Comparison of the EQ-5D and SF-12 health surveys in a general population survey in Alberta, Canada. *Medical Care.* 2000 Jan; 38, (1): 115-21.

Keller S D; Ware Jr J E; Hatoum H T, and Kong S X. The SF-36 Arthritis-Specific Health Index (ASHI): II. Tests of validity in four clinical trials. *Medical Care.* 1999 May; 37, (5 Suppl): MS51-60.

Kosinski M; Keller S D; Ware Jr J E; Hatoum H T, and Kong S X. The SF-36 Health Survey as a generic outcome measure in clinical trials of patients with osteoarthritis and rheumatoid arthritis: relative validity of scales in relation to clinical measures of arthritis severity. *Medical Care.* 1999 May; 37, (5 Suppl): MS23-39.

Mahomed N N; Spellmann M, and Goldberg M J. Functional health status of adults with achondroplasia. *American Journal of Medical Genetics.* 1998 Jun 16; 78, (1): 30-5.

Mancuso C A; Peterson M G, and Charlson M E. Effects of depressive symptoms on health-related quality of life in asthma patients [In Process Citation]. *Journal of Genetics, Intern Med.* 2000 May; 15, (5): 301-10.

---. Effects of depressive symptoms on health-related quality of life in asthma patients [see comments]. 2000 May; 15, (5): 301-10. Notes: Comment in: *J Gen Intern Med* 2000 May,15(5):344-5

Mant J W; Jenkinson C; Murphy M F; Clipsham K; Marshall P, and Vessey M P. Use of the Short Form-36 to detect the influence of upper gastrointestinal disease on self-reported health status. *Quality of Life Research.* 1998 Apr; 7, (3): 221-6.

McEwen S; Mayo N, and Wood-Dauphinee S. Inferring quality of life from performance-based assessments [In Process Citation]. *Disability Rehabilitation.* 2000 Jul 10; 22, (10): 456-63.

McGuire M S; Grimaldi G; Grotas J, and Russo P. The type of urinary diversion after radical cystectomy significantly impacts on the patient's quality of life [see comments]. 2000 Jan-2000 Feb 28; 7, (1): 4-8.
Notes: Comment in: *Ann Surg Oncol* 2000 Jan-Feb,7(1):1

Miners A H; Sabin C A; Tolley K H; Jenkinson C; Ebrahim S, and Lee C A. Assessing Health-Related Quality-of-Life In Patients With Severe Haemophilia A And B. *Psychology, Health & Medicine.* 1999; 4, (1): 5-15.

Morrin L; Black S, and Reid R. Impact of duration in a cardiac rehabilitation program on coronary risk profile and health-related quality of life outcomes. *Journal of Cardiopulminary Rehabilitation.* 2000 Mar-2000 Apr 30; 20, (2): 115-21.

Mouton C P; Rovi S; Furniss K, and Lasser N L. The associations between health and domestic violence in older women: results of a pilot study. *Journal of Women's Health, Gender-Based Medicine.* 1999 Nov; 8, (9): 1173-9.

Murray M; Lefort S, and Ribeiro V. The SF-36: Reliable And Valid For The Institutionalized Elderly? *Aging & Mental Health.* 1998; 2, (1): 24-7.

Nash I S; Curtis L H, and Rubin H. Predictors of patient-reported physical and mental health 6 months after percutaneous coronary revascularization. *American Heart Journal.* 1999 Sep; 138, (3 Pt 1): 422-9.

Pickard A S; Johnson J A; Penn A; Lau F, and Noseworthy T. Replicability of SF-36 summary scores by the SF-12 in stroke patients. *Stroke.* 1999 Jun; 30, (6): 1213-7.

Rai G S; Kiniorns M, and Burns W. New Handicap Scale For Elderly In Hospital. *Arch Gerontol Geriatr.* 1999; 28, ( ): 99-104.

Revicki D A; Genduso L A; Hamilton S H; Ganoczy D, and Beasley Jr C M. Olanzapine versus haloperidol in the treatment of schizophrenia and other psychotic disorders: quality of life and clinical outcomes of a randomized clinical trial. *Quality of Life Research.* 1999 Aug; 8, (5): 417-26.

Revicki Dennis A; Genduso Laura A; Hamilton Susan H; Ganoczy Dara, and Beasley Charles M. Olanzapine versus haloperidol in the treatment of schizophrenia and other psychotic disorders: Quality of life and clinical outcomes of a randomized clinical trial. *Quality of Life Research.* 1999 Aug; 8, (5): 417-26. Notes: Entered manually- no medline citation

Salsberry P J; Nickel J T; Polivka B J; Kuthy R A; Slack C, and Shapiro N. Self-reported health status of low-income mothers. *Image J Nurse Sch.* 1999; 31, (4): 375-80.

Schofield Margot J; Lattimore-Foot Glenda, and Sanson-Fisher Rob. SF-36 Health Profiles of Recently Discharged Hospital Patients In Australia. 1998; 3, (4): 551-63.

Shmueli A. Subjective health status and health values in the general population. *Journal of Health Psychology.* 1999 Apr-1999 Jun 30; 19, (2): 122-7.

Sullivan Marianne; Karlsson Jan; Franz Bo, and Nyth Anna-Lena. Lakares Halsa. SF-36 Halsoenkat (Physicians' Health. SF-36 Health Survey). 1996; ( ): .

Ware Jr J; Kosinski M, and Keller S D. A 12-Item Short-Form Health Survey: construction of scales and preliminary tests of reliability and validity. *Med Care.* 1996 Mar; 34, (3): 220-33.

Ware Jr J E; Gandek B; Kosinski M; Aaronson N K; Apolone G; Brazier J; Bullinger M; Kaasa S; Leplege A; Prieto L; Sullivan M, and Thunedborg K. The equivalence of SF-36 summary health scores estimated using standard and country-specific algorithms in 10 countries: results from the IQOLA Project. International Quality of Life Assessment. *Journal of Clinical Epidemiology.* 1998 Nov; 51, (11): 1167-70.

Ware Jr J E; Kosinski M; Bayliss M S; McHorney C A; Rogers W H, and Raczek A. Comparison of methods for the scoring and statistical analysis of SF-36 health profile and summary measures: summary of results from the Medical Outcomes Study. *Medical Care.* 1995 Apr; 33, (4 Suppl): AS264-79.

Appendix A: Definitions Of Criterion Variables

TABLE A.1 DEFINITIONS OF CHRONIC CONDITIONS FROM THE GENERAL U.S. POPULATION SURVEY

CONDITION	DEFINITION
Hypertension	Has doctor ever told you that you have hypertension (sometimes called high blood pressure).
Congestive Heart Failure	Has doctor ever told you that you have congestive heart failure (heart failure or enlarged heart).
MI (Recent)	Has doctor ever told you that you had a heart attack in the last year (myocardial infarction).
Diabetes	Has doctor ever told you that you have diabetes (high blood sugar).
Angina	Has doctor ever told you that you have angina.
Cancer	Has doctor ever told you that you have cancer (except skin cancer).
Allergies	Do you now have chronic allergies or sinus trouble.
Arthritis	Do you now have arthritis of any kind or rheumatism.
Back Pain/Sciatica	Do you now have sciatica or chronic back problems.
Vision Impairment	Do you now have blindness or trouble seeing with one or both eyes, even when wearing glasses.
Chronic Lung Disease	Do you now have chronic lung disease (like chronic bronchitis, asthma, or emphysema).
Dermatitis	Do you now have dermatitis or other chronic skin rash.
Hearing Impairment	Do you now have deafness or other trouble hearing with one or both ears.
Limitations in use of arm(s)/leg(s)	Do you now have limitation in the use of an arm or leg (missing, paralyzed, or weakness).
Depression Screener	Two or more weeks in the past year feeling sad, blue, or depressed; or two years or more feeling depressed or sad most days; or feeling depressed or sad much of the time in the past year?
Stress	Quite a bit or a great deal of stress or pressure experienced in daily living in the past four weeks
Life Satisfaction	Extremely or very happy with personal life during the past four weeks

TABLE A.2 DEFINITIONS OF CRITERIA USED IN FIRST ROUND OF EMPIRICAL VALIDATION

CRITERION	DEFINITION
MOS Tracer Conditions (Table 8.1, column 1; Appendix B1)	
Hypertension	Physician report of current hypertension (or independently derived probability of hypertension if physician report missing or questionable).
Congestive Heart Failure	Physician report of current congestive heart failure (or independently derived probability of CHF if physician report missing or questionable).
MI (Recent)	Physician report of MI within the past year (or independently derived probability of MI if physician report of MI missing or questionable).
Diabetes, Type II	Physician report of diabetes with age at onset 30 years or older (or independently derived probability of diabetes and age at onset if actual information missing or questionable).
Severity of MOS Tracer Conditions (Table 8.1, column 2-4; Appendix B2)	
Hypertension	Severity defined by diastolic blood pressure above 100 mm Hg (2 levels)
Congestive Heart Failure	Severity defined by the presence of dyspnea on one-block exertion or while lying flat (2 levels)
MI (Recent)	Severity defined by the presence of premature ventricular contractions and/or angina (2 levels)
Diabetes, Type II	Severity defined by the presence of complications and duration of diabetes (4 levels: 1-free of complications and duration less than 10 years; 2-free of complications and duration 10 or more years; 3-complications of eye of foot only; 4-complications of diabetic heart and/or kidney disease)
MOS Comorbid Conditions* (Table 8.1, column 5; Appendix B3)	
Asthma	Had any asthma attacks in past six months
COPD	Now have lung disease ever diagnosed by physician as chronic obstructive pulmonary disease (like chronic bronchitis or emphysema) in past six months.
Angina - ever**	Ever told by physician have angina.
Angina, recent - no MI	Symptoms of angina in past six months in the absence of an MI within one year.
MI, past	Ever had a heart attack diagnosed by physician, more than one year ago.
Other lung disease	Any other lung disease such as tuberculosis or pneumonia in past six months.
Back pain/sciatica	Attacks of back pain or sciatica last six months.
Hip impairments	Ever told by physician have hip impairments
Rheumatoid arthritis	Now have active condition physician ever diagnosed as arthritis and physician labeled it rheumatoid arthritis and morning stiffness.

TABLE A.2 DEFINITIONS OF CRITERIA USED IN FIRST ROUND OF EMPIRICAL VALIDATION (CONTINUED)

CRITERION	DEFINITION
Osteoarthritis	Now have active condition physician ever diagnosed as arthritis and physician labeled it osteoarthritis or degenerative arthritis and patient is · 55 years old.
Musculoskeletal complaints	Active condition physician ever diagnosed as arthritis but criteria for osteoarthritis or rheumatoid arthritis not met.
Other rheumatic disease**	Now have active rheumatic disease other than arthritis physician ever diagnosed (e.g., systemic lupus erythematosus, scleroderma, or gout).
Colitis**	Now have active disease physician ever diagnosed as Crohn's disease or ulcerative colitis (severe bowel irritation).
Diverticulitis**	Now have active disease physician ever diagnosed as diverticulitis.
Fistulas**	Now have active disease physician ever diagnosed as intestinal fistulas.
Gallbladder disease**	Now have active disease physician ever diagnosed as chronic gallbladder disease.
Irritable bowel disease	Ever told by physician have irritable bowel syndrome, functional bowel disease, or chronic bowel disease.
Liver disease**	Now have active disease physician ever diagnosed as chronic hepatitis or cirrhosis.
Diabetes, Type I**	Physician report of diabetes with age at onset younger than 30 years (or independently derived probability of diabetes and age at onset if actual information missing or questionable).
Ulcer	Now have active disease physician ever diagnosed as an ulcer (peptic, gastric, stomach, or duodenal).
Kidney disease**	Disease physician ever diagnosed as serious kidney disease in last six months.
Benign Prostatic Hypertrophy**	Male, age · 50 years, history of nocturia in past six months, no serious kidney disease ever diagnosed, and no report of prostatic cancer.
UTI	Kidney or bladder infection diagnosed by physician in past six months.
Varicosities**	Now have condition physician ever diagnosed as varicose veins/deep varicosities.
Cancer**	Ever had cancer.
Dermatitis	Repeated episodes of dermatitis or skin rash in past six months.
Anemia	Told by doctor have anemia (past six months.)

(continued)

TABLE A.2   DEFINITIONS OF CRITERIA USED IN FIRST ROUND OF EMPIRICAL VALIDATION (CONTINUED)

CRITERION	DEFINITION
Symptom Clusters (Table 8.1, column 6-9; Appendix B4)	
Ear, nose & throat	Patient reported frequency of blurred vision, dry mouth or lump in throat in the past four weeks.
Central Nervous system	Patient report of fainting, drowsiness or dizziness, shortness of breath, chest pain heart palpitations, or frequent headaches in the past four weeks.
Musculoskeletal	Patient report of stiffness or soreness in the joints, backache, heavy feeling in arm or legs, or numbness in the feet in the past four weeks.
GI/GU	Patient report of acid indigestion, heartburn, nausea, or trouble passing urine in the past four weeks.
Clinical Depression Groups (Table 8.1, column 15-16; Appendix B8 & B9)	
Cross-sectional	NIMH (DIS) criteria met for major depression and/or dysthymia at baseline assessment.
Longitudinal	Major depression and/or dysthymia present at one-year follow-up but *not present* at two-year follow-up.
Age Groups (Table 8.1, column 10-11; Appendix B5 & B6)	
Age 18-44	Uncomplicated hypertensives (patients with hypertension and no other major medical conditions), age 18 - 44
Age 45-64	Uncomplicated hypertensives (patients with hypertension and no other major medical conditions), age 45 - 64
Age 65 or older	Uncomplicated hypertensives (patients with hypertension and no other major medical conditions), age >= 65
Self-Reported Transition Groups (Table 8.1, column 12-14; Appendix B7)	
Physical	Patient report at two-year follow-up of change in physical health over two years: (a lot more limited now, a little more limited now, about the same, somewhat less limited now or a lot less limited now.
General	Patient report at two-year follow-up of change in general health over two years: (a lot more limited now, a little more limited now, about the same, somewhat less limited now or a lot less limited now.
Mental	Patient report at two-year follow-up of change in mental health over two years: (a lot more limited now, a little more limited now, about the same, somewhat less limited now or a lot less limited now.

* Information regarding the comorbid medical conditions was obtained from the patient during a structured medical history interview conducted by a trained clinician. If information regarding a condition (or conditions) was missing, an independently derived probability of each diagnosis was substituted.

** Because of very low prevalence, the following conditions are incorporated into an index of eleven comorbid conditions: angina-ever, other rheumatic disease, colitis, diverticulitis, intestinal fistulas, gallbladder disease, liver disease, benign prostatic hypertrophy, varicosities, cancer, and type I diabetes.

Appendix B: Table Of Results

TABLE B.1 ADJUSTED MEAN SCORES FOR SF-36 SCALES AND COMPONENT SUMMARIES FOR FOUR CHRONIC CONDITIONS

	PF	RP	BP	GH	VT	SF	RE	MH	PCS	MCS
Hypertension	78.27 (0.8)	65.90 (1.4)	75.08 (1.0)	66.79 (0.7)	61.63 (0.8)	90.08 (0.6)	79.85 (1.3)	80.39 (0.6)	45.94 (0.3)	53.40 (0.3)
Congestive Heart Failure	59.47 (2.8)	46.33 (4.3)	69.57 (3.4)	50.18 (2.3)	47.15 (2.7)	78.59 (3.3)	69.02 (3.9)	78.54 (1.7)	38.29 (1.2)	51.43 (1.0)
Myocardial Infarction, Recent	72.40 (2.8)	50.60 (4.7)	76.15 (2.66)	61.59 (2.4)	56.10 (2.0)	87.68 (2.4)	73.04 (4.5)	76.33 (1.5)	43.51 (1.2)	51.69 (0.9)
Type II Diabetes	74.36 (2.1)	63.32 (3.7)	73.55 (2.2)	59.16 (1.9)	59.11 (1.7)	86.54 (1.9)	80.63 (2.8)	78.83 (1.4)	43.86 (1.0)	53.04 (0.8)
F for Four Conditions	21.76[a]	10.75[a]	3.05[c]	25.10[a]	11.48[a]	5.19[a]	2.99[c]	2.93[c]	17.85[a]	1.59
Adjusted R^2 (Four Conditions)	0.252	0.113	0.074	0.084	0.105	0.058	0.036	0.078	0.186	0.069
N	1413	1413	1413	1413	1413	1413	1413	1413	1413	1413

[a] $p < 0.001$, two-tailed test
[b] $p < 0.01$, two-tailed test
[c] $p > 0.05$, two-tailed test

Note: Mean scores were estimated from linear regression models that controlled for demographics and MOS design variables.

SF-36™ Summary Measures Manual

TABLE B.2 ADJUSTED SF-36 AND COMPONENT SCALE SCORES FOR THE SEVERITY OF FOUR CHRONIC CONDITIONS

	PF	RP	BP	GH	VT	SF	RE	MH	PCS	MCS
Hypertension										
Severity 1	77.72 (0.8)	64.64 (1.6)	75.06 (1.1)	66.60 (0.8)	61.51 (0.9)	89.88 (0.7)	78.78 (1.6)	80.13 (0.7)	45.78 (0.4)	53.25 (0.4)
Severity 2	81.12 (2.9)	73.98 (5.2)	74.92 (3.6)	67.89 (3.0)	62.13 (2.0)	91.43 (1.9)	86.75 (3.4)	82.11 (2.1)	46.81 (1.4)	54.41 (0.9)
Congestive Heart Failure										
Severity 1	67.55 (3.5)	57.18 (5.4)	74.83 (3.9)	56.59 (3.0)	53.49 (3.2)	84.33 (3.9)	76.99 (4.3)	81.51 (1.7)	41.60 (1.7)	53.25 (1.2)
Severity 2	47.95 (2.7)	29.36 (4.8)	61.31 (4.8)	40.76 (2.8)	36.99 (3.6)	69.54 (4.9)	57.27 (5.0)	73.99 (2.9)	33.31 (1.2)	48.58 (1.7)
Type II Diabetes										
Severity 1	77.16 (2.7)	64.73 (4.8)	73.70 (2.6)	61.53 (2.3)	59.85 (1.9)	87.59 (2.5)	82.92 (3.7)	79.09 (1.9)	44.66 (1.3)	53.27 (1.0)
Severity 2	75.47 (2.8)	65.80 (4.9)	76.83 (2.9)	57.79 (3.1)	60.07 (3.1)	87.49 (2.5)	74.82 (4.6)	78.09 (2.4)	45.06 (1.3)	51.94 (1.3)
Severity 3	73.36 (5.0)	63.88 (8.1)	73.25 (5.0)	58.69 (5.8)	57.91 (3.2)	85.09 (3.9)	88.17 (5.8)	81.51 (2.9)	42.83 (2.3)	54.51 (1.2)
Severity 4	63.55 (3.8)	54.69 (7.3)	66.93 (5.2)	52.96 (2.8)	55.36 (3.2)	80.83 (4.3)	76.01 (5.2)	76.72 (2.1)	39.86 (1.7)	52.41 (1.2)
F for Severity of Hypertension	1.20	2.62	0.00	0.15	0.07	0.55	3.59	0.69	0.43	1.13
F for Severity of CHF	21.08[a]	17.31[a]	5.34[c]	17.25[a]	13.50[a]	5.65[c]	7.24[b]	5.32[c]	17.33[a]	5.01[c]
F for Severity of Diabetes	3.61[c]	0.89	1.42	2.61	0.70	0.94	1.44	1.02	2.94[c]	0.93
Adjusted R^2 (Severity)	0.270	0.126	0.080	0.098	0.113	0.069	0.047	0.081	0.202	0.074
N	1413	1413	1413	1413	1413	1413	1413	1413	1413	1413

[a] $p < 0.001$, two-tailed test
[b] $p < 0.01$, two-tailed test
[c] $p < 0.05$, two-tailed test

Note: Mean scores were estimated from linear regression models that controlled for demographics and MOS design variables.

TABLE B.3 ESTIMATES OF DIFFERENCES IN MEAN SCORES FOR SF-36 SCALES AND COMPONENT SUMMARIES FOR PATIENTS WITH COMORBIDITIES VS PATIENTS WITHOUT COMORBIDITIES

	PF	RP	BP	VT	GH	SF	RE	MH	PCS	MCS
Asthma	-10.72	-16.61[c]	-5.59	-4.50	-7.92	-2.98	-38.18[a]	-4.71	-2.21	-5.63[c]
	(6.2)	(8.2)	(5.8)	(5.4)	(4.8)	(5.3)	(11.0)	(4.2)	(2.8)	(2.8)
COPD	-5.86	-10.68[c]	-5.98	-9.01[b]	-4.57[c]	-4.80	-.82	-4.37	-3.65[c]	-1.06
	(3.4)	(5.4)	(3.1)	(2.8)	(2.2)	(3.6)	(4.1)	(2.2)	(1.5)	(1.2)
Angina (no MI)	-5.69[b]	-13.78[a]	-6.41[b]	-7.96[a]	-7.67[a]	-.89	-4.19	-2.06	-3.94[a]	-.72
	(2.1)	(3.4)	(2.2)	(1.7)	(2.0)	(1.9)	(3.2)	(1.5)	(.9)	(.9)
Past MI	-7.33[b]	-5.04	-3.59	-5.66[b]	-3.85	-4.32	-3.16	-1.61	-2.70[b]	-.71
	(2.5)	(3.9)	(2.7)	(2.1)	(2.1)	(2.5)	(4.7)	(2.2)	(1.0)	(1.3)
Other Lung Disease	.84	2.11	2.91	-2.02	9.25[c]	-4.15	-3.59	-6.00	1.68	-1.83
	(4.4)	(6.3)	(5.0)	(6.1)	(4.6)	(3.5)	(10.4)	(5.4)	(2.6)	(1.9)
Back Complaints	-4.92[b]	-11.71[a]	-13.39[a]	-3.04[c]	-4.86[b]	-5.50[a]	-2.46	-2.54[c]	-3.95[a]	-.71
	(1.7)	(3.1)	(1.9)	(1.2)	(1.7)	(1.6)	(3.0)	(1.2)	(.7)	(.7)
Hip Impairment	-15.41[a]	-6.29	-9.97[a]	-12.30[a]	-7.89[c]	-10.97[b]	-9.45	-4.07	-5.36[a]	-2.28
	(3.5)	(5.8)	(2.8)	(2.7)	(3.1)	(4.2)	(6.4)	(2.7)	(1.5)	(1.7)
Rheumatoid Arthritis	-7.90	-13.92[c]	-20.50[a]	-5.94	-.61	-2.99	4.56	-.33	-6.42[a]	2.20
	(5.8)	(6.9)	(4.6)	(3.7)	(3.5)	(3.8)	(7.0)	(4.1)	(1.9)	(2.0)
Osteoarthritis	-8.38[c]	-12.73[c]	-13.61[a]	-2.64	-5.36	-2.78	3.47	-1.33	-5.17[a]	1.05
	(3.3)	(5.1)	(3.1)	(2.2)	(3.0)	(2.7)	(5.0)	(2.0)	(1.3)	(1.1)
Musculoskeletal	-4.23[c]	-6.62[c]	-6.20[a]	-3.62[c]	-3.50	.36	-1.82	-3.48[b]	-2.29[b]	-.58
	(1.7)	(3.4)	(1.9)	(1.7)	(1.8)	(1.5)	(3.4)	(1.3)	(.8)	(.8)
Irritable Bowel	-6.08	-3.25	-5.04	-6.39	-6.16	-5.46	-3.40	-2.03	-2.57	-1.37
	(3.5)	(6.0)	(3.6)	(3.8)	(3.3)	(2.8)	(6.6)	(2.5)	(1.5)	(1.5)
Ulcers	-12.07[c]	-14.03[c]	-7.17	-9.29[c]	-6.01	-6.71	-7.42	-4.62	-4.87	-1.61
	(6.0)	(6.8)	(4.5)	(4.3)	(4.3)	(4.2)	(6.8)	(3.6)	(2.9)	(1.9)
Kidney Disease	-3.95	-13.78	-12.09†	-5.73	-7.24	-9.34	4.17	-7.35	-3.90	-1.93
	(5.7)	(12.2)	(7.2)	(5.9)	(4.2)	(7.5)	(13.9)	(6.7)	(3.0)	(4.1)
UTI	-.93	-1.77	-4.26	-1.83	-5.35	.07	-1.72	.42	-1.13	-.40
	(2.1)	(4.7)	(2.5)	(2.4)	(2.8)	(3.4)	(5.4)	(1.9)	(1.0)	(1.3)
Dermatitis	3.22[c]	-6.50	-3.00	-2.04	-.88	-.76	-2.60	-1.37	-.40	-.86
	(1.6)	(3.9)	(2.6)	(1.6)	(1.9)	(1.9)	(3.5)	(1.9)	(.8)	(.9)
Anemia	-6.03	-4.66	-4.61	-9.51[b]	-5.62	-.62	5.25	-.69	-3.65[c]	.81
	(4.1)	(6.5)	(3.4)	(3.1)	(3.5)	(3.9)	(4.2)	(3.0)	(1.7)	(1.4)
Intercept	89.21[a]	6.22[a]	90.05[a]	72.92[a]	67.14[a]	95.66[a]	79.59[a]	83.02[a]	51.53[a]	53.27[a]
	(2.2)	(4.3)	(2.6)	(2.1)	(2.1)	(.21)	(3.5)	(1.8)	(1.1)	(1.0)
F for Significance of Comorbidities	8.48[a]	6.71[a]	14.50[a]	12.44[a]	6.98[a]	4.06[a]	1.72[c]	4.07[a]	12.63[a]	1.93[c]
Adjusted R^2	0.3436	0.1972	0.2451	0.1973	0.1852	0.1182	0.0670	0.1129	0.3310	0.0866
	n = 1413	n = 1413	n = 1413	n = 1413	n = 1413	n = 1413	n = 1413	n = 1413	n = 1413	n = 1413

[a] $p < 0.001$ [b] $p < 0.01$ [c] $p < 0.05$

Note: Each entry is the difference in score between having and not having the comorbid condition.
Note: These estimates are from a fully loaded model that controlled for demographics, diagnosis, and severity of diagnosis.
Note: Clinically depressed patients were excluded from the models.

TABLE B.4   CORRELATIONS[1] BETWEEN SYMPTOM FREQUENCY AND SF-36 SCALES AND COMPONENT SUMMARIES (N = 1,397)

Symptoms[2]	Mean	SD	\multicolumn{8}{c}{SF-36 Scales}	\multicolumn{2}{c}{Component Summaries}
			PF	RP	BP	GH	VT	SF	RE	MH	PCS[3]	MCS[4]
Ears, Nose, and Throat												
Blurred Vision	1.55	1.01	-.31	-.32	-.29	-.30	-.30	-.29	-.20	-.21	-.32	-.18
Dry Mouth	2.13	1.27	-.38	-.37	-.34	-.35	-.37	-.31	-.22	-.29	-.37	-.22
Lump in throat	1.29	0.73	-.20	-.23	-.24	-.21	-.30	-.24	-.23	-.31	-.18	-.27
Under Central Nervous System Control												
Fainting or passing out	1.03	0.25	-.09	-.09	-.08	-.07	-.11	-.12	-.13	-.12	-.06	-.13
Shortness of breath (lying down)	1.38	0.89	-.43	-.37	.35	-.36	-.40	-.32	-.20	-.25	-.41	-.19
Feeling drowsy or sedated	1.88	1.09	-.33	-.41	-.36	-.38	-.52	-.40	-.34	-.40	-.34	-.38
Feeling dizzy when standing up	1.64	0.93	-.23	-.32	-.29	-.28	-.38	-.31	-.27	-.34	-.25	-.31
Chest pain relieved by nitroglycerin	1.31	0.79	-.36	-.31	-.25	-.33	-.28	-.23	-.17	-.12	-.35	-.10
Heart pounding or palpitations	1.50	0.88	-.31	-.32	-.32	-.31	-.36	-.30	-.28	-.33	-.29	-.28
Headaches more than usual	1.66	1.01	-.16	-.25	-.34	-.24	-.36	-.35	-.30	-.42	-.18	-.39
Musculoskeletal/Extremities												
Backaches or lower back pains	2.47	1.37	-.34	-.37	-.53	-.30	-.34	-.29	-.22	-.22	-.41	-.17
Pins and needles in your feet	1.75	1.17	-.35	-.34	.38	-.34	-.32	-.26	-.20	-.20	-.38	-.15
Heavy Feeling in arms and legs	1.51	0.97	-.40	-.42	-.45	-.36	-.42	-.39	-.31	-.29	-.42	-.26
Stiffness, pain in muscles	2.95	1.36	-.46	-.48	-.64	-.38	-.41	-.32	-.21	-.19	-.55	-.11
GI/GU												
Acid indigestion after meals	2.20	1.18	-.24	-.28	-.34	-.29	-.34	-.26	-.23	-.26	-.27	-.23
Trouble passing urine	1.23	0.69	-.12	-.19	-.16	-.18	-.19	-.16	-.15	-.18	-.15	-.16
Nausea (upset stomach)	1.57	0.92	-.21	-.31	-.34	-.29	-.33	-.35	-.29	-.37	-.24	-.33
Other												
Waking up early, not able to go back to sleep	2.21	1.22	-.35	-.39	-.34	-.33	-.38	-.31	-.28	-.31	-.34	-.26
Coughing producing sputum	1.81	1.19	-.22	-.28	-.25	-.29	-.27	-.23	-.16	-.20	-.26	-.16
F for All Symptoms			53.71[a]	53.93[a]	91.44[a]	38.83[a]	59.78[a]	34.94[a]	21.12[a]	36.11[a]	68.43[a]	29.33[a]
F for Significance of Ears, Nose & Throat			10.48[a]	3.36[c]	0.70	4.43[b]	2.44	1.63	0.96	9.53[a]	8.99[a]	5.36[a]
F for Significance of CNS Symptoms			25.76[a]	12.28[a]	7.77[a]	13.20[a]	34.37[a]	15.10[a]	13.02[a]	28.91[a]	20.98[a]	29.56[a]
F for Significance of GI/GU Symptoms			3.20[c]	1.05	2.89[c]	1.48	1.30	6.10[a]	2.72[c]	7.48[a]	0.38	6.94[a]
F for Significance of Musculoskeletal			56.55[a]	54.17[a]	199.61[a]	22.51[a]	25.54[a]	15.95[a]	6.45[a]	2.06	110.43[a]	5.93[a]
Adjusted R^2			0.418	0.418	0.552	0.340	0.444	0.316	0.215	0.323	0.479	0.278

[1]   The F-statistics summarized here and in the text are based on comparison of means for the symptom clusters. Those statistics are summarized as correlations coefficients so as to make the results presentable.
[2]   Reported frequency in the past four weeks scored as followed: 1=never, 2=once or twice, 3=a few times, 4=fairly often, 5=very often
[3]   Physical Component Summary Scale
[4]   Mental Component Summary Scale

Note: All correlations between symptom frequency and scales are significant at $p < .001$, except underlined correlations. Underlined correlations are not significant at $p < .05$.

TABLE B.5 CROSS SECTIONAL AGE RELATED DIFFERENCES IN HEALTH STATUS: UNCOMPLICATED HYPERTENSION[1]

	Age 18-44 (n=206)	Age 45-54 (n=193)	Age 55-64 (n=311)	Age 65+ (n=362)	(F)	RV
Physical Functioning (PF)	88.7 (1.3)	83.3 (1.5)	79.0 (1.3)	70.0 (1.3)	33.7[a]	1.00
Role-Physical (RP)	77.2 (2.4)	75.1 (2.5)	68.0 (2.2)	57.8 (2.2)	14.2[a]	0.42
Bodily Pain (BP)	78.4 (1.4)	75.3 (1.6)	75.1 (1.3)	70.9 (1.3)	4.8[b]	0.14
General Health (GH)	69.6 (1.3)	63.7 (2.7)	64.7 (1.1)	65.6 (1.0)	2.8[c]	0.08
Vitality (VT)	63.1 (1.4)	59.0 (1.6)	62.9 (1.1)	59.9 (1.2)	2.4	0.07
Social Functioning (SF)	90.6 (1.2)	90.6 (1.2)	91.5 (0.9)	90.1 (0.9)	0.4	0.01
Role Emotional (RE)	82.1 (2.4)	83.5 (2.3)	85.3 (1.8)	81.7 (1.8)	0.8	0.02
Mental Health (MH)	78.8 (1.1)	79.4 (1.1)	82.5 (0.8)	83.9 (0.8)	6.7[a]	0.19
Physical Component (PCS)	49.9 (0.6)	47.6 (0.7)	45.5 (0.6)	42.3 (0.6)	27.7[a]	0.82
Mental Component (MCS)	51.9 (0.6)	52.5 (0.6)	54.8 (0.4)	55.6 (0.4)	11.9[a]	0.35

[a] $p < .001$
[b] $p < .01$
[c] $p < .05$
[1] Uncomplicated Hypertension is defined as patients with hypertension and classified as "Minor Medical" in previous sickgroup comparisons.

TABLE B.6 DETECTING TWO YEAR CHANGE IN HEALTH STATUS AMONG UNCOMPLICATED HYPERTENSIVE PATIENTS[1] (N = 591)

	Baseline	Exit[1]	Mean Difference	(F)	RV
Physical Functioning (PF)	78.8 (0.9)	75.3 (1.0)	-3.5 (0.7)	23.04[a]	0.62
Role Physical (RP)	66.4 (1.6)	67.4 (1.7)	1.0 (1.7)	0.36	0.01
Bodily Pain (BP)	74.0 (0.9)	75.5 (0.9)	1.5 (0.9)	2.89	0.08
General Health (GH)	67.0 (0.7)	62.7 (0.9)	-4.3 (0.7)	37.21[a]	1.00
Vitality (VT)	61.5 (0.8)	64.4 (0.8)	2.9 (0.7)	15.21[a]	0.41
Social Functioning (SF)	91.6 (0.6)	90.2 (0.7)	-1.4 (0.8)	3.24	0.09
Role Emotional (RE)	85.2 (1.3)	84.7 (1.3)	-0.5 (1.5)	0.09	0.00
Mental Health (MH)	81.8 (0.6)	82.1 (0.6)	0.3 (0.5)	0.49	0.01
Physical Component (PCS)	45.6 (0.4)	44.6 (0.4)	-1.0 (0.4)	7.29[b]	0.20
Mental Component (MCS)	54.4 (0.3)	54.9 (0.3)	0.5 (0.3)	2.56	0.07

[a] $p < .001$
[b] $p < .01$
[c] $p < .05$
[1] Uncomplicated Hypertension is defined as patients with hypertension and classified as "Minor Medical" in previous sickgroup comparisons.

TABLE B.7 SF-36 AND COMPONENT ONE-YEAR DIFFERENCE SCORES BY SELF-REPORTED PHYSICAL, MENTAL, AND GENERAL HEALTH TRANSITIONS

	N	\multicolumn{16}{c}{SF-36 Scales}				

		PF		RP		BP		GH		VT		SF		RE		MH		PCS		MCS	
	N	x̄	SE	x̄	SE	x̄	SE	x̄	SE	x̄	SE	x̄	SE	x̄	SE	x̄	SE	x̄	SE	x̄	SE
Physical																					
Lot more	97	-12.4	2.6	-12.4	3.8	-11.1	3.8	-13.9	2.1	-3.9	2.2	-14.7	3.0	-1.4	4.7	-2.9	1.7	-6.6	1.2	-0.9	1.1
Some more	200	-8.4	1.2	-9.0	3.0	-2.7	2.3	-11.3	1.1	-3.9	1.2	-7.6	1.8	-0.2	3.3	-2.2	1.0	-3.9	0.6	-0.7	0.7
Same	888	-0.3	0.5	5.1	1.2	1.6	0.9	-2.4	0.5	2.1	0.6	0.1	.7	5.5	1.2	1.5	0.5	-0.1	0.2	1.2	0.3
Some less	161	3.8	1.6	8.5	3.2	4.7	2.4	-1.8	1.4	3.9	1.6	2.0	1.9	0.8	3.3	2.6	1.2	1.4	0.7	0.7	0.8
Lot less	167	11.0	1.6	18.1	3.1	8.5	2.1	5.1	1.3	.6	1.6	11.7	1.9	16.4	3.4	7.5	1.3	3.8	0.7	4.4	0.8
F		44.7[a]		17.5[a]		9.0[a]		33.1[a]		13.5[a]		27.5[a]		5.1[a]		11.8[a]		35.5[a]		7.9[a]	
RV		1.00		0.39		0.20		0.74		0.30		0.61		0.11		0.26		0.79		0.18	
Mental																					
Lot more	51	-6.6	3.9	-12.2	5.9	-12.5	4.9	-15.9	3.1	-9.6	3.4	-18.1	4.1	-17.0	5.7	-13.0	3.1	-3.4	1.7	-7.3	1.6
Some more	146	-6.9	1.6	-5.8	3.3	-2.5	2.7	-9.5	1.5	-4.5	1.6	-11.5	2.1	-12.1	3.9	-6.2	1.2	-2.1	0.8	-4.0	0.8
Same	728	-0.8	0.6	4.2	1.3	1.1	1.0	-4.5	0.6	0.9	0.6	-1.0	0.7	1.7	1.3	0.1	0.4	-0.2	0.3	0.1	0.3
Some less	248	0.8	1.1	6.2	2.4	4.2	1.8	-0.1	1.0	4.5	1.1	0.6	1.4	11.7	2.6	3.0	1.0	0.2	0.5	2.5	0.6
Lot less	269	3.9	1.2	9.4	2.5	4.5	1.7	2.4	1.2	8.4	1.3	9.7	1.6	22.4	2.6	11.5	1.1	-0.1	0.6	7.2	0.7
F		11.2[a]		6.4[a]		5.2[a]		22.9[a]		20.1[a]		31.3[a]		27.2[a]		58.4[a]		3.3[b]		59.8[a]	
RV		0.19		0.11		0.09		0.39		0.34		0.53		0.46		1.00		0.05		1.02	
General																					
Lot more	24	-20.9	6.7	-28.1	8.0	-17.5	9.1	-28.0	5.2	-18.1	5.6	-29.7	7.0	-11.1	8.7	-8.8	3.8	-11.1	3.1	-5.3	2.4
Some more	188	-8.7	.3	-12.9	2.6	-8.8	2.4	-14.1	1.2	-5.2	1.2	-10.0	1.9	-3.7	3.3	-4.3	1.1	-5.3	0.6	-1.6	0.7
Same	827	-1.5	0.5	4.6	1.2	2.5	0.9	-3.4	0.5	1.5	0.6	-1.2	0.7	4.8	1.3	0.7	0.5	-0.2	0.2	0.7	0.3
Some less	251	3.4	1.1	7.5	2.5	3.0	1.8	0.6	1.1	4.8	1.1	3.8	1.4	8.4	2.8	5.0	1.0	0.6	0.5	2.7	0.7
Lot less	164	11.1	1.6	19.7	3.2	8.3	2.2	7.2	1.4	11.1	1.6	12.5	2.0	15.4	3.3	9.5	1.5	4.2	0.7	5.1	0.9
F		43.6[a]		22.2[a]		11.9[a]		55.3[a]		26.9[a]		36.0[a]		6.3[a]		25.7[a]		40.9[a]		16.3[a]	
RV		0.79		0.40		0.21		1.00		0.49		0.65		0.11		0.46		0.74		0.29	

The MANOVA F for the one year self-reported physical health transition was F = 9.35, p < .00001, df 32,1508.
The MANOVA F for the one year self-reported mental health transition was F = 10.97, p < .00001, df 32,1437.
The MANOVA F for the one year self-reported general health transition was F = 12.91, p < .00001, df 32,1449.

TABLE B.8 SUMMARY OF RESULTS FOR CROSS-SECTIONAL TESTS FOR DETECTING DIFFERENCES IN MENTAL HEALTH BETWEEN PATIENTS WITH CLINICAL DEPRESSION AND PATIENTS WITH MINOR MEDICAL CONDITIONS

Measures	Clinical Depression (N = 263)	Minor Medical (N = 999)	Mean Diff. (S.E.)	F	RV	ES²
Physical Functioning (PF)	79.2 (1.4)	79.5 (0.7)	-0.3 (1.6)	.03	--[e]	--[e]
Role Physical (RP)	48.7 (2.5)	68.4 (1.2)	-19.7 (2.7)	57.00[a]	.05	.58
Bodily Pain (BP)	62.9 (1.6)	75.3 (0.7)	-12.4 (1.7)	60.37[a]	.06	.52
General Health (GH)	55.2 (1.4)	66.5 (0.6)	-11.3 (1.5)	70.73[a]	.07	.55
Vitality (VT)	39.8 (1.3)	61.8 (0.6)	-22.0 (1.4)	251.85[a]	.24	1.05
Social Functioning (SF)	59.2 (1.6)	90.9 (0.5)	-31.7 (1.7)	603.19[a]	.58	1.39
Role Emotional (RE)	36.4 (2.4)	83.5 (1.0)	-47.1 (2.6)	430.15[a]	.41	1.42
Mental Health (MH)	46.0 (1.2)	81.7 (0.5)	-35.7 (1.3)	<u>1043.93[a]</u>	1.00[d]	1.98
Physical Component Summary (PCS)	47.8 (0.7)	46.0 (0.3)	1.8 (0.8)	5.95[c]	.01	.18
Mental Component Summary (MCS)	33.4 (0.7)	54.1 (0.3)	-20.7 (0.8)	1072.56[a]	1.03	2.07

[a] $p < .001$;   [b] $p < .01$;   [c] $p < .05$
[1] Effect size determined by dividing the U.S. general population sd for each scale into each difference score.
[d] Best validity (highest F-ratio) among eight SF-36 scales is underlined
[e] Not statistically significant, RV and ES not estimated
RV = relative validity (see text)
ES = effect size (see text)

TABLE B.9 SUMMARY OF RESULTS FOR LONGITUDINAL TESTS FOR DIFFERENCES IN MENTAL HEALTH AFTER RECOVERY FROM CLINICAL DEPRESSION

Measures	Baseline	Follow-Up[1]	Average Change (SE)	F	RV	ES[2]
Physical Functioning (PF)	86.3 (1.6)	85.7 (1.8)	-0.5 (1.8)	.09	--[e]	--[e]
Role Physical (RP)	56.4 (4.1)	72.6 (3.9)	16.2 (4.7)	11.97[a]	.23	.48
Bodily Pain (BP)	69.7 (2.3)	69.9 (2.2)	0.2 (2.6)	.01	--[e]	--[e]
General Health (GH)	61.8 (2.1)	67.5 (2.4)	5.7 (2.0)	8.06[b]	.16	.28
Vitality (VT)	46.4 (1.9)	60.5 (1.9)	14.1 (1.9)	<u>51.84[a]</u>	1.00[d]	.67
Social Functioning (SF)	68.1 (2.4)	82.2 (1.9)	14.1 (2.7)	26.83[a]	.52	.62
Role Emotional (RE)	42.6 (4.2)	73.7 (3.7)	31.1 (4.7)	43.69[a]	.84	.93
Mental Health (MH)	51.9 (1.8)	67.3 (1.7)	15.4 (2.1)	50.98[a]	.98	.85
Physical Component Summary (PCS)	50.6 (1.0)	49.3 (1.1)	-1.3 (1.1)	1.34	--[e]	--[e]
Mental Component Summary (MCS)	36.3 (1.3)	47.2 (1.0)	10.9 (1.3)	74.99[a]	1.45	1.09

[a] $p < .001$;   [b] $p < .01$;   [c] $p < .05$
[1] Note: Follow-up scores were obtained two years after baseline (N = 94).
[2] Effect size determined by dividing the U.S. general population sd for each scale into each difference score.
[d] Best validity (highest F-ratio) among eight SF-36 scales is underlined
[e] Not statistically significant, RV and ES not estimated
RV = relative validity (see text)
ES = effect size (see text)

Appendix C: Scoring Algorithms: 1990 Norm-Based Scoring of PCS and MCS

This appendix reproduces Chapter 4 from the first edition of this manual (Ware et al., 1994) to document SF-36 (v1) PCS and MCS scoring based on 1990 norms.

Scoring of the Physical (PCS) and Mental (MCS) Component Summary measures involves three steps. First, the eight SF-36 scales are standardized using means and standard deviations from the general U.S. population. Second, they are aggregated using weights (factor score coefficients) from the general U.S. population. Finally aggregate PCS and MCS scores are standardized using a linear T-score transformation to have a mean of 50 and a standard deviation of 10, in the general U.S. population.

General U.S. population statistics used in the standardization and in the aggregation of SF-36 scale scores are presented in Table C.1. Detailed information including formulas for scale aggregation and transformation of scores are presented below. Formal checks using a test dataset recorded on a computer diskette (see back cover) can be performed to confirm the successful reproduction of PCS and MCS scales, as discussed later in the chapter. We strongly recommend these tests.

TABLE C.1 GENERAL U.S. POPULATION MEANS, STANDARD DEVIATIONS AND FACTOR SCORE COEFFICIENTS USED TO DERIVE PCS AND MCS SCALE SCORES

SF-36 SCALES	MEAN	SD	FACTOR SCORE COEFFICIENTS	
			PCS	MCS
Physical Functioning (PF)	84.52404	22.89490	0.42402	-0.22999
Role Physical (RP)	81.19907	33.79729	0.35119	-0.12329
Bodily Pain (BP)	75.49196	23.55879	0.31754	-0.09731
General Health (GH)	72.21316	20.16964	0.24954	-0.01571
Vitality (VT)	61.05453	20.86942	0.02877	0.23534
Social Functioning (SF)	83.59753	22.37642	-0.00753	0.26876
Role Emotional (RE)	81.29467	33.02717	-0.19206	0.43407
Mental Health (MH)	74.84212	18.01189	-0.22069	0.48581

Importance Of Standardization

As with the scoring of SF-36 items and scales, which are aggregated to score the summary measures, standardization of the scoring of the PCS and MCS scales is vital to their interpretation. Any changes in scoring of the SF-36 items, scales, or the algorithms for the summary scales may compromise their reliability and validity. Changes in scoring have also been shown to invalidate normative comparisons, and changes are likely to complicate or prevent meaningful comparisons of results across studies. (Ware, Snow, Kosinski, et al., 1993)

Norm-Based Scoring

The PCS and MCS scales are scored using norm-based methods. The means, standard deviations, and factor score coefficients used in scoring come from the general U.S. population. A linear T-score transformation method is used so that both the PCS and MCS have a mean of 50 and a standard deviation of 10 in the general U.S. population. This transformation is in contrast to the 0-100 scoring used to date for the eight SF-36 scales. The eight SF-36 scales have means ranging from 61 to 84 and standard deviations ranging from 18 to 34 in the general U.S. population (Ware, Snow, Kosinski, et al., 1993).

The advantage of the standardization and norm-based scoring of the PCS and MCS is that results for one can be meaningfully compared with the other and their scores have a direct interpretation in relation to the distribution of scores in the general U.S. population. Specifically, all scores above and below 50 are above and below the average, respectively, in the general U.S. population. Because the standard deviation is 10 for both PCS and MCS measures, each one point difference in scores also has a direct interpretation. A one point difference is one-tenth of a standard deviation. Advantages of norm-based scoring are illustrated in Appendix D.

Steps In Scoring

Following the scoring of the eight scales according to the standard SF-36 scoring algorithms (Medical Outcomes Trust, 1991, 1994; Ware, Snow, Kosinski, et al., 1993), PCS and MCS are scored in three steps as explained below.

Standardization Of Scales

First, each SF-36 *scale* is standardized using a z-score transformation and SF-36 scale means and standard deviations from the general U.S. population as given in Table 1. A z-score for each scale is computed by subtracting the general U.S. population mean from each SF-36 scale score and dividing the difference by the corresponding scale standard deviation from the general U.S. population. Formulas are listed below.

Step 1. Formulas for z-score standardizations of SF-36 scales.

$$PF\_Z = (PF - 84.52404) / 22.89490$$
$$RP\_Z = (RP - 81.19907) / 33.79729$$
$$BP\_Z = (BP - 75.49196) / 23.55879$$
$$GH\_Z = (GH - 72.21316) / 20.16964$$
$$VT\_Z = (VT - 61.05453) / 20.86942$$
$$SF\_Z = (SF - 83.59753) / 22.37642$$
$$RE\_Z = (RE - 81.29467) / 33.02717$$
$$MH\_Z = (MH - 74.84212) / 18.01189$$

Aggregation Of Scale Scores

After a z-score has been computed for each SF-36 scale, the second step involves computation of aggregate scores for the physical and mental components using the physical and mental factor score coefficients from the general U.S. population as given in Table C.1.

Computation of an aggregate physical component score consists of multiplying each SF-36 scale z-score by its respective physical factor score coefficient and summing the eight products, as shown below. Similarly, an aggregate mental component score is obtained by multiplying each SF-36 scale z-score by its respective mental factor score coefficient and summing the eight products.

Step 2. Formulas for aggregating standardized scales in estimating aggregate physical and mental component scores:

AGG_PHYS = (PF_Z * .42402) + (RP_Z * .35119) + BP_Z * .31754 + (GH_Z * .24954) + (VT_Z * .02877) + (SF_Z * -.00753) + (RE_Z * -.19206) + (MH_Z * -.22069)

AGG_MENT = (PF_Z * .22999) + (RP_Z * .12329) + BP_Z * .09731 + (GH_Z * .01571) + (VT_Z * .23534) + (SF_Z * -.26876) + (RE_Z * -.43407) + (MH_Z * -.48581)

Pending results from ongoing evaluations of other options, it is recommended that component scale scores be set to missing if the respondent is missing any one of the eight SF-36 scales. To minimize the number of component scores missing, we recommend estimating each of the eight scale scores if half or more of the items are complete, as documented elsewhere (Medical Outcomes Trust, 1991, 1994; Ware, Snow, Kosinski, et al., 1993)

Transformation Of Summary Scores

The third step involves transforming each component score to the norm-based (50, 10) scoring. This is accomplished by multiplying each aggregate component scale score by 10 and adding the resulting product to 50. Formulas are listed below.

Step 3. Formulas for T-score transformation of component scores.

Transformed Physical (PCS) = 50 + (AGG_PHYS * 10)
Transformed Mental (MCS) = 50 + (AGG_MENT * 10)

Features Of PCS And MCS Scores

The PCS and MCS were constructed and scored to achieve a number of advantages, in addition to reducing the SF-36 from an eight-scale profile to two summary measures without substantial loss of information. Features of the PCS and MCS scores, including their reliability, confidence intervals (CI), skewness (percent ceiling and floor), and number of levels observed in the general U.S. population, are summarized in Table C.2. These results confirm some of the theoretical advantages of the two summary measures as compared to the eight SF-36 scales, including a very large increase in the number of levels defined, smaller confidence intervals relative to each of the eight scales, as well as the elimination of both floor and ceiling effects. Reliability estimates and confidence intervals are discussed further in Chapter 7. Tradeoffs between the two summary measures versus the eight-scale profile are evaluated in Chapter 8.

TABLE C.2 COMPARISON OF FEATURES OF SF-36 SCALES AND SUMMARY MEASURES, GENERAL U.S. POPULATION

	SF-36 SCALES[b]	SUMMARY MEASURES	
		PCS	MCS
Reliability	.78 - .93	.93	.88
95% CI ($\pm$)	13 – 33	5.7	6.3
% Floor	1 – 24	0	0
% Ceiling	1 – 56	0	0
# of Levels[a]	4 – 26	567	493

[a] Scores rounded to first decimal place
[b] Statistics are presented as the range of results found across the eight SF-36 scales in the general U.S. population.

Scoring Checks

Because errors can lead to inaccurate scale scores, we strongly recommend formal scoring checks of SF-36 scales prior to computing the SF-36 component summary scales. These formal scoring checks are explained in full detail in the SF-36 Scoring Exercise available through the Medical Outcomes Trust (Medical Outcomes Trust, 1994).

The following scoring checks are also strongly recommended for the SF-36 component summary scales. Any discrepancies should be investigated for scoring errors:

1. Check correlations between the eight SF-36 scales and the PCS and MCS scales. The PF, RP, and BP scales should correlate highest with the PCS and lowest with the MCS. The MH, RE, and SF scales should correlate highest with the MCS and lowest with the PCS. The GH and VT scales should correlate moderately with both physical and mental component scales.

2. Check the correlation between the physical and mental component summary scales. The correlation should be very low.

Tables and text in Chapter 3 summarize results from these tests across numerous studies.

Scoring Exercise

A test dataset and SAS code for scoring the SF-36 PCS and MCS scales has been provided on a computer diskette (see inside back cover). The SAS code begins with algorithms for scoring the eight SF-36 scales and finishes with the computation of the component summary scales. The purpose of this scoring exercise is to help SF-36 users evaluate results from each step in the process of calculating SF-36 component summary scale scores. The test dataset for this scoring exercise contains 100 administrations of the SF-36 Health Survey. The test dataset is called "RAWDATA." The SAS code for scoring the scales is called "SF36SUMM.SCR."

Table C.3 presents descriptive statistics for the eight SF-36 scales and the physical and mental component scales from the test dataset. After scoring the test data set, you should observe the same means, standard deviations, and minimum and maximum observed values as those presented in Table C.3.

Note that the missing data rate for each scale is artificially large as part of the scoring exercise. Consequently, the missing data rate for the two component summary scales is uncharacteristically high.

TABLE C.3 TEST DATASET DESCRIPTIVE STATISTICS SF-36 SCALES AND SUMMARY MEASURES (N = 100)

SF-36 SCALES	# OF CASES	MEAN	STANDARD DEVIATION	MINIMUM OBSERVED VALUE	MAXIMUM OBSERVED VALUE
Physical Functioning (PF)	99	75.8	25.1	5	100
Role Physical (RP)	93	58.0	40.5	0	100
Bodily Pain (BP)	98	69.5	25.1	0	100
General Health (GH)	96	59.6	22.9	5	100
Vitality (VT)	100	56.2	19.3	15	95
Social Functioning (SF)	100	83.5	24.7	12.5	100
Role Emotional (RE)	92	72.5	39.1	0	100
Mental Health (MH)	100	74.4	19.5	8	100
PCS	88	45.2	10.9	17.2	66.7
MCS	88	49.7	11.2	12.7	64.2

Appendix D: 1990 Norms for PCS and MCS

This appendix reproduces Chapter 10 from the first edition of this manual (Ware et al., 1994) to document the 1990 norms for the SF-36 Health Survey (v1).

Normative data makes it possible to interpret PCS and MCS scores for an individual respondent or the average for a group of respondents by comparing them with the distribution of scores for other individuals. Because the PCS and MCS scoring is "norm-based," they have the advantage of a direct interpretation in this regard (see Appendix C). For example, it is useful to know that all scores above and below 50 are above and below the mean, respectively, in the general U.S. population for both the PCS and MCS. However, because physical and mental health scores vary with age, gender, and with the presence of diseases and other conditions, specific norms for subgroups of the population will greatly facilitate the interpretation of PCS and MCS scores. (Some background information about the norming of the SF-36 and other health status surveys in the U.S. and other countries, is briefly summarized in the original *SF-36 User's Manual* (Ware, Snow, Kosinski, et al., 1993), which presents norms for the eight SF-36 scales.)

Organization Of This Chapter

This chapter presents norms for the PCS and MCS for both the general U.S. population and for specific groups of patients participating in the Medical Outcomes Study (MOS). In addition to 21 tables that mirror those previously published for the eight SF-36 scales, 29 additional tables of normative data are presented here, including:

1. Norms for "well" adults sampled from the general U.S. population and for those reporting each of 14 chronic physical conditions and for those in the general population who screened positive for possibility of clinical depression.
2. Norms for changes observed in PCS and MCS scores over a one-year period for MOS patients with various chronic conditions.

This chapter is divided into seven sections, beginning with background information about the use of norm-based interpretations and a summary of the methods used to collect normative data. Most of the chapter is devoted to tables of norms for the general U.S. population in total and by age and gender, norms for self-reported chronic conditions in the general U.S. population and for more rigorously determined "tracer" conditions and "comorbid" conditions, based on the MOS. The last section presents norms for one-year change scores for groups of patients with chronic conditions and comorbidities in the MOS.

To make it easier to find normative data among the 50 data tables presented at the end of this chapter, Figure D.1 lists all tables with their page numbers.

FIGURE D.1 GUIDE TO TABLES PRESENTING NORMATIVE DATA

NORMS FOR THE GENERAL U.S. POPULATION	TABLE	PAGE
Total Sample, Descriptive Statistics	D.1	185
Total Sample, Frequency Distributions	D.2	186
Males	D.3	187
Females	D.3	187
Males & Females, by Age Group	D.4	188
Males, by Age Group	D.5	190
Females, by Age Group	D.6	192

NORMS FOR CHRONIC CONDITIONS IN THE GENERAL U.S. POPULATION	TABLE	PAGE
No Chronic Conditions	D.7	194
Allergies	D.8	195
Angina	D.9	196
Arthritis	D.10	197
Back Pain/Sciatica	D.11	198
Cancer (except skin cancer)	D.12	199
Congestive Heart Failure	D.13	200
Depression Screener	D.14	201
Dermatitis	D.15	202
Diabetes	D.16	203
Hearing Impairment	D.17	204
Hypertension	D.18	205
Limitations in Use of Arm(s)/Leg(s)	D.19	206
Lung Disease, Chronic	D.20	207
Myocardial Infarction, Recent	D.21	208
Vision Impairment	D.22	209

NORMS FOR CHRONIC CONDITIONS IN THE MOS	TABLE	PAGE
Total MOS Sample, All Conditions	D.23	210
Clinical Depression	D.24	211
Congestive Heart Failure	D.25	212
Diabetes, Type II	D.26	213
Hypertension	D.27	214
Myocardial Infarction, Recent Acute	D.28	215

NORMS FOR MOS PATIENTS WITH HYPERTENSION AND COMORBID CONDITIONS	TABLE	PAGE
Angina (Recent) without Myocardial Infarction	D.29	216
Back Pain/Sciatica	D.30	217
Benign Prostatic Hypertrophy Symptoms	D.31	218
Chronic Obstructive Pulmonary Disease	D.32	219
Dermatitis	D.33	220
Musculoskeletal Complaints	D.34	221
Osteoarthritis	D.35	222
Varicosities	D.36	223

NORMS FOR ONE-YEAR CHANGE SCORES, FIVE MOS CONDITIONS	TABLE	PAGE
Total MOS Sample	D.37	224
Clinical Depression	D.38	225
Congestive Heart Failure	D.39	226
Diabetes, Type II	D.40	227
Hypertension	D.41	228
Myocardial Infarction, Recent Acute	D.42	229

NORMS FOR ONE-YEAR CHANGE SCORES, MOS PATIENTS WITH HYPERTENSION AND COMORBID CONDITIONS	TABLE	PAGE
Angina (Recent) without Myocardial Infarction	D.43	230
Back Pain/Sciatica	D.44	231
Benign Prostatic Hypertrophy Symptoms	D.45	232
Chronic Obstructive Pulmonary Disease	D.46	233
Dermatitis	D.47	234
Musculoskeletal Complaints	D.48	235
Osteoarthritis	D.49	236
Varicosities	D.50	237

General Population Database

Norms for the general U.S. population were estimated from responses to the 1990 National Survey of Functional Health Status (NSFHS), a survey that included the SF-36 conducted by the National Opinion Research Center (NORC) (Thalji, Haggerty, Rubin, et al., 1991). Respondents were drawn from the sample frame of the 1989 and 1990 General Social Survey (GSS), an annual interview survey of the noninstitutionalized adult U.S. population. The GSS consisted of a two-stage probability sample design. In the first stage, quota sampling was used based on age, gender, and employment status at the block level. The primary sampling units were Standard Metropolitan Statistical Areas or non-metropolitan counties. Primary sampling units were stratified according to region, age, and race before selection. The unit for selection at the second stage were blocked groups stratified by race and income. The sample frame was 1,537 households from the 1989 GSS and 1,372 households from the 1990 GSS, for a base sample of 2,909 households. Two categories of respondents were drawn from the 2,909 households. The first category of respondents drawn were single members (core sample) of each household who had been previously interviewed in the 1989 or 1990 GSS (N = 2,909). The second category of respondents was an oversample of elderly (age 65 years or older) members residing in the 2,909 GSS households (N = 342). The total designated sample was 3,251 persons residing in 2,909 households.

All core sample respondents were randomly assigned to a self-administered mail survey (80%) or to a telephone interview (20%), with oversampled respondents assigned to the same survey mode as their core household member. The mail survey was conducted in two waves, with a postcard prompt occurring in between each wave. A $2.00 incentive was provided to respondents assigned to the mail survey. Follow-up of nonresponders consisted of a telephone interview.

The telephone survey was a computer-assisted telephone interview. An advanced letter describing the purpose of the survey and how each respondent was selected was sent out prior to the interview. No incentive was provided to respondents assigned to the telephone survey. The first follow-up of nonresponders consisted of a personalized letter explaining the importance of the survey. Trained interviewers followed the letters five days later with a second attempt to administer the survey by telephone. On average, six calls were placed to each nonresponder. After all attempts to reach nonresponders by telephone failed, a self-administered mail survey was sent.

The data collection period was 10 weeks for the mail survey and eight weeks for the telephone survey between October 15, 1990 and December 22, 1990. At least 50% of the data collection period for both surveys overlapped. The locating protocol for both surveys followed three steps: (1) a call to directory assistance; (2) a check of returned envelopes for forwarding addresses; and (3) a review of the GSS case locator page and call record for possible locating leads. The unlocatable rate was 10% for respondents assigned to the mail survey and 12% for respondents assigned to the telephone survey. The overall response rate for the 1990 NSFHS was 77.1% using a combination of mail and telephone survey methods.

MOS Database

The MOS is an observational study of variations in functional status and well-being among adult patients sampled from various systems of care, as documented in detail elsewhere (Stewart, Greenfield, Hays, et al., 1989; Tarlov, Ware, Greenfield, et al., 1989; Greenfield, Nelson, Zubkoff, et al., 1992; Kravitz, Greenfield, Rogers, et al., 1992; Stewart & Ware, 1992).

Ages at baseline ranged from 18-98 with a mean of 58 years; 54% were female, 16% black, and 6% other minorities. One in five (22%) had household incomes below 200% of the poverty line and 42% were educated beyond the high school level. The characteristics of patients in each diagnostic group are documented in the lower panel of the table of norms for that group.

Effects Of Data Collection Method

A randomized trial of data collection methods conducted during the norming of the SF-36 confirmed the practical advantages of mail-out/mail-back (MO/MB) surveys and documented the lack of equivalence between responses to MO/MB surveys and those from telephone interviews (McHorney, Kosinski, & Ware, 1994). As shown in Figure E.2, average scores for the SF-36 Mental Component Summary (MCS) measure were 2.43 (±0.3, $p < 0.001$) points higher for those interviewed by telephone, in comparison with those who self-administered the SF-36 by MO/MB. This difference is nearly one-fourth of a standard deviation (SD = 10 for the MCS) and appears to be noteworthy in clinical terms. The effect of data collection method on MCS scores is approximately one-fourth the impact of depressive disorder. Underlying this difference in MCS scores were significant differences for seven of the eight SF-36 scales (all but General Health). There was no effect on the Physical Component Summary (PCS); a difference greater than half a point would have been significant. For this reason, data collection methods should be standardized for surveys of health status.

When results are compared across methods, the effect of methods should be taken into account and results should be interpreted with caution, as discussed elsewhere for the eight SF-36 scales (McHorney, Kosinski, & Ware, 1994). For the two summary measures, which take into account the correlations among the eight scales, the effect is apparent only in the MCS. When comparing results from telephone interviews with norms reported in this manual, it is important to consider the possibility that MCS scores from telephone interviews may be *inflated* by about two to three points, which is two to three tenths of an SD unit.

FIGURE D.2 EFFFECTS OF TELEPHONE INTERVIEWS ON THE PCS AND MCS SCORES IN THE GENERAL U.S. POPULATION

[Bar chart showing PCS: Mail 48.4, Phone 48.5; MCS: Mail 49.4, Phone 51.8*; * p < .001]

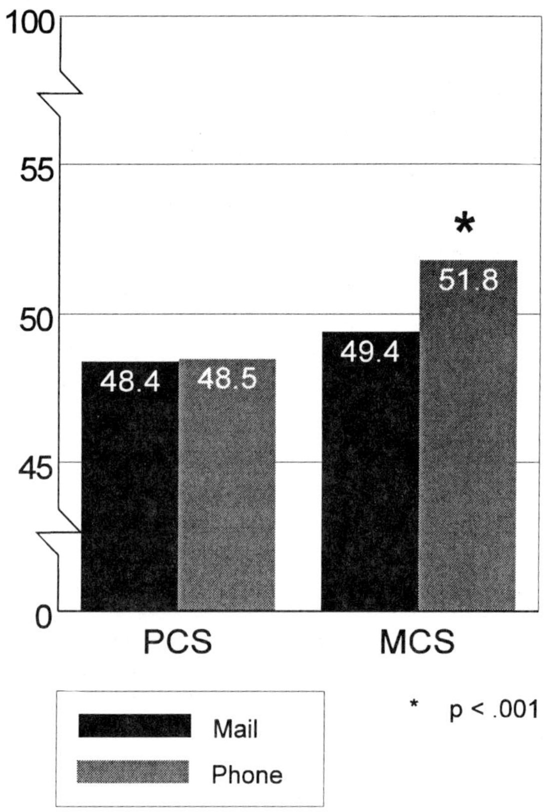

Note: Mean scores are adjusted for differences in demographics and chronic conditions between mail and telephone surveys

U.S. General Population Norms

Table D.1 presents descriptive statistics for PCS and MCS scores in the general U.S. population. Descriptive statistics include the mean, median (50th percentile), 25th and 75th percentiles, standard deviation, and the observed range of scores. The descriptive statistics presented in Table D.1 are for the total population (N = 2,474). Complete frequency distributions, including cumulative percentages (percentile ranks), for the PCS and MCS in the total population are presented in Table D.2. Note that among the advantages of the PCS and MCS, evidenced in Tables D.1 and D.2, are many more scale levels, standardized variances, and much less skewness of score distributions, relative to the eight SF-36 scales.

The mean and standard deviation are 50 and 10, respectively, for the PCS and MCS because of the scoring method used in their standardization (see Chapter 13). The medians for the PCS and MCS are higher than the means — 51.84 and 53.15, respectively. This reflects some skewness of the score distributions in the general U.S. population, with more respondents (about 60%) scoring above the mean. The frequency distributions in Table D.2 can be used to determine more precisely where in the distribution a score falls.

Norms For Males And Females

Table D.3 presents general U.S. population norms separately for males and females (all age groups combined). This table is useful in determining whether a score for males or females is above or below the average score for males and females in the general U.S. population. From Table D.3, it is evident that males score higher (more favorably) on average than females.

Norms For Age Groups

Table D.4 presents PCS and MCS general U.S. population norms for seven different age groups for males and females combined. These age groups were selected to: (1) be large enough to satisfy minimum standards for precision; and (2) correspond with standard practices for defining age-specific groups, which were also followed in norming the eight SF-36 scales. Consistent with the literature linking physical and mental health to age (Bergner, Bobbitt, Carter, et al., 1981; Stoller, 1984; Okun & Stock, 1987; Wood, Rhodes & Whelan, 1989; Idler & Angel, 1990; Idler, Kasl, & Lemke, 1990), the data in Table D.4 demonstrates a roughly linear decline in PCS scale scores for older age groups, while MCS scale scores do not.

Norms For Age Groups, Males And Females

Tables D.5 - D.6 present PCS and MCS norms by age group separately for males and females in the general U.S. population. These tables differ from the previous tables in that the oldest two age groups have been collapsed into one group (ages 65 and older) in order to have adequate sample sizes to maintain precision. Note that across all age groups, and consistent with previous literature linking gender to differences in health status, PCS and MCS scale scores are higher for males than for females (Bergner, Bergner, Hallstrom, et al., 1984; Wood, Rhodes, & Whelan, 1989; Liang, Wu, Krause, et al., 1992).

Norms For Chronic Conditions, General U.S. Population

Normative data for individuals with various chronic conditions were estimated using data from the general U.S. population (self-reported) and the MOS patient population (physician-reported).

The norms presented in these tables are not adjusted for differences in sociodemographic characteristics and comorbid conditions. Information about these variables is presented in the lower panel of each table to facilitate their consideration. It is important to keep these differences in mind when making comparisons. For example, in interpreting general population norms for those who report CHF (Table D.13), note that the majority also suffered from arthritis and that the average age was nearly 66 years. Analyses presented in Chapter 9, which used these data to estimate the unique effects of each condition on PCS and MCS scores, included statistical adjustments for differences in sociodemographic characteristics and comorbid conditions.

Norms For Chronic Conditions: General U.S. Population

The U.S. general population survey included a checklist of 14 chronic conditions. This checklist asked respondents if their doctor had ever told them they had any of the following conditions: hypertension, a heart attack in the last year, congestive heart failure (CHF), diabetes, angina, and cancer (except skin). In addition, respondents were asked if they now have any of the following conditions: chronic allergies or sinus troubles, arthritis of any kind or rheumatism, sciatica or chronic back problems, blindness or other trouble seeing with one or both eyes, chronic lung disease (like bronchitis, asthma, or emphysema), dermatitis or other chronic skin rash, deafness or other trouble hearing, and limitation in the use of an arm or leg (missing, paralyzed, or weakness). Definitions are repeated in footnotes to the data tables.

Tables D.7 - D.22 present PCS and MCS norms for the 14 chronic conditions and for a "well" group of respondents who did not report any of the 14 chronic conditions on the checklist and made no response to the "other" condition categories. At the bottom of each table is a sample description for each chronic condition, including the five most prevalent comorbidities. From comparing results for the chronic conditions to the "well" respondents in Tables D.7-D.22, it is clear that the "well" respondents have substantially better physical and mental health status than those respondents who reported chronic conditions. It is also evident that respondents with the more physically morbid conditions, such as CHF, MI, diabetes, and limitations in the use of an arm or leg, have lower PCS scale scores, on average, than those reporting other conditions. As expected, respondents who screened positive for possibility of depression have the lowest MCS scores on average.

Norms For Chronic Conditions: MOS

Norms for the PCS and MCS scales were estimated for patients with five "tracer" conditions selected for the longitudinal follow-up in the MOS: hypertension, CHF, recent MI, type II diabetes, and clinical depression (Tarlov, Ware, Greenfield, et al., 1989; Stewart & Ware, 1992). Diagnosis of the first four conditions (all but depression) was determined by standardized forms completed by the MOS physicians. The prevalence of these conditions among the 18,762 patients screened from doctor offices varied: 30.2% had hypertension, 3.2% had CHF, 1.5% had a recent MI, and 9.2% had diabetes, type II. An additional 4,335 patients were screened from offices of formally trained mental health providers. All patients were screened for clinical depression using a two-stage process with the CES-D and telephone administered diagnostic interview schedule (Wells, Hays, Burnam, et al., 1989).

Tables D.23 - D.28 present norms for the PCS and MCS for the total sample and for each of the five tracer groups. In addition, sample descriptions, including sociodemographic characteristics and the five most prevalent comorbid conditions, are presented for each tracer group.

Norms For Comorbid Conditions: MOS

To estimate the effects of comorbid conditions on PCS and MCS scores, we focused on the group of patients with hypertension because this represented the largest group of MOS patients and was the least morbid of the five MOS tracer conditions (Stewart, Greenfield, Hays, et al., 1989). Tables D.29 - D.36 present norms for MOS patients with hypertension and each of eight comorbid conditions including: angina (recent with no history of MI) (N = 256), back pain/sciatica (N = 481), benign prostatic hypertrophy symptoms (N = 184), chronic obstructive pulmonary disease (COPD) (N = 85), dermatitis (N = 231), musculoskeletal complaints (N = 341), osteoarthritis (N = 175), and varicosities (N = 222).

Norms for one-year change scores for the PCS and MCS were based on data from patients in the MOS with data at baseline and one-year follow-up assessments. These data are presented for five "tracer" conditions (hypertension, congestive heart failure, myocardial infarction, type II diabetes, and clinical depression) and for patients with conditions comorbid to hypertension. These norms can provide a basis for comparing the degree of change exhibited by a particular sample of patients. For example, one can assess whether there was more or less change in a sample of people, on average, than is typical of people with a condition such as hypertension. These norms can also be used for comparisons within a disease to see if the distribution of change scores exhibited in a particular sample is typical.

Computation Of Change Scores

One-year change scores were computed by subtracting baseline PCS and MCS scores from the PCS and MCS scores collected at the 12-month follow-up assessment. Change scores were categorized as "better," "stayed the same," or "worse" by comparing them with amounts equal to two standard errors of measurement (SEM) (see Chapter 7), which is approximately the 95% confidence interval for an individual score. Patients were classified as "worse" if their change score for the PCS/MCS indicated a decline greater than two SEMs. Patients were classified as "stayed the same" if their change score for the PCS/MCS was within two SEMs. Patients were classified as "better" if their change score for the PCS/MCS improved greater than two SEMs. The SEM for the PCS and MCS was computed by taking the square root of one minus the reliability for the PCS and MCS scales and multiplying the result by the standard deviation for the PCS/MCS in the general U.S. population (SD=10) (Nunnally, 1978). For the PCS, two SEM's equal ±5.42. For the MCS, two SEM's equal ±6.33. Thus, for example, patients with change scores showing improvement greater than 5.42 for the PCS and above 6.33 for the MCS were classified as "better." Patients with change scores below -5.42 for the PCS and below -6.33 for the MCS were classified as "worse." Patients with change scores between ±5.42 for the PCS and ±6.33 for the MCS were classified as "stayed the same."

The number of patients who died between baseline assessment and the one-year follow-up for each tracer group is noted at the bottom of each table. Those who died were not included in the analyses. They could be added to the "worse" category in the "better-same-worse" analysis.

Chronic Conditions: MOS

Tables D.37 - D.42 present norms for one-year change scores and the percent "better," "stayed the same," and "worse" for the total MOS sample and for the five MOS tracer conditions. Appendix A presents frequency distributions of one-year change scores for the total MOS sample and for the four most prevalent tracer conditions.

As shown in these tables, average one-year change scores ranged from -1.69 (for CHF) to .44 (for MI survivors) for the PCS and from .18 (for diabetes) to 3.86 (for depression) for the MCS for the five MOS tracer conditions. Underlying these average changes in physical and mental health status are noteworthy variations in the percentages who were classified as "better," "same," or "worse." Clinically depressed patients were more than twice as likely to improve (42.3%) than to decline (17.9%) in mental health. For patients with CHF, diabetes type II, and hypertension, a slightly greater percentage declined in physical health status than improved.

Tables D.43 - D.50 present one-year change score norms and the percent classified as "better," "stayed the same," or "worse" for the patients with hypertension and one of eight comorbid conditions. The results presented in each of these tables clearly demonstrates the added burden associated with a comorbidity on physical and mental health status for patients with hypertension.

TABLE D.1 NORMS FOR GENERAL U.S. POPULATION, TOTAL SAMPLE

Total Sample (N = 2,474)	PCS	MCS
Mean	50.00	50.00
25th Percentile	42.83	45.03
50th Percentile (Median)	51.84	53.15
75th Percentile	56.01	57.44
Standard Deviation	10.00	10.00
Range	8-73	10-74

TABLE D.2 FREQUENCY DISTRIBUTION FOR PCS AND MCS SCALE SCORES: GENERAL U.S. POPULATION (N = 2,474)

PCS Scale				MCS Scale			
Score	f	%	Cum. %	Score	f	%	Cum. %
66-73	14	0.5	100.0	66-74	20	0.8	100.0
65	14	0.6	99.5	65	6	0.2	99.3
64	19	0.8	98.9	64	19	0.8	99.1
63	13	0.5	98.1	63	28	1.2	98.3
62	30	1.2	97.6	62	57	2.4	97.2
61	39	1.6	96.3	61	49	2.0	94.8
60	78	3.2	94.7	60	72	3.0	92.8
59	133	5.5	91.5	59	133	5.5	89.8
58	152	6.3	86.0	58	160	6.6	84.3
57	157	6.5	79.7	57	191	7.9	77.7
56	160	6.6	73.2	56	149	6.2	69.8
55	184	7.6	66.5	55	120	5.0	63.6
54	164	6.8	58.9	54	137	5.7	58.6
53	155	6.4	52.1	53	124	5.2	53.0
52	107	4.4	45.7	52	91	3.8	47.8
51	86	3.6	41.3	51	95	4.0	44.1
50	88	3.6	37.7	50	89	3.7	40.1
49	72	3.0	34.1	49	83	3.4	36.4
48	65	2.7	31.1	48	46	1.9	33.0
47	55	2.3	28.4	47	82	3.4	31.0
46	50	2.1	26.1	46	48	2.0	27.6
45	44	1.8	24.1	45	52	2.1	25.6
44	49	2.0	22.3	44	42	1.7	23.5
43	31	1.3	20.2	43	36	1.5	21.8
42	23	1.0	18.9	42	30	1.2	20.3
41	30	1.3	18.0	41	40	1.7	19.0
40	30	1.2	16.7	40	44	1.8	17.4
39	32	1.3	15.5	39	43	1.8	15.5
38	31	1.3	14.2	38	32	1.3	13.8
37	28	1.2	12.9	37	22	0.9	12.4
36	26	1.1	11.7	36	28	1.2	11.5
35	21	0.9	10.7	35	27	1.1	10.3
34	17	0.7	9.8	34	24	1.0	9.2
33	20	0.8	9.1	33	19	0.8	8.2
32	21	0.9	8.2	32	18	0.7	7.4
31	21	0.9	7.4	31	16	0.7	6.7
30	17	0.7	6.5	30	20	0.8	6.0
29	18	0.8	5.8	29	11	0.5	5.2
28	11	0.5	5.0	28	17	0.7	4.7
27	14	0.6	4.5	27	9	0.4	4.0
26	13	0.5	4.0	26	9	0.4	3.6
25	15	0.6	3.4	25	14	0.6	3.2
24	5	0.2	2.8	24	9	0.4	2.7
23	14	0.6	2.6	23	8	0.3	2.3
22	7	0.3	2.0	22	9	0.4	1.9
21	6	0.3	1.7	21	6	0.3	1.6
8-20	35	1.5	1.5	9-20	31	1.3	1.3
	2415				2415		

Frequency missing = 59

TABLE D.3   NORMS FOR MALES AND FEMALES: GENERAL U.S. POPULATION

Males (N = 1,055)		PCS	MCS
	Mean	51.05	50.73
	25th Percentile	44.95	46.89
	50th Percentile (Median)	52.91	53.81
	75th Percentile	56.38	57.78
	Standard Deviation	9.39	9.57
	Range	9-73	12-74
Females (N = 1,412)		**PCS**	**MCS**
	Mean	49.07	49.33
	25th Percentile	41.10	43.40
	50th Percentile (Median)	51.03	52.80
	75th Percentile	55.86	57.19
	Standard Deviation	10.42	10.32
	Range	8-69	10-71

TABLE D.4 NORMS FOR SEVEN AGE GROUPS, MALES AND FEMALES COMBINED: GENERAL U.S. POPULATION

Ages 18-24 (N = 173)		PCS	MCS
	Mean	53.44	49.11
	25th Percentile	51.36	44.82
	50th Percentile (Median)	55.03	51.13
	75th Percentile	58.32	56.48
	Standard Deviation	7.59	10.16
	Range	22-73	13-67

Ages 25-34 (N = 474)		PCS	MCS
	Mean	53.72	48.64
	25th Percentile	50.55	43.33
	50th Percentile (Median)	55.24	51.68
	75th Percentile	58.11	56.16
	Standard Deviation	7.13	10.22
	Range	18-69	13-64

Ages 35-44 (N = 503)		PCS	MCS
	Mean	52.15	49.91
	25th Percentile	49.22	45.13
	50th Percentile (Median)	53.98	52.32
	75th Percentile	57.21	56.64
	Standard Deviation	7.75	9.26
	Range	10-67	17-65

Ages 45-54 (N = 338)		PCS	MCS
	Mean	49.64	50.53
	25th Percentile	45.70	46.35
	50th Percentile (Median)	52.60	53.55
	75th Percentile	55.89	57.20
	Standard Deviation	9.67	10.02
	Range	14-67	10-68

TABLE D.4 NORMS FOR SEVEN AGE GROUPS, MALES AND FEMALES COMBINED: GENERAL U.S. POPULATION (CONTINUED)

Age 55-64 (N = 269)		PCS	MCS
	Mean	45.90	51.05
	25th Percentile	38.66	46.71
	50th Percentile (Median)	49.86	54.35
	75th Percentile	54.32	57.90
	Standard Deviation	11.25	9.69
	Range	13-62	13-65

Age 65-74 (N = 442)		PCS	MCS
	Mean	43.33	52.68
	25th Percentile	35.04	48.34
	50th Percentile (Median)	46.18	55.67
	75th Percentile	52.50	59.13
	Standard Deviation	11.16	9.29
	Range	8-59	21-74

Ages 75 & Over (N = 264)		PCS	MCS
	Mean	37.89	50.44
	25th Percentile	28.99	41.36
	50th Percentile (Median)	38.16	53.69
	75th Percentile	47.35	59.36
	Standard Deviation	11.16	11.66
	Range	13-59	18-71

TABLE D.5 NORMS FOR MALES BY AGE GROUP: GENERAL U.S. POPULATION

Ages 18-24 Males (N = 71)		PCS	MCS
	Mean	53.50	50.89
	25th Percentile	51.12	45.91
	50th Percentile (Median)	54.70	51.40
	75th Percentile	57.80	57.17
	Standard Deviation	6.22	8.66
	Range	35-73	13-62
Ages 25-34 Males (N = 199)		**PCS**	**MCS**
	Mean	54.98	48.93
	25th Percentile	52.74	45.12
	50th Percentile (Median)	55.82	51.87
	75th Percentile	58.64	56.67
	Standard Deviation	6.33	10.32
	Range	18-67	14-63
Ages 35-44 Males (N = 239)		**PCS**	**MCS**
	Mean	52.95	51.00
	25th Percentile	49.99	46.89
	50th Percentile (Median)	54.61	53.63
	75th Percentile	57.34	57.27
	Standard Deviation	7.03	8.90
	Range	22-67	21-65
Ages 45-54 Males (N = 145)		**PCS**	**MCS**
	Mean	50.40	51.03
	25th Percentile	48.14	48.10
	50th Percentile (Median)	53.36	53.94
	75th Percentile	56.13	57.51
	Standard Deviation	9.68	9.86
	Range	13-67	17-67

TABLE D.5 NORMS FOR MALES BY AGE GROUP: GENERAL U.S. POPULATION (CONTINUED)

Ages 55-64 Males (N = 105)	PCS	MCS
Mean	46.90	51.60
25th Percentile	40.57	48.46
50th Percentile (Median)	49.80	54.63
75th Percentile	54.99	57.58
Standard Deviation	10.82	9.11
Range	16-58	25-63

Ages 65 & Over Males (N = 293)	PCS	MCS
Mean	41.95	52.51
25th Percentile	33.48	47.95
50th Percentile (Median)	43.84	54.83
75th Percentile	51.64	59.44
Standard Deviation	11.35	9.78
Range	9-59	19-74

TABLE D.6 NORMS FOR FEMALES BY AGE GROUP: GENERAL U.S. POPULATION

Ages 18-24 Females (N = 102)		PCS	MCS
	Mean	53.39	47.37
	25th Percentile	51.81	43.40
	50th Percentile (Median)	55.15	50.26
	75th Percentile	58.58	55.70
	Standard Deviation	8.74	11.18
	Range	22-68	15-67

Ages 25-34 Females (N = 275)		PCS	MCS
	Mean	52.46	48.34
	25th Percentile	49.32	41.90
	50th Percentile (Median)	54.43	51.31
	75th Percentile	57.67	55.22
	Standard Deviation	7.66	10.12
	Range	18-69	13-64

Ages 35-44 Females (N = 264)		PCS	MCS
	Mean	51.36	48.84
	25th Percentile	46.99	43.23
	50th Percentile (Median)	52.58	51.16
	75th Percentile	57.09	55.75
	Standard Deviation	8.34	9.49
	Range	10-65	17-64

Ages 45-54 Females (N = 193)		PCS	MCS
	Mean	48.95	50.07
	25th Percentile	43.40	45.55
	50th Percentile (Median)	51.61	53.48
	75th Percentile	55.79	56.99
	Standard Deviation	9.64	10.18
	Range	20-65	10-68

Ages 55-64 Females (N = 164)		PCS	MCS
	Mean	45.03	50.56
	25th Percentile	38.18	44.61
	50th Percentile (Median)	49.91	53.71
	75th Percentile	54.14	57.94
	Standard Deviation	11.57	10.16
	Range	13-62	13-65

TABLE D.6   NORMS FOR FEMALES BY AGE GROUP: GENERAL U.S. POPULATION (CONTINUED)

Ages 65 & Over Females (N = 413)		PCS	MCS
	Mean	41.02	51.44
	25th Percentile	32.00	43.43
	50th Percentile (Median)	42.93	55.08
	75th Percentile	49.83	58.96
	Standard Deviation	11.52	10.54
	Range	8-59	19-71

TABLE D.7 NORMS FOR "HEALTHY" GROUP WITH NO CHRONIC CONDITIONS: GENERAL U.S. POPULATION

	PCS	MCS
Mean	55.26	53.43
25th Percentile	53.69	50.33
50th Percentile (median)	55.85	54.74
75th Percentile	58.44	57.74
Standard Deviation	5.10	6.33
Range	23-67	23-67

Same Description

N	465	**Five Most Prevalent Comorbidities**
Mean Age	35.8	N/A[a]
Percent Over 65	3.9	N/A
Percent Female	45.2	N/A
Mean Education	13.7	N/A
Percent Nonwhite	18.1	N/A

[a] Not applicable, these respondents have no chronic conditions.

TABLE D.8 NORMS FOR ALLERGIES: GENERAL U.S. POPULATION

	PCS	MCS
Mean	47.44	48.23
25th Percentile	41.36	40.89
50th Percentile (median)	50.82	51.16
75th Percentile	55.36	56.67
Standard Deviation	10.81	10.74
Range	8-69	10-74

Same Description

N	818	**Five Most Prevalent Comorbidities**	
Mean Age	45.3	Depression Screener	46.2%
Percent Over 65	16.5	Arthritis	39.4%
Percent Female	59.2	Hypertension	30.8%
Mean Education	12.7	Back Pain/Sciatica	29.9%
Percent Nonwhite	15.5	Hearing Impairment	16.6%

Definition of allergies: Self-report of chronic allergies or sinus trouble

TABLE D.9 NORMS FOR ANGINA: GENERAL U.S. POPULATION

	PCS	MCS
Mean	36.36	48.04
25th Percentile	27.12	39.41
50th Percentile (median)	35.88	50.05
75th Percentile	46.95	58.33
Standard Deviation	12.38	12.42
Range	8-59	18-68

Sample Description

N	107	Five Most Prevalent Comorbidities	
Mean Age	62.6	Arthritis	75.1%
Percent Over 65	51.3	Hypertension	66.3%
Percent Female	58.6	Depression Screener	47.4%
Mean Education	11.7	Allergies	44.4%
Percent Nonwhite	12.6	Back Pain/Sciatica	39.1%

Definition of angina: Self-report of angina.

TABLE D.10 NORMS FOR ARTHRITIS: GENERAL U.S. POPULATION

	PCS	MCS
Mean	43.15	48.81
25th Percentile	34.71	41.40
50th Percentile (median)	45.83	51.74
75th Percentile	52.52	57.25
Standard Deviation	11.62	11.11
Range	8-65	13-74

Sample Description

N	826	**Five Most Prevalent Comorbidities**	
Mean Age	56.0	Allergies	45.9
Percent Over 65	34.5	Depression Screener	42.7
Percent Female	57.5	Hypertension	39.3
Mean Education	11.9	Back Pain/Sciatica	38.2
Percent Nonwhite	14.2	Hearing Impairment	25.0

Definition of arthritis: Self-report of arthritis or any kind of rheumatism.

TABLE D.11 NORMS FOR BACK PAIN/SCIATICA: GENERAL U.S. POPULATION

	PCS	MCS
Mean	43.14	46.88
25th Percentile	35.01	38.88
50th Percentile (median)	45.46	49.34
75th Percentile	52.41	56.67
Standard Deviation	11.56	11.73
Range	8-66	14-74

Sample Description

		Five Most Prevalent Comorbidities	
N	519		
Mean Age	49.1	Arthritis	55.7%
Percent Over 65	23.1	Allergies	50.8%
Percent Female	55.9	Depression Screener	49.1%
Mean Education	12.3	Hypertension	35.2%
Percent Nonwhite	13.8	Hearing Impairment	23.2%

Definition of back pain/sciatica: Self-report of sciatica or chronic back problems.

TABLE D.12   NORMS FOR CANCER: GENERAL U.S. POPULATION

	PCS	MCS
Mean	45.12	48.82
25th Percentile	36.61	43.40
50th Percentile (median)	47.39	53.03
75th Percentile	54.06	57.91
Standard Deviation	11.60	11.07
Range	16-64	13-65

Sample Description

		Five Most Prevalent Comorbidities	
N	105		
Mean Age	53.2	Depression Screener	45.7
Percent Over 65	40.4	Arthritis	43.6
Percent Female	57.3	Hypertension	36.3
Mean Education	12.2	Allergies	34.5
Percent Nonwhite	10.1	Back Pain/Sciatica	26.9

Definition of cancer: Self-report of cancer (except skin cancer).

TABLE D.13 NORMS FOR CONGESTIVE HEART FAILURE: GENERAL U.S. POPULATION

	PCS	MCS
Mean	31.02	45.65
25th Percentile	23.13	35.16
50th Percentile (median)	29.17	45.55
75th Percentile	35.79	56.03
Standard Deviation	10.64	12.49
Range	8-57	18-71

Sample Description

N	83	**Five Most Prevalent Comorbidities**	
Mean Age	65.6	Arthritis	64.8%
Percent Over 65	49.7	Hypertension	63.1%
Percent Female	59.1	Depression Screener	61.3%
Mean Education	10.5	Allergies	51.7%
Percent Nonwhite	28.3	Back Pain/Sciatica	39.8%

Definition of congestive heart failure: Self-report of congestive heart failure.

TABLE D.14   NORMS FOR DEPRESSION SCREENER: GENERAL U.S. POPULATION

	PCS	MCS
Mean	47.92	43.46
25th Percentile	39.78	35.41
50th Percentile (median)	51.37	45.24
75th Percentile	56.72	52.82
Standard Deviation	11.62	11.42
Range	13-73	13-67

Sample Description

N	881	**Five Most Prevalent Comorbidities**	
Mean Age	43.1	Allergies	41.3%
Percent Over 65	14.6	Arthritis	32.7%
Percent Female	57.7	Hypertension	28.8%
Mean Education	12.4	Back Pain/Sciatica	25.8%
Percent Nonwhite	21.1	Limited Use in Arm/Leg	16.3%

Definition of depression screener: Self-report of two weeks or more feeling sad, blue or depressed in the past year; or two years or more feeling sad or blue most days; or feeling depressed or sad much of the time in the past year.

TABLE D.15 NORMS FOR DERMATITIS: GENERAL U.S. POPULATION

	PCS	MCS
Mean	46.88	46.16
25th Percentile	38.99	38.48
50th Percentile (median)	50.96	50.16
75th Percentile	55.87	56.63
Standard Deviation	11.49	12.06
Range	13-65	18-64

Sample Description

N	214	**Five Most Prevalent Comorbidities**	
Mean Age	46.4	Allergies	52.5
Percent Over 65	19.2	Depression Screener	48.3
Percent Female	53.7	Arthritis	42.0
Mean Education	13.0	Back Pain/Sciatica	29.1
Percent Nonwhite	13.9	Hypertension	25.4

Definition of dermatitis: Self-report of dermatitis or other chronic skin rash.

TABLE D.16   NORMS FOR DIABETES: GENERAL U.S. POPULATION

	PCS	MCS
Mean	39.30	47.90
25th Percentile	29.97	39.52
50th Percentile (median)	38.22	49.92
75th Percentile	49.35	56.66
Standard Deviation	11.32	11.37
Range	13-60	18-67

Sample Description

N	145	**Five Most Prevalent Comorbidities**	
Mean Age	56.2	Hypertension	57.6
Percent Over 65	38.0	Depression Screener	53.1
Percent Female	59.2	Arthritis	51.9
Mean Education	11.2	Allergies	38.2
Percent Nonwhite	34.7	Back Pain/Sciatica	33.4

Definition of diabetes: Self-report of diabetes.

TABLE D.17   NORMS FOR HEARING IMPAIRMENT: GENERAL U.S. POPULATION

	PCS	MCS
Mean	43.70	48.74
25th Percentile	34.61	42.72
50th Percentile (median)	46.74	51.09
75th Percentile	53.31	56.80
Standard Deviation	12.18	10.62
Range	8-67	13-74

Sample Description

N	387	**Five Most Prevalent Comorbidities**	
Mean Age	57.6	Arthritis	55.7%
Percent Over 65	43.8	Allergies	43.3%
Percent Female	37.1	Depression Screener	39.8%
Mean Education	11.8	Hypertension	39.2%
Percent Nonwhite	11.3	Back Pain/Sciatica	35.6%

Definition of hearing impairment: Self-report of deafness or other trouble hearing with one or both ears.

TABLE D.18　NORMS FOR HYPERTENSION: GENERAL U.S. POPULATION

	PCS	MCS
Mean	44.57	49.24
25th Percentile	37.29	42.15
50th Percentile (median)	47.39	52.07
75th Percentile	53.51	57.15
Standard Deviation	11.29	10.55
Range	9-67	14-74

Sample Description

N	670	Five Most Prevalent Comorbidities	
Mean Age	54.0	Arthritis	46.7%
Percent Over 65	32.4	Depression Screener	44.7%
Percent Female	51.4	Allergies	42.6%
Mean Education	12.1	Back Pain/Sciatica	28.7%
Percent Nonwhite	25.7	Hearing Impairment	20.8%

Definition of hypertension: Self-report of current hypertension.

TABLE D.19   NORMS FOR LIMITATION IN THE USE OF AN ARM(S) OR LEG(S): GENERAL U.S. POPULATION

	PCS	MCS
Mean	37.74	45.89
25th Percentile	26.61	38.42
50th Percentile (median)	37.36	46.50
75th Percentile	49.37	56.03
Standard Deviation	12.99	11.57
Range	8-60	18-71

Sample Description

N	263	**Five Most Prevalent Comorbidities**	
Mean Age	51.5	Arthritis	64.2
Percent Over 65	28.2	Depression Screener	63.3
Percent Female	54.1	Allergies	48.2
Mean Education	11.8	Back Pain/Sciatica	43.6
Percent Nonwhite	18.9	Hypertension	39.4

Definition of limitations in use of arm(s)/leg(s): Self-report of limitations in the use of an arm or leg (missing, paralyzed or weakness).

TABLE D.20 NORMS FOR CHRONIC LUNG DISEASE: GENERAL U.S. POPULATION

	PCS	MCS
Mean	42.31	44.47
25th Percentile	30.09	35.30
50th Percentile (median)	45.71	46.59
75th Percentile	53.74	55.06
Standard Deviation	14.08	12.28
Range	8-69	10-68

Sample Description

		Five Most Prevalent Comorbidities	
N	182		
Mean Age	48.9	Allergies	63.3%
Percent Over 65	24.1	Depression Screener	55.4%
Percent Female	59.4	Arthritis	52.1%
Mean Education	11.7	Back Pain/Sciatica	33.8%
Percent Nonwhite	14.6	Hypertension	32.7%

Definition of chronic lung disease: Self-report of chronic lung disease (like chronic bronchitis, asthma, or emphysema).

TABLE D.21 NORMS FOR MYOCARDIAL INFARCTION: GENERAL U.S. POPULATION

	PCS	MCS
Mean	35.97	45.73
25th Percentile	25.17	35.65
50th Percentile (median)	34.51	47.34
75th Percentile	46.95	56.28
Standard Deviation	12.10	12.41
Range	13-59	18-68

Sample Description

N	62	**Five Most Prevalent Comorbidities**	
Mean Age	60.6	Hypertension	67.9
Percent Over 65	42.8	Arthritis	63.1
Percent Female	46.5	Depression Screener	56.1
Mean Education	11.0	Allergies	54.0
Percent Nonwhite	21.2	Angina	39.7

Definition of myocardial infarction: Self-report of myocardial infarction in the past year.

TABLE D.22 NORMS FOR VISION IMPAIRMENT: GENERAL U.S. POPULATION

	PCS	MCS
Mean	41.86	45.21
25th Percentile	31.50	35.65
50th Percentile (median)	44.22	47.06
75th Percentile	53.26	56.86
Standard Deviation	12.52	12.81
Range	8-62	16-68

Sample Description

N	259	**Five Most Prevalent Comorbidities**	
Mean Age	54.6	Arthritis	54.3
Percent Over 65	34.2	Depression Screener	53.2
Percent Female	59.0	Allergies	51.2
Mean Education	11.6	Hypertension	48.7
Percent Nonwhite	20.0	Back Pain/Sciatica	38.0

Definition of vision impairment: Self-report of blindness or other trouble seeing with one or both eyes, even when wearing glasses.

TABLE D.23 NORMS FOR ALL CONDITIONS COMBINED, MOS PARTICIPANTS

	PCS	MCS
Mean	44.92	48.44
25th Percentile	38.01	41.08
50th Percentile (Median)	47.18	52.01
75th Percentile	53.42	57.35
Standard Deviation	10.94	11.77
Range	8-71	3-71

Sample Description

N	**3445**	**Prevalence of Five MOS Conditions**	
Mean Age	54.3	Hypertension	60.6%
Percent Over 65	28.6	Myocardial Infarction	3.1%
Percent Female	61.7	Congestive Heart Failure	6.3%
Mean Education	13.3	Diabetes Type II	15.7%
Percent Poverty	19.8	Clinical Depression	14.6%

TABLE D.24   NORMS FOR CLINICAL DEPRESSION, MOS PARTICIPANTS

	PCS	MCS
Mean	44.96	34.84
25th Percentile	36.57	25.60
50th Percentile (median)	45.63	33.30
75th Percentile	54.29	43.63
Standard Deviation	12.05	12.17
Range	14-71	3-64

Sample Description

N	502	**Five Most Prevalent Comorbidities**	
Mean Age	41.6	Back Pain/Sciatica	45.9%
Percent Over 65	6.0	Angina-Recent	25.0%
Percent Female	75.8	Hypertension	20.9%
Mean Education	13.4	Musculoskeletal Complaints	17.6%
Percent Poverty	23.3	Dermatitis	17.5%

Definition of clinical depression: NIMH (DIS) criteria met for major depression and/or dysthymia.

SF-36™ Summary Measures Manual — Appendix D: 1990 Norms for PCS and MCS

TABLE D.25 NORMS FOR CONGESTIVE HEART FAILURE, MOS PARTICIPANTS

	PCS	MCS
Mean	34.50	50.43
25th Percentile	25.44	45.03
50th Percentile (median)	34.03	52.88
75th Percentile	43.54	58.37
Standard Deviation	12.08	11.13
Range	12-64	15-69

Sample Description

		Five Most Prevalent Comorbidities	
N	216		
Mean Age	67.4	Hypertension	52.8%
Percent Over 65	59.7	Back Pain/Sciatica	47.4%
Percent Female	52.3	Past MI	45.7%
Mean Education	12.2	Angina-Recent	40.2%
Percent Poverty	27.3	Musculoskeletal Complaints	32.7%

Definition of congestive heart failure: Physician report of current congestive heart failure.

TABLE D.26 NORMS FOR DIABETES TYPE II, MOS PARTICIPANTS

	PCS	MCS
Mean	41.52	51.90
25th Percentile	33.38	48.07
50th Percentile (median)	43.72	54.56
75th Percentile	50.41	58.43
Standard Deviation	11.27	9.55
Range	8-64	19-71

Sample Description

N	541	**Five Most Prevalent Comorbidities**	
Mean Age	60.2	Hypertension	64.3%
Percent Over 65	37.7	Back Pain/Sciatica	31.0%
Percent Female	55.6	Musculoskeletal Complaints	25.6%
Mean Education	12.5	Angina-Recent	18.6%
Percent Poverty	22.7	Dermatitis	17.0%

Definition of diabetes type II: Physician report of diabetes with age of onset 30 years or older.

TABLE D.27 NORMS FOR HYPERTENSION, MOS PARTICIPANTS

	PCS	MCS
Mean	44.31	52.22
25th Percentile	37.75	47.20
50th Percentile (Median)	47.00	54.95
75th Percentile	52.77	58.81
Standard Deviation	10.76	9.28
Range	8-67	19-71

Sample Description

N	2089	Five Most Prevalent Comorbidities	
Mean Age	59.1	Back Pain/Sciatica	34.0%
Percent Over 65	35.7	Musculoskeletal Complaints	24.6%
Percent Female	58.5	Recent Angina	16.3%
Mean Education	12.5	Diabetes Type II	16.2%
Percent Poverty	19.2	Varicosities	15.1%

Definition of hypertension: Physician report of current hypertension.

TABLE D.28 NORMS FOR MYOCARDIAL INFARCTION, MOS PARTICIPANTS

	PCS	MCS
Mean	42.64	51.67
25th Percentile	36.32	47.39
50th Percentile (median)	43.57	53.14
75th Percentile	49.81	57.51
Standard Deviation	10.02	8.19
Range	17-58	23-65

Sample Description

N	107	**Five Most Prevalent Comorbidities**	
Mean Age	59.2	Angina-Ever	55.8%
Percent Over 65	29.0	Angina-Recent	50.7%
Percent Female	30.8	Hypertension	42.5%
Mean Education	12.8	Back Pain/Sciatica	28.7%
Percent Poverty	14.0	Diabetes Type II	24.3%

Definition of myocardial infarction: Physician report of myocardial infarction within the past year.

TABLE D.29   NORMS FOR COMORBID RECENT ANGINA WITHOUT MYOCARDIAL INFARCTION, WITH HYPERTENSION, MOS PARTICIPANTS

	PCS	MCS
Mean	38.63	50.43
25th Percentile	29.57	44.89
50th Percentile (median)	40.64	52.82
75th Percentile	48.00	57.96
Standard Deviation	11.04	9.68
Range	14-58	19-68

Sample Description

N	256	**Five Most Prevalent Comorbidities**	
Mean Age	59.7	Back Pain/Sciatica	50.0%
Percent Over 65	39.4	Musculoskeletal Complaints	29.0%
Percent Female	55.1	Past MI	24.0%
Mean Education	12.7	Dermatitis	21.0%
Percent Poverty	23.0	Osteoarthritis	18.0%

Definition of angina: Symptoms of angina in past six months in the absence of an MI within one year.

TABLE D.30 NORMS FOR COMORBID BACK PAIN/SCIATICA WITH HYPERTENSION, MOS PARTICIPANTS

	PCS	MCS
Mean	39.67	51.37
25th Percentile	29.57	45.01
50th Percentile (median)	41.22	54.39
75th Percentile	49.54	59.11
Standard Deviation	11.71	10.32
Range	12-71	21-71

Sample Description

N	481	**Five Most Prevalent Comorbidities**	
Mean Age	60.4	Musculoskeletal Complaints	30.0%
Percent Over 65	35.8	Angina-Recent	28.0%
Percent Female	64.2	Angina-No MI	27.0%
Mean Education	12.2	Varicosities	21.0%
Percent Poverty	20.6	Osteoarthritis	21.0%

Definition of back pain/sciatica: Attacks of back pain or sciatica in last six months.

TABLE D.31 NORMS FOR COMORBID BENIGN PROSTATIC HYPERTROPHY (BPH) SYMPTOMS WITH HYPERTENSION, MOS PARTICIPANTS

	PCS	MCS
Mean	43.84	54.13
25th Percentile	38.28	49.95
50th Percentile (median)	46.44	57.04
75th Percentile	51.63	59.67
Standard Deviation	10.28	8.72
Range	13-59	26-70

Sample Description

N	184	**Five Most Prevalent Comorbidities**	
Mean Age	67.1	Back Pain/Sciatica	32.0%
Percent Over 65	53.8	Musculoskeletal Complaints	27.0%
Percent Female	0.0	Past MI	22.0%
Mean Education	12.5	Angina-Recent	22.0%
Percent Poverty	17.6	Angina-No MI	21.0%

Definition of BPH symptoms: Male, age 50 years or older, history of nocturia in past six months, no serious kidney disease ever diagnosed, and no report of prostatic cancer.

TABLE D.32   NORMS FOR COMORBID CHRONIC OBSTRUCTIVE PULMONARY DISEASE (COPD) WITH HYPERTENSION, MOS PARTICIPANTS

	PCS	MCS
Mean	35.90	47.73
25th Percentile	25.71	39.97
50th Percentile (median)	36.51	50.93
75th Percentile	43.52	57.79
Standard Deviation	10.38	11.44
Range	17-55	19-68

Sample Description

N	85	**Five Most Prevalent Comorbidities**	
Mean Age	62.4	Back Pain/Sciatica	55.0%
Percent Over 65	42.4	Angina-Recent	38.0%
Percent Female	63.5	Angina-No MI	36.0%
Mean Education	11.6	Varicosities	31.0%
Percent Poverty	34.7	Musculoskeletal Complaints	27.0%

Definition of COPD: Lung disease diagnosed by physician as COPD in past six months.

TABLE D.33 NORMS FOR COMORBID DERMATITIS WITH HYPERTENSION, MOS PARTICIPANTS

	PCS	MCS
Mean	43.06	51.94
25th Percentile	35.43	46.72
50th Percentile (median)	46.19	54.00
75th Percentile	52.22	58.64
Standard Deviation	10.89	8.93
Range	10-63	24-68

Sample Description

N	231	**Five Most Prevalent Comorbidities**	
Mean Age	57.6	Back Pain/Sciatica	55.0%
Percent Over 65	35.5	Musculoskeletal Complaints	38.0%
Percent Female	48.5	Angina-Recent	36.0%
Mean Education	13.2	Angina-No MI	31.0%
Percent Poverty	16.3	Varicosities	27.0%

Definition of dermatitis: Now have condition that physician ever diagnosed chronic skin rash.

TABLE D.34 NORMS FOR COMORBID MUSCULOSKELETAL COMPLAINTS WITH HYPERTENSION, MOS PARTICIPANTS

	PCS	MCS
Mean	41.60	52.79
25th Percentile	33.43	46.17
50th Percentile (median)	43.15	55.80
75th Percentile	50.26	59.71
Standard Deviation	10.42	9.76
Range	13-59	21-70

Sample Description

N	341	**Five Most Prevalent Comorbidities**	
Mean Age	61.4	Back Pain/Sciatica	43.0%
Percent Over 65	41.6	Angina-Recent	23.0%
Percent Female	63.0	Varicosities	22.0%
Mean Education	12.0	Angina-No MI	21.0%
Percent Poverty	22.7	Dermatitis	18.0%

Definition of musculoskeletal complaints: Now have active condition physician ever diagnosed arthritis, but criteria for osteoarthritis or rheumatoid arthritis not met.

TABLE D.35 NORMS FOR COMORBID OSTEOARTHRITIS WITH HYPERTENSION, MOS PARTICIPANTS

	PCS	MCS
Mean	36.10	53.43
25th Percentile	25.71	48.68
50th Percentile (median)	37.00	56.30
75th Percentile	46.83	60.50
Standard Deviation	12.00	10.20
Range	12-57	24-71

Sample Description

N	175	**Five Most Prevalent Comorbidities**	
Mean Age	67.8	Back Pain/Sciatica	57.0%
Percent Over 65	58.9	Varicosities	27.0%
Percent Female	74.3	Angina-No MI	26.0%
Mean Education	11.9	Angina-Recent	26.0%
Percent Poverty	23.8	Musculoskeletal Complaints	19.0%

Definition of osteoarthritis: Now have active condition physician ever diagnosed as arthritis, and physician ever labeled it osteoarthritis or degenerative arthritis, and patient is 55 years or older.

TABLE D.36   NORMS FOR COMORBID VARICOSITIES WITH HYPERTENSION, MOS PARTICIPANTS

	PCS	MCS
Mean	41.91	51.66
25th Percentile	34.47	46.15
50th Percentile (median)	43.54	54.60
75th Percentile	49.97	58.85
Standard Deviation	10.64	10.01
Range	10-64	20-70

Sample Description

N	222	**Five Most Prevalent Comorbidities**	
Mean Age	61.8	Back Pain/Sciatica	45.0%
Percent Over 65	39.2	Musculoskeletal Complaints	34.0%
Percent Female	72.1	Angina-Recent	25.0%
Mean Education	12.2	Angina-No MI	25.0%
Percent Poverty	20.1	Osteoarthritis	21.0%

Definition of varicosities: Now have condition that physician ever diagnosed as varicose veins/deep varicosities.

TABLE D.37   NORMS FOR ONE-YEAR CHANGE SCORES, TOTAL MOS SAMPLE (N = 1539)

	PCS	MCS
Mean	-0.53	1.10
25th Percentile	-4.96	-3.76
50th Percentile (Median)	-0.39	0.36
75th Percentile	4.15	5.46
Standard Deviation	8.91	9.53
Range	-43 to 34	-37 to 45

Percent Classified as Better, Stayed the Same, or Worse:[1] *Total*

PCS	f	%	MCS	f	%
Worse	358	23.3	Worse	238	15.5
Stayed the Same	874	56.8	Stayed the Same	960	62.4
Better	307	19.9	Better	341	22.1

Note: 39 patients who died between baseline and one-year follow-up are excluded.

TABLE D.38   NORMS FOR ONE YEAR CHANGE SCORES, CLINICAL DEPRESSION (N = 279)

	PCS	MCS
Mean	-0.64	3.86
25th Percentile	-6.04	-4.33
50th Percentile (median)	-0.25	2.78
75th Percentile	4.70	11.30
Standard Deviation	9.31	13.18
Range	-36 to 34	-35 to 45

Percent Classified as Better, Stayed the Same, or Worse:[1] Depression

PCS	f	%	MCS	f	%
Worse	81	29.0	Worse	50	17.9
Stayed the Same	139	49.8	Stayed the Same	111	39.8
Better	59	21.2	Better	118	42.3

Note: 5 patients who died between baseline and one-year follow-up are excluded.

[1] Patients classified as "better" or "worse" have one year difference scores greater than ±2 SEMs. Patients with one year difference scores within ±2 SEMs are classified as "stayed the same" (2 SEMs: PCS = ±5.42; MCS = ±6.33).

TABLE D.39 NORMS FOR ONE YEAR CHANGE SCORES, CONGESTIVE HEART FAILURE (N = 131)

	PCS	MCS
Mean	-1.69	0.91
25th Percentile	-5.50	-2.83
50th Percentile (median)	-0.80	0.64
75th Percentile	2.55	4.53
Standard Deviation	8.82	9.44
Range	-34 to 20	-27 to 33

Percent Classified as Better, Stayed the Same, or Worse:[1] *CHF*

PCS	f	%	MCS	f	%
Worse	35	26.7	Worse	23	17.6
Stayed the Same	66	50.4	Stayed the Same	80	61.1
Better	30	22.9	Better	28	21.3

Note: 15 patients who died between baseline and one-year follow-up are excluded.

TABLE D.40   NORMS FOR ONE YEAR CHANGE SCORES, DIABETES TYPE II (N = 291)

	PCS	MCS
Mean	0.22	0.18
25th Percentile	-4.52	-3.87
50th Percentile (median)	0.11	-0.29
75th Percentile	5.39	4.32
Standard Deviation	8.83	8.60
Range	-28 to 27	-22 to 24

Percent Classified as Better, Stayed the Same, or Worse:[1] Diabetes

PCS	f	%	MCS	f	%
% Worse	71	24.4	% Worse	48	16.5
% Stayed the Same	159	54.6	% Stayed the Same	191	65.6
% Better	61	21.0	% Better	52	17.9

Note: 11 patients who died between baseline and one-year follow-up are excluded.

[1] Patients classified as "better" or "worse" have one year differences scores greater than ±2 SEMs of the mean. Patients with one year difference scores within ±2 SEM's are classified as "stayed the same" (2 SEM's: PCS = ±5.42; MCS = ±6.33).

TABLE D.41 NORMS FOR ONE YEAR CHANGE SCORES, HYPERTENSION (N = 895)

	PCS	MCS
Mean	-0.40	0.20
25th Percentile	-4.86	-3.73
50th Percentile (Median)	-0.40	-0.02
75th Percentile	4.16	4.08
Standard Deviation	9.09	8.07
Range	-43 to 33	-37 to 33

Percent Classified as Better, Stayed the Same, or Worse:[1] *Hypertension*

PCS	%	MCS	%
Worse	21.6	Worse	14.3
Stayed the Same	58.5	Stayed the Same	69.5
Better	19.9	Better	16.2

Note: 25 patients who died between baseline and one-year follow-up are excluded.

TABLE D.42   NORMS FOR ONE YEAR CHANGE SCORES, MYOCARDIAL INFARCTION (N = 67)

	PCS	MCS
Mean	0.44	1.57
25th Percentile	-1.92	-2.54
50th Percentile (median)	0.98	1.94
75th Percentile	3.64	5.34
Standard Deviation	6.05	7.79
Range	-17 to 19	-20 to 26

Percent Classified as Better, Stayed the Same, or Worse:[1] MI

PCS	f	%	MCS	f	%
Worse	10	14.9	Worse	7	10.5
Stayed the Same	44	65.7	Stayed the Same	50	74.6
Better	13	19.4	Better	10	14.9

Note: 4 patients who died between baseline and one-year follow-up are excluded.

[1] Patients classified as "better" or "worse have one year difference greater than ±2 SEMs. Patients with one year difference scores within ±2 SEM's are classified as "stayed the same" (2 SEMs for PCS = ±5.42, MCS = ±6.33).

TABLE D.43 NORMS FOR ONE YEAR CHANGE SCORES, RECENT ANGINA WITHOUT MYOCARDIAL INFARCTION, WITH HYPERTENSION (N = 133)

	PCS	MCS
Mean	-1.24	0.10
25th Percentile	-4.91	-5.03
50th Percentile (median)	-1.04	-0.05
75th Percentile	3.44	6.11
Standard Deviation	8.82	9.65
Range	-23 to 21	-36 to 26

Percent Classified as Better, Stayed the Same, or Worse:[1] Angina

PCS	f	%	MCS	f	%
Worse	33	24.9	Worse	25	18.8
Stayed the Same	80	60.1	Stayed the Same	77	57.9
Better	20	15.0	Better	31	23.3

Note: 6 patients who died between baseline and one-year follow-up are excluded.

TABLE D.44 NORMS FOR ONE YEAR CHANGE SCORES, BACK PAINS / SCIATICA WITH HYPERTENSION (N = 241)

	PCS	MCS
Mean	-0.33	-0.56
25th Percentile	-5.65	-5.04
50th Percentile (median)	-0.62	-0.34
75th Percentile	5.13	4.40
Standard Deviation	10.15	9.42
Range	-33 to 29	-36 to 29

Percent Classified as Better, Stayed the Same, or Worse:[1] Back Pain/Sciatica

PCS	f	%	MCS	f	%
Worse	61	25.3	Worse	47	19.5
Stayed the Same	129	53.5	Stayed the Same	150	62.2
Better	51	21.2	Better	44	18.3

Note: 6 patients who died between baseline and one-year follow-up are excluded.

1 Patients classifies as "better" or "worse" have one year difference scores greater than ±2 SEMs. Patients with one year difference scores within ±2 SEMs are classified as "stayed the same" (2 SEM: PCS = ±5.42; MCS = ±6.33).

TABLE D.45   NORMS FOR ONE YEAR CHANGE SCORES, BENIGN PROSTATIC HYPERTROPHY SYMPTOMS (BPH) WITH HYPERTENSION (N = 101)

	PCS	MCS
Mean	-0.70	-0.13
25th Percentile	-4.34	-4.21
50th Percentile (median)	-1.67	0.35
75th Percentile	3.44	3.99
Standard Deviation	7.83	6.95
Range	-23 to 25	-21 to 26

Percent Classified as Better, Stayed the Same, or Worse[1] BPH

PCS	f	%	MCS	f	%
Worse	25	24.8	Worse	16	15.8
Stayed the Same	56	55.4	Stayed the Same	70	69.3
Better	20	19.8	Better	15	14.9

Note: 5 patients who died between baseline and one-year follow-up are excluded.

TABLE D.46 NORMS FOR ONE YEAR CHANGE SCORES, CHRONIC OBSTRUCTIVE PULMONARY DISEASE (COPD) WITH HYPERTENSION (N = 46)

	PCS	MCS
Mean	-3.30	4.78
25th Percentile	-7.78	-2.81
50th Percentile (median)	-3.23	3.49
75th Percentile	0.44	9.40
Standard Deviation	6.84	11.48
Range	-18 to 19	-20 to 33

Percent Classified as Better, Stayed the Same, or Worse:[1] COPD

PCS	f	%	MCS	f	%
Worse	14	30.5	Worse	7	15.2
Stayed the Same	26	56.5	Stayed the Same	26	56.5
Better	6	13.0	Better	13	28.3

Note: 1 patient who died between baseline and one-year follow-up is excluded.

1 Patients classifies as "better" or "worse" have one year difference scores greater then ±2 SEMs. Patients with one year difference scores within ±2 SEMs are classified as "stayed the same" (2 SEMs: PCS = ±5.42; MCS = ±6.33).

TABLE D.47   NORMS FOR ONE YEAR CHANGE SCORES, DERMATITIS WITH HYPERTENSION (N = 125)

	PCS	MCS
Mean	-0.20	-0.17
25th Percentile	-5.49	-4.60
50th Percentile (median)	-1.54	0.21
75th Percentile	3.86	4.68
Standard Deviation	8.90	7.73
Range	-20 to 31	-29 to 17

Percent Classified as Better, Stayed the Same, or Worse:[1] Dermatitis

PCS	f	%	MCS	f	%
Worse	29	23.2	Worse	21	16.8
Stayed the Same	71	56.8	Stayed the Same	85	68.0
Better	25	20.0	Better	19	15.2

Note: 4 patients who died between baseline and one-year follow-up are excluded.

TABLE D.48   NORMS FOR ONE YEAR CHANGE SCORES, MUSCULOSKELETAL COMPLAINTS WITH HYPERTENSION (N = 168)

	PCS	MCS
Mean	-1.96	-1.24
25th Percentile	-5.87	-5.11
50th Percentile (median)	-1.22	-0.29
75th Percentile	4.15	3.16
Standard Deviation	10.89	9.09
Range	-38 to 32	-36 to 33

Percent Classified as Better, Stayed the Same, or Worse:[1]
Musculoskeletal Comp.

PCS	f	%	MCS	f	%
Worse	38	22.6	Worse	27	16.1
Stayed the Same	93	55.4	Stayed the Same	116	69.0
Better	37	22.0	Better	25	14.9

Note: 7 patients who died between baseline and one-year follow-up are excluded.
1   Patients classified as "better" or "worse" have one year difference scores greater than ±2 SEMs. Patients with one year difference scores within ±2 SEMs are classified as "stayed the same" (2 SEMs: PCS = ±5.42; MCS = ±6.33).

TABLE D.49 NORMS FOR ONE YEAR CHANGE SCORES, OSTEOARTHRITIS WITH HYPERTENSION (N = 102)

	PCS	MCS
Mean	-2.09	-0.21
25th Percentile	-8.98	-4.82
50th Percentile (median)	-3.34	-0.20
75th Percentile	3.20	3.81
Standard Deviation	8.13	8.53
Range	-23 - 19	-20 - 26

Percent Classified as Better, Stayed the Same, or Worse:[1] *Osteoarthritis*

PCS	f	%	MCS	f	%
Worse	31	30.4	Worse	23	22.6
Stayed the Same	52	51.0	Stayed the Same	65	63.7
Better	19	18.6	Better	14	13.7

Note: 2 patients who died between baseline and one-year follow-up are excluded.

TABLE D.50 NORMS FOR ONE YEAR CHANGE SCORES, VARICOSITIES WITH HYPERTENSION (N = 130)

	PCS	MCS
Mean	-0.70	-1.00
25th Percentile	-4.89	-6.38
50th Percentile (median)	-0.40	0.76
75th Percentile	3.75	3.63
Standard Deviation	8.31	8.02
Range	-28 to 32	-26 to 19

Percent Classified as Better, Stayed the Same, or Worse:[1] Varicosities

PCS	f	%	MCS	f	%
Worse	30	23.1	Worse	30	23.1
Stayed the Same	73	56.1	Stayed the Same	83	63.8
Better	27	20.8	Better	17	13.1

[1] Patients classified as "better" or "worse" have one year difference scores greater than ±2 SEMs. Patients with one year difference scores within ±2 SEMs are classified as "stayed the same" (2 SEMs: PCS = ±5.42; MCS = ±6.33).